Scientific Computation

Editorial Board

J.-J. Chattot, Davis, CA, USA
P. Colella, Berkeley, CA, USA
R. Glowinski, Houston, TX, USA
M. Holt, Berkeley, CA, USA
Y. Hussaini, Tallahassee, FL, USA
P. Joly, Le Chesnay, France
H. B. Keller, Pasadena, CA, USA
D. I. Meiron, Pasadena, CA, USA
O. Pironneau, Paris, France
A. Quarteroni, Lausanne, Switzerland
J. Rappaz, Lausanne, Switzerland
R. Rosner, Chicago, IL, USA
J. H. Seinfeld, Pasadena, CA, USA
A. Szepessy, Stockholm, Sweden
M. F. Wheeler, Austin, TX, USA

Springer
*Berlin
Heidelberg
New York
Hong Kong
London
Milan
Paris
Tokyo*

G. N. Milstein   M. V. Tretyakov

Stochastic Numerics for Mathematical Physics

With 48 Figures and 28 Tables

 Springer

Professor Grigori N. Milstein
Weierstrass Institute
for Applied Analysis and Stochastics
Mohrenstrasse 39
10117 Berlin, Germany
and
Ural State University
Department of Mathematics
Lenin Str. 51
620083 Ekaterinburg, Russia

Dr. Michael V. Tretyakov
University of Leicester
Department of Mathematics
Leicester LE1 7RH
United Kingdom

Library of Congress Control Number: 2004103618

Bibliographic information published by Die Deutsche Bibliothek
Die Deutsche Bibliothek lists this publication in the Deutsche Nationalbibliografie; detailed bibliographic data is available in the Internet at <http://dnb.ddb.de>.

ISBN 3-540-21110-1 Springer-Verlag Berlin Heidelberg New York

This work is subject to copyright. All rights are reserved, whether the whole or part of the material is concerned, specifically the rights of translation, reprinting, reuse of illustrations, recitation, broadcasting, reproduction on microfilm or in any other way, and storage in data banks. Duplication of this publication or parts thereof is permitted only under the provisions of the German Copyright Law of September 9, 1965, in its current version, and permission for use must always be obtained from Springer-Verlag. Violations are liable to prosecution under the German Copyright Law.

Springer-Verlag is a part of Springer Science+Business Media

springeronline.com

© Springer-Verlag Berlin Heidelberg 2004
Printed in Germany

The use of general descriptive names, registered names, trademarks, etc. in this publication does not imply, even in the absence of a specific statement, that such names are exempt from the relevant protective laws and regulations and therefore free for general use.

Typesetting: Data prepared by the authors using a Springer TeX macro package
Production: PTP-Berlin Protago-TeX-Production GmbH, Germany
Cover design: *design & production* GmbH, Heidelberg
Printed on acid-free paper   55/3141/Yu   5 4 3 2 1 0

Preface

Using stochastic differential equations (SDEs), we can successfully model systems that function in the presence of random perturbations. Such systems are among the basic objects of modern control theory, signal processing and filtering, physics, biology and chemistry, economics and finance, to mention just a few. However the very importance acquired by SDEs lies, to a large extent, in the strong connection they have with the equations of mathematical physics. It is well known that problems of mathematical physics involve "damned dimensions", often leading to severe difficulties in solving boundary value problems. A way out is provided by the employment of probabilistic representations together with Monte Carlo methods. As a result, a complex multi-dimensional problem for partial differential equations (PDEs) reduces to the Cauchy problem for a system of SDEs. The last system can naturally be regarded as one-dimensional, since it contains only one independent variable; it arises as a characteristic system of the considered problem for PDEs. The importance of this approach, while enabling the reduction of a multi-dimensional boundary value problem to the one-dimensional Cauchy problem, cannot be underestimated for computational mathematics.

Two books:

> Milstein G.N., Numerical Integration of Stochastic Differential Equations. Kluwer, 1995 (English translation from Russian 1988),
> Kloeden P.E. and Platen E., Numerical Solution of Stochastic Differential Equations. Springer, 1992,

present a systematic treatment of mean-square and weak numerical schemes for SDEs. These approximations represent two fundamental aspects in the contemporary theory of numerical integration of SDEs. Mean-square methods are useful for direct simulation of stochastic trajectories which, for instance, can give information on general behavior of a stochastic model. They are the basis for construction of weak methods which are important for many practical applications. Weak methods are sufficient for evaluation of mean values and solving problems of mathematical physics by the Monte Carlo technique, and they are simpler than mean-square ones.

In the present book, numerical integration of SDEs receives a large developmental effort in two directions: a lot of new special schemes are constructed for a number of stochastic systems which are important for applications and for the first time numerical methods for SDEs in bounded domains are proposed. The second part of the book is devoted to construction of stochastic numerical algorithms for solving complicated problems for PDEs, both linear and nonlinear.

The first two chapters contain essentially revised material of the previously mentioned book by the first author, with broad supplements. For instance, a number of effective numerical methods for systems with colored noise are included in these chapters. Another example of the new material is construction of fully implicit mean-square schemes for SDEs with multiplicative noise.

Many difficulties arise with realizing numerical methods for SDEs of a general form. At the same time methods adapted to specific systems can be more efficient than general methods. Very often fluctuations, which affect a physical system, are small. Fortunately, as shown in Chap. 3, in the case of stochastic systems with small noise, it is possible to construct special numerical methods. The errors of these methods are estimated in terms of products $h^i \varepsilon^j$, where h is the step-size of discretization and ε is a small noise parameter. Usually, the global errors in these methods have the form $O(h^j + \varepsilon^k h^l)$, where $j > l$, $k > 0$. Thanks to the fact that the accuracy order l of such methods is comparatively small, they are not too complicated, while due to the large j and the small factor ε^k at h^l, their errors are fairly low. This allows us to construct effective (high-exactness) methods with low time-step order which nevertheless have small errors.

In Chap. 4, specific methods for stochastic Hamiltonian systems and Langevin type equations are proposed. Stochastic Hamiltonian systems, like deterministic Hamiltonian systems, possess the property of preserving symplectic structure (symplecticness). For instance, Hamiltonian systems with additive noise are a rather wide and important class of equations having this property. It is well known from deterministic numerical analysis that an effective numerical solution of deterministic Hamiltonian systems on long time intervals requires symplectic methods. It turns out that symplectic methods for stochastic Hamiltonian systems, which are proposed in the first part of Chap. 4, have significant advantages over standard schemes for SDEs.

In the second part of Chap. 4 we construct special numerical methods (we call them quasi-symplectic) for Langevin type equations which have widespread occurrence in models from physics, chemistry, and biology. The proposed methods are such that they degenerate to symplectic methods when the system degenerates to a Hamiltonian one and their law of phase volume contractivity is close to the exact one. The presented numerical tests of both symplectic and quasi-symplectic methods clearly demonstrate superiority of the proposed methods over very long time intervals in comparison with standard ones.

Our probabilistic methods for solving boundary value problems ensure that the proposed approximations of solutions of the corresponding SDEs belong to a bounded domain. Such mean-square approximations are considered in Chap. 5. A numerical method for simulation of an autonomous diffusion process in a space bounded domain is based on a space discretization using a random walk over small spheres. The algorithm gives points which are close to the points of the real phase trajectory for SDEs. To realize the algorithm, the exit point of the Wiener process from a d-dimensional ball has to be constructed at each step. Due to independence of the first exit time and the first exit point of the Wiener process from the ball, they can be simulated separately. It is known that the exit point is distributed uniformly on the sphere, but simulation of the exit time is a fairly laborious problem. Consequently, the algorithm gives only the phase component of the approximate trajectory without modelling the corresponding time component. The space-time point lies on the d-dimensional lateral surface of a semicylinder with spherical base in the $(d+1)$-dimensional semispace $[t_0, \infty) \times R^d$. The algorithm ensures smallness of the phase increments at each step, but the nonsimulated time increments can take arbitrary large values with some probability. "Ordinary" mean-square methods from Chap. 1, intended to solve SDEs on a finite time interval, are based on a time discretization. The space-time point, corresponding to an "ordinary" one-step approximation constructed at a time point t_k, lies on the d-dimensional plane $t = t_k$, which belongs to the $(d+1)$-dimensional semispace $[t_0, \infty) \times R^d$. The "ordinary" mean-square methods give both time and phase components of the approximate trajectory. They ensure smallness of time increments at each step, but space increments can take arbitrary large values with some probability. In Chap. 5 we also introduce mean-square approximations which control boundedness of both space increments and time increments. In addition they give approximate values for both phase and time components of the space-time diffusion in the space-time bounded domain. The space-time point lies on a bounded d-dimensional manifold. This problem is solved in a constructive manner by the implementation of a space-time discretization with a random walk over boundaries of small space-time parallelepipeds.

Chapter 6 is devoted to random walks related to linear Dirichlet and Neumann boundary value problems for PDEs of elliptic and parabolic type. These random walks are Markov chains. Using them together with the Monte Carlo technique, complex multi-dimensional problems for linear PDEs can be solved. The random walks are constructed on the basis of mean-square and weak approximations for the characteristic system of SDEs due to the corresponding probabilistic representations of the solution to the considered boundary value problem. As in Chap. 5, a certain boundedness of the simulated increments of the Markov chains is necessary here. The proposed algorithms are accompanied by convergence theorems and numerical tests.

Nonlinear PDEs are suggested as mathematical models of problems in many fields such as fluid dynamics, combustion, biochemistry, dynamics of populations, finance, etc. They are mostly investigated by numerical methods, which are traditionally based on deterministic approaches. A probabilistic approach to construction of new numerical methods for solving initial and boundary value problems for nonlinear parabolic PDEs is developed in Chaps. 7 and 8. The approach is based on making use of the well-known probabilistic representations of solutions of linear PDEs and the idea of SDE numerical integration in the weak sense. Despite their probabilistic nature these methods are nevertheless deterministic. The probabilistic approach takes into account a coefficient dependence on the space variables and a relationship between diffusion and advection in an intrinsic manner. In particular, the layer methods derived allow us to avoid difficulties stemming from essentially changing coefficients and strong advection. A lot of computer experiments were made using the numerical algorithms proposed in Chaps. 7 and 8. Among them are numerical tests on the Burgers equation with small viscosity and on the FKPP-equation. Their results are in a good agreement with the theory. We also present a comparison analysis of the layer methods and the well known finite-difference schemes demonstrating some of the advantages of the proposed methods.

Chapter 9 is devoted to applications of stochastic numerics. Among lots of possibilities, we select applications of constructed stochastic simulation algorithms to such models of stochastic dynamics as systems with stochastic resonance and stochastic ratchets. We demonstrate here both mean-square methods for simulating trajectories of the considered models and weak methods for solving a number of boundary value problems.

An overwhelming majority of the methods proposed in this book are brought to numerical algorithms. Then it only remains to write a computer program, which is usually not complicated, and to use the method in practice. We give some illustrations in the Appendix how our methods can be implemented.

The field of stochastic numerics and its applications is too broad for one book. For example, such important topics as numerical integration of stochastic partial differential equations and of backward stochastic differential equations, being close to interests of the authors, are not covered in the book. The authors do not aim to provide an exhaustive review of literature. As a rule, the references are cited in the course of presentation. Some references are given without comments. On the whole, the content of the book is mainly based on the results obtained by the authors.

Throughout the book we use the hierarchical numbering system: the k-th equation (or the k-th theorem, figure, table, etc) in Sect. j of Chap. i is labelled at the place, where it occurs, and is cited within Chap. i as $(j.k)$ (or Theorem $j.k$, etc); it is cited as $(i.j.k)$ (or Theorem $i.j.k$, etc) outside Chap. i. The equation (theorem, figure, etc) counter is reset at the beginning of each section. The only exception is listings in the Appendix: the k-th listing is labelled and cited as Listing A.k.

We would like to thank the institutions which have made this work possible – the Ural State University and the Institute of Mathematics and Mechanics of the Russian Academy of Sciences (Ekaterinburg, Russia), the Weierstrass Institute for Applied Analysis and Stochastics (Berlin, Germany), and the University of Leicester (Leicester, UK).

Berlin and Leicester,   *Grigori N. Milstein*
December 2003   *Michael V. Tretyakov*

Table of Contents

**1   Mean-square approximation
    for stochastic differential equations**   1
   1.1   Fundamental theorem on the mean-square order of convergence   3
         1.1.1   Statement of the theorem   3
         1.1.2   Proof of the fundamental theorem   5
         1.1.3   The fundamental theorem for equations
                 in the sense of Stratonovich.......................   9
         1.1.4   Discussion 10
         1.1.5   The explicit Euler method 12
         1.1.6   Nonglobally Lipschitz conditions 16
   1.2   Methods based on Taylor-type expansion 18
         1.2.1   Taylor expansion of solutions
                 of ordinary differential equations 18
         1.2.2   Wagner–Platen expansion of solutions
                 of stochastic differential equations 19
         1.2.3   Construction of explicit methods 23
   1.3   Implicit mean-square methods 29
         1.3.1   Construction of drift-implicit methods 29
         1.3.2   The balanced method 33
         1.3.3   Implicit methods for SDEs
                 with locally Lipschitz vector fields 37
         1.3.4   Fully implicit mean-square methods: The main idea .. 38
         1.3.5   Convergence theorem for fully implicit methods 41
         1.3.6   General construction of fully implicit methods 43
   1.4   Modeling of Ito integrals 45
         1.4.1   Ito integrals depending on a single noise
                 and methods of order 3/2 and 2 45
         1.4.2   Modeling Ito integrals by the rectangle
                 and trapezium methods 51
         1.4.3   Modeling Ito integrals by the Fourier method 54
   1.5   Explicit and implicit methods of order 3/2 for systems
         with additive noise 60
         1.5.1   Explicit methods based on Taylor-type expansion..... 60
         1.5.2   Implicit methods based on Taylor-type expansion..... 62

		1.5.3	Stiff systems of stochastic differential equations with additive noise. A-stability	66
		1.5.4	Runge–Kutta type methods........................	71
		1.5.5	Two-step difference methods	75
	1.6	Numerical schemes for equations with colored noise		77
		1.6.1	Explicit schemes of orders 2 and 5/2	78
		1.6.2	Runge–Kutta schemes	80
		1.6.3	Implicit schemes	81

2 Weak approximation for stochastic differential equations 83

	2.1	One-step approximation		87
		2.1.1	Properties of remainders and Ito integrals	88
		2.1.2	One-step approximations of third order	92
		2.1.3	The Taylor expansion of mathematical expectations ..	98
	2.2	The main theorem on convergence of weak approximations and methods of order two		99
		2.2.1	The general convergence theorem...................	100
		2.2.2	Runge–Kutta type methods.......................	104
		2.2.3	The Talay–Tubaro extrapolation method	105
		2.2.4	Implicit method	109
	2.3	Weak methods for systems with additive and colored noise...		112
		2.3.1	Second-order methods	113
		2.3.2	Main lemmas for third-order methods...............	113
		2.3.3	Construction of a method of order three.............	116
		2.3.4	Weak schemes for systems with colored noise	121
	2.4	Variance reduction..		123
		2.4.1	The method of important sampling	123
		2.4.2	Variance reduction by control variates and combining method	126
		2.4.3	Variance reduction for boundary value problems	129
	2.5	Application of weak methods to the Monte Carlo computation of Wiener integrals...........................		130
		2.5.1	The trapezium, rectangle, and other methods of second order..................................	133
		2.5.2	A fourth-order Runge–Kutta method for computing Wiener integrals of functionals of exponential type....	135
		2.5.3	Explicit Runge–Kutta method of order four for conditional Wiener integrals of exponential-type functionals	137
		2.5.4	Theorem on one-step error.......................	140
		2.5.5	Implicit Runge–Kutta methods for conditional Wiener integrals of exponential-type functionals	145
		2.5.6	Numerical experiments	148
	2.6	Random number generators...............................		159

		2.6.1	Some uniform random number generators 160

        2.6.1 Some uniform random number generators 160
        2.6.2 A specific test for SDE integration 163
        2.6.3 Generation of Gaussian random numbers 166
        2.6.4 Parallel implementation 168

3 Numerical methods for SDEs with small noise 171
   3.1 Mean-square approximations and estimation of their errors .. 173
        3.1.1 Construction of one-step mean-square approximation . 173
        3.1.2 Theorem on mean-square global estimate 175
        3.1.3 Selection of time increment h
               depending on parameter ε 177
        3.1.4 (h,ε)-approach versus (ε,h)-approach 177
   3.2 Some concrete mean-square methods
       for systems with small noise 178
        3.2.1 Taylor-type numerical methods..................... 179
        3.2.2 Runge–Kutta methods 180
        3.2.3 Implicit methods 182
        3.2.4 Stratonovich SDEs with small noise................. 183
        3.2.5 Mean-square methods for systems
               with small additive noise 184
   3.3 Numerical tests of mean-square methods 185
        3.3.1 Simulation of Lyapunov exponent of a linear system
               with small noise.................................. 185
        3.3.2 Stochastic model of a laser 188
   3.4 The main theorem on error estimation and general approach
       to construction of weak methods 190
   3.5 Some concrete weak methods 193
        3.5.1 Taylor-type methods.............................. 193
        3.5.2 Runge–Kutta methods 196
        3.5.3 Weak methods for systems with small additive noise .. 199
   3.6 Expansion of the global error in powers of h and ε 202
   3.7 Reduction of the Monte Carlo error 203
   3.8 Simulation of the Lyapunov exponent of a linear system
       with small noise by weak methods 206

**4 Stochastic Hamiltonian systems
and Langevin-type equations** 211
   4.1 Preservation of symplectic structure 213
   4.2 Mean-square symplectic methods
       for stochastic Hamiltonian systems 216
        4.2.1 General stochastic Hamiltonian systems 216
        4.2.2 Explicit methods in the case of separable Hamiltonians 220
   4.3 Mean-square symplectic methods for Hamiltonian systems
       with additive noise 224
        4.3.1 The case of a general Hamiltonian 224
        4.3.2 The case of separable Hamiltonians 231

- 4.3.3 The case of Hamiltonian
  $H(t,p,q) = \frac{1}{2}p^\top M^{-1}p + U(t,q)$ 234
- 4.4 Numerical tests of mean-square symplectic methods......... 237
  - 4.4.1 Kubo oscillator 237
  - 4.4.2 A model for synchrotron oscillations of particles in storage rings 239
  - 4.4.3 Linear oscillator with additive noise................ 240
- 4.5 Liouvillian methods for stochastic systems preserving phase volume................................. 246
  - 4.5.1 Liouvillian methods for partitioned systems with multiplicative noise 248
  - 4.5.2 Liouvillian methods for a volume-preserving system with additive noise 250
- 4.6 Weak symplectic methods for stochastic Hamiltonian systems 251
  - 4.6.1 Hamiltonian systems with multiplicative noise 251
  - 4.6.2 Hamiltonian systems with additive noise 255
  - 4.6.3 Numerical tests 257
- 4.7 Quasi-symplectic mean-square methods for Langevin-type equations............................. 261
  - 4.7.1 Langevin equation: Linear damping and additive noise 262
  - 4.7.2 Langevin-type equation: Nonlinear damping and multiplicative noise 270
- 4.8 Quasi-symplectic weak methods for Langevin-type equations . 273
  - 4.8.1 Langevin equation: Linear damping and additive noise 273
  - 4.8.2 Langevin-type equation: Nonlinear damping and multiplicative noise 275
  - 4.8.3 Numerical examples 276

5 Simulation of space and space-time bounded diffusions ... 283
- 5.1 Mean-square approximation for autonomous SDEs without drift in a space bounded domain 286
  - 5.1.1 Local approximation of diffusion in a space bounded domain 287
  - 5.1.2 Global algorithm for diffusion in a space bounded domain 291
  - 5.1.3 Simulation of exit point $X_x(\tau_x)$ 301
- 5.2 Systems with drift in a space bounded domain 302
- 5.3 Space-time Brownian motion............................. 306
  - 5.3.1 Auxiliary knowledge 306
  - 5.3.2 Some distributions for one-dimensional Wiener process 308
  - 5.3.3 Simulation of exit time and exit point of Wiener process from a cube 313
  - 5.3.4 Simulation of exit point of the space-time Brownian motion from a space-time parallelepiped with cubic base 316

		Table of Contents XIII

     5.4 Approximations for SDEs in a space-time bounded domain .. 317
           5.4.1 Local mean-square approximation in a space-time
                  bounded domain 318
           5.4.2 Global algorithm in a space-time bounded domain 322
           5.4.3 Approximation of exit point $(\tau, X(\tau))$ 325
           5.4.4 Simulation of space-time Brownian motion with drift . 328
     5.5 Numerical examples 329
     5.6 Mean-square approximation of diffusion with reflection 337

6 Random walks for linear boundary value problems 339
     6.1 Algorithms for solving the Dirichlet problem
          based on time-step control 339
           6.1.1 Theorems on one-step approximation 341
           6.1.2 Numerical algorithms and convergence theorems...... 348
     6.2 The simplest random walk for the Dirichlet problem
          for parabolic equations 353
           6.2.1 The algorithm of the simplest random walk 353
           6.2.2 Convergence theorem 356
           6.2.3 Other random walks 359
           6.2.4 Numerical tests 364
     6.3 Random walks for the elliptic Dirichlet problem 365
           6.3.1 The simplest random walk for elliptic equations 366
           6.3.2 Other methods for elliptic problems................. 370
           6.3.3 Numerical tests 372
     6.4 Specific random walks for elliptic equations
          and boundary layer 374
           6.4.1 Conditional expectation of Ito integrals
                  connected with Wiener process in the ball 376
           6.4.2 Specific one-step approximations for elliptic equations . 380
           6.4.3 The average number of steps 384
           6.4.4 Numerical algorithms and convergence theorems...... 388
     6.5 Methods for elliptic equations with small parameter
          at higher derivatives 392
     6.6 Methods for the Neumann problem for parabolic equations .. 397
           6.6.1 One-step approximation for boundary points 399
           6.6.2 Convergence theorems 403

7 Probabilistic approach to numerical solution
of the Cauchy problem for nonlinear parabolic equations... 407
     7.1 Probabilistic approach to linear parabolic equations......... 408
     7.2 Layer methods for semilinear parabolic equations 415
           7.2.1 The construction of layer methods 415
           7.2.2 Convergence theorem for a layer method 419
           7.2.3 Numerical algorithms 422
     7.3 Multi-dimensional case 427

- 7.3.1 Multidimensional parabolic equation 427
- 7.3.2 Probabilistic approach to reaction-diffusion systems ... 429
- 7.4 Numerical examples 431
- 7.5 Probabilistic approach to semilinear parabolic equations with small parameter 438
  - 7.5.1 Implicit layer method and its convergence 440
  - 7.5.2 Explicit layer methods 442
  - 7.5.3 Singular case 443
  - 7.5.4 Numerical algorithms based on interpolation 445
- 7.6 High-order methods for semilinear equation with small constant diffusion and zero advection 446
  - 7.6.1 Two-layer methods 447
  - 7.6.2 Three-layer methods 448
- 7.7 Numerical tests 451
  - 7.7.1 The Burgers equation with small viscosity 452
  - 7.7.2 The generalized FKPP-equation with a small parameter 455

8 Numerical solution of the nonlinear Dirichlet and Neumann problems based on the probabilistic approach .. 461
- 8.1 Layer methods for the Dirichlet problem for semilinear parabolic equations 461
  - 8.1.1 Construction of a layer method of first order 463
  - 8.1.2 Convergence theorem 467
  - 8.1.3 A layer method with a simpler approximation near the boundary 468
  - 8.1.4 Numerical algorithms and their convergence 474
- 8.2 Extension to the multi-dimensional Dirichlet problem 476
- 8.3 Numerical tests of layer methods for the Dirichlet problems .. 479
  - 8.3.1 The Burgers equation 479
  - 8.3.2 Comparison analysis 482
  - 8.3.3 Quasilinear equation with power law nonlinearities ... 485
- 8.4 Layer methods for the Neumann problem for semilinear parabolic equations 488
  - 8.4.1 Construction of layer methods 489
  - 8.4.2 Convergence theorems 493
  - 8.4.3 Numerical algorithms 497
  - 8.4.4 Some other layer methods 499
- 8.5 Extension to the multi-dimensional Neumann problem 501
- 8.6 Numerical tests for the Neumann problem 503
  - 8.6.1 Comparison of various layer methods 503
  - 8.6.2 A comparison analysis of layer methods and finite-difference schemes 505

9 Application of stochastic numerics to models with stochastic resonance and to Brownian ratchets 509
9.1 Noise-induced regular oscillations in systems with stochastic resonance 510
9.1.1 Sufficient conditions for regular oscillations 512
9.1.2 Comparison with the approach based on Kramers' theory of diffusion over a potential barrier 517
9.1.3 High-frequency regular oscillations in systems with multiplicative noise 518
9.1.4 Large-amplitude regular oscillations in monostable system 523
9.1.5 Regular oscillations in a system of two coupled oscillators 525
9.2 Noise-induced unidirectional transport 526
9.2.1 Systems with state-dependent diffusion 528
9.2.2 Forced thermal ratchets 533

A Appendix: Practical guidance to implementation of the stochastic numerical methods 541
A.1 Mean-square methods 541
A.2 Weak methods and the Monte Carlo technique 544
A.3 Algorithms for bounded diffusions 550
A.4 Random walks for linear boundary value problems 558
A.5 Nonlinear PDEs 560
A.6 Miscellaneous 565

References 571

Index 587

Frequently Used Notation

PDE	partial differential equation
SDE	stochastic differential equation
MC	Monte Carlo
RK	Runge–Kutta
PRK	partitioned Runge–Kutta
RKN	Runge–Kutta–Nyström
RNG	random number generator
LCG	linear congruential generator
LFSR	linear feedback shift register generator
PRNG	parallel random number generator
FKPP	Fisher–Kolmogorov–Petrovskii–Piskunov
SR	stochastic resonance
a.s.	almost surely
i.i.d.	independent identically distributed
$:=$	equal by definition
$\doteq$ or $\simeq$ or $\approx$	approximately equal
$\cdots \circ dw(t)$	Stratonovich SDE
δ_{ij}	Kronecker delta symbol
$\chi_A(\omega)$ or $I_A(x)$	indicator function of the set A
$\mathbf{R}^d$	d-dimensional Euclidean space
$x = (x^1, \ldots, x^d)^\mathsf{T}$	column vector $x \in \mathbf{R}^d$ with ith-component x^i
$x^\mathsf{T} y$ or (x, y)	scalar product of vectors $x, y \in \mathbf{R}^d$
$\lvert x \rvert$	Euclidean norm of vector $x \in \mathbf{R}^d$
$A = \{a^{ij}\}$	matrix A with ijth-component a^{ij}
$\lvert \sigma \rvert = (tr\sigma\sigma^\mathsf{T})^{1/2}$	Euclidean norm of matrix σ
$O(\rho)$	expression being divided by ρ remains bounded as $\rho \to 0$
$(\Omega, \mathcal{F}, P)$	probability space
$\mathcal{F}_t$	nondecreasing family of σ-subalgebras of $\mathcal{F}$
$E(X)$ or EX	expectation of X
$Var(X)$ or $D(X)$	variance of X

Frequently Used Notation

$\bar{D}_M$	sample variance
$E(X\|\mathcal{F}_t)$	conditional expectation of X under $\mathcal{F}_t$
$\mathcal{N}(\mu,\sigma^2)$	Gaussian distribution with mean μ and variance σ^2
$w_r(t)$ or $W_r(t)$	standard Wiener process
$I_{i\ldots j}$, $J_{i\ldots j}$, $I_{i_1,\ldots,i_j}$	Ito integrals
$X_{t,x}(s)$, $s \geq t$	trajectory such that $X_{t,x}(t) = x$
h or h_k or Δt	step of time discretization
$\Delta_k w_r(h) = w_r(t_k+h) - w_r(t_k)$	increment of $w_r(t)$ on $[t_k, t_k+h]$
$\bar{X}(t_k)$ or $\bar{X}_k$ or X_k	approximation of $X(t_k)$
$\bar{X}_{t,x}(t+h)$	one-step approximation
$\Delta = X_{t,x}(t+h) - x$	increment of solution
$\bar{\Delta} = \bar{X}_{t,x}(t+h) - x$	increment of approximation of solution
$\bar{X}^{(m)}$, $m=1,\ldots,M$	independent realizations
$\mathbf{F}$	class of functions with polynomial growth
$C^k(\bar{D})$	space of k times continuously differentiable functions on $\bar{D}$
$C_{0,0}^d$	set of all d-dimensional continuous vector-functions $x(t)$ satisfying the condition $x(0) = 0$
$C_{0,a;T,b}^d$	set of all d-dimensional continuous vector-functions $x(t)$ satisfying the conditions $x(0) = a$, $x(T) = b$
$\mu_{0,0}(x)$	Wiener measure corresponding to Brownian paths with fixed initial point $(0,0)$
$\mu_{0,a}^{T,b}(x)$	conditional Wiener measure corresponding to Brownian paths $X_{0,a}^{T,b}(t)$ with fixed initial and final points
L, Λ_r	differential operators
Δ	Laplace operator
$H(t,p,q)$, $H_r(t,p,q)$	Hamiltonians
G	bounded domain in $\mathbf{R}^d$
∂G	boundary of domain G
Γ_δ	interior of δ-neighborhood of ∂G belonging to G
$Q = [t_0, T) \times G$	cylinder in $\mathbf{R}^{d+1}$
$\Gamma = \bar{Q}\backslash Q$	part of boundary of Q consisting of upper base and lateral surface
$U_r \subset \mathbf{R}^d$	open sphere of radius r with center at the origin of $\mathbf{R}^d$
$U_r^\sigma(x)$	open ellipsoid obtained from U_r by linear transformation $\sigma(x)$ and shift x

$C_r \subset \mathbf{R}^d$	cube with center at the origin of $\mathbf{R}^d$ and edges of the length $2r$ being parallel to the coordinate axes
$\Pi_{r,l} = [0, lr^2) \times C_r$	space-time parallelepiped
$\bar{\nu}_x(D)$	first exit time of approximate trajectory from D
$\tau,\ \theta,\ \bar{\theta}$	Markov moments
$\nu,\ \bar{\nu},\ \varkappa$	random steps

1 Mean-square approximation for stochastic differential equations

The simplest approximate method for solving the Ito system

$$dX = a(t,X)dt + \sum_{r=1}^{q} \sigma_r(t,X)dw_r(t) \tag{0.1}$$

is *Euler's method*:

$$X_{k+1} = X_k + \sum_{r=1}^{q} \sigma_{rk}\Delta_k w_r(h) + a_k h, \tag{0.2}$$

where $\Delta_k w_r(h) = w_r(t_{k+1}) - w_r(t_k)$, and the index k at σ_r and a indicates that these functions are evaluated at the point (t_k, X_k).

G. Marujama [167] showed the mean-square convergence of this method, while I.I. Gichman and A.V. Skorochod [82] proved that the order of accuracy of Euler's method is $1/2$, i.e.,

$$(E(X(t_k) - X_k)^2)^{1/2} \leq Ch^{1/2}, \tag{0.3}$$

where C is a constant not depending on k and h.

If for some method we would have

$$(E(X(t_k) - X_k)^2)^{1/2} \leq Ch^p, \tag{0.4}$$

then we say that *the mean-square order of accuracy* of the method is p.

A method of first order of accuracy was first constructed in [175]. It is as follows:

$$X_{k+1} = X_k + \sum_{r=1}^{q} \sigma_{rk}\Delta_k w_r(h) + a_k h$$
$$+ \sum_{i=1}^{q}\sum_{r=1}^{q}(\Lambda_i \sigma_r)_k \int_{t_k}^{t_{k+1}}(w_i(\theta) - w_i(t_k))dw_r(\theta), \tag{0.5}$$

where $\Lambda_i = (\sigma_i, \dfrac{\partial}{\partial x})$.

For a single noise ($q = 1$), the integral in (0.5) can be expressed in terms of $\Delta_k w$ and the formula takes the following form, e.g. in the scalar case:

$$X_{k+1} = X_k + \sigma_k \Delta_k w + a_k h + \frac{1}{2}(\sigma \frac{\partial \sigma}{\partial x})_k \Delta_k^2 w - \frac{1}{2}(\sigma \frac{\partial \sigma}{\partial x})_k h. \qquad (0.6)$$

The same paper [175] also contains methods of higher order of accuracy for the scalar case.

The authors of [226, 231, 244, 255, 280, 281] and many others obtained results that border on those of [175].

The technique by which a number of formulas were obtained in [175] is somewhat laborious. It uses the theory of Markov operator semigroups in combination with the method of undetermined coefficients. W. Wagner and E. Platen [308] gave a very simple proof (using only Ito's formula) of the expansion of the solution $X_{t,x}(t+h)$ of the system (0.1) in powers of h and in integrals depending on the increments $w_r(\theta) - w_r(t)$, $t \leq \theta \leq t+h$, $r = 1, \ldots, q$. This expansion generalizes (0.5). In the deterministic situation it comes down to Taylor's formula for $X_{t,x}(t+h)$ in powers of h in a neighborhood of the point (t, x). Theoretically, this expansion allows one to construct methods of arbitrary high order of accuracy (of course, with corresponding conditions on the coefficients of the system (0.1)).

We begin with the fundamental convergence theorem which establishes the mean-square order of convergence of a method resting on properties of its one-step approximation only [179, 180]. The conditions of this theorem use both properties of mean and mean-square deviation of one-step approximation. The theorem asserts that if p_1 and p_2 are orders of accuracy of the one-step approximation for the deviation of mathematical expectation and the mean-square deviation, respectively, and if $p_2 > 1/2$, $p_1 \geq p_2 + 1/2$, then the method converges and the order of the method is $p_2 - 1/2$. The proof of mean-square convergence of overwhelming majority of the methods considered in this book is based on this theorem. To derive a one-step approximation, which determines a method, many approaches can be used. In the deterministic theory, an expansion of the solution by Taylor's formula underlies all one-step methods, both implicit and explicit. The analogue of such an expansion for stochastic systems – Wagner–Platen expansion – is considered in Sect. 1.2. Here, on the basis of the same convergence theorem, we establish the order of a method in dependence of components in the expansion. Section 1.3 is devoted to implicit methods. Recall that in distinction to *explicit methods*, *implicit methods* have far better stability properties. In particular, they are necessary for integrating stiff systems of differential equations (see, e.g. [98, 99, 243]). Let us also note that in the case of general Hamiltonian systems symplectic Runge–Kutta methods are all implicit ([96, 97, 261]). We start Sect. 1.3 considering methods with general implicitness in drift terms (drift-implicit methods) and methods with implicitness of a special kind in stochastic terms. Drift-implicit methods were proposed in many papers. Stochastic implicit methods of the special kind (balanced implicit methods) were first considered in [189]. They are very far from being a sufficiently reach class of implicit methods. In particular, the balanced methods are linear with respect to the one-step approximation, and they cannot be used for constructing symplectic methods which we introduce for stochastic Hamiltonian systems (see Chap. 4). General stochastic implicit (fully implicit) methods [195] are also considered in Sect. 1.3. The increments of Wiener processes are substituted by some truncated random variables in these fully implicit schemes. In Sect. 1.4 we consider means for both exact and approximate modeling of Ito integrals depending on single or several noises. It turns out that the possibilities for exact modeling are very limited, especially in the case of several noises. Therefore, our main interest is in methods of approximate modelling. We develop such methods in great detail for those Ito integrals that are necessary for construction of mean-square approximations with first order of accuracy in the general case and with orders 3/2 and 2 for systems with one noise. Stochastic differential equations with additive and colored noise are met very often in applications and that is why they are of great independent interest. In Sects. 1.5 – 1.6, we construct efficient high-order methods because of some special properties of such systems.

Note that the wish to construct a large amount of methods, both explicit and implicit, is related to the fact that distinct methods have various properties as regards accuracy, stability, time and labor expenses, etc.

1.1 Fundamental theorem on the mean-square order of convergence

1.1.1 Statement of the theorem

Let $(\Omega, \mathcal{F}, P)$ be a probability space, $\mathcal{F}_t$, $t_0 \leq t \leq T$, be a nondecreasing family of σ-subalgebras of $\mathcal{F}$, and $(w_r(t), \mathcal{F}_t)$, $r = 1, \ldots, q$, be independent Wiener processes. Consider the system of SDEs in the sense of Ito

$$dX = a(t, X)dt + \sum_{r=1}^{q} \sigma_r(t, X)dw_r(t), \qquad (1.1)$$

where X, a, σ_r are vectors of dimension d.

Assume that the functions $a(t, x)$ and $\sigma_r(t, x)$ are defined and continuous for $t \in [t_0, T]$, $x \in \mathbf{R}^d$, and satisfy a *globally Lipschitz condition*: for all $t \in [t_0, T]$, $x \in \mathbf{R}^d$, $y \in \mathbf{R}^d$ there is an inequality

$$|a(t, x) - a(t, y)| + \sum_{r=1}^{q} |\sigma_r(t, x) - \sigma_r(t, y)| \leq K|x - y|. \qquad (1.2)$$

Here and below we denote by $|x|$ the Euclidean norm of the vector x and by $x^\mathsf{T} y$ or by (x, y) the scalar (inner) product of two vectors x and y.

Let $(X(t), \mathcal{F}_t)$, $t_0 \leq t \leq T$, be a solution of the system (1.1) with $E|X(t_0)|^2 < \infty$. The *one-step approximation* $\bar{X}_{t,x}(t+h)$, $t_0 \leq t < t+h \leq T$, depends on x, t, h, and $\{w_1(\theta) - w_1(t), \ldots w_q(\theta) - w_q(t),\ t \leq \theta \leq t+h\}$ and is defined as follows:

$$\bar{X}_{t,x}(t+h) = x + A(t, x, h; w_i(\theta) - w_i(t),\ i = 1, \ldots, q,\ t \leq \theta \leq t+h). \quad (1.3)$$

Using the one-step approximation, we recurrently construct the approximations $(\bar{X}_k, \mathcal{F}_{t_k})$, $k = 0, \ldots, N$, $t_{k+1} - t_k = h_{k+1}$, $T_N = T$:

$$\bar{X}_0 = X_0 = X(t_0),$$
$$\bar{X}_{k+1} = \bar{X}_{t_k, \bar{X}_k}(t_{k+1}) = \bar{X}_k$$
$$+ A(t_k, \bar{X}_k, h_{k+1}; w_i(\theta) - w_i(t_k),\ i = 1, \ldots, q,\ t_k \leq \theta \leq t_{k+1}). \quad (1.4)$$

We will use the following notation. An approximation of $X(t_k)$ will be denoted by $\bar{X}(t_k)$, $\bar{X}_k$, or simply by X_k. Everywhere below we put $\bar{X}_0 = X(t_0)$. Further, let X be an $\mathcal{F}_{t_k}$-measurable random variable with $E|X|^2 < \infty$; as usual, $X_{t_k,X}(t)$ denotes the solution of the system (1.1) for $t_k \leq t \leq T$ satisfying the following initial condition at $t = t_k$: $X(t_k) = X$. By $\bar{X}_{t_k,X}(t_i)$, $t_i \geq t_k$, we denote an approximation of the solution at step i such that $\bar{X}_k = X$. Clearly,

$$\bar{X}_{k+1} = \bar{X}_{t_k, \bar{X}_k}(t_{k+1}) = \bar{X}_{t_0, \bar{X}_0}(t_{k+1}).$$

For simplicity, we assume that $t_{k+1} - t_k = h = (T - t_0)/N$.

Theorem 1.1. *Suppose the one-step approximation $\bar{X}_{t,x}(t+h)$ has order of accuracy p_1 for the mathematical expectation of the deviation and order of accuracy p_2 for the mean-square deviation; more precisely, for arbitrary $t_0 \leq t \leq T - h$, $x \in \mathbf{R}^d$ the following inequalities hold:*

$$|E(X_{t,x}(t+h) - \bar{X}_{t,x}(t+h))| \leq K(1 + |x|^2)^{1/2} h^{p_1}, \quad (1.5)$$

$$\left[E|X_{t,x}(t+h) - \bar{X}_{t,x}(t+h)|^2\right]^{1/2} \leq K(1 + |x|^2)^{1/2} h^{p_2}. \quad (1.6)$$

Also let

$$p_2 \geq \frac{1}{2},\ p_1 \geq p_2 + \frac{1}{2}. \quad (1.7)$$

Then for any N and $k = 0, 1, \ldots, N$ the following inequality holds:

$$\left[E|X_{t_0,X_0}(t_k) - \bar{X}_{t_0,X_0}(t_k)|^2\right]^{1/2} \leq K(1 + E|X_0|^2)^{1/2} h^{p_2 - 1/2}, \quad (1.8)$$

i.e., the order of accuracy of the method constructed using the one-step approximation $\bar{X}_{t,x}(t+h)$ is $p = p_2 - 1/2$.

We note that all the constants K mentioned above, as well as the ones that will appear in the sequel, depend in the final analysis on the system (1.1) and the approximation (1.3) only and do not depend on X_0 and N.

Remark 1.2. Often the notion of *strong order of accuracy* is used (see, e.g., [131]): if for some method

$$E\,|X(t_k) - X_k| \leq Kh^p,$$

where K is a positive constant independent of k and h, then we say that the strong order of accuracy of the method is equal to p. Clearly, if the mean-square order of a method is p, then the method has the same strong order. We prefer to use the notions of accuracy and convergence in the mean-square sense.

1.1.2 Proof of the fundamental theorem

Lemma 1.3. *There is a representation*

$$X_{t,x}(t+h) - X_{t,y}(t+h) = x - y + Z \tag{1.9}$$

for which

$$E|X_{t,x}(t+h) - X_{t,y}(t+h)|^2 \leq |x-y|^2(1+Kh), \tag{1.10}$$

$$EZ^2 \leq K|x-y|^2 h. \tag{1.11}$$

Proof. Ito's formula readily implies that for $0 \leq \theta \leq h$:

$$E|X_{t,x}(t+\theta) - X_{t,y}(t+\theta)|^2 = |x-y|^2$$

$$+ 2E \int_t^{t+\theta} (X_{t,x}(s) - X_{t,y}(s))^\mathsf{T} (a(s, X_{t,x}(s)) - a(s, X_{t,y}(s))) ds$$

$$+ E \int_t^{t+\theta} \sum_{r=1}^q |\sigma_r(s, X_{t,x}(s)) - \sigma_r(s, X_{t,y}(s))|^2 ds.$$

It follows from here and (1.2) that

$$E|X_{t,x}(t+\theta) - X_{t,y}(t+\theta)|^2 \leq |x-y|^2 + K \int_t^{t+\theta} E|X_{t,x}(s) - X_{t,y}(s)|^2 ds. \tag{1.12}$$

In turn, this implies

$$E|X_{t,x}(t+\theta) - X_{t,y}(t+\theta)|^2 \leq |x-y|^2 \times e^{Kh}, \ 0 \leq \theta \leq h, \tag{1.13}$$

from which (1.10) follows. Further, noting that

$$Z = \int_t^{t+h} \sum_{r=1}^q (\sigma_r(s, X_{t,x}(s)) - \sigma_r(s, X_{t,y}(s))) dw_r(s)$$

$$+ \int_t^{t+h} (a(s, X_{t,x}(s)) - a(s, X_{t,y}(s))) ds,$$

and using (1.13), it is not difficult to obtain (1.11). □

Remark 1.4. In the sequel we will use a *conditional version of the inequalities* (1.5), (1.6), (1.10), (1.11). In this conditional version the deterministic variables x, y are replaced by $\mathcal{F}_t$-measurable random variables X, Y. For example, the conditional version of (1.5) reads:

$$|E(X_{t,X}(t+h) - \bar{X}_{t,X}(t+h)|\mathcal{F}_t)| \leq K(1+|X|^2)^{1/2} h^{p_1}. \tag{1.14}$$

We will also use simple consequences of these inequalities. For example, (1.14) implies

$$E|E(X_{t,X}(t+h) - \bar{X}_{t,X}(t+h)|\mathcal{F}_t)|^2 \leq K(1+E|X|^2) h^{2p_1}. \tag{1.15}$$

The proof of these conditional versions rests on an assertion of the following kind: if ζ is $\tilde{\mathcal{F}}$-measurable, $\tilde{\mathcal{F}} \subset \mathcal{F}$, $f(x,\omega)$ does not depend on $\tilde{\mathcal{F}}$, $\omega \in \Omega$, and $Ef(x,\omega) = \phi(x)$, then $E(f(\zeta,\omega)|\tilde{\mathcal{F}}) = \phi(\zeta)$ (see [82, 138]). In the case under consideration *the increments of the Wiener processes* $w_1(\theta) - w_1(t), \ldots, w_q(\theta) - w_q(t)$, $t \leq \theta \leq t + h$, do not depend on $\mathcal{F}_t$ (see, e.g. [82]), and neither does $\bar{X}_{t,x}(t+h)$, which is formed so that it depends on x, t, h, and the mentioned increments only.

Lemma 1.5. *For all natural N and all $k = 0, \ldots, N$ the following inequality holds:*

$$E|\bar{X}_k|^2 \leq K(1 + E|X_0|^2). \tag{1.16}$$

Proof. Suppose that $E|\bar{X}_k|^2 < \infty$. Then using the conditional version of (1.6), we obtain

$$E|X_{t_k,\bar{X}_k}(t_{k+1}) - \bar{X}_{t_k,\bar{X}_k}(t_{k+1})|^2 \leq K(1 + E|\bar{X}_k|^2) h^{2p_2}. \tag{1.17}$$

It is well known [82] that if an $\mathcal{F}_t$-measurable random variable X has bounded second moment, then the solution $X_{t,X}(t+\vartheta)$ also has bounded second moment. Therefore, $E|X_{t_k,\bar{X}_k}(t_{k+1})|^2 < \infty$. This and (1.17) readily imply that $E|\bar{X}_{k+1}|^2 < \infty$ (recall that $\bar{X}_{t_k,\bar{X}_k}(t_{k+1}) = \bar{X}_{k+1}$). Since $E|\bar{X}_0|^2 < \infty$, we have proved the existence of all $E|\bar{X}_k|^2 < \infty$, $k = 0, \ldots, N$.

Consider the equation

$$\begin{aligned} E|\bar{X}_{k+1}|^2 &= E|\bar{X}_k|^2 + E|X_{t_k,\bar{X}_k}(t_{k+1}) - \bar{X}_k|^2 \\ &\quad + E|X_{t_k,\bar{X}_k}(t_{k+1}) - \bar{X}_{t_k,\bar{X}_k}(t_{k+1})|^2 + 2E\bar{X}_k^\mathsf{T}(X_{t_k,\bar{X}_k}(t_{k+1}) - \bar{X}_k) \\ &\quad + 2E\bar{X}_k^\mathsf{T}(\bar{X}_{t_k,\bar{X}_k}(t_{k+1}) - X_{t_k,\bar{X}_k}(t_{k+1})) \\ &\quad + 2E(X_{t_k,\bar{X}_k}(t_{k+1}) - \bar{X}_k)^\mathsf{T}(\bar{X}_{t_k,\bar{X}_k}(t_{k+1}) - X_{t_k,\bar{X}_k}(t_{k+1})). \end{aligned} \tag{1.18}$$

We have (see [82])

$$E|X_{t_k,\bar{X}_k}(t_{k+1}) - \bar{X}_k|^2 \leq K(1 + E|\bar{X}_k|^2) h. \tag{1.19}$$

1.1 Fundamental theorem on the mean-square order of convergence

Further, we obtain from (1.17) and (1.19):
$$2|E(X_{t_k,\bar{X}_k}(t_{k+1}) - \bar{X}_k)^\mathsf{T}(\bar{X}_{t_k,\bar{X}_k}(t_{k+1}) - X_{t_k,\bar{X}_k}(t_{k+1}))|$$
$$\leq 2(E|X_{t_k,\bar{X}_k}(t_{k+1}) - \bar{X}_k|^2)^{1/2}(E|X_{t_k,\bar{X}_k}(t_{k+1}) - \bar{X}_{t_k,\bar{X}_k}(t_{k+1})|^2)^{1/2}$$
$$\leq K(1 + E|\bar{X}_k|^2)h^{p_2+1/2} . \qquad (1.20)$$

It is not difficult to prove the inequality
$$E|E(X_{t_k,\bar{X}_k}(t_{k+1}) - \bar{X}_k|\mathcal{F}_{t_k})|^2 \leq K(1 + E|\bar{X}_k|^2)h^2 . \qquad (1.21)$$

Therefore
$$2|E\bar{X}_k^\mathsf{T}(X_{t_k,\bar{X}_k}(t_{k+1}) - \bar{X}_k)| = 2|E\bar{X}_k^\mathsf{T} E(X_{t_k,\bar{X}_k}(t_{k+1}) - \bar{X}_k|\mathcal{F}_{t_k})|$$
$$\leq 2(E|\bar{X}_k|^2)^{1/2}(E|E(X_{t_k,\bar{X}_k}(t_{k+1}) - \bar{X}_k|\mathcal{F}_{t_k})|^2)^{1/2} \leq K(1 + E|\bar{X}_k|^2)h .$$
$$(1.22)$$

Similarly, but referring to (1.15) instead of (1.21), we obtain
$$2|E\bar{X}_k^\mathsf{T}(\bar{X}_{t_k,\bar{X}_k}(t_{k+1}) - X_{t_k,\bar{X}_k}(t_{k+1}))| \leq K(1 + E|\bar{X}_k|^2)h^{p_1} . \qquad (1.23)$$

Applying the inequalities (1.19), (1.17), (1.22), (1.23), and (1.20) to the equality (1.18) and recalling that $p_1 \geq 1$, $p_2 \geq 1/2$, we arrive at the inequality (taking, without loss of generality, $h \leq 1$):
$$E|\bar{X}_{k+1}|^2 \leq E|\bar{X}_k|^2 + K(1 + E|\bar{X}_k|^2)h = (1 + Kh)E|\bar{X}_k|^2 + Kh . \qquad (1.24)$$

Hence, using the well-known result which, for the sake of reference, is stated as Lemma 1.6 below, we obtain (1.16). □

Lemma 1.6. *Suppose that for arbitrary N and $k = 0, \ldots, N$ we have*
$$u_{k+1} \leq (1 + Ah)u_k + Bh^p , \qquad (1.25)$$
where $h = T/N$, $A \geq 0$, $B \geq 0$, $p \geq 1$, $u_k \geq 0$, $k = 0, \ldots, N$. Then
$$u_k \leq e^{AT}u_0 + \frac{B}{A}(e^{AT} - 1)h^{p-1} \qquad (1.26)$$
(where for $A = 0$ we put $(e^{AT} - 1)/A$ equal to zero).

Let us proceed to the proof of Theorem 1.1 itself. We have
$$X_{t_0,X_0}(t_{k+1}) - \bar{X}_{t_0,X_0}(t_{k+1}) = X_{t_k,X(t_k)}(t_{k+1}) - \bar{X}_{t_k,\bar{X}_k}(t_{k+1})$$
$$= (X_{t_k,X(t_k)}(t_{k+1}) - X_{t_k,\bar{X}_k}(t_{k+1})) + (X_{t_k,\bar{X}_k}(t_{k+1}) - \bar{X}_{t_k,\bar{X}_k}(t_{k+1})) .$$
$$(1.27)$$

The first difference in the right-hand side of (1.27) is the error of the solution arising due to the error in the initial data at time t_k, accumulated at the k-th step. The second difference is the one-step error at the $(k+1)$-step. Taking the square of both sides of the equation, we obtain

$$E|X_{t_0,X_0}(t_{k+1}) - \bar{X}_{t_0,X_0}(t_{k+1})|^2$$
$$= EE(|X_{t_k,X(t_k)}(t_{k+1}) - X_{t_k,\bar{X}_k}(t_{k+1})|^2|\mathcal{F}_{t_k})$$
$$+ EE(|X_{t_k,\bar{X}_k}(t_{k+1}) - \bar{X}_{t_k,\bar{X}_k}(t_{k+1})|^2|\mathcal{F}_{t_k})$$
$$+ 2EE((X_{t_k,X(t_k)}(t_{k+1}) - X_{t_k,\bar{X}_k}(t_{k+1}))^\mathsf{T}$$
$$\times (X_{t_k,\bar{X}_k}(t_{k+1}) - \bar{X}_{t_k,\bar{X}_k}(t_{k+1}))|\mathcal{F}_{t_k}) \,. \tag{1.28}$$

By the conditional version of Lemma 1.3, we get

$$EE(|X_{t_k,X(t_k)}(t_{k+1}) - X_{t_k,\bar{X}_k}(t_{k+1})|^2|\mathcal{F}_{t_k})$$
$$\leq E|X(t_k) - \bar{X}_k|^2 \times (1 + Kh) \,. \tag{1.29}$$

By the conditional version of (1.6) and Lemma 1.5, we have

$$EE(|X_{t_k,\bar{X}_k}(t_{k+1}) - \bar{X}_{t_k,\bar{X}_k}(t_{k+1})|^2|\mathcal{F}_{t_k})$$
$$\leq K(1 + E|\bar{X}_k|^2)h^{2p_2} \leq K(1 + E|X_0|^2)h^{2p_2} \,. \tag{1.30}$$

The difference $X_{t_k,X(t_k)}(t_{k+1}) - X_{t_k,\bar{X}_k}(t_{k+1})$ in the last summand in (1.28) can be treated using Lemma 1.3:

$$X_{t_k,X(t_k)}(t_{k+1}) - X_{t_k,\bar{X}_k}(t_{k+1}) := X(t_k) - \bar{X}_k + Z \,.$$

Then we obtain two terms each of which can be estimated individually. Using (1.15) and Lemma 1.5, we get

$$|EE((X(t_k) - \bar{X}_k)^\mathsf{T}(X_{t_k,\bar{X}_k}(t_{k+1}) - \bar{X}_{t_k,\bar{X}_k}(t_{k+1}))|\mathcal{F}_{t_k})|$$
$$= |E((X(t_k) - \bar{X}_k)^\mathsf{T} E(X_{t_k,\bar{X}_k}(t_{k+1}) - \bar{X}_{t_k,\bar{X}_k}(t_{k+1})|\mathcal{F}_{t_k}))|$$
$$\leq (E|X(t_k) - \bar{X}_k|^2)^{1/2} K(1 + E|X_0|^2)^{1/2} h^{p_1} \,. \tag{1.31}$$

Finally, using Lemma 1.5 and the inequality (1.6), we obtain

$$|E(Z^\mathsf{T}(X_{t_k,\bar{X}_k}(t_{k+1}) - \bar{X}_{t_k,\bar{X}_k}(t_{k+1})))|$$
$$\leq (EE(|Z|^2|\mathcal{F}_{t_k}))^{1/2}(EE(|X_{t_k,\bar{X}_k}(t_{k+1}) - \bar{X}_{t_k,\bar{X}_k}(t_{k+1})|^2|\mathcal{F}_{t_k}))^{1/2}$$
$$\leq K(E|X(t_k) - \bar{X}_k|^2)^{1/2}(1 + E|X_0|^2)^{1/2} h^{p_2+1/2} \,. \tag{1.32}$$

Introduce the notation $\varepsilon_k^2 := E|X(t_k) - \bar{X}_k|^2$. The relations (1.28) – (1.32) and the condition $p_1 \geq p_2 + 1/2$ then lead to the inequality (we take $h < 1$):

$$\varepsilon_{k+1}^2 \leq \varepsilon_k^2(1 + Kh) + K(1 + E|X_0|^2)^{1/2}\varepsilon_k h^{p_2+1/2} + K(1 + E|X_0|^2)h^{2p_2} \,.$$

Using the elementary relation

$$(1+E|X_0|^2)^{1/2}\varepsilon_k h^{p_2+1/2} \le \frac{\varepsilon_k^2 h}{2} + \frac{1+E|X_0|^2}{2}h^{2p_2},$$

we get

$$\varepsilon_{k+1}^2 \le \varepsilon_k^2(1+Kh) + K(1+E|X_0|^2)h^{2p_2}.$$

The inequality (1.8) follows from this, taking into account Lemma 1.6 and the fact that $\varepsilon_0^2 = 0$. This proves Theorem 1.1. □

In [180] a strengthening of the main convergence theorem (Theorem 1.1) is proved. Here we give this result without proof.

Theorem 1.7. *In addition to the conditions of Theorem 1.1, suppose that also*

$$\left[E|X_{t,x}(t+h) - \bar{X}_{t,x}(t+h)|^4\right]^{1/4} \le K(1+|x|^4)^{1/4}h^{p_2-1/4},$$

and let $p_2 \ge 3/4$. Then

$$\left[E \max_{0 \le k \le N} |X_{t_0,X_0}(t_k) - \bar{X}_{t_0,X_0}(t_k)|^2\right]^{1/2} \le K(1+E|X_0|^4)^{1/4}h^{p_2-1/2}.$$

1.1.3 The fundamental theorem for equations in the sense of Stratonovich

Consider the following *system in the sense of Stratonovich*:

$$dX = a(t,X)dt + \sum_{r=1}^{q} \sigma_r(t,X) \circ dw_r(t), \qquad (1.33)$$

where we use the sign ∘ in distinction to (1.1). It is well known (see, e.g., [110]) that this system is equivalent to the following system in the sense of Ito:

$$dX = \left(a(t,X) + \frac{1}{2}\sum_{r=1}^{q} \frac{\partial \sigma_r}{\partial x}(t,X)\sigma_r(t,X)\right)dt + \sum_{r=1}^{q} \sigma_r(t,X)dw_r(t). \quad (1.34)$$

In this system, $\partial \sigma_r/\partial x$ is the matrix with entry $\partial \sigma_r^i/\partial x^j$ at the intersection of the i-th row and j-th column. Here, of course, we assume the σ_r, $r = 1,\ldots,q$, to be not only differentiable but also require that the vectors $(\partial \sigma_r/\partial x)\sigma_r$ satisfy a global Lipschitz condition with respect to $x \in \mathbf{R}^d$, i.e., that system (1.34) satisfies the condition (1.2).

By the above it is not difficult to see that Theorem 1.1 remains true for solutions of equations understood in the sense of Stratonovich. Incidentally, we note that systems with additive noise (let us recall that one distinguishes between multiplicative and additive noise depending on whether the diffusion coefficients σ_r depend on state X or not) have evidently the same form for systems in both sense of Ito and Stratonovich. Another example of systems for which these forms coincide is given by the second-order differential equation with noise

$$\ddot{X} = a(t, X, \dot{X})dt + \sum_{r=1}^{q} \gamma_r(t, X) \circ \dot{w}_r(t), \quad (1.35)$$

which can be written as $2d$-dimensional system

$$dX = Y\,dt$$
$$dY = a(t, X, Y)dt + \sum_{r=1}^{q} \gamma_r(t, X) \circ dw_r(t). \quad (1.36)$$

Clearly, for any $2d$-dimensional column-vector σ_r corresponding to this system, d former components are equal to zero and the other d components compose γ_r. From here the $2d \times 2d$-matrix, which plays the role of $\partial \sigma_r/\partial x$, has nonzero entries at the intersection of last d rows and former d columns only, whence the correction term for (1.36) is equal to zero, i.e., the Ito form of (1.36) is the same.

1.1.4 Discussion

The rule:

"if, in a single step, the mean-square error has order h^{p_2} (i.e., inequality (1.6) holds), then it has order $h^{p_2-1/2}$ on the whole interval"

is not true without the additional condition $p_1 \geq p_2 + 1/2$. A simple example in which this can be seen is given by the method

$$\bar{X}_{k+1} = \bar{X}_k + \sigma(t_k, \bar{X}_k)\Delta_k w(h)$$

for the scalar version of system (1.1). It is easy to see that here $p_2 = 1$ while the method diverges for $a \neq 0$.

The rule:

"if, in a single step, the mean-square deviation has order h^{p_2}, then it has order h^{p_2-1} on the whole interval"

is true, but a bit rough. Following this rule, we cannot prove the convergence of the Euler method, in which $p_2 = 1$. Moreover, a more efficient rule for the mean-square deviation cannot be found if we are guided only by the mean-square characteristic of the one-step approximation. The rule following from Theorem 1.1 is based on properties of both the *mean* and the *mean-square deviation* of the one-step approximation. In particular, for Euler's method $p_1 = 2$, $p_2 = 1$ (see Sect. 1.1.5), and so it follows from Theorem 1.1 that Euler's method has order of accuracy $1/2$.

Properties of the mean are used at a single place in the proof of the theorem (and at a "delicate" place indeed), more precisely, when deriving the

1.1 Fundamental theorem on the mean-square order of convergence

inequality (1.31). If the left-hand side of this inequality is roughly estimated, without taking into account (1.5), using the Bunyakovsky–Schwarz inequality, we obtain

$$|EE((X(t_k) - \bar{X}_k)^\mathsf{T}(X_{t_k,\bar{X}_k}(t_{k+1}) - \bar{X}_{t_k,\bar{X}_k}(t_{k+1}))|\mathcal{F}_{t_k})|$$
$$\leq (E|X(t_k) - \bar{X}_k|^2)^{1/2}(E|X_{t_k,\bar{X}_k}(t_{k+1}) - \bar{X}_{t_k,\bar{X}_k}(t_{k+1})|^2)^{1/2}$$
$$\leq (E|X(t_k) - \bar{X}_k|^2)^{1/2}K(1 + E|\bar{X}_k|^2)^{1/2}h^{p_2}.$$

While at the right-hand side of (1.31) we had the factor h^{p_1}, here we only have h^{p_2}, which is not sufficient for concluding the theorem.

Let us consider two simple examples which warn against a noncritical use of "sufficiently natural" methods.

Example 1.8. Take a piecewise linear interpolation of the Wiener process, $w^h(t) = w(t_k) + \Delta_k w(h)(t - t_k)/h$, $t_k \leq t \leq t_{k+1}$, and consider, instead of the equation

$$dX = a(t, X)dt + \sigma(t, X)dw(t),$$

on each interval $t_k \leq t \leq t_{k+1}$ the equation

$$dX^h = a(t, X^h)dt + \sigma(t, X^h)dw^h(t)$$

and its solution $X_k = X^h(t_k)$ at the nodes, which is $\mathcal{F}_{t_k}$-measurable. For the equation

$$dX = aX dt + \sigma X dw, \ 0 \leq t \leq T, \tag{1.37}$$

we obtain the one-step approximation:

$$\bar{X}_{t,x}(t + h) = x \exp(ah + \sigma(w(t + h) - w(t))). \tag{1.38}$$

At the same time,

$$X_{t,x}(t + h) = x \exp((a - \frac{\sigma^2}{2})h + \sigma(w(t + h) - w(t))). \tag{1.39}$$

Since

$$Ee^{\sigma(w(t+h)-w(t))} = e^{(\sigma^2/2)h},$$

we have

$$E(X_{t,x}(t + h) - \bar{X}_{t,x}(t + h)) = (e^{ah} - e^{(a+\sigma^2/2)h})x = O(h),$$

i.e., $p_1 = 1$ and Theorem 1.1 cannot be used to ensure convergence of the approximations. Actually, there is no convergence, since

$$E X_{k+1} = x \exp(ah + \sigma(w(t + h) - w(t))) \times E X_k = e^{(a+\sigma^2/2)h} E X_k,$$
$$E \bar{X}(T) = E X_N = e^{(a+\sigma^2/2)T} E X_0,$$

while
$$EX(T) = e^{aT} EX_0.$$

Note that if, instead of (1.37), we consider an equation in the sense of Stratonovich,
$$dX = aX\,dt + \sigma X \circ dw, \tag{1.40}$$
then the approximation (1.38) coincides with the solution of (1.40) (see the relation between (1.33) and (1.34)), i.e., the approximation (1.38) for (1.40) has infinite order of accuracy.

Example 1.9. We consider the implicit method (trapezium method) for the equation (1.37):
$$X_{k+1} = X_k + a\frac{X_k + X_{k+1}}{2}h + \sigma\frac{X_k + X_{k+1}}{2}\Delta_k w(h). \tag{1.41}$$
So,
$$X_{k+1} = X_k + X_k\frac{ah + \sigma\Delta_k w(h)}{1 - ah/2 - \sigma\Delta_k w(h)/2}. \tag{1.42}$$
Since the mathematical expectation of the right-hand side does not exist, it is clear that the approximation (1.41) cannot converge in mean-square to the solution of (1.37).

Consider the following one-step approximation:
$$\bar{X}_{t,x}(t+h) = x + axh + \sigma x \Delta w(h) + \frac{\sigma^2}{2}x\Delta^2 w(h),$$
$$\Delta w(h) = w(t+h) - w(t), \tag{1.43}$$
which can be thought of as corresponding to (1.42). We have
$$E\bar{X}_{t,x}(t+h) = (1 + ah + \frac{\sigma^2}{2}h)x,$$
which implies that $p_1 = 1$. It is not difficult to prove that also in this case there is no convergence of the approximation (1.43) to the solution of the equation (1.37). At the same time, an explicit computation gives that for the approximation (1.43) to the solution of (1.40) we have $p_1 = 2$, $p_2 = 3/2$, i.e., the method based on (1.43) has first order of accuracy for the equation (1.40). The computation can be performed by expanding the solution of (1.40) $X_{t,x}(t+h) = x\exp(ah + \sigma\Delta w(h))$ in powers of h and $\Delta w(h)$ and retaining the powers up to h^2 and $(\Delta w(h))^4$ inclusive in all the relations.

1.1.5 The explicit Euler method

For the system (1.1) we consider the one-step approximation (1.3) of the form
$$\bar{X}_{t,x}(t+h) = x + a(t,x)h + \sum_{r=1}^{q}\sigma_r(t,x)\Delta_t w_r(h), \tag{1.44}$$
where $\Delta_t w_r(h) = w_r(t+h) - w_r(t)$.

1.1 Fundamental theorem on the mean-square order of convergence

By (1.4), this approximation generates the *Euler method*:

$$X_0 = X(t_0), \quad X_{k+1} = X_k + a_k h + \sum_{r=1}^{q} \sigma_{rk} \Delta_k w_r(h), \tag{1.45}$$

where a_k, σ_{rk} are the values of the coefficients a and σ_r at the point (t_k, X_k), and $\Delta_k w_r(h) = w_r(t_k + h) - w_r(t_k)$.

For (1.44) we can find p_1 and p_2 satisfying the estimates (1.5) and (1.6). To this end we termwise subtract (1.45) from the identity

$$X_{t,x}(t+h) = x + \int_t^{t+h} a(s, X_{t,x}(s))ds + \sum_{r=1}^{q} \sigma_r(s, X_{t,x}(s))dw_r(s). \tag{1.46}$$

We obtain

$$X_{t,x}(t+h) - \bar{X}_{t,x}(t+h) = \int_t^{t+h} (a(s, X_{t,x}(s)) - a(t, x))ds$$

$$+ \sum_{r=1}^{q} \int_t^{t+h} (\sigma_r(s, X_{t,x}(s)) - \sigma_r(t, x))dw_r(s). \tag{1.47}$$

Calculation of the expectation leads to

$$E(X_{t,x}(t+h) - \bar{X}_{t,x}(t+h)) = E \int_t^{t+h} (a(s, X_{t,x}(s)) - a(t, x))ds. \tag{1.48}$$

Taking the square of both sides of (1.47) and then evaluating mathematical expectation, we obtain

$$E|X_{t,x}(t+h) - \bar{X}_{t,x}(t+h)|^2$$

$$\leq 2E| \int_t^{t+h} (a(s, X_{t,x}(s)) - a(t, x))ds|^2$$

$$+ 2 \sum_{r=1}^{q} \int_t^{t+h} E|\sigma_r(s, X_{t,x}(s)) - \sigma_r(t, x)|^2 ds. \tag{1.49}$$

We assume that, in addition to (1.2), the functions $a(t, x)$ and $\sigma_r(t, x)$ have partial derivatives with respect to t that grow at most as linear functions of x as $|x| \to \infty$. Then a and σ_r satisfy an inequality of the form

$$|a(s, X_{t,x}(s)) - a(t, x)| \leq |a(s, X_{t,x}(s)) - a(s, x)| + |a(s, x) - a(t, x)|$$

$$\leq K|X_{t,x}(s) - x| + K(1 + |x|^2)^{1/2}(s - t). \tag{1.50}$$

By the Bunyakovsky–Schwarz inequality we have

$$|\int_t^{t+h} (a(s, X_{t,x}(s)) - a(t, x))ds|^2 \leq h \int_t^{t+h} |a(s, X_{t,x}(s)) - a(t, x)|^2 ds,$$

and due to (1.50) and (1.19) the first term of the right-hand side of (1.49) is bounded by $K(1+|x|^2)h^3$. Using the inequality (1.50) for σ_r and (1.19), we see that the second term of the right-hand side of (1.49) is bounded by $K(1+|x|^2)h^2$. Thus $p_2 = 1$. Note that this value of p_2 is determined by the second term at the right-hand side of (1.49).

To find p_1, we turn to (1.48). We assume in addition that the derivatives $\partial a^i/\partial x^j$ and $\partial^2 a^i/\partial x^j \partial x^k$, $i,j,k = 1,\ldots,d$, are uniformly bounded. We write

$$a(s, X_{t,x}(s)) - a(t, x)$$
$$= a(t, X_{t,x}(s)) - a(t, x) - (a(t, X_{t,x}(s)) - a(s, X_{t,x}(s)))$$
$$= \frac{\partial a}{\partial x}(t, x)(X_{t,x}(s) - x) + \rho_1 + \rho_2, \qquad (1.51)$$

where $\partial a/\partial x$ is the $d \times d$-matrix with entries $\partial a^i/\partial x^j$ at the intersection of the i-th row and j-th column, and

$$|\rho_1| \leq K|X_{t,x}(s) - x|^2, \quad |\rho_2| \leq K(1 + |X_{t,x}(s)|^2)^{1/2}(s-t). \qquad (1.52)$$

It follows from

$$E(X_{t,x}(s) - x) = E\int_t^s a(\theta, X_{t,x}(\theta))d\theta \qquad (1.53)$$

and (1.2) for $y = 0$ that for all $t_0 \leq \theta \leq T$

$$|a(\theta, X_{t,x}(\theta))| \leq |a(\theta, 0)| + |a(\theta, X_{t,x}(\theta)) - a(\theta, 0)| \leq K(1 + |X_{t,x}(\theta)|).$$

Then it is not difficult to obtain the estimate

$$\left|\int_t^{t+h} E(X_{t,x}(s) - x)ds\right| \leq K(1 + |x|^2)^{1/2}h^2.$$

This and (1.48), (1.51), (1.52) imply (1.5) with $p_1 = 2$.

By Theorem 1.1, under the given assumptions regarding the coefficients a and σ_r, *Euler's method is a method of order* $p = p_2 - 1/2 = 1/2$.

Now we consider Euler's method for the system with additive noise:

$$dX = a(t,X)dt + \sum_{r=1}^q \sigma_r(t)dw_r(t). \qquad (1.54)$$

It can be readily seen that in this case the second term at the right-hand side of (1.49) can be bounded by $K(1+|x|^2)h^3$. As a result, we find $p_2 = 3/2$. Since, as before, $p_1 = 2 \geq p_2 + 1/2$, Theorem 1.1 implies that *the order of Euler's method for systems with additive noise is* $p = p_2 - 1/2 = 1$.

Remark 1.10. The formula (1.45) approximates the solution $X(t)$ of (1.1) at the nodes t_k. Consider a piecewise linear Euler approximation on the entire interval $t_0 \leq t \leq T$:

$$\bar{X}(t) = X_k + a_k(t - t_k) + \sum_{r=1}^{q} \sigma_{rk} \Delta_k w_r(h) \frac{t - t_k}{h}, \quad t_k \leq t \leq t_{k+1}. \quad (1.55)$$

Note that now $\bar{X}(t)$ is not $\mathcal{F}_t$-measurable since $w_r(t_{k+1})$, $r = 1, \ldots, q$, participate in (1.55). Equation (1.55) immediately implies

$$E|\bar{X}(t) - X_k|^2 = O(h), \quad t_k \leq t \leq t_{k+1}.$$

Therefore, we have on the interval $t_0 \leq t \leq T$:

$$(E|\bar{X}(t) - X(t)|^2)^{1/2} = O(h^{1/2}),$$

i.e., a piecewise linear approximation has the same order of accuracy on the entire interval as at the nodes. It is necessary to emphasize that this fact only holds for methods whose order of accuracy does not exceed $1/2$. In fact, consider the scalar equation

$$dX = dw, \quad X(0) = 0.$$

Its Euler approximation

$$X_0 = 0, \quad X_{k+1} = X_k + \Delta_k w(h)$$

is exact at nodes, i.e., it has infinite order of accuracy at nodes. At the same time, a piecewise linear approximation, e.g. for $t = (t_k + t_{k+1})/2$, has overall error of order $1/2$:

$$E(X(\frac{t_k + t_{k+1}}{2}) - \bar{X}(\frac{t_k + t_{k+1}}{2}))^2$$
$$= E(w(\frac{t_k + t_{k+1}}{2}) - \frac{1}{2}(w(t_{k+1}) + w(t_k)))^2 = \frac{1}{4}h.$$

Remark 1.11. Consider the system of linear differential equations with additive noise

$$dX = A(t)X dt + \sum_{r=1}^{q} \sigma_r(t) dw_r(t). \quad (1.56)$$

In [190] it is constructed a method that uses simulation of discrete processes $\Delta_k w_r(h)$ in an optimal manner. Already for the scalar equation

$$dX = aX dt + \sigma dw(t)$$

with constant coefficients $a \neq 0$, $\sigma \neq 0$, this method is precisely of first order of accuracy. Thus, there is no numerical integration method for (1.56) that uses only information about $w_r(t)$, $r = 1, \ldots, q$, at discrete time moments t_k and would have order of accuracy exceeding $O(h)$ (see also [38]). We note that though the optimal method has the same order as the Euler method, it gives essentially more refined results in a number of cases (see [190] and also Sect. 4.4.3).

1.1.6 Nonglobally Lipschitz conditions

The conditions of Theorem 1.1 on the coefficients are rather restrictive. Nevertheless this theorem is very useful since it allows us to derive various methods of numerical integration. Most likely, the methods or some their modifications obtained on the basis of this theorem can be applied more widely. For practice, globally Lipschitz conditions (1.2) are most restrictive. The point is that the equations with nonglobally Lipschitz conditions and at the same time such that there exist unique extendable solutions for them make up a very broad and important class in applications. There are papers on the Euler method (see e.g. [92, 94, 102, 171]), where its convergence is studied under more relaxed conditions. Let us observe briefly (without proof) some results of [102, 171]. Consider the autonomous system of SDEs

$$dX = a(X)dt + \sum_{r=1}^{q} \sigma_r(X) dw_r(t). \qquad (1.57)$$

Assume the local Lipschitz condition: for each $R > 0$ there exists a constant C_R such that

$$|a(x) - a(y)| + \sum_{r=1}^{q} |\sigma_r(x) - \sigma_r(y)| \leq C_R |x - y| \qquad (1.58)$$

for any $x, y \in \mathbf{R}^d$, $|x| \leq R$, $|y| \leq R$. Besides, assume the boundedness of the p-th moments (for some $p > 2$) of both the exact solution $X(t)$ of (1.57) and its Euler approximation X_k: there is a positive constant A such that

$$E \sup_{0 \leq t \leq T} |X(t)|^p \leq A, \ E \sup_{1 \leq k \leq N} |X_k|^p \leq A. \qquad (1.59)$$

It is proved in [102] that under (1.58) and (1.59)

$$\lim_{h \to 0} E \sup_{1 \leq k \leq N} |X(t_k) - X_k|^2 = 0. \qquad (1.60)$$

In the same paper [102] a rate of convergence is established as well. The following equality is proved

$$E \sup_{1 \leq k \leq N} |X(t_k) - X_k|^2 = O(h) \qquad (1.61)$$

under some further assumptions. The most essential additional assumptions are a global Lipschitz condition for $\sigma_r(x)$ and a one-sided Lipschitz condition for $a(x)$: there exists a positive $C > 0$ such that

$$(x - y)^\mathsf{T}(a(x) - a(y)) \leq C|x - y|^2. \qquad (1.62)$$

The condition (1.62) occurs very often in applications (see [98, 274] and references therein). However, the second inequality in (1.59) is broken for

1.1 Fundamental theorem on the mean-square order of convergence

many interesting systems satisfying (1.62). In [171, 285] this is shown for the equation

$$dX = -X^3 dt + dw(t). \tag{1.63}$$

As a result, the explicit Euler method for (1.63) diverges. It can be heuristically explained in the following way. Let us start, for instance, at $X(0) = X_0 = 1/h$ (we can do so without loss of generality since we always reach a similar position with a positive probability). Then

$$X_1 = X_0 - X_0^3 h + \Delta_0 w(h) = \frac{1}{h} - \frac{1}{h^2} + \Delta_0 w(h)$$

with a large probability is a negative number which can be approximately considered as $\approx -1/h^2$. Then

$$X_2 = -\frac{1}{h^2} + \frac{1}{h^5} + \Delta_1 w(h) \approx \frac{1}{h^5}, \ X_3 \approx -\frac{1}{h^{14}},$$

and so on. In experiments one can observe that the explicit Euler method leads to computer overflows. It is intuitively clear that under fixed h the local error increases with increasing $|x|$. And one of the natural modifications of the usual Euler method consists in using a variable step-size. For example, let us consider the explicit Euler method for (1.63), which is based on the following one-step approximation

$$\bar{X}_{t,x}(t+h_x) = x - x^3 h_x + \Delta_t w(h_x), \tag{1.64}$$

where $\Delta_t w(h_x) = w(t+h_x) - w(t)$ and

$$h_x = \begin{cases} h, & |x| \leq 1/\sqrt{h}, \\ 1/x^2, & |x| > 1/\sqrt{h}. \end{cases}$$

The numerical experiments show that a typical trajectory of the explicit Euler method for (1.63) blows up at times t between 2.5 and 800 for the time step $h = 0.36$, $X_0 = 0$. A decrease of the time step improves the situation, e.g., for $h = 0.25$ a typical trajectory blows up at times from 33 to 8×10^4. At the same time, the approximation (1.64) with $h = 0.36$ remains stable on an arbitrary time interval.

Another possibility of constructing convergence modifications of Euler methods consists in using implicit schemes (see [102, 286] and Sect. 1.3). One can say that convergence of most of methods takes place under global Lipschitz conditions on the coefficients and their derivatives. If these conditions are violated, one needs special and comparatively complicated investigations. We pay attention that depending on considered class of systems many different demands are imposed on a method connected with such properties as rate of convergence, complexity, efficiency, behavior on a long time interval and so on. It is usual that a method which is good in a certain sense (i.e., it is good for a class of problems) can be very bad in other respects. Therefore, we need both derivation of various methods and additional investigation of properties of derived methods. When we derive a method, it is natural to use some convenient and at the same time sufficiently broad conditions which allow us to prove convergence of the method. The proof of convergence under less restrictive conditions and the investigation of its properties are further problems which have to be considered both theoretically and experimentally.

1.2 Methods based on Taylor-type expansion

1.2.1 Taylor expansion of solutions of ordinary differential equations

As is well known, the expansion just mentioned lies at the basis of all one-step methods, both explicit and implicit (e.g., Runge–Kutta type methods). Here we will give it in a form convenient for our subsequent construction of its stochastic counterpart.

Consider the system of ordinary differential equations

$$\frac{dX}{dt} = a(t, X). \tag{2.1}$$

The right-hand side of (2.1) is assumed to be such that all the subsequent constructions can be performed. It suffices for this that the function $a(t, x)$, $t_0 \le t \le T$, $x \in \mathbf{R}^d$, is sufficiently smooth and that $a(t, x)$ grows at most as a linear function of $|x|$ as $|x| \to \infty$.

Let $f(t, x)$ be a scalar or vector function (of course, sufficiently smooth). We have along a solution $X(t)$ of (2.1):

$$\begin{aligned}\frac{d}{dt} f(t, X(t)) &= \frac{\partial f}{\partial t}(t, X(t)) + \frac{\partial f}{\partial x}(t, X(t)) a(t, X(t)) \\ &= \frac{\partial f}{\partial t}(t, X(t)) + \sum_{i=1}^{d} \frac{\partial f}{\partial x^i}(t, X(t)) a^i(t, X(t)).\end{aligned} \tag{2.2}$$

Define by L the operator:

$$L = \frac{\partial}{\partial t} + a^\mathsf{T} \frac{\partial}{\partial x} = \frac{\partial}{\partial t} + (a, \frac{\partial}{\partial x}) = \frac{\partial}{\partial t} + \sum_{i=1}^{d} a^i \frac{\partial}{\partial x^i}.$$

Assuming $X(t) = x$, (2.2) implies

$$f(s, X(s)) = f(t, x) + \int_t^s L f(\theta, X(\theta)) d\theta. \tag{2.3}$$

Let $f(t, x) = x$. Then $L f(t, x) = a(t, x)$, $L^2 f(t, x) = L a(t, x)$, etc. Therefore, (2.3) for $s = t + h$ implies

$$X(t+h) = x + \int_t^{t+h} a(s, X(s))ds \qquad (2.4)$$

(we have given these arguments for deriving the obvious identity (2.4) for uniformity with the subsequent computations). Further, using (2.3) for $a(s, X(s))$, we obtain

$$X(t+h) = x + \int_t^{t+h} \left(a(t,x) + \int_t^s La(\theta, X(\theta))d\theta \right) ds$$

$$= x + a(t,x)h + \int_t^{t+h} (t+h-s)La(s, X(s))ds. \qquad (2.5)$$

Again we use (2.3), but now for $La(s, X(s))$. We find

$$X(t+h) = x + a(t,x)h + La(t,x)\frac{h^2}{2}$$

$$+ \int_t^{t+h} \frac{(t+h-s)^2}{2} L^2 a(s, X(s))ds. \qquad (2.6)$$

Continuing in this way, we obtain the well-known *Taylor expansion* in powers of h in a neighborhood of t for the solution of (2.1). This expansion lies at the basis of creating explicit methods of various orders of accuracy and reads:

$$X(t+h) = x + a(t,x)h + La(t,x)\frac{h^2}{2} + \cdots$$

$$+ L^{m-1}a(t,x)\frac{h^m}{m!} + \int_t^{t+h} \frac{(t+h-s)^m}{m!} L^m a(s, X(s))ds. \qquad (2.7)$$

By (2.7), the *one-step approximation*

$$\bar{X}_{t,x}(t+h) = x + a(t,x)h + La(t,x)\frac{h^2}{2} + \cdots + L^{m-1}a(t,x)\frac{h^m}{m!} \qquad (2.8)$$

has error of order $m+1$ at a single step, and the method based on (2.8) has the m-th global order of accuracy.

1.2.2 Wagner–Platen expansion of solutions of stochastic differential equations

Let $X_{t,x}(s) = X(s)$ be the solution of the system (1.1), and let $f(t,x)$ be a sufficiently smooth (scalar or vector) function. By Ito's formula, we have for $t_0 \leq t \leq \theta \leq T$:

$$f(\theta, X(\theta)) = f(t,x) + \sum_{r=1}^q \int_t^\theta \Lambda_r f(\theta_1, X(\theta_1))dw_r(\theta_1)$$

$$+ \int_t^\theta Lf(\theta_1, X(\theta_1))d\theta_1, \qquad (2.9)$$

where the operators Λ_r, $r = 1, \ldots, q$, and L are given by:

$$\Lambda_r = \sigma_r^\mathsf{T} \frac{\partial}{\partial x} = (\sigma_r, \frac{\partial}{\partial x}) = \sum_{i=1}^{d} \sigma_r^i \frac{\partial}{\partial x^i},$$

$$L = \frac{\partial}{\partial t} + a^\mathsf{T} \frac{\partial}{\partial x} + \frac{1}{2} \sum_{r=1}^{q} \sum_{i=1}^{d} \sum_{j=1}^{d} \sigma_r^i \sigma_r^j \frac{\partial^2}{\partial x^i \partial x^j}.$$

The formula (2.9) is an analogue of formula (2.3).

Apply (2.9) to the functions $\Lambda_r f$ and Lf and then insert the expressions obtained for $\Lambda_r f(\theta_1, X(\theta_1))$ and $Lf(\theta_1, X(\theta_1))$ into (2.9). We find

$$f(s, X(s)) = f + \sum_{r=1}^{q} \Lambda_r f \int_t^s dw_r(\theta) + Lf \int_t^s d\theta$$

$$+ \sum_{r=1}^{q} \int_t^s (\sum_{i=1}^{q} \int_t^\theta \Lambda_i \Lambda_r f(\theta_1, X(\theta_1)) dw_i(\theta_1)) dw_r(\theta)$$

$$+ \sum_{r=1}^{q} \int_t^s (\int_t^\theta L\Lambda_r f(\theta_1, X(\theta_1)) d\theta_1) dw_r(\theta)$$

$$+ \sum_{r=1}^{q} \int_t^s (\int_t^\theta \Lambda_r Lf(\theta_1, X(\theta_1)) dw_r(\theta_1)) d\theta$$

$$+ \int_t^s (\int_t^\theta L^2 f(\theta_1, X(\theta_1)) d\theta_1) d\theta, \qquad (2.10)$$

where f, $\Lambda_r f$, and Lf are computed at (t, x).

Continuing in this way, we obtain an expansion for $f(t+h, X(t+h))$. As proved in Sect. 1.2.1, in the deterministic situation this expansion is the Taylor expansion in powers of h with remainder of integral type. In the stochastic situation the role of powers is played by random variables of the form (they are independent of $\mathcal{F}_t$):

$$I_{i_1, \ldots, i_j}(h) = \int_t^{t+h} dw_{i_j}(\theta) \int_t^\theta dw_{i_{j-1}}(\theta_1) \int_t^{\theta_1} \cdots \int_t^{\theta_{j-2}} dw_{i_1}(\theta_{j-1}), \quad (2.11)$$

where $i_1, \ldots, i_j$ take values in the set $\{0, 1, \ldots, q\}$, and $dw_0(\theta_i)$ is understood to mean $d\theta_i$.

It is obvious that $EI_{i_1,\ldots,i_j}(h) = 0$ if at least one $i_k \neq 0$, $k = 1, \ldots, j$, while $EI_{i_1,\ldots,i_j}(h) = O(h^j)$ if all $i_k = 0$, $k = 1, \ldots, j$.

Let us evaluate $E(I_{i_1,\ldots,i_j})^2$.

Lemma 2.1. *We have*

$$(E(I_{i_1,\ldots,i_j})^2)^{1/2} = O(h^{\sum_{k=1}^{j}(2-i_k')/2}), \qquad (2.12)$$

1.2 Methods based on Taylor-type expansion

where

$$i'_k = \begin{cases} 0, & i_k = 0, \\ 1, & i_k \neq 0. \end{cases}$$

In other words, when computing the *order of smallness of the integral* (2.11) we should be guided by the following rule: $d\theta$ contributes one to the order of smallness and $dw_r(\theta)$, $r = 1, \ldots, q$, contributes one half.

Proof. Suppose $i_j \neq 0$. Then (we can put $t = 0$ in (2.11) without loss of generality)

$$E(I_{i_1,\ldots,i_j})^2 = \int_0^h E(I_{i_1,\ldots,i_{j-1}}(\theta))^2 d\theta. \qquad (2.13)$$

If $i_j = 0$, i.e. $dw_{i_j}(\theta) = d\theta$, then

$$E(I_{i_1,\ldots,i_j})^2 = E(\int_0^h I_{i_1,\ldots,i_{j-1}}(\theta)d\theta)^2 \leq h\int_0^h E(I_{i_1,\ldots,i_{j-1}}(\theta))^2 d\theta. \qquad (2.14)$$

Denote by $p(i_1, \ldots, i_j)$ the order of smallness of $E(I_{i_1,\ldots,i_j})^2$. Then the formulas (2.13) and (2.14) give the recurrence relation

$$p(i_1, \ldots, i_j) = p(i_1, \ldots, i_{j-1}) + (2 - i'_j),$$

which proves (2.12). □

To clarify the general rule for establishing expansions of the form (2.10) using the integrals (2.11), we give the following formula, which can be obtained by a series of direct substitutions (we take s equal to $t + h$):

$$f(t+h, X(t+h)) = f + \sum_{r=1}^q \Lambda_r f \int_t^{t+h} dw_r(\theta) + Lf \int_t^{t+h} d\theta$$
$$+ \sum_{r=1}^q \sum_{i=1}^q \Lambda_i \Lambda_r f \int_t^{t+h} dw_r(\theta) \int_t^\theta dw_i(\theta_1)$$
$$+ \sum_{r=1}^q \sum_{i=1}^q \sum_{s=1}^q \Lambda_s \Lambda_i \Lambda_r f \int_t^{t+h} dw_r(\theta) \int_t^\theta dw_i(\theta_1) \int_t^{\theta_1} dw_s(\theta_2)$$
$$+ \sum_{r=1}^q \Lambda_r Lf \int_t^{t+h} d\theta \int_t^\theta dw_r(\theta_1) + \sum_{r=1}^q L\Lambda_r f \int_t^{t+h} dw_r(\theta) \int_t^\theta d\theta_1$$
$$+ L^2 f \int_t^{t+h} d\theta \int_t^\theta d\theta_1 + \rho, \qquad (2.15)$$

where

$$\rho = \sum_{r=1}^q \sum_{i=1}^q \sum_{s=1}^q \sum_{j=1}^q \int_t^{t+h} (\int_t^\theta (\int_t^{\theta_1} (\int_t^{\theta_2} \Lambda_j \Lambda_s \Lambda_i \Lambda_r f(\theta_3, X(\theta_3))$$
$$\times dw_j(\theta_3))dw_s(\theta_2))dw_i(\theta_1))dw_r(\theta)$$
$$+ \sum_{r=1}^q \sum_{i=1}^q \int_t^{t+h} (\int_t^\theta (\int_t^{\theta_1} L\Lambda_i \Lambda_r f(\theta_2, X(\theta_2))d\theta_2)dw_i(\theta_1))dw_r(\theta)$$

$$+ \sum_{r=1}^{q} \sum_{i=1}^{q} \int_{t}^{t+h} (\int_{t}^{\theta} (\int_{t}^{\theta_1} \Lambda_i L \Lambda_r f(\theta_2, X(\theta_2)) dw_i(\theta_2)) d\theta_1) dw_r(\theta)$$

$$+ \sum_{r=1}^{q} \sum_{i=1}^{q} \int_{t}^{t+h} (\int_{t}^{\theta} (\int_{t}^{\theta_1} \Lambda_i \Lambda_r L f(\theta_2, X(\theta_2)) dw_i(\theta_2)) dw_r(\theta_1)) d\theta$$

$$+ \sum_{r=1}^{q} \sum_{i=1}^{q} \sum_{s=1}^{q} \int_{t}^{t+h} (\int_{t}^{\theta} (\int_{t}^{\theta_1} (\int_{t}^{\theta_2} L\Lambda_s \Lambda_i \Lambda_r f(\theta_3, X(\theta_3)) d\theta_3)$$
$$\times dw_s(\theta_2)) dw_i(\theta_1)) dw_r(\theta)$$

$$+ \sum_{r=1}^{q} \int_{t}^{t+h} (\int_{t}^{\theta} (\int_{t}^{\theta_1} L^2 \Lambda_r f(\theta_2, X(\theta_2)) d\theta_2) d\theta_1) dw_r(\theta)$$

$$+ \sum_{r=1}^{q} \int_{t}^{t+h} (\int_{t}^{\theta} (\int_{t}^{\theta_1} L \Lambda_r L f(\theta_2, X(\theta_2)) d\theta_2) dw_r(\theta_1)) d\theta$$

$$+ \sum_{r=1}^{q} \int_{t}^{t+h} (\int_{t}^{\theta} (\int_{t}^{\theta_1} \Lambda_r L^2 f(\theta_2, X(\theta_2)) dw_r(\theta_2)) d\theta_1) d\theta$$

$$+ \int_{t}^{t+h} (\int_{t}^{\theta} (\int_{t}^{\theta_1} L^3 f(\theta_2, X(\theta_2)) d\theta_2) d\theta_1) d\theta. \qquad (2.16)$$

The right-hand side of (2.15) consists of:

(a) a term of zero order of smallness, the term f;
(b) terms of order of smallness $1/2$, they make up the sum of all possible integrals of the form (2.11) of order $1/2$ with corresponding coefficients, each of these terms is $\Lambda_r f \int_t^{t+h} dw_r(\theta)$;
(c) terms of order of smallness one, they make up the sum of all possible integrals of the form (2.11) of order one with corresponding coefficients, here the terms are of two kinds: $Lf \int_t^{t+h} d\theta$ and $\Lambda_i \Lambda_r f \int_t^{t+h} dw_r(\theta) \int_t^{\theta} dw_i(\theta_1)$;
(d) all possible terms of order of smallness $3/2$;
(e) one term of order of smallness two, the term $L^2 f \int_t^{t+h} d\theta \int_t^{\theta} d\theta_1$;
(f) the remainder ρ.

It is easy to see that the coefficient at the integral $I_{i_1,\ldots,i_j}$, whose order of smallness is $\sum_{k=1}^{j}(2-i'_k)/2$ by Lemma 2.1, is equal to $\Lambda_{i_j} \cdots \Lambda_{i_1} f$, where Λ_0 is taken to mean L.

In (2.15) (discarding ρ) all terms of order of smallness up to $3/2$ have been included, as well as one term of order of smallness two. This term is characteristic in that the integral in it does not involve Wiener processes, and so its expectation is not equal to zero (of course, if $L^2 f \neq 0$). The term $L^2 f \int_t^{t+h} d\theta \int_t^{\theta} d\theta_1 = L^2 f(h^2/2)$ has been included in the main part of (2.15) for convenience of reference; the true reason for its inclusion will be revealed later.

Lemma 2.2. *Suppose*

$$|\Lambda_{i_j}\cdots\Lambda_{i_1}f(t,x)|\leq K(1+|x|^2)^{1/2}. \qquad (2.17)$$

Then the quantity

$$I_{i_1,\ldots,i_j}(f,h)=\int_t^{t+h}dw_{i_j}(\theta)\int_t^\theta dw_{i_{j-1}}(\theta_1)\int_t^{\theta_1}\cdots$$

$$\cdots\int_t^{\theta_{j-2}}\Lambda_{i_j}\cdots\Lambda_{i_1}f(\theta_{j-1},X(\theta_{j-1}))dw_{i_1}(\theta_{j-1}) \quad (2.18)$$

satisfies the inequality

$$E|I_{i_1,\ldots,i_j}(f,h)|^2\leq K(1+E|X(t)|^2)h^{\sum_{k=1}^j(2-i_k')}, \qquad (2.19)$$

i.e., in particular, its order of smallness is the same as that of $I_{i_1,\ldots,i_j}(h)$. Furthermore, if at least one index i_k, $k=1,\ldots,q$, is not equal to zero, then

$$EI_{i_1,\ldots,i_j}(f,h)=0,\ \sum_{k=1}^j i_k\neq 0. \qquad (2.20)$$

Proof. The proof of this lemma does, in essence, not differ from that of Lemma 2.1. We only have to estimate $E|\Lambda_{i_j}\cdots\Lambda_{i_1}f(\theta_{j-1},X(\theta_{j-1}))|^2$ after the last step. Using (2.17) we obtain

$$E|\Lambda_{i_j}\cdots\Lambda_{i_1}f(\theta_{j-1},X(\theta_{j-1}))|^2\leq K(1+E|X(\theta_{j-1})|^2),$$

whence follows (2.19) in view of the inequality $E|X(\theta_{j-1})|^2\leq K(1+E|X(t)|^2)$ for $t<\theta_{j-1}$. □

Corollary 2.3. *Lemma 2.2 implies that each term in the remainder ρ (see (2.15) and (2.16)) has order of smallness at most two. Moreover, the expectation of all the terms of order of smallness two and 5/2 from ρ vanishes by (2.20). So,*

$$|E\rho|=O(h^3).$$

This is true if all the integrands in ρ satisfy, e.g., (2.17).

1.2.3 Construction of explicit methods

Now we substitute x for $f(t,x)$ in (2.15) and (2.16). Note that in this case

$$\Lambda_r f=\sigma_r,\ Lf=a.$$

Therefore

$$X_{t,x}(t+h) = x + \sum_{r=1}^{q} \sigma_r \int_t^{t+h} dw_r(\theta) + ah$$

$$+ \sum_{r=1}^{q} \sum_{i=1}^{q} \Lambda_i \sigma_r \int_t^{t+h} (w_i(\theta) - w_i(t)) dw_r(\theta)$$

$$+ \sum_{r=1}^{q} L\sigma_r \int_t^{t+h} (\theta - t) dw_r(\theta)$$

$$+ \sum_{r=1}^{q} \Lambda_r a \int_t^{t+h} (w_r(\theta) - w_r(t)) d\theta$$

$$+ \sum_{r=1}^{q} \sum_{i=1}^{q} \sum_{s=1}^{q} \Lambda_s \Lambda_i \sigma_r \int_t^{t+h} (\int_t^{\theta} (w_s(\theta_1) - w_s(t)) dw_i(\theta_1)) dw_r(\theta)$$

$$+ La \frac{h^2}{2} + \rho. \qquad (2.21)$$

In this formula all the coefficients σ_r, a, $\Lambda_i \sigma_r$, $L\sigma_r$, $\Lambda_r a$, $\Lambda_s \Lambda_i \sigma_r$, and La are computed at the point (t, x), while the remainder ρ is equal to

$$\rho = \sum_{r=1}^{q} \sum_{i=1}^{q} \sum_{s=1}^{q} \sum_{j=1}^{q} \int_t^{t+h} (\int_t^{\theta} (\int_t^{\theta_1} (\int_t^{\theta_2} \Lambda_j \Lambda_s \Lambda_i \sigma_r(\theta_3, X(\theta_3))$$

$$\times dw_j(\theta_3)) dw_s(\theta_2)) dw_i(\theta_1)) dw_r(\theta)$$

$$+ \sum_{r=1}^{q} \sum_{i=1}^{q} \int_t^{t+h} (\int_t^{\theta} (\int_t^{\theta_1} L\Lambda_i \sigma_r(\theta_2, X(\theta_2)) d\theta_2) dw_i(\theta_1)) dw_r(\theta)$$

$$+ \sum_{r=1}^{q} \sum_{i=1}^{q} \int_t^{t+h} (\int_t^{\theta} (\int_t^{\theta_1} \Lambda_i L\sigma_r(\theta_2, X(\theta_2)) dw_i(\theta_2)) d\theta_1) dw_r(\theta)$$

$$+ \sum_{r=1}^{q} \sum_{i=1}^{q} \int_t^{t+h} (\int_t^{\theta} (\int_t^{\theta_1} \Lambda_i \Lambda_r a(\theta_2, X(\theta_2)) dw_i(\theta_2)) dw_r(\theta_1)) d\theta$$

$$+ \sum_{r=1}^{q} \sum_{i=1}^{q} \sum_{s=1}^{q} \int_t^{t+h} (\int_t^{\theta} (\int_t^{\theta_1} (\int_t^{\theta_2} L\Lambda_s \Lambda_i \sigma_r(\theta_3, X(\theta_3)) d\theta_3)$$

$$\times dw_s(\theta_2)) dw_i(\theta_1)) dw_r(\theta)$$

$$+ \sum_{r=1}^{q} \int_t^{t+h} (\int_t^{\theta} (\int_t^{\theta_1} L^2 \sigma_r(\theta_2, X(\theta_2)) d\theta_2) d\theta_1) dw_r(\theta)$$

$$+ \sum_{r=1}^{q} \int_t^{t+h} (\int_t^{\theta} (\int_t^{\theta_1} L\Lambda_r a(\theta_2, X(\theta_2)) d\theta_2) dw_r(\theta_1)) d\theta$$

$$+ \sum_{r=1}^{q} \int_t^{t+h} (\int_t^{\theta} (\int_t^{\theta_1} \Lambda_r La(\theta_2, X(\theta_2)) dw_r(\theta_2)) d\theta_1) d\theta$$

1.2 Methods based on Taylor-type expansion

$$+ \int_t^{t+h} (\int_t^\theta (\int_t^{\theta_1} L^2 a(\theta_2, X(\theta_2)) d\theta_2) d\theta_1) d\theta .$$

$$+ \sum_{r=1}^q \int_t^{t+h} (\int_t^\theta (\int_t^{\theta_1} \Lambda_r L a(\theta_2, X(\theta_2)) dw_r(\theta_2)) d\theta_1) d\theta$$

$$+ \int_t^{t+h} (\int_t^\theta (\int_t^{\theta_1} L^2 a(\theta_2, X(\theta_2)) d\theta_2) d\theta_1) d\theta . \quad (2.22)$$

In connection with the formulas (2.21), (2.22) we consider the following *one-step approximations*:

$$\bar{X}_{t,x}^{(1)}(t+h) = x + \sum_{r=1}^q \sigma_r (w_r(t+h) - w_r(t)), \quad (2.23)$$

$$\bar{X}_{t,x}^{(2)}(t+h) = \bar{X}_{t,x}^{(1)}(t+h) + ah, \quad (2.24)$$

$$\bar{X}_{t,x}^{(3)}(t+h) = \bar{X}_{t,x}^{(2)}(t+h) + \sum_{r=1}^q \sum_{i=1}^q \Lambda_i \sigma_r \int_t^{t+h} (w_i(\theta) - w_i(t)) dw_r(\theta), \quad (2.25)$$

$$\bar{X}_{t,x}^{(4)}(t+h) = \bar{X}_{t,x}^{(3)}(t+h) + \sum_{r=1}^q L\sigma_r \int_t^{t+h} (\theta - t) dw_r(\theta)$$

$$+ \sum_{r=1}^q \Lambda_r a \int_t^{t+h} (w_r(\theta) - w_r(t)) d\theta$$

$$+ \sum_{r=1}^q \sum_{i=1}^q \sum_{s=1}^q \Lambda_s \Lambda_i \sigma_r$$

$$\times \int_t^{t+h} (\int_t^\theta (w_s(\theta_1) - w_s(t)) dw_i(\theta_1)) dw_r(\theta), \quad (2.26)$$

$$\bar{X}_{t,x}^{(5)}(t+h) = \bar{X}_{t,x}^{(4)}(t+h) + La\frac{h^2}{2} . \quad (2.27)$$

To each of these approximations, we associate the error

$$\rho^{(i)} = X_{t,x}^{(i)}(t+h) - \bar{X}_{t,x}^{(i)}(t+h), \ i = 1, \ldots, 5 .$$

Using Lemmas 2.1 and 2.2, it can be readily shown that (under the condition (2.17) on the corresponding functions)

$$|E\rho^{(1)}| = O(h), \ E|\rho^{(1)}|^2 = O(h^2) ,$$

i.e., $p_1 = 1$, $p_2 = 1$. Therefore, in order to satisfy the conditions of Theorem 1.1 (more precisely, the condition $p_1 \geq p_2 + 1/2$), we should take $p_2 = 1/2$. As a result, Theorem 1.1 does not guarantee convergence and, as it is already noted earlier, the method (2.23) clearly does not converge.

We have for $\rho^{(2)}$:

$$|E\rho^{(2)}| = O(h^2), \ E|\rho^{(2)}|^2 = O(h^2),$$

i.e., $p_1 = 2$, $p_2 = 1$. Therefore, the second method (Euler's method) has order of convergence equal to $1/2$.

We have for $\rho^{(3)}$:

$$|E\rho^{(3)}| = O(h^2), \ E|\rho^{(3)}|^2 = O(h^3),$$

i.e., $p_1 = 2$, $p_2 = 3/2$. Therefore, the third method has order of convergence equal to one.

We have for $\rho^{(4)}$:

$$|E\rho^{(4)}| = O(h^2), \ E|\rho^{(4)}|^2 = O(h^4),$$

i.e., $p_1 = 2$, $p_2 = 2$. But to satisfy the conditions of Theorem 1.1, we have to put $p_2 = 3/2$. As a result, the fourth method is also of order one.

Finally, we have for $\rho^{(5)}$:

$$|E\rho^{(5)}| = O(h^3), \ E|\rho^{(5)}|^2 = O(h^4),$$

i.e., $p_1 = 3$, $p_2 = 2$. Therefore, the fifth method has order of convergence equal to $3/2$.

The results concerning (2.24), (2.25), and (2.27) are stated as the following theorem.

Theorem 2.4. *Suppose the conditions (2.17) on the corresponding functions hold. Then the mean-square order of accuracy of the methods based on the approximations (2.24), (2.25), and (2.27) are equal to $1/2$, 1, and $3/2$, respectively.*

These examples of methods (which are of independent significance) readily give support to and enable the formulation of a general result.

Suppose that an expansion of the type under consideration includes all the terms of order m. Then the remainder ρ includes terms of half-integer order $m+1/2$ and of integer order $m+1$. Since the mathematical expectation of any term of half-integer order vanishes, we have $|E\rho| = O(h^{m+1})$, i.e., $p_1 = m+1$. At the same time, $E|\rho|^2 = O(h^{2m+1})$, i.e., $p_2 = m + 1/2$. By Theorem 1.1, the order of accuracy of such a method is m.

Let an expansion contain only all terms of half-integer order $m+1/2$. Then among terms of order $m+1$ in the remainder there is, in general, one term having nonzero mathematical expectation; to be precise, it is $\int_t^{t+h} d\theta \int_t^\theta d\theta_1 \cdots \int_t^{\theta_{m-1}} L^m a(\theta_m, X(\theta_m)) d\theta_m$. So, $|E\rho| = O(h^{m+1})$, $(E|\rho|^2)^{1/2} = O(h^{m+1})$. Hence Theorem 1.1 can be applied with $p_2 = m + 1/2$ only and the order of accuracy of such a method is again m. So, if we add all terms of order $m + 1/2$ to all terms up to order m inclusive, then the order of accuracy of the method does not increase. However, if we add the single term of order $m+1$ mentioned above to all terms up to order $m+1/2$ inclusive, then the order of accuracy of the method increases by $1/2$. In fact, the mathematical expectation of all the remaining terms of order $m+1$ is zero, and so $|E\rho| = O(h^{m+2})$, i.e., $p_1 = m+2$, $p_2 = m+1$, and $p = m+1/2$. Thus, the following theorem holds (see also [236, 238, 308]).

Theorem 2.5. *Suppose that $\bar{X}_{t,x}(t+h)$ includes all terms of the form $\Lambda_{i_1} \cdots \Lambda_{i_j} f I_{i_1,\ldots,i_j}$, where $f \equiv x$, up to order m inclusive. Let all functions $\Lambda_{i_1} \cdots \Lambda_{i_j} f(t,x)$, where $f \equiv x$, $\sum_{k=1}^{j}(2-i'_k)/2 \leq m+1$, satisfy the inequality (2.17). Then the mean-square order of accuracy of the method based on this approximation is equal to m.*

Suppose that $\bar{X}_{t,x}(t+h)$ includes all terms of the form $\Lambda_{i_1} \cdots \Lambda_{i_j} f I_{i_1,\ldots,i_j}$, where $f \equiv x$, up to order $m+1/2$ inclusive, as well as the term

$$L^m a \int_t^{t+h} d\theta \int_t^{\theta} d\theta_1 \cdots \int_t^{\theta_{m-1}} d\theta_m = L^m a \frac{h^{m+1}}{(m+1)!}.$$

Suppose that all functions $\Lambda_{i_1} \cdots \Lambda_{i_j} f(t,x)$, where $f \equiv x$, $\sum_{k=1}^{j}(2-i'_k)/2 \leq m+2$, satisfy the inequality (2.17). Then the mean-square order of accuracy of the method based on this approximation is equal to $m+1/2$.

Remark 2.6. The sufficient conditions on the drift and diffusion coefficients in Theorems 2.4 and 2.5 consist in existence and boundedness of their derivatives up to a certain order. They are rather restrictive. Nevertheless these theorems are very useful since they allow us to derive various methods of numerical integration. Most likely the methods derived can be applied more widely. Of course, as in the case of the Euler method, the issues of convergence, study of different properties of methods (for example, their relation to stiffness, ergodicity and so on) require some additional investigations and developing of corresponding recommendations concerning their practical applications (see also Sect. 1.1.6).

Example 2.7. Consider the linear system of stochastic differential equations

$$dX = A(t)X dt + \sum_{r=1}^{q} B_r(t)X dw_r(t), \quad t_0 \leq t \leq T. \tag{2.28}$$

Here, $A(t)$ and $B_r(t)$ are $d \times d$-matrices with entries that are smooth on $[t_0, T]$, and $a(t,x) = A(t)x$, $\sigma_r(t,x) = B_r(t)x$. Therefore, the conditions of Theorem 2.5 hold if $A(t)$ and $B_r(t)$ are sufficiently smooth.

Let us take the method of first order of accuracy given by the approximation (2.25). Since

$$\Lambda_i \sigma_r(t,x) = (B_i(t)x, \frac{\partial}{\partial x})B_r(t)x = B_r(t)B_i(t)x,$$

this method has the form

$$X_{k+1} = X_k + \sum_{r=1}^{q} B_r(t_k)X_k\Delta_k w_r(h) + A(t_k)X_k h$$
$$+ \sum_{r=1}^{q}\sum_{i=1}^{q} B_r(t_k)B_i(t_k)X_k \int_{t_k}^{t_k+h} (w_i(\theta) - w_i(t_k))dw_r(\theta). \quad (2.29)$$

It is easy to verify the formula

$$\int_{t_k}^{t_k+h} (w_i(\theta) - w_i(t_k))dw_r(\theta) = \Delta_k w_i(h)\Delta_k w_r(h)$$
$$- \int_{t_k}^{t_k+h} (w_r(\theta) - w_r(t_k))dw_i(\theta).$$

If all $B_r(t)$ commute, then we have for $i \neq r$:

$$B_r B_i X \int_{t_k}^{t_k+h} (w_i(\theta) - w_i(t_k))dw_r(\theta)$$
$$= B_i B_r X \Delta_k w_i(h)\Delta_k w_r(h) - B_i B_r X \int_{t_k}^{t_k+h} (w_r(\theta) - w_r(t_k))dw_i(\theta).$$

Then the formula (2.29) takes the form

$$X_{k+1} = X_k + \sum_{r=1}^{q} B_r(t_k)X_k\Delta_k w_r(h) + A(t_k)X_k h$$
$$+ \frac{1}{2}\sum_{r=1}^{q} B_i^2(t_k)X_k(\Delta_k^2 w_i(h) - h)$$
$$+ \sum_{r=2}^{q}\sum_{i=1}^{r-1} B_i(t_k)B_r(t_k)X_k\Delta_k w_i(h)\Delta_k w_r(h). \quad (2.30)$$

Thus, in the *commutative situation* we can construct a method of first order of accuracy by modeling the increments of the Wiener processes only.

We also note that in the nonlinear case the approximation (2.25) can be simplified in a similar manner if

$$\Lambda_i \sigma_r(t,x) = \Lambda_r \sigma_i(t,x). \quad (2.31)$$

Example 2.8. Consider the following system which is a generalization of (1.36):

$$dX = a(t,X,Y)dt$$
$$dY = b(t,X,Y)dt + \sum_{r=1}^{q} \gamma_r(t,X)dw_r(t). \quad (2.32)$$

For this system, the first d components of σ_r are zeros, the last d components of σ_r are γ_r which do not depend on y, and, consequently, $\Lambda_i \sigma_r = 0$. The approximation (2.25) gives

$$\bar{X}_{t,x,y}(t+h) = x + a(t,x,y)h$$

$$\bar{Y}_{t,x,y}(t+h) = y + b(t,x,y)h + \sum_{r=1}^{q} \gamma_r(t,x) \left(w_r(t+h) - w_r(t) \right), \quad (2.33)$$

i.e., the approximation (2.25) coincides with the Euler method which is of order one in the case of the system (2.32).

1.3 Implicit mean-square methods

1.3.1 Construction of drift-implicit methods

To clarify the matter, we start with construction of implicit methods for ordinary differential equations. Rearranging (2.3), we can write the formula

$$f(s, X(s)) = f(t+h, X(t+h)) - \int_s^{t+h} Lf(\theta, X(\theta)) d\theta. \quad (3.1)$$

Using (3.1), we replace $a(s, X(s))$ in (2.4) and obtain

$$X(t+h) = x + a(t+h, X(t+h))h - \int_t^{t+h} (s-t) La(s, X(s)) ds. \quad (3.2)$$

Continuing in this way, we find

$$X(t+h) = x + a(t+h, X(t+h))h - La(t+h, X(t+h))\frac{h^2}{2} + \cdots$$

$$+ (-1)^{m-1} L^{m-1} a(t+h, X(t+h)) \frac{h^m}{m!}$$

$$+ \int_t^{t+h} (-1)^m \frac{(s-t)^m}{m!} L^m a(s, X(s)) ds. \quad (3.3)$$

If we discard the integral in this formula, then we obtain the implicit one-step approximation on which we can base a method of m-th order of accuracy.

Using a simple trick, we can obtain a whole class of implicit methods. We illustrate this trick by deriving a class of implicit methods of second order of accuracy.

Introducing a parameter α, we write (2.4) in the following form:

$$X(t+h) = x + \alpha \int_t^{t+h} a(s, X(s)) ds + (1-\alpha) \int_t^{t+h} a(s, X(s)) ds. \quad (3.4)$$

Now we replace $a(s, X(s))$ in the integral at α by (2.3) and in the integral at $(1-\alpha)$ by (3.1). We obtain

$$X(t+h) = x + \alpha a(t,x)h + (1-\alpha)a(t+h, X(t+h))h$$
$$+ \int_t^{t+h}(t + \alpha h - s)La(s, X(s))ds. \qquad (3.5)$$

Further, we introduce a parameter β and rewrite the integral in (3.5) as the sum of two integrals:

$$\int_t^{t+h}(t+\alpha h - s)La(s,X(s))ds = \beta \int_t^{t+h}(t+\alpha h - s)La(s,X(s))ds$$
$$+ (1-\beta)\int_t^{t+h}(t+\alpha h - s)La(s,X(s))ds.$$

Substituting $La(s, X(s)) = La(t, x) + O(h)$ in the first integral and $La(s, X(s)) = La(t+h, X(t+h)) + O(h)$ in the second integral, we obtain the following *implicit one-step approximation*

$$\bar{X}_{t,x}(t+h) = x + \alpha a(t,x)h + (1-\alpha)a(t+h, \bar{X}_{t,x}(t+h))h$$
$$+ \beta(2\alpha - 1)\, La(t,x)\frac{h^2}{2}$$
$$+ (1-\beta)(2\alpha - 1)La(t+h, \bar{X}_{t,x}(t+h))\frac{h^2}{2}. \qquad (3.6)$$

Using this approximation, we construct a two-parameter family of implicit methods of second order of accuracy:

$$X_{k+1} = X_k + \alpha a_k h + (1-\alpha)a_{k+1}h$$
$$+ \beta(2\alpha - 1)\,(La)_k\frac{h^2}{2} + (1-\beta)(2\alpha - 1)(La)_{k+1}\frac{h^2}{2}, \qquad (3.7)$$

where the functions with index k are computed at (t_k, X_k), while those with index $k+1$ are computed at (t_{k+1}, X_{k+1}).

Now we turn to construction of drift-implicit methods for SDEs. One of the simplest and popular *drift-implicit methods* has the form

$$X_{k+1} = X_k + \alpha a_k h + (1-\alpha)a_{k+1}h + \sum_{r=1}^q \sigma_{rk}\Delta_k w_r(h). \qquad (3.8)$$

This method (in fact, the one-parameter family of methods) is of mean-square order $1/2$ for general systems and of order one for systems with additive noise. For $\alpha = 1$ it coincides with the explicit Euler method.

To construct more accurate methods, let us consider, next to (2.9), the formula ($t \leq \theta_1 \leq \theta$)

1.3 Implicit mean-square methods

$$Lf(\theta_1, X(\theta_1)) = Lf(\theta, X(\theta)) - \sum_{r=1}^{q} \int_{\theta_1}^{\theta} \Lambda_r Lf(\theta_2, X(\theta_2)) dw_r(\theta_2)$$

$$- \int_{\theta_1}^{\theta} L^2 f(\theta_2, X(\theta_2)) d\theta_2. \quad (3.9)$$

As in the deterministic case, we substitute (3.9) in (2.9). Then putting $\theta = t + h$, we obtain

$$f(t+h, X(t+h)) = f(t,x) + \sum_{r=1}^{q} \int_{t}^{t+h} \Lambda_r f(\theta_1, X(\theta_1)) dw_r(\theta_1)$$

$$+ Lf(t+h, X(t+h))h$$

$$- \sum_{r=1}^{q} \int_{t}^{t+h} (\int_{\theta_1}^{t+h} \Lambda_r Lf(\theta_2, X(\theta_2)) dw_r(\theta_2)) d\theta_1$$

$$- \int_{t}^{t+h} (\int_{\theta_1}^{t+h} L^2 f(\theta_2, X(\theta_2)) d\theta_2) d\theta_1. \quad (3.10)$$

It would be rather unwise to use a formula of the form (3.9) in order to represent $\Lambda_r f(\theta_1, X(\theta_1))$ with its subsequent substitution in (3.10). Indeed, although the function $\Lambda_r f(\theta_1, X(\theta_1))$ does not depend itself on the future, all the terms in its representation of the form (3.9) do depend on the future and, e.g., the integral $\int_{t}^{t+h} (\int_{\theta_1}^{t+h} L\Lambda_r f(\theta_2, X(\theta_2)) d\theta_2) dw_r(\theta_1)$ does not make sense without some additional clarification. This complication can be overcome using the following trick. Consider, e.g., the formula (2.15) and represent the coefficient $\Lambda_r f$ in it as follows:

$$\Lambda_r f(t, x) = \Lambda_r f(t+h, X(t+h))$$

$$- \sum_{i=1}^{q} \int_{t}^{t+h} \Lambda_i \Lambda_r f(\theta, X(\theta)) dw_i(\theta) - \int_{t}^{t+h} L\Lambda_r f(\theta, X(\theta)) d\theta. \quad (3.11)$$

After substitution of (3.11) in (2.15), the right-hand side contains a dependency on $X(t+h)$ which is a factor not at h, as in (3.10), but at $\Delta w_r(h)$. The appearance of this implicitness may lead to a method which is not acceptable a priori. We clarify this considering, e.g. the following equation:

$$dX = aX dt + \sigma X dw(t).$$

We have
$$X_{t,x}(t+h) = x + \sigma x \Delta w(h) + axh + \rho,$$
where $E\rho = O(h^2)$, $E\rho^2 = O(h^2)$.

Further, as in (3.11), σx can be written as

$$\sigma x = \sigma X(t+h) - \int_t^{t+h} \sigma X(\theta) dw(\theta) - \int_t^{t+h} aX(\theta) d\theta.$$

As a result, we find

$$X(t+h) = x + \sigma X(t+h)\Delta w(h) + axh + \rho_1. \qquad (3.12)$$

Omitting ρ_1 (at this moment we are not concerned with justifying this), we obtain the method

$$X_{k+1} = X_k + \sigma X_{k+1}\Delta w(h) + aX_k h, \qquad (3.13)$$

which gives X_{k+1} with infinite second moment like the method (1.42) in Example 1.9. Generally, if the right-hand side of an equation contains terms with factors $\Delta w(h)$ (i.e., quantities taking arbitrary large values) and coefficients depending on X_{k+1}, then the solvability with respect to X_{k+1} of this relation is questionable. At the same time, if the expressions containing X_{k+1} involve a factor h with positive power, then for sufficiently small h the solvability is guaranteed under natural assumptions. Therefore, a certain caution is necessary when an implicitness is introduced by expressions occurring in stochastic integrals. However, we may hope that if an implicitness is introduced owing to expressions occurring in nonstochastic integrals then it is possible to get a suitable stability of the methods which is actually the reason for constructing implicit methods. Let us return to (2.15) again and write the coefficient Lf as

$$\begin{aligned} Lf &= Lf(t+h, X(t+h)) \\ &\quad - \sum_{r=1}^q \int_t^{t+h} \Lambda_r Lf(\theta, X(\theta)) dw_r(\theta) - \int_t^{t+h} L^2 f(\theta, X(\theta)) d\theta \\ &= Lf(t+h, X(t+h)) - \sum_{r=1}^q \Lambda_r Lf \int_t^{t+h} dw_r(\theta) \\ &\quad - L^2 f \int_t^{t+h} d\theta + \rho_1, \end{aligned} \qquad (3.14)$$

where, as can be readily shown,

$$|E\rho_1 h| \le K(1+|x|^2)^{1/2} h^3, \quad E\rho_1^2 h^2 \le K(1+|x|^2)^{1/2} h^4.$$

The other coefficients of (2.15) can be treated in the same manner (e.g., the above reasoning is immediately applicable to the coefficient $L^2 f$). Moreover, similar transformations can be performed repeatedly over individual terms containing, say, Lf, since a formula of the type (3.14) can be written for any smooth function. As a result, we can obtain a large amount of various representations for $f(t+h, X(t+h))$ using the integrals $I_{i_1,\ldots,i_j}$ or using products of such integrals with coefficients depending on the points (t,x) and $(t+h, X(t+h))$. As in the deterministic situation, the amount of such representations can be increased by considering splittings of, e.g., the term $Lf \int_t^{t+h} d\theta$ into the sum of two integrals with coefficients α and $(1-\alpha)$ such that the term $\alpha Lf \int_t^{t+h} d\theta$ remains unchanged while the term Lf in $(1-\alpha)Lf \int_t^{t+h} d\theta$ is replaced by (3.14). In Sect. 1.5 we give a number of concrete implicit methods obtained on the basis of such representations for systems with additive noise.

Here we have introduced implicitness in deterministic terms only. It is natural to call such methods as *drift-implicit* (or deterministically implicit) methods. Drift-implicit methods are well adapted, for instance, for stiff systems with additive noise (see Sect. 1.5). But when the stochastic part plays an essential role, the application of fully implicit (stochastically implicit) methods, which also involve implicit stochastic terms, is unavoidable. A good illustration of a situation when fully implicit methods should be applied is given in Sect. 1.3.2 below.

1.3.2 The balanced method

First we consider the following one-dimensional Ito equation with multiplicative noise:

$$dX = \sigma X dw, \ X(0) = x. \tag{3.15}$$

The solution of (3.15) decreases rapidly to zero for $|\sigma| \gg 1$ because its Lyapunov exponent $\lambda = -\sigma^2/2$ is negative. The one-dimensional equation (3.15) cannot be simply called stiff, but it can be an equation for one component in a stiff multidimensional problem. For large parameter $|\sigma|$ in (3.15), one observes that mean-square explicit methods work unreliable and have large errors for not too small time step sizes. They even lead to computer overflow. On the other hand, using very small time step sizes may require too much computational time. In stiff situation this is the crucial point where one has to look for other more suitable methods. For example, these difficulties occur in the estimation of the Lyapunov exponent where the long-time behavior of the numerical solution is decisive for the calculations.

Obviously, one cannot apply drift-implicit schemes to improve the stability of the numerical solution of the stochastic equation (3.15), which does not contain any drift component. Thus, we have to construct fully implicit methods which involve implicitness in the deterministic as well as in the stochastic terms.

The Euler method for (3.15) has the form

$$X_{k+1} = X_k + \sigma X_k \Delta_k w(h), \ X_0 = x.$$

There is no simple stochastic counterpart of the deterministic implicit Euler method, i.e., the method

$$X_{k+1} = X_k + \sigma X_{k+1}\Delta_k w(h)$$

fails because we have $E|(1-\sigma\Delta_k w)^{-1}| = \infty$. Nevertheless, a way to introduce implicitness in the numerical treatment for this special equation could be to look at a higher-order explicit mean-square method and try to introduce implicitness there. For this purpose, we start from the scheme

$$X_{k+1} = X_k + \sigma X_k \Delta_k w(h) + \frac{1}{2}\sigma^2 X_k(\Delta_k^2 w(h) - h), \quad X_0 = x,$$

which represents a numerical method of mean-square order one. Again, the introduction of implicitness in $\sigma X_k \Delta_k w(h)$ as above fails, but one can analyze the term $\sigma^2 X_k(\Delta_k^2 w(h) - h)/2$ and introduce a partial implicitness. This leads to the scheme

$$X_{k+1} = X_k + (\sigma \Delta_k w(h) + \frac{1}{2}\sigma^2 \Delta_k^2 w(h))X_k - \frac{1}{2}\sigma^2 X_{k+1} h. \qquad (3.16)$$

The numerical approximation described by the scheme (3.16) converges to the exact solution with order one. This statement can be verified by Theorem 1.1. We note that no random term in (3.16) is implicit.

We derive a method with random implicit terms using heuristic reasoning. Let us denote the result due to the Euler method by $X = x + \sigma x \Delta w$. Let $x > 0$ for definiteness. If $\Delta w < 0$, then X is very often much less than the value $X(h)$ of the exact solution. Therefore, X needs a correction which has to be positive. It is natural to take this correction proportionally to σ, $|\Delta w|$, and the difference $x - X(h)$, i.e., to take the balanced term of the form $\sigma(x - X(h))|\Delta w|$. We come to the same conclusion if $\Delta w > 0$. Taking X_k instead of x and X_{k+1} instead of $X(h)$, we propose the following scheme

$$X_{k+1} = X_k + \sigma X_k \Delta_k w(h) + \sigma(X_k - X_{k+1})|\Delta_k w(h)|. \qquad (3.17)$$

The method (3.17) belongs to the class of balanced methods. We will prove that the balanced method converges with the same order $1/2$ as the Euler method does.

Several numerical experiments for the linear equation (3.15), which has the explicit solution

$$X(t) = x \exp\{\sigma w(t) - \frac{\sigma^2}{2}t\}, \quad t \geq 0,$$

are performed in [189]. They demonstrate superiority of the method (3.17) over very long time in comparison with the Euler method and the method (3.16). The method (3.16) works slightly better than the Euler method but in spite of the fact that its order of convergence is higher than the order of (3.17), it works worse on long time intervals in comparison with (3.17). This can be clarified on a heuristic level as follows. The stochastic implicit term in (3.16) is of order one and in (3.17) is of order $1/2$, i.e., the level of implicitness in (3.16) is insufficient for an adequate behavior on long time intervals. Besides, we note that the scheme (3.17) preserves the positiveness property of solutions of the equation (3.15) in contrast to (3.16) and to the Euler method.

Now we consider the d-dimensional system of SDEs (see (1.1))

$$dX = a(t,X)dt + \sum_{r=1}^{q} \sigma_r(t,X)dw_r(t), \qquad (3.18)$$

under the conditions of Theorem 1.1. Let us introduce the family of balanced methods. A balanced method applied to (3.18) can be written in the general form:

$$X_{k+1} = X_k + a(t_k, X_k)h + \sum_{r=1}^{q} \sigma_r(t_k, X_k)\Delta_k w_r(h) + C_k(X_k - X_{k+1}), \quad (3.19)$$

where

$$C_k = c_0(t_k, X_k)h + \sum_{r=1}^{q} c_r(t_k, X_k)|\Delta_k w_r(h)|. \qquad (3.20)$$

Here $c_0, \ldots, c_r$ represent $d \times d$-matrix-valued functions. We assume that for any sequence of real numbers (α_i) with $\alpha_0 \in [0, \bar{\alpha}]$, $\alpha_1 \geq 0, \ldots, \alpha_r \geq 0$, where $\bar{\alpha} \geq h$ for all step sizes h considered and $(t, x) \in [0, \infty) \times \mathbf{R}^d$, the matrix

$$M(t,x) := I + \alpha_0 c_0(t,x) + \sum_{r=1}^{q} \alpha_r c_r(t,x)$$

has inverse and satisfies the condition

$$|(M(t,x))^{-1}| \leq K < \infty. \qquad (3.21)$$

Here I is the unit matrix. Obviously, (3.21) can easily be fulfilled by keeping $c_0, \ldots, c_r$ all positive semidefinite. Thus, under these conditions one directly obtains the one-step increment $X_{k+1} - X_k$ of the balanced method via the solution of a system of linear algebraic equations. Furthermore, we suppose that the components of the matrices $c_0, \ldots, c_r$ are uniformly bounded. The latter condition will be necessary to prove the convergence of the balanced method.

We remark that in the purely deterministic case the method (3.19)–(3.20) covers, for instance, the implicit Euler method with one or more Newton iteration steps.

Now we are able to state the corresponding convergence theorem for the general balanced method.

Theorem 3.1. *Under the above assumptions the balanced method* (3.19) – (3.20) *converges with mean-square order* 1/2.

Proof. First we show that the estimate (1.5) holds for the balanced method with $p_1 = 3/2$. For this purpose, let us introduce the local Euler approximation step

$$X^E = x + a(t,x)h + \sum_{r=1}^{q} \sigma_r(t,x)\Delta w_r(h)$$

and the local approximation of method (3.19)–(3.20)

$$\bar{X} = x + a(t,x)h + \sum_{r=1}^{q} \sigma_r(t,x)\Delta w_r(h) + C(t,x)(x - \bar{X}),$$

where

$$C(t,x) = c_0(t,X)h + \sum_{r=1}^{q} c_r(t,X)|\Delta w_r(h)|.$$

Since $p_1 = 2$ for the Euler method, we have

$$|E(X_{t,x}(t+h) - \bar{X}_{t,x}(t+h))| = |E(X_{t,x}(t+h) - X^E) + E(X^E - \bar{X})|$$
$$\leq K(1+|x|^2)^{1/2}h^2 + |E(X^E - \bar{X})|.$$

Further,

$$|E(X^E - \bar{X})| = |E(I - (I+C)^{-1})(a(t,x)h + \sum_{r=1}^{q} \sigma_r(t,x)\Delta w_r(h))|$$
$$= |E(I+C)^{-1}C(a(t,x)h + \sum_{r=1}^{q} \sigma_r(t,x)\Delta w_r(h))|.$$

Exploiting above the symmetry property of $\Delta w_r(h)$, $r = 1, \ldots, q$, in those expressions involving these zero-mean Gaussian variables, we get

$$|E(X^E - \bar{X})| = |E(I+C)^{-1}Ca(t,x)h|.$$

Now using (3.21) and boundedness of the components of the matrices $c_0, \ldots, c_r$, we obtain

$$|E(X^E - \bar{X})| \leq KE|Ca(t,x)h| \leq K(1+|x|^2)^{1/2}h^{3/2}.$$

Hence the assumption (1.5) of Theorem 1.1 is satisfied with $p_1 = 3/2$ for the balanced method.

Similarly, we check the assumption (1.6) by standard arguments:

$$\left[E|X_{t,x}(t+h) - \bar{X}_{t,x}(t+h)|^2\right]^{1/2} \leq (E|X_{t,x}(t+h) - X^E|^2)^{1/2}$$
$$+ (E|X^E - \bar{X}|^2)^{1/2} \leq K(1+|x|^2)^{1/2}h.$$

Thus we can choose the exponents p_2 and p_1 in Theorem 1.1 equal to 1 and 3/2, respectively, and apply Theorem 1.1 to finally prove the mean-square order $p_2 - 1/2 = 1/2$ of the balanced method, as is claimed in Theorem 3.1. □

Remark 3.2. Theorem 3.1 remains evidently true for a more general method. Namely, the matrix C_k in (3.20) can be taken in the form

$$C_k = c_0(t_k, X_k)h + \sum_{r=1}^{q} c_r(t_k, X_k)|\Delta_k w_r(h)| + c_{q+1}(t_k, X_k)h^{1/2}. \quad (3.22)$$

Some numerical experiments for two-dimensional systems are presented in [189]. They show that numerical methods, which also involve implicit random terms, can be successfully implemented. For the balanced methods, the type and degree of implicitness can be chosen by appropriate weights. An appropriate choice of these weights depends on the underlying dynamics and requires further investigations. This problem is closely connected with the problem of determining a suitable test equation for such methods.

1.3.3 Implicit methods for SDEs with locally Lipschitz vector fields

As in Sect. 1.1.6, we can note that the convergence conditions for methods considered are rather restricted. Again, the globally Lipschitz condition is the most restrictive. We saw that without this condition (for example, in the case of one-sided Lipschitz condition for a) the property of boundedness of moments for Euler approximations can be broken and the explicit Euler method, as a rule, diverges. Fortunately, the convergence properties for implicit methods are much better. Let us consider the autonomous system of SDEs (see (1.57))

$$dX = a(X)dt + \sum_{r=1}^{q} \sigma_r(X)dw_r(t). \quad (3.23)$$

We assume that σ_r satisfy a global Lipschitz condition and a satisfies a *one-sided Lipschitz condition* (see (1.62)):

$$(x-y)^\mathsf{T}(a(x) - a(y)) \leq C|x-y|^2. \quad (3.24)$$

Now we consider the *split-step Euler method* introduced in [171] and [102]:

$$X_k^* = X_k + a(X_k^*)h, \quad (3.25)$$

$$X_{k+1} = X_k^* + \sum_{r=1}^{q} \sigma_r(X_k^*)\Delta_k w_r(h). \quad (3.26)$$

The authors of [102] prove that under (3.24) and for a sufficiently small h the equation (3.25) is resolved with respect to X_k^* for any X_k. Thus, the method (3.25)-(3.26) is well-defined. The following theorem is valid (see [102]).

Theorem 3.3. *Consider the split-step backward Euler method* (3.25)–(3.26) *applied to the SDE* (3.23). *Let* $\sigma_r(x)$, $r = 1 \ldots, q$, *satisfy the global Lipschitz condition and* $a(x)$ *satisfy the one-sided Lipschitz condition* (3.24). *Then*

$$\lim_{h \to 0} E \sup_{1 \leq k \leq N} |X(t_k) - X_k|^2 = 0. \tag{3.27}$$

If additionally $a(x)$ *behaves polynomially, more precisely there exist constants* $K > 0$ *and* $m > 0$ *such that for all* $x, y \in \mathbf{R}^d$

$$|a(x) - a(y)| \leq K(1 + |x|^m + |y|^m)|x - y|^2, \tag{3.28}$$

then

$$E \sup_{1 \leq k \leq N} |X(t_k) - X_k|^2 = O(h). \tag{3.29}$$

Analogous results are obtained in [102] for the drift-implicit Euler method (see also [286]):

$$X_{k+1} = X_k + a(X_{k+1})h + \sum_{r=1}^{q} \sigma_r(X_k) \Delta_k w_r(h). \tag{3.30}$$

1.3.4 Fully implicit mean-square methods: The main idea

Construction of implicit methods for stochastic systems with additive noise does not cause any difficulties in principle. However, as we saw in the previous subsections, all is much more intricate in the case of stochastic systems with multiplicative noise. The balanced methods from Sect. 1.3.2 are of a very special form. In particular, this form does not allow us to construct symplectic methods for stochastic Hamiltonian systems with multiplicative noise (see Chap. 4). In this section we construct a sufficiently large class of fully implicit methods of mean-square order $1/2$ for general stochastic systems.

Let us start with an example. Consider the Ito scalar equation

$$dX = \sigma X dw(t). \tag{3.31}$$

The one-step approximation of the Euler method $\hat{X}$ for (3.31) is

$$\hat{X} = x + \sigma x \Delta w(h). \tag{3.32}$$

We can represent this approximation in the form

$$\hat{X} = x + \sigma \hat{X} \Delta w + \sigma(x - \hat{X}) \Delta w = x - \sigma^2 x (\Delta w)^2 + \sigma \hat{X} \Delta w.$$

As h is small, $(\Delta w)^2 \sim h$ and we obtain the following "natural" implicit method

$$\tilde{X} = x - \sigma^2 x h + \sigma \tilde{X} \Delta w(h). \tag{3.33}$$

1.3 Implicit mean-square methods

However, this method cannot be realized since $1 - \sigma \Delta w(h)$ can vanish for any small h. Further, for the formal value of $\tilde{X}$ from (3.33):

$$\tilde{X} = \frac{x(1 - \sigma^2 h)}{1 - \sigma \Delta w(h)},$$

we have $E|\tilde{X}| = \infty$ (see also Example 1.9). Clearly, the method (3.33) is not suitable. The reason for this is the unboundedness of the random variable $\Delta w(h)$ for any arbitrarily small h.

Our basic idea consists in replacement of $\Delta w(h) = \xi \sqrt{h}$, where ξ is an $\mathcal{N}(0,1)$-distributed random variable, by another random variable $\zeta \sqrt{h} = \zeta_h \sqrt{h}$ such that $\zeta \sqrt{h}$ is bounded and the Euler type method

$$\check{X} = x + \sigma x \zeta \sqrt{h} \qquad (3.34)$$

is of the mean-square order $1/2$ as well. To achieve this, it is sufficient to require:

$$E(\check{X} - \hat{X}) = O(h^{3/2}), \ E(\check{X} - \hat{X})^2 = O(h^2). \qquad (3.35)$$

We take a symmetric ζ. Then $E(\check{X} - \hat{X}) = 0$. To satisfy the second equation in (3.35), the condition $E(\zeta_h - \xi)^2 = O(h)$ is sufficient.

We shall require a stronger inequality

$$E(\zeta_h - \xi)^2 \leq h^k, \ k \geq 1. \qquad (3.36)$$

Let for $A_h > 0$

$$\zeta_h = \begin{cases} \xi, & |\xi| \leq A_h, \\ A_h, & \xi > A_h, \\ -A_h, & \xi < -A_h. \end{cases} \qquad (3.37)$$

Since

$$E(\zeta_h - \xi)^2 = \frac{2}{\sqrt{2\pi}} \int_{A_h}^{\infty} (x - A_h)^2 e^{-x^2/2} dx$$

$$= \frac{2}{\sqrt{2\pi}} e^{-A_h^2/2} \int_0^{\infty} y^2 e^{-y^2/2} e^{-A_h y} dy < e^{-A_h^2/2},$$

the inequality (3.36) is fulfilled if $e^{-A_h^2/2} \leq h^k$, i.e. $A_h^2 \geq 2k|\ln h|$. Thus, if

$$A_h = \sqrt{2k|\ln h|}, \ k \geq 1,$$

then the method based on the one-step approximation (3.34) has the mean-square order $1/2$.

Lemma 3.4. *Let $A_h = \sqrt{2k|\ln h|}$, $k \geq 1$, and ζ_h be defined by (3.37). Then the following inequality holds:*

$$0 \leq E(\xi^2 - \zeta_h^2) = 1 - E\zeta_h^2 \leq (1 + 2\sqrt{2k|\ln h|}) h^k. \qquad (3.38)$$

Proof. We have

$$1 - E\zeta_h^2 = \frac{2}{\sqrt{2\pi}} \int_{A_h}^{\infty} (x^2 - A_h^2) e^{-x^2/2} dx$$

$$= \frac{2}{\sqrt{2\pi}} \int_{A_h}^{\infty} \left[(x - A_h)^2 + 2A_h(x - A_h) \right] e^{-x^2/2} dx$$

$$\leq e^{-A_h^2/2} + \frac{4A_h}{\sqrt{2\pi}} \int_{A_h}^{\infty} x e^{-x^2/2} dx$$

$$= e^{-A_h^2/2} \left(1 + \frac{4A_h}{\sqrt{2\pi}} \right) \leq (1 + 2A_h) e^{-A_h^2/2},$$

whence (3.38) follows. □

Now consider the following implicit method (for definiteness we put $k = 1$ and $A_h = \sqrt{2|\ln h|}$):

$$\bar{X} = x - \sigma^2 x h + \sigma \bar{X} \zeta_h \sqrt{h},$$

$$\bar{X} = \frac{x(1 - \sigma^2 h)}{1 - \sigma \zeta_h \sqrt{h}}. \quad (3.39)$$

Since $|\zeta_h| \leq \sqrt{2|\ln h|}$, this method is realizable for all h satisfying the inequality

$$2h|\ln h| < \frac{1}{\sigma^2}. \quad (3.40)$$

Proposition 3.5. *The method (3.39) is of the mean-square order $1/2$.*

Proof. Let us compare the method (3.39) with the Euler method (3.32). We get

$$E\bar{X} = x(1 - \sigma^2 h) E \sum_{m=0}^{\infty} \sigma^m \zeta_h^m h^{m/2} = x(1 - \sigma^2 h) E \sum_{m=0}^{\infty} \sigma^{2m} \zeta_h^{2m} h^m.$$

It is obvious from here that the principal term in the expansion of $E(\bar{X} - \hat{X})$ is equal to $x\sigma^2 h(E\zeta_h^2 - 1)$. Due to Lemma 3.4, we obtain for all sufficiently small h:

$$|E(\bar{X} - \hat{X})| \leq C|x|\sigma^2(1 + 2\sqrt{2|\ln h|}) h^2, \quad (3.41)$$

where C is a positive constant.

Further

$$E(\bar{X} - \hat{X})^2 = E(-\sigma^2 x h + \sigma \bar{X} \zeta_h \sqrt{h} - \sigma x \xi \sqrt{h})^2$$

$$\leq 2\sigma^4 x^2 h^2 + 2E(\sigma \bar{X} \zeta_h \sqrt{h} - \sigma x \xi \sqrt{h})^2$$

$$= 2\sigma^4 x^2 h^2 + 2E(\sigma \cdot (x - \sigma^2 x h + \sigma \bar{X} \zeta_h \sqrt{h}) \zeta_h \sqrt{h} - \sigma x \xi \sqrt{h})^2$$

$$\leq 2\sigma^4 x^2 h^2 + 2\sigma^2 x^2 h E(\zeta_h - \xi)^2 + C_1 x^2 h^2 \leq C_2 x^2 h^2 \quad (3.42)$$

for all sufficiently small h and some positive constants C_1 and C_2. The inequalities (3.41) and (3.42) imply the mean-square convergence of implicit method (3.39) with order $1/2$. $\square$

Introduction of implicitness in the stochastic term leads to appearance of the compensating term $-\sigma^2 xh$ in (3.39). This can be explained in the following way. Since $\bar{X}$ must be close to $x + \sigma x \zeta_h \sqrt{h}$, the expression $x + \sigma \bar{X} \zeta_h \sqrt{h}$ is close to $x + \sigma x \zeta_h \sqrt{h} + \sigma^2 x \zeta_h^2 h$. Consequently, making use of the compensating term results in $x + \sigma \bar{X} \zeta_h \sqrt{h} - \sigma^2 xh = x + \sigma x \zeta_h \sqrt{h} + \sigma^2 x (\zeta_h^2 - 1)h \approx x + \sigma x \zeta_h \sqrt{h}$, i.e., we get the correct result.

Now let us consider the expression $\sigma((1-\beta)x + \beta \bar{X})\zeta_h \sqrt{h}$ which introduces implicitness in the stochastic term with the parameter $0 \le \beta \le 1$. Clearly, the compensating term in this case is equal to $-\sigma^2 \beta xh$. Thus, we derive the method:

$$\bar{X} = x - \sigma^2 \beta xh + \sigma((1-\beta)x + \beta \bar{X})\zeta_h \sqrt{h},\ 0 \le \beta \le 1. \qquad (3.43)$$

The following proposition can be proved analogously to Proposition 3.5.

Proposition 3.6. *The method* (3.43) *as well as the methods*

$$\bar{X} = x - \sigma^2 \beta x \zeta_h^2 h + \sigma((1-\beta)x + \beta \bar{X})\zeta_h \sqrt{h},\ 0 \le \beta \le 1, \qquad (3.44)$$

$$\bar{X} = x - \sigma^2 \beta((1-\alpha)x + \alpha \bar{X})h + \sigma((1-\beta)x + \beta \bar{X})\zeta_h \sqrt{h},\ 0 \le \alpha, \beta \le 1, \quad (3.45)$$

are of mean-square order $1/2$.

1.3.5 Convergence theorem for fully implicit methods

Now we are in position to introduce fully implicit methods for general systems of stochastic differential equations. For simplicity in writing, we deal here with the scalar Ito SDE:

$$dX = a(t, X)dt + \sigma(t, X)dw(t). \qquad (3.46)$$

We suppose that $a(t, x)$, $\sigma(t, x)$, $\dfrac{\partial \sigma}{\partial x}(t, x)$ are continuous for $t_0 \le t \le T$, $x \in \mathbf{R}$, and there exists a positive constant L such that

$$|a(t, y) - a(t, x)| \le L|y - x|,\ \left|\frac{\partial \sigma}{\partial x}(t, x)\right| \le L,\ t_0 \le t \le T,\ x, y \in \mathbf{R}. \qquad (3.47)$$

Recall that the same letter L (or K, or C) is used for various constants.

Consider the implicit one-step approximation (cf. (3.39))

$$\bar{X} = x + a(t, \bar{X})h - \sigma(t, x)\frac{\partial \sigma}{\partial x}(t, x)h + \sigma(t, \bar{X})\zeta_h \sqrt{h}, \qquad (3.48)$$

where ζ_h is defined by (3.37) with $A_h = \sqrt{2|\ln h|}$ for definiteness.

Lemma 3.7. *There exist constants $K > 0$ and $h_0 > 0$ such that for any $h \leq h_0$, $t_0 \leq t \leq T$, $x \in \mathbf{R}$ the equation (3.48) has a unique solution $\bar{X}$ which satisfies the inequality*

$$|\bar{X} - x| \leq K(1 + |x|)(|\zeta_h|\sqrt{h} + h). \tag{3.49}$$

The solution $\bar{X}$ of equation (3.48) can be found by the method of simple iteration with x as the initial approximation.

Proof. For any fixed t, x, and h, let us introduce the function

$$\varphi(z) = x + a(t,z)h - \sigma(t,x)\frac{\partial \sigma}{\partial x}(t,x)h + \sigma(t,z)\zeta_h\sqrt{h}.$$

Then (3.48) can be written as

$$\bar{X} = \varphi(\bar{X}).$$

There is a positive constant C such that for any $z \in \mathbf{R}$

$$|\varphi(z) - x| \leq |a(t,x)|h + |a(t,z) - a(t,x)|h + |\sigma(t,x)||\zeta_h|\sqrt{h}$$
$$+ |\sigma(t,z) - \sigma(t,x)||\zeta_h|\sqrt{h} + |\sigma(t,x)\frac{\partial \sigma}{\partial x}(t,x)|h$$
$$\leq C(1+|x|)(|\zeta_h|\sqrt{h} + h) + L|z-x|(|\zeta_h|\sqrt{h} + h).$$

Further, for any $z_1, z_2 \in \mathbf{R}$

$$|\varphi(z_2) - \varphi(z_1)| \leq L|z_2 - z_1|\,(|\zeta_h|\sqrt{h} + h).$$

Clearly, there exist positive constants K and h_0 such that for any $h \leq h_0$, $x \in \mathbf{R}$

$$L(|\zeta_h|\sqrt{h} + h) < 1,$$

and

$$|\varphi(z) - x| \leq K(1+|x|)(|\zeta_h|\sqrt{h} + h)$$

if

$$|z - x| \leq K(1+|x|)(|\zeta_h|\sqrt{h} + h).$$

Let us note that the constants K in the last two inequalities are the same. Now the lemma follows from the contraction mapping principle. □

In addition to (3.47) we suppose that there exist continuous $\partial a/\partial t$, $\partial \sigma/\partial t$, and $\partial^2 \sigma/\partial x^2$ and the inequalities

$$\left|\frac{\partial a}{\partial t}(t,x)\right| \leq L(1+|x|), \quad \left|\frac{\partial \sigma}{\partial t}(t,x)\right| \leq L(1+|x|), \quad t_0 \leq t \leq T, \ x \in \mathbf{R}, \tag{3.50}$$

hold.

Theorem 3.8. *Assume* (3.47) *and* (3.50). *Let there exist* $\delta > 0$ *such that if* $|y - x| \leq \delta(1 + |x|)$, *the inequality*

$$|\sigma(t,x)\frac{\partial^2 \sigma}{\partial x^2}(t,y)| \leq L, \ t_0 \leq t \leq T, \quad (3.51)$$

holds.

Then the implicit method based on the one-step approximation (3.48) *converges in mean-square with the order* $1/2$.

We omit the proof of this theorem, see it in [195].

Remark 3.9. The condition (3.51) is satisfied if, for instance,

$$|\sigma(t,x)| \leq L, \ |\frac{\partial^2 \sigma}{\partial x^2}(t,x)| \leq L, \ t_0 \leq t \leq T, \ x \in \mathbf{R}, \quad (3.52)$$

or

$$|\frac{\partial^2 \sigma}{\partial x^2}(t,x)| \leq \frac{L}{1+|x|}, \ t_0 \leq t \leq T, \ x \in \mathbf{R}, \quad (3.53)$$

holds.

Let us underline that the conditions of Theorem 3.8 are not necessary and the method is applicable more widely.

Remark 3.10. Let the function $c(t,x) := \sigma(t,x)\frac{\partial \sigma}{\partial x}(t,x)$ satisfy the condition

$$|c(t,y) - c(t,x)| \leq L|y-x|. \quad (3.54)$$

Consider the implicit one-step approximation

$$\bar{X} = x + a(t, \bar{X})h - \sigma(t, \bar{X})\frac{\partial \sigma}{\partial x}(t, \bar{X})h + \sigma(t, \bar{X})\zeta_h\sqrt{h}. \quad (3.55)$$

It is not difficult to prove that Theorem 3.8 is valid for the implicit method based on (3.55) provided (3.54) is fulfilled.

1.3.6 General construction of fully implicit methods

Let

$$dX^i = a^i(t, X)dt + \sum_{r=1}^{m} \sigma^i_r(t, X)dw_r(t), \ i = 1, \ldots, d. \quad (3.56)$$

Introduce the one-step approximation:

$$\bar{X}^i = x^i$$
$$+ \sum_{k=1}^{l} \lambda^i_k a^i(t + \nu^i_k h, (1-\alpha^i_{k1})x^1 + \alpha^i_{k1}\bar{X}^1, \ldots, (1-\alpha^i_{kd})x^d + \alpha^i_{kd}\bar{X}^d)h$$
$$+ \sum_{r=1}^{m}\sum_{k=1}^{l} \mu^i_{rk}\sigma^i_r(t + \nu^i_{rk}h, (1-\beta^i_{rk1})x^1 + \beta^i_{rk1}\bar{X}^1,$$
$$\ldots, (1-\beta^i_{rkd})x^d + \beta^i_{rkd}\bar{X}^d)\zeta_{rh}\sqrt{h} + A^i, \quad (3.57)$$

where $0 \le \nu, \alpha, \beta \le 1$, $\lambda, \mu \ge 0$, $\sum_{k=1}^{l} \lambda_k^i = 1$, $\sum_{k=1}^{l} \mu_{rk}^i = 1$, $i = 1, \ldots, d$, l is a positive integer, and A^i are some expressions to be found. Substituting the Euler-type approximation

$$\hat{X}^j = x^j + a^j(t,x)h + \sum_{s=1}^{m} \sigma_s^j(t,x)\zeta_{sh}\sqrt{h}$$

instead of $\bar{X}^j$, $j = 1, \ldots, d$, in σ_r^i, we obtain

$$\sigma_r^i(t + \nu_{rk}^i h, (1-\beta_{rk1}^i)x^1 + \beta_{rk1}^i \bar{X}^1, \ldots, (1-\beta_{rkd}^i)x^d + \beta_{rkd}^i \bar{X}^d)$$

$$\approx \sigma_r^i(t,x) + \sum_{j=1}^{d} \frac{\partial \sigma_r^i}{\partial x^j}(t,x)\beta_{rkj}^i \sum_{s=1}^{m} \sigma_s^j(t,x)\zeta_{sh}\sqrt{h}.$$

It is clear from here that either

$$A^i = -\sum_{r=1}^{m}\sum_{k=1}^{l} \mu_{rk}^i \sum_{j=1}^{d} \frac{\partial \sigma_r^i}{\partial x^j}(t,x)\beta_{rkj}^i \sum_{s=1}^{m} \sigma_s^j(t,x)\zeta_{sh}\sqrt{h}\zeta_{rh}\sqrt{h} \qquad (3.58)$$

or

$$A^i = -\sum_{r=1}^{m}\sum_{k=1}^{l} \mu_{rk}^i \sum_{j=1}^{d} \frac{\partial \sigma_r^i}{\partial x^j}(t,x)\beta_{rkj}^i \sigma_r^j(t,x)h \qquad (3.59)$$

can be put in (3.57).

Substituting one of these expressions in (3.57), we obtain a multi-parameter family of implicit methods. Here we will not precisely indicate assumptions on the coefficients a and σ_r assuming that appropriate conditions on the coefficients hold.

Theorem 3.11. *Under appropriate conditions of smoothness and boundedness on the coefficients of (3.56) the method based on the one-step approximation (3.57) with A^i as in (3.58) or (3.59) is of mean-square order $1/2$.*

We omit the proof here (see details in [195]).

Remark 3.12. It is also possible to introduce implicitness in A^i by changing t, x as it was done in the terms connecting with a^i. Moreover, the family can be extended if some a^i or σ_r^i are represented as sums of terms. In this case the coefficients λ, ν, α, μ, β can differ for different terms.

Let us give an example of fully implicit methods:

$$\bar{X} = x + a(t,\bar{X})h - \sum_{r=1}^{m}\sum_{j=1}^{d} \frac{\partial \sigma_r}{\partial x^j}(t,\bar{X})\sigma_r^j(t,\bar{X})h + \sum_{r=1}^{m} \sigma_r(t,\bar{X})\zeta_{rh}\sqrt{h}.$$

Further, in the case of SDEs in the sense of Stratonovich

$$dX = a(t, X)dt + \sum_{r=1}^{m} \sigma_r(t, X) \circ dw_r(t) \qquad (3.60)$$

we construct the derivative-free fully implicit method (midpoint method):

$$X_{k+1} = X_k + a(t_k + \frac{h}{2}, \frac{X_k + X_{k+1}}{2})h + \sum_{r=1}^{m} \sigma_r(t_k, \frac{X_k + X_{k+1}}{2})(\zeta_{rh})_k \sqrt{h}. \qquad (3.61)$$

For $\sigma_r^i = 0$, this method coincides with the well-known deterministic midpoint scheme, which has the second order of convergence.

In the general case the method (3.61) is of the mean-square order $1/2$. In the commutative case, i.e., when $\Lambda_i \sigma_r = \Lambda_r \sigma_i$ (here the operator $\Lambda_r := (\sigma_r, \partial/\partial x)$) or in the case of a system with one noise (i.e., $m = 1$) the midpoint method (3.61) has the first mean-square order of convergence which is stated in the next theorem [195].

Theorem 3.13. *Suppose that the commutative conditions $\Lambda_i \sigma_r = \Lambda_r \sigma_i$, $i, r = 1, \ldots, m$, are fulfilled. Let ζ_{rh} be defined by (3.37) with $A_h = \sqrt{4|\ln h|}$. Then the method (3.61) for the system (3.60) has the first mean-square order of convergence.*

1.4 Modeling of Ito integrals

Such integrals arise in numerical integration formulas when the system of equations has the form

$$dX = a(t, X)dt + \sigma(t, X)dw, \qquad (4.1)$$

where $w(t)$ is a scalar process ($q = 1$). We restrict ourselves to the integrals which are needed for methods of orders $3/2$ and 2 (more details see in [180]). We consider here explicit methods. Clearly, the integrals can be used for implicit methods as well.

1.4.1 Ito integrals depending on a single noise and methods of order 3/2 and 2

Due to Theorem 2.4, the method (2.27) is of order $3/2$. For (4.1) it acquires the form

$$\bar{X}_{t,x}(t+h) = x + \sigma(w(t+h) - w(t)) + \Lambda\sigma \int_t^{t+h} (w(\theta) - w(t))dw(\theta)$$

$$+ ah + L\sigma \int_t^{t+h} (\theta - t)dw(\theta) + \Lambda a \int_t^{t+h} (w(\theta) - w(t))d\theta$$

$$+ \Lambda^2 \sigma \int_t^{t+h} (\int_t^{\theta} (w(\theta_1) - w(t))dw(\theta_1))dw(\theta) + La\frac{h^2}{2}, \qquad (4.2)$$

where all the coefficients σ, a, $\Lambda\sigma$, $L\sigma$, Λa, $\Lambda^2\sigma$, La are calculated at (t, x).

Since the distribution of random variables in (4.2) does not depend on t, we have to be able to model together with $w(h)$, $\int_0^h w(\theta)d\theta$ the following variables

$$\int_0^h w(\theta)dw(\theta) = \frac{1}{2}(w^2(h) - h),$$

$$\int_0^h \theta dw(\theta) = hw(h) - \int_0^h w(\theta)d\theta,$$

$$\int_0^h (\int_0^\theta w(\theta_1)dw(\theta_1))dw(\theta) = \frac{1}{6}w^3(h) - \frac{1}{2}hw(h),$$

i.e., to construct the method of order $3/2$ it suffices at each step to model $w(h)$ and $\int_0^h w(\theta)d\theta$. The problem of modeling these variables can be solved very simply. In fact, their joint distribution is Gaussian. Write the integral $\int_0^h w(\theta)d\theta$ as

$$\int_0^h w(\theta)d\theta = \alpha w(h) + (\int_0^h w(\theta)d\theta - \alpha w(h))$$

and choose α such that $w(h)$ and $\int_0^h w(\theta)d\theta - \alpha w(h)$ are independent. Clearly, α can be found from the condition that the mathematical expectation of their product should vanish. Since $E(w(h)\int_0^h w(\theta)d\theta) = h^2/2$, we have $\alpha = h/2$. As a result, the integral can be written as a sum of two independent normally distributed random variables:

$$\int_0^h w(\theta)d\theta = \frac{1}{2}hw(h) + (\int_0^h w(\theta)d\theta - \frac{1}{2}hw(h)). \qquad (4.3)$$

The first term at the right-hand side of (4.3) is $\mathcal{N}(0, h^3/4)$-distributed, and the second term is $\mathcal{N}(0, h^3/12)$-distributed. Thus, from the point of view of modeling the random variables involved, the method (4.2) for the system (4.1) with a single noise is rather simple.

It is not difficult to prove that to construct a method of order two, it suffices to model at each step three random variables $w(h)$, $\int_0^h w(\theta)d\theta$, and

$\int_0^h w^2(\theta)d\theta$. In [180], the characteristic function of these random variables is found. However, it is very complicated and cannot be useful in practice. Thus, the exact modeling has bad perspectives, and therefore we need to be able to model these variables approximately.

Lemma 4.1. *Suppose that the one-step approximation* (see (1.3))

$$\bar{X}_{t,x}(t+h) = x + A(t, x, h; w_i(\theta) - w_i(t), i = 1, \ldots, q, t \leq \theta \leq t+h) \quad (4.4)$$

generates a method with order of accuracy m. *Suppose* A *contains terms of the form* $P(t,x) \cdot \xi(w_i(\theta) - w_i(t), i = 1, \ldots, q, t \leq \theta \leq t+h)$, *where* $|P(t,x)| \leq K(1 + |x|^2)^{1/2}$ *and* ξ *is a random variable, depending on the Wiener processes on the interval* $[t, t+h]$, *as indicated between the brackets. Let* $\xi = \eta + \varsigma$, *where* η *and* ς *are random variables depending on the same Wiener processes on the same interval. Finally, suppose*

$$|E\varsigma| \leq Kh^{m+1}, \; (E\varsigma^2)^{1/2} \leq Kh^{m+1/2}. \quad (4.5)$$

Then the method based on the one-step approximation (4.4) *and with* $P \cdot \xi$ *replaced by* $P \cdot \eta$ *has order of accuracy equal to* m.

The proof of this lemma easily follows from fundamental Theorem 1.1.

This lemma makes it possible to replace random variables, which participate in the method and are difficult in modeling, by random variables for which the modeling is simpler. A sufficiently general approach to the approximate modeling of random variables is as follows. Make up the system of SDEs whose solution at time h is given by the set of random variables to be modeled. If this system is integrable over the interval $[0, h]$ with sufficiently high accuracy (at the expense of a small integration step h_1, this can be done by a "rough" method in which simpler random variables are modeled), then we can construct the required approximations for the variables to be modeled. We note one important particular point that the needed random variables themselves have different orders of smallness with respect to h. And so we need approximations for them of a different order of accuracy. We clarify this using the variables $w(h)$, $\int_0^h w(\theta)d\theta$, and $\int_0^h w^2(\theta)d\theta$ as an example.

Introduce the new process

$$v(s) = \frac{w(sh)}{\sqrt{h}}, \; 0 \leq s \leq 1. \quad (4.6)$$

It is obvious that $v(s)$ is a standard Wiener process. We have

$$w(h) = h^{1/2}v(1), \; \int_0^h w(\theta)d\theta = h^{3/2}\int_0^1 v(s)ds, \; \int_0^h w^2(\theta)d\theta = h^2\int_0^1 v^2(s)ds. \quad (4.7)$$

Thus, the problem of modeling the variables $w(h)$, $\int_0^h w(\theta)d\theta$, and $\int_0^h w^2(\theta)d\theta$ can be reduced to that of modeling the variables $v(1)$, $\int_0^1 v(s)ds$, and $\int_0^1 v^2(s)ds$. These variables are the solution of the system of equations

$$dx = dv(s), \; x(0) = 0,$$
$$dy = x\,ds, \; y(0) = 0,$$
$$dz = x^2 ds, \; z(0) = 0, \qquad (4.8)$$

at the moment $s = 1$.

Let $\bar{x}(s_k)$, $\bar{y}(s_k)$, $\bar{z}(s_k)$, $0 = s_0 < s_1 < \cdots < s_N = 1$, $s_{k+1} - s_k = h_1 = 1/N$, be an approximate solution of (4.8). If we are guided by Lemma 4.1 when constructing a method of order two, then the random variables $w(h) = h^{1/2} x(1)$, $\int_0^h w(\theta)d\theta = h^{3/2} y(1)$, and $\int_0^h w^2(\theta)d\theta = h^2 z(1)$ participating in it can be replaced by $h^{1/2} \bar{x}(1)$, $h^{3/2} \bar{y}(1)$, and $h^2 \bar{z}(1)$, if only the conditions

$$|E(v(1) - \bar{x}(1))| = O(h^{5/2}), \; (E(v(1) - \bar{x}(1))^2)^{1/2} = O(h^2), \qquad (4.9)$$

$$|E(\int_0^1 v(s)ds - \bar{y}(1))| = O(h^{3/2}), \; (E(\int_0^1 v(s)ds - \bar{y}(1))^2)^{1/2} = O(h), \qquad (4.10)$$

$$|E(\int_0^1 v^2(s)ds - \bar{z}(1))| = O(h), \; (E(\int_0^1 v^2(s)ds - \bar{z}(1))^2)^{1/2} = O(h^{1/2}) \qquad (4.11)$$

are fulfilled.

First we integrate (4.8) by the Euler method with step h_1:

$$x_{k+1} = x_k + \Delta_k v(h_1), \; x_0 = 0,$$
$$y_{k+1} = y_k + x_k h_1, \; y_0 = 0,$$
$$z_{k+1} = z_k + x_k^2 h_1, \; z_0 = 0. \qquad (4.12)$$

Since

$$x_k = v(s_k), \; x_N = \bar{x}(1) = v(1),$$
$$y_k = h_1 \sum_{i=0}^{k-1} v(s_i), \; \bar{y}(1) = h_1 \sum_{i=0}^{N-1} v(s_i),$$
$$z_k = h_1 \sum_{i=0}^{k-1} v^2(s_i), \; \bar{z}(1) = h_1 \sum_{i=0}^{N-1} v^2(s_i),$$

we have

1.4 Modeling of Ito integrals

$$Ev(1) = E\bar{x}(1) = 0,$$

$$E\int_0^1 v(s)ds = E\bar{y}(1) = 0,$$

$$E\int_0^1 v^2(s)ds = \frac{1}{2}, \quad E\bar{z}(1) = \frac{1}{2} - \frac{h_1}{2}. \tag{4.13}$$

Further, since the system (4.8) is a system with additive noise, Euler's method has order $O(h_1)$. Therefore, taking $h_1 = h$ we find that all the relations (4.9)-(4.11) hold (note that in our case $E(v(1) - \bar{x}(1))^2 = 0$). Indeed, direct computations (of quite some length in the case of the third equation) give (for $h_1 = h$):

$$E(\int_0^1 v(s)ds - \bar{y}(1))^2 = \frac{h^2}{3}, \quad |E(\int_0^1 v^2(s)ds - \bar{z}(1))| = \frac{h}{2},$$

$$E(\int_0^1 v^2(s)ds - \bar{z}(1))^2 = \frac{11}{12}h^2 - \frac{h^3}{3}. \tag{4.14}$$

Thus, if we approximately model the random variables, needed for a method of order two in the case of a system with single noise, by applying Euler's method to (4.8), then at each step we have to model $\approx 1/h$ normally distributed random variables (here h is the integration step in both the original system and the auxiliary system (4.8)).

We will now use a method of order 3/2 (see (4.2)) to integrate (4.8). Since

$$a = \begin{bmatrix} 0 \\ x \\ x^2 \end{bmatrix}, \quad \sigma = \begin{bmatrix} 1 \\ 0 \\ 0 \end{bmatrix}, \quad \Lambda\sigma = \Lambda^2\sigma = L\sigma = 0, \quad \Lambda a = \begin{bmatrix} 0 \\ 1 \\ 2x \end{bmatrix}, \quad La = \begin{bmatrix} 0 \\ 0 \\ 1 \end{bmatrix},$$

we get

$$x_{k+1} = x_k + \Delta_k v(h_1),$$

$$y_{k+1} = y_k + x_k h_1 + \int_{s_k}^{s_{k+1}} (v(\theta) - v(s_k))d\theta,$$

$$z_{k+1} = z_k + x_k^2 h_1 + 2x_k \int_{s_k}^{s_{k+1}} (v(\theta) - v(s_k))d\theta + \frac{h_1^2}{2}. \tag{4.15}$$

The method (4.15) has the following properties. First x_k and y_k are equal to $v(s_k)$ and $\int_0^{s_k} v(\theta)d\theta$, respectively (this is obvious) and, secondly, $E\bar{z}(1) =$

$Ez(1)$ (this will be proved below). As a result, only the second relation in (4.11) among all the relations (4.9)-(4.11) has to be satisfied. However, since the method (4.15) is of order 3/2, this relation reduces to the requirement

$$(E(z(1) - \bar{z}(1))^2)^{1/2} = O(h_-^{3/2}) = O(h^{1/2}). \tag{4.16}$$

Thus, if we choose h_1 such that $h_1 = h^{1/3}$, then the conditions of Lemma 4.1 hold.

Let us we compare computational costs of two procedures proposed above to approximately model the needed random variables. Suppose that we integrate the original system by a method of order two with the step $h = 0.001$. Then, if we use the first procedure to model the needed random variables, we have $h_1 = h$ and we need $\approx 1/h$, i.e., ≈ 1000 normally distributed random variables. In the second procedure we have $h_1 = h^{1/3} = 0.1$; however at each step we have to model not one but two random variables, $\Delta_k v(h_1)$ and $\int_{s_k}^{s_{k+1}} (v(\theta) - v(s_k))d\theta$. As a result, we need only 20 variables here instead of 1000 random variables. Of course, the second procedure is much more economical. Now we give a theorem (see its proof in [180, p. 88]) on second-order methods for the system (4.1) related to approximate modeling by (4.15).

Theorem 4.2. *Suppose we solve the system (4.1) by a method of second order of accuracy with step h (for example, in accordance with Theorem 2.5), in which at each step the integrals $\int_{t_j}^{t_{j+1}} dw(\theta)$, $\int_{t_j}^{t_{j+1}} (w(\theta) - w(t_j))d\theta$, $\int_{t_j}^{t_{j+1}} (w(\theta) - w(t_j))^2 d\theta$ participate. If these random variables are replaced, independently at each step, by the random variables $h^{1/2}x_N$, $h^{3/2}y_N$, $h^2 z_N$, where x_N, y_N, z_N, can be found recurrently from (4.15) with step $h_1 = O(h^{1/3})$, then the order of accuracy of the obtained method remains the same, i.e., it is equal to 2.*

Remark 4.3. Application of the usual numerical integration formulas for modeling the integrals does not lead to a success. We will convince ourselves of this by the example of modeling the integral $\int_0^1 v(s)ds$ using the trapezium formula. It is well known that the trapezium formula has error $O(h^2)$. This is true for integrands having bounded second derivative. Here, however, $v(s)$ is a Wiener process with nonsmooth trajectories. Of course, we are interested in accuracy in the mean sense. More precisely, we are interested in mean and mean-square deviation of the approximation of the integral by the trapezium formula. Applying the trapezium formula to the integral $\int_0^1 v(s)ds$, we get

$$\int_0^1 v(s)ds \doteq \frac{h}{2}(v(0) + 2v(s_1) + \cdots + 2v(s_{N-1}) + v(s_N)), \quad h = \frac{1}{N}. \tag{4.17}$$

Here the mean deviation is zero, since the mathematical expectation of both sides of (4.17) are zero. It is easy to compute that

$$E(\int_0^1 v(s)ds - \sum_{k=0}^{N-1} \frac{v(s_k)+v(s_{k+1})}{2}h)^2$$

$$= \sum_{k=0}^{N-1} E(\int_{s_k}^{s_{k+1}} (v(s)-v(s_k)) - \frac{v(s_{k+1})-v(s_k)}{2})ds)^2 = \frac{h^2}{12}.$$

Thus, the mean-square deviation in the trapezium method is $O(h)$, which is of lower order than expected.

1.4.2 Modeling Ito integrals by the rectangle and trapezium methods

A method of first order of accuracy for the system (1.1) has the form

$$X_0 = X(t_0),\ X_{n+1} = X_n + \sum_{j=1}^q (\sigma_j)_n \Delta_n w_j(h) + a_n h$$

$$+ \sum_{i=1}^q \sum_{j=1}^q (\Lambda_i \sigma_j)_n \int_{t_n}^{t_{n+1}} (w_i(s) - w_i(t_n))dw_j(s), \quad (4.18)$$

where $t_0 < t_1 < \cdots < t_N = T$, $h = t_{n+1} - t_n = (T-t_0)/N$. To realize the method (4.18), we have to model the set of random variables $\Delta_n w_j$, $\int_{t_n}^{t_{n+1}} (w_i(s) - w_i(t_n))dw_j(s)$, $i,j = 1,\ldots,q$, at each step. Since at different steps these sets are independent, the problem reduces to modeling the variables $w_j(h)$, $I_{ij} = \int_0^h w_i(s)dw_j(s)$, $i,j = 1,\ldots,q$. Due to

$$\int_0^h w_j(s)dw_i(s) = w_i(h)w_j(h) - \int_0^h w_i(s)dw_j(s),\ i \neq j, \quad (4.19)$$

$$\int_0^h w_i(s)dw_i(s) = \frac{w_i^2(h)}{2} - \frac{h}{2}, \quad (4.20)$$

it suffices to model the set of variables

$$w_j(h),\ j = 1,\ldots,q,$$

$$I_{ij} = \int_0^h w_i(s)dw_j(s), \ i=1,\ldots,q, \ j=i+1,\ldots,q. \tag{4.21}$$

Consider the *rectangle method*. We write the integral $\int_0^h w_i(s)dw_j(s)$ as a sum of l integrals:

$$\int_0^h w_i(s)dw_j(s) = \sum_{k=1}^l \int_{s_{k-1}}^{s_k} w_i(s)dw_j(s), \tag{4.22}$$

where $s_0 = 0$, $s_k - s_{k-1} = h/l$, $k=1,\ldots,l$. We replace each of these integrals using the left rectangle formula and obtain

$$\int_0^h w_i(s)dw_j(s) \doteq \sum_{k=1}^l w_i(s_{k-1})(w_j(s_k) - w_j(s_{k-1})). \tag{4.23}$$

For the error

$$\Delta_{ij} = \sum_{k=1}^l \int_{s_{k-1}}^{s_k} (w_i(s) - w_i(s_{k-1}))dw_j(s)$$

in the approximate identity (4.23), we have

$$E\Delta_{ij} = 0, \ E\Delta_{ij}^2 = \sum_{k=1}^l E(\int_{s_{k-1}}^{s_k} (w_i(s) - w_i(s_{k-1}))dw_j(s))^2$$

$$= \sum_{k=1}^l \int_{s_{k-1}}^{s_k} E(w_i(s) - w_i(s_{k-1}))^2 ds = \frac{h^2}{2l}. \tag{4.24}$$

Thus, to approximately represent the variables (4.21) by (4.23), we have to model ql independent $\mathcal{N}(0,1)$-distributed random variables ξ_{ik}, $i=1,\ldots,q$, $k=1,\ldots,l$ ($\sqrt{h/l}\xi_{ik} = w_j(s_k) - w_j(s_{k-1})$) and put for $i=1,\ldots,q$, $j=i+1,\ldots,q$:

$$w_i(h) = \sqrt{\frac{h}{l}} \sum_{r=1}^l \xi_{ir}, \ \int_0^h w_i(s)dw_j(s) \doteq \frac{h}{l} \sum_{k=1}^l \sum_{r=1}^{k-1} \xi_{ir}\xi_{jk}. \tag{4.25}$$

Note that if the integral $\int_0^h w_i(s)dw_i(s)$ is modeled according to (4.25), then the error involved is $h^2/(2l)$, while this integral can be computed exactly by (4.20). Further, the integrals for $i > j$ can be approximately computed either by using (4.19) after having approximately modeled (4.21), or by modeling them according to (4.23). In both cases the error will be the same. It is easy to see that the use of (4.19) is equivalent to the use of the right rectangle formula.

We now consider the *trapezium method*. Applying to each integral in the sum (4.22) the trapezium formula, we find

$$\int_0^h w_i(s)dw_j(s) \doteq \sum_{k=1}^{l} \frac{1}{2}(w_i(s_{k-1}) + w_i(s_k))(w_j(s_k) - w_j(s_{k-1})). \quad (4.26)$$

For the error

$$\Delta_{ij} = \frac{1}{2} \sum_{k=1}^{l} \int_{s_{k-1}}^{s_k} ((w_i(s) - w_i(s_{k-1})) - (w_i(s_{k-1}) - w_i(s)))dw_j(s)$$

in the approximate identity (4.26), we have

$$E\Delta_{ij} = 0,$$

$$E\Delta_{ij}^2 = \frac{1}{4} \sum_{k=1}^{l} \int_{s_{k-1}}^{s_k} E((w_i(s) - w_i(s_{k-1})) - (w_i(s_{k-1}) - w_i(s)))^2 ds$$

$$= \frac{1}{4} \sum_{k=1}^{l}((s - s_{k-1}) + (s_k - s))ds = \frac{1}{4}\frac{h^2}{l}. \quad (4.27)$$

Here, having modeled ql independent $\mathcal{N}(0,1)$-distributed random variables ξ_{ik}, we can set

$$w_i(h) = \sqrt{\frac{h}{l}} \sum_{r=1}^{l} \xi_{ir}, \quad \int_0^h w_i(s)dw_j(s) \doteq \frac{h}{l} \sum_{k=1}^{l} (\sum_{r=1}^{k-1} \xi_{ir} + \frac{1}{2}\xi_{ik})\xi_{jk}, \quad i < j. \quad (4.28)$$

It can be readily seen that (4.19) gives the same result as (4.28) for $i > j$. If we take $l \approx 1/h$, then (4.24), (4.27) imply that the mean-square error of the approximations of the integrals is $O(h^{3/2})$ in both the rectangle and the trapezium method, while the mean error is zero. By Lemma 4.1 this implies the following result.

Theorem 4.4. *If we replace the random variables at each step of the method (4.18) using either the rectangle or the trapezium formula with $l \approx 1/h$, then the order of accuracy of the obtained method remains the same, i.e., it is equal to 1.*

1.4.3 Modeling Ito integrals by the Fourier method

We now turn to a method which can be naturally called the Fourier method.

Consider the Fourier coefficients of the process $w_i(t) - (t/h)w_i(h)$, $i = 1,\ldots,q$, on the interval $0 \leq t \leq h$ with respect to the trigonometric system of functions 1, $\cos 2k\pi t/h$, $\sin 2k\pi t/h$, $k = 1, 2, \ldots$ (see [113, 229] for the *Wiener construction* of Brownian motion). We have

$$a_{ik} = \frac{2}{h} \int_0^h (w_i(s) - \frac{s}{h}w_i(h)) \cos \frac{2\pi ks}{h} ds, \; k = 0, 1, 2, \ldots,$$

$$b_{ik} = \frac{2}{h} \int_0^h (w_i(s) - \frac{s}{h}w_i(h)) \sin \frac{2\pi ks}{h} ds, \; k = 1, 2, \ldots. \quad (4.29)$$

The distribution of these coefficients is clearly Gaussian.

Lemma 4.5. *The following equalities hold:*

$$Ew_i(h)a_{ik} = 0, \; k = 0, 1, 2, \ldots, \; Ew_i(h)b_{ik} = 0, \; k = 1, 2, \ldots, \quad (4.30)$$

$$Ea_{i0}^2 = \frac{h}{3}, \; Ea_{ik}^2 = Eb_{ik}^2 = \frac{h}{2k^2\pi^2}, \; k = 1, 2, \ldots, \quad (4.31)$$

$$Ea_{i0}a_{ik} = -\frac{h}{k^2\pi^2}, \; k = 1, 2, \ldots, \quad (4.32)$$

$$Ea_{i0}b_{ik} = Ea_{ik}a_{im} = Ea_{ik}b_{im}$$
$$= Eb_{ik}b_{im} = 0, \; k, m = 1, 2, \ldots, \; k \neq m. \quad (4.33)$$

Proof. All these formulas can be obtained by direct computation. We give, as an example, a detailed proof of one of the formulas in (4.31). We have for $k \neq 0$:

$$Ea_{ik}^2 = \frac{4}{h^2} \int_0^h \int_0^h E((w(t) - \frac{t}{h}w(h))(w(s) - \frac{s}{h}w(h))) \cos \frac{2\pi kt}{h} \cos \frac{2\pi ks}{h} dt ds.$$

We evaluate:

$$E((w(t) - \frac{t}{h}w(h))(w(s) - \frac{s}{h}w(h))) = \begin{cases} t - ts/h, \; t \leq s, \\ s - ts/h, \; t > s. \end{cases}$$

Therefore

$$Ea_{ik}^2 = \frac{4}{h^2} \int_0^h (\int_0^s t \cos \frac{2\pi kt}{h} dt) \cos \frac{2\pi ks}{h} ds$$
$$+ \frac{4}{h^2} \int_0^h (\int_s^h \cos \frac{2\pi kt}{h} dt) s \cos \frac{2\pi ks}{h} ds$$
$$- \frac{4}{h^2} \int_0^h \int_0^h \frac{st}{h} \cos \frac{2\pi kt}{h} \cos \frac{2\pi ks}{h} dt ds. \quad (4.34)$$

Further,
$$\int_0^s t\cos\frac{2\pi kt}{h}dt = s\sin\frac{2\pi ks}{h}\cdot\frac{h}{2\pi k} + (\frac{h}{2\pi k})^2(\cos\frac{2\pi ks}{h} - 1).$$
Then, in particular,
$$\int_0^h t\cos\frac{2\pi kt}{h}dt = 0.$$
Hence the last term in (4.34) vanishes. Using these equations, we obtain
$$Ea_{ik}^2 = \frac{4}{h^2}\int_0^h (s\sin\frac{2\pi ks}{h}\cdot\frac{h}{2\pi k} + (\frac{h}{2\pi k})^2(\cos\frac{2\pi ks}{h} - 1))\cos\frac{2\pi ks}{h}ds$$
$$+\frac{4}{h^2}\int_0^h (-\frac{h}{2\pi k}\sin\frac{2\pi ks}{h})s\cos\frac{2\pi ks}{h}ds$$
$$= \frac{4}{h^2}\int_0^h (\frac{h}{2\pi k})^2(\cos\frac{2\pi ks}{h} - 1)\cos\frac{2\pi ks}{h}ds = \frac{h}{2k^2\pi^2}.$$

The other formulas can be proved similarly. □

We replace the integrand $w_i(s)$ in the integral $I_{ij} = \int_0^h w_i(s)dw_j(s)$ by an expression containing a part of its Fourier series:

$$I_{ij} = \int_0^h w_i(s)dw_j(s) \doteq \bar{I}_{ij}$$
$$:= \int_0^h (\frac{s}{h}w_i(h) + \frac{a_{i0}}{2} + \sum_{k=1}^m (a_{ik}\cos\frac{2\pi ks}{h} + b_{ik}\sin\frac{2\pi ks}{h}))dw_j(s). \quad (4.35)$$

Lemma 4.6. *For the error Δ_{ij} in the approximate identity (4.35), the following relations hold:*

$$E\Delta_{ij} = 0, \quad E\Delta_{ij}^2 = \frac{h^2}{12} - \frac{h^2}{2\pi^2}\sum_{k=1}^m \frac{1}{k^2}, \quad i \neq j. \quad (4.36)$$

Proof. The relation $E\Delta_{ij} = 0$ is evident. We have for $i \neq j$:

$$E\Delta_{ij}^2 = E(\int_0^h (w_i(t) - \frac{t}{h}w_i(h) - \frac{a_{i0}}{2}$$
$$- \sum_{k=1}^m (a_{ik}\cos\frac{2\pi kt}{h} + b_{ik}\sin\frac{2\pi kt}{h}))dw_j(t))^2$$
$$= E\int_0^h (w_i(t) - \frac{t}{h}w_i(h) - \frac{a_{i0}}{2} - \sum_{k=1}^m (a_{ik}\cos\frac{2\pi kt}{h} + b_{ik}\sin\frac{2\pi kt}{h}))^2 dt.$$

Since the sum

$$\frac{a_{i0}}{2} + \sum_{k=1}^{m}(a_{ik}\cos\frac{2\pi kt}{h} + b_{ik}\sin\frac{2\pi kt}{h})$$

is a part of the Fourier series of the function $w_i(t) - \frac{t}{h}w_i(h)$, we have

$$\int_0^h (w_i(t) - \frac{t}{h}w_i(h) - \frac{a_{i0}}{2} - \sum_{k=1}^{m}(a_{ik}\cos\frac{2\pi kt}{h} + b_{ik}\sin\frac{2\pi kt}{h}))^2 dt$$

$$= \int_0^h (w_i(t) - \frac{t}{h}w_i(h))^2 dt - \frac{1}{4}ha_{i0}^2 - \sum_{k=1}^{m}\frac{h}{2}(a_{ik}^2 + b_{ik}^2).$$

Then using Lemma 4.5, we find

$$E\Delta_{ij}^2 = \int_0^h E(w_i(t) - \frac{t}{h}w_i(h))^2 dt - \frac{h}{4}Ea_{i0}^2 - \sum_{k=1}^{m}\frac{h}{2}E(a_{ik}^2 + b_{ik}^2)$$

$$= \int_0^h (t - \frac{2t^2}{h} + \frac{t^2}{h})dt - \frac{1}{12}h^2 - \sum_{k=1}^{m}\frac{h}{2}\cdot\frac{h}{k^2\pi^2}.$$

This implies the second relation of (4.36). □

Lemma 4.7. *The following formula is valid:*

$$\bar{I}_{ij} = \int_0^h (\frac{t}{h}w_i(h) + \frac{a_{i0}}{2} + \sum_{k=1}^{m}(a_{ik}\cos\frac{2\pi kt}{h} + b_{ik}\sin\frac{2\pi kt}{h}))dw_j(t)$$

$$= \frac{1}{2}w_i(h)w_j(h) + \frac{a_{i0}}{2}w_j(h) - \frac{a_{j0}}{2}w_i(h) - \pi\sum_{k=1}^{m}k(a_{ik}b_{jk} - a_{jk}b_{ik}). \quad (4.37)$$

Proof. We have

$$\int_0^h t\,dw_j(t) = hw_j(h) - \int_0^h w_j(t)dt,$$

$$\int_0^h \cos\frac{2\pi kt}{h}dw_j(t) = w_j(h) + \frac{2\pi k}{h}\int_0^h w_j(t)\sin\frac{2\pi kt}{h}dt,$$

$$\int_0^h \sin\frac{2\pi kt}{h}dw_j(t) = -\frac{2\pi k}{h}\int_0^h w_j(t)\cos\frac{2\pi kt}{h}dt. \quad (4.38)$$

1.4 Modeling of Ito integrals

Further, we get (see the formulas (4.29) defining the Fourier coefficients):

$$a_{j0} = \frac{2}{h}\int_0^h w_j(t)dt - w_j(h),$$

$$a_{jk} = \frac{2}{h}\int_0^h w_j(t)\cos\frac{2\pi kt}{h}dt, \ k \neq 0,$$

$$b_{jk} = \frac{2}{h}\int_0^h w_j(t)\sin\frac{2\pi kt}{h}dt + \frac{w_j(h)}{k\pi}. \qquad (4.39)$$

Transforming first the expression for $\bar{I}_{ij}$ by using (4.38) and then by using (4.39), we come to (4.37). This proves the lemma. □

In the right-hand side of (4.37) the coefficient a_{i0} depends on a_{ik}, the coefficient a_{j0} depends on a_{jk}, and all the remaining coefficients are mutually independent (see Lemma 4.5). Introduce the new random variable $a_{i0}^{(m)}$ by

$$a_{i0}^{(m)} = -\frac{a_{i0}}{2} - \sum_{k=1}^m a_{ik}.$$

We show that $a_{i0}^{(m)}$ does not depend on a_{ik}, $k = 1, \ldots, m$. Indeed, by Lemma 4.5 we have

$$Ea_{i0}^{(m)}a_{ik} = -\frac{1}{2}Ea_{i0}a_{ik} - Ea_{ik}^2 = 0, \ k = 1, \ldots, m.$$

Then the independence follows from the fact that all the variables under consideration are Gaussian.

We can directly compute that

$$E(a_{i0}^{(m)})^2 = \frac{h}{12} - \frac{h}{2\pi^2}\sum_{k=1}^m \frac{1}{k^2}. \qquad (4.40)$$

Substituting $a_{i0} = -2a_{i0}^{(m)} - 2\sum_{k=1}^m a_{ik}$ in (4.37), we obtain

$$\bar{I}_{ij} = \frac{1}{2}w_i(h)w_j(h) + a_{j0}^{(m)}w_i(h) - a_{i0}^{(m)}w_j(h)$$

$$+ \sum_{k=1}^m (a_{jk}w_i(h) - a_{ik}w_j(h)) + \pi\sum_{k=1}^m k(a_{ik}b_{jk} - a_{jk}b_{ik}). \qquad (4.41)$$

In (4.41) all $w_i(h)$, $w_j(h)$, $a_{i0}^{(m)}$, $a_{j0}^{(m)}$, a_{ik}, a_{jk}, b_{ik}, b_{jk} are independent Gaussian random variables.

We gather the results concerning the approximate modeling of the variables $w_i(h)$ and $I_{ij} = \int_0^h w_i(s)dw_j(s)$ obtained above in the following theorem.

Theorem 4.8. *Making up* $\bar{I}_{ij}$, $i,j = 1,\ldots,q$, *reduces to modeling* $2(m+1)q$ *independent* $\mathcal{N}(0,1)$-*distributed random variables* ξ_i, ξ_{ik}, $k=0,\ldots,m$, η_{ik}, $k=1,\ldots,m$. *Here*

$$\xi_i = h^{-1/2}w_i(h), \ \xi_{i0} = \left(\frac{h}{12} - \frac{h}{2\pi^2}\sum_{k=1}^{m}\frac{1}{k^2}\right)^{-1/2} a_{i0}^{(m)},$$

$$\xi_{ik} = \sqrt{2}\pi k h^{-1/2} a_{ik}, \ \eta_{ik} = \sqrt{2}\pi k h^{-1/2} b_{ik}.$$

The quantities $w_i(h)$ *and* $I_{ij}(h)$ *can be expressed in terms of these variables as*

$$w_i(h) = h^{1/2}\xi_i, \ I_{ii} = \bar{I}_{ii} = \frac{h}{2}(\xi_i^2 - 1), \ i = 1,\ldots,q,$$

$$I_{ij} \doteq \bar{I}_{ij} = \frac{h}{2}\xi_i\xi_j + h\left(\frac{1}{12} - \frac{1}{2\pi^2}\sum_{k=1}^{m}\frac{1}{k^2}\right)^{1/2}(\xi_{j0}\xi_i - \xi_{i0}\xi_j)$$

$$+ \frac{h}{\pi\sqrt{2}}\sum_{k=1}^{m}\frac{1}{k}(\xi_{jk}\xi_i - \xi_{ik}\xi_j)$$

$$+ \frac{h}{2\pi}\sum_{k=1}^{m}\frac{1}{k}(\xi_{ik}\eta_{jk} - \xi_{jk}\eta_{ik}), \ i \neq j. \tag{4.42}$$

The error $I_{ij} - \bar{I}_{ij}$ *of the approximate modeling is characterized by the relations*

$$E\Delta_{ij} = 0, \ E\Delta_{ij}^2 = \frac{h^2}{12} - \frac{h^2}{2\pi^2}\sum_{k=1}^{m}\frac{1}{k^2}, \ i \neq j. \tag{4.43}$$

If $m \approx 1/h$, *then the following method (which is constructive from the point of view of modeling random variables):*

$$X_{n+1} = X_n + \sum_{j=1}^{q}(\sigma_j)_n \xi_j^{(n)} h^{1/2} + a_n h + \sum_{i=1}^{q}\sum_{j=1}^{q}(\Lambda_i \sigma_j)_n \bar{I}_{ij}^{(n)}, \tag{4.44}$$

where the index n *indicates that the random variables are modeled according to* (4.42) *independently at each step, is a method of the first order of accuracy for integrating* (1.1).

Proof. We only need to prove the last assertion, since all the previous ones follow from Lemmas 4.5 – 4.7. However its proof does not differ at all from the proof of Theorem 4.4 if we take into account that (since $\sum_{k=1}^{\infty} 1/k^2 = \pi^2/6$)

$$\frac{1}{12} - \frac{1}{2\pi^2}\sum_{k=1}^{m}\frac{1}{k^2} = \frac{1}{2\pi^2}\sum_{k=m+1}^{\infty}\frac{1}{k^2} \leq \frac{1}{2\pi^2}\int_{m}^{\infty}\frac{dx}{x^2} = \frac{1}{2\pi^2 m}.$$

The theorem is proved. □

Table 4.1. Coefficients of the error in the trapezium and the Fourier method.

m	1	2	3	4	5	10	20
$\dfrac{1}{8(m+1)}$	0.0625	0.0417	0.0312	0.0250	0.0208	0.0114	0.0060
$\dfrac{1}{12} - \dfrac{1}{2\pi^2}\sum_{k=1}^{m}\dfrac{1}{k^2}$	0.0327	0.0200	0.0144	0.0112	0.0092	0.0048	0.0025

We will now compare results of modeling by the rectangle method, the trapezium method, and the Fourier method. It is clear from (4.24) and (4.27) that to achieve the same accuracy, the rectangle method requires twice as many independent $\mathcal{N}(0,1)$-distributed random variables as does the trapezium method. We compare the Fourier method and the trapezium method from this point of view. To this end we put $l = 2(m+1)$ in (4.27). Then for identical costs as regards forming random variables (since in the trapezium method we have to model $ql = 2(m+1)q$ independent $\mathcal{N}(0,1)$-distributed random variables in order to approximately form all the necessary Ito integrals by (4.28)) the error of the trapezium method can be computed by (4.27) and that of the Fourier method by (4.43). We give some values of the coefficients at h^2 in the error $E\Delta_{ij}^2$ (Table 4.1): the second row corresponds to the coefficients in the trapezium method; the third row corresponds to those in the Fourier method. For m large the Fourier method is 2.5 times more economical than the trapezium method, since the error of the Fourier method is close to $1/(20m)$ while that of the trapezium method is close to $1/(8m)$.

Remark 4.9. A rather large literature is devoted to theoretical and practical points of modeling Ito integrals (along with [180] and [131] see e.g., [75, 133, 142, 310] and references therein). We recall (see [180]) that the integrals $I_{i_1,\ldots,i_j}(h)$ (see (2.11)) needed for a method can be simulated approximately by numerical integration of a system of SDEs. Indeed, consider all Ito integrals of orders $m - 1/2$ and m. If $I_{i_1,\ldots,i_j}(\theta)$ is an Ito integral of order $m - 1/2$, then $I_{i_1,\ldots,i_k,i_{k+1}}$, where $i_{k+1} = 0$, is an Ito integral of order $m + 1/2$, and it satisfies the equation

$$dI_{i_1,\ldots,i_k,i_{k+1}} = I_{i_1,\ldots,i_k}\,d\theta\,. \tag{4.45}$$

If $I_{i_1,\ldots,i_l}$ is an Ito integral of order m, then $I_{i_1,\ldots,i_l,i_{l+1}}$, $i_{l+1} = 1,\ldots,q$, satisfies the equation

$$dI_{i_1,\ldots,i_l,i_{l+1}} = I_{i_1,\ldots,i_l}\,dw_{i_{l+1}}(\theta)\,. \tag{4.46}$$

As a result, we have accounted for all Ito integrals of order $m+1/2$. Adjoining the equations of the type (4.45) and (4.46) to the system of equations for the Ito integrals up to order m, we obtain a system of equations for the Ito integrals up to order $m+1/2$ inclusively. It can be readily seen that this system is a linear autonomous system of SDEs. The initial data for each variable is zero. It is clear that the dimension of this system can be substantially reduced because of relations between the integrals.

1.5 Explicit and implicit methods of order 3/2 for systems with additive noise

From the point of view of numerical integration, the distinguishing mark of systems with additive noise is the absence of random variables of the form I_{i_1,i_2} and I_{i_1,i_2,i_3}, $i_1, i_2, i_3 \neq 0$, in the Taylor-type expansions, and therefore we are able to construct various constructive (with respect to modeling of random variables) methods with order of accuracy reaching 3/2.

1.5.1 Explicit methods based on Taylor-type expansion

Consider the system of stochastic differential equations with additive noise

$$dX = a(t,X)dt + \sum_{r=1}^{q} \sigma_r(t) dw_r(t). \tag{5.1}$$

We use the formulas (2.21) and (2.22). We have

$$\Lambda_i \sigma_r = 0, \ L\sigma_r = \frac{d\sigma_r}{dt}(t), \ \Lambda_r a = (\sigma_r, \frac{\partial}{\partial x}) a(t,x), \ \Lambda_s \Lambda_i \sigma_r = 0,$$

$$La = \frac{\partial a}{\partial t}(t,x) + (a, \frac{\partial}{\partial x}) a(t,x) + \frac{1}{2} \sum_{r=1}^{q} \sum_{i=1}^{d} \sum_{j=1}^{d} \sigma_r^i \sigma_r^j \frac{\partial^2 a}{\partial x^i \partial x^j}(t,x).$$

By (2.21), we can write down the following numerical integration formula

$$\begin{aligned} X_{k+1} = X_k &+ \sum_{r=1}^{q} \sigma_{r_k} \Delta_k w_r(h) + a_k h \\ &+ \sum_{r=1}^{q} (\Lambda_r a)_k \int_{t_k}^{t_{k+1}} (w_r(\theta) - w_r(t_k)) d\theta \\ &+ \sum_{r=1}^{q} \sigma'_{r_k} \int_{t_k}^{t_{k+1}} (\theta - t_k) dw_r(\theta) + (La)_k \frac{h^2}{2}. \end{aligned} \tag{5.2}$$

Suppose the functions $a(t,x)$ and $\sigma_r(t)$ satisfy for $t_0 \leq t \leq T$, $x \in \mathbf{R}^d$ the following conditions:

(a) the function a and all its first- and second-order partial derivatives as well as the partial derivatives $\partial^3 a/\partial t \partial x^i \partial x^j$, $\partial^3 a/\partial x^i \partial x^j \partial x^k$, and $\partial^4 a/\partial x^i \partial x^j \partial x^k \partial x^l$ are continuous;

1.5 Explicit and implicit methods of order 3/2

(b) the functions $\sigma_r(t)$ are twice continuously differentiable;

(c) the first-order partial derivatives of a with respect to x are uniformly bounded (so that the global Lipschitz condition is satisfied), while its remaining partial derivatives listed above, regarded as functions of x, grow at most as a linear function of $|x|$ as $|x| \to \infty$.

Then Theorem 2.4 holds (see (2.22) and Remark 2.6). Thus, the method (5.2) has order of accuracy equal to 3/2.

In (5.2) we have the following random variables:

$$\Delta_k w_r(h), \quad \int_{t_k}^{t_{k+1}} (w_r(\theta) - w_r(t_k))d\theta, \quad \int_{t_k}^{t_{k+1}} (\theta - t_k)dw_r(\theta) .$$

Since

$$\int_{t_k}^{t_{k+1}} (\theta - t_k)dw_r(\theta) = h\Delta_k w_r(h) - \int_{t_k}^{t_{k+1}} (w_r(\theta) - w_r(t_k))d\theta, \quad (5.3)$$

to use (5.2) at the $(k+1)$-st step it suffices to model the random variables $\Delta_k w_r(h)$ and $\int_{t_k}^{t_{k+1}} (w_r(\theta) - w_r(t_k))d\theta$, $r = 1,\ldots,q$. Each of them has a Gaussian distribution. We can directly compute that

$$E[\Delta_k w_r(h)(\int_{t_k}^{t_{k+1}} (w_r(\theta) - w_r(t_k))d\theta - \frac{1}{2}h\Delta_k w_r(h))] = 0, \quad (5.4)$$

$$E[\int_{t_k}^{t_{k+1}} (w_r(\theta) - w_r(t_k))d\theta - \frac{1}{2}h\Delta_k w_r(h)]^2 = \frac{1}{12}h^3. \quad (5.5)$$

Introduce the following independent $\mathcal{N}(0,1)$-distributed random variables ξ_{rk} and η_{rk}:

$$\xi_{rk} = h^{-1/2}\Delta_k w_r(h),$$

$$\eta_{rk} = \sqrt{12}h^{-3/2}(\int_{t_k}^{t_{k+1}} (w_r(\theta) - w_r(t_k))d\theta - \frac{1}{2}h\Delta_k w_r(h)). \quad (5.6)$$

Using these random variables, we obtain

$$\Delta_k w_r(h) = h^{1/2}\xi_{rk},$$

$$\int_{t_k}^{t_{k+1}} (w_r(\theta) - w_r(t_k))d\theta = h^{3/2}(\frac{1}{2}\xi_{rk} + \frac{1}{\sqrt{12}}\eta_{rk}). \quad (5.7)$$

As a result, the formula (5.2) takes the following concrete form:

$$X_{k+1} = X_k + \sum_{r=1}^{q} \sigma_r(t_k)\xi_{rk}h^{1/2} + a(t_k, X_k)h$$

$$+ \sum_{r=1}^{q} \Lambda_r a(t_k, X_k)(\frac{1}{2}\xi_{rk} + \frac{1}{\sqrt{12}}\eta_{rk})h^{3/2}$$

$$+ \sum_{r=1}^{q} \frac{d\sigma_r}{dt}(t_k)(\frac{1}{2}\xi_{rk} - \frac{1}{\sqrt{12}}\eta_{rk})h^{3/2} + La(t_k, X_k)\frac{h^2}{2}. \quad (5.8)$$

We state the result in the next theorem. Here and below, as a rule, we will not precisely indicate the conditions on the coefficients a and σ_r (see, for instance, conditions (a) - (c) above), and will be satisfied by saying that these coefficients satisfy appropriate smoothness and boundedness conditions.

Theorem 5.1. *Suppose the coefficients $a(t,x)$ and $\sigma_r(t)$ of (5.1) satisfy appropriate smoothness and boundedness conditions. Then the method (5.8) has mean-square order of accuracy $3/2$.*

Remark 5.2. In (5.2) and (5.8) the terms containing $(d\sigma_r/dt)(t_k)$ appear because of the approximate representation of the integral $\int_{t_k}^{t_{k+1}} \sigma_r(\theta) dw_r(\theta)$ in the form

$$\int_{t_k}^{t_{k+1}} \sigma_r(\theta) dw_r(\theta) = \sigma_r(t_k)\Delta_k w_r(h) + \frac{d\sigma_r}{dt}(t_k) \int_{t_k}^{t_{k+1}} (\theta - t_k) dw_r(\theta) + \rho, \quad (5.9)$$

where ρ satisfies the relations $E\rho = 0$, $E\rho^2 = O(h^5)$.

The random variables $\int_{t_k}^{t_{k+1}} \sigma_r(\theta) dw_r(\theta)$, $r = 1, \ldots, q$, have a Gaussian distribution. If we model the integrals $\int_{t_k}^{t_{k+1}} \sigma_r(\theta) dw_r(\theta)$ exactly, then we can avoid the computation of $(d\sigma_r/dt)(t_k)$ and drop the requirement on smoothness of $\sigma_r(t)$ (see details in [180, pp. 40-41]).

1.5.2 Implicit methods based on Taylor-type expansion

First, we note that in the case of system (5.1) the formulas (2.21) and (2.22) take the simpler form:

$$X(t+h) = x + \sum_{r=1}^{q} \int_{t}^{t+h} \sigma_r(\theta) dw_r(\theta) + ah$$

$$+ \sum_{r=1}^{q} \Lambda_r a \int_{t}^{t+h} (w_r(\theta) - w_r(t)) d\theta + La\frac{h^2}{2} + \rho, \quad (5.10)$$

$$\rho = \sum_{r=1}^{q}\sum_{i=1}^{q} \int_{t}^{t+h} (\int_{t}^{\theta} (\int_{t}^{\theta_1} \Lambda_i \Lambda_r a(\theta_2, X(\theta_2)) dw_i(\theta_2)) dw_r(\theta_1)) d\theta$$

$$+ \sum_{r=1}^{q} \int_{t}^{t+h} (\int_{t}^{\theta} (\int_{t}^{\theta_1} L\Lambda_r a(\theta_2, X(\theta_2)) d\theta_2) dw_r(\theta_1)) d\theta$$

$$+ \sum_{r=1}^{q} \int_{t}^{t+h} (\int_{t}^{\theta} (\int_{t}^{\theta_1} \Lambda_r La(\theta_2, X(\theta_2)) dw_r(\theta_2)) d\theta_1) d\theta$$

$$+ \int_{t}^{t+h} (\int_{t}^{\theta} (\int_{t}^{\theta_1} L^2 a(\theta_2, X(\theta_2)) d\theta_2) d\theta_1) d\theta. \qquad (5.11)$$

Following the recipe of Sect. 1.3.1, we write the term a in (5.10) as the sum $\alpha a + (1-\alpha)a$. In the second term of this sum we replace a by

$$a(t, x) = a(t+h, X(t+h))$$

$$- \sum_{r=1}^{q} \int_{t}^{t+h} \Lambda_r a(\theta, X(\theta)) dw_r(\theta) - \int_{t}^{t+h} La(\theta, X(\theta)) d\theta$$

$$= a(t+h, X(t+h)) - \sum_{r=1}^{q} \Lambda_r a \int_{t}^{t+h} dw_r(\theta) - La \cdot h + \rho_1, \quad (5.12)$$

where

$$\rho_1 = - \sum_{r=1}^{q} \sum_{i=1}^{q} \int_{t}^{t+h} (\int_{t}^{\theta} \Lambda_i \Lambda_r a(\theta_1, X(\theta_1)) dw_i(\theta_1)) dw_r(\theta)$$

$$- \sum_{r=1}^{q} \int_{t}^{t+h} (\int_{t}^{\theta} L\Lambda_r a(\theta_1, X(\theta_1)) d\theta_1) dw_r(\theta)$$

$$- \sum_{r=1}^{q} \int_{t}^{t+h} (\int_{t}^{\theta} \Lambda_r La(\theta_1, X(\theta_1)) dw_r(\theta_1)) d\theta$$

$$- \int_{t}^{t+h} (\int_{t}^{\theta} L^2 a(\theta_1, X(\theta_1)) d\theta_1) d\theta. \qquad (5.13)$$

Substitute (5.12) in (5.10). The relation obtained will contain a term $(2\alpha - 1)La \cdot h^2/2$. Again we write La as $\beta La + (1-\beta)La$ and replace La in the second term by

$$La(t, x) = La(t+h, X(t+h)) + \rho_2, \qquad (5.14)$$

where

$$\rho_2 = - \int_{t}^{t+h} L^2 a(\theta, X(\theta)) d\theta - \sum_{r=1}^{q} \int_{t}^{t+h} \Lambda_r La(\theta, X(\theta)) dw_r(\theta). \qquad (5.15)$$

Combining all these expressions, we obtain

$$X(t+h) = x + \sum_{r=1}^{q} \int_{t}^{t+h} \sigma_r(\theta)dw_r(\theta)$$

$$+\alpha ah + (1-\alpha)a(t+h, X(t+h))h$$

$$-(1-\alpha)h\sum_{r=1}^{q} \Lambda_r a \int_{t}^{t+h} dw_r(\theta) + \sum_{r=1}^{q} \Lambda_r a \int_{t}^{t+h} (w_r(\theta) - w_r(t))d\theta$$

$$+\beta(2\alpha-1)La\frac{h^2}{2} + (1-\beta)(2\alpha-1)La(t+h, X(t+h))\frac{h^2}{2}$$

$$+\rho + (1-\alpha)\rho_1 h + (1-\beta)(2\alpha-1)\rho_2\frac{h^2}{2}, \qquad (5.16)$$

where ρ is defined by (5.11), ρ_1 by (5.13), and ρ_2 by (5.15).

It can be readily seen that $R = \rho + (1-\alpha)\rho_1 h + (1-\beta)(2\alpha-1)\rho_2 h^2/2$ satisfies the inequalities (of course, under appropriate conditions on a; the σ_r are assumed to be continuous):

$$|ER| \leq K(1+|x|^2)^{1/2}h^3, \ (ER^2)^{1/2} \leq K(1+|x|^2)^{1/2}h^2. \qquad (5.17)$$

If we omit the term R in (5.16), we obtain an implicit one-step approximation whose realization requires modeling of the integrals $\int_{t}^{t+h} \sigma_r(\theta)dw_r(\theta)$. If $\sigma_r''(t)$ exist and are bounded, we can use the representation (5.9). Substituting this in (5.16), we obtain a new remainder, R_1, which, as can readily be seen, satisfies the same inequality as R (see above). Thus we obtain the one-step approximation

$$\bar{X}(t+h) = x + \sum_{r=1}^{q} \sigma_r(t)(w_r(t+h) - w_r(t))$$

$$+\alpha a(t,x)h + (1-\alpha)a(t+h, \bar{X}(t+h))h$$

$$-(1-\alpha)h\sum_{r=1}^{q} \Lambda_r a(t,x)(w_r(t+h) - w_r(t))$$

$$+\sum_{r=1}^{q} \Lambda_r a(t,x) \int_{t}^{t+h} (w_r(\theta) - w_r(t))d\theta$$

$$+\sum_{r=1}^{q} \sigma_r'(t)((w_r(t+h) - w_r(t))h - \int_{t}^{t+h} (w_r(\theta) - w_r(t))d\theta)$$

$$+\beta(2\alpha-1)La(t,x)\frac{h^2}{2}$$

$$+(1-\beta)(2\alpha-1)La(t+h, \bar{X}(t+h))\frac{h^2}{2}. \qquad (5.18)$$

1.5 Explicit and implicit methods of order 3/2

The following two-parameter implicit method corresponds to the approximation (5.18):

$$X_{k+1} = X_k + \sum_{r=1}^{q} \sigma_r(t_k)\xi_{rk}h^{1/2}$$
$$+ \alpha a(t_k, X_k)h + (1-\alpha)a(t_{k+1}, X_{k+1})h$$
$$+ \sum_{r=1}^{q} \Lambda_r a(t_k, X_k)\left(\frac{2\alpha-1}{2}\xi_{rk} + \frac{1}{\sqrt{12}}\eta_{rk}\right)h^{3/2}$$
$$+ \sum_{r=1}^{q} \frac{d\sigma_r}{dt}(t_k)\left(\frac{1}{2}\xi_{rk} - \frac{1}{\sqrt{12}}\eta_{rk}\right)h^{3/2}$$
$$+ \beta(2\alpha-1)La(t_k, X_k)\frac{h^2}{2}$$
$$+ (1-\beta)(2\alpha-1)La(t_{k+1}, X_{k+1})\frac{h^2}{2}, \tag{5.19}$$

where ξ_{rk} and η_{rk} are the same as in the method (5.8).

Theorem 5.3. *Suppose that the coefficients $a(t,x)$ and $\sigma_r(t)$ of (5.1) satisfy appropriate smoothness and boundedness conditions (in particular, a and La satisfy a global Lipschitz condition). Then the implicit one-step approximation (5.18) satisfies the conditions of Theorem 1.1 with $p_1 = 3$, $p_2 = 2$, and so the order of accuracy of the method (5.19) is equal to 3/2. The order of accuracy of the method based on one-step approximation according to formula (5.16) by omitting R is also equal to 3/2.*

Proof. Denote the right-hand side of (5.18) by $F(\bar{X}(t+h))$ regarding all other variables as parameters and rewrite (5.18) as

$$\bar{X}(t+h) = F(\bar{X}(t+h)). \tag{5.20}$$

Then we can write the equation

$$X(t+h) = F(X(t+h)) + R_1, \tag{5.21}$$

where R_1 satisfies inequalities of the same type as R in (5.17).
Evaluate the difference $X(t+h) - \bar{X}(t+h)$ using (5.18), (5.20) – (5.21):

$$X(t+h) - \bar{X}(t+h)$$
$$= (1-\alpha)(a(t+h, X(t+h)) - a(t+h, \bar{X}(t+h)))h$$
$$+ (1-\beta)(2\alpha-1)(La(t+h, X(t+h))$$
$$- La(t+h, \bar{X}(t+h)))\frac{h^2}{2} + R_1. \tag{5.22}$$

Since a and La satisfy a global Lipschitz condition, we have

$$|X(t+h) - \bar{X}(t+h)|$$
$$\leq |1-\alpha| \cdot h \cdot K |X(t+h) - \bar{X}(t+h)|$$
$$+ |1-\beta| \cdot |2\alpha - 1| \cdot K \frac{h^2}{2} |X(t+h) - \bar{X}(t+h)| + |R_1|.$$

Hence, we get for sufficiently small h:

$$|X(t+h) - \bar{X}(t+h)| \leq 2|R_1|,$$

and so (since $X(t) = \bar{X}(t) = x$)

$$E|X(t+h) - \bar{X}(t+h)|^2 \leq K(1+|x|^2)h^4. \tag{5.23}$$

Further, (5.22) implies

$$|E(X(t+h) - \bar{X}(t+h))|$$
$$\leq |1-\alpha| \cdot h \cdot KE|X(t+h) - \bar{X}(t+h)|$$
$$+ |1-\beta| \cdot |2\alpha - 1| \cdot K \frac{h^2}{2} E|X(t+h) - \bar{X}(t+h)| + |ER_1|,$$

whence, by (5.23), we have

$$|E(X(t+h) - \bar{X}(t+h))| \leq K(1+|x|^2)^{1/2} h^3. \tag{5.24}$$

The inequalities (5.23), (5.24) and Theorem 1.1 imply that the method (5.19) has order of accuracy equal to 3/2. The second part of the theorem can be proved in a similar manner. □

Example 5.4. For $\alpha = 1/2$ the method (5.19) becomes

$$X_{k+1} = X_k + \sum_{r=1}^{q} \sigma_r(t_k) \xi_{rk} h^{1/2} + \frac{a(t_k, X_k) + a(t_{k+1}, X_{k+1})}{2} h$$
$$+ \sum_{r=1}^{q} \Lambda_r a(t_k, X_k) \frac{1}{\sqrt{12}} \eta_{rk} h^{3/2} + \sum_{r=1}^{q} \frac{d\sigma_r}{dt}(t_k) (\frac{1}{2} \xi_{rk} - \frac{1}{\sqrt{12}} \eta_{rk}) h^{3/2}. \tag{5.25}$$

In certain respects this method is even simpler than the explicit method (5.8) (it does not contain La). Moreover, as can be seen from the proof of the theorem, in this case it is not necessary to require La to satisfy a Lipschitz condition.

1.5.3 Stiff systems of stochastic differential equations with additive noise. A-stability

As is well known, one often meets deterministic systems for which the application of explicit methods (e.g., Runge–Kutta methods of various orders of accuracy, Adams methods) requires the use of a very small step h on the whole interval of integration. In such a situation, on a relatively small interval of rapid change of the solution the choice of h is dictated by interpolation conditions, while on a substantially larger part of the interval there is no objective necessity for choosing a very small step, and this small step arises only as a consequence of some instability properties of the method itself. We clarify this by an example. Consider the two-dimensional system of linear differential equations with constant coefficients

$$\frac{dX}{dt} = AX, \ X(0) = X_0, \qquad (5.26)$$

where $\lambda_1 \ll \lambda_2 < 0$ are the eigenvalues of the matrix A with eigenvectors a_1 and a_2 (of course, in simulations λ_1, λ_2, a_1, a_2 are not known).

We apply Euler's method to the problem (5.26):

$$X_{k+1} = X_k + AX_k h. \qquad (5.27)$$

Let $X_0 = \alpha^1 a_1 + \alpha^2 a_2$. Then the solution of (5.26) has the form

$$X(t) = \alpha^1 e^{\lambda_1 t} a_1 + \alpha^2 e^{\lambda_2 t} a_2. \qquad (5.28)$$

Since the first component of the solution decreases rapidly because of the condition $\lambda_1 \ll \lambda_2 < 0$, on an initial interval of small length $\sim 1/|\lambda_1|$ we have to choose a very small step in (5.27). On the rest of the interval (whose length we can compare in a natural way with $1/|\lambda_2|$), $X(t)$ changes slowly, and here the interpolation conditions do not require us to choose a small step h. We consider in more detail the nature of the method (5.27). We have

$$X_1 = (I + Ah)X_0 = (I + Ah)(\alpha^1 a_1 + \alpha^2 a_2)$$
$$= (1 + \lambda_1 h)\alpha^1 a_1 + (1 + \lambda_2 h)\alpha^2 a_2.$$

It can be readily seen that

$$X_{k+1} = (1 + \lambda_1 h)^{k+1} \alpha^1 a_1 + (1 + \lambda_2 h)^{k+1} \alpha^2 a_2. \qquad (5.29)$$

The computation by (5.29) will be suitable if h is chosen to satisfy $|1+\lambda_1 h| \leq 1$, i.e.,

$$h \leq \frac{2}{|\lambda_1|}. \qquad (5.30)$$

Moreover, the condition (5.30) must be satisfied on the whole interval of length $\sim 1/|\lambda_2|$. Even when the first component in (5.28) has practically damped out, a step choice $h > 2/|\lambda_1|$ would, because of inevitable errors in the computations, again catch it, and a sharp increase in the error would result. Thus, using the method (5.27), we have to choose the step very small (in accordance with (5.30)) on the whole interval of integration (of course, on the initial interval of length $\approx 1/|\lambda_1|$ the step must even be smaller, but this is because of natural causes and, in view of the smallness of $1/|\lambda_1|$, does not lead to any complications). The necessity of choosing the integration step small not only implies that the amount of computations increases, but also, more importantly, that the computational error increases. As a result, for appropriate λ_1, λ_2 this error may become so large that the method (5.27) becomes inapplicable for solving (5.26).

For $\lambda_1 \ll \lambda_2 < 0$, the system (5.26) belongs to the class of so-called stiff systems [98, 99, 243]. There is no unique generally accepted notion of stiffness, and different authors have proposed various definitions. Here it is better to talk about the *phenomenon of stiffness*, which is characterized, from the point of view of physics, by the presence of both fast and slow processes described by the system of differential equations. When we solve such systems by explicit numerical integration methods, there arises a mismatch between the necessity of choosing a very small integration step on the whole interval and the objective possibility of interpolating the solution on a large part of the interval with a large step (since the solution changes slowly). Moreover, when we use explicit methods, a small increase of the integration step within definite bounds leads to an explosion of the computational error.

Consider the implicit Euler method applied to the system (5.26):

$$X_{k+1} = X_k + AX_{k+1}h. \tag{5.31}$$

We have

$$X_{k+1} = (I - Ah)^{-1} X_k = (I - Ah)^{-(k+1)} X_0$$
$$= \frac{1}{(1 - \lambda_1 h)^{k+1}} \alpha^1 a_1 + \frac{1}{(1 - \lambda_2 h)^{k+1}} \alpha^2 a_2. \tag{5.32}$$

It is clear from (5.32) that the method (5.31) does not have the property of instability, even for arbitrary large h, i.e., choosing h in (5.31) we need only worry about the error of the method. It is clear that in the end the major differences in properties between the methods (5.27) and (5.31) are related to the various means of interpolating the exponents $e^{\lambda_1 t}$. Of course, implicit methods are far more laborious than explicit methods, since in general they require to solve at each step a system of nonlinear equations in X_{k+1}.

A system of *linear equations*

$$\frac{dX}{dt} = AX + b(t), \tag{5.33}$$

with a constant matrix A, which eigenvalues λ_k, $k = 1, \ldots, d$, have negative real parts, is called *stiff* (see [98, 99, 243]) if the following condition holds:

$$\frac{\max_k |\operatorname{Re} \lambda_k|}{\min_k |\operatorname{Re} \lambda_k|} \gg 1. \tag{5.34}$$

A system of *nonlinear equations* is said to belong to the class of *stiff* systems if in a neighborhood of each point (t, x) in the domain under consideration its system of first approximation is stiff.

1.5 Explicit and implicit methods of order 3/2

The quality, with respect to some measure, of a method is conveniently judged by the action of the method on some test system having a small number of parameters and a simple form. To clarify stability properties of a method, one chooses as a test system the equation

$$\frac{dX}{dt} = \lambda X, \qquad (5.35)$$

where λ is a complex parameter with $\operatorname{Re}\lambda < 0$. The choice of (5.35) is related to the fact that every homogeneous system with constant coefficients having distinct eigenvalues with negative real parts can be decomposed into equations of the form (5.35).

The result of applying some method to (5.35) is a difference equation. For example, application of the explicit Euler method leads to the difference equation

$$X_{k+1} = X_k + \lambda h X_k, \qquad (5.36)$$

while application of the implicit Euler method leads to

$$X_{k+1} = X_k + \lambda h X_{k+1}. \qquad (5.37)$$

In such difference equations λh is a parameter. It is natural to require for computational purposes that the trivial solution of such systems be stable. And we have to require stability for all λh belonging to the left halfplane of the complex λh-plane in order to ensure applicability of the method for equations with arbitrary λ (requiring only $\operatorname{Re}\lambda < 0$) as h does not tend to zero.

Definition 5.5. *The region of stability of a method is the set of values of λh satisfying the condition of asymptotic stability of the trivial solution of the difference equation arising when the test equation (5.35) is integrated by this method. A method is called A-stable (absolutely stable) if the halfplane $\operatorname{Re}\lambda h < 0$ belongs to its region of stability.*

There are other definitions of stability, but we will not be concerned with them. It is well known that no explicit Runge–Kutta or Adams method is A-stable. Hence, numerical integration of stiff systems leads to the necessity of constructing implicit methods and of investigating their stability.

It can be seen from (5.37) that the implicit Euler method is A-stable, while (5.36) implies that the region of stability of the explicit Euler method is the interior of the disk with radius one and center at the point $\lambda h = -1$ in the complex λh-plane.

We now turn to the stochastic system with additive noise (5.1). It is natural to say that it is stiff if its deterministic part is stiff; as test equation we can naturally take

$$dX = \lambda X dt + \sigma dw, \qquad (5.38)$$

where λ is a complex parameter with $\operatorname{Re}\lambda < 0$ and σ is an arbitrary real parameter.

The method (5.19) applied to equation (5.38) takes the form

$$X_{k+1} = (1 + \alpha\lambda h + \beta(2\alpha - 1)\frac{(\lambda h)^2}{2})X_k$$
$$+ ((1-\alpha)\lambda h + (1-\beta)(2\alpha-1)\frac{(\lambda h)^2}{2})X_{k+1} + \sigma\xi_k h^{1/2}$$
$$+ \lambda\sigma(\frac{2\alpha-1}{2}\xi_k + \frac{1}{\sqrt{12}}\eta_k)h^{3/2}. \tag{5.39}$$

The equation (5.39) is a difference equation with additive noise. If for $\sigma = 0$ the trivial solution of (5.39) is asymptotically stable, then, in particular, for any σ any solution of (5.39) with $E|X_0|^2 < \infty$ has second-order moments that are uniformly bounded in k. It is readily verified that in the opposite case the second-order moments tend to infinity as $k \to \infty$. Therefore the properties of the method (5.19) can be judged from the stability properties of (5.39) for $\sigma = 0$. Thus, e.g., to clarify the region of stability of a stochastic method, we have to apply the method to equation (5.38) with $\sigma = 0$, i.e., to (5.35), and then clarify the stability properties of the difference equation obtained; the latter is clearly deterministic. In relation with this, Definition 5.5, concerning the region of stability and A-stability of a method, can be transferred without modifications to stochastic numerical integration methods.

Example 5.6. Consider the method (5.25). Applying it to (5.35), we get the difference equation

$$X_{k+1} = (1 + \frac{\lambda h}{2})X_k + \frac{\lambda h}{2}X_{k+1},$$

i.e.,

$$X_{k+1} = \frac{1 + \lambda h/2}{1 - \lambda h/2}X_k.$$

One can see that if $\operatorname{Re}\lambda h < 0$ then

$$\left|\frac{1 + \lambda h/2}{1 - \lambda h/2}\right| < 1,$$

i.e., the region of stability includes the whole left halfplane of the complex λh-plane, and so the method (5.25) is A-stable.

Example 5.7. Consider the method (5.19) for $\alpha = \beta = 0$. It can be readily computed that the region of stability of this method is given by the inequality

$$\frac{1}{|1 - \lambda h + (\lambda h)^2/2|} < 1,$$

or, setting $\lambda h = \mu + i\nu$, by

$$\left|1 - (\mu + i\nu) + \frac{(\mu + i\nu)^2}{2}\right|^2 > 1.$$

The left-hand side of this inequality can be rewritten as follows:

$$(1 - \mu + \frac{\mu^2 - \nu^2}{2})^2 + \nu^2(-1 + \mu)^2$$

$$= (1 - \mu)^2 + (\frac{\mu^2 - \nu^2}{2})^2 + \mu^2(1 - \mu) + \nu^2((1 - \mu)^2 - (1 - \mu)),$$

which clearly exceeds 1 for $\mu < 0$. Hence the method under consideration is A-stable.

Remark 5.8. The question of stability of (implicit or explicit) methods in the case of systems with diffusion coefficients depending on x is far more complicated. A linear autonomous stochastic system

$$dX = AX\,dt + \sum_{r=1}^{q} B_r X\,dw_r(t) \tag{5.40}$$

can be regarded as stiff if, first, its trivial solution is asymptotically stable, e.g. in mean-square, and, secondly, among the negative eigenvalues for the system of second-order moments for (5.40) there are eigenvalues with large as well as small modulus. We may hope that precisely such systems have, in a certain sense, both fast and slow processes. Apparently, the test system in this case is a second-order system of the form (5.40) with one-two noises and a small number of parameters. The application of a method to such a test system leads to a stochastic linear difference system having a trivial solution. As a result, there arises a possibility of judging the quality of the method by investigating the mean-square stability of the trivial solution. All of the above represents not more than mere assumptions. It is clear that in this direction to find appropriate estimates requires a large amount of numerical experiments and serious theoretical investigations.

1.5.4 Runge–Kutta type methods

The most complicated term in the method (5.8) is usually $La(t_k, X_k)$. Using the idea of recalculation, we will construct a method in which $La(t_k, X_k)$ does not occur.

Introduce $\bar{X}^{(1)}(t + h)$ by Euler's method:

$$\bar{X}^{(1)}(t + h) = x + \sum_{r=1}^{q} \sigma_r(t)\xi_r h^{1/2} + a(t, x)h. \tag{5.41}$$

Under the conditions of Theorem 5.1 we have for $\rho_1 = X_{t,x}(t+h) - \bar{X}^{(1)}(t+h)$ (recall that we are considering a system with additive noise for which Euler's method has order of accuracy one):

$$|E\rho_1| \le K(1+|x|^2)^{1/2}h^2, \ (E\rho_1^2)^{1/2} \le K(1+|x|^2)^{1/2}h^{3/2}. \qquad (5.42)$$

Further,

$$a(t+h, X_{t,x}(t+h)) = a(t,x) + \sum_{r=1}^{q} \Lambda_r a(t,x)\xi_r h^{1/2} + La(t,x)h + \rho_2, \qquad (5.43)$$

where

$$|E\rho_2| \le K(1+|x|^2)^{1/2}h^2, \ (E\rho_2^2)^{1/2} \le K(1+|x|^2)^{1/2}h. \qquad (5.44)$$

Put $\rho_3 = a(t+h, X_{t,x}(t+h)) - a(t+h, \bar{X}^{(1)}(t+h))$. Since a satisfies a Lipschitz condition, by the second relation in (5.42) we successively have

$$(E\rho_3^2)^{1/2} \le K(1+|x|^2)^{1/2}h^{3/2}, \ |E\rho_3| \le E|\rho_3| \le (E\rho_3^2)^{1/2}. \qquad (5.45)$$

We can write

$$La(t,x)h = a(t+h, \bar{X}^{(1)}(t+h)) - a(t,x) - \sum_{r=1}^{q} \Lambda_r a(t,x)\xi_r h^{1/2} + \rho_4, \qquad (5.46)$$

where $\rho_4 = \rho_3 - \rho_2$.

By (5.44) and (5.45) we have

$$|E\rho_4| \le K(1+|x|^2)^{1/2}h^{3/2}, \ (E\rho_4^2)^{1/2} \le K(1+|x|^2)^{1/2}h. \qquad (5.47)$$

Consider the one-step approximation that is obtained from (5.8) by replacing $La(t,x)h$ in (5.8) by the right-hand side of (5.46) without the term ρ_4 (of course, in our context we also have to replace (t_k, X_k) by (t,x), ξ_{rk} by ξ_r, and η_{rk} by η_r):

$$\bar{X}(t+h) = x + \sum_{r=1}^{q} \sigma_r(t)\xi_r h^{1/2} + (a(t,x) + a(t+h, \bar{X}^{(1)}(t+h)))\frac{h}{2}$$
$$+ \sum_{r=1}^{q} \Lambda_r a(t,x)\frac{1}{\sqrt{12}}\eta_r h^{3/2} + \sum_{r=1}^{q} \sigma'_r(t)(\frac{1}{2}\xi_r - \frac{1}{\sqrt{12}}\eta_r)h^{3/2}. \qquad (5.48)$$

The value $\bar{X}(t+h)$ computed by (5.48) differs from $\bar{X}(t+h)$ of the one-step approximation (5.8) by $\rho_4 h/2$. Therefore, we have for the $\bar{X}(t+h)$ in (5.48):

$$X(t+h) - \bar{X}(t+h) = \rho + \rho_4 \frac{h}{2},$$

where ρ satisfies the relations (see the proof of (5.8) and of its order of accuracy):

$$|E\rho| \le K(1+|x|^2)^{1/2}h^3, \ (E\rho^2)^{1/2} \le K(1+|x|^2)^{1/2}h^2.$$

1.5 Explicit and implicit methods of order 3/2   73

By (5.47) we have

$$|E(\rho + \rho_4 \frac{h}{2})| = O(h^{5/2}), \ (E(\rho + \rho_4 \frac{h}{2})^2)^{1/2} = O(h^2).$$

Since $p_1 = 5/2$, $p_2 = 2$, Theorem 1.1 implies that the method corresponding to the one-step approximation (5.48) has order of accuracy 3/2. We state the result in the next theorem.

Theorem 5.9. *Suppose that the coefficients $a(t,x)$ and $\sigma_r(t)$ of (5.1) satisfy appropriate smoothness and boundedness conditions. Then the method*

$$X_{k+1} = X_k + \sum_{r=1}^{q} \sigma_r(t_k)\xi_{rk}h^{1/2}$$

$$+ (a(t_k, X_k) + a(t_{k+1}, X_k + \sum_{r=1}^{q} \sigma_r(t_k)\xi_{rk}h^{1/2} + a(t_k, X_k)h))\frac{h}{2}$$

$$+ \sum_{r=1}^{q} \Lambda_r a(t_k, X_k) \frac{1}{\sqrt{12}} \eta_{rk} h^{3/2}$$

$$+ \sum_{r=1}^{q} \frac{d\sigma_r}{dt}(t_k)(\frac{1}{2}\xi_{rk} - \frac{1}{\sqrt{12}}\eta_{rk})h^{3/2}, \tag{5.49}$$

where ξ_{rk} and η_{rk} are the same as in the method (5.8), has order of accuracy 3/2.

Remark 5.10. In (5.48), $\Lambda_r a(t,x)$ contains first-order derivatives of a with respect to x:

$$\Lambda_r a^i(t,x) = \sum_{j=1}^{d} \sigma_r^j(t)(\partial a^i/\partial x^j)(t,x).$$

Therefore, the method (5.49), which is based on (5.48), is not a fully Runge–Kutta method. In fact, getting rid of the computation of the derivatives is not difficult. For example,

$$\frac{\partial a^i}{\partial x^j} \approx \frac{a^i(x^1, \ldots, x^j + \Delta x^j, \ldots, x^d) - a^i(x^1, \ldots, x^j - \Delta x^j, \ldots, x^d)}{2\Delta x^j}.$$

This is an equality up to $O((\Delta x^j)^2)$ by the assumptions made with respect to the function a^i. Then replacement of all $\partial a^i/\partial x^j$ in (5.49) by their difference relations preserves the order of accuracy. However, this approach requires a large amount of recalculations. In the deterministic theory Runge–Kutta methods use a minimal amount of recalculations. Here we can also compute $\Lambda_r a(t,x)$ using only a single recalculation of vector a. For this it suffices to use the identity

$$\Lambda_r a(t,x) = \frac{a(t, x + \sigma_r(t)h) - a(t,x)}{h} + O(h). \tag{5.50}$$

Note that $x + \sigma_r(t)h$ is the Euler approximation for the Cauchy problem

$$\frac{dY_r}{dt} = \sigma_r(t), \quad Y_r(t) = x. \tag{5.51}$$

As a result, we write the following Runge–Kutta method instead of (5.49):

$$X^{(1)}_{k+1} = X_k + \sum_{r=1}^{q} \sigma_r(t_k)\xi_{rk}h^{1/2} + c(t_k, X_k)h,$$

$$Y^{(1)}_{r,k+1} = X_k + \sigma_r(t_k)h,$$

$$X_{k+1} = X_k + \sum_{r=1}^{q} \sigma_r(t_k)\xi_{rk}h^{1/2} + (a(t_k, X_k) + a(t_k, X^{(1)}_{k+1}))\frac{h}{2}$$

$$+ \sum_{r=1}^{q}(a(t_k, Y^{(1)}_{r,k+1}) - a(t_k, X_k))\frac{1}{\sqrt{12}}\eta_{rk}h^{1/2}$$

$$+ \sum_{r=1}^{q}\frac{d\sigma_r}{dt}(t_k)(\frac{1}{2}\xi_{rk} - \frac{1}{\sqrt{12}}\eta_{rk})h^{3/2}. \tag{5.52}$$

The idea of invoking other systems of differential equations along with the original system, in the spirit of (5.50)-(5.51), to economize the amount of recalculations may turn out to be also useful in substantially more general situations. However, here we restrict ourselves to the remark.

Remark 5.11. It is also possible to construct fully Runge–Kutta schemes by the method of undetermined coefficients demonstrated in Sect. 4.3.1. Using this approach, we construct the following explicit Runge–Kutta method for the system (5.1):

$$X_{k+1} = X_k + \sum_{r=1}^{q} \sigma_r(t_k)\xi_{rk}h^{1/2}$$

$$+ \frac{h}{2}a(t_k, X_k + \sum_{r=1}^{q}\sigma_r(t_k)[(\frac{1}{2} + \frac{1}{\sqrt{6}})\xi_{rk} + \frac{1}{\sqrt{12}}\eta_{rk}]h^{1/2})$$

$$+ \frac{h}{2}a(t_{k+1}, X_k + ha(t_k, X_k))$$

$$+ \sum_{r=1}^{q}\sigma_r(t_k)[(\frac{1}{2} - \frac{1}{\sqrt{6}})\xi_{rk} + \frac{1}{\sqrt{12}}\eta_{rk}]h^{1/2})$$

$$+ \sum_{r=1}^{q}\frac{d\sigma_r}{dt}(t_k)(\frac{1}{2}\xi_{rk} - \frac{1}{\sqrt{12}}\eta_{rk})h^{3/2}. \tag{5.53}$$

One can prove that under appropriate assumptions on the coefficients of (5.1) the method (5.53) is of mean-square order 3/2.

The method (5.49) is an explicit Runge–Kutta method. We can construct implicit Runge–Kutta methods by writing the implicit versions of formulas (5.41) and (5.43) and preserve all the remaining derivations. We can also substitute the right-hand side of (5.46) without ρ_4 in (5.19) putting $\beta = 1$ in (5.19). Having done, for example, the latter, we obtain a one-parameter family of implicit Runge–Kutta methods:

$$X_{k+1} = X_k + \sum_{r=1}^{q} \sigma_r(t_k)\xi_{rk}h^{1/2} + a(t_k, X_k)\frac{h}{2}$$

$$+ (1-\alpha)a(t_{k+1}, X_{k+1})h + (\alpha - \frac{1}{2})a(t_{k+1}, X_{k+1}^{(1)})h$$

$$+ \sum_{r=1}^{q} \Lambda_r a(t_k, X_k)\frac{1}{\sqrt{12}}\eta_{rk}h^{3/2}$$

$$+ \sum_{r=1}^{q} \frac{d\sigma_r}{dt}(t_k)(\frac{1}{2}\xi_{rk} - \frac{1}{\sqrt{12}}\eta_{rk})h^{3/2}, \tag{5.54}$$

where $X_{k+1}^{(1)}$ is the same as in (5.52). Following the proofs of Theorems 5.3 and 5.9, it is not difficult to prove that the method (5.54) has order of accuracy 3/2 (here the additional assumption that $La(t,x)$ has to satisfy a uniform Lipschitz condition in x can be dropped).

Another implicit Runge–Kutta method of order 3/2 for (5.1) is constructed in Sect. 4.3.1 (see (4.3.11)-(4.3.14)).

1.5.5 Two-step difference methods

In (5.19) we put $\beta = 0$, $\alpha = \alpha_1 \neq 1/2$, and we express $La(t_{k+1}, X_{k+1})$ in terms of X_k, X_{k+1}, ξ_{rk}, and η_{rk}. Then we take $k+1$ instead of k in (5.19), put $\beta = 1$, $\alpha = \alpha_2$, and replace $La(t_{k+1}, X_{k+1})$ by the expression just found. As a result, we obtain

$$X_{k+2} = \frac{1 - 2\alpha_2}{2\alpha_1 - 1}X_k + \frac{2(\alpha_1 + \alpha_2 - 1)}{2\alpha_1 - 1}X_{k+1}$$

$$+ \frac{1 - 2\alpha_2}{2\alpha_1 - 1}\sum_{r=1}^{q} \sigma_r(t_k)\xi_{rk}h^{1/2}$$

$$+ \sum_{r=1}^{q} \sigma_r(t_{k+1})\xi_{r(k+1)}h^{1/2} + \frac{\alpha_1(1 - 2\alpha_2)}{2\alpha_1 - 1}a(t_k, X_k)h$$

$$+ \frac{4\alpha_1\alpha_2 - \alpha_1 - 3\alpha_2 + 1}{2\alpha_1 - 1}a(t_{k+1}, X_{k+1})h + (1 - \alpha_2)a(t_{k+2}, X_{k+2})h$$

$$+\frac{1-2\alpha_2}{2\alpha_1-1}\sum_{r=1}^{q}\Lambda_r a(t_k,X_k)(\frac{2\alpha_1-1}{2}\xi_{rk}+\frac{1}{\sqrt{12}}\eta_{rk})h^{3/2}$$

$$+\frac{1-2\alpha_2}{2\alpha_1-1}\sum_{r=1}^{q}\frac{d\sigma_r}{dt}(t_k)(\frac{1}{2}\xi_{rk}-\frac{1}{\sqrt{12}}\eta_{rk})h^{3/2}$$

$$+\sum_{r=1}^{q}\Lambda_r a(t_{k+1},X_{k+1})(\frac{2\alpha_2-1}{2}\xi_{r(k+1)}+\frac{1}{\sqrt{12}}\eta_{r(k+1)})h^{3/2}$$

$$+\sum_{r=1}^{q}\frac{d\sigma_r}{dt}(t_{k+1})(\frac{1}{2}\xi_{r(k+1)}-\frac{1}{\sqrt{12}}\eta_{r(k+1)})h^{3/2}. \tag{5.55}$$

For $\alpha_2 = 1/2$ the method (5.55) coincides with the implicit one-step method (5.25) from Example 5.4 (with index k increased by one). For $\alpha_2 = 1$, $\alpha_1 \neq 1/2$ this is a one-parameter family of explicit two-step difference methods. For other α_2 and $\alpha_1 \neq 1/2$ this is a two-parameter family of implicit two-step difference methods. The order of accuracy of the method (5.55) is stated in Theorem 5.12, a proof of which is available in [180]. Note that we cannot use Theorem 1.1 here, since it is highly accommodated to one-step methods only.

Theorem 5.12. *Suppose that the coefficients $a(t,x)$ and $\sigma_r(t)$ of (5.1) satisfy appropriate smoothness and boundedness conditions. Suppose*

$$0 \leq \frac{1-2\alpha_2}{2\alpha_1-1} \leq 1. \tag{5.56}$$

Then the method (5.55) has order of accuracy $3/2$ (of course, under the assumptions that $X_0 = X(t_0)$, $X_1 = X(t_1)$).

Remark 5.13. The method (5.55) has the same features as difference methods in the deterministic situation. We do not compute La in it, while in comparison with the Runge–Kutta method it does not require recalculations. At the same time, to use it one has to find a value X_1 that is sufficiently close to $X(t_1)$. To this end, as in the deterministic situation, X_1 has to be found beforehand by using a one-step method that integrates the system (5.1) on the interval $[t_0, t_0 + h]$ with a small auxiliary step.

Example 5.14. Consider the method (5.55) with $\alpha_1 = -1/2$, $\alpha_2 = 1$ (it is explicit). We investigate its A-stability. Applying it to the test equation (5.35), we get the difference equation

$$X_{k+2} = (\frac{1}{2}-\frac{1}{4}\lambda h)X_k + (\frac{1}{2}-\frac{7}{4}\lambda h)X_{k+1}.$$

It is easy to convince oneself that negative λh with sufficiently large absolute values do not belong to the region of stability. Therefore this method is not A-stable.

Consider now the method (5.55) with $\alpha_1 = 1$, $\alpha_2 = 0$. The corresponding difference equation has the form

$$X_{k+2} = (1 + \lambda h)X_k + \lambda h X_{k+2}.$$

Its trivial solution is asymptotically stable for all λh in the left halfplane. Therefore this method is A-stable.

1.6 Numerical schemes for equations with colored noise

The simplest approximation of real fluctuations that affect a physical system is Gaussian white noise. However, Gaussian white noise, or a Gaussian delta-correlated random process, is a stochastic process with zero correlation time and infinite variance, so it is an unreal process. Such a random process may be considered only as the first approximation of real fluctuation with a short correlation time. This shortcoming is overcome by colored noise (finite-bandwidth noise) [78, 106].

Herein we consider differential equations with exponentially correlated colored noise

$$dY = f(Y)dt + G(Y)Zdt$$
$$dZ = AZdt + \sum_{r=1}^{q} b_r dw_r(t), \qquad (6.1)$$

where Y and f are l-dimensional vectors, Z and b_r are m-dimensional vectors, A is an $m \times m$ matrix, G is an $l \times m$ matrix, and w_r are uncorrelated standard Wiener processes. In the one-dimensional case equations (6.1) are rewritten in the form

$$dy = f(y)dt + g(y)zdt$$
$$dz = -azdt + bdw, \qquad (6.2)$$

where z is the well-known Ornstein–Uhlenbeck process (a is supposed to be a positive number), or exponentially correlated colored noise, with the properties

$$Ez(t) = 0, \quad Ez(t)z(s) = \frac{b^2}{2a} \exp(-a|t-s|).$$

The system (6.1) is simpler than the general system of SDEs by two reasons: (1) (6.1) is a system with additive Gaussian white noise, (2) equations (6.1) are linear with respect to Z. That is why comparatively simple high-order methods may be constructed for such a system. It is also possible to consider nonautonomous systems, however, here we restrict ourselves to the system (6.1). In the earlier works (see, e.g. [69] and references therein), efficient explicit algorithms up to the second order were obtained. For the first time various methods for the system (6.1) were easily derived and justified on the basis of the general theory in [199]. Moreover, we present efficient implicit and Runge–Kutta schemes. From the point of view of numerical integration, the special features of the system (6.1) consist in the absence of random variables of the form I_{i_1,i_2}, I_{i_1,i_2,i_3}, and $I_{i_1,i_2,0}$, $i_1,i_2,i_3 \neq 0$, in the Taylor-type expansions, and therefore we are able to construct various constructive (with respect to modeling of random variables) methods with order of accuracy reaching 5/2. Some numerical tests of methods given in this section are presented in [199].

1.6.1 Explicit schemes of orders 2 and 5/2

For the system (6.1), the coefficients a, σ_r and the operators L and Λ_r take the form

$$x = \begin{bmatrix} y \\ z \end{bmatrix}, \quad a = \begin{bmatrix} f(y) + G(y)z \\ Az \end{bmatrix}, \quad \sigma_r = \begin{bmatrix} 0 \\ b_r \end{bmatrix},$$

$$L = (f(y) + G(y)z, \frac{\partial}{\partial y}) + (Az, \frac{\partial}{\partial z}) + \frac{1}{2}\sum_{r=1}^{q}\sum_{i,j=1}^{m} b_r^i b_r^j \frac{\partial^2}{\partial z^i \partial z^j},$$

$$\Lambda_r = (b_r, \frac{\partial}{\partial z}). \tag{6.3}$$

We have

$$\Lambda_r a = \begin{bmatrix} G(y)b_r \\ Ab_r \end{bmatrix}, \quad La = \begin{bmatrix} [f'_y + (Gz)'_y](f + Gz) + GAz \\ A^2 z \end{bmatrix}, \tag{6.4}$$

where f'_y is the Jacobian matrix, $(Gz)'_y = [G'_{y^1}z \; G'_{y^2}z \cdots G'_{y^l}z]$ is an $l \times l$ matrix the columns of which are $G'_{y^1}z, G'_{y^2}z, \ldots, G'_{y^l}z$.

The method (5.8) in the case of the considered system (6.1) acquires the form

$$Y_{k+1} = Y_k + (f + Gz)_k h + \frac{1}{2}G_k \sum_{r=1}^{q} b_r(\xi_{rk} + \frac{1}{\sqrt{3}}\eta_{rk})h^{3/2}$$

$$+ \frac{h^2}{2}([f'_y + (Gz)'_y]_k(f + Gz)_k + G_k AZ_k),$$

$$Z_{k+1} = Z_k + \sum_{r=1}^{q} b_r \xi_{rk} h^{1/2} + AZ_k h$$

$$+ \frac{1}{2}A\sum_{r=1}^{q} b_r(\xi_{rk} + \frac{1}{\sqrt{3}}\eta_{rk})h^{3/2} + \frac{h^2}{2}A^2 Z_k, \tag{6.5}$$

where, for example, $(f + Gz)_k = f(Y_k) + G(Y_k)Z_k$.

Recall that due to Theorem 5.1 the method (5.8) for the general system with additive noise is of order 3/2. It turns out that for the system (6.1) this method is of order 2.

Theorem 6.1. *Suppose the coefficients $f(y)$ and $G(y)$ of (6.1) satisfy appropriate smoothness and boundedness conditions. Then the method (6.5) has mean-square order of accuracy equal to 2.*

Proof. Since $\Lambda_r a$ depends on y only (see (6.3)), all the $\Lambda_i \Lambda_r a$ are equal to zero. Consequently, the remainder ρ (see formula (5.11)) for method (6.5) does not contain the integrals of order 2, and we obtain

$$|E\rho| = O(h^3), \ (E|\rho|^2)^{1/2} = O(h^{5/2}),$$

i.e., $p_1 = 3$, $p_2 = 5/2$. Now application of the fundamental theorem proves the result. □

Further expansion of the integrals of ρ in the formula (5.11) for the system (6.1) gives the following scheme

$$Y_{k+1} = \tilde{Y}_{k+1} + \sum_{r=1}^{q}[(Gb_r)'_y]_k(f+Gz)_k(I_{0r0})_k$$

$$+ \sum_{r=1}^{q}[f'_y Gb_r + GAb_r + \Lambda_r\{(Gz)'_y(f+Gz)\}]_k(I_{r00})_k$$

$$+ \frac{h^3}{6}[L^2(f+Gz)]_k,$$

$$Z_{k+1} = \tilde{Z}_{k+1} + \sum_{r=1}^{q} A^2 b_r (I_{r00})_k + \frac{h^3}{6} A^3 Z_k. \quad (6.6)$$

In (6.6) $\tilde{Y}_{k+1}$ and $\tilde{Z}_{k+1}$ are the right-hand sides of (6.5), $(Gb_r)'_y = [G'_{y^1} b_r\ G'_{y^2} b_r \cdots G'_{y^l} b_r]$ is an $l \times l$ matrix the columns of which are $G'_{y^1} b_r, G'_{y^2} b_r, \ldots, G'_{y^l} b_r$, I_{0r0} and I_{r00} are the known integrals (see the notation in Sect. 1.2.2). We have

$$I_{0r0} = 2J_r - hI_{r0}, \ I_{r00} = hI_{r0} - J_r, \ J_r = \int_0^h \theta w_r(\theta) d\theta,$$

These integrals can be simulated according to relations

$$I_{r0} = \frac{1}{2}h^{3/2}(\xi_r + \frac{1}{\sqrt{3}}\eta_r), \ J_r = h^{5/2}(\frac{1}{3}\xi_r + \frac{1}{4\sqrt{3}}\eta_r + \frac{1}{12\sqrt{5}}\zeta_r),$$

where ξ_r, η_r, and ζ_r are independent random variables with standard Gaussian distribution $\mathcal{N}(0,1)$ which are independently simulated at each step.

It is not difficult to prove that for this method $p_1 = 4$, $p_2 = 3$. Therefore the following theorem is true.

Theorem 6.2. *Suppose the coefficients $f(y)$ and $G(y)$ of (6.1) satisfy appropriate smoothness and boundedness conditions. Then the method (6.6) has mean-square order of accuracy equal to 5/2.*

Remark 6.3. As mentioned above, for a general system mean-square methods of order 1/2 only may be obtained with easily simulated random variables. The higher-order methods need numerical solution of a special system of SDEs at each step for the simulation of the Ito integrals or some approximation of repeated Ito integrals in the case of the first-order scheme. However, for the system with colored noise (6.1) efficient mean-square methods up to the 5/2 order are derived according to special properties of the system (6.1). By the way, third-order schemes for (6.1) require calculation of repeated Ito integrals, and in the case of nonlinear functions f and G it is impossible to obtain an efficient third-order mean-square method with easily simulated random variables.

1.6.2 Runge–Kutta schemes

To reduce calculations of derivatives, we propose the explicit second-order Runge–Kutta scheme

$$Y_{k+1} = Y_k + \frac{1}{2}h(f(Y_k) + G(Y_k)Z_k) + \frac{1}{2}h(f(\tilde{Y}_k) + G(\tilde{Y}_k)\tilde{Z}_k)$$

$$+ \sum_{r=1}^{q} G(Y_k) b_r h^{3/2} \eta_{rk}/\sqrt{12},$$

$$Z_{k+1} = Z_k + \sum_{r=1}^{q} b_r \xi_{rk} h^{1/2} + \frac{1}{2}h(Z_k + \tilde{Z}_k) + \sum_{r=1}^{q} A b_r h^{3/2} \eta_{rk}/\sqrt{12}, \quad (6.7)$$

where

$$\tilde{Y}_k = Y_k + (f(Y_k) + G(Y_k)Z_k)h,$$

$$\tilde{Z}_k = Z_k + \sum_{r=1}^{q} b_r \xi_{rk} h^{1/2} + A Z_k h. \quad (6.8)$$

This algorithm has been derived by the substitution of the expansions

$$\frac{1}{2}(f(\tilde{Y}_k) + G(\tilde{Y}_k)\tilde{Z}_k) = \frac{1}{2}(f(Y_k) + G(Y_k)Z_k) + \frac{1}{2}\sum_{r=1}^{q} G(Y_k) b_r h^{1/2} \xi_{rk}$$

$$+ \frac{h}{2}([f'_y + (Gz)'_y]_k(f + Gz)_k + G(Y_k)AZ_k) + \rho_1,$$

$$\frac{1}{2}A\tilde{Z}_k = \frac{1}{2}AZ_k + \frac{1}{2}\sum_{r=1}^{q} A b_r h^{1/2} \xi_{rk} + \frac{1}{2}A^2 Z_k h,$$

$$|E\rho_1| = O(h^2), \quad (E|\rho_1|^2)^{1/2} = O(h^{3/2}), \quad (6.9)$$

in the second-order scheme (6.5).

The Runge–Kutta scheme (6.7)-(6.8) does not include any derivatives. Thanks to the special properties of the system (6.1), it is a fully Runge–Kutta algorithm. The 5/2-order explicit method (6.6) can be simplified by the idea of attracting a subsidiary system of deterministic equations (see Remark 5.10). A method obtained in this way is available in [199].

1.6.3 Implicit schemes

A family of the first-order implicit methods (implicit Euler schemes) has the form (cf. (3.8)):

$$Y_{k+1} = Y_k + \alpha h(f + Gz)_k + (1 - \alpha)h(f + Gz)_{k+1}$$

$$Z_{k+1} = Z_k + \sum_{r=1}^{q} b_r \xi_{rk} h^{1/2} + \alpha h A Z_k + (1 - \alpha) h A Z_{k+1}, \qquad (6.10)$$

where ξ_{rk} are independent normally distributed $\mathcal{N}(0, 1)$ random variables, and $0 \leq \alpha \leq 1$.

We present the two-parameter family of second-order implicit schemes (cf. (5.19)):

$$Y_{k+1} = Y_k + \alpha h(f + Gz)_k + (1 - \alpha)h(f + Gz)_{k+1}$$

$$+ h^{3/2} \sum_{r=1}^{q} G(Y_k) b_r ((2\alpha - 1)\xi_{rk}/2 + \eta_{rk}/\sqrt{12})$$

$$+ \beta(2\alpha - 1)h^2 [L(f + Gz)]_k/2 + (1 - \beta)(2\alpha - 1)h^2 [L(f + Gz)]_{k+1}/2,$$

$$Z_{k+1} = Z_k + \sum_{r=1}^{q} b_r \xi_{rk} h^{1/2} + \alpha h A Z_k + (1 - \alpha) h A Z_{k+1}$$

$$+ h^{3/2} \sum_{r=1}^{q} A b_r ((2\alpha - 1)\xi_{rk}/2 + \eta_{rk}/\sqrt{12})$$

$$+ \beta(2\alpha - 1)h^2 A^2 Z_k/2 + (1 - \beta)(2\alpha - 1)h^2 A^2 Z_{k+1}/2, \quad (6.11)$$

where $0 \leq \alpha, \beta \leq 1$. The family (6.11) is derived by representing the terms $(f + Gz)_k$ and AZ_k of (6.5) in the form

$$(f + Gz)_k = \alpha(f + Gz)_k + (1 - \alpha)(f + Gz)_{k+1}$$

$$- (1 - \alpha)(\sum_{r=1}^{q} G(Y_k) b_r h^{1/2} \xi_{rk} + h[L(f + Gz)]_k + \rho_1),$$

$$AZ_k = \alpha AZ_k + (1 - \alpha)(AZ_{k+1} - \sum_{r=1}^{q} A b_r h^{1/2} \xi_{rk} + h A^2 Z_k + \rho_2),$$

$$|E\rho_i| = O(h^2), \ (E|\rho_i|^2)^{1/2} = O(h^{3/2}), \ i = 1, 2, \qquad (6.12)$$

and with the expressions

$$[L(f+Gz)]_k = \beta[L(f+Gz)]_k + (1-\beta)[L(f+Gz)]_{k+1} + \rho_3,$$
$$A^2 Z_k = \beta A^2 Z_k + (1-\beta) A^2 Z_{k+1} + \rho_4,$$
$$|E\rho_i| = O(h),\ (E|\rho_i|^2)^{1/2} = O(h^{1/2}),\ i = 3, 4. \qquad (6.13)$$

If we choose $\alpha = 1/2$ in (6.11), we obtain the simplest scheme of the family (6.11), which is called the trapezoidal method (cf. (5.25)):

$$Y_{k+1} = Y_k + \frac{1}{2}h(f+Gz)_k + \frac{1}{2}h(f+Gz)_{k+1}$$
$$+ h^{3/2} \sum_{r=1}^{q} G(Y_k) b_r \eta_{rk}/\sqrt{12},$$
$$Z_{k+1} = Z_k + \sum_{r=1}^{q} b_r \xi_{rk} h^{1/2} + \frac{1}{2} hA(Z_k + Z_{k+1})$$
$$+ h^{3/2} \sum_{r=1}^{q} A b_r \eta_{rk}/\sqrt{12}. \qquad (6.14)$$

2 Weak approximation for stochastic differential equations

Using probabilistic representations together with Monte Carlo methods, a complex multi-dimensional problem for partial differential equations can be reduced to the Cauchy problem for a system of SDEs. The last system, which contains one independent variable only, arises as a characteristic system of the considered problems for PDEs.

In its simplest form, the method of characteristics is as follows. Consider a system of d ordinary differential equations

$$dX = a(X)dt. \tag{0.1}$$

Let $X_x(t)$ be the solution of this system satisfying the initial condition $X_x(0) = 0$. Then we have for an arbitrary continuously differentiable function $u(x)$:

$$u(X_x(t)) - u(x) = \int_0^t (a(X_x(t)), \frac{\partial u}{\partial x}(X_x(t)))dt. \tag{0.2}$$

Consider the Cauchy problem for the first-order linear partial differential equation

$$(a(x), \frac{\partial u}{\partial x}) = 0, \tag{0.3}$$

$$u_{|\gamma} = f(x), \tag{0.4}$$

where γ is a curve in the d-dimensional space of the variable x. Let u be a solution of equation (0.3). Then (0.2) implies

$$u(x) = u(X_x(t)). \tag{0.5}$$

The formula (0.5) indicates the following way for solving the problem (0.3)-(0.4): starting at x, draw the trajectory $X_x(t)$ of the system (0.1) up to the moment τ of its intersection with γ. By (0.4), u is known on γ. Therefore

$$u(x) = u(X_x(\tau)) = f(X_x(\tau)). \tag{0.6}$$

Now we consider the system of stochastic differential equations

$$dX = a(X)dt + \sum_{r=1}^{q} \sigma_r(X)dw_r(t). \tag{0.7}$$

Applying Ito's formula to a sufficiently smooth function $u(x)$, we get the following analogue of (0.2):

$$u(X_x(\tau)) - u(x) = \int_0^\tau Lu(X_x(t))dt + \sum_{r=1}^q \int_0^\tau \Lambda_r u(X_x(t))dw_r(t). \quad (0.8)$$

In this formula, τ is a Markov moment and

$$L = (a, \frac{\partial}{\partial x}) + \frac{1}{2}\sum_{r=1}^q (\sigma_r, \frac{\partial}{\partial x})^2 = \sum_{i=1}^d a^i \frac{\partial}{\partial x^i} + \frac{1}{2}\sum_{r=1}^q \sum_{i,j=1}^d \sigma_r^i \sigma_r^j \frac{\partial^2}{\partial x^i \partial x^j},$$

$$\Lambda_r = (\sigma_r, \frac{\partial}{\partial x}) = \sum_{i=1}^d \sigma_r^i \frac{\partial}{\partial x^i},$$

where a^i, σ_r^i are the components of the vectors a, σ_r.

For an elliptic-type equation

$$Lu = 0, \quad (0.9)$$

we consider the Dirichlet problem in a domain G with boundary condition

$$u|_{\partial G} = f(x). \quad (0.10)$$

Let u be a solution of (0.9). Then (0.8) implies

$$u(x) = u(X_x(\tau)) - \sum_{r=1}^q \int_0^\tau \Lambda_r u(X_x(t))dw_r(t). \quad (0.11)$$

Taking τ as the time at which the trajectory $X_x(t)$ hits the boundary ∂G and averaging (0.11), we arrive at a probabilistic representation of the solution of (0.9)-(0.10):

$$u(x) = Eu(X_x(\tau)) = Ef(X_x(\tau)). \quad (0.12)$$

Using the Monte Carlo approach, we obtain

$$u(x) \simeq \frac{1}{M}\sum_{m=1}^M f(X_x^{(m)}(\tau_x^{(m)})), \quad (0.13)$$

where $X_x^{(m)}(t)$, $m = 1, \ldots, M$, are independent realizations of the process $X_x(t)$ defined by the system (0.7).

Thus, the multi-dimensional boundary value problem (0.9)-(0.10) reduces to the Cauchy problem for the system (0.7). This system can be naturally regarded as one-dimensional, since it contains one independent variable only. The system (0.7) comes about as characteristic system of differential equations for the problem (0.9)-(0.10). This approach, which enables reduction of

2 Weak approximation for stochastic differential equations

a multi-dimensional boundary value problem to a one-dimensional Cauchy problem, is of considerable importance for computational mathematics.

Let us give the well-known probabilistic representation to the solution of the Cauchy problem for the heat equation

$$\frac{\partial u}{\partial t} + \sum_{i=1}^{d} a^i(t,x)\frac{\partial u}{\partial x^i} + \frac{1}{2}\sum_{r=1}^{q}\sum_{i,j=1}^{d} \sigma_r^i(t,x)\sigma_r^j(t,x)\frac{\partial^2 u}{\partial x^i \partial x^j} = 0, \quad (0.14)$$

$$u(T,x) = f(x), \quad (0.15)$$

where $t_0 \leq t \leq T$, $x \in \mathbf{R}^d$.

The value of the unknown function u at a point (s,x) can be expressed as a mathematical expectation:

$$u(s,x) = Ef(X_{s,x}(T)), \quad (0.16)$$

where $X_{s,x}(t)$ is the solution of the following system of SDEs (which is not autonomous in distinction to (0.7)):

$$dX = a(t,X)dt + \sum_{r=1}^{q} \sigma_r(t,X)dw_r(t),$$

$$X_{s,x}(s) = x, \ s \leq t \leq T. \quad (0.17)$$

We note that the representations (0.12) and (0.16) are the well-known *Feynman–Kac formula*.

Application of the Monte Carlo technique gives

$$u(s,x) = Ef(X_{s,x}(T)) \simeq \frac{1}{M}\sum_{m=1}^{M} f(X_{s,x}^{(m)}(T)), \quad (0.18)$$

where $X_{s,x}^{(m)}(T)$, $m = 1, \ldots, M$, are independent realizations of the random variable $X_{s,x}(T)$.

To be able to use (0.18) (see also (0.13)), we have to model the random variable $X_{s,x}(T)$. The exact computation of $X_{s,x}(T)$ is impossible by and large even in the deterministic situation. Therefore, we have to replace $X_{s,x}(T)$ by a nearly random variable $\bar{X}_{s,x}(T)$ that can be modeled. Instead of (0.18) we obtain

$$u(s,x) = Ef(X_{s,x}(T)) \simeq Ef(\bar{X}_{s,x}(T)) \simeq \frac{1}{M}\sum_{m=1}^{M} f(X_{s,x}^{(m)}(T)). \quad (0.19)$$

The first approximate equality in (0.19) involves an error brought about by replacing X by $\bar{X}$ (an error related to the approximate integration of the system (0.17)); in the second approximate equality the error comes from the Monte Carlo technique.

2 Weak approximation for stochastic differential equations

While modeling the solution of a system of SDEs is a prerequisite for using the Monte Carlo technique, it is not necessary at all to solve the very complicated problem of constructing mean-square approximations. Let $X(t)$ be the exact and $\bar{X}(t)$ be an approximate solution. In many problems of mathematical physics it is only required that the expectation $Ef(\bar{X}(T))$ is close to $Ef(X(T))$ for a sufficiently large class of functions f, i.e., that $\bar{X}(t)$ is close to $X(t)$ in a weak sense. If an approximation $\bar{X}$ is such that

$$|Ef(\bar{X}(T)) - Ef(X(T))| \leq Ch^p \tag{0.20}$$

for f from a sufficiently large class of functions, then we say that the *weak order of accuracy* of the approximation $\bar{X}$ (the method $\bar{X}$) is p. We can prove, for example, that the weak order of accuracy of Euler's method is one. Note that numerical integration in the mean-square sense with some order of accuracy guarantees an approximation in the weak sense with the same order of accuracy, since if $(E|\bar{X}(t) - X(t)|^2)^{1/2} = O(h^p)$ then for every function f satisfying a Lipschitz condition we have $E(f(\bar{X}(T) - f(X(T))) = O(h^p)$. Moreover, an increase in the order of accuracy in the mean-square sense does not, in general, imply an increase of the weak order of accuracy. For example, the method (1.0.5) has first weak order of accuracy as Euler's method does. At the same time, a crude method like (we give the formula for a scalar equation):

$$X_{k+1} = X_k + a_k h + \sigma_k \alpha_k h^{1/2}, \tag{0.21}$$

where α_k, $k = 0, \ldots, N-1$, are independent random variables taking the values $+1$ and -1 with probabilities $1/2$, also has first order of accuracy in the sense of weak approximation. The main interest in weak approximations lies in the hope to obtain simpler methods and, in particular, methods not requiring modeling of complicated random variables. We recall that, e.g., the mean-square method (1.0.5), which is of the first order only, requires to solve the difficult problem of modeling complicated random variables of the type $\int_0^h w_i(\theta) dw_j(\theta)$. These problems of modeling complicated random variables can be avoided by integrating in the weak sense, which gives an impetus for the development of methods for constructing weak approximations. In addition we note that while in the deterministic theory the one-dimensional case differs but little from the multi-dimensional one, for the numerical integration of stochastic differential equations the multi-dimensional case, especially when several noises are involved, is essentially more complicated than the one-dimensional case. Weak approximations were introduced for the first time ever in [176] (see also [178,231,282]). The fact that weak approximations suffice for the equations of mathematical physics shows that precisely such approximations are of the most interest in applications and should thus be in the center of investigations on numerical integration of SDEs. At the same time, it must be stressed that the construction of weak approximations uses the general theory of mean-square approximations in an essential way.

In this chapter we construct various methods of second order of accuracy in the weak sense for general systems of stochastic differential equations as well as methods of third order of accuracy for systems with additive and colored noise. These methods use random variables that are simple to model. They are also simpler than mean-square methods from another point of view: for the same mean-square and weak orders of accuracy the weak methods require calculating substantially fewer operators of coefficients of the considered system. In Sect. 2.1 we give a detailed construction of a one-step approximation of third order of accuracy. It is the basis for constructing methods of second order of accuracy for stochastic systems of general type. In Sect. 2.2 we prove a theorem stating that if a one-step approximation has $(p+1)$-th order of accuracy then approximation on a finite interval has p-th order of accuracy. This theorem plays the same role in the theory of weak approximation as the main convergence theorem does in the theory of mean-square approximation. Some Runge–Kutta type and implicit methods are constructed using this theorem. The Talay–Tubaro extrapolation method is considered in this section as well. In the next section we derive weak methods for systems with additive and colored noise. Section 2.4 is devoted to the important question on error reduction of the Monte Carlo method. Both the method of important sampling and the method of control variates are considered here. In addition a combining method is proposed.

The weak approximation of SDEs for solving PDEs has numerous applications. The Monte Carlo simulation of option prices and its derivatives is a typical instance of such an application. Many works in financial mathematics are devoted to this approach (see among them [28,31,46,67,68,192,197,214,251]).

Here we restrict ourselves to the application to Wiener integrals. On the basis of the relation between some important classes of Wiener integrals and stochastic systems of equations we develop new approximate methods for computing Wiener integrals in Sect. 2.5. For Wiener integrals of functionals of integral type, we find methods of second order of accuracy. For integrals of a more particular, but often encountered, form (integrals of functionals of exponential type), we succeed in constructing efficient methods of fourth order of accuracy. Finally, we give numerical results that fit in nicely with the theoretical results. Implementation of numerical methods for integrating SDEs requires a source of random numbers. On a computer random numbers are usually generated via iterative deterministic algorithms known as random number generators. They are discussed in Sect. 2.6.

In this chapter the word "weak" will be omitted if this does not lead to misunderstanding.

2.1 One-step approximation

The one-step weak approximation $\bar{X}_{t,x}(t+h)$ of the solution $X_{t,x}(t+h)$ can be constructed by computing moments (from the first up to the r-th inclusively) of the vector $\bar{X}_{t,x}(t+h) - x$ and the corresponding moments of the vector $X_{t,x}(t+h)-x$. In this case the order of accuracy of the one-step approximation depends on both the order of the moments under consideration and on the order of closeness of those. To construct the one-step approximations of third order of accuracy considered in this section, we have to take into account all moments up to order six inclusively. In Sect. 2.1.3 we give an expansion formula for $Ef(t+h, X_{t,x}(t+h))$ in powers of h.

2.1.1 Properties of remainders and Ito integrals

As before, we consider the system

$$dX = a(t,X)dt + \sum_{r=1}^{q} \sigma_r(t,X)dw_r(t), \qquad (1.1)$$

where X, a, and σ are vectors of dimension d with components X^i, a^i, σ_r^i. We assume that the functions $a(t,x)$ and $\sigma_r(t,x)$ are sufficiently smooth with respect to the variables t, x and satisfy a global Lipschitz condition with respect to x: for all $t \in [t_0, T]$, $x \in \mathbf{R}^d$, $y \in \mathbf{R}^d$ the following inequality holds:

$$|a(t,x) - a(t,y)| + \sum_{r=1}^{q} |\sigma_r(t,x) - \sigma_r(t,y)| \leq K|x-y|. \qquad (1.2)$$

Here and below $|x|$ denotes the Euclidean norm of the vector x, and we denote by $x^\mathsf{T} y$ or by (x,y) the scalar (inner) product of two vectors x and y. We introduce the operators

$$\Lambda_r f = (\sigma_r, \frac{\partial}{\partial x})f = \sum_{i=1}^{d} \sigma_r^i \frac{\partial f}{\partial x^i},$$

$$Lf = (\frac{\partial}{\partial t} + a^\mathsf{T} \frac{\partial}{\partial x} + \frac{1}{2}\sum_{r=1}^{q}\sum_{i=1}^{d}\sum_{j=1}^{d} \sigma_r^i \sigma_r^j \frac{\partial^2}{\partial x^i \partial x^j})f,$$

where f may be a scalar function or a vector-function.

In the course of exposition we will impose additional conditions on a and σ_r. Note that the conditions on a and σ_r given in Theorem 1.2.5 are sufficient for all results in this section to hold. We recall that these conditions are related to the growth of functions of the form $\Lambda_{i_j} \cdots \Lambda_{i_1} f(t,x)$ for $f \equiv x$ as $|x| \to \infty$ (see (1.2.17)); more precisely, these functions grow with respect to x at most as a linear function of $|x|$ as $|x| \to \infty$. The indices $i_1, \ldots, i_j$ take the values $0, 1, \ldots, q$, and $\Lambda_0 = L$. We rewrite (1.2.21):

2.1 One-step approximation

$$X_{t,x}(t+h) = x + \sum_{r=1}^{q} \sigma_r \int_t^{t+h} dw_r(\theta) + ah$$

$$+ \sum_{r=1}^{q} \sum_{i=1}^{q} \Lambda_i \sigma_r \int_t^{t+h} (w_i(\theta) - w_i(t)) dw_r(\theta)$$

$$+ \sum_{r=1}^{q} L\sigma_r \int_t^{t+h} (\theta - t) dw_r(\theta) + \sum_{r=1}^{q} \Lambda_r a \int_t^{t+h} (w_r(\theta) - w_r(t)) d\theta$$

$$+ \sum_{r=1}^{q} \sum_{i=1}^{q} \sum_{s=1}^{q} \Lambda_s \Lambda_i \sigma_r \int_t^{t+h} (\int_t^{\theta} (w_s(\theta_1) - w_s(t)) dw_i(\theta_1)) dw_r(\theta)$$

$$+ La\frac{h^2}{2} + \rho, \tag{1.3}$$

where the coefficients σ_r, a, $\Lambda_i \sigma_r$, $L\sigma_r$, $\Lambda_r a$, $\Lambda_s \Lambda_i \sigma_r$, La are calculated at the point (t,x), while the remainder ρ is given in (1.2.22) (we do not write it here).

Definition 1.1. *We say that a function $f(x)$ belongs to the class $\mathbf{F}$, written as $f \in \mathbf{F}$, if we can find constants $K > 0$, $\kappa > 0$ such that for all $x \in \mathbf{R}^d$ the following inequality holds:*

$$|f(x)| \leq K(1 + |x|^\kappa). \tag{1.4}$$

If a function $f(s,x)$ depends not only on $x \in \mathbf{R}^d$ but also on a parameter $s \in S$, then we say that $f(s,x)$ belongs to $\mathbf{F}$ (with respect to the variable x) if an inequality of the type (1.4) holds uniformly in $s \in S$.

In the sequel we need that σ_r, a, $\Lambda_i \sigma_r$, $L\sigma_r$, $\Lambda_r a$, $\Lambda_s \Lambda_i \sigma_r$, La, etc. belong to the class $\mathbf{F}$. For example, in the proof of Lemma 1.2 (see below) we use the fact that all integrands participating in the remainder ρ as well as all functions obtained by applying the operators Λ_l, $l = 1, \ldots, q$, and L to the functions $\Lambda_j \Lambda_s \Lambda_i \sigma_r$, $L\Lambda_i \sigma_r$, $\Lambda_i L\sigma_r$, $\Lambda_i \Lambda_r a$ belong to the class $\mathbf{F}$. Clearly, it is sufficient for this to require that all partial derivatives up to order five, inclusively, of the coefficients a, σ_r with respect to t and x belong to $\mathbf{F}$. In such cases we assert that the coefficients a, σ_r, $r = 1, \ldots, q$, together with their partial derivatives of sufficiently high order belong to $\mathbf{F}$.

Lemma 1.2. *Suppose that the Lipschitz condition (1.2) holds and the functions a, σ_r, $r = 1, \ldots, q$, together with their partial derivatives of a sufficiently high order belong to $\mathbf{F}$. Then the following inequalities hold:*

$$|E\rho| \leq K(x)h^3, \ K(x) \in \mathbf{F}, \tag{1.5}$$

$$E|\rho|^2 \leq K(x)h^4, \ K(x) \in \mathbf{F}, \tag{1.6}$$

$$|E\rho \int_t^{t+h} dw_r(\theta)| \leq K(x)h^3, \ K(x) \in \mathbf{F}. \tag{1.7}$$

90   2 Weak approximation for stochastic differential equations

Proof. The form of the remainder ρ (see (1.2.22)) and the fact that $L^2 a \in \mathbf{F}$ imply that we can find an even number $2m$ and a number $K > 0$ such that

$$|E\rho| = |E \int_t^{t+h} (\int_t^\theta (\int_t^{\theta_1} L^2 a(\theta_2, X(\theta_2)) d\theta_2) d\theta_1) d\theta|$$

$$\leq |\int_t^{t+h} (\int_t^\theta (\int_t^{\theta_1} K(1 + E|X(\theta_2)|^{2m}) d\theta_2) d\theta_1) d\theta|. \qquad (1.8)$$

But according to [82], $E|X(\theta_2)|^{2m}$ is bounded by a quantity $K(1 + |x|^{2m})$. Hence (1.8) implies (1.5). In the proof of (1.6) we use the fact that each term (more precisely, the mathematical expectation of the norm of each term in (1.2.22)) is, in any case, of second order of smallness with respect to h. To prove (1.7), we have to treat each integral in the first four sums in (1.2.22) by Ito's formula. For example,

$$\int_t^{t+h} (\int_t^\theta (\int_t^{\theta_1} (\int_t^{\theta_2} \Lambda_j \Lambda_s \Lambda_i \sigma_r(\theta_3, X(\theta_3)) dw_j(\theta_3)) dw_s(\theta_2)) dw_i(\theta_1)) dw_r(\theta)$$

$$= \Lambda_j \Lambda_s \Lambda_i \sigma_r(t, x) \int_t^{t+h} (\int_t^\theta (\int_t^{\theta_1} (\int_t^{\theta_2} dw_j(\theta_3)) dw_s(\theta_2)) dw_i(\theta_1)) dw_r(\theta)$$

$$+ \sum_{l=1}^q \int_t^{t+h} (\int_t^\theta (\int_t^{\theta_1} (\int_t^{\theta_2} (\int_t^{\theta_3} \Lambda_l \Lambda_j \Lambda_s \Lambda_i \sigma_r(\theta_4, X(\theta_4)) dw_l(\theta_4)) dw_j(\theta_3))$$

$$\times dw_s(\theta_2)) dw_i(\theta_1)) dw_r(\theta)$$

$$+ \int_t^{t+h} (\int_t^\theta (\int_t^{\theta_1} (\int_t^{\theta_2} (\int_t^{\theta_3} L\Lambda_j \Lambda_s \Lambda_i \sigma_r(\theta_4, X(\theta_4)) d\theta_4) dw_j(\theta_3))$$

$$\times dw_s(\theta_2)) dw_i(\theta_1)) dw_r(\theta).$$

As a result, ρ can be written as a sum of terms of second, or higher, order of smallness with respect to h. Moreover, the terms of second order look like one of the integrals

$$I_{risj} = \int_t^{t+h} (\int_t^\theta (\int_t^{\theta_1} (\int_t^{\theta_2} dw_j(\theta_3)) dw_s(\theta_2)) dw_i(\theta_1)) dw_r(\theta),$$

$$I_{r0i} = \int_t^{t+h} (\int_t^\theta (\int_t^{\theta_1} dw_i(\theta_2)) d\theta_1) dw_r(\theta),$$

$$I_{ri0} = \int_t^{t+h} (\int_t^\theta (\int_t^{\theta_1} d\theta_2) dw_i(\theta_1)) dw_r(\theta),$$

$$I_{0ri} = \int_t^{t+h} (\int_t^\theta (\int_t^{\theta_1} dw_i(\theta_2)) dw_r(\theta_1)) d\theta,$$

with nonrandom coefficients $\Lambda_j \Lambda_s \Lambda_i \sigma_r$, $\Lambda_i L \sigma_r$, $L\Lambda_i \sigma_r$, $\Lambda_i \Lambda_r a$, respectively. It is easy to prove that the expectation of the product of $\int_t^{t+h} dw_l(\theta)$ with any term of second order of smallness is zero. For example, let us show that

$$E(I_{risj} \cdot I_l) = E(\int_t^{t+h} (\int_t^\theta (\int_t^{\theta_1} (\int_t^{\theta_2} dw_j(\theta_3))dw_s(\theta_2)) \\ \times dw_i(\theta_1))dw_r(\theta) \cdot \int_t^{t+h} dw_l(\theta)). \tag{1.9}$$

Indeed, by changing variables

$$v_k = -w_k, \ k = 1,\ldots,q,$$

the v_k are independent Wiener processes. Since the Wiener processes participating in (1.9) are odd, we have

$$E(I_{risj} \cdot I_l)$$
$$= -E(\int_t^{t+h} (\int_t^\theta (\int_t^{\theta_1} (\int_t^{\theta_2} dv_j(\theta_3))dv_s(\theta_2))dv_i(\theta_1))dv_r(\theta) \cdot \int_t^{t+h} dv_l(\theta))$$
$$= -E(I_{risj} \cdot I_l),$$

which implies (1.9).

The other terms in ρ have order of smallness at least $5/2$. Using the Bunyakovsky–Schwarz inequality, we can readily show that the mathematical expectation of absolute value of the product of each of such terms with $\int_t^{t+h} dw_l(\theta)$ is smaller than or equal to $K(x)h^3$ with $K(x) \in \mathbf{F}$. This proves (1.7) and hence the lemma. □

Using the identity

$$\int_t^{t+h} (\theta - t)dw_r(\theta) = h \int_t^{t+h} dw_r(\theta) - \int_t^{t+h} (w_r(\theta) - w_r(t))d\theta$$

and the notation

$$I_j = \int_t^{t+h} dw_j(\theta) = w_j(t+h) - w_j(t), \quad I_{jp} = \int_t^{t+h} (w_j(\theta) - w_j(t))dw_p(\theta),$$

$$I_{sir} = \int_t^{t+h} (\int_t^\theta (w_s(\theta_1) - w_s(t))dw_i(\theta_1))dw_r(\theta), \quad J_r = \int_t^{t+h} (w_r(\theta) - w_r(t))d\theta,$$

we introduce $\tilde{X}$ by the formula

$$\tilde{X} = x + \sum_{r=1}^{q} \sigma_r I_r + ah + \sum_{r=1}^{q}\sum_{i=1}^{q} \Lambda_i \sigma_r I_{ir}$$

$$+ \sum_{r=1}^{q} L\sigma_r \cdot I_r h + \sum_{r=1}^{q} (\Lambda_r a - L\sigma_r) \cdot J_r + La \cdot \frac{h^2}{2} \qquad (1.10)$$

and rewrite (1.3) as follows:

$$X = \tilde{X} + \sum_{r=1}^{q}\sum_{i=1}^{q}\sum_{s=1}^{q} \Lambda_s \Lambda_i \sigma_r \cdot I_{sir} + \rho. \qquad (1.11)$$

Lemma 1.3. *The following identities hold:*

$$EI_{sir} = 0,\ EI_{sir}I_j = 0,\ EI_{sir}I_jI_p = 0,$$
$$EI_{sir}I_{jp} = 0,\ i,j,p,r,s = 1,\ldots,q. \qquad (1.12)$$

Proof. The first, third, and fourth identities in (1.12) are obvious because of oddness. We prove the second identity. Without loss of generality we may put $t = 0$. For $j \neq s$, $j \neq i$, $j \neq r$ this identity follows from the independence of I_j and I_{sir}. To consider the other cases, we introduce the system of equations

$$dx(\theta) = w_s(\theta)dw_i(\theta),\ x(0) = 0,$$
$$dy(\theta) = x(\theta)dw_r(\theta),\ y(0) = 0.$$

Then $I_{sir} = y(h)$ and $EI_{sir}I_j = Ey(h)w_j(h)$. Let, e.g., $j = s$. Then by Ito's formula,

$$d(y(\theta)w_s(\theta)) = w_s(\theta)x(\theta)dw_r(\theta) + y(\theta)dw_s(\theta) + x(\theta)\delta_{sr}d\theta,$$

where δ_{sr} is the Kronecker symbol. Hence,

$$dE(y(\theta)w_s(\theta)) = Ex(\theta)\delta_{sr}d\theta.$$

Since $Ex(\theta) = 0$ and $E(y(0)w_s(0)) = 0$, we have $EI_{sir}I_s = Ey(h)w_s(h) = 0$. The cases $j = i$ and $j = r$ can be treated in a similar way. □

2.1.2 One-step approximations of third order

We introduce the notation $X = X(t+h)$, $\Delta = X - x$, $\tilde{\Delta} = \tilde{X} - x$, $\bar{\Delta} = \bar{X} - x$, and denote by x^i the i-th coordinate of the vector x. Our nearest goal is to form a random vector $\bar{X}$ such that the difference of all moments up to order five, inclusively, of the coordinates of the vectors Δ and $\bar{\Delta}$ would have third order of smallness with respect to h. More precisely,

$$|E(\prod_{j=1}^{s} \Delta^{i_j} - \prod_{j=1}^{s} \bar{\Delta}^{i_j})| \leq K(x)h^3, \ i_j = 1,\ldots,d, \tag{1.13}$$

$$s = 1,\ldots,5, \ K(x) \in \mathbf{F},$$

and, moreover,

$$E\prod_{j=1}^{s}|\bar{\Delta}^{i_j}| \leq K(x)h^3, \ i_j = 1,\ldots,d, \ s = 6, \ K(x) \in \mathbf{F}. \tag{1.14}$$

First of all we state the following lemma.

Lemma 1.4. *Under the conditions of Lemma 1.2 the following inequalities hold:*

$$|E(\prod_{j=1}^{s} \Delta^{i_j} - \prod_{j=1}^{s} \tilde{\Delta}^{i_j})| \leq K(x)h^3, \ s = 1,\ldots,5, \ K(x) \in \mathbf{F}. \tag{1.15}$$

Proof. The proof of this lemma is based on Lemmas 1.2 and 1.3. In fact, by (1.11) each component Δ^{i_j} of Δ differs from the corresponding component $\tilde{\Delta}^{i_j}$ of $\tilde{\Delta}$ by a sum made up from the corresponding components of the vectors ρ and $\Lambda_s\Lambda_i\sigma_r \cdot I_{sir}$. Therefore the difference $\prod_{j=1}^{s} \Delta^{i_j} - \prod_{j=1}^{s} \tilde{\Delta}^{i_j}$ consists of terms each of which must have a component of ρ or a component of the integral I_{sir} as at least one of its factors. If $s = 1$, i.e., we are considering first moments, then these terms do not have other factors and (1.15) follows from (1.5) and the first identity in (1.12). If $s = 2$ then the terms containing ρ either have I_r as a factor or they have a factor whose order is at least one. In the first case we use the estimate (1.7), and in the second case we use the Bunyakovsky–Schwarz inequality and, subsequently, (1.6). In other words, in the second case the order of a term is at least three because one factor has order at least one and the other factor (which is a component of ρ) has, by (1.6), order two, i.e., by the Bunyakovsky–Schwarz inequality we may sum the orders of the factors. This already makes clear that for $s = 3, 4, 5$ the terms containing at least one component of ρ as a factor have order of smallness at least three with respect to h. For $s = 2$ we return to the terms containing I_{sir} as a factor; I_{sir} has order of smallness $3/2$ with respect to h. Such terms either contain an expression of the form $I_{sir}I_j$, $I_{sir}h$, $I_{sir}I_{jp}$ as a factor and the expectation of such terms is zero (by the first, second, and fourth identity in (1.12)), or they have at least third order of smallness with respect to h. For $s = 3$, the terms containing I_{sir} and having a mean-square order of smallness with respect to h which is less than three are easily seen to contain an expression of the form $I_{sir}I_jI_p$, and their expectation is zero (see the third identity in (1.12)). For $s = 4, 5$, all terms containing I_{sir} are at least of order three. These considerations imply that the inequality (1.15) holds. □

94   2 Weak approximation for stochastic differential equations

We will form a random vector $\bar{X}$ such that the inequalities

$$|E(\prod_{j=1}^{s}\tilde{\Delta}^{i_j} - \prod_{j=1}^{s}\bar{\Delta}^{i_j})| \le K(x)h^3,\ s=1,\ldots,5,\ K(x) \in \mathbf{F}, \qquad (1.16)$$

as well as (1.14) hold. Since (1.15) and (1.16) imply (1.13), we will reach our aim in constructing the vector $\bar{X}$ satisfying the inequalities (1.13) and (1.14).

We construct $\bar{X}$ similar to $\tilde{X}$ as follows:

$$\bar{X} = x + \sum_{r=1}^{q}\sigma_r\xi_r h^{1/2} + ah + \sum_{r=1}^{q}\sum_{i=1}^{q}\Lambda_i\sigma_r\xi_{ir}h$$
$$+ \sum_{r=1}^{q}L\sigma_r\xi_r h^{3/2} + \sum_{r=1}^{q}(\Lambda_r a - L\sigma_r)\eta_r h^{3/2} + La\frac{h^2}{2}. \qquad (1.17)$$

Lemma 1.5. *Suppose the conditions of Lemma 1.2 are satisfied. Then the inequalities (1.13) and (1.14) hold if the random variables ξ_r, ξ_{ir}, η_r in (1.17) have finite moments up to order six, inclusively, and the following relations hold:*

$$E\xi_r h^{1/2} = EI_r = 0,\ E\xi_{ir}h = EI_{ir} = 0,\ E\eta_r h^{3/2} = EJ_r = 0; \qquad (1.18)$$

$$E\xi_i\xi_r h = EI_i I_r = \delta_{ir}h,\ E\xi_i\xi_{rj}h^{3/2} = EI_i I_{rj} = 0,$$
$$E\xi_r\eta_j h^2 = EI_r J_j = \delta_{rj}\frac{h^2}{2},$$
$$E\xi_{ir}\xi_{js}h^2 = EI_{ir}I_{js} = \begin{cases} h^2/2 & \text{if } i=j,\ r=s,\\ 0 & \text{otherwise}, \end{cases}$$
$$E\xi_{ir}\eta_j h^{5/2} = EI_{ir}J_j = 0; \qquad (1.19)$$

$$E\xi_i\xi_r\xi_j h^{3/2} = EI_i I_r I_j = 0,$$
$$E\xi_i\xi_r\xi_{js}h^2 = EI_i I_r I_{js}$$
$$= \begin{cases} h^2/2 & \text{if } j \ne s \text{ and either } i=j,\ r=s \text{ or } i=s,\ r=j,\\ h^2 & \text{if } i=r=j=s,\\ 0 & \text{otherwise}, \end{cases}$$
$$E\xi_i\xi_r\eta_j h^{5/2} = EI_i I_r J_j = 0,\ E\xi_i\xi_{jr}\xi_{sl}h^{5/2} = EI_i I_{jr}I_{sl} = 0; \qquad (1.20)$$

$$E\xi_i\xi_r\xi_j\xi_s h^2 = EI_i I_r I_j I_s$$
$$= \begin{cases} h^2 & \text{if } \{i,r,j,s\} \text{ consists of two pairs of equal numbers},\\ 3h^2 & \text{if } i=r=j=s,\\ 0 & \text{otherwise}, \end{cases}$$
$$E\xi_i\xi_r\xi_j\xi_{sl}h^{5/2} = EI_i I_r I_j I_{sl} = 0; \qquad (1.21)$$

$$E\xi_i\xi_r\xi_j\xi_s\xi_l h^{5/2} = EI_i I_r I_j I_s I_l = 0. \qquad (1.22)$$

Proof. The inequality (1.14) for sixth moments of the absolute values of the coordinates of the vector $\bar{\Delta} = \bar{X} - x$ evidently follows from (1.17), since each term in $\bar{\Delta}$ has at least order of smallness $1/2$ with respect to h. Further, all the identities (1.18)-(1.22) consist of two parts: the right part and the left part. We prove the right parts below. The left parts of (1.18) are clearly sufficient to get (1.16) and (1.13) for $s = 1$, i.e., to prove that the first moments of the coordinates of the vectors Δ and $\bar{\Delta}$ coincide up to $O(h^3)$. The left parts of (1.18)-(1.19) suffice for the second moments; (1.18)-(1.20) suffice for the third moments; (1.18)-(1.21) suffice for the fourth moments; and (1.18)-(1.22) suffice for the fifth moments.

Almost all the right parts of (1.18)-(1.22) can be easily derived taking into account oddness and independence; only evaluation of the expectations $EI_{ir}I_{js}$ and $EI_iI_rI_{js}$ causes some difficulties. Without loss of generality, we set $t = 0$. To evaluate

$$EI_{ir}I_{js} = E\int_0^h w_i(\theta)dw_r(\theta) \int_0^h w_j(\theta)dw_s(\theta),$$

we introduce the system of equations

$$dx(\theta) = w_i(\theta)dw_r(\theta),\ x(0) = 0,$$
$$dy(\theta) = w_j(\theta)dw_s(\theta),\ y(0) = 0.$$

It is evident that $EI_{ir}I_{js} = Ex(h)y(h)$. By Ito's formula, we have

$$dxy = yw_idw_r + w_jdw_s + w_iw_j\delta_{rs}d\theta.$$

Therefore

$$dE(x(\theta)y(\theta)) = \delta_{rs}E(w_i(\theta)w_j(\theta))d\theta,$$

which immediately implies the last of the identities (1.19).

To evaluate

$$EI_iI_rI_{js} = Ew_i(h)w_r(h)\int_0^h w_j(\theta)dw_s(\theta),$$

we introduce the equation

$$dy(\theta) = w_j(\theta)dw_s(\theta),\ y(0) = 0.$$

It is obvious that $EI_iI_rI_{js} = Ew_i(h)w_r(h)\,y(h)$. By Ito's formula,

$$d(w_iw_ry) = w_rydw_i + w_iydw_r + w_iw_rw_jdw_s$$
$$+y\delta_{ir}d\theta + w_rw_j\delta_{is}d\theta + w_iw_j\delta_{rs}d\theta.$$

In view of $Ey(\theta) = 0$ we obtain

$$dE(w_iw_ry) = \delta_{is}E(w_rw_j)d\theta + \delta_{rs}E(w_iw_j)d\theta,$$

which immediately implies the second identity in (1.20). □

Theorem 1.6. *Suppose the conditions of Lemma 1.2 hold. Let a function $f(x)$ and all its partial derivatives up to order six inclusively belong to the class* **F**. *Let ξ_i, η_i, ξ_{ij} be chosen such that (1.18) – (1.22) hold. Then $\bar{X}$ from (1.17) satisfies the inequality (recall that $X = X(t+h)$):*

$$|Ef(X) - Ef(\bar{X})| \le K(x)h^3, \ K(x) \in \mathbf{F}, \tag{1.23}$$

i.e., the method (1.17) has the third order of accuracy on a single step in the sense of weak approximations.

Proof. Lemmas 1.2 - 1.5 imply the inequalities (1.13) and (1.14). Moreover, similarly to the proof of (1.14) in Lemma 1.5, we can prove the inequality

$$E \prod_{j=1}^{s} |\Delta^{i_j}| \le K(x)h^3, \ i_j = 1, \ldots, d, \ s = 6, \ K(x) \in \mathbf{F}. \tag{1.24}$$

Now we write the Taylor expansion of $f(X)$ with respect to powers of $\Delta^i = X^i - x^i$ in a neighborhood of x and with Lagrange remainder term containing terms of order six. We similarly expand $f(\bar{X})$ with respect to the $\bar{\Delta}^i = \bar{X}^i - x^i$. Using (1.13), (1.14), and (1.24), we arrive at (1.23). □

Let us proceed to modeling of random variables and constructive formation of a one-step approximations of third order of accuracy. There are various methods that satisfy the relations (1.18)-(1.22). For example (see [282]), let ξ_i, $i = 1, \ldots, q$, ζ_{ij}, $i = 2, \ldots, q$, $j = 1, \ldots, i-1$, be mutually independent random variables where ξ_i are distributed by the law $P(\xi = 0) = 2/3$, $P(\xi = -\sqrt{3}) = P(\xi = \sqrt{3}) = 1/6$, and ζ_{ij} are distributed by the law $P(\zeta = -1) = P(\zeta = 1) = 1/2$. Then if $\zeta_{ii} := -1$, $\zeta_{ij} = -\zeta_{ji}$, $j > i$, and

$$\eta_i = \frac{1}{2}\xi_i, \ \xi_{ij} = \frac{1}{2}\xi_i\xi_j + \frac{1}{2}\zeta_{ij},$$

the relations (1.18)-(1.22) are satisfied. This is easy to verify. This method requires modeling $q(q+1)/2$ random variables. We propose a method (see [178]) which requires $2q$ random variables only.

Consider mutually independent random variables ξ_i and ζ_j, $i, j = 1, \ldots, q$, and put

$$\eta_i = \frac{1}{2}\xi_i, \ \xi_{ij} = \frac{1}{2}\xi_i\xi_j - \frac{1}{2}\gamma_{ij}\zeta_i\zeta_j, \ \gamma_{ij} = \begin{cases} -1, & i < j, \\ 1 & i \ge j. \end{cases} \tag{1.25}$$

We will assume that ξ_i and ζ_j have all moments needed. Below we will verify that if, in addition to the above-said, we require that

$$E\xi_i = E\xi_i^3 = E\xi_i^5 = 0, \ E\xi_i^2 = 1, \ E\xi_i^4 = 3,$$
$$E\zeta_i = E\zeta_i^3 = 0, \ E\zeta_i^2 = E\zeta_i^4 = 1, \tag{1.26}$$

then all the relations (1.18)-(1.22) are satisfied.

For example, we can model ξ_i by the law $\mathcal{N}(0,1)$ and the ζ_i by the law $P(\zeta=-1)=P(\zeta=1)=1/2$. But ξ_i can be modeled by a much simpler law as above: $P(\xi=0)=2/3$, $P(\xi=\pm\sqrt{3})=1/6$. In practice (see the end of Sect. 2.6.2), the following modeling of ξ_i is of interest too: $P(\xi=\pm\sqrt{1-\sqrt{6}/3})=3/8$, $P(\xi=\pm\sqrt{1+\sqrt{6}})=1/8$.

We turn to the direct verification of (1.18)-(1.22) for the random variables (1.25). Many of them can be verified rather simply. Therefore we verify more complicated ones only.

Lemma 1.7. Let ξ_i, ζ_j, $i,j=1,\ldots,q$, be mutually independent random variables such that (1.25) – (1.26) hold. Then

$$E\xi_{ir}\xi_{js} = \begin{cases} 1/2 & \text{if } i=j,\ r=s, \\ 0 & \text{otherwise}. \end{cases} \quad (1.27)$$

Proof. We have

$$E\xi_{ir}\xi_{js} = \frac{1}{4}(E\xi_i\xi_r\xi_j\xi_s - \gamma_{js}E\xi_i\xi_r\zeta_j\zeta_s - \gamma_{ir}E\xi_j\xi_s\zeta_i\zeta_r + \gamma_{ir}\gamma_{js}E\zeta_i\zeta_r\zeta_j\zeta_s). \quad (1.28)$$

Let $i \neq j$. It is obvious that if r is not equal to j, then all expectations at the right-hand side of (1.28) are zero. So, for $i \neq j$ the right-hand side can be nonzero only if $r=i$ or $r=j$. Consider the case $i \neq j$, $r=i$. If also $s \neq j$, then the right-hand side is zero. Now we evaluate (1.28) for $i \neq j$, $r=i$, $s=j$. We have $\gamma_{ir}=\gamma_{js}=1$ and each of the four expectations at the right-hand side of (1.28) is equal to 1. As a result, for $i \neq j$, $r=i$ the right-hand side vanishes. Consider now the case $i \neq j$, $r=j$. In this case the right-hand side of (1.28) can be nonzero only if $s=i$. So, let $i \neq j$, $r=j$, $s=i$. We have $E\xi_i\xi_r\xi_j\xi_s = E\zeta_i\zeta_r\zeta_j\zeta_s = 1$ but $E\xi_i\xi_r\zeta_j\zeta_s = E\xi_j\xi_s\zeta_i\zeta_r = 0$. For $i \neq j$, $r=j$, $s=i$ the product $\gamma_{ir}\gamma_{js}$ is always -1. Indeed, if $i<j$ then $i<r$ and $j>s$, since $r=j$, $s=i$. But $\gamma_{ir}=-1$ for $i<r$ and $\gamma_{js}=1$ for $j>s$. Hence $\gamma_{ir}\gamma_{js}=-1$. The case $i>j$ can be treated similarly. So, if $i \neq j$, the right-hand side of (1.28) is always zero.

Now let $i=j$. Then the right-hand side of (1.28) can be nonzero for $r=s$ only. We distinguish three cases. In the first case $i<r$. Then $j<s$, $\gamma_{ir}=\gamma_{js}=-1$, $E\xi_i\xi_r\xi_j\xi_s = E\zeta_i\zeta_r\zeta_j\zeta_s = 1$, $E\xi_i\xi_r\zeta_j\zeta_s = E\xi_j\xi_s\zeta_i\zeta_r = 0$, and hence $E\xi_{ir}\xi_{js}=1/2$. The second case, $i>r$, can be treated similarly. It differs by the relations $\gamma_{ir}=\gamma_{js}=1$. The third case, $i=r$, gives $i=j=r=s$, $E\xi_i\xi_r\xi_j\xi_s=3$, $E\xi_i\xi_r\zeta_j\zeta_s = E\xi_j\xi_s\zeta_i\zeta_r = E\zeta_i\zeta_r\zeta_j\zeta_s = 1$, $\gamma_{ir}=\gamma_{js}=1$, and hence $E\xi_{ir}\xi_{js}=1/2$. $\square$

We need this lemma to substantiate the next to last relation in (1.19). The other relations in (1.19) as well as (1.18) can be verified in an obvious manner. In (1.20) the second relation presents some difficulty. To verify it, we prove the following lemma.

Lemma 1.8. *Let ξ_i, ζ_j, $i,j = 1,\ldots,q$, be mutually independent random variables such that (1.25) − (1.26) hold. Then*

$$E\xi_i\xi_r\xi_{js} = \begin{cases} 1/2 & \text{if } j \neq s \text{ and either } i = j,\ r = s \text{ or } i = s,\ r = j, \\ 1 & \text{if } i = r = j = s, \\ 0 & \text{otherwise}. \end{cases} \quad (1.29)$$

Proof. We have

$$E\xi_i\xi_r\xi_{js} = \frac{1}{2}(E\xi_i\xi_r\xi_j\xi_s - \gamma_{js}E\xi_i\xi_r\zeta_j\zeta_s). \quad (1.30)$$

For $j \neq s$ the right-hand side of (1.30) can be nonzero only for $i = j$, $r = s$ or $i = s$, $r = j$. In both these cases $E\xi_i\xi_r\xi_j\xi_s = 1$, $E\xi_i\xi_r\zeta_j\zeta_s = 0$, which proves (1.29) for $j \neq s$. If $j = s$ but $i \neq j$, then the right-hand side of (1.30) can be nonzero for $i = r$ only. But in this case $\gamma_{js} = 1$, $E\xi_i\xi_r\xi_j\xi_s = E\xi_i\xi_r\zeta_j\zeta_s = 1$ and, hence, the right-hand side of (1.30) is zero. Let $j = s$, $i = j$. Then (1.30) can be nonzero only if $i = r$, i.e. $i = r = j = s$. If $i = r = j = s$, then $E\xi_i\xi_r\xi_j\xi_s = 3$, $E\xi_i\xi_r\zeta_j\zeta_s = 1$, $\gamma_{js} = 1$, i.e. $E\xi_i\xi_r\xi_{js} = 1$. □

The other relations (1.18)-(1.22) can be verified in a simple way. As a result, we can write the one-step approximation (1.17) as

$$\bar{X} = x + \sum_{r=1}^{q} \sigma_r \xi_r h^{1/2} + ah - \sum_{r=1}^{q}\sum_{i=1}^{q} \Lambda_i \sigma_r \xi_{ir} h$$

$$+ \frac{1}{2}\sum_{r=1}^{q}(\Lambda_r a + L\sigma_r)\xi_r h^{3/2} + La\frac{h^2}{2}, \quad (1.31)$$

where ξ_{ir} satisfy (1.25), and ξ_i, ζ_j are independent random variables satisfying (1.26). We recall that ξ_i can be modeled by, e.g., the law $P(\xi = 0) = 2/3$, $P(\xi = \pm\sqrt{3}) = 1/6$ or $P(\xi = \pm\sqrt{1 - \sqrt{6}/3}) = 3/8$, $P(\xi = \pm\sqrt{1 + \sqrt{6}}) = 1/8$ and ζ_j can be modeled by $P(\zeta = \pm 1) = 1/2$. The one-step approximation (1.31) has the third order of accuracy in the sense of weak approximation.

2.1.3 The Taylor expansion of mathematical expectations

Let us derive some expansions of $Ef(t + h, X(t + h))$ with respect to powers of h. By Ito's formula, we have for $u \geq t$:

$$f(u, X(u)) = f(t, X(t)) + \int_t^u Lf(\theta, X(\theta))d\theta$$

$$+ \sum_{r=1}^{q} \int_t^u \Lambda_r \tilde{f}(\theta, X(\theta))dw_r(\theta). \quad (1.32)$$

Applying (1.32) to $Lf(\theta, X(\theta))$, substituting the obtained expression in $\int_t^u Lf(\theta, X(\theta))d\theta$, and making a few simple transformations, we get

$$f(u, X(u)) = f(t, X(t)) + Lf(t, X(t))(u-t) + \int_t^u (u-\theta)L^2 f(\theta, X(\theta))d\theta$$

$$+ \sum_{r=1}^q \int_t^u (\Lambda_r f(\theta, X(\theta)) + (u-\theta)\Lambda_r Lf(\theta, X(\theta)))dw_r(\theta).$$

Proceeding further in this way, we find

$$f(u, X(u)) = f(t, X(t)) + Lf(t, X(t))(u-t) + \cdots$$

$$+ \frac{1}{m!} L^m f(t, X(t))(u-t)^m + \int_t^u \frac{(u-\theta)^m}{m!} L^{m+1} f(\theta, X(\theta))d\theta$$

$$+ \int_t^u \sum_{r=1}^q (\Lambda_r f(\theta, X(\theta)))$$

$$+ \cdots + \frac{(u-\theta)^m}{m!} \Lambda_r L^m f(\theta, X(\theta)))dw_r(\theta). \quad (1.33)$$

Lemma 1.9. *Suppose that the following mathematical expectations exist and are continuous with respect to θ:*

$$EL^k f(\theta, X(\theta)), \ k = 0, 1, \ldots, m+1,$$
$$E(\Lambda_r L^k f(\theta, X(\theta)))^2, \ k = 0, 1, \ldots, m, \ r = 1, \ldots, q.$$

Then the following formulas hold for $t \leq s \leq t+h$:

$$E(f(t+h, X_{t,x}(t+h))|\mathcal{F}_s) = f(s, X_{t,x}(s)) + (t+h-s)Lf(s, X_{t,x}(s))$$

$$+ \cdots + \frac{(t+h-s)^m}{m!} L^m f(s, X_{t,x}(s))$$

$$+ \int_s^{t+h} \frac{(t+h-\theta)^m}{m!} E(L^{m+1} f(\theta, X_{t,x}(\theta))|\mathcal{F}_s)d\theta, \quad (1.34)$$

$$Ef(t+h, X_{t,x}(t+h)) = f(t,x) + hLf(t,x) + \cdots + \frac{h^m}{m!} L^m f(t,x)$$

$$+ \int_t^{t+h} \frac{(t+h-\theta)^m}{m!} EL^{m+1} f(\theta, X_{t,x}(\theta))d\theta, \quad (1.35)$$

The proof clearly follows from (1.33).

The formula (1.35) is related to the Taylor expansion of semigroups [103]. It is more convenient than the Taylor expansion of semigroups because, in particular, it is also applicable to unbounded functions f. Clearly, in (1.34) and (1.35) the remainders of integral type are $O(h^{m+1})$.

2.2 The main theorem on convergence of weak approximations and methods of order two

The mentioned theorem is the fundamental convergence theorem for weak approximations. It establishes the weak order of convergence of a method resting on properties of its one-step approximation only. Using this theorem, we prove, in particular, that the method based on the one-step approximation of third order constructed in the previous section is of weak order 2. Acting analogously, it is not difficult, in principle, to construct weak methods of any orders (see e.g. the next section where methods of weak order 3 are constructed for systems with additive noise).

2.2.1 The general convergence theorem

Along with the system (1.1), we consider the approximation

$$\bar{X}_{t,x}(t+h) = x + A(t, x, h; \xi), \qquad (2.1)$$

where ξ is a random variable (in general, a vector) having moments of a sufficiently high order, and A is a vector function of dimension d. Partition the interval $[t_0, T]$ into N equal parts with step $h = (T - t_0)/N$: $t_0 < t_1 < \cdots < t_N = T$, $t_{k+1} - t_k = h$. According to (2.1), we construct the sequence

$$\bar{X}_0 = X_0 = X(t_0), \ \bar{X}_{k+1} = \bar{X}_k + A(t, \bar{X}_k, h; \xi_k), \ k = 0, \ldots, N-1, \qquad (2.2)$$

where ξ_0 is independent of $\bar{X}_0$, while ξ_k for $k > 0$ is independent of $\bar{X}_0, \ldots, \bar{X}_k, \xi_0, \ldots, \xi_{k-1}$. As before, we write $\Delta = X - x = X_{t,x}(t+h) - x$, $\bar{\Delta} = \bar{X} - x = \bar{X}_{t,x}(t+h) - x$. Let $X(t) = X_{t_0, X_0}(t)$ be a solution of (1.1) and $\bar{X}_{t_0, X_0}(t_k) = \bar{X}_k$.

Theorem 2.1. *Suppose that*
(a) the coefficients of equation (1.1) are continuous, satisfy a Lipschitz condition (1.2) and together with their partial derivatives with respect to x of order up to $2p+2$, inclusively, belong to $\mathbf{F}$;
(b) the method (2.1) is such that

$$|E(\prod_{j=1}^{s}\Delta^{i_j} - \prod_{j=1}^{s}\bar{\Delta}^{i_j})| \le K(x)h^{p+1}, \ s = 1, \ldots, 2p+1, \ K(x) \in \mathbf{F}, \qquad (2.3)$$

$$E\prod_{j=1}^{2p+2}|\bar{\Delta}^{i_j}| \le K(x)h^{p+1}, \ K(x) \in \mathbf{F}; \qquad (2.4)$$

(c) the function $f(x)$ together with its partial derivatives of order up to $2p+2$, inclusively, belong to $\mathbf{F}$;
(d) for a sufficiently large m (specified below) the expectations $E|\bar{X}_k|^{2m}$ exist and are uniformly bounded with respect to N and $k = 0, 1, \ldots, N$.
Then, for all N and all $k = 0, 1, \ldots, N$ the following inequality holds:

$$|Ef(X_{t_0, X_0}(t_k)) - Ef(\bar{X}_{t_0, X_0}(t_k))| \le Kh^p, \qquad (2.5)$$

i.e., the method (2.2) has order of accuracy p in the sense of weak approximations.

2.2 The main theorem on convergence of weak approximations

Proof. First of all we note that the Lipschitz condition (1.2) implies that for any $m > 0$ the expectations $E|X(\theta)|^{2m}$ exist and are uniformly bounded with respect to $\theta \in [t_0, T]$ if only $E|X(t_0)|^{2m} < \infty$ (see [82]). Moreover, the same (1.2) implies

$$E \prod_{j=1}^{2p+2} |\Delta^{i_j}| \leq K(x) h^{p+1}, \quad K(x) \in \mathbf{F}. \tag{2.6}$$

Further, suppose that $u(x)$ is a function that together with its partial derivatives of order up to $2p+2$, inclusively, belong to $\mathbf{F}$. Then

$$|Eu(X_{t,x}(t+h)) - Eu(\bar{X}_{t,x}(t+h))| \leq K(x) h^{p+1}, \quad K(x) \in \mathbf{F}. \tag{2.7}$$

Thanks to (2.3), (2.4), (2.6), the proof of (2.7) is completely similar to the proof of Theorem 1.6.

We introduce the function

$$u(s, x) = Ef(X_{s,x}(t_{k+1})).$$

By the requirements (a) and (c), u has partial derivatives with respect to x of order up to $2p+2$, inclusively; moreover, these derivatives belong to F (see [82]). Therefore the function $u(s, x)$ satisfies an estimate of the form (2.7) uniformly with respect to $s \in [t_0, t_{k+1}]$.

Further, since $\bar{X}_0 = X_0$, $X_{t_0, X_0}(t_1) = X(t_1)$, $X_{t_1, X_{t_0, \bar{X}_0}(t_1)}(t_{k+1}) = X(t_{k+1})$, we have

$$Ef(X(t_{k+1})) = Ef(X_{t_1, X_{t_0, \bar{X}_0}(t_1)}(t_{k+1})) - Ef(X_{t_1, \bar{X}_1}(t_{k+1}))$$
$$+ Ef(X_{t_1, \bar{X}_1}(t_{k+1})). \tag{2.8}$$

Similarly, since $X_{t_1, \bar{X}_1}(t_{k+1}) = X_{t_2, X_{t_1, \bar{X}_1}(t_2)}(t_{k+1})$, we have

$$Ef(X_{t_1, \bar{X}_1}(t_{k+1})) = Ef(X_{t_2, X_{t_1, \bar{X}_1}(t_2)}(t_{k+1})) - Ef(X_{t_2, \bar{X}_2}(t_{k+1}))$$
$$+ Ef(X_{t_2, \bar{X}_2}(t_{k+1})). \tag{2.9}$$

Now (2.8) and (2.9) imply

$$Ef(X(t_{k+1})) = Ef(X_{t_1, X_{t_0, \bar{X}_0}(t_1)}(t_{k+1})) - Ef(X_{t_1, \bar{X}_1}(t_{k+1}))$$
$$+ Ef(X_{t_2, X_{t_1, \bar{X}_1}(t_2)}(t_{k+1})) - Ef(X_{t_2, \bar{X}_2}(t_{k+1}))$$
$$+ Ef(X_{t_2, \bar{X}_2}(t_{k+1})).$$

Proceeding further, we obtain

$$Ef(X(t_{k+1})) = \sum_{i=0}^{k-1} Ef(X_{t_{i+1}, X_{t_i, \bar{X}_i}(t_{i+1})}(t_{k+1}))$$
$$- \sum_{i=0}^{k-1} Ef(X_{t_{i+1}, \bar{X}_{i+1}}(t_{k+1})) + Ef(X_{t_k, \bar{X}_k}(t_{k+1})). \tag{2.10}$$

This immediately implies the identity (recall that $\bar{X}_{i+1} = \bar{X}_{t_i,\bar{X}_i}(t_{i+1})$)

$$Ef(X(t_{k+1})) - Ef(\bar{X}_{k+1})$$
$$= \sum_{i=0}^{k-1} (EE(f(X_{t_{i+1},X_{t_i,\bar{X}_i}(t_{i+1})}(t_{k+1}))|X_{t_i,\bar{X}_i}(t_{i+1}))$$
$$- EE(f(X_{t_{i+1},\bar{X}_{t_i,\bar{X}_i}(t_{i+1})}(t_{k+1}))|\bar{X}_{t_i,\bar{X}_i}(t_{i+1}))$$
$$+ Ef(X_{t_k,\bar{X}_k}(t_{k+1})) - Ef(\bar{X}_{t_k,\bar{X}_k}(t_{k+1})). \quad (2.11)$$

According to the definition of $u(s,x)$, (2.11) implies

$$|Ef(X(t_{k+1})) - Ef(\bar{X}_{k+1})|$$
$$= |\sum_{i=0}^{k-1} (Eu(t_{i+1}, X_{t_i,\bar{X}_i}(t_i + h)) - Eu(t_{i+1}, \bar{X}_{t_i,\bar{X}_i}(t_i + h)))$$
$$+ (Ef(X_{t_k,\bar{X}_k}(t_{k+1})) - Ef(\bar{X}_{t_k,\bar{X}_k}(t_{k+1})))|$$
$$\leq \sum_{i=0}^{k-1} E|E(u(t_{i+1}, X_{t_i,\bar{X}_i}(t_i + h)) - u(t_{i+1}, \bar{X}_{t_i,\bar{X}_i}(t_i + h))|\bar{X}_i)|$$
$$+ E|E(f(X_{t_k,\bar{X}_k}(t_{k+1})) - f(\bar{X}_{t_k,\bar{X}_k}(t_{k+1}))|\bar{X}_k)|. \quad (2.12)$$

We note that the functions $u(s,x)$ and $f(x)$, which belong to F and so satisfy an inequality of the form (2.7), also satisfy the conditional version of this inequality. Suppose that for both $u(s,x)$ and $f(x)$ we have a function $K(x)$ in this inequality with $\kappa = 2m$. Then (2.12) implies

$$|Ef(X(t_{k+1})) - Ef(\bar{X}_{k+1})|$$
$$\leq \sum_{i=0}^{k-1} K(1 + E|\bar{X}_i|^{2m})h^{p+1} + K(1 + E|\bar{X}_k|^{2m})h^{p+1}.$$

Assuming that the requirement (d) holds for precisely this $2m$, we arrive at (2.5). □

We will now give a sufficient condition for requirement (d) in Theorem 2.1 which is convenient in practice.

Lemma 2.2. *Suppose that for $h < 1$,*

$$|EA(t_k, x, h; \xi_k)| \leq K(1 + |x|)h, \quad (2.13)$$
$$|A(t_k, x, h; \xi_k)| \leq M(\xi_k)(1 + |x|)h^{1/2}, \quad (2.14)$$

where $M(\xi_k)$ has moments of all orders.
Then for every even number $2m$ the mathematical expectations $E|\bar{X}_k|^{2m}$ exist and are uniformly bounded with respect to N and $k = 1, \ldots, N$, if only $E|\bar{X}_0|^{2m}$ exists.

2.2 The main theorem on convergence of weak approximations

Proof. For the i-th coordinate of the vector $\bar{X}_{k+1}$ we have

$$(\bar{X}_{k+1}^i)^{2m} = (\bar{X}_k^i + A^i(t_k, \bar{X}_k, h; \xi_k))^{2m}$$
$$= (\bar{X}_k^i)^{2m} + C_{2m}^1 (\bar{X}_k^i)^{2m-1} A^i(t_k, \bar{X}_k, h; \xi_k)$$
$$+ \sum_{j=2}^{2m} C_{2m}^j (\bar{X}_k^i)^{2m-j} (A^i(t_k, \bar{X}_k, h; \xi_k))^j. \quad (2.15)$$

Using (2.13), we obtain

$$|E(\bar{X}_k^i)^{2m-1} A^i(t_k, \bar{X}_k, h; \xi_k)|$$
$$= |E((\bar{X}_k^i)^{2m-1} E(A^i(t_k, \bar{X}_k, h; \xi_k)|\bar{X}_k))|$$
$$\leq |E|\bar{X}_k^i|^{2m-1} K(1 + |\bar{X}_k|) h \leq K(1 + E|\bar{X}_k|^{2m}) h. \quad (2.16)$$

By (2.14), we obtain for $h < 1$ and $j = 2, \ldots, 2m$:

$$|E|\bar{X}_k^i|^{2m-j} (A^i(t_k, \bar{X}_k, h; \xi_k))^j|$$
$$\leq E(|\bar{X}_k^i|^{2m-j} (M(\xi_k))^j (1 + |\bar{X}_k|)^j h^{j/2}) \leq K(1 + E|\bar{X}_k|^{2m}) h. \quad (2.17)$$

Because of (2.15)-(2.17) and the inequality $|x|^{2m} \leq K \sum_{i=1}^d (x^i)^{2m}$, where the constant K depends on d and m only, we obtain

$$E \sum_{i=1}^d (\bar{X}_{k+1}^i)^{2m} \leq E \sum_{i=1}^d (\bar{X}_k^i)^{2m} + K(1 + E \sum_{i=1}^d (\bar{X}_k^i)^{2m}) h.$$

Using Lemma 1.1.6, this concludes the proof of the lemma. □

Theorem 2.1 and Lemma 2.2 imply a theorem on the order of accuracy of the method

$$X_{k+1} = X_k + \sum_{r=1}^q \sigma_{rk} \xi_{rk} h^{1/2} + a_k h + \sum_{r=1}^q \sum_{i=1}^q (\Lambda_i \sigma_r)_k \xi_{irk} h$$
$$+ \frac{1}{2} \sum_{r=1}^q (\Lambda_r a + L \sigma_r)_k \xi_{rk} h^{3/2} + (La)_k \frac{h^2}{2}, \quad (2.18)$$

which is constructed according to (1.31).

In (2.18) the coefficients σ_{rk}, a_k, $(\Lambda_i \sigma_r)_k$, etc. are calculated at the point (t_k, X_k), and the sets of random variables ξ_{rk}, ξ_{irk} are independent and can be modeled for each k as in (1.31).

Theorem 2.3. *Suppose the conditions of Lemma 1.2 hold. Suppose also that the functions $\Lambda_i \sigma_r$, $\Lambda_r a$, $L \sigma_r$, and La grow at most as a linear function in $|x|$ as $|x|$ grows (the functions a and σ_r satisfy this requirement thanks to the Lipschitz condition (1.2)), i.e., (2.13) – (2.14) hold for (2.18). Then the method (2.18) has order of accuracy 2 in the sense of weak approximations,* i.e., for a sufficiently large class of functions f we have (2.5) with $p = 2$ (under the conditions of this theorem, this class of functions contains the functions that belong, together with their partial derivatives with respect to x up to order 6, inclusively, to $\mathbf{F}$).

The proof of this theorem clearly follows from the properties of the one-step approximation (1.31) proved in Sect. 2.1, Lemma 2.2, and Theorem 2.1.

Example 2.4. Consider the one-dimensional equation (1.1) with a single noise, i.e. $q = 1$. In this case

$$\xi_{11} = \frac{1}{2}(\xi^2 - 1),$$

where ξ is, e.g., $\mathcal{N}(0,1)$-distributed or distributed by the law $P(\xi = 0) = 2/3$, $P(\xi = -\sqrt{3}) = P(\xi = \sqrt{3}) = 1/6$. The formula (2.18) takes the form

$$\begin{aligned}X_{k+1} = X_k &+ \sigma_k \xi_k h^{1/2} + a_k h + \frac{1}{2}(\sigma \frac{\partial \sigma}{\partial x})_k (\xi_k^2 - 1)h \\ &+ \frac{1}{2}(\frac{\partial \sigma}{\partial t} + a\frac{\partial \sigma}{\partial x} + \frac{1}{2}\sigma^2 \frac{\partial^2 \sigma}{\partial x^2} + \sigma \frac{\partial a}{\partial x})_k \xi_k h^{3/2} \\ &+ (\frac{\partial a}{\partial t} + a\frac{\partial a}{\partial x} + \frac{1}{2}\sigma^2 \frac{\partial^2 a}{\partial x^2})_k \frac{h^2}{2}.\end{aligned} \quad (2.19)$$

This formula was derived in [176] using Taylor expansions of the characteristic functions of the variables $\Delta = X_{t,x}(t+h) - x$ and $\bar{\Delta} = \bar{X}_{t,x}(t+h) - x$.

2.2.2 Runge–Kutta type methods

The method (2.19) may present some difficulties because of the necessity of computing the derivatives of the coefficients a and σ at each step. Using the idea of the Runge–Kutta method, one can propose a number of ways in which by recalculation one can obtain a method not including all the derivatives participating in (2.19). We give a concrete, sufficiently simple method of this kind (it was proposed in [176]):

$$\begin{aligned}X_{k+1} = X_k &+ \frac{1}{2}\sigma_k \xi_k h^{1/2} + \frac{1}{2}(a - \sigma\frac{\partial \sigma}{\partial x})_k h + \frac{1}{2}(\sigma \frac{\partial \sigma}{\partial x})_k \xi_k^2 h \\ &+ \frac{1}{2}a(t_k + h, X_k + \sigma_k \xi_k h^{1/2} + a_k h)h \\ &+ \frac{1}{4}\sigma(t_k + h, X_k + \sigma_k \xi_k (\frac{h}{3})^{1/2} + a_k h)\xi_k h^{1/2} \\ &+ \frac{1}{4}\sigma(t_k + h, X_k - \sigma_k \xi_k (\frac{h}{3})^{1/2} + a_k h)\xi_k h^{1/2},\end{aligned} \quad (2.20)$$

where ξ_k are the same variables as in (2.19).

To get convinced of the fact that the method (2.20) is a method of order two, we note that

2.2 The main theorem on convergence of weak approximations

$$\frac{1}{2}a(t+h, x+\sigma\xi h^{1/2}+ah)h$$
$$=\frac{1}{2}(a+\frac{\partial a}{\partial t}h+\frac{\partial a}{\partial x}ah+\frac{\partial a}{\partial x}\sigma\xi h^{1/2}+\frac{1}{2}\frac{\partial^2 a}{\partial x^2}\sigma^2\xi^2 h)h$$
$$+\frac{1}{2}\frac{\partial^2 a}{\partial x^2}a\sigma\xi h^{5/2}+\frac{1}{2}\frac{\partial^2 a}{\partial t\partial x}\sigma\xi h^{5/2}+O(h^3)\,,\tag{2.21}$$

$$\frac{1}{4}\sigma(t+h, x+\sigma\xi(\frac{h}{3})^{1/2}+ah)\xi h^{1/2}+\frac{1}{4}\sigma(t+h, x-\sigma\xi(\frac{h}{3})^{1/2}+ah)\xi h^{1/2}$$
$$=\frac{1}{4}(2\sigma+2\frac{\partial\sigma}{\partial t}h+2\frac{\partial\sigma}{\partial x}ah+\frac{1}{3}\frac{\partial^2\sigma}{\partial x^2}\sigma^2\xi^2 h)\xi h^{1/2}+\frac{1}{4}\frac{\partial^2\sigma}{\partial t^2}\xi h^{5/2}$$
$$+\frac{1}{2}\frac{\partial^2\sigma}{\partial t\partial x}a\xi h^{5/2}+\frac{1}{12}\frac{\partial^3\sigma}{\partial t\partial x^2}\sigma^2\xi^3 h^{5/2}+\frac{1}{4}\frac{\partial^2\sigma}{\partial x^2}a^2\xi h^{5/2}$$
$$+\frac{1}{12}\frac{\partial^3\sigma}{\partial x^3}a\sigma^2\xi^3 h^{5/2}+\frac{1}{432}\frac{\partial^4\sigma}{\partial x^4}\sigma^4\xi^5 h^{5/2}+O(h^3)\,.\tag{2.22}$$

Substituting (2.21) and (2.22) in (2.20), we observe that X_{k+1} in (2.20) differs from X_{k+1} in (2.19), first by the sum

$$s_1 = (\frac{1}{2}\frac{\partial^2 a}{\partial x^2}a\sigma\xi_k + \frac{1}{2}\frac{\partial^2 a}{\partial t\partial x}\sigma\xi_k + \frac{1}{4}\frac{\partial^2\sigma}{\partial t^2}\xi_k + \frac{1}{2}\frac{\partial^2\sigma}{\partial t\partial x}a\xi_k + \frac{1}{12}\frac{\partial^3\sigma}{\partial t\partial x^2}\sigma^2\xi_k^3$$
$$+\frac{1}{4}\frac{\partial^2\sigma}{\partial x^2}a^2\xi_k + \frac{1}{12}\frac{\partial^3\sigma}{\partial x^3}a\sigma^2\xi_k^3 + \frac{1}{432}\frac{\partial^4\sigma}{\partial x^4}\sigma^4\xi_k^5)h^{5/2}+O(h^3)\,,$$

secondly, the term $\frac{1}{4}\sigma^2\frac{\partial^2\sigma}{\partial x^2}\xi_k h^{3/2}$ in (2.19) is replaced by $\frac{1}{12}\sigma^2\frac{\partial^2\sigma}{\partial x^2}\xi_k^3 h^{3/2}$, and, thirdly, the term $\frac{1}{4}\sigma^2\frac{\partial^2 a}{\partial x^2}h^2$ in (2.19) is replaced by $\frac{1}{4}\sigma^2\frac{\partial^2 a}{\partial x^2}\xi_k^2 h^2$. It is easy to see that these differences have no influence on the fulfillment of the conditions of the main Theorem 2.1 with $p=2$.

Thus, the method (2.20) has order of accuracy two. At each step it requires two recalculations of the function a, three recalculations of the function σ, one calculation of the function $\partial\sigma/\partial x$, and modeling of the single random variable ξ.

Some other Runge–Kutta type methods see in [159, 231, 232, 282, 295].

2.2.3 The Talay–Tubaro extrapolation method

Talay and Tubaro proved in [287] that it is possible to expand the global errors of weak methods for stochastic systems in powers of time increment h. Their approach is analogous to the Richardson–Runge extrapolation method for ordinary differential equations and allows us to estimate the global error as well as to improve the accuracy of the method. In particular, we can construct a method of order two applying the Euler method twice with different time steps.

Here we suppose that the coefficients of (1.1) are sufficiently smooth and all their derivatives up to a sufficiently large order are bounded.

Theorem 2.5. *Let a one-step weak approximation $\bar{X}_{t,x}(t+h)$ of the solution $X_{t,x}(t+h)$ of (1.1) generate a method of order p. Then the global error*

$$R := Ef(X_{t_0,X_0}(T)) - Ef(\bar{X}_{t_0,X_0}(T))$$

of the method has the following expansion

$$R = C_0 h^p + \cdots + C_n h^{p+n} + O(h^{p+n+1}), \quad (2.23)$$

where the constants $C_0, \ldots, C_n$ are independent of h, and n is an integer, $n \geq 0$ (n can be anyhow large if the coefficients of (1.1) belong to C^∞ and their derivatives of any order are bounded).

In particular, for the Euler method

$$R = C_0 h + \cdots + C_n h^{1+n} + O(h^{n+2}). \quad (2.24)$$

Proof. We begin with the proof of the formula

$$R = Ch + O(h^2) \quad (2.25)$$

for the Euler method

$$X_{k+1} = X_k + \sum_{r=1}^{q} \sigma_{rk} \xi_{rk} h^{1/2} + a_k h, \quad (2.26)$$

where all the ξ_{rk} are independent and are modeled by the law $P(\xi = \pm 1) = 1/2$. As in the proof of the convergence Theorem 2.1, we obtain the equality (see (2.11) and (2.12))

$$R = \sum_{i=0}^{N-1} E(u(t_{i+1}, X_{t_i, \bar{X}_i}(t_{i+1})) - u(t_{i+1}, \bar{X}_{t_i, \bar{X}_i}(t_{i+1})))$$

$$= E \sum_{i=0}^{N-1} E(u(t_{i+1}, X_{t_i, \bar{X}_i}(t_{i+1})) - u(t_{i+1}, \bar{X}_{t_i, \bar{X}_i}(t_{i+1}))|\bar{X}_i), \quad (2.27)$$

where

$$u(s, x) = Ef(X_{s,x}(T)).$$

By Lemma 1.9 we get

$$Eu(t+h, X_{t,x}(t+h)) = u(t,x) + hLu(t,x) + \frac{1}{2} h^2 L^2 u(t,x) + O(h^3). \quad (2.28)$$

Expanding $Eu(t+h, \bar{X}_{t,x}(t+h))$ in powers of h by the usual Taylor formula, we obtain

$$Eu(t+h, \bar{X}_{t,x}(t+h)) = u(t,x) + hLu(t,x) + \frac{1}{2}h^2 A(t,x) + O(h^3). \quad (2.29)$$

Two first terms in the right-hand sides of (2.28) and (2.29) coincide since the one-step order for the Euler method is equal to two. The direct computation gives

$$A(t,x) = \frac{\partial^2 u}{\partial t^2} + 2\sum_{i=1}^{d} a^i \frac{\partial^2 u}{\partial t \partial x^i} + \sum_{r=1}^{q}\sum_{i,j=1}^{d} \sigma_r^i \sigma_r^j \frac{\partial^3 u}{\partial t \partial x^i \partial x^j}$$

$$+ \sum_{i,j=1}^{d} a^i a^j \frac{\partial^2 u}{\partial x^i \partial x^j} + \sum_{r=1}^{q}\sum_{i,j,l=1}^{d} a^i \sigma_r^j \sigma_r^l \frac{\partial^3 u}{\partial x^i \partial x^j \partial x^l}$$

$$+ \frac{1}{12}\sum_{r,s=1}^{q}\sum_{i,j,l,m=1}^{d} \sigma_r^i \sigma_r^j \sigma_s^l \sigma_s^m \frac{\partial^4 u}{\partial x^i \partial x^j \partial x^l \partial x^m},$$

where all the coefficients and the derivatives of the function u are calculated at the point (t,x). We should underline that in fact we need not the explicit form of the function $A(t,x)$. We have given it for definiteness only. A little later we make use of this observation in the proof of the general assertion.

It follows from (2.28) and (2.29) that

$$Eu(t+h, X_{t,x}(t+h)) - Eu(t+h, \bar{X}_{t,x}(t+h)) = h^2 B(t,x) + O(h^3), \quad (2.30)$$

where $B(t,x) = (L^2 u(t,x) - A(t,x))/2$. The formula (2.27) can be rewritten in the form

$$R = E\sum_{i=0}^{N-1} h^2 B(t, \bar{X}_i) + O(h^2). \quad (2.31)$$

Consider now the $(d+1)$-dimensional system

$$dX = a(t,X)dt + \sum_{r=1}^{q} \sigma_r(t,X)dw_r(t), \quad X(t_0) = X_0,$$
$$dY = B(t,X)dt, \quad Y(t_0) = 0. \quad (2.32)$$

Solving (2.32) by the Euler method, we get

$$E\sum_{i=0}^{N-1} B(t,\bar{X}_i)h = E\bar{Y}(T) = EY(T) + O(h) = C + O(h), \quad (2.33)$$

where the constant C is equal to

$$C = EY(T) = E\int_{t_0}^{T} B(\theta, X(\theta))d\theta. \quad (2.34)$$

The formula (2.25) follows from (2.31) and (2.33).

Let us prove (2.24). Now, instead of (2.30), we use the formula

$$Eu(t+h, X_{t,x}(t+h)) - Eu(t+h, \bar{X}_{t,x}(t+h))$$
$$= h^2 B_0(t,x) + \cdots + h^{n+2} B_n(t,x) + O(h^{n+3}),$$

with

$$B_j(t,x) = \frac{1}{(j+2)!}(L^{2+j}u(t,x) - A_j(t,x)),$$

where $A_j(t,x)$ are coefficients of the corresponding expansion of $Eu(t+h, \bar{X}_{t,x}(t+h))$. As a result, we get instead of (2.31):

$$R = E \sum_{i=0}^{N-1} \sum_{j=0}^{n} h^{2+j} B_j(t, \bar{X}_i) + O(h^{n+2}). \tag{2.35}$$

For each $j = 0, \ldots, n$ consider the $(d+1)$-dimensional system like to (2.32). Due to (2.33) we can write

$$R = \sum_{j=0}^{n} K_j h^{1+j} + h R_1 + O(h^{n+2}), \tag{2.36}$$

where $K_j = E Y_j(T)$ and

$$R_1 = E \sum_{j=0}^{n} h^j (\bar{Y}_j - Y_j).$$

Clearly, R_1 has a representation analogous to (2.36) and we obtain (2.24) in a finite number of such steps. The formula (2.23) is proved analogously. □

Due to Theorem 2.5, we obtain an extension of the well known in the case of deterministic differential equations extrapolation methods to the stochastic case. For example, simulating $u = Ef(X_{t_0,X_0}(T))$ twice by the Euler scheme but with varying time steps $h_1 = h$, $h_2 = \alpha h$, $\alpha > 0$, $\alpha \neq 1$, we obtain $\bar{u}^{h_1} = Ef(\bar{X}_{t_0,X_0}^{h_1}(T))$ and $\bar{u}^{h_2} = Ef(\bar{X}_{t_0,X_0}^{h_2}(T))$. We can expand (see (2.25))

$$u = \bar{u}^{h_1} + C h_1 + O(h^2),$$
$$u = \bar{u}^{h_2} + C h_2 + O(h^2), \tag{2.37}$$

whence

$$C = -\frac{\bar{u}^{h_2} - \bar{u}^{h_1}}{h_2 - h_1} + O(h). \tag{2.38}$$

By (2.37) and (2.38) we get the improved value with error $O(h^2)$:

$$\bar{u}_{imp} = \bar{u}^{h_1} \frac{h_2}{h_2 - h_1} - \bar{u}^{h_2} \frac{h_1}{h_2 - h_1}, \quad u = \bar{u}_{imp} + O(h^2). \tag{2.39}$$

Thus, the obtained method has an accuracy of order two. In the same spirit, using three recalculations of $u = Ef(X_{t_0,X_0}(T))$ by the Euler method with varying time-steps, one can find C_0 and C_1 from (2.24) and, as a consequence, a method of order three can be constructed, and so on. The formula (2.23) can be used in the same way. We see that extrapolation procedures make possible to construct methods of higher orders much simpler to implement than procedures based on one-step approximations of high order. But the extrapolation procedures have certain deficiencies, they are not general-purpose. For example, they do not allow us to use schemes with variable step. So that, similarly to deterministic case, the one-step high-order approximations are of independent significance.

2.2.4 Implicit method

The following formula holds (cf. Sect. 1.2.2):

$$a(t+h, X(t+h)) = a + \sum_{r=1}^{q} \Lambda_r a \int_t^{t+h} dw_r(\theta) + La \cdot h + \rho_1, \qquad (2.40)$$

where

$$\rho_1 = \sum_{r=1}^{q} \sum_{i=1}^{q} \Lambda_i \Lambda_r a \int_t^{t+h} (\int_t^{\theta} dw_i(\theta_1)) dw_r(\theta)$$

$$+ \sum_{r=1}^{q} \sum_{i=1}^{q} \sum_{s=1}^{q} \int_t^{t+h} (\int_t^{\theta} (\int_t^{\theta_1} \Lambda_s \Lambda_i \Lambda_r a(\theta_2, X(\theta_2))$$
$$\times dw_s(\theta_2)) dw_i(\theta_1)) dw_r(\theta)$$

$$+ \sum_{r=1}^{q} \sum_{i=1}^{q} \int_t^{t+h} (\int_t^{\theta} (\int_t^{\theta_1} L\Lambda_i \Lambda_r a(\theta_2, X(\theta_2)) d\theta_2) dw_i(\theta_1)) dw_r(\theta)$$

$$+ \sum_{r=1}^{q} \int_t^{t+h} (\int_t^{\theta} L\Lambda_r a(\theta_1, X(\theta_1)) d\theta_1) dw_r(\theta)$$

$$+ \sum_{r=1}^{q} \int_t^{t+h} (\int_t^{\theta} \Lambda_r La(\theta_1, X(\theta_1)) dw_r(\theta_1)) d\theta. \qquad (2.41)$$

As in Lemma 1.2, we can show that

$$|E\rho_1| \leq K(x)h^2, \ K(x) \in \mathbf{F}, \qquad (2.42)$$

$$E|\rho_1|^2 \leq K(x)h^2, \ K(x) \in \mathbf{F}, \qquad (2.43)$$

$$|E\rho_1 \int_t^{t+h} dw_r(\theta)| \leq K(x)h^2, \ K(x) \in \mathbf{F}. \qquad (2.44)$$

Further,
$$La(t+h, X(t+h)) = La + \rho_2, \quad (2.45)$$
where
$$\rho_2 = \sum_{r=1}^{q} \int_t^{t+h} \Lambda_r La(\theta, X(\theta)) dw_r(\theta) + \int_t^{t+h} L^2 a(\theta, X(\theta)) d\theta. \quad (2.46)$$

We can readily prove that
$$|E\rho_2| \le K(x)h, \ K(x) \in \mathbf{F}, \quad (2.47)$$
$$E|\rho_2|^2 \le K(x)h, \ K(x) \in \mathbf{F}, \quad (2.48)$$
$$|E\rho_2 \int_t^{t+h} dw_r(\theta)| \le K(x)h, \ K(x) \in \mathbf{F}. \quad (2.49)$$

Using the relations (1.3), (2.40), and (2.45), it is easy to obtain the following formula, involving the arbitrary constants α, β:

$$X(t+h) = x + \sum_{r=1}^{q} \sigma_r I_r + \alpha a h + (1-\alpha)a(t+h, X(t+h))h$$
$$+ \sum_{r=1}^{q} \sum_{i=1}^{q} \Lambda_i \sigma_r I_{ir} + \sum_{r=1}^{q} (L\sigma_r - (1-\alpha)\Lambda_r a) I_r h$$
$$+ \sum_{r=1}^{q} (\Lambda_r a - L\sigma_r) J_r + \beta(2\alpha - 1) La \frac{h^2}{2}$$
$$+ (1-\beta)(2\alpha - 1) La(t+h, X(t+h)) \frac{h^2}{2}$$
$$+ \sum_{r=1}^{q} \sum_{i=1}^{q} \sum_{s=1}^{q} \Lambda_s \Lambda_i \sigma_r I_{sir}$$
$$+ \rho - (1-\alpha)\rho_1 h - (1-\beta)(2\alpha - 1)\rho_2 h^2. \quad (2.50)$$

Introduce the two-parameter family of implicit methods
$$\bar{X} = x + \sum_{r=1}^{q} \sigma_r \xi_r h^{1/2} + \alpha a h + (1-\alpha)a(t+h, \bar{X})h$$
$$+ \sum_{r=1}^{q} \sum_{i=1}^{q} \Lambda_i \sigma_r \xi_{ir} h + \sum_{r=1}^{q} (\frac{2\alpha - 1}{2}\Lambda_r a + \frac{1}{2}L\sigma_r)\xi_r h^{3/2}$$
$$+ \beta(2\alpha - 1) La \frac{h^2}{2} + (1-\beta)(2\alpha - 1) La(t+h, \bar{X}) \frac{h^2}{2}, \quad (2.51)$$

where ξ_r, ξ_{ir} can be modeled as in Sect. 2.1 (see (1.31)).

We show that (under certain natural assumptions) the method (2.51) has order of accuracy two. To this end we consider the equation

2.2 The main theorem on convergence of weak approximations 111

$$X - (1-\alpha)a(t+h,X)h - (1-\beta)(2\alpha-1)La(t+h,X)\frac{h^2}{2} = Z. \quad (2.52)$$

We assume that for sufficiently small h and all Z this equation can be solved for X:

$$X = \varphi(t+h, Z). \quad (2.53)$$

Introduce vectors Y, $\bar{Y}$:

$$Y = x + \sum_{r=1}^{q} \sigma_r I_r + \alpha a h + \sum_{r=1}^{q}\sum_{i=1}^{q} \Lambda_i \sigma_r I_{ir}$$

$$+ \sum_{r=1}^{q}(L\sigma_r - (1-\alpha)\Lambda_r a)I_r h + \sum_{r=1}^{q}(\Lambda_r a - L\sigma_r)J_r + \beta(2\alpha-1)La\frac{h^2}{2}$$

$$+ \sum_{r=1}^{q}\sum_{i=1}^{q}\sum_{s=1}^{q} \Lambda_s \Lambda_i \sigma_r I_{sir}$$

$$+ \rho - (1-\alpha)\rho_1 h - (1-\beta)(2\alpha-1)\rho_2 h^2, \quad (2.54)$$

$$\bar{Y} = x + \sum_{r=1}^{q} \sigma_r \xi_r h^{1/2} + \alpha a h + \sum_{r=1}^{q}\sum_{i=1}^{q} \Lambda_i \sigma_r \xi_{ir} h$$

$$+ \sum_{r=1}^{q}(\frac{2\alpha-1}{2}\Lambda_r a + \frac{1}{2}L\sigma_r)\xi_r h^{3/2} + \beta(2\alpha-1)La\frac{h^2}{2}. \quad (2.55)$$

Then by (2.50), (2.52), and (2.54):

$$X(t+h) = \varphi(t+h, Y),$$

and by (2.51), (2.52), and (2.55):

$$\bar{X} = \bar{X}(t+h) = \varphi(t+h, \bar{Y}).$$

Assume that the function $\varphi(t+h, y)$ has partial derivatives with respect to y up to order six, inclusively, and that they together with φ belong to **F**. For $s = 1, \ldots, 5$ we write

$$|E(\prod_{j=1}^{s} \Delta^{i_j} - \prod_{j=1}^{s} \bar{\Delta}^{i_j})|$$

$$= |E(\prod_{j=1}^{s}(X^{i_j} - x^{i_j}) - \prod_{j=1}^{s}(\bar{X}^{i_j} - x^{i_j}))|$$

$$= |E(\prod_{j=1}^{s}(\varphi^{i_j}(t+h, Y) - x^{i_j}) - \prod_{j=1}^{s}(\varphi^{i_j}(t+h, \bar{Y}) - x^{i_j}))|. \quad (2.56)$$

The right-hand side of (2.56) is $O(h^3)$ if (see the proof of Theorem 1.6):

$$|E(\prod_{j=1}^{s}(Y^{r_j} - x^{r_j}) - \prod_{j=1}^{s}(\bar{Y}^{r_j} - x^{r_j}))| \leq K(x)h^3, \ s = 1, \ldots, 5, \quad (2.57)$$

and if

$$E\prod_{j=1}^{s}|Y^{r_j} - x^{r_j}| \leq K(x)h^3, \ E\prod_{j=1}^{s}|\bar{Y}^{r_j} - x^{r_j}| \leq K(x)h^3, \ s = 6. \quad (2.58)$$

Taking into account the properties (2.42)-(2.44) and (2.47)-(2.49) of the remainders ρ_1 and ρ_2, the relations (2.57)-(2.58) can be proved as in Sect. 2.1 we proved the analogous relations for the differences $X^{i_j} - x^{i_j}$ and $\bar{X}^{i_j} - x^{i_j}$. So, we have proved that $\bar{X}$, which is implicitly defined by (2.51), satisfies

$$|E(\prod_{j=1}^{s}\Delta^{i_j} - \prod_{j=1}^{s}\bar{\Delta}^{i_j})| \leq K(x)h^3, \ s = 1, \ldots, 5, \ K(x) \in \mathbf{F}. \quad (2.59)$$

We will prove the inequality

$$E\prod_{j=1}^{s}|\bar{\Delta}^{i_j}| \leq K(x)h^3, \ s = 6, \ K(x) \in \mathbf{F}. \quad (2.60)$$

In fact, the solvability of (2.51) for $\bar{X}$ in the form $\bar{X} = \varphi(t+h, \bar{Y})$ with $\varphi \in \mathbf{F}$, implies existence of all sufficiently high moments of $\bar{X}$ if only ξ_r and ξ_{ir} (which participate in the formula for $\bar{Y}$) have sufficiently high moments. Further, since $a \in \mathbf{F}$ and $La \in \mathbf{F}$, moments (up to order six, inclusively) for $a(t+h, \bar{X})$ and $La(t+h, \bar{X})$ exist. Now (2.60) immediately follows from (2.51).

Finally, assume that $\varphi(t+h, \bar{Y})$ grows at most linearly as $|x|$ goes to infinity. Then the subsequent application of Lemma 2.2 and Theorem 2.1 leads to the result which we state as the theorem.

Theorem 2.6. *Suppose that for sufficiently small h the relation (2.52) is solvable for $Z: X = \varphi(t+h, Z)$. Suppose that the function $\varphi(t+h, y)$ has partial derivatives with respect to y up to order six, inclusively, that together with φ belong to $\mathbf{F}$. Finally, assume that the superposition $\varphi(t+h, \bar{Y})$, with $\bar{Y}$ defined by (2.55), grows at most linearly as $|x| \to \infty$. Then the implicit method based on (2.51) has order of accuracy two in the sense of weak approximation.*

2.3 Weak methods for systems with additive and colored noise

Consider the system of SDEs with additive noise

$$dX = a(t, X)dt + \sum_{r=1}^{q}\sigma_r(t)dw_r(t). \quad (3.1)$$

Since the σ_r do not depend on x, numerical methods are essentially simpler for such systems.

2.3 Weak methods for systems with additive and colored noise

2.3.1 Second-order methods

Since in the case of (3.1) the $\Lambda_i \sigma_r$ vanish, the terms ξ_{irk} in the method (2.18) (which has order of accuracy two) are absent, and consequently, we only have to model the random variables ξ_{rk} at each step. The method (2.18) takes the following form for the system (3.1):

$$X_{k+1} = X_k + \sum_{r=1}^{q} \sigma_r(t_k)\xi_{rk}h^{1/2} + a_k h$$

$$+ \frac{1}{2}\sum_{r=1}^{q}(\sigma'_r + (\sigma_r, \frac{\partial}{\partial x})a)_k \xi_{rk} h^{3/2} + (La)_k \frac{h^2}{2}. \quad (3.2)$$

Moreover, it is not difficult to obtain the fully Runge–Kutta method of order two for (3.1):

$$X_{k+1} = X_k + \sum_{r=1}^{q} \sigma_r(t_k + \frac{h}{2})\xi_{rk}h^{1/2}$$

$$+ \frac{h}{2}[a_k + a(t_{k+1}, X_k + \sum_{r=1}^{q}\sigma_r(t_k)\xi_{rk}h^{1/2} + a_k h)]. \quad (3.3)$$

While for systems of a general form the attempt to arrive at a method of order of accuracy three meets with extremely awkward constructions, for systems with additive noise the problem of constructing such a method can be solved relatively simply. This problem is considered in the next two subsections.

2.3.2 Main lemmas for third-order methods

To construct a method of order of accuracy three we write down the following formula for the solution $X_{t,x}(\theta) = X(\theta)$ of (3.1):

$$X(t+h) = x + \sum_{r=1}^{q}\sigma_r \int_t^{t+h} dw_r(\theta) + ah$$

$$+ \sum_{r=1}^{q}\Lambda_r a \int_t^{t+h}(w_r(\theta) - w_r(t))d\theta + \sum_{r=1}^{q}\sigma'_r \int_t^{t+h}(\theta - t)dw_r(\theta)$$

$$+ La\frac{h^2}{2} + \sum_{r=1}^{q}\sum_{i=1}^{q}\Lambda_i\Lambda_r a \int_t^{t+h}(\int_t^{\theta}(w_i(\theta_1) - w_i(t))dw_r(\theta_1))d\theta$$

$$+ \sum_{r=1}^{q}\sum_{i=1}^{q}\sum_{s=1}^{q}\Lambda_s\Lambda_i\Lambda_r a \int_t^{t+h}(\int_t^{\theta}(\int_t^{\theta_1}(w_s(\theta_2)$$

$$- w_s(t))dw_i(\theta_2))dw_r(\theta_1))d\theta$$

$$+ \sum_{r=1}^{q} \sigma_r'' \int_t^{t+h} (\int_t^{\theta} (\theta_1 - t) d\theta_1) dw_r(\theta)$$

$$+ \sum_{r=1}^{q} L\Lambda_r a \int_t^{t+h} (\int_t^{\theta} (\theta_1 - t) dw_r(\theta_1)) d\theta$$

$$+ \sum_{r=1}^{q} \Lambda_r L a \int_t^{t+h} (\int_t^{\theta} (w_r(\theta_1) - w_r(t)) d\theta_1) d\theta + L^2 a \frac{h^3}{6} + \rho, \quad (3.4)$$

where all coefficients σ_r, $\Lambda_r a$, σ_r', La, $\Lambda_i \Lambda_r a$, $\Lambda_s \Lambda_i \Lambda_r a$, σ_r'', $L\Lambda_r a$, $\Lambda_r La$, $L^2 a$ are calculated at the point (t, x).

Lemma 3.1. *The remainder ρ in (3.4) satisfies the relations*

$$|E\rho| = O(h^4), \qquad (3.5)$$

$$E|\rho|^2 = O(h^6), \qquad (3.6)$$

$$|E\rho \int_t^{t+h} dw_r(\theta)| = O(h^4). \qquad (3.7)$$

This lemma can be proved similarly to Lemma 1.2. To shorten the exposition, we have not listed all assumptions on the coefficients a and σ_r in detail, they are similar to those in Lemma 1.2. For brevity here and below we use, e.g., (3.5) instead of (1.5).

We introduce the notation

$$I_r = \int_t^{t+h} dw_r(\theta), \quad J_r = \int_t^{t+h} (w_r(\theta) - w_r(t)) d\theta,$$

$$G_r = \int_t^{t+h} (w_r(\theta) - w_r(t))(\theta - t) d\theta,$$

$$J_{ir} = \int_t^{t+h} (\int_t^{\theta} (w_i(\theta_1) - w_i(t)) dw_r(\theta_1)) d\theta,$$

$$J_{sir} = \int_t^{t+h} (\int_t^{\theta} (\int_t^{\theta_1} (w_s(\theta_2) - w_s(t)) dw_i(\theta_2)) dw_r(\theta_1)) d\theta.$$

Lemma 3.2. *The following identities hold:*

$$\int_t^{t+h} (\theta - t) dw_r(\theta) = hI_r - J_r, \qquad (3.8)$$

$$\int_t^{t+h} (\int_t^{\theta} (\theta_1 - t) dw_r(\theta_1)) d\theta = 2G_r - hJ_r, \qquad (3.9)$$

2.3 Weak methods for systems with additive and colored noise

$$\int_t^{t+h}(\int_t^\theta (w_r(\theta_1)-w_r(t))d\theta_1)d\theta = hJ_r - G_r, \tag{3.10}$$

$$\int_t^{t+h}(\int_t^\theta (\theta_1-t)d\theta_1)dw_r(\theta) = \frac{1}{2}h^2 I_r - G_r. \tag{3.11}$$

Proof. We give a proof of (3.9). We have:

$$d(\int_t^\theta (\theta_1-t)dw_r(\theta_1) \cdot (\theta-t)) = \int_t^\theta (\theta_1-t)dw_r(\theta_1) \cdot d\theta + (\theta-t)^2 dw_r(\theta).$$

Integration of this identity from t to $t+h$ gives

$$\int_t^{t+h}(\int_t^\theta (\theta_1-t)dw_r(\theta_1))d\theta = h\int_t^{t+h}(\theta-t)dw_r(\theta) - \int_t^{t+h}(\theta-t)^2 dw_r(\theta). \tag{3.12}$$

Further,

$$d((\theta-t)^2(w_r(\theta)-w_r(t))) = 2(\theta-t)(w_r(\theta)-w_r(t))d\theta + (\theta-t)^2 dw_r(\theta),$$

whence

$$\int_t^{t+h}(\theta-t)^2 dw_r(\theta) = h^2(w_r(t+h)-w_r(t))$$

$$-2\int_t^{t+h}(w_r(\theta)-w_r(t))(\theta-t)d\theta = h^2 I_r - 2G_r. \tag{3.13}$$

The formula (3.8) for the integral $\int_t^{t+h}(\theta-t)dw_r(\theta)$ on the right-hand side of (3.12) can be obtained in a similar way. Substituting (3.8) and (3.13) in (3.12), we obtain (3.9). Thus, (3.9) has been proved. The derivation of (3.10) and (3.11) is even simpler. □

By Lemma 3.2, formula (3.4) can be written as

$$X(t+h) = x + \sum_{r=1}^q \sigma_r I_r + ah + \sum_{r=1}^q \Lambda_r a J_r + \sum_{r=1}^q \sigma'_r(hI_r - J_r)$$

$$+ La\frac{h^2}{2} + \sum_{r=1}^q \sum_{i=1}^q \Lambda_i \Lambda_r a J_{ir} + \sum_{r=1}^q L\Lambda_r a(2G_r - hJ_r)$$

$$+ \sum_{r=1}^q \Lambda_r La(hJ_r - G_r) + \sum_{r=1}^q \sigma''_r(\frac{1}{2}h^2 I_r - G_r)$$

$$+L^2 a \frac{h^3}{6} + \sum_{r=1}^{q}\sum_{i=1}^{q}\sum_{s=1}^{q} \Lambda_s \Lambda_i \Lambda_r a J_{sir} + \rho. \tag{3.14}$$

Lemma 3.3. *We have*

$$E J_{sir} = 0, \ E J_{sir} I_j = 0, \ E J_{sir} I_j I_l = 0. \tag{3.15}$$

Proof. The first and last identities in (3.15) can be proved using oddness. The second identity can be proved using arguments similar to the ones in the proofs of Lemmas 1.3 and 1.5. □

As in Sect. 2.1, we introduce an auxiliary vector $\tilde{X}$ equal to the right-hand side of (3.14) without the last two terms:

$$\tilde{X} = x + \sum_{r=1}^{q} \sigma_r I_r + ah + \sum_{r=1}^{q} \Lambda_r a J_r + \sum_{r=1}^{q} \sigma'_r(hI_r - J_r)$$

$$+ La\frac{h^2}{2} + \sum_{r=1}^{q}\sum_{i=1}^{q} \Lambda_i \Lambda_r a J_{ir} + \sum_{r=1}^{q} L\Lambda_r a(2G_r - hJ_r)$$

$$+ \sum_{r=1}^{q} \Lambda_r La(hJ_r - G_r) + \sum_{r=1}^{q} \sigma''_r(\frac{1}{2}h^2 I_r - G_r) + L^2 a \frac{h^3}{6}. \tag{3.16}$$

Lemma 3.4. *The following relations hold:*

$$|E(\prod_{j=1}^{s} \Delta^{i_j} - \prod_{j=1}^{s} \tilde{\Delta}^{i_j})| = O(h^4), \ s = 1, \ldots, 7. \tag{3.17}$$

The proof of this lemma is based on Lemmas 3.1 and 3.3 and differs not essentially from the proof of Lemma 1.4.

2.3.3 Construction of a method of order three

We construct the one-step approximation $\bar{X}$ on the basis of $\tilde{X}$ as follows:

$$\bar{X} = x + \sum_{r=1}^{q} \sigma_r \xi_r h^{1/2} + ah + \sum_{r=1}^{q} \Lambda_r a \eta_r h^{3/2} + \sum_{r=1}^{q} \sigma'_r(\xi_r - \eta_r)h^{3/2}$$

$$+ La\frac{h^2}{2} + \sum_{r=1}^{q}\sum_{i=1}^{q} \Lambda_i \Lambda_r a \eta_{ir} h^2 + \sum_{r=1}^{q} L\Lambda_r a(2\mu_r - \eta_r)h^{5/2}$$

$$+ \sum_{r=1}^{q} \Lambda_r La(\eta_r - \mu_r)h^{5/2} + \sum_{r=1}^{q} \sigma''_r(\frac{1}{2}\xi_r - \mu_r)h^{5/2} + L^2 a \frac{h^3}{6}. \tag{3.18}$$

To construct a method of order of accuracy three (the one-step order of accuracy of such a method equals 4), we need fulfillment of the relations

2.3 Weak methods for systems with additive and colored noise   117

$$|E(\prod_{j=1}^{s} \tilde{\Delta}^{i_j} - \prod_{j=1}^{s} \bar{\Delta}^{i_j})| = O(h^4), \ s = 1, \ldots, 7. \tag{3.19}$$

Indeed, Lemma 3.4 in this case implies

$$|E(\prod_{j=1}^{s} \Delta^{i_j} - \prod_{j=1}^{s} \bar{\Delta}^{i_j})| = O(h^4), \ s = 1, \ldots, 7. \tag{3.20}$$

Then according to Theorem 2.1, a method based on the one-step approximation $\bar{X}$ with the properties (3.20) and (3.21) will have order of accuracy three. Of course, the above said is valid under the standard assumptions on the coefficients of the system (3.1) which, in particular for the approximation (3.18), ensure the relation

$$E\prod_{j=1}^{s} |\bar{\Delta}^{i_j}| = O(h^4), \ s = 8. \tag{3.21}$$

Thus, we should consider the relations (3.19) only.

Lemma 3.5. *The following seven groups of identities ensure the relations (3.19):*

$$E\xi_r h^{1/2} = EI_r = 0, \ E\eta_r h^{3/2} = EJ_r = 0,$$
$$E\eta_{ir} h^2 = EJ_{ir} = 0, \ E\mu_r h^{5/2} = EG_r = 0; \tag{3.22}$$

$$E\xi_i \xi_r h = EI_i I_r = \delta_{ir} h, \ E\xi_r \eta_j h^2 = EI_r J_j = \delta_{rj}\frac{h^2}{2},$$
$$E\xi_i \eta_{jr} h^{5/2} = EI_i J_{jr} = 0, \ E\xi_i \mu_r h^3 = EI_i G_r = \delta_{ir}\frac{h^3}{3},$$
$$E\eta_i \eta_j h^3 = EJ_i J_j = \delta_{ij}\frac{h^3}{3}, \ E\eta_i \eta_{jr} h^{7/2} = EJ_i J_{jr} = 0; \tag{3.23}$$

$$E\xi_i \xi_r \eta_{js} h^3 = EI_i I_r J_{js}$$
$$= \begin{cases} h^3/6 \text{ if } j \neq s \text{ and either } i = j, \ r = s \text{ or } i = s, \ r = j, \\ h^3/3 \text{ if } i = r = j = s, \\ 0 \quad \text{otherwise}, \end{cases}$$
$$E\xi_i \xi_r \mu_j h^{7/2} = EI_i I_r G_j = 0, \ E\xi_i \eta_j \eta_r h^{7/2} = EI_i J_j J_r = 0; \tag{3.24}$$

118   2 Weak approximation for stochastic differential equations

$$E\xi_i\xi_r\xi_j\xi_s h^2 = EI_iI_rI_jI_s = \begin{cases} h^2 & \text{if } \{i,r,j,s\} \text{ consists of two pairs} \\ & \text{of equal numbers,} \\ 3h^2 & \text{if } i = r = j = s, \\ 0 & \text{otherwise,} \end{cases}$$

$$E\xi_i\xi_r\xi_j\eta_s h^3 = EI_iI_rI_jJ_s = \begin{cases} h^3/2 & \text{if } \{i,r,j,s\} \text{ consists of two pairs} \\ & \text{of equal numbers,} \\ 3h^3/2 & \text{if } i = r = j = s, \\ 0 & \text{otherwise,} \end{cases}$$

$$E\xi_i\xi_r\xi_j\eta_{sl} h^{7/2} = EI_iI_rI_jJ_{sl} = 0; \tag{3.25}$$

$$E\xi_i\xi_r\xi_j\xi_s\xi_l h^{5/2} = EI_iI_rI_jI_sI_l = 0,$$
$$E\xi_i\xi_r\xi_j\xi_s\eta_l h^{7/2} = EI_iI_rI_jI_sJ_l = 0; \tag{3.26}$$

in the following identities we assume, without loss of generality, that $i_1 \leq i_2 \leq i_3 \leq i_4 \leq i_5 \leq i_6$:

$$E\prod_{j=1}^{6}\xi_{i_j} h^3 = E\prod_{j=1}^{6} I_{i_j}$$

$$= \begin{cases} h^3, & i_1 = i_2 < i_3 = i_4 < i_5 = i_6, \\ 3h^3, & i_1 = i_2 < i_3 = i_4 = i_5 = i_6, \\ 3h^3, & i_1 = i_2 = i_3 = i_4 < i_5 = i_6, \\ 15h^3, & i_1 = i_2 = i_3 = i_4 = i_5 = i_6, \\ 0, & \text{otherwise;} \end{cases} \tag{3.27}$$

$$E\prod_{j=1}^{7}\xi_{i_j} h^{7/2} = E\prod_{j=1}^{7} I_{i_j} = 0. \tag{3.28}$$

Proof. The proof of this lemma repeats the proof of Lemma 1.5 in many respects. Here we will consider the proof of the identity for $I_iI_rJ_{js}$ in (3.24) only. Without loss of generality, we may put $t = 0$. We introduce the equations

$$dx = w_j dw_s(\theta), \quad x(0) = 0,$$
$$dy = x d\theta, \quad y(0) = 0.$$

Then

$$y(h) = \int_0^h x(\theta) d\theta = \int_0^h (\int_0^\theta w_j(\theta_1) dw_s(\theta_1)) d\theta = J_{js}(h),$$

$$EI_iI_rJ_{js} = E(w_i(h)w_r(h)y(h)).$$

We have

$$d(w_iw_ry) = w_r y dw_i + w_i y dw_r + w_iw_r x d\theta + y\delta_{ir} d\theta.$$

Hence
$$dE(w_i w_r y) = E(w_i w_r x) d\theta, \qquad (3.29)$$
since, obviously, Ey vanishes. We turn to evaluation of $E(w_i w_r x)$. We have
$$d(w_i w_r x) = w_r x dw_i + w_i x dw_r + w_i w_r w_j dw_s$$
$$+ x \delta_{ir} d\theta + w_i w_j \delta_{rs} d\theta + w_r w_j \delta_{is} d\theta,$$
whence
$$dE(w_i w_r x) = \delta_{ir} Ex d\theta + E(w_i w_j)\delta_{rs} d\theta + E(w_r w_j)\delta_{is} d\theta. \qquad (3.30)$$

Since $Ex(t) \equiv 0$, the right-hand side of (3.30) does not vanish in three cases only.

The first case: $i = j \neq s = r$. We have
$$dE(w_i w_r x) = \theta d\theta, \ E(w_i w_r x) = \frac{\theta^2}{2},$$
and (3.29) implies
$$EI_i I_r J_{js} = E(w_i(h) w_r(h) y(h)) = \frac{h^3}{6}.$$

The second case: $r = j \neq s = i$. This can be considered in a similar way and leads to the same result.

The third case: $i = j = r = s$. This gives
$$dE(w_i w_r x) = 2\theta d\theta, \ EI_i I_r J_{js} = \frac{h^3}{3}.$$

So, the identity for $EI_i I_r J_{js}$ in (3.24) has been proved completely. The remaining identities in (3.22)-(3.28) can be proved in a way that is definitely not more complicated. This proves Lemma 3.5. □

To finish construction of the one-step approximation $\bar{X}$ (see (3.18)), it remains to choose the random variables ξ_i, η_r, η_{ir}, μ_r so that the relations (3.22)-(3.28) hold. This can be done by modeling these random variables in various ways. Here we can choose them to be even simpler than in Sect. 2.1: although we are constructing a method of higher order of accuracy, we are doing it for systems of a less general form.

We will look for these variables in the following way. Consider symmetric random variables ξ_i, ν_j, ζ_r, $i, j, r = 1, \ldots, q$, that are all mutually independent (the condition of symmetry can be replaced by the weaker condition of vanishing of the corresponding odd moments), and put

$$\eta_i = \frac{\xi_i}{2} + \nu_i, \ \eta_{ij} = \frac{1}{6}(\xi_i \xi_j - \zeta_i \zeta_j), \ \mu_i = \frac{\xi_i}{3}. \qquad (3.31)$$

Let ξ_i, ν_j, ζ_r have the following moments:

$$E\xi_i = E\xi_i^3 = E\xi_i^5 = E\xi_i^7 = 0, \ E\nu_j = 0, \ E\zeta_r = 0, \quad (3.32)$$

$$E\xi_i^2 = 1, \ E\xi_i^4 = 3, \ E\xi_i^6 = 15, \ E\nu_j^2 = \frac{1}{12}, \ E\zeta_i^2 = 1. \quad (3.33)$$

Lemma 3.6. *Suppose that ξ_i, ν_j, ζ_r are independent random variables with moments satisfying (3.32) – (3.33). Then the variables (3.31) satisfy the relations (3.22) – (3.28).*

The proof of this lemma consists of a simple verification of the relations (3.22)-(3.28).

For the identities (3.32)-(3.33) to be satisfied, the simplest modeling of the random variables ν_j and ζ_r is by the laws $P(\nu = \pm 1/\sqrt{12}) = 1/2$, $P(\zeta = \pm 1) = 1/2$, while ξ_i can be modeled by the law $\mathcal{N}(0,1)$. However, for ξ_i we can also choose a simpler law. For example, $P(\xi = 0) = 1/3$, $P(\xi = \pm 1) = 3/10$, $P(\xi = \pm\sqrt{6}) = 1/30$.

Since (3.20)-(3.21) hold, the one-step approximation (3.18) with random variables (3.31) has order of accuracy four. We summarize the obtained result in the following theorem.

Theorem 3.7. *Suppose the coefficient $a(t,x)$ in the system (3.1) satisfies the Lipschitz condition*

$$|a(t,x) - a(t,y)| \leq K|x-y|.$$

Suppose that $a(t,x)$ together with its partial derivatives up to a sufficiently high order (at least up to order seven, inclusively) belongs to $\mathbf{F}$, and suppose that the coefficients $\sigma_r(t)$ are three time continuously differentiable with respect to $t \in [t_0, T]$. Assume that the functions a, $\Lambda_r a$, La, $\Lambda_i \Lambda_r a$, $L\Lambda_r a$, $\Lambda_r La$, and $L^2 a$ grow at most linearly as $|x|$ goes to infinity. Let random variables ξ_{ik}, ν_{ik}, ζ_{ik} be independent and such that the relations (3.32) – (3.33) hold. Then the method

$$X_{k+1} = X_k + \sum_{r=1}^{q} \sigma_r(t_k)\xi_{rk}h^{1/2} + a_k h$$

$$+ \sum_{r=1}^{q}(\Lambda_r a)_k(\frac{\xi_{rk}}{2} + \nu_{rk})h^{3/2} + \sum_{r=1}^{q} \sigma'_r(t_k)(\frac{\xi_{rk}}{2} - \nu_{rk})h^{3/2} + (La)_k\frac{h^2}{2}$$

$$+ \frac{1}{6}\sum_{r=1}^{q}\sum_{i=1}^{q}(\Lambda_i \Lambda_r a)_k(\xi_{ik}\xi_{rk} - \zeta_{ik}\zeta_{rk})h^2$$

$$+ \sum_{r=1}^{q}(L\Lambda_r a)_k(\frac{\xi_{rk}}{6} - \nu_{rk})h^{5/2} + \sum_{r=1}^{q}(\Lambda_r La)_k(\frac{\xi_{rk}}{6} + \nu_{rk})h^{5/2}$$

$$+ \sum_{r=1}^{q}\sigma''_r(t_k)\xi_{rk}h^{5/2} + (L^2 a)_k\frac{h^3}{6}, \quad (3.34)$$

has order of accuracy three in the sense of weak approximation (i.e., the relation (2.5) with $p = 3$ holds for all functions f belonging together with their partial derivatives up to order eight, inclusively, to the class $\mathbf{F}$).

Thanks to the lemmas proved in this section, the proof of this theorem follows immediately from Lemma 2.2 and Theorem 2.1.

Remark 3.8. It is not difficult to see that omission of the random variables ν_{rk} in the terms at $h^{5/2}$ does not change the order of the method (3.34).

2.3.4 Weak schemes for systems with colored noise

In this subsection we present several weak methods for differential equations with colored noise (see Sect. 1.6):

$$dY = f(Y)dt + G(Y)Zdt$$
$$dZ = AZdt + \sum_{r=1}^{q} b_r dw_r(t), \qquad (3.35)$$

where Y and f are l-dimensional vectors, Z and b_r are m-dimensional vectors, A is an $m \times m$ matrix, G is an $l \times m$ matrix, and $w_r(t)$ are independent standard Wiener processes.

Here we restrict ourselves to presentation of some explicit Runge–Kutta (RK) methods and implicit schemes. See other methods for (3.35) as well as some numerical experiments in [199]. The methods can be derived using the corresponding mean-square schemes of Sect. 1.6 and the results of the previous subsections. We note that the methods can easily be carried over to a non-autonomous system with colored noise.

Obviously, the first-order weak method coincides with the Euler scheme. Applying the RK method (3.3) to (3.35), we obtain the second-order RK method:

$$Y_{k+1} = Y_k + \frac{h}{2}[f(Y_k) + G(Y_k)Z_k] + \frac{h}{2}[f(\tilde{Y}_k) + G(\tilde{Y}_k)\tilde{Z}_k]$$
$$Z_{k+1} = Z_k + \sum_{r=1}^{q} b_r \xi_{rk} h^{1/2} + \frac{h}{2}(Z_k + \tilde{Z}_k), \qquad (3.36)$$

where

$$\tilde{Y}_k = Y_k + h[f(Y_k) + G(Y_k)Z_k]$$
$$\tilde{Z}_k = Z_k + \sum_{r=1}^{q} b_r \xi_{rk} h^{1/2} + AZ_k h, \qquad (3.37)$$

and ξ_{rk} are independent random variables with standard normal distribution $\mathcal{N}(0,1)$ or distributed according to the law $P(\xi=0) = 2/3$, $P(\xi=\pm\sqrt{3}) = 1/6$.

Due to specific features of the system (3.35), we succeeded in construction of the fully third-order RK scheme:

$$Y_{k+1} = Y_k + \frac{1}{6}(k_1 + 4k_2 + k_3)$$

$$Z_{k+1} = Z_k + \sum_{r=1}^{q} b_r \xi_{rk} h^{1/2} + \frac{1}{6}(l_1 + 4l_2 + l_3), \qquad (3.38)$$

where

$$\mathbf{f}(y,z) := f(y) + G(y)z,$$

$$k_1 = h\mathbf{f}_k, \quad k_2 = h\mathbf{f}(Y_k + \frac{k_1}{2}, Z_k + \frac{l_1}{2} + \frac{1}{2}\sum_{r=1}^{q} b_r \xi_{rk} h^{1/2}),$$

$$k_3 = h\mathbf{f}(Y_k - k_1 + 2k_2, Z_k - l_1 + 2l_2 + \sum_{r=1}^{q} b_r(\xi_{rk} + 6\nu_{rk})h^{1/2}),$$

$$l_1 = hAZ_k, \quad l_2 = hA(Z_k + \frac{l_1}{2} + \frac{1}{2}\sum_{r=1}^{q} b_r \xi_{rk} h^{1/2}),$$

$$l_3 = hA(Z_k - l_1 + 2l_2 + \sum_{r=1}^{q} b_r(\xi_{rk} - 6\nu_{rk})h^{1/2})$$

and the random variables ξ_{rk} and ν_{rk} are independent and can be simulated either as $\mathcal{N}(0,1)$ and $\mathcal{N}(0,1/\sqrt{12})$, respectively, or by the laws

$$P(\xi = 0) = 1/3, \quad P(\xi = \pm 1) = 3/10, \quad P(\xi = \pm\sqrt{6}) = 1/30,$$
$$P(\nu = \pm 1/\sqrt{12}) = 1/2.$$

Now we present weak implicit schemes. The first-order implicit weak methods coincide with the Euler mean-square schemes (1.6.10), but independent random variables ξ_{rk} can be simulated as $P(\xi = \pm 1) = 1/2$.

The two-parameter family of second-order implicit weak schemes has the form

$$Y_{k+1} = Y_k + \alpha h(f + Gz)_k + (1-\alpha)h(f + Gz)_{k+1}$$
$$+ h^{3/2} \sum_{r=1}^{q} G(Y_k) b_r (2\alpha - 1) \xi_{rk}/2$$
$$+ \beta(2\alpha - 1)h^2 [L(f + Gz)]_k/2 + (1-\beta)(2\alpha - 1)h^2 [L(f + Gz)]_{k+1}/2,$$

$$Z_{k+1} = Z_k + \sum_{r=1}^{q} b_r \xi_{rk} h^{1/2} + \alpha h A Z_k + (1-\alpha)h A Z_{k+1}$$
$$+ h^{3/2} \sum_{r=1}^{q} A b_r (2\alpha - 1) \xi_{rk}/2$$
$$+ \beta(2\alpha - 1)h^2 A^2 Z_k/2 + (1-\beta)(2\alpha - 1)h^2 A^2 Z_{k+1}/2. \quad (3.39)$$

The random variables ξ_{rk} here are the same as in the scheme (3.36)-(3.37), and $0 \le \alpha,\ \beta \le 1$.

If the parameter α in (3.39) is equal to 1/2, we obtain the trapezoidal weak method which is the simplest one among the family (3.39).

2.4 Variance reduction

If we compute $Ef(X(T))$ by the Monte Carlo method, using an approximate method for integrating the system

$$dX = a(t,X)dt + \sum_{r=1}^{q} \sigma_r(t,X)dw_r(t) \qquad (4.1)$$

to find $X(T)$, two errors arise. One of them is the numerical integration error:

$$Ef(X(T)) = Ef(\bar{X}(T)) + O(h^p).$$

The other is the error of the Monte Carlo method:

$$Ef(\bar{X}(T)) = \frac{1}{M}\sum_{i=1}^{M} Ef(\bar{X}^{(i)}(T)) \pm c\frac{(Varf(\bar{X}(T)))^{1/2}}{M^{1/2}}, \qquad (4.2)$$

where, e.g. the values $c = 1, 2, 3$ correspond to the fiducial probabilities 0.68, 0.95, 0.997, respectively.

Since $Varf(\bar{X}(T))$ is close to $Varf(X(T))$, we may assume that the error of the Monte Carlo method can be estimated by $(Varf(X(T))/M)^{1/2}$. If $Varf(X(T))$ is large, then to achieve a satisfactory accuracy we have to simulate a very large number of trajectories. If it were possible to change $f(X(T))$ by a variable Z such that $EZ = Ef(X(T))$ but with $VarZ$ substantially smaller than $Varf(X(T))$, then the modeling of Z instead of $f(X(T))$ would make it possible to obtain more accurate results for the same computational costs.

Two variance reduction methods are known: the method of important sampling (see [84, 180, 222, 223, 305]) and the method of control variates (see [222, 223]). A combining method is given in [197, 198].

2.4.1 The method of important sampling

Along with (4.1), we consider the system

$$dX = a(t,X)dt - \sum_{r=1}^{q}\mu_r(t,X)\sigma_r(t,X)dt + \sum_{r=1}^{q}\sigma_r(t,X)dw_r(t),$$

$$dY = \sum_{r=1}^{q}\mu_r(t,X)Y\,dw_r(t), \qquad (4.3)$$

where μ_r and Y are scalars, μ_r are rather arbitrary functions, however, with good analytical properties (for example, they are sufficiently smooth and have bounded derivatives).

By Girsanov's theorem, we have for any μ_r:

$$yEf(X_{s,x}(T))|_{(4.1)} = EY_{s,x,y}(T)f(X_{s,x}(T))|_{(4.3)}.$$

Putting $Z = Y_{s,x,y}(T)f(X_{s,x}(T))$, we see that EZ does not depend on the choice of the μ_r, while for $y = 1$ it equals to the desired quantity. At the same time, $VarZ$ does depend on the μ_r. Then it is natural to regard $\mu_1, \ldots, \mu_q$ as controls and to choose them by the condition that the variance $VarZ = EZ^2 - (EZ)^2$ is minimal. Since EZ is independent of $\mu_1, \ldots, \mu_q$, this choice reduces to solving the following problem from the optimal control theory: it is required to choose the controls $\mu_1, \ldots, \mu_q$ constituting a minimum of the functional

$$I = EY^2_{s,x,y}(T)f^2(X_{s,x}(T))$$

with respect to (4.3).

The function $u(s,x) = Ef(X_{s,x}(T))|_{(4.1)}$ satisfies the equation

$$Lu \equiv \frac{\partial u}{\partial s} + \sum_{i=1}^{d} a^i \frac{\partial u}{\partial x^i} + \frac{1}{2} \sum_{r=1}^{q} \sum_{i=1}^{d} \sum_{j=1}^{d} \sigma_r^i \sigma_r^j \frac{\partial^2 u}{\partial x^i \partial x^j} = 0 \qquad (4.4)$$

with the condition

$$u(T,x) = f(x) \qquad (4.5)$$

at the end of the time interval.

Introduce the function

$$v(s,x)y^2 = \min_{\mu_1,\ldots,\mu_q} I = \min_{\mu_1,\ldots,\mu_q} EY^2_{s,x,y}(T)f^2(X_{s,x}(T))$$

(it is clearly homogeneous of order two in y, which is already reflected in the notation). We write the Bellman equation for this function:

$$\min_{\mu_1,\ldots,\mu_q}\left(Lvy^2 + \sum_{r=1}^{q}(\sigma_r, \frac{\partial v}{\partial x})\mu_r y^2 + v\sum_{r=1}^{q}\mu_r^2 y^2\right) = 0. \qquad (4.6)$$

The minimization condition in (4.6) implies (if $v \neq 0$):

$$\mu_r = -\frac{1}{2v}(\sigma_r, \frac{\partial v}{\partial x}). \qquad (4.7)$$

Thus, v satisfies the equation

$$Lv - \frac{1}{4v}\sum_{r=1}^{q}(\sigma_r, \frac{\partial v}{\partial x})^2 = 0. \qquad (4.8)$$

Moreover, it is clear that
$$v(T,x) = f^2(x). \tag{4.9}$$

Let $f > 0$. Then $v > 0$. By some simple computations, we are readily verify that $\sqrt{v}$ is a solution of the problem (4.4)-(4.5). Thus, $v = u^2$. By (4.7), this implies

$$\mu_r = -\frac{1}{u}(\sigma_r, \frac{\partial u}{\partial x}). \tag{4.10}$$

Further, if we write the relation $v = u^2$ in the form

$$EZ^2 = (EZ)^2,$$

then we find that $Var Z = 0$ for μ_r from (4.10), i.e., the variable $Y_{s,x,y}(T) \times f(X_{s,x}(T))$ with X and Y from (4.3), (4.10) is deterministic.

Of course, the controls μ_r, $r = 1, \ldots, q$, cannot be constructed without knowing the function u. Nevertheless, the result obtained establishes that, in principle, it is possible to arbitrarily reduce the variance $Var Z$ by conveniently choosing the functions μ_r.

Note that the reasoning above is not completely rigorous. However, using its results, it is not difficult to prove the following theorem.

Theorem 4.1. *Let $f > 0$ and suppose there is a solution $u > 0$ of the problem (4.4)-(4.5). Suppose there is a solution of the system (4.3), (4.10) for $t_0 \le s < T$ and $x \in \mathbf{R}^d$. Then $Z = Y_{s,x,y}(T)f(X_{s,x}(T))$ computed according to (4.3), (4.10) is a deterministic variable.*

Proof. Let $u > 0$ be a solution of (4.4)-(4.5) and μ_r in (4.3) be such that there is a solution of the system (4.3). Using Ito's formula, we obtain (taking into account that $Lu = 0$):

$$d(u(t, X_{s,x}(t)) \cdot Y_{s,x,y}(t))$$
$$= Lu \cdot Y dt - \sum_{r=1}^{q} \mu_r(\sigma_r, \frac{\partial u}{\partial x}) Y dt + \sum_{r=1}^{q} (\sigma_r, \frac{\partial u}{\partial x}) Y dw_r(t)$$
$$+ u \sum_{r=1}^{q} \mu_r Y dw_r(t) + \sum_{r=1}^{q} (\sigma_r, \frac{\partial u}{\partial x}) \mu_r Y dt$$
$$= \sum_{r=1}^{q} ((\sigma_r, \frac{\partial u}{\partial x}) + \mu_r u) Y dw_r(t),$$

whence

$$u(t, X_{s,x}(t)) \cdot Y_{s,x,y}(t) = u(s,x)y + \int_t^s \sum_{r=1}^{q} ((\sigma_r, \frac{\partial u}{\partial x}) + \mu_r u) Y dw_r. \tag{4.11}$$

For the μ_r from (4.10), the relation (4.11) reduces to

$$u(t, X_{s,x}(t)) \cdot Y_{s,x,y}(t) = u(s,x)y,$$

i.e., for each t (so, in particular, for $t = T$) the quantity $u(t, X_{s,x}(t)) \times Y_{s,x,y}(t)$ is deterministic. By (4.5), this quantity for $t = T$ is equal to $Y_{s,x,y}(T)f(X_{s,x}(T))$. □

The results obtained can be used in, e.g., the following situation. Let f be a function close to a function f_0, and let the solution of the problem (4.4)-(4.5) for $f = f_0$ be known and be equal to u_0. If we take μ_r in (4.3) equal to

$$\mu_r = -\frac{1}{u_0}(\sigma_r, \frac{\partial u_0}{\partial x}),$$

the variance $Var(Y_{s,x,y}(T)f(X_{s,x}(T))$, although not zero, is small.

An illustration of the method of important sampling is given in Example 5.14 from Sect. 2.5.6. See another example in Sect. 3.7.

Remark 4.2. If the condition $f > 0$ in Theorem 4.1 is not satisfied, but if, e.g., $f > -C$, $C > 0$, then for $f + C$ the solution of the problem (4.4)-(4.5) is $u + C$, and the dependence

$$\mu_r = -\frac{1}{u+C}(\sigma_r, \frac{\partial u}{\partial x})$$

in (4.3) leads to $Z = Y_{s,x,y}(T)(f(X_{s,x}(T)+C)$ being a deterministic variable (as in Theorem 4.1). If f is neither bounded from below nor above but $f = g - h$ with $g > 0$ and $h > 0$ and for each of the functions g, h the conditions of Theorem 4.1 hold, then Theorem 4.1 can be used for g and h separately to compute Ef.

2.4.2 Variance reduction by control variates and combining method

Consider the Cauchy problem for linear parabolic equation

$$\frac{\partial u}{\partial s} + \frac{1}{2}\sum_{i,j=1}^{d} a^{ij}(s,x)\frac{\partial^2 u}{\partial x^i \partial x^j} + \sum_{i=1}^{d} b^i(s,x)\frac{\partial u}{\partial x^i} + c(s,x)u + g(s,x) = 0, \quad (4.12)$$

$$t_0 \leq s < T, \ x \in \mathbf{R}^d,$$

with the initial condition

$$u(T,x) = f(x). \tag{4.13}$$

The matrix $a(s,x) = \{a^{ij}(s,x)\}$ is supposed to be symmetric and positive semidefinite.

Let $\sigma(s,x)$ be a matrix obtained from the equation
$$a(s,x) = \sigma(s,x)\sigma^\top(s,x).$$
This equation is solvable with respect to σ (for instance, by a lower triangular matrix) at least for a positively definite a.

The solution to the problem (4.12)-(4.13) has the following probabilistic representations (see [48]):
$$u(s,x) = E(f(X_{s,x}(T))Y_{s,x,1}(T) + Z_{s,x,1,0}(T)), \quad (4.14)$$
$$s \leq T, \ x \in \mathbf{R}^d,$$
where $X_{s,x}(t), \ Y_{s,x,y}(t), \ Z_{s,x,y,z}(t), \ t \geq s$, is the solution of the Cauchy problem for the system of SDEs
$$dX = b(t,X)dt + \sigma(t,X)dw(t), \ X(s) = x, \quad (4.15)$$
$$dY = c(t,X)Y dt, \ Y(s) = y, \quad (4.16)$$
$$dZ = g(t,X)Y dt, \ Z(s) = z. \quad (4.17)$$

Here $w(t) = (w^1(t), \ldots, w^d(t))^\top$ is a d-dimensional standard Wiener process, Y and Z are scalars.

Let $u(s,x)$ be a solution of the problem (4.12)-(4.13). Introduce the process
$$\varphi_F(t) = u(t,X_{s,x}(t))Y_{s,x,1}(t) + Z_{s,x,1,0}(t)$$
$$+ \int_s^t Y_{s,x,1}(s')F^\top(s', X_{t,x}(s'))dw(s'), \quad (4.18)$$
where $F(s,x)$ is a d-dimensional vector-function with good analytical properties but arbitrary otherwise. Clearly, $\varphi_F(s) = u(s,x)$ and
$$\varphi_F(T) = f(X_{s,x}(T))Y_{s,x,1}(T) + Z_{s,x,1,0}(T)$$
$$+ \int_s^T Y_{s,x,1}(s')F^\top(s', X_{t,x}(s'))dw(s').$$

Further, the expectation $E\varphi_F(T)$ is equal to $u(s,x)$ and does not depend on the choice of F. At the same time, the variance $Var\varphi_F(T)$ does depend on F. In this situation it also turns out that the variance can be reduced to zero.

Theorem 4.3. *For*
$$F^j(t,x) = -\sum_{i=1}^d \sigma^{ij}(t,x)\frac{\partial u}{\partial x^i}(t,x), \ j = 1,\ldots,d, \quad (4.19)$$
the variable $\varphi_F(T)$ is deterministic, i.e., $Var\varphi_F(T) = 0$.

Proof. Taking into account that $u(s,x)$ is the solution of (4.12)-(4.13) and using Ito's formula, we obtain

$$d\varphi_F(t) = Y_{s,x,1}(t) \sum_{j=1}^{d} \sum_{i=1}^{d} \sigma^{ij} \frac{\partial u}{\partial x^i} dw^j(t) + Y_{s,x,1}(t) \sum_{j=1}^{d} F^j dw^j(t),$$

where σ^{ij}, F^j, u, and $\partial u/\partial x^i$ have $t, X_{s,x}(t)$ as their arguments. From here

$$\varphi_F(T) = u(s,x) + \int_s^T Y_{s,x,1}(t) \sum_{j=1}^{d} (\sum_{i=1}^{d} \sigma^{ij} \frac{\partial u}{\partial x^i} + F^j) dw^j(t).$$

Clearly,

$$Var \varphi_F(T) = E \int_s^T Y_{s,x,1}^2(t) \sum_{j=1}^{d} (\sum_{i=1}^{d} \sigma^{ij} \frac{\partial u}{\partial x^i} + F^j)^2 dt \qquad (4.20)$$

which is equal to zero for F^j according to (4.19). □

The method of reducing variance by a suitable choice of F is known as the method of control variates (see [222, 223]). See its application in Examples 5.17 and 5.18 from Sect. 2.5.6.

The combining method is proposed in [197] (see also [198]). We combine the method of important sampling and the method of control variates by introducing the system:

$$\begin{aligned} dX &= b(t,X)dt - \sigma(t,X)\mu(t,X)dt + \sigma(t,X)dw(t), \ X(s) = x, \\ dY &= c(t,X)Y dt + \mu^\top(t,X) Y dw(t), \ Y(s) = y, \\ dZ &= g(t,X)Y dt + F^\top(t,X) Y dw(t), \ Z(s) = z, \end{aligned} \qquad (4.21)$$

where $\mu(t,x)$ is a column-vector of dimension d, and the random variable

$$\varphi_{\mu,F}(t) = u(t, X_{s,x}(t)) Y_{s,x,1}(t) + Z_{s,x,1,0}(t). \qquad (4.22)$$

The solution to the problem (4.12)-(4.13) has various probabilistic representations:

$$u(s,x) = E\varphi_{\mu,F}(T) = E(f(X_{s,x}(T)) Y_{s,x,1}(T) + Z_{s,x,1,0}(T)), \qquad (4.23)$$

where $X_{s,x}(t)$, $Y_{s,x,y}(t)$, $Z_{s,x,y,z}(t)$, $t \geq s$, is the solution of the Cauchy problem for the system (4.21).

Let $F = 0$. Then the usual representation (the well-known Feynman–Kac formula) can be seen if $\mu = 0$; the others rest on Girsanov's theorem. For $F \neq 0$, the representation (4.23) is evidently true as well. We see that the expectation $E\varphi_{\mu,F}(T)$ does not depend on a choice of both μ and F.

The following theorem can be proved analogously to Theorem 4.3.

Theorem 4.4. *Let μ and F be such that there is a solution of the system (4.21) on the interval $[s,T]$. Then the variance $Var\varphi_{\mu,F}(T)$ is equal to*

$$Var\varphi_{\mu,F}(T) = E \int_s^T Y^2_{s,x,1}(t) \sum_{j=1}^d \left(\sum_{i=1}^d \sigma^{ij} \frac{\partial u}{\partial x^i} + u\mu^j + F^j \right)^2 dt \quad (4.24)$$

provided that the expectation in (4.24) exists.
  In particular, if μ and F are such that

$$\sum_{i=1}^d \sigma^{ij} \frac{\partial u}{\partial x^i} + u\mu^j + F^j = 0 \, , \; j = 1,\ldots,d \, , \quad (4.25)$$

then $Var\varphi_{\mu,F}(T) = 0$ and $\varphi_{\mu,F}(t) \equiv u(s,x)$, $s \leq t \leq T$, i.e., $\varphi_{\mu,F}(t)$ is deterministic and independent of $t \in [s,T]$.

2.4.3 Variance reduction for boundary value problems

Let G be a bounded domain in $\mathbf{R}^d$ and $Q = [t_0,T) \times G$ be a cylinder in $\mathbf{R}^{d+1}$, $\Gamma = \bar{Q}\backslash Q$ be the part of the cylinder's boundary consisting of the upper base and lateral surface. Consider the Dirichlet problem for the parabolic equation:

$$Lu + g = 0, \; (s,x) \in Q \, , \quad (4.26)$$

$$u\,|_\Gamma = f(s,x) \, , \quad (4.27)$$

where

$$Lu := \frac{\partial u}{\partial s} + \frac{1}{2} \sum_{i,j=1}^d a^{ij}(s,x) \frac{\partial^2 u}{\partial x^i \partial x^j} + \sum_{i=1}^d b^i(s,x) \frac{\partial u}{\partial x^i} + c(s,x)u.$$

We assume that conditions hold which guarantee existence of a sufficiently smooth classical solution $u(s,x)$ of the problem (4.26)-(4.27).

The solution of the problem (4.26)-(4.27) has various probabilistic representations:

$$u(s,x) = \mathbf{E}\left[f(\tau, X_{s,x}(\tau))Y_{s,x,1}(\tau) + Z_{s,x,1,0}(\tau) \right] \, , \quad (4.28)$$

where $X_{s,x}(t)$, $Y_{s,x,y}(t)$, $Z_{s,x,y,z}(t)$, $s \leq t \leq \tau$, is the solution of the Cauchy problem for the system of SDEs

$$\begin{aligned}
dX &= \chi_{\{\tau > t\}}[(b(t,X) - \sigma(t,X)\mu(t,X))\,dt + \sigma(t,X)\,dw(t)] \, , \\
dY &= \chi_{\{\tau > t\}}[c(t,X)Y\,dt + \mu^\intercal(t,X)Y\,dw(t)] \, , \\
dZ &= \chi_{\{\tau > t\}}[g(t,X)Y\,dt + F^\intercal(t,X)Y\,dw(t)] \, , \\
X(s) &= x, \;\; Y(s) = y, \;\; Z(s) = z,
\end{aligned} \quad (4.29)$$

$(s,x) \in Q$, and $\tau = \tau_{s,x}$ is the first exit time of the trajectory $(t, X_{s,x}(t))$ to the boundary Γ:

$$\tau = \tau_{s,x} = T \wedge \inf\{t : X_{s,x}(t) \in \Gamma\}.$$

Introduce the process

$$\varphi_{\mu,F}(t) = u(t \wedge \tau, X_{s,x}(t \wedge \tau))Y_{s,x,1}(t \wedge \tau) + Z_{s,x,1,0}(t \wedge \tau).$$

Clearly

$$\varphi_{\mu,F}(s) = u(s,x), \ \varphi_{\mu,F}(\tau) = f(\tau, X_{s,x}(\tau))Y_{s,x,1}(\tau) + Z_{s,x,1,0}(\tau) = \varphi_{\mu,F}(T),$$

and consequently $u(s,x) = E\varphi_{\mu,F}(T)$.

Theorem 4.5. *The variance $Var\varphi_{\mu,F}(T)$ is equal to*

$$Var\varphi_{\mu,F}(T) = E \int_s^T \chi_{\{\tau>t\}} Y_{s,x,1}^2(t) \sum_{j=1}^d (\sum_{i=1}^d \sigma^{ij} \frac{\partial u}{\partial x^i} + u\mu^j + F^j)^2 dt. \quad (4.30)$$

In particular, if μ and F are such that

$$\sum_{i=1}^d \sigma^{ij} \frac{\partial u}{\partial x^i} + u\mu^j + F^j = 0, \ j = 1, \ldots, d, \quad (4.31)$$

then $Var\varphi_{\mu,F}(T) = 0$ and $\varphi_{\mu,F}(t) \equiv u(s,x)$, $s \leq t \leq \tau$, i.e., $\varphi_{\mu,F}(t)$ is deterministic and independent of $t \in [s, \tau]$.

Proof. Using the Ito formula and taking into account that $Lu + g = 0$, we derive that $d\varphi_F(t)$ contains again martingale terms only:

$$d\varphi_F(t) = \chi_{\{\tau>t\}} Y_{s,x,1}(t) \sum_{j=1}^d (\sum_{i=1}^d \sigma^{ij} \frac{\partial u}{\partial x^i} + u\mu^j + F^j) dw^j(t),$$

whence the theorem follows. □

We note that an analogous result is true for elliptic boundary value problems as well.

2.5 Application of weak methods to the Monte Carlo computation of Wiener integrals

Consider Wiener integrals

$$I = \int_{C_{0,0}^d} F(x(\cdot))\, d\mu_{0,0}(x), \quad (5.1)$$

2.5 Application of weak methods to the computation of Wiener integrals

where $\mu_{0,0}(x)$ is a Wiener measure corresponding to Brownian paths with the fixed initial point $(0,0)$ and

$$F(x(\cdot)) = \varphi(x(T), \int_0^T a(t, x(t))dt). \tag{5.2}$$

The integral (5.1) is understood in the sense of Lebesgue integral with respect to the measure $\mu_{0,0}(x)$ and is taken over the set $C_{0,0}^d$ of all d-dimensional continuous vector-functions $x(t)$ satisfying the condition $x(0) = 0$ (see, e.g. [80]). A relation of such integrals with quantum physics and some equations of mathematical physics can be found, e.g., in [50, 61, 80, 129, 250].

Numerical evaluation of Wiener integrals is an important and difficult task. Many approaches are proposed for solving this problem (see, e.g. [50, 53, 306] and references therein). As a rule, the known numerical methods reduce a path integral to a high dimensional integral which is then approximated using either classical or Monte Carlo methods. The high order of these integrals makes calculation of the Wiener integrals extremely difficult.

In this section we consider Monte Carlo methods for computing Wiener integrals of functionals of integral type (5.1)-(5.2) based on the relation between such integrals and stochastic differential equations.

Let $w(t) = (w^1(t), \ldots, w^d(t))$ be a d-dimensional Wiener process. We introduce the system of SDEs

$$dX^1(t) = dw^1(t)$$
$$\ldots\ldots\ldots\ldots$$
$$dX^d(t) = dw^d(t) \tag{5.3}$$
$$dZ(t) = a(t, X^1(t), \ldots, X^d(t))dt, \ t \geq s, \tag{5.4}$$

with initial conditions

$$X^1(s) = x^1, \ldots, X^d(s) = x^d, \ Z(s) = z. \tag{5.5}$$

We will denote the solution of the system (5.3)-(5.5) by either $X_{s,x}(t)$, $Z_{s,x,z}(t)$, or, if this does not lead to confusion, simply by $X(t), Z(t)$.

The Wiener integral (5.1) of the functional (5.2) is equal to

$$I = E\varphi(X_{0,0}(T), Z_{0,0,0}(T)). \tag{5.6}$$

According to the Monte Carlo method, the mathematical expectation $E\varphi$ can be estimated by the sum

$$I_M = \frac{1}{M} \sum_{m=1}^M \varphi(X_{0,0}^{(m)}(T), Z_{0,0,0}^{(m)}(T)), \tag{5.7}$$

where $X^{(m)}(T), Z^{(m)}(T), m = 1, \ldots, M$, are independent realizations of the random variables $X(t), Z(t)$.

An efficiency of this approach is due to the fact that the system (5.3)-(5.5) has the fixed dimension d and the corresponding accuracy is reached by means of a choice of a method for (5.3)-(5.4) and a step of numerical integration h and a number M of Monte Carlo simulations. Thus, the problem of calculating the infinite-dimensional Wiener integral I is reduced to the Cauchy problem (5.3)-(5.5). This problem can naturally be regarded as one-dimensional since it contains one independent variable only. We underline that in other methods the path integral is reduced to a high dimensional Riemann integral and the accuracy is reached on account of increasing its dimension. Of course, for the numerical computation of I it suffices to construct weak approximations of the solution of the system (5.3)-(5.4) and to use the general methods developed in this chapter. However, in view of the specificity of the problem under consideration we can construct more effective methods. The effectiveness of the constructed algorithms allows us to evaluate integrals (5.1)-(5.2) for a large dimension d.

Since (5.3)-(5.5) implies

$$X_{0,0}(t) = w(t), \ Z_{0,0,0}(t) = \int_0^t a(\theta, x(\theta))d\theta, \qquad (5.8)$$

and $w(t)$ can exactly be modeled at any moment t, finding an approximation $\bar{Z}_{0,0,0}(t)$ reduces to approximately computing the integral (5.8) for $t = T$. Because there are many efficient quadrature formulas, at first glance this problem does not seem difficult. At the same time we have to keep in mind that quadrature formulas have a high order of accuracy only for integrands that are sufficiently smooth with respect to t. In view of nonregularity of $w(t)$, the integrand $a(t, w(t))$ does not satisfy the usual conditions of smoothness. Below we show that the trapezium formula applied to the integral (5.8) has the second weak order. In view of above said, this requires a separate proof, of course. For Wiener integrals of functionals of exponential type

$$F(x(\cdot)) = \exp\left[\int_0^T f(t, x(t))\, dt\right] \qquad (5.9)$$

we derive a method of order four (see Sect. 2.5.2).

In Sects. 2.5.3-2.5.5 we consider *conditional* Wiener integrals of exponential-type functionals. The corresponding probabilistic representation contains a more complicated system than (5.3). The solution of this system gives a Markov representation of the Brownian bridge. The system is singular and this circumstance stipulates a certain complexity of theoretical proofs. Nevertheless the constructed fourth-order Runge–Kutta algorithms (see Sect. 2.5.3) are equally simple and effective as in the case of the Wiener integral (5.1), (5.9). Its one-step error is analyzed in Sect. 2.5.4. Implicit methods of order two for conditional Wiener integrals are derived in Sect. 2.5.5.

Some numerical tests of the proposed methods for both Wiener integrals (5.1) and conditional Wiener integrals are presented in Sect. 2.5.6.

2.5.1 The trapezium, rectangle, and other methods of second order

We introduce the one-step approximation for the system (5.3)-(5.4):

$$\bar{X}_{t,x}(t+h) = x + \xi h^{1/2},$$
$$\bar{Z}_{t,x,z}(t+h) = z + \frac{h}{2}(a(t,x) + a(t+h, \bar{X}_{t,x}(t+h))), \quad (5.10)$$

where $\xi = (\xi^1, \ldots, \xi^d)$ is a d-dimensional random variable with independent coordinates ξ^i, $i = 1, \ldots, d$, such that $E\xi^i = E(\xi^i)^3 = E(\xi^i)^5 = 0$, $E(\xi^i)^2 = 1$, $E(\xi^i)^4 = 3$. We divide the interval $[0, T]$ into N equal parts with step $h = T/N$: $0 = t_0 < t_1 < \cdots < t_N = T$, $t_{k+1} - t_k = h$. Using (5.10) we construct the approximate solution

$$X_0 = 0, \quad X_{k+1} = X_k + \xi_k h^{1/2}, \quad Z_0 = 0,$$
$$Z_{k+1} = Z_k + \frac{h}{2}(a(t_k, X_k) + a(t_{k+1}, X_{k+1})), \quad k = 0, \ldots, N-1, \quad (5.11)$$

where ξ_k, $k = 0, \ldots, N-1$, are independent d-dimensional random variable distributed like ξ. We can take, e.g., normally distributed random variables as such ξ_k. However, as in the previous sections, in (5.11) we may also use random variables that are more convenient for computing purposes, e.g. with the coordinates taking the values 0, $\sqrt{3}$, $-\sqrt{3}$ with probabilities $2/3$, $1/6$, $1/6$.

Theorem 5.1. *Suppose the function a satisfies a global Lipschitz condition. Suppose also that a and φ together with their partial derivatives of order up to six, inclusively, belong to $\mathbf{F}$. Then*

$$E\varphi(\bar{X}_{0,0}(T), \bar{Z}_{0,0,0}(T)) - E\varphi(X_{0,0}(T), Z_{0,0,0}(T)) = O(h^2). \quad (5.12)$$

Proof. The proof rests on Theorem 2.1. The conditions (a), (c), (d) of that theorem (see also Lemma 2.2) are obviously fulfilled. Therefore it remains to verify the condition (b). It is not difficult to see that in our case the inequality (2.4) holds. It can be shown (see details in [180]) that $E\prod_{j=1}^{s} \bar{\Delta}^{i_j}$ for $s = 1, \ldots, 5$ coincides with $E\prod_{j=1}^{s} \Delta^{i_j}$ up to $O(h^3)$. This proves the inequality (2.3). □

We can prove that not only the trapezium formula, but any interpolation formula of third order of accuracy with respect to h applied to the integral in (5.8) leads to a method of second order of accuracy for Wiener integrals. In particular, an application of the midpoint formula gives

$$I = E\varphi(X_{0,0}(T), Z_{0,0,0}(T)) = E\varphi(w(T), \frac{1}{N}\sum_{k=1}^{N} a(t_{k-1/2}, w(t_{k-1/2}))) + O(h^2), \quad (5.13)$$

where $t_{k-1/2} = t_k - h/2$.

It is clear that $E\varphi(w(T), \frac{1}{N}\sum_{k=1}^{N} a(t_{k-1/2}, w(t_{k-1/2})))$ can be realized in the form $E\varphi(\bar{X}_{0,0}(T), \bar{Z}_{0,0,0}(T))$ where

$$X_0 = 0, \ X_{k+1} = X_k + (\xi_k + \eta_k)\left(\frac{h}{2}\right)^{1/2}, \ Z_0 = 0,$$

$$Z_{k+1} = Z_k + ha\left(t_{k+1/2}, X_k + \eta_k\left(\frac{h}{2}\right)^{1/2}\right), \ k = 0, \ldots, N-1, \quad (5.14)$$

and ξ_k and η_k are independent d-dimensional random variables whose coordinates, in turn, are independent $\mathcal{N}(0,1)$-distributed random variables. As in the trapezium method, we can also use simpler random variables.

Remark 5.2. We stress that because of the nonregularity of Brownian trajectories separate proofs are required for the result that the considered quadrature formulas for (5.8) give the same accuracy as in the deterministic case. To confirm this, we consider the following system of two equations:

$$dX(s) = dw(s), \ X(0) = 0,$$
$$dZ(s) = X(s)ds, \ Z(0) = 0,$$

as well as the function $\varphi(z) = z^2$. Here, $EZ^2(h)$ can readily be evaluated exactly: $EZ^2(h) = h^3/3$. We compute the approximation $\bar{Z}(h)$ by Simpson's formula:

$$\bar{Z}(h) = \frac{h}{6}\left(X(0) + 4X\left(\frac{h}{2}\right) + X(h)\right) = \frac{h}{6}\left(4X\left(\frac{h}{2}\right) + X(h)\right).$$

We are immediately convinced that $E\bar{Z}^2(h) = 13h^3/36$, and hence that $E\bar{Z}^2(h) - EZ^2(h) = O(h^3)$ instead of the expected $O(h^5)$, since to compute the integral (5.8) for $t = h$ we have used Simpson's formula, which has order of accuracy five in the deterministic case. Moreover, in general, we can prove that there is no "natural" method of third order of accuracy (see details in [180]).

Remark 5.3. Define the step random process $w^h(t)$, $0 \leq t \leq T$, to be equal to $w(t_k)$ for $t \in [t_k - h/2, t_k + h/2] \cap [0, T]$. The approximation of the integral $\int_0^T a(w(t))dt$ by the trapezium method coincides with $\int_0^T a(w^h(t))dt$.

In this subsection we have proved that deviation of $E\varphi(\int_0^T a(w(t))dt)$ from the expectation of the same expression with the integrand replaced by its approximation has order $O(h^2)$ if only φ and a are sufficiently regular functions. Then one can conjecture that this might also be true for functionals $F(x(\cdot))$ of a more general form than $F(x(\cdot)) = \varphi(\int_0^T a(x(t))dt)$. In [300] (see also [180]), it is shown that this conjecture is true for a very large class of functionals.

2.5 Application of weak methods to the computation of Wiener integrals

Remark 5.4. Note that the piecewise linear approximation $w_h(t)$ of a Wiener process defined as

$$w_h(t) = w(t_k)\frac{t_{k+1} - t}{h} + w(t_{k+1})\frac{t - t_k}{h}, \quad t_k \leq t \leq t_{k+1},$$

and differing from the piecewise constant approximation $w^h(t)$ used in the trapezium method, gives an error of order $O(h)$. The simplest way to confirm this is to compute

$$E\int_0^1 w^2(t)dt = \frac{1}{2}, \quad E\int_0^1 (w^h(t))^2 dt = \frac{1}{2}, \quad E\int_0^1 w_h^2(t)dt = \frac{1}{2} - \frac{h}{6}.$$

2.5.2 A fourth-order Runge–Kutta method for computing Wiener integrals of functionals of exponential type

This subsection is devoted to the computation of Wiener integrals (5.1) of often encountered functionals of exponential type

$$F(x(\cdot)) = \exp(\int_0^T a(x(t))dt), \quad x(t) \in \mathbf{R}^d. \tag{5.15}$$

The functional (5.15) is a particular case of the functional (5.2), and therefore the results of the previous subsection can be applied here too. In Remark 5.2 we have noted that there is no "natural" method of order of accuracy exceeding two for integrating the system (5.3)-(5.4). However, thanks to the special form of the functional (5.15), the computation of the integral (5.1) with (5.15) can be done by using another system, for which we can successfully develop a method of order four. This system has the form

$$dX^1(t) = dw^1(t), \ X^1(0) = x^1,$$
$$\cdots\cdots\cdots\cdots$$
$$dX^d(t) = dw^d(t), \ X^d(0) = x^d,$$
$$dY(t) = Y(t)a(t, X^1(t), \ldots, X^d(t))dt, \ Y(0) = 1. \tag{5.16}$$

It can readily be seen that the Wiener integral of the functional (5.15) is equal to

$$I = EY(T),$$

where $x^1 = \cdots = x^d = 0$.

Thus, evaluation of the Wiener integral (5.1), (5.15) leads to the problem of numerical integration of the system (5.16). For our purposes, the approximate solution is better, if the difference $EY(t) - E\bar{Y}(t)$ is smaller. As in the previous subsection, despite the facts that the $X(t)$ in (5.16) can be found exactly and that the equation for $Y(t)$ does not have stochastic components, we need special proofs for using methods that are well known in the deterministic case.

Let $h = T/N$, $t_k = kh$, $k = 0, \ldots, N$, $t_{k-1/2} = (k-1/2)h$, $k = 1, \ldots, N$. Consider the following approximation:

$$X_0 = 0, \ X_{k-1/2} = X_{k-1} + \frac{1}{\sqrt{2}}\xi_{k-1/2}h^{1/2},$$

$$X_k = X_{k-1/2} + \frac{1}{\sqrt{2}}\xi_k h^{1/2},$$

$$Y_0 = 1, \ Y_k = Y_{k-1} + \frac{1}{6}(k_1 - 2k_2 + 2k_3 + k_4), \tag{5.17}$$

where the $\xi_{k-1/2}$, ξ_k are mutually independent d-vectors of independent $\mathcal{N}(0,1)$-distributed components and

$$k_1 = ha(t_{k-1}, X_{k-1})Y_{k-1},$$

$$k_2 = ha(t_{k-1/2}, X_{k-1/2})(Y_{k-1} + \frac{k_1}{2}),$$

$$k_3 = ha(t_{k-1/2}, X_{k-1/2})(Y_{k-1} + \frac{k_2}{2}),$$

$$k_4 = ha(t_k, X_k)(Y_{k-1} + k_3). \tag{5.18}$$

Clearly, the method (5.17)-(5.18) is a Runge–Kutta method of fourth order of accuracy for integrating the last equation in the system (5.16) if we consider $a(t, w^1(t), \ldots, w^d(t))$ to be a sufficiently smooth function of t. Since this function is random and nonsmooth in t, we need a separate proof. This proof is rather long. It rests on Theorem 2.1 and Lemma 1.9 and can be found in [84] and [180]. The following result is true under some natural assumptions on the function $a(t,x)$ (e.g., a is non-positive and sufficiently smooth).

Theorem 5.5. *The method* (5.17)-(5.18) *for the Wiener integral* (5.1), (5.15) *is of order four, i.e.,*

$$E\bar{Y}(t) - EY(t) = O(h^4). \tag{5.19}$$

Remark 5.6. We can also obtain Runge–Kutta methods of second or third order of accuracy. Note that the method (5.17)-(5.18) requires only a double modeling of a normally-distributed random vector and a double computation of the function a at each step. The method of third order of accuracy requires the same amount of computations as does the method (5.17)-(5.18), while the method of second order of accuracy, which is already quite inferior as regards accuracy, requires only a somewhat less amount of computations. Therefore the Runge–Kutta methods of third and second order of accuracy are of limited interest in this case.

2.5.3 Explicit Runge–Kutta method of order four for conditional Wiener integrals of exponential-type functionals

Consider *conditional* Wiener integrals

$$J = \int_{C^d_{0,a;T,b}} F(x(\cdot))\, d\mu^{T,b}_{0,a}(x) \qquad (5.20)$$

of the exponential-type functionals

$$F(x(\cdot)) = \exp\left[\int_0^T f(t, x(t))\, dt\right]. \qquad (5.21)$$

Here $\mu^{T,b}_{0,a}(x)$ is a conditional Wiener measure which corresponds to the Brownian paths $X^{T,b}_{0,a}(t)$ with fixed initial and final points, i.e., it corresponds to the d-dimensional Brownian bridge from a at the time $t = 0$ into b at the time $t = T$. The integral (5.20) is understood in the sense of Lebesgue integral with respect to the measure $\mu^{T,b}_{0,a}(x)$ and is taken over the set $C^d_{0,a;T,b}$ of all d-dimensional continuous vector-functions $x(t)$ satisfying the conditions $x(0) = a$, $x(T) = b$ (see, e.g. [80]). We note [61,80,129,250] that the Feynman path integral of the form

$$J = \int \exp\left(\int_0^T \left[\frac{m\dot{x}^2(t)}{2} - V(x(t))\right] dt\right) \mathcal{D}x(t)$$

is another writing of the integral (5.20)-(5.21) with $f = -V$.

As it is known [110, 120], the d-dimensional Brownian bridge $X(t) = X_{0,a}(t) = X^{T,b}_{0,a}(t)$, $0 \le t \le T$, from a to b can be characterized as the pathwise unique solution of the system of SDEs

$$dX = \frac{b - X}{T - t}\, dt + dw(t),\ 0 \le t < T,\ X(0) = a, \qquad (5.22)$$

with

$$X(T) = b. \qquad (5.23)$$

Let us also introduce the scalar equation

$$dY = f(t, X(t))\, Y\, dt,\ 0 \le t \le T,\ Y(0) = 1, \qquad (5.24)$$

where $X(t)$ is defined by (5.22)-(5.23) and $f(t, x)$ is the same as in (5.21). Then the Wiener integral (5.20)-(5.21) is equal to

$$J = EY(T). \qquad (5.25)$$

Thus, evaluation of the Wiener integral (5.20)-(5.21) is reduced to the problem of numerical integration of the system (5.22)-(5.24).

Introduce a discretization of the time interval $[0, T]$, for definiteness the equidistant one with a time step $h > 0$:

$$t_k = kh, \quad k = 0, \ldots, N, \quad t_N = T,$$

and let $t_{k+1/2} := t_k + h/2$.

To get a higher order method for (5.22)-(5.24), we need to simulate the solution of (5.22) exactly. The solution of (5.22) is

$$X(t) = a\frac{T-t}{T} + b\frac{t}{T} + (T-t)\int_0^t \frac{dw(s)}{T-s}.$$

Hence

$$X(t+h) = X(t) + h\frac{b - X(t)}{T-t} + (T-t-h)\int_t^{t+h} \frac{dw(s)}{T-s}.$$

We have

$$E\left[(T-t-h)\int_t^{t+h} \frac{dw(s)}{T-s} \Big| X(t)\right] = 0,$$

$$E\left[(T-t-h)\int_t^{t+h} \frac{dw(s)}{T-s} \Big| X(t)\right]^2 = \left(1 - \frac{h}{T-t}\right)h. \quad (5.26)$$

We can exactly simulate the solution of (5.22) by a simple recurrent procedure based on the formula

$$X(t+h) = X(t) + h\frac{b - X(t)}{T-t} + h^{1/2}\sqrt{\frac{T-t-h}{T-t}}\,\xi, \quad t+h \leq T, \quad (5.27)$$

where ξ is a random vector which components are Gaussian random variables with zero mean and unit variance and they are independent of $X(t)$.

Now let us formally apply a standard deterministic explicit fourth-order Runge–Kutta method to the equation (5.24) assuming that $X(t)$ is a known function. Then, taking into account (5.27), we obtain the following algorithm for integrating the system (5.22)-(5.24):

$$X(0) = a,$$

$$X(t_{k+1/2}) = X(t_k) + \frac{h}{2}\frac{b - X(t_k)}{T - t_k} + \frac{h^{1/2}}{\sqrt{2}}\sqrt{\frac{T - t_{k+1/2}}{T - t_k}}\,\xi_{k+1/2},$$

$$k = 0, \ldots, N-1,$$

$$X(t_{k+1}) = X(t_{k+1/2}) + \frac{h}{2}\frac{b - X(t_{k+1/2})}{T - t_{k+1/2}} + \frac{h^{1/2}}{\sqrt{2}}\sqrt{\frac{T - t_{k+1}}{T - t_{k+1/2}}}\,\xi_{k+1},$$

$$k = 0, \ldots, N-2, \quad X(t_N) = b, \quad (5.28)$$

2.5 Application of weak methods to the computation of Wiener integrals

$$Y_0 = 1,$$
$$k_1 = f(t_k, X(t_k))Y_k, \quad k_2 = f(t_{k+1/2}, X(t_{k+1/2}))\left[Y_k + hk_1/2\right],$$
$$k_3 = f(t_{k+1/2}, X(t_{k+1/2}))\left[Y_k + hk_2/2\right],$$
$$k_4 = f(t_{k+1}, X(t_{k+1}))\left[Y_k + hk_3\right],$$
$$Y_{k+1} = Y_k + \frac{h}{6}(k_1 + 2k_2 + 2k_3 + k_4), \quad k = 0, \ldots, N-1, \quad (5.29)$$

where $\xi_{k+1/2}$, ξ_{k+1} are d-dimensional random vectors which components are mutually independent random variables with standard normal distribution $\mathcal{N}(0,1)$.

Since the function $X(t)$ is non-smooth, the deterministic result on the accuracy order of the involved Runge–Kutta method is not applicable here and a separate convergence theorem is needed.

Introduce the operator

$$L = \frac{\partial}{\partial t} + \sum_{i=1}^{d} \frac{b^i - x^i}{T-t} \frac{\partial}{\partial x^i} + \frac{1}{2}\sum_{i=1}^{d} \frac{\partial^2}{(\partial x^i)^2}, \quad 0 \leq t < T. \quad (5.30)$$

We observe that this operator contains singularity since the denominator $T-t$ tends to zero as t goes to T.

Consider the function

$$u(t,x) = EY_{t,x,1}(T). \quad (5.31)$$

It satisfies the Cauchy problem

$$Lu + fu = 0, \quad 0 \leq t < T, \ x \in \mathbf{R}^d, \quad (5.32)$$
$$u(T,x) = 1.$$

We assume that the function $f(t,x)$ is sufficiently smooth, belongs to the class **F** together with its partial derivatives of a sufficiently high order and is such that the problem (5.32) has a unique solution which is sufficiently smooth and belongs to the class **F** together with its partial derivatives of a sufficiently high order. In addition, we suppose that $EY^2(t)$ exists and bounded on $[0,T]$ and that for all sufficiently small h the second moments EY_k^2 are uniformly bounded with respect to h. For instance, the latter conditions are satisfied when the function $f(t,x)$ is bounded. The following theorem is proved under these assumptions on the function $f(t,x)$.

Theorem 5.7. *The method (5.28)-(5.29) applied to evaluation of the conditional Wiener integral (5.25) is of fourth order of accuracy, i.e.,*

$$|J - EY_N| = |EY(T) - EY_N| \leq Kh^4, \quad (5.33)$$

where the constant K is independent of h.

Theoretically, we can use Theorem 5.7, approximating $f(t,x)$ (if it is unbounded) by an appropriate bounded function. The proof of Theorem 5.7 is based on a thorough analysis of the one-step error which is made in the next section. Complete proofs are available in [215].

2.5.4 Theorem on one-step error

In this subsection we consider a one-step error of the method (5.28)-(5.29).

It is convenient to introduce the additional notation for the approximation defined by (5.29): $\bar{Y}_{0,a,1}(t_k) = Y_k$ and also $\bar{Y}_{t_k,x,y}(t_i)$, $t_i \geq t_k$, by which we mean the approximation of (5.24) started from y at $t = t_k$ with $X(t_k) = x$.

It is not difficult to see that

$$Y_{t,x,y}(t+t') = yY_{t,x,1}(t+t'), \quad \bar{Y}_{t_k,x,y}(t_{k+k'}) = y\bar{Y}_{t_k,x,1}(t_{k+k'}),$$
$$EY_{t,x,y}(T) = yEY_{t,x,1}(T) = yu(t,x), \tag{5.34}$$

where $u(t,x)$ is the solution of the problem (5.32).

Recall that $t_0 = 0$, $X_0 = a$, $Y_0 = 1$. Using (5.31) and (5.34) and the fact that we simulate $X_k = X(t_k)$ exactly, we can represent the *global error* of the method (5.28)-(5.29) (cf. (5.33)) in the form

$$\begin{aligned}\left|EY_{0,a,1}(T) - E\bar{Y}_{0,a,1}(T)\right| &= |EY_{t_0,X_0,Y_0}(T) - EY_N| \\ &= |u(t_0,X_0)Y_0 - Eu(t_N,X_N)Y_N| \\ &= \left|\sum_{k=0}^{N-1}\left[Eu(t_k,X(t_k))Y_k - Eu(t_{k+1},X(t_{k+1}))\bar{Y}_{t_k,X_k,Y_k}(t_{k+1})\right]\right| \\ &= \left|\sum_{k=0}^{N-1}EY_k\left[u(t_k,X(t_k)) - u(t_{k+1},X(t_{k+1}))\bar{Y}_{t_k,X_k,1}(t_{k+1})\right]\right| \\ &\leq \sum_{k=0}^{N-1}\left|EY_k\left[u(t_k,X(t_k)) - u(t_{k+1},X_{t_k,X_k}(t_{k+1}))\bar{Y}_{t_k,X_k,1}(t_{k+1})\right]\right|. \end{aligned} \tag{5.35}$$

We have

$$\begin{aligned}R_k :&= \left|EY_k\left[u(t_k,X(t_k)) - u(t_{k+1},X_{t_k,X_k}(t_{k+1}))\bar{Y}_{t_k,X_k,1}(t_{k+1})\right]\right| \\ &= \left|EY_k E\left[u(t_k,X_k) - u(t_{k+1},X_{t_k,X_k}(t_{k+1}))\bar{Y}_{t_k,X_k,1}(t_{k+1})|\mathcal{F}_{t_k}\right]\right|. \end{aligned} \tag{5.36}$$

First, we analyze R_k for $k = 0, \ldots, N-2$. To this end, we consider the *one-step error* for $0 \leq t < T - h$:

$$r(t,x) := Eu(t+h, X_{t,x}(t+h))\bar{Y}_{t,x,1}(t+h) - u(t,x). \tag{5.37}$$

We rewrite (5.29) on a single step in the form:

$$\bar{Y}_{t,x,1}(t+h) = 1 + \frac{h}{6}\left(f_0 + 4f_{1/2} + f_1\right)$$
$$+\frac{h^2}{6}\left(f_0 f_{1/2} + f_{1/2}^2 + f_{1/2}f_1\right) + \frac{h^3}{12}\left(f_0 f_{1/2}^2 + f_{1/2}^2 f_1\right) + \frac{h^4}{24} f_0 f_{1/2}^2 f_1, \tag{5.38}$$

2.5 Application of weak methods to the computation of Wiener integrals

where $f_0 := f(t,x)$, $f_{1/2} := f(t + h/2, X_{t,x}(t + h/2))$, and $f_1 := f(t + h, X_{t,x}(t + h))$.

Using (1.35), we get

$$Eu(t+h, X_{t,x}(t+h)) = u(t,x) + hLu(t,x) + \frac{h^2}{2}L^2u(t,x) + \frac{h^3}{6}L^3u(t,x)$$
$$+ \frac{h^4}{24}L^4u(t,x) + \int_t^{t+h} \frac{(t+h-\theta)^4}{24} EL^5u(\theta, X_{t,x}(\theta))\, d\theta, \quad (5.39)$$

$$Ef_0 u(t+h, X_{t,x}(t+h)) = f_0 Eu(t+h, X_{t,x}(t+h))$$
$$= f_0 \left[u(t,x) + hLu(t,x) + \frac{h^2}{2}L^2u(t,x) + \frac{h^3}{6}L^3u(t,x) \right.$$
$$\left. + \int_t^{t+h} \frac{(t+h-\theta)^3}{6} EL^4u(\theta, X_{t,x}(\theta))\, d\theta \right], \quad (5.40)$$

$$Ef_1 u(t+h, X_{t,x}(t+h)) = f_0 u(t,x) + hL(fu)(t,x) + \frac{h^2}{2}L^2(fu)(t,x)$$
$$+ \frac{h^3}{6}L^3(fu)(t,x) + \int_t^{t+h} \frac{(t+h-\theta)^3}{6} EL^4(fu)(\theta, X_{t,x}(\theta))\, d\theta. \quad (5.41)$$

Further,

$$Ef_{1/2} u(t+h, X_{t,x}(t+h)) = E\left(f_{1/2} E\left[u(t+h, X_{t,x}(t+h))|\mathcal{F}_{t+h/2}\right]\right),$$

and by (1.34) we obtain

$$E\left[u(t+h, X_{t,x}(t+h))|\mathcal{F}_{t+h/2}\right] = u(t+h/2, X_{t,x}(t+h/2))$$
$$+ \frac{h}{2}Lu(t+h/2, X_{t,x}(t+h/2)) + \frac{h^2}{8}L^2u(t+h/2, X_{t,x}(t+h/2))$$
$$+ \frac{h^3}{48}L^3u(t+h/2, X_{t,x}(t+h/2))$$
$$+ \int_{t+h/2}^{t+h} \frac{(t+h-\theta)^3}{6} E\left[L^4u(\theta, X_{t,x}(\theta))|\mathcal{F}_{t+h/2}\right] d\theta,$$

then

$$Ef_{1/2} u(t+h, X_{t,x}(t+h)) = f_0 u(t,x) + \frac{h}{2}L(fu)(t,x)$$
$$+ \frac{h^2}{8}L^2(fu)(t,x) + \frac{h^3}{48}L^3(fu)(t,x)$$

$$+ \int_t^{t+h/2} \frac{(t+h/2-\theta)^3}{6} EL^4(fu)(\theta, X_{t,x}(\theta))d\theta + \frac{h}{2} f_0 Lu(t,x)$$

$$+ \frac{h^2}{4} L(fLu)(t,x) + \frac{h^3}{16} L^2(fLu)(t,x)$$

$$+ \frac{h}{2} \int_t^{t+h/2} \frac{(t+h/2-\theta)^2}{2} EL^3(fLu)(\theta, X_{t,x}(\theta))d\theta$$

$$+ \frac{h^2}{8} f_0 L^2 u(t,x) + \frac{h^3}{16} L(fL^2 u)(t,x)$$

$$+ \frac{h^2}{8} \int_t^{t+h/2} (t+h/2-\theta) EL^2(fL^2 u)(\theta, X_{t,x}(\theta))d\theta$$

$$+ \frac{h^3}{48} f_0 L^3 u(t,x) + \frac{h^3}{48} \int_t^{t+h/2} EL(fL^3 u)(\theta, X_{t,x}(\theta))d\theta$$

$$+ Ef_{1/2} \int_{t+h/2}^{t+h} \frac{(t+h-\theta)^3}{6} L^4 u(\theta, X_{t,x}(\theta))\, d\theta. \tag{5.42}$$

Analogously, we get

$$Ef_0 f_{1/2} u(t+h, X_{t,x}(t+h)) = f_0^2 u(t,x) + \frac{h}{2} f_0 L(fu)(t,x)$$

$$+ \frac{h^2}{8} f_0 L^2(fu)(t,x) + f_0 \int_t^{t+h/2} \frac{(t+h/2-\theta)^2}{2} EL^3(fu)(\theta, X_{t,x}(\theta))d\theta$$

$$+ \frac{h}{2} f_0^2 Lu(t,x) + \frac{h^2}{4} f_0 L(fLu)(t,x)$$

$$+ \frac{h}{2} f_0 \int_t^{t+h/2} (t+h/2-\theta) EL^2(fLu)(\theta, X_{t,x}(\theta))d\theta + \frac{h^2}{8} f_0^2 L^2 u(t,x)$$

$$+ \frac{h^2}{8} f_0 \int_t^{t+h/2} EL(fL^2 u)(\theta, X_{t,x}(\theta))d\theta$$

$$+ f_0 Ef_{1/2} \int_{t+h/2}^{t+h} \frac{(t+h-\theta)^2}{2} L^3 u(\theta, X_{t,x}(\theta))\, d\theta, \tag{5.43}$$

$$Ef_{1/2}^2 u(t+h, X_{t,x}(t+h)) = f_0^2 u(t,x) + \frac{h}{2} L(f^2 u)(t,x)$$

$$+ \frac{h^2}{8} L^2(f^2 u)(t,x) + \int_t^{t+h/2} \frac{(t+h/2-\theta)^2}{2} EL^3(f^2 u)(\theta, X_{t,x}(\theta))d\theta$$

$$+ \frac{h}{2} f_0^2 Lu(t,x) + \frac{h^2}{4} L(f^2 Lu)(t,x)$$

$$+ \frac{h}{2} \int_t^{t+h/2} (t+h/2-\theta) EL^2(f^2 Lu)(\theta, X_{t,x}(\theta))d\theta + \frac{h^2}{8} f_0^2 L^2 u(t,x)$$

2.5 Application of weak methods to the computation of Wiener integrals

$$+\frac{h^2}{8}\int_t^{t+h/2} EL\left(f^2L^2u\right)(\theta,X_{t,x}(\theta))d\theta$$

$$+Ef_{1/2}^2\int_{t+h/2}^{t+h}\frac{(t+h-\theta)^2}{2}L^3u(\theta,X_{t,x}(\theta))\,d\theta\,, \tag{5.44}$$

$$Ef_{1/2}f_1u(t+h,X_{t,x}(t+h)) = f_0^2u(t,x) + \frac{h}{2}L\left(f^2u\right)(t,x)$$

$$+\frac{h^2}{8}L^2\left(f^2u\right)(t,x) + \int_t^{t+h/2}\frac{(t+h/2-\theta)^2}{2}EL^3\left(f^2u\right)(\theta,X_{t,x}(\theta))d\theta$$

$$+\frac{h}{2}f_0L\left(fu\right)(t,x) + \frac{h^2}{4}L\left(fL\left(fu\right)\right)(t,x)$$

$$+\frac{h}{2}\int_t^{t+h/2}(t+h/2-\theta)EL^2\left(fL\left(fu\right)\right)(\theta,X_{t,x}(\theta))d\theta + \frac{h^2}{8}f_0L^2\left(fu\right)(t,x)$$

$$+\frac{h^2}{8}\int_t^{t+h/2} EL\left(fL^2\left(fu\right)\right)(\theta,X_{t,x}(\theta))d\theta$$

$$+Ef_{1/2}\int_{t+h/2}^{t+h}\frac{(t+h-\theta)^2}{2}L^3\left(fu\right)(\theta,X_{t,x}(\theta))\,d\theta\,, \tag{5.45}$$

$$Ef_0f_{1/2}^2u(t+h,X_{t,x}(t+h)) = f_0^3u(t,x) + \frac{h}{2}f_0L\left(f^2u\right)(t,x)$$

$$+f_0\int_t^{t+h/2}(t+h/2-\theta)EL^2\left(f^2u\right)(\theta,X_{t,x}(\theta))d\theta + \frac{h}{2}f_0^3Lu(t,x)$$

$$+\frac{h}{2}f_0\int_t^{t+h/2} EL\left(f^2Lu\right)(\theta,X_{t,x}(\theta))d\theta$$

$$+f_0Ef_{1/2}^2\int_{t+h/2}^{t+h}(t+h-\theta)L^2u(\theta,X_{t,x}(\theta))\,d\theta\,, \tag{5.46}$$

$$Ef_{1/2}^2f_1u(t+h,X_{t,x}(t+h)) = f_0^3u(t,x) + \frac{h}{2}L\left(f^3u\right)(t,x)$$

$$+\int_t^{t+h/2}(t+h/2-\theta)EL^2\left(f^3u\right)(\theta,X_{t,x}(\theta))d\theta + \frac{h}{2}f_0^2L(fu)(t,x)$$

$$+\frac{h}{2}\int_t^{t+h/2} EL\left(f^2L(fu)\right)(\theta,X_{t,x}(\theta))d\theta$$

$$+Ef_{1/2}^2\int_{t+h/2}^{t+h}(t+h-\theta)L^2\left(fu\right)(\theta,X_{t,x}(\theta))\,d\theta\,, \tag{5.47}$$

$$Ef_0 f_{1/2}^2 f_1 u(t+h, X_{t,x}(t-h)) = f_0^4 u(t,x)$$
$$+ f_0 \int_t^{t+h/2} EL\left(f^3 u\right)(\theta, X_{t,x}(\theta)) d\theta + f_0 E f_{1/2}^2 \int_{t+h/2}^{t+h} L\left(fu\right)(\theta, X_{t,x}(\theta)) d\theta.$$
(5.48)

Substituting (5.38)-(5.48) in (5.37), we obtain

$$r = h\left[Lu + fu\right] + \frac{h^2}{2}\left[L^2 u + L(fu) + fLu + f^2 u\right]$$
$$+ \frac{h^3}{6}[L^3 u + L^2(fu) + fL^2 u + fL(fu) + L(fLu) + L\left(f^2 u\right)$$
$$+ f^2 Lu + f^3 u] + \frac{h^4}{24}[L^4 u + L^3(fu) + fL^3 u + fL^2(fu)$$
$$+ f^2 L^2 u + f^2 L(fu) + f^3 Lu + f^4 u + L^2(fLu) + L^2(f^2 u) + fL(fLu)$$
$$+ fL(f^2 u) + L(fL^2 u) + L(fL(fu)) + L(f^2 Lu) + L(f^3 u)] + \tilde{r}, \quad (5.49)$$

where all the operators and functions are evaluated at the point (t,x) and $\tilde{r}$ accumulates all the integrals present in (5.39)-(5.48) multiplied by h to the corresponding power. Taking into account that $u(t,x)$ satisfies the equation from (5.32), we get

$$r(t,x) = \tilde{r}(t,x). \quad (5.50)$$

If the terms in the one-step error $r(t,x)$ of the method (5.28)-(5.29) (i.e., the terms in $\tilde{r}$) were bounded by $K(x)h^5$, $K(x) \in \mathbf{F}$, for all $t \leq T - h$, the relations (5.35)-(5.37) would imply that $\sum_{k=0}^{N-2} R_k \leq Ch^4$, where C is independent of h. But we see that the one-step error consists of integrals with integrands containing terms of the form $A(t,x) = L^n\left(q_1 L^l q_2\right)(t,x)$, where $q_1(t,x)$ and $q_2(t,x)$ are some functions from the class $\mathbf{F}$. The functions $A(t,x)$ belong to the class $\mathbf{F}$ for $t \in [0, T_*]$, where $T_* < T$ is a fixed (independent of h) time moment. Then $|r(t,x)| \leq K(x)h^5$, $K(x) \in \mathbf{F}$, $t \in [0, T_*]$, with $K(x)$ depending on T_*. However, the functions $A(t,x)$ do not belong to the class $\mathbf{F}$ for $t \in [0, T)$ due to the singularity in L (see (5.30)). Consequently, $r(t,x)$ can not be bounded by $K(x)h^5$, $K(x) \in \mathbf{F}$, for all $t < T$, and a more detailed analysis of the one-step error is required to prove the convergence theorem. In particular, we need to consider the structure of the functions $A(t,x)$ in detail. We always assume that L^0 is an identity operator. The following lemma is proved by induction.

Lemma 5.8. *Let $q_1(t,x)$ and $q_2(t,x)$ be sufficiently smooth functions belonging to the class $\mathbf{F}$ together with their partial derivatives of a sufficiently high order. Then for $0 \leq t < T$:*

$$L^n(q_1 L^l q_2)(t,x) = g_0(t,x) + \sum_{j=1}^m \sum_{\alpha_j} g_{\alpha_j}(t,x)\psi^{\alpha_j}(t,x), \quad (5.51)$$

$$l, n = 0, 1, \ldots, \quad m = l + n,$$

2.5 Application of weak methods to the computation of Wiener integrals

where α_j is a multi-index such that $\alpha_j = (i_1, \ldots, i_j)$ and each i_k is from $\{1, \ldots, d\}$, the summation in (5.51) is over all possible values of α_j, g_0 and g_{α_j} are some functions from the class $\mathbf{F}$, and

$$\psi^r = \frac{b^r - x^r}{T-t}, \ r = 1, \ldots, d,$$

$$\psi^{\alpha_{j+1}} = \frac{b^{i_{j+1}} - x^{i_{j+1}}}{T-t} \psi^{\alpha_j} + \frac{\partial}{\partial x^r} \psi^{\alpha_j}, \ \alpha_j = (i_1, \ldots, i_j),$$

$$\alpha_{j+1} = (i_1, \ldots, i_j, i_{j+1}), \ j = 1, 2, \ldots,$$

and for all α_j

$$L\psi^{\alpha_j} = 0.$$

Using specific properties of the functions ψ^{α_j}, the following theorem on one-step error is proved in [215].

Theorem 5.9. *The one-step error of the method* (5.28)-(5.29) *can be written in the form*

$$r(t,x) = \tilde{r}(t,x) = h^5 S(t,x) + E\rho(t,x;h), \tag{5.52}$$

where $S(t,x)$ *is a linear combination of the functions* $\psi^{\alpha_2}(t,x)$, $\psi^{\alpha_3}(t,x)$, $\psi^{\alpha_4}(t,x)$, $(T-t)\psi^{\alpha_4}(t,x)$, $h\psi^{\alpha_4}(t,x)$, $(T-t)\psi^{\alpha_5}(t,x)$, $h\psi^{\alpha_5}(t,x)$, $(T-t)^2\psi^{\alpha_6}(t,x)$, $(T-t)h\psi^{\alpha_6}(t,x)$, $h^2\psi^{\alpha_6}(t,x)$, *coefficients in this linear combination are independent of* t, x, *and* h; $\rho(t,x;h)$ *is such that*

$$\left(E\left[\rho(t, X_{0,a}(t); h)\right]^{2n}\right)^{1/2n} \leq \frac{Ch^5}{\sqrt{T-t-h}}, \ t+h < T,$$

with a constant C *independent of* t *and* h.

We should emphasize that the most important part of this theorem consists in the equality $r(t,x) = \tilde{r}(t,x)$ which is due to equations (5.37)-(5.49). Theorem 5.9 is a basis for the proof of Theorem 5.7 on the global error of the method (5.28)-(5.29) (see [215]).

2.5.5 Implicit Runge–Kutta methods for conditional Wiener integrals of exponential-type functionals

From the point of view of possible applications, the most interesting case is when the function f is bounded from above, for example, when f is negative. In this case the explicit Runge–Kutta method from Sect. 2.5.3 may cause some computational problems since, for instance, Y_{k+1} in (5.29) can become a large negative number while the exact $Y(t)$ is always positive. Apparently, this may occasionally lead to some instabilities and require a very small time step to achieve a reasonable accuracy. In such a situation an implicit method can behave better.

Let us formally apply the deterministic midpoint method to (5.24) provided $X(t)$ is a known function. As a result, we obtain

$$X(h/2) = a + \frac{h}{2}\frac{b-a}{T} + \sqrt{\frac{h}{2}}\sqrt{\frac{T-h/2}{T}}\xi_{1/2},$$

$$X(t_{k+1/2}) = X(t_{k-1/2}) + h\frac{b - X(t_{k-1/2})}{T - t_{k-1/2}} + \sqrt{h}\sqrt{\frac{T - t_{k+1/2}}{T - t_{k-1/2}}}\xi_{k+1/2},$$

$$k = 1, \ldots, N-1, \qquad (5.53)$$

$$Y_0 = 1,$$
$$Y_{k+1} = Y_k + hf(t_{k+1/2}, X(t_{k+1/2}))\frac{Y_k + Y_{k+1}}{2}, \ k = 0, \ldots, N-1, \quad (5.54)$$

where $\xi_{k+1/2}, \ k = 0, \ldots, N-1,$ are d-dimensional random vectors which components are mutually independent random variables with standard normal distribution $\mathcal{N}(0,1)$.

Resolving the implicitness in (5.54), we get

$$Y_{k+1} = Y_k \frac{1 + \dfrac{h}{2}f(t_{k+1/2}, X(t_{k+1/2}))}{1 - \dfrac{h}{2}f(t_{k+1/2}, X(t_{k+1/2}))}. \qquad (5.55)$$

To ensure that the denominator in (5.55) does not vanish for all sufficiently small h, we should require that the function $f(t,x)$ is bounded from above, i.e., that $f(t,x) \leq c$ for all (t,x), c is a constant. In this case for all sufficiently small h the denominator in (5.55) is positive. If $f(t,x) \leq 0$, then $-1 \leq Y_k \leq 1$ for all k.

We prove the convergence theorem for the method (5.53)-(5.54) under the same assumptions as in Sect. 2.5.3 (see them before Theorem 5.7). Note that in the case of $f(t,x) \leq 0$, the condition $EY_k^2 \leq C$ is satisfied due to the uniform boundedness of the random variables Y_k.

Theorem 5.10. *The method (5.53)-(5.54) applied to evaluation of the conditional Wiener integral (5.25) is of second accuracy order, i.e.,*

$$|J - EY_N| = |EY(T) - EY_N| \leq Kh^2, \qquad (5.56)$$

where the constant K is independent of h.

The proof of this theorem is given in [215].

If we formally apply the deterministic Gauss method of order four (see, e.g., [98, p. 71]) to (5.24), assuming that $X(t)$ is a known function, we obtain

2.5 Application of weak methods to the computation of Wiener integrals

$$X(\gamma h) = a + \gamma h \frac{b-a}{T} + \sqrt{\gamma h} \sqrt{\frac{T-\gamma h}{T}} \xi_\gamma,$$

$$X((1-\gamma)h) = X(\gamma h) + (1-2\gamma)h \frac{b - X(\gamma h)}{T - \gamma h}$$

$$+ \sqrt{(1-2\gamma)h} \sqrt{\frac{T-(1-\gamma)h}{T-\gamma h}} \xi_{1-\gamma},$$

$$X(t_k + \gamma h) = X(t_{k-1} + (1-\gamma)h) + 2\gamma h \frac{b - X(t_{k-1} + (1-\gamma)h)}{T - t_k + \gamma h}$$

$$+ \sqrt{2\gamma h} \sqrt{\frac{T-t_k-\gamma h}{T-t_k+\gamma h}} \xi_{k+\gamma},$$

$$X(t_k + (1-\gamma)h) = X(t_k + \gamma h) + (1-2\gamma)h \frac{b - X(t_k + \gamma h)}{T - t_k - \gamma h}$$

$$+ \sqrt{(1-2\gamma)h} \sqrt{\frac{T-t_{k+1}+\gamma h}{T-t_k-\gamma h}} \xi_{k+1-\gamma}, \quad k = 1, \ldots, N-1, \quad (5.57)$$

$$Y_0 = 1,$$

$$k_1 = f(t_k + \gamma h, X(t_k + \gamma h)) \left[Y_k + \frac{h}{4} k_1 + \left(\frac{1}{4} - \frac{\sqrt{3}}{6} \right) h k_2 \right],$$

$$k_2 = f(t_k + (1-\gamma)h, X(t_k + (1-\gamma)h)) \left[Y_k + \left(\frac{1}{4} + \frac{\sqrt{3}}{6} \right) h k_1 + \frac{h}{4} k_2 \right],$$

$$Y_{k+1} = Y_k + \frac{h}{2}(k_1 + k_2), \quad k = 0, \ldots, N-1, \quad (5.58)$$

where $\gamma = \frac{1}{2} - \frac{\sqrt{3}}{6}$ and $\xi_{k+\gamma}, \xi_{k+1-\gamma}, k = 0, \ldots, N-1$, are d-dimensional random vectors which components are mutually independent random variables with standard normal distribution $\mathcal{N}(0,1)$.

Resolving (5.58) with respect to k_1 and k_2, we get

$$Y_{k+1} = Y_k \frac{1 + \frac{h}{4}(f_1 + f_2) + \frac{h^2}{12} f_1 f_2}{1 - \frac{h}{4}(f_1 + f_2) + \frac{h^2}{12} f_1 f_2}, \quad (5.59)$$

where $f_1 := f(t_k + \gamma h, X(t_k + \gamma h))$ and $f_2 := f(t_k + (1-\gamma)h, X(t_k + (1-\gamma)h))$.

The denominator in (5.59) does not vanish for all sufficiently small h for functions $f(t,x)$ being bounded from above. And if $f(t,x) \leq 0$, then $-1 \leq Y_k \leq 1$ for all k.

The intuition built on the previous analysis of the methods (5.28)-(5.29) and (5.53)-(5.54) tells us that the method (5.57)-(5.58) should be of order four. But this assertion turned out to be wrong, the method is of order two only just as the method (5.53)-(5.54). We have not found an implicit method for (5.25) that satisfies the condition $|Y_k| \leq 1$ for $f(t,x) \leq 0$ and has the fourth order of accuracy. In this search it was natural to restrict ourselves to standard fourth-order deterministic implicit methods for ordinary differential equations as a basis for potentially higher-order implicit methods for (5.25).

The following convergence theorem is valid.

Theorem 5.11. *The method* (5.57)-(5.58) *applied to evaluation of the conditional Wiener integral* (5.25) *is of second order of accuracy, i.e.,*

$$|J - EY_N| = |EY(T) - EY_N| \leq Kh^2, \qquad (5.60)$$

where the constant K is independent of h.

Although the methods (5.53)-(5.54) and (5.57)-(5.58) are of the same order of convergence, in our numerical tests (see Sect. 2.5.6) the method (5.57)-(5.58) gives more accurate results. Apparently, this is due to the fact that the constant K in (5.60) is, in general, less than its counterpart in (5.56). At the same time, the method (5.53)-(5.54) requires one evaluation of f per step, while (5.57)-(5.58) requires two evaluations of f per step.

2.5.6 Numerical experiments

The first part of this subsection (Examples 5.12-5.15) deals with testing the proposed methods for Wiener integrals (5.1) with respect to the "usual" Wiener measure while in the second part (Examples 5.16-5.18) methods for conditional Wiener integrals (5.20)-(5.21) are tested.

In Examples 5.12-5.15 numerical experiments are mainly related to the computation of the Wiener integral

$$I = \int_{C_{0,0}^d} \exp(\alpha \int_0^T x^2(s)ds) \, d\mu_{0,0}(x). \qquad (5.61)$$

In this case the system (5.16) with initial conditions at a moment $0 \leq t < T$ is written as the system of two equations for $t \leq s \leq T$:

$$dX(s) = dw(s), \ X(t) = x,$$
$$dY(s) = \alpha Y(s) X^2(s) ds, \ Y(t) = y. \qquad (5.62)$$

Recall that

$$I = EY_{0,0,1}(T) = E\exp(\alpha \int_0^T X_{0,0}^2(s)ds), \qquad (5.63)$$

where $X_{0,0}(s)$, $Y_{0,0,1}(s)$ is the solution of the system (5.62) for $t = 0$, $x = 0$, $y = 1$.

2.5 Application of weak methods to the computation of Wiener integrals

Introduce the function

$$u(t,x) = EY_{t,x,1}(T) = E\exp(\alpha \int_t^T X_{t,x}^2(s)ds). \tag{5.64}$$

This function satisfies the Cauchy problem

$$\frac{\partial u}{\partial t} + \frac{1}{2}\frac{\partial^2 u}{\partial x^2} + \alpha x^2 u = 0, \ u(T,x) = 1. \tag{5.65}$$

The solution of (5.65) for $\alpha = \lambda^2/2$, $0 \le \lambda < \pi/2T$, has the form

$$u(t,x) = \exp[\frac{1}{2}\lambda x^2 \tan\lambda(T-t) - \frac{1}{2}\ln\cos(\lambda(T-t))]. \tag{5.66}$$

Since $I = u(0,0)$, this leads to the well-known result

$$I = (\cos\lambda T)^{-1/2}, \ \alpha = \frac{\lambda^2}{2}, \ 0 \le \lambda < \frac{\pi}{2T}. \tag{5.67}$$

Further, the solution of (5.65) for $\alpha = -\lambda^2/2$ has the form

$$u(t,x) = \exp\left[\frac{1}{2}\lambda\frac{1-\exp 2\lambda(T-t)}{1+\exp 2\lambda(T-t)}x^2\right.$$
$$\left. + \frac{\lambda(T-t)}{2} + \frac{1}{2}\ln\frac{2}{1+\exp 2\lambda(T-t)}\right], \tag{5.68}$$

and, consequently,

$$I = \left(\frac{2\exp\lambda T}{1+\exp 2\lambda T}\right)^{1/2}, \ \alpha = -\frac{\lambda^2}{2}. \tag{5.69}$$

Now we look for the variance of the variable $Y_{0,0,1}(T)$. For $\alpha = \lambda^2/2$:

$$VarY_{0,0,1}(T) = EY_{0,0,1}^2(T) - (EY_{0,0,1}(T))^2$$
$$= E\exp\left(2\alpha\int_0^T X_{0,0}^2(s)ds\right) - (\cos\lambda T)^{-1}, \tag{5.70}$$

$$0 \le \lambda < \frac{\pi}{2\sqrt{2}T}.$$

For $\pi/2\sqrt{2}T \le \lambda < \pi/2T$, the variance is equal to infinity.
For $\alpha = -\lambda^2/2$:

$$VarY_{0,0,1}(T) = \left(\frac{2\exp\sqrt{2}\lambda T}{1+\exp 2\sqrt{2}\lambda T}\right)^{1/2} - \frac{2\exp\lambda T}{1+\exp 2\lambda T}. \tag{5.71}$$

To reduce variance, let us use Theorem 4.1 with $s = 0$, $x = 0$. Note that $u(T, x) = 1$ and take $\alpha = \lambda^2/2$. Then, if (see (4.10))

$$\mu(t, x) = -\frac{1}{u}\frac{\partial u}{\partial x} = -\lambda x \tan \lambda(T - t),$$

the variable $Y_{0,0,1}(T)$ computed along the system

$$dX = \lambda(\tan \lambda(T - s))X ds + dw(s), \quad X(0) = 0,$$

$$dY = \frac{\lambda^2}{2}X^2 Y ds - \lambda(\tan \lambda(T - s))XY dw(s), \quad Y(0) = 1, \quad (5.72)$$

is deterministic.

For $\alpha = -\lambda^2/2$, the variable $Y_{0,0,1}(T)$ becomes deterministic as the solution of the following system:

$$dX = \lambda \frac{1 - \exp 2\lambda(T - s)}{1 + \exp 2\lambda(T - s)} X + dw(s), \quad X(0) = 0,$$

$$dY = -\frac{\lambda^2}{2}X^2 Y ds - \lambda \frac{1 - \exp 2\lambda(T - s)}{1 + \exp 2\lambda(T - s)} XY dw(s), \quad Y(0) = 1. \quad (5.73)$$

For completeness of exposition, we give the derivation of, e.g., (5.68). To this end we change variables in (5.65) for $\alpha = -\lambda^2/2$:

$$u = \exp v.$$

We get

$$\frac{\partial v}{\partial t} + \frac{1}{2}\left(\frac{\partial v}{\partial x}\right)^2 + \frac{1}{2}\frac{\partial^2 v}{\partial x^2} - \frac{\lambda^2}{2}x^2 = 0, \quad v(T, x) = 0. \quad (5.74)$$

We look for a solution of the problem (5.74) in the form

$$v(t, x) = \frac{1}{2}p(t)x^2 + r(t), \quad p(T) = r(T) = 0.$$

For $p(t)$ and $r(t)$ we obtain the Cauchy problem

$$p' + p^2 - \lambda^2 = 0, \quad p(T) = 0,$$

$$r' + \frac{1}{2}p = 0, \quad r(T) = 0,$$

which solution leads to formula (5.68).

Example 5.12. In Table 5.1 we give the results of integrating the system (5.62) over the interval $[0, 1]$ with initial data $X(0) = 0$, $Y(0) = 1$ for the α's indicated, by the method of first (M_I), second (M_{II}), and third (M_{III}) orders of accuracy constructed in Sect. 2.3 for systems with additive noise. The system (5.62) is a system with single noise. Therefore the methods of first (Euler's method) and second (see (3.2)) orders of accuracy require the

2.5 Application of weak methods to the computation of Wiener integrals   151

Table 5.1. Computation of the Wiener integral I by methods from Sect. 2.3. The first part of the table: $\alpha = -1$, $T = 1$, $I = EY(1) \doteq 0.6776$; the second part: $\alpha = -0.5$, $T = 1$, $I = EY(1) \doteq 0.8050$; the third part: $\alpha = 0.5$, $T = 1$, $I = EY(1) \doteq 1.3604$; $\gamma := 2\sqrt{VarY(1)/M}$.

h	M	M_I	M_{II}	M_{III}	γ
0.2	100	0.6576 ± 0.0547	0.6316 ± 0.0494	0.6330 ± 0.0494	0.0475
0.1	100	0.6810 ± 0.0532	0.6688 ± 0.0510	0.6695 ± 0.0510	0.0475
0.01	100	0.6650 ± 0.0442	0.6634 ± 0.0441	0.6640 ± 0.0441	0.0475
0.2	10^4	0.6955 ± 0.0052	0.6713 ± 0.0049	0.6749 ± 0.0048	0.0048
0.1	10^4	0.6841 ± 0.0049	0.6743 ± 0.0048	0.6749 ± 0.0048	0.0048
0.2	10^5	0.6973 ± 0.0016	0.6733 ± 0.0015	0.6769 ± 0.0015	0.0015
0.2	100	0.8008 ± 0.0376	0.7734 ± 0.0369	0.7749 ± 0.0365	0.0344
0.1	100	0.8086 ± 0.0371	0.7961 ± 0.0375	0.7967 ± 0.0373	0.0344
0.01	100	0.8018 ± 0.0302	0.8007 ± 0.0329	0.8007 ± 0.0303	0.0344
0.2	10^4	0.8254 ± 0.0034	0.8011 ± 0.0035	0.8030 ± 0.0035	0.0034
0.1	10^4	0.8135 ± 0.0034	0.8028 ± 0.0035	0.8032 ± 0.0034	0.0034
0.2	100	1.2759 ± 0.0763	1.4222 ± 0.1455	1.4443 ± 0.1663	0.1651
0.1	100	1.3093 ± 0.0971	1.3928 ± 0.1352	1.3999 ± 0.1397	0.1651
0.01	100	1.3048 ± 0.0654	1.3105 ± 0.1342	1.3111 ± 0.1344	0.1651
0.2	10^4	1.2356 ± 0.0068	1.3453 ± 0.0124	1.3598 ± 0.0143	0.0165
0.1	10^4	1.2865 ± 0.0087	1.3524 ± 0.0126	1.3572 ± 0.0132	0.0165

modeling of one random variable per step. In our case, the method of third order of accuracy (see (3.34)) requires the modeling of two random variables per step. This is related to the fact that i and r in (3.34) can take only the unit value (since $q = 1$) and ζ_{1k}^2 can therefore be replaced by unit (see (3.33)). To obtain the results, the variable ξ is modeled (in all three methods) by the $\mathcal{N}(0,1)$-distribution, and the variable ν is modeled by the law $P(\nu = \pm 1/\sqrt{12}) = 1/2$ (see (3.32)-(3.33)).

The numbers presented in Table 5.1 are approximations of $E\bar{Y}(1)$ computed by

$$E\bar{Y}(1) \simeq \frac{1}{M} \sum_{m=1}^{M} \bar{Y}^{(m)}(1)$$

$$\pm \frac{2}{M^{1/2}} \left[\frac{1}{M} \sum_{m=1}^{M} (\bar{Y}^{(m)}(1))^2 - \left(\frac{1}{M} \sum_{m=1}^{M} \bar{Y}^{(m)}(1) \right)^2 \right]^{1/2}, \quad (5.75)$$

i.e., under the natural assumption that the sample variance is sufficiently close to $Var\bar{Y}(1)$, the quantity $E\bar{Y}(1)$ lies between the given limits with probability 0.95. It is obvious that the true value of the required Wiener

152   2 Weak approximation for stochastic differential equations

integral (5.61) at $T = 1$, which is equal to $EY(1)$, differs from $E\bar{Y}(1)$ by $O(h)$ for Euler's method, by $O(h^2)$ for the method (3.2), and by $O(h^3)$ for the method (3.34). It is also obvious that with increasing M the error of the Monte Carlo method reduces, and if $M \to \infty$, the difference between the tabulated values and the true value of $EY(1)$ tends to the error of the numerical integration. For $\alpha = -1$ and $h = 0.1, 0.2$, it can be seen from the table that this error for Euler's method is one-two units of the second position after the point, for the method of second order it is several units of the third position while for the method of third order it is even less. As α increases, the efficiency of the methods of higher order becomes even more evident.

The value $\gamma = (2/\sqrt{M})(VarY(1))^{1/2}$ differs from the sample values of $(2/\sqrt{M}) \times (Var\bar{Y}(1))^{1/2}$, i.e., from the component in (5.75):

$$\frac{2}{M^{1/2}} \left[\frac{1}{M} \sum_{m=1}^{M} \left(\bar{Y}^{(m)}(1) \right)^2 - \left(\frac{1}{M} \sum_{m=1}^{M} \bar{Y}^{(m)}(1) \right)^2 \right]^{1/2},$$

first because of the estimation error, and secondly because of the numerical integration error. The components in the columns M_I, M_{II}, and M_{III} cannot become smaller as the order of accuracy increases. With an increase of the order of accuracy these components can become close to γ, because the numerical error increases. Moreover, they may also increase (see, e.g., the data for $\alpha = 0.5$).

Table 5.2. Computation of the Wiener integral I by the Euler method and the Runge–Kutta method of fourth order.

α	h	M	$EY(1)$	M_E	M_{R-K}	γ
-1	0.2	100	0.6776	0.7299 ± 0.0441	0.7025 ± 0.0436	0.0475
-1	0.2	10^4	0.6776	0.6987 ± 0.0051	0.6770 ± 0.0048	0.0048
-1	0.1	100	0.6776	0.7035 ± 0.0483	0.6883 ± 0.0477	0.0475
-1	0.1	10^4	0.6776	0.6876 ± 0.0048	0.6779 ± 0.0047	0.0048
-0.5	0.2	100	0.8050	0.8502 ± 0.0273	0.8253 ± 0.0295	0.0344
-0.5	0.2	10^4	0.8050	0.8279 ± 0.0034	0.8050 ± 0.0034	0.0034
-0.5	0.1	100	0.8050	0.8259 ± 0.0231	0.8122 ± 0.0339	0.0344
-0.5	0.1	10^4	0.8050	0.8160 ± 0.0034	0.8055 ± 0.0034	0.0034
0.5	0.2	100	1.3604	1.1864 ± 0.0429	1.2685 ± 0.0688	0.1651
0.5	0.2	10^4	1.3604	1.2297 ± 0.0065	1.3536 ± 0.0151	0.0165
0.5	0.1	100	1.3604	1.2611 ± 0.0762	1.3322 ± 0.1126	0.1651
0.5	0.1	10^4	1.3604	1.2824 ± 0.0089	1.3540 ± 0.0134	0.0165

2.5 Application of weak methods to the computation of Wiener integrals

We should also stress that the numbers given in the table include the numerical integration error, and therefore the indicated region of variation of them need not cover the true value, especially for methods of low order (see, e.g., the data M_I for $h = 0.2$, $N = 10000$).

Example 5.13. In Table 5.2 we give the results of integrating the system (5.62) by Euler's method (M_E) and the Runge–Kutta method of fourth order of accuracy (M_{R-K}). It can be seen from this table that, e.g., for $\alpha = 0.5$ the numerical integration error of Euler's method is more than one unit of the first position after the point, while in the Runge–Kutta method it moves to the unit of the third position. We draw attention to the fact that the sample variance for $\alpha = 0.5$ is less for the Euler method than for the Runge–Kutta method. As already noted in the previous example, the "large" variance of the Runge–Kutta method can in this case be explained by its greater accuracy.

Example 5.14. For relatively small ε, we compute the integral

$$I(\lambda, \varepsilon) = I = \int_{C_{0,0}^1} \exp[\frac{\lambda^2}{2} \int_0^T (x^2(s) + \varepsilon g(x(s)))ds] \, d\mu_{0,0}(x). \qquad (5.76)$$

To reduce the error of the Monte Carlo method, it is natural to integrate the system (see (5.72))

$$dX = p(s)X ds + dw(s), \; X(0) = 0,$$
$$dY = Y\frac{\lambda^2}{2}(X^2 + \varepsilon g(X))ds - p(s)XY dw(s), \; Y(0) = 1, \qquad (5.77)$$

where $p(s) = \lambda \tan \lambda(T-s)$, $T = 1$.

By Euler's method, we integrate the system (5.77) and the system

$$dX = dw(s), \; X(0) = 0,$$
$$dY = Y\frac{\lambda^2}{2}(X^2 + \varepsilon g(X))ds, \; Y(0) = 1. \qquad (5.78)$$

We obtain for $\lambda = 1$, $\varepsilon g(x) = 0.1x^2 \cos x$, $h = 0.0005$, $N = 100$:

for (5.77) we have $E\bar{Y}(1) = 1.3599 \pm 0.0053$,
for (5.78) we have $E\bar{Y}(1) = 1.3747 \pm 0.0630$.

We see that the error of the Monte Carlo method for the system (5.77) is approximately 10 times smaller than the error of the Monte Carlo method for the system (5.78).

Example 5.15. By the Cameron–Martin formula (see [156]), the value of the Wiener integral of the functional

$$F(x(\cdot)) = \exp(-\int_0^1 (x_1^2 + 2x_2^2 + 2x_3^2 + x_4^2 + x_1x_2 + x_2x_3 + x_3x_4)ds) \quad (5.79)$$

is equal to

$$I = \int_{C_{0,0}^4} F(x(\cdot))\mu_{0,0}(x) = \exp(\frac{1}{2}\int_0^1 tr\Gamma(s)ds),$$

where the matrix $\Gamma(s)$ can be found from the Cauchy problem

$$\frac{d\Gamma(s)}{ds} = 2Q - \Gamma^2(s), \; Q = \begin{bmatrix} 1 & 1/2 & 0 & 0 \\ 1/2 & 2 & 1/2 & 0 \\ 0 & 1/2 & 2 & 1/2 \\ 0 & 0 & 1/2 & 1 \end{bmatrix}, \; \Gamma(1) = 0.$$

Numerical integration of this system gives the above-mentioned Wiener integral: $I \simeq 0.1285$.

Table 5.3. Computation of the multi-dimensional Wiener integral of the functional (5.79).

h	M	M_E	M_{R-K}
0.2	25	0.1142 ± 0.0790	0.1410 ± 0.0536
0.2	100	0.0616 ± 0.0503	0.1230 ± 0.0231
0.2	1000	0.0848 ± 0.0179	0.1785 ± 0.0711
0.1	100	0.0997 ± 0.0262	0.1094 ± 0.0239
0.1	1000	0.1126 ± 0.0089	0.1220 ± 0.0081
0.1	10^4	0.1166 ± 0.0029	0.1270 ± 0.0026
0.05	10^4	0.1215 ± 0.0027	0.1264 ± 0.0026

Computations of this integral via integration of the system (5.16) by the Euler and Runge–Kutta methods are presented in Table 5.3. As in Table 5.1 and Table 5.2, along with the Monte Carlo error there arises the error $O(h)$ for the Euler method and $O(h^4)$ for the Runge–Kutta method. As can be seen from the table, this error is essential for the Euler method (see, e.g., the last row of Table 5.3, where the exact value 0.1285 of this integral is not covered by the values 0.1285 ± 0.0027), but it is not essential in the case of the Runge–Kutta method for the taken number of trajectories. Recall that the volume of calculations for the considered problems required by the Runge–Kutta method for given h and M is only twice as much as in the case of the Euler method.

Example 5.16. We consider the *conditional* Wiener integral (5.20)-(5.21) with the function $f(t,x)$ of the form

2.5 Application of weak methods to the computation of Wiener integrals

$$f(t,x) = (A(t)x, x) + (a_1(t), x) + a_0(t), \tag{5.80}$$

where $A(t)$ is a $d \times d$ symmetric matrix, $a_1(t)$ is a d-dimensional vector, and $a_0(t)$ is a scalar function.

Let $u(t, x)$ be the solution of (5.32) with f from (5.80). Introduce the function $P(t, x)$:

$$u(t, x) = \exp(P(t, x)). \tag{5.81}$$

This function satisfies the problem

$$LP + (A(t)x, x) + (a_1(t), x) + a_0(t) + \frac{1}{2} \sum_{i=1}^{d} \left(\frac{\partial P}{\partial x^i} \right)^2 = 0, \quad x \in \mathbf{R}^d, \ t < T,$$

$$P(T, x) = 0. \tag{5.82}$$

We look for a solution of (5.82) in the form

$$P(t, x) = \frac{1}{2} (P(t)x, x) + (p(t), x) + q(t), \tag{5.83}$$

where $P(t)$ is a $d \times d$ symmetric matrix, $p(t)$ is a d-dimensional vector, and $q(t)$ is a scalar function.

Substituting (5.83) in (5.82) and collecting terms $(\cdot x, x)$, $(\cdot, x)$ and terms independent of x separately, we arrive at the system for $P(t)$, $p(t)$, and $q(t)$:

$$P'(t) - \frac{2}{T-t}P + 2A(t) + P^2(t) = 0, \quad P(T) = 0, \tag{5.84}$$

$$p'(t) - \frac{1}{T-t}p + \frac{1}{T-t}P(t)b + P(t)p + a_1(t) = 0, \quad p(T) = 0, \tag{5.85}$$

$$q'(t) + \frac{1}{T-t}(p(t), b) + \frac{1}{2}\text{tr}\, P(t) + \frac{1}{2}(p(t), p(t)) + a_0(t) = 0, \quad q(T) = 0. \tag{5.86}$$

Note that if $a_1(t) \equiv 0$ and $b = 0$, then $p(t) \equiv 0$. And if in addition $a_0(t) \equiv 0$, then

$$q(t) = \frac{1}{2} \int_t^T \text{tr}\, P(s)\, ds.$$

The solution of (5.84) can be expanded in (positive) powers of $T - t$. If $A(t)$ is a constant matrix A, then this formal expansion starts with the terms:

$$P(t) = \frac{2}{3}A \cdot (T-t) + \frac{4}{45}A^2 \cdot (T-t)^3 + \cdots. \tag{5.87}$$

For test purposes, it is convenient to have an exact solution of (5.84)-(5.86) in a closed analytical form. To this end, we choose a variable matrix $A(t)$ such that

$$A(t) = A - \frac{2}{9}A^2 \cdot (T-t)^2, \tag{5.88}$$

Table 5.4. The results of simulaton of the conditional Wiener integral (5.20)-(5.21) for f from (5.80) with $a_0 = 0$, $a_1 = 0$, $A(t)$ from (5.88), (5.91) and for $a = b = 0$, $T = 1$ by the explicit Runge–Kutta method (5.28)-(5.29) and the implicit Runge–Kutta methods (5.53), (5.55) and (5.57), (5.59). The exact solution is 1.

h	M	(5.28)-(5.29)	(5.53), (5.55)	(5.57), (5.59)
0.2	10^6	0.9994 ± 0.0013	1.0176 ± 0.0044	1.0040 ± 0.0013
0.1	10^8	1.00002 ± 0.00013	1.00361 ± 0.00015	1.00093 ± 0.00013
0.05	10^8	0.99996 ± 0.00013	1.00089 ± 0.00013	1.00019 ± 0.00013

where A is a constant symmetric matrix. Then the exact solution of the system (5.84)-(5.86) with $b = 0$, $a_0(t) \equiv 0$, and $a_1(t) \equiv 0$ has the form

$$P(t) = \frac{2}{3}(T-t)A, \quad p(t) = 0, \quad q(t) = \frac{(T-t)^2}{6} tr A. \tag{5.89}$$

Consequently, the solution of (5.83) is

$$P(t, x) = \frac{T-t}{3}(Ax, x) + \frac{(T-t)^2}{6} tr A. \tag{5.90}$$

Then the conditional Wiener integral (5.20)–(5.21) for f from (5.80) with $a_0 = 0$, $a_1 = 0$, $A(t)$ from (5.88) and for $a = b = 0$ is equal to

$$J = u(0,0) = \exp\left(\frac{T^2}{6} tr A\right).$$

In our experiments we take the dimension $d = 4$ and the following matrix A:

$$A = \begin{bmatrix} -1 & -0.5 & 0 & 0 \\ -0.5 & 2 & -0.5 & 0 \\ 0 & -0.5 & -2 & -0.5 \\ 0 & 0 & -0.5 & 1 \end{bmatrix}, \tag{5.91}$$

for which $tr A = 0$.

In Table 5.4 we give results of simulation of the conditional Wiener integral (5.20)-(5.21) for f from (5.80) with $a_0 = 0$, $a_1 = 0$, $A(t)$ from (5.88), (5.91) and for $a = b = 0$, $T = 1$ by the explicit Runge–Kutta method (5.28)-(5.29) and the implicit Runge–Kutta methods (5.53), (5.55) and (5.57), (5.59). We have two types of errors in numerical simulations here: the error of a method used and the Monte Carlo error. The results in the table are approximations of $E\bar{Y}(1)$ calculated as in (5.75). Note that the "$\pm$" reflects the Monte Carlo error only and it does not reflect the error of a method. The results obtained are in agreement with the proved convergence theorems (see also Table 5.5). Recall that the implicit methods (5.53)-(5.54) and (5.57)-(5.58) are both of order two. In our tests the method (5.57)-(5.58) performs

2.5 Application of weak methods to the computation of Wiener integrals   157

better. Apparently, this is due to the fact that the constant K in (5.60) is, in general, less than its counterpart in (5.56).

We also note that for the considered test problem we do not have any numerical instabilities and the explicit method is computationally effective. As has been discussed at the beginning of Sect. 2.5.5, implicit methods should be used in practice when explicit methods are affected by instabilities.

Example 5.17. To reduce the Monte Carlo error, variance reduction techniques from Sect. 2.4 can be used. In the case of evaluating Wiener integrals (5.20)-(5.21) application of method of important sampling changes the linear system (5.22) for X to a system with, in general, a nonlinear drift. As a result, we lose the advantage of simulating $X(t)$ exactly and of approximating the conditional Wiener integral by higher-order numerical integrators from Sects. 2.5.3 and 2.5.5. This shortcoming does not arise in the case of the method of control variates from Sect. 2.4.2. That is why, we restrict ourselves here to this method only.

In connection with the evaluation of the Wiener integral (5.20)-(5.21) consider the following system of Ito SDEs (cf. (5.22)-(5.24)):

$$dX = \frac{b-X}{T-t} dt + dw(t), \ X(s) = x, \tag{5.92}$$

$$dY = f(t, X(t)) Y \, dt, \ Y(s) = y, \tag{5.93}$$

$$dZ = G^\top(t, X) Y \, dw(t), \ Z(s) = z. \tag{5.94}$$

Here Z is a scalar and $G(t, x)$ is a column-vector of dimension d with good analytical properties, the other notation is the same as it was in this section before.

It is clear that

$$u(s, x) = EY_{s,x,1}(T) = E\left[Y_{s,x,1}(T) + Z_{s,x,1,0}(T)\right].$$

Theorem 4.3 implies that by choosing $G(t, x)$ as

$$G^i = -\frac{\partial u}{\partial x^i}, \quad i = 1, \ldots, d, \tag{5.95}$$

we obtain that the variance of $Y_{s,x,1}(T) + Z_{s,x,1,0}(T)$ is equal to zero.

Applying a numerical method to (5.92)-(5.94), we get the approximate $\bar{Y}_{s,x,1}(T)$ and $\bar{Z}_{s,x,1,0}(T)$. The variance $Var\left[\bar{Y}_{s,x,1}(T) + \bar{Z}_{s,x,1,0}(T)\right]$ is close to $Var\left[Y_{s,x,1}(T) + Z_{s,x,1,0}(T)\right]$, i.e., it is small in the case of G from (5.95), and, consequently, a smaller number of independent realizations M is needed to have a satisfactory accuracy.

For f from (5.80) with $a_0 = 0$, $a_1 = 0$, $A(t)$ from (5.88), (5.91) and for $b = 0$, the solution $u(t, x)$ of (5.32) has the form (5.81), (5.90). Therefore, in this case the vector function G defined in (5.95) is equal to

$$G^i(t,x) = -\frac{2}{3}(T-t)\,\exp(P(t,x))\sum_{j=1}^{d} A^{ij}x^j\,, \quad i=1,\ldots,d, \qquad (5.96)$$

where $P(t,x)$ is from (5.90) and A is from (5.91).

Applying the Euler method to the equation (5.94), we get

$$Z_0 = 0,$$
$$Z_{k+1} = Z_k + G^\top(t_k, X)Y_k\,\Delta w_k, \quad k=1,\ldots,N-1. \qquad (5.97)$$

If we approximate (5.92)-(5.93) using the explicit fourth-order Runge–Kutta method (5.28)-(5.29), then Y_k in (5.97) is from (5.29) and the Wiener increment is

$$\Delta w_k := w(t_{k+1}) - w(t_k) = \frac{h^{1/2}}{\sqrt{2}}\left(\xi_{k+1/2} + \xi_{k+1}\right),$$

where $\xi_{k+1/2}$ and ξ_{k+1} are the same as in (5.28)-(5.29).

It is clear that $EZ_{k+1} = 0$. This implies that the method (5.28)-(5.29), (5.97) applying to (5.92)-(5.94) to approximate the Wiener integral $J = EY(T)$ is of order four, i.e., the above realization of the variance reduction technique does not affect the accuracy of the numerical method. The variance $Var\,Y(T)$ is approximated with accuracy $O(h)$. Consequently, for a fixed number of realizations M the Monte Carlo error in simulations using the variance reduction technique is $\sim 1/\sqrt{h}$ times less than in simulations without variance reduction. In other words, in the case of variance reduction the Monte Carlo error is proportional to $\sqrt{h}/\sqrt{M}$. This is illustrated in Table 5.5. In particular, we see for $h = 0.05$ that to produce results of the same quality we need $M = 10^8$ independent trajectories without variance reduction and $M = 10^7$ independent realizations in the variance reduction case (compare Tables 5.4 and 5.5).

Table 5.5. The results of simulaton of the conditional Wiener integral (5.20)-(5.21) for f from (5.80) with $a_0 = 0$, $a_1 = 0$, $A(t)$ from (5.88), (5.91) and for $a = b = 0$, $T = 1$ by the explicit Runge–Kutta method (5.28)-(5.29) and the implicit Runge–Kutta methods (5.53), (5.55) and (5.57), (5.59) using the variance reduction technique. The exact solution is 1.

h	M	(5.28)-(5.29)	(5.53), (5.55)	(5.57), (5.59)
0.1	10^7	0.99977 ± 0.00024	1.00396 ± 0.00050	1.00103 ± 0.00023
0.05	10^7	0.99992 ± 0.00017	1.00098 ± 0.00017	1.00023 ± 0.00016
0.05	10^8	0.99999 ± 0.00005	1.00088 ± 0.00005	1.00027 ± 0.00005
0.01	10^7	1.00003 ± 0.00007	1.00001 ± 0.00007	1.00003 ± 0.00007

Example 5.18. Of course, in practice the solution $u(t,x)$ is not known. However, an approximate solution $\tilde{u}$ to the problem (5.32) can be known. In this

case we can take $G(t,x)$ in the form of (5.95) with $\tilde{u}$ instead of u and we may expect a variance reduction. To illustrate this assertion, we take the function $f(t,x)$ in the form (5.80) with the constant matrix $A(t) \equiv A$ from (5.91) and $a_0 = 0$, $a_1 = 0$. We also put $b = 0$. In this case we do not know the exact solution $u(t,x)$ of (5.32). But for the variance reduction we can use an approximation $\tilde{u}(t,x)$ of the solution based on the formal expansion (5.87):

$$\tilde{u}(t,x) = \exp\left(\frac{1}{2}\left(\tilde{P}(t)x, x\right)\right), \tag{5.98}$$

where

$$\tilde{P}(t) = \frac{2}{3} A \cdot (T-t).$$

Deriving (5.98), we take into account that $tr\tilde{P}(t) = 0$ because of the specific choice of the matrix A which is from (5.91).

Then we take the function G in (5.94) of the form

$$G^i(t,x) = -\frac{\partial \tilde{u}}{\partial x^i}, \quad i = 1, \ldots, d.$$

Putting $a = 0$ and $T = 1$, we evaluate the corresponding conditional Wiener integral (5.20)-(5.21) by the fourth-order explicit Runge–Kutta method (5.28)-(5.29) with time step $h = 0.01$ and we simulate $M = 10^5$ independent realizations. Without variance reduction, we get: $J \doteq 1.1536 \pm 0.0093$, while applying the variance reduction technique (i.e., using the method (5.28)-(5.29), (5.97) for (5.92)-(5.94)) we obtain $J \doteq 1.1482 \pm 0.0018$. We see that the Monte Carlo error is 5 times less when we use the variance reduction technique.

2.6 Random number generators

Implementation of a numerical method for integrating SDEs requires a source of random numbers. Typically, random numbers are approximated by pseudorandom ones which are generated via iterative deterministic algorithms. These algorithms produce a sequence of numbers which are in fact not random at all. The program that produces a sequence of pseudorandom numbers is called pseudorandom number generator or, for simplification, random number generator (RNG). A generator should satisfy the following natural requirements. First, the sequence of random numbers must have appropriate statistical properties, i.e., a RNG is constructed in such a way that it produces the deterministic sequence of numbers which shares enough of the statistical properties of a true random sequence. On the other hand, the RNG has to be as quick as possible because in the case of integrating SDEs a sufficient part of the simulation time is spent by generating random numbers. Besides, the sequence of random numbers should be reproducible. Finally, RNG is

160   2 Weak approximation for stochastic differential equations

preferable to be portable, i.e., the same on any computer. One can find a lot of various RNGs in literature including their theoretical backgrounds, empirical testing, implementations, recommendations for their use, etc. (see, e.g., [40, 81, 100, 135, 147, 225] and references therein).

Random variables with any distribution can be obtained by transforming independent random variables distributed uniformly over the interval [0, 1]. Then, first of all, we are interested in a source of randomness which will supply us with a realization of independent uniformly distributed random variables. In particular, it is very simple to simulate discrete random variables needed for weak methods of this chapter using uniform random numbers (see also [81, 135, 254]). In Sect. 2.6.1, we give definitions and some properties of the following uniform RNGs: linear congruential generators, additive lagged Fibonacci generators, and Tausworthe (shift register) generators. It is known that in addition to standard statistical tests of RNGs, it is useful to apply application-specific tests which are more relevant to a concrete type of applications. A comparative analysis of the RNGs from Sect. 2.6.1 in the context of SDEs integration is given in Sect. 2.6.2. Generation of Gaussian random numbers is considered in Sect. 2.6.3. Parallel RNGs are discussed in Sect. 2.6.4. See also implementation of some RNGs in Appendix A.6.

2.6.1 Some uniform random number generators

The most important requirements for uniform RNGs are the following ones. First, the produced sequence should pass statistical tests on uniformity and independence (see details, e.g., in [135, 147, 225]). Further, as a rule, RNGs produce periodic sequences of numbers with a period ρ. Weak numerical integration of SDEs usually requires a large amount M of random numbers. Moreover, due to common practice in Monte Carlo simulations, the period ρ of taken RNG should be essentially greater than M (there are recommendations not to use more than 5% of a RNG's period in a single calculation [239]). So, the period of the generated sequence has to be large. At the same time, a RNG should be efficient. Besides, as it was noticed before, the sequence of random numbers has to be reproducible and the RNG has to be portable. The last requirements are essential for testing and development purposes. A parallel implementation puts additional requirements on RNGs. They are discussed in Sect. 2.6.4.

In this subsection we deal with three types of RNGs: linear congruential generators (LCG), additive lagged Fibonacci generators, and linear feedback shift register generators (LFSR).

The most popular and well studied RNGs are linear congruential generators [135, 147, 153]. They are also known as Lehmer generators. For instance, some standard Unix generators (e.g., rand(), drand48(), ranf()) are of this type. A LCG is defined by the recurrence relation:

$$x_n = (a\, x_{n-1} + c) \bmod m, \qquad (6.1)$$

where the multiplier a and the modulus m are positive integers, the additive constant c is a non-negative integer.

The normalized sequence

$$u_n = x_n/m \qquad (6.2)$$

is used as a source of random numbers in $[0, 1]$.

We shall use the notation $\text{LCG}(m, a, c)$ for an LCG generator with the parameters m, a, c.

If $c > 0$, maximum possible period of $\text{LCG}(m, a, c)$ is equal to m, and if $c = 0$, its maximum possible period is equal to $m-1$. LCGs can be effectively implemented on 32-bit computers with $m \leq 2^{32}$ (to avoid using slow multi-precision arithmetic). Therefore, their periods do not exceed $2^{32} \sim 10^9$ which may be too small for SDEs integration. However, LCGs with $m \sim 2^{64}$ can be effectively implemented on 64-bit computers which are becoming increasingly common. Periods of such LCGs are about $2^{64} \sim 10^{19}$ that is quite enough for the most of numerical experiments today.

For drawbacks of LCGs, see, e.g. [100, 135, 147]. The parameters of a LCG should be selected carefully. As is known (see, e.g., [40, 100, 147] and references therein), LCGs with properly chosen parameters work very well for most applications. Examples of LCGs with good parameters are given in [148].

To get generators with larger periods as well as to improve quality of RNGs, it is possible to combine LCGs. The two most widely known combination methods are shuffling one sequence with another or with itself (see, e.g., [135, 147]) and modular addition of two or more sequences (see, e.g., [145, 151]). A large number of empirical investigations performed over the past 30 years strongly support shuffling and some generators available in software libraries use it (e.g., the procedures ran1() and ran2() from [239] and see also the procedure rng_lcgs.c in Appendix A.6). However, it has two drawbacks: the effect of shuffling is not well understood from theoretical point of view and one does not know how to jump ahead quickly in the sequence with shuffling. The last one does not allow us to use generators with shuffling in parallel implementations. The second class of combination method, by modular addition, is generally better understood theoretically and good jump-ahead techniques are known.

Additive lagged Fibonacci generators [40, 135, 164] are another type of quite popular generators. For instance, the standard Unix generator random() is of this type. They have very long periods and are very fast. An additive lagged Fibonacci generator is defined by

$$x_n = (x_{n-p} \pm x_{n-q}) \bmod m, \qquad (6.3)$$

where p and q are the lags, $p > q$. The addition (or subtraction) is done modulo any large integer m or with $m = 1$ when x_i are presented as floating point numbers in the interval $[0, 1)$. This method requires storing the p

previous values of x_i in an array called a lag table. We denote the generator (6.3) by $F(p,q,m)$ for $m > 1$ and $F(p,q)$ for $m = 1$. If $m > 1$, the normalized sequence is obtained as in (6.2).

Again, it is important that the parameters be carefully chosen in order to provide good randomness properties and a large period. Tables of suitable parameters are given, e.g., in [135, 164]. An advantage of additive lagged Fibonacci generators (as well as of other Fibonacci generators) is that the period can be made arbitrary large by just increasing the lag p. For instance, for k-bit precision x_i (e.g., when $m = 2^k$) and correctly chosen lags p and q, it is possible to obtain period of the generator $\rho = 2^{k-1}(2^p - 1)$. Increase of the lag also improves its randomness properties. To improve quality of RNGs, it is also possible to combine a lagged Fibonacci generator with an LCG or to combine three or more previous elements of the sequence rather than two.

It is commonly recommended (see, e.g., [40, 164]) to take a sufficiently large lag p. Naturally, the larger p the larger amount of memory is needed. Moreover, a large lag may affect the speed of generator [40]. For instance, we observe in our numerical experiments that $F(97, 33)$ is faster than $F(1279, 418)$ (see Table 6.2). An implementation of $F(1279, 418)$ is given in Listing A.15 from Appendix A.6.

Note that additive lagged Fibonacci generators are not so convenient as LCGs and Tausworthe generators for parallel implementations.

Now consider Tausworthe generators. Define

$$x_n = \sum_{j=1}^{L} y_{ns+j-1} 2^{-j}, \qquad (6.4)$$

where s and L are positive integers and y_n is a sequence of bits (zeros and ones):

$$y_n = (y_{n-r} + y_{n-k}) \bmod 2 \qquad (6.5)$$

with $s \leq r$ and $2r > k$. This is called linear feedback shift register (LFSR) or Tausworthe generator [135, 147, 290, 292].

The normalized sequence u_n is obtained as

$$u_n = x_n \, 2^{-L} \qquad (6.6)$$

We denote the generator (6.4)-(6.6) by $T(L, k, r, s)$. Under some conditions, its period is equal to $2^k - 1$. An efficient computer code of Tausworthe generator (QuickTaus algorithm) can be found in [146, 292]. For fast implementation, L is taken equal to 32 (64) on 32-bit (64-bit) computers. As is known [135, 147, 292], the simple recurrences of this type lead to generators with bad properties (as an example, see TAUS63 in our tests in Sect. 2.6.2 below). But combining several recurrences with carefully chosen parameters can give quite good and effective generators [146, 149, 292].

Remark 6.1. There are other types of RNGs, e.g., multiple Fibonacci generators, generalized feedback shift register generators and their modifications, multiple recursive generators, nonlinear generators (see [135, 147, 225] and references therein for description of these generators).

Remark 6.2. It was shown in [105] that sequences of low discrepancy, which are superior in the case of classical quasi-Monte Carlo methods [225], can lead to wrong results in the case of numerical integration of SDEs.

2.6.2 A specific test for SDE integration

As is known (see, e.g. [40, 100, 135, 147, 225] and references therein), in addition to standard statistical tests of RNGs, it is useful to apply application-specific tests which are more relevant to a concrete type of applications. Here we test generators from the previous subsection using the model problem

$$dX^i = \sigma dw_i(t), \ X^i(0) = x, \ i = 1, 2, \tag{6.7}$$

where $w_i(t)$, $i = 1, 2$, are independent standard Wiener processes, σ and x are constants.

It is not difficult to check that

$$E \cos X^i(t) = \cos x \, e^{-\sigma^2 t/2}, \tag{6.8}$$

$$E \cos \left[(X^1(t) - x)(X^2(t) - x) \right] = \frac{1}{\sqrt{1 + \sigma^4 t^2}}. \tag{6.9}$$

In our tests we evaluate the expectations

$$E \cos X^i(t) \quad \text{and} \quad E \cos \left[(X^1(t) - x)(X^2(t) - x) \right]$$

by simulation of the system (6.7) using the weak method (3.34) together with the Monte Carlo technique. As is known, in such simulations two types of errors arise: an error of numerical integration and the Monte Carlo error. We have

$$Ef(X(T)) \doteq Ef(\bar{X}(T)) \doteq \frac{1}{M} \sum_{m=1}^{M} f(\bar{X}^{(m)}(T)) \pm \frac{c\sqrt{\bar{D}_M}}{\sqrt{M}},$$

where

$$\bar{D}_M = \frac{1}{M} \sum_{m=1}^{M} \left(f(\bar{X}^{(m)}(T)) \right)^2 - \left(\frac{1}{M} \sum_{m=1}^{M} f(\bar{X}^{(m)}(T)) \right)^2,$$

M is the number of independent realizations $\bar{X}^{(m)}$ simulated by a weak method, and c is a constant. If the values $c = 1, 2, 3$, then the fiducial probabilities are equal to 0.68, 0.95, 0.997, respectively.

In our tests we take sufficiently small time steps h such that for a fixed M the error of numerical integration is essentially less than the Monte Carlo error. Then we are able to analyze the Monte Carlo error and test RNGs in such a fashion. We simulate the value $\frac{1}{M}\sum_{m=1}^{M} f(\bar{X}^{(m)}(T))$ N times using consequent non-overlapped parts of a sequence generated by a RNG. We say that the RNG works properly for our application if results of not less than $A\%$ of the N experiments belongs to the confidential interval:

$$\left|\frac{1}{M}\sum_{m=1}^{M} f(\bar{X}^{(m)}(T)) - Ef(X(T))\right| \leq \frac{c\sqrt{\bar{D}_M}}{\sqrt{M}} ,$$

where $A = 68, 95, 99.7$ for $c = 1, 2, 3$, respectively.

The following RNGs are used in the tests:

1. DRAND48 - the standard UNIX generator, which is LCG($2^{48}, 25214903917, 11$); its period length $\sim 10^{13}$.
2. RANECU - L'Ecuyer's combined by addition LCG with the components LCG(2147483563, 40014, 0) and LCG(2147483399, 40692, 0) [145]; its period length $\sim 10^{18}$.
3. RNG_LCGS from Appendix A.6 - the RNG combined by shuffling LCGs with the components LCG(2147483563, 40014, 0) and LCG(2147483399, 40692, 0); its period length $\sim 10^{18}$. It is analogous to ran2() of [239].
4. $F(97, 33)$, the seed table of this additive lagged Fibonacci generator is initialized using ran1() of [239]; its period length $\sim 10^{47}$.
5. $F(1279, 418)$, the seed table of this additive lagged Fibonacci generator is initialized using ran1() of [239]; its period length $\sim 10^{403}$.
6. TAUS57 - the combined Tausworthe generator with the components $T(32, 29, 27, 18)$ and $T(32, 28, 19, 14)$ [146]; its period length $\sim 10^{17}$.
7. TAUS63 - the Tausworthe generator $T(64, 63, 62, 10)$ [239, 309]; its period length $\sim 10^{19}$.
8. TAUS88 - the combined Tausworthe generator with the components $T(32, 31, 18, 12)$, $T(32, 29, 27, 4)$, and $T(32, 28, 25, 17)$ [146], its period length $\sim 10^{26}$.

The results of the tests are given in Table 6.1. We can conclude that all the generators except TAUS63 are quite acceptable. Table 6.2 gives an indication of the relative speeds of the generators. Our tests show that additive lagged Fibonacci generators are preferable.

The simulations were performed on Dec AlphaStation with CPU Alpha-21164, 533MHz; OS Digital Unix 4.0D; using the standard C translator with the command line: cc -newc -non_shared -O4 -inline speed -fast. Evidently, speed of a RNG depends on the architecture of a computer. Some comparison analysis of timing of various RNGs made on Dec AlphaStation 250 and SUN Ultra-2 can be found in, e.g., [150]. Further, the speed of a RNG can usually be increased by allowing the routine to return an array of values, rather than

Table 6.1. Percent of experiments belonging to confidential intervals for various RNGs. The top position corresponds to evaluation of (6.8) with $i = 1$ and the bottom one - to (6.9). The system (6.7) is simulated by the weak method (3.34) with $\sigma = 1$, $t = 1$, $x = 0$, $h = 0.01$, $M = 10^5$, $N = 10^4$.

c	DRAND48	RANECU	RNG_LCGS	$F(97,33)$	$F(1279,418)$	TAUS57	TAUS63	TAUS88
1	69.0	68.5	68.8	68.6	68.3	69.2	4.6	67.8
	68.1	68.4	68.0	68.2	68.3	68.4	15.4	68.4
2	95.3	95.5	95.5	95.4	95.8	95.1	25.5	95.3
	95.6	95.4	95.6	95.4	95.4	95.7	48.5	95.0
3	99.8	99.7	99.8	99.7	99.7	99.8	62.7	99.7
	99.8	99.7	99.7	99.7	99.7	99.7	83.4	99.8

Table 6.2. Time required to simulate (6.8) with $i = 1$ and (6.9) by the weak method (3.34) with $\sigma = 1$, $t = 1$, $x = 0$, $h = 0.01$, $M = 10^5$ using various random generators and percentage of the time spent on generation of uniform random numbers.

generator	time, s	%
DRAND48	7.6	78
RANECU	5.5	70
RNG_LCGS	6.4	74
$F(97, 33)$	2.7	39
$F(1279, 418)$	3.0	42
TAUS57	3.3	50
TAUS63	2.6	37
TAUS88	3.7	54

a single value [40, 147]. This is clearly advantageous for a vector or parallel implementations, and it is also true for a sequential implementation.

We do not discuss LCGs with small periods $\sim 10^9$ here. However, we did some tests with ran1() of [239] (i.e., LCG(2147483647,16807,0) with shuffling) and got quite good results. But this short period generator is slower than, e.g., an additive lagged Fibonacci generator with an essentially larger period. Let us note in passing that modern computers are able to simulate 10^9 numbers in a few minutes and in a lot of practical applications we need generators with periods $\sim 10^{18}$ or larger.

In a majority of numerical experiments presented in this book we used the generators rng_lcgs() (see Listing A.14 in Appendix A.6) and $F(1279, 418)$ (see its realization rng_fiba.c in Listing A.15). In general, we prefer the additive lagged Fibonacci generator $F(1279, 418)$ since it is fast and of a good quality. At the same time, it is strongly recommended to run all simulations with two (or even more) generators, and the results compared to check whether the RNG is introducing a bias.

Remark 6.3. In the experiments presented in this book, we simulate discrete random numbers, needed for weak schemes, as follows (see also the procedure rng_disc.c in Appendix A.6). Suppose we would like to simulate the discrete random variable ξ which takes values $x_1, \ldots, x_n$ with probabilities $p_1, \ldots, p_n$. To this end, we can generate a uniform number U and then simulate ξ as

$$\xi = \begin{cases} x_1 \text{ if } 0 \leq U < p_1; \\ x_2 \text{ if } p_1 \leq U < p_1 + p_2; \\ \ldots\ldots\ldots\ldots \\ x_n \text{ if } p_1 + \cdots + p_{n-1} \leq U < 1. \end{cases}$$

Weak methods given in this chapter use discrete random variables for which n is small: 2, 3, 4, or 5. Some of these random variables can be simulated much more effectively than in the way described above. For instance, consider i.i.d. discrete random variables ξ_i taking two values ± 1 with probability $1/2$ which are needed for Euler-type methods. These random variables can be effectively generated using "random bit" generators (see, e.g. [135]). At the same time, uniform random numbers can be exploited for our purposes in a very efficient way as it is described below (see also [135, p. 101]).

Let an ideal uniform RNG produces 32-bit numbers. Then the sequence of bits in these numbers (with the change of 0 by -1) gives us a sequence ξ_i required for Euler-type methods. These bits should be extracted from the most-significant (left-hand) part of the computer word (e.g., 24 first bits of 32) since in existing RNGs the least significant bits can be not sufficiently random. As a result, producing just one number by a uniform 32-bit RNG, we get 24 numbers ξ_i, i.e., in this way we simulate the i.i.d. discrete random variables ξ_i with the law $P(\xi = \pm 1) = 1/2$ very economically. Moreover, such an approach can be applied to generating some other discrete random variables. Consider, for example, the random variable with the law

$$P\left(\xi = \pm\sqrt{1 - \sqrt{6}/3}\right) = \frac{3}{8}, \quad P\left(\xi = \pm\sqrt{1 + \sqrt{6}}\right) = \frac{1}{8}, \qquad (6.10)$$

which is used in the weak second order methods (see (2.18) and (1.31)). It is clear that such a number can be generated by three bits of a uniform random number, i.e., we again have the very economical way of generating the needed discrete random numbers.

2.6.3 Generation of Gaussian random numbers

A source of Gaussian random numbers is needed for all the methods of Chap. 1. There are many choices of algorithms for generation of Gaussian random numbers based on applying some transformations to uniform random numbers (see, e.g. [81, 135, 254] and references therein).

1. First, there are "universal" methods of transforming uniform random variables, which are applicable to almost any distribution and, in particular, to the Gaussian one. The inverse transform algorithm and the general rejection method [81, 135, 254] are among the "universal" methods.

The inverse transform algorithm is based on the following assertion. If U is a uniform $(0,1)$ random variable, then for any continuous distribution function F the random variable X defined by $X = F^{-1}(U)$ has the distribution F. Thus the problem reduces to evaluating the inverse function. Specialized algorithms for inverting normal distribution function are considered, e.g. in [166]. The "universal" methods are usually rather slow and special algorithms are preferable in general.

2. Using the Central Limit Theorem, it is possible to propose some approximate methods. As an example, consider the following one. If U_i, $i = 1, \ldots, 12$, are independent uniform $(0,1)$ random variables, then

$$X = \sum_{i=1}^{12} U_i - 6$$

has an approximate normal distribution $\mathcal{N}(0,1)$. This method is not only approximate, but it is also rather slow.

3. Ones of the most popular algorithms for generating Gaussian random numbers are polar methods. The well-known Box–Muller transformation [27] (see also [81, 254]) belongs to this type of methods. The Box–Muller transformation itself is computationally not very efficient since it requires to compute the sine and cosine trigonometric functions. But it is possible to implement the Box–Muller transformation via rejection method and avoid costly evaluations of sine and cosine [163] (see also [81, 135, 254]).

The rejection polar method can be described as follows. First, generate two independent uniform $(-1,1)$ random numbers U_1 and U_2 and set $R^2 = U_1^2 + U_2^2$. Second, if $R^2 \geq 1$ then repeat the first step, otherwise

$$X_1 = U_1\sqrt{-2\log R^2/R^2}, \qquad X_2 = U_2\sqrt{-2\log R^2/R^2}. \tag{6.11}$$

The obtained X_1 and X_2 are independent normally distributed $\mathcal{N}(0,1)$ random numbers.

On average, the rejection polar method requires 2.546 uniform random numbers, 1 logarithm, 1 square-root, 1 division, and 4.546 multiplications to generate two independent Gaussian random numbers. This method is very popular and quite fast. In Appendix A.6 it is realized as the procedure rng_gau.c. This generator uses the procedure rng_lcgs.c from Appendix A.6 as a source of uniform random numbers. The procedure rng_gau.c was used in a majority of experiments in the book where the source of Gaussian random numbers was needed.

The (statistical) quality of random numbers generated by, e.g., the rejection polar method ultimately depends only on the quality of the underlying uniform generator (see Sect. 2.6.1). A particular method for generating Gaussian random numbers, however, may exacerbate some fault in the uniform generator (see [81] and references therein for further discussion).

4. The fastest algorithms for generating Gaussian random numbers are based on the ratio-of-uniforms method [81,135]. The idea of ratio-of-uniforms method [126] in application to Gaussian random numbers is to generate a random point (U,V) uniformly distributed in the region of the plane $\{(u,v) : u > 0, \ v^2 \leq -4u^2 \log u\}$ and then to output the ratio $X = V/U$ which is normally $\mathcal{N}(0,1)$ distributed. The point (U,V) can be conveniently generated using an acceptance-rejection technique. A faster version of the ratio-of-uniforms algorithm is proposed in [155], where its FORTRAN implementation and comparison analysis with other methods for generating normal random numbers are also given.

2.6.4 Parallel implementation

Monte Carlo simulations are well suited to parallel computers. A common way of parallelizing Monte Carlo simulations is to run identical procedures but with different random number sequences on the various processors. A communication between the processors is needed only to start the simulation and to make final averaging and output. Then the use of p independent processors should reduce computational costs of the simulation in p times. Of course, this is true only if the results obtained on each processor are statistically independent, i.e., the random number sequences generated in the processors have to be independent. Thus, the problem of parallelization of Monte Carlo simulation consists in finding a good parallel random number generator (PRNG) (see, e.g. [30,40,168] and references therein).

Note that running a RNG (a centralized RNG) on a particular processor to supply all the other processors with random numbers is not an appropriate solution for parallelizing Monte Carlo simulations because inter-processor communication is very expensive and the desirable speed-up of calculations cannot be achieved on this way. Besides, in this case the requirement of reproducibility of calculations is often difficult to satisfy since the different processors may request random numbers in different orders in different runs of the program depending on the network traffic and implementation of the communication software.

Thus, each processor has to have its own RNG which produces a sequence of random numbers independent of the sequences on the other processors. To parallelize RNGs, the following main techniques are used:

(i) producing parallel streams of random numbers by taking subsequences from a single, long-period RNG [30,40,81];
(ii) parametrization of RNGs [168].

The first one is subdivided into (ia) leapfrog method and (ib) sequence splitting. In the leapfrog method the sequence is partitioned among the pro-

2.6 Random number generators

cessors in a cyclic fashion, like a desk of cards dealt to card players. In the sequence splitting, the sequence is partitioned among processors in a block fashion, by splitting it into non-overlapping contiguous sections.

The leapfrog method can effectively be realized for LCGs, combined LCGs and LFSRs (see, e.g. [40, 81] and references therein). For instance, in the case of LCG (6.1) it works as follows. Let

$$x_{1,0} = x_0$$
$$x_{j,0} = (ax_{j-1,0} + c) \bmod m, \quad j = 2, \ldots, p,$$

and

$$x_{j,n} = (a_p x_{j,n-1} + c_p) \bmod m, \quad j = 1, \ldots, p, \quad n = 1, 2, \ldots, \quad (6.12)$$

where $a_p = a^p \bmod m$ and $c_p = c(a^p - 1)/(a-1) \bmod m$. It is not difficult to show that $x_{j,n} = x_{j+np}$, where x_i is the sequence of pseudorandom numbers obtained by the original LCG (6.1), i.e., $x_{j,n}$, $j = 1, \ldots, p$, are subsequences of the sequence x_i. Here p is called the step size and j selects one of the p parallel streams.

If the original generator (6.1) has good statistical properties, then it is natural to think that the subsequences $x_{j,n}$ also possess good statistical properties. But in general this is not true [56]. Although the parameters a, c, and m may be chosen so that the LCG (6.1) performs well in the statistical tests, there is no guarantee that the LCG (6.1) with $a = a_p$ and $c = c_p$ will also have good statistical properties for an arbitrary p. See further discussion in [30, 40, 81].

The sequence splitting into consecutive blocks of a given length can effectively be realized for LCGs, combined LCGs, LFSRs, and additive lagged Fibonacci generators (see, e.g. [30, 40, 81] and reference therein). As an example, consider again the LCG (6.1). Define the p subsequences $x_{j,n}$, $j = 1, \ldots, p$, as

$$x_{1,0} = x_0$$
$$x_{j,0} = (a_L x_{j-1,0} + c_L) \bmod m, \quad j = 2, \ldots, p,$$

where $a_L = a^L \bmod m$ and $c_L = c(a^L - 1)/(a-1) \bmod m$ and L is a sufficiently large positive integer but it is smaller than $1/p$ of the length of the original LCG (6.1), and

$$x_{j,n} = (ax_{j,n-1} + c) \bmod m, \quad j = 1, \ldots, p, \quad n = 1, 2, \ldots.$$

The obtained $x_{j,n} = x_{(j-1)L+n}$, where x_i is generated by the original LCG (6.1). Each subsequence $x_{j,n}$ has the length L.

The well known problem with this PRNG is that although the subsequences $x_{j,n}$ are disjoint, this does not necessarily mean that they are uncorrelated. See further discussion in [30, 40, 44, 152] and references therein.

Finally, we briefly consider the parametrization method (for a review see [168]). This method identifies a parameter in the underlying recursion of a serial RNG that can be varied. Each valid value of this parameter leads to a recursion that produces a unique, full-period steam of random numbers. The exact meaning of parametrization depends on the type of RNG. For instance, the additive lagged Fibonacci generator (see (6.3)) can be parameterized through its initial values [40, 168, 169]. This parametrization based on the fact [165] that the lagged Fibonacci generator (6.3) has many disjoint full-period cycles. As a result, different seed tables may produce completely different non-overlapping periodic sequences of numbers. For suitably chosen parameters, the number of such disjoint cycles is $2^{(p-1)(k-1)}$ provided $m = 2^k$ [165, 169]. This type of PRNG is quite popular and has been implemented for a number of parallel computers and parallel languages.

See also parametrization of LCGs, linear matrix generators, LFSRs, and inverse congruential generators in [168].

Remark 6.4. If the number of processors is different in different runs of the program then to ensure the reproducibility in parallel Monte Carlo simulation it is necessary to write the program in terms of "virtual processors", each virtual processor having its own RNG.

A discussion of RNG implementations for computers with vector processors can be found in, e.g. [2, 30].

3 Numerical methods for SDEs with small noise

In the general case many difficulties arise with realizing numerical methods for SDEs. But we know (see, e.g., Chaps. 1-2 where, in particular, such specific systems as systems with additive and colored noises are treated) that numerical methods adapted to specific systems can be more efficient and easier than general methods. An important instance of a stochastic system is given by differential equations with small noise, since often fluctuations, which affect a dynamical system, are sufficiently small.

The system of Ito stochastic differential equations with *small noise* can be written in the form

$$dX = a(t,X)\,dt + \varepsilon^2 b(t,X)\,dt + \varepsilon \sum_{r=1}^{q} \sigma_r(t,X)\,dw_r\,,\ X(t_0) = X_0\,, \quad (0.1)$$

$$t \in [t_0, T],\ 0 \le \varepsilon \le \varepsilon_0\,,$$

where ε is a small parameter, ε_0 is a positive number, $X = (X^1, \ldots, X^d)^\mathsf{T}$, $a(t,x) = (a^1(t,x), \ldots, a^d(t,x))^\mathsf{T}$, $b(t,x) = (b^1(t,x), \ldots, b^d(t,x))^\mathsf{T}$, $\sigma_r(t,x) = (\sigma_r^1(t,x), \ldots, \sigma_r^d(t,x))^\mathsf{T}$, $r = 1, \ldots, q$, are d-dimensional vectors, $w_r(t)$, $r = 1, \ldots, q$, are independent standard Wiener processes, and X_0 does not depend on $w_r(t) - w_r(t_0)$, $t_0 < t \le T$, $r = 1, \ldots, q$.

If the parameter ε is equal to zero, we have a deterministic system for which various effective numerical methods exist. One can believe that if parameter ε is sufficiently small, i.e., the system (0.1) is sufficiently close to the deterministic one, it is also possible to obtain effective methods taking into account that ε is small.

Introduce the equidistant discretization of the interval $[t_0, T]$: $\{t_k : k = 0, 1, \ldots, N;\ t_0 < t_1 < \cdots < t_N = T\}$ and the time increment $h = t_{k+1} - t_k$. The errors of the methods proposed in this chapter are estimated in terms of products $h^i \varepsilon^j$, and they are usually of the form $O(h^p + \varepsilon^k h^q)$, $q < p$. The order of such a method is equal to q which may be low, e.g. $1/2$ or 1. For small ε the product $\varepsilon^k h^q$ also becomes small and, consequently, so does the error. This allows us to construct effective methods with low order but which nevertheless have small errors. For instance, the methods of this chapter are effectively applied to evaluation of the signal-to-noise ratio in systems with stochastic resonance in [296] (cf. Chap. 9).

In the first part of this chapter (Sects. 3.1-3.3) we derive specific *mean-square* methods for systems with small noise [201]. In these sections, we put $b=0$ for simplicity, i.e., we consider the system

$$dX = a(t,X)\,dt + \varepsilon \sum_{r=1}^{q} \sigma_r(t,X)\,dw_r, \ X(t_0) = X_0. \tag{0.2}$$

In Sect. 3.1 we propose our approach to constructing one-step mean-square approximations for solutions of the system (0.2) and prove the theorem on mean-square estimate of method error. Various efficient mean-square schemes for systems with small noise are presented in Sect. 3.2. Numerical tests of the proposed mean-square methods are presented in Sect. 3.3.

The second part of this chapter (Sects. 3.4-3.8) is devoted to *weak* approximations of the system with small noise (0.1) [202]. In Sect. 3.4 we state the theorem on error estimate of weak methods and illustrate our approach to construction of weak schemes for the system (0.1). Various specific weak schemes for systems with small noise are proposed in Sect. 3.5. Section 3.6 deals with the expansion of the global error of weak methods in powers of h and ε. The method of important sampling from Sect. 2.4.1 allows us to effectively reduce the Monte Carlo error in the case of a system with small noise that is demonstrated in Sect. 3.7. Numerical tests of the proposed weak methods are presented in Sect. 3.8.

We note that in the case of the system (0.1) the operators L and Λ_r from Chaps. 1 and 2 take the form

$$L = L_1 + \varepsilon^2 L_2, \ L_1 = \frac{\partial}{\partial t} + (a, \frac{\partial}{\partial x}) = \frac{\partial}{\partial t} + \sum_{i=1}^{d} a^i \frac{\partial}{\partial x^i},$$

$$L_2 = (b, \frac{\partial}{\partial x}) + \frac{1}{2}\sum_{r=1}^{q}(\sigma_r, \frac{\partial}{\partial x})^2 = \sum_{i=1}^{d} b^i \frac{\partial}{\partial x^i} + \frac{1}{2}\sum_{r=1}^{q}\sum_{i,j=1}^{d} \sigma_r^i \sigma_r^j \frac{\partial^2}{\partial x^i \partial x^j},$$

$$\Lambda_r = (\sigma_r, \frac{\partial}{\partial x}) = \sum_{i=1}^{d} \sigma_r^i \frac{\partial}{\partial x^i}.$$

We will use the following notation for Ito integrals:

$$I_{i_1,\ldots,i_j}(F,t,h)$$
$$= \int_t^{t+h} dw_{i_j}(\theta) \int_t^{\theta} dw_{i_{j-1}}(\theta_1) \int_t^{\theta_1} \cdots \int_t^{\theta_{j-2}} F(\theta_{j-1})\,dw_{i_1}(\theta_{j-1}),$$

where $i_1,\ldots,i_j$ are from the set of numbers $\{0,1,\ldots,q\}$ and $dw_0(\theta_r)$ designates $d\theta_r$, $F(\theta)$ is a deterministic (for simplicity continuous) function; $I_{i_1,i_2,\ldots,i_j}(t,h) \equiv I_{i_1,i_2,\ldots,i_j}(1(\cdot),t,h)$ where $1(\theta)$ is the function which is everywhere equal to one. Properties of Ito integrals were established in Lemma 1.2.1.

3.1 Mean-square approximations and estimation of their errors

3.1.1 Construction of one-step mean-square approximation

Consider the one-step approximation with the local order two (cf. (1.2.27)):

$$\bar{X}(t+h) = X(t) + \varepsilon \sum_{r=1}^{q} \sigma_r(t, X(t)) I_r(t, h) + a(t, X(t)) h$$

$$+ \varepsilon^2 \sum_{i,r=1}^{q} \Lambda_r \sigma_i(t, X(t)) I_{ri}(t, h) + \varepsilon \sum_{r=1}^{q} L_1 \sigma_r(t, X(t)) I_{0r}(t, h)$$

$$+ \varepsilon^3 \sum_{r=1}^{q} L_2 \sigma_r(t, X(t)) I_{0r}(t, h) + \varepsilon \sum_{r=1}^{q} \Lambda_r a(t, X(t)) I_{r0}(t, h)$$

$$+ \varepsilon^3 \sum_{s,i,r=1}^{q} \Lambda_s \Lambda_i \sigma_r(t, X(t)) I_{sir}(t, h) + L_1 a(t, X(t)) h^2/2$$

$$+ \varepsilon^2 L_2 a(t, X(t)) h^2/2. \tag{1.1}$$

The remainder $\rho = X(t+h) - \bar{X}(t+h)$ of this approximation has the form (cf. (1.2.22)):

$$\rho = \varepsilon^4 \sum_{r,i,s,j=1}^{q} I_{risj}(\Lambda_r \Lambda_i \Lambda_s \sigma_j, t, h) + \varepsilon^2 \sum_{i,r=1}^{q} I_{0ir}(L_1 \Lambda_i \sigma_r, t, h)$$

$$+ \varepsilon^4 \sum_{i,r=1}^{q} I_{0ir}(L_2 \Lambda_i \sigma_r, t, h) + \varepsilon^2 \sum_{i,r=1}^{q} I_{i0r}(\Lambda_i L_1 \sigma_r, t, h)$$

$$+ \varepsilon^4 \sum_{i,r=1}^{q} I_{i0r}(\Lambda_i L_2 \sigma_r, t, h) + \varepsilon^2 \sum_{i,r=1}^{q} I_{ir0}(\Lambda_i \Lambda_r a, t, h)$$

$$+ \varepsilon^3 \sum_{r,i,s=1}^{q} I_{0sir}(L_1 \Lambda_s \Lambda_i \sigma_r, t, h) + \varepsilon^5 \sum_{r,i,s=1}^{q} I_{0sir}(L_2 \Lambda_s \Lambda_i \sigma_r, t, h)$$

$$+ \varepsilon \sum_{r=1}^{q} I_{00r}(L_1^2 \sigma_r, t, h) + \varepsilon^3 \sum_{r=1}^{q} I_{00r}((L_1 L_2 + L_2 L_1) \sigma_r, t, h)$$

$$+ \varepsilon^5 \sum_{r=1}^{q} I_{00r}(L_2^2 \sigma_r, t, h) + \varepsilon \sum_{r=1}^{q} I_{0r0}(L_1 \Lambda_r a, t, h)$$

$$+ \varepsilon^3 \sum_{r=1}^{q} I_{0r0}(L_2 \Lambda_r a, t, h) + \varepsilon \sum_{r=1}^{q} I_{r00}(\Lambda_r L_1 a, t, h)$$

$$+\varepsilon^3 \sum_{r=1}^{q} I_{r00}(\Lambda_r L_2 a, t, h) + I_{000}(L_1^2 a, t, h)$$
$$+\varepsilon^2 I_{000}((L_1 L_2 + L_2 L_1)a, t, h) + \varepsilon^4 I_{000}(L_2^2 a, t, h). \quad (1.2)$$

It is not difficult to obtain
$$E\rho = O(h^3), \ (E\rho^2)^{1/2} = O(h^3 + \varepsilon^2 h^2).$$

The Ito integrals I_{ri} and I_{sir} of the method (1.1) cannot be easily simulated. But these integrals are multiplied by ε^2 and ε^3, respectively. That is why, they may be transferred to the remainder and the error of the approximation would still not be large. Further, if we transfer from (1.1) not only the terms with complicated Ito integrals but also the terms which are sufficiently small, we obtain the reduced one-step approximation

$$\bar{X}(t+h) = X(t) + \varepsilon \sum_{r=1}^{q} \sigma_r(t, X(t)) I_r(t, h) + a(t, X(t))h$$
$$+\varepsilon \sum_{r=1}^{q} L_1 \sigma_r(t, X(t)) I_{0r}(t, h) + \varepsilon \sum_{r=1}^{q} \Lambda_r a(t, X(t)) I_{r0}(t, h)$$
$$+L_1 a(t, X(t)) h^2/2, \quad (1.3)$$

the remainder ρ_1 of which is equal to

$$\rho_1 = \rho + \varepsilon^2 \sum_{i,r=1}^{q} \Lambda_r \sigma_i(t, X(t)) I_{ri}(t, h) + \varepsilon^3 \sum_{r=1}^{q} L_2 \sigma_r(t, X(t)) I_{0r}(t, h)$$
$$+\varepsilon^3 \sum_{s,i,r=1}^{q} \Lambda_s \Lambda_i \sigma_r(t, X(t)) I_{sir}(t, h) + \varepsilon^2 L_2 a(t, X(t)) h^2/2, \quad (1.4)$$

where ρ is taken from (1.2).

One can obtain
$$E\rho_1 = O(h^3 + \varepsilon^2 h^2),$$
$$(E\rho_1^2)^{1/2} = O(h^3 + \varepsilon h^{5/2} + \varepsilon^2 h^2 + \varepsilon^3 h^{3/2} + \varepsilon^2 h) = O(h^3 + \varepsilon^2 h). \quad (1.5)$$

The terms $\varepsilon h^{5/2}$, $\varepsilon^2 h^2$, and $\varepsilon^3 h^{3/2}$ of the second expression are omitted because they are not greater than $O(h^3 + \varepsilon^2 h)$. Of course, the order of the approximation (1.3) is less (the order is equal to one due to the term $\varepsilon^2 \sum_{i,r=1}^{q} \Lambda_r \sigma_i I_{ri}$ in ρ_1) than the order of the approximation (1.1), but the error of the approximation (1.3) has the small factor ε^2 at h. Thus, we obtain the one-step approximation (1.3) which has sufficiently small mean-square local error and is efficient as to simulation of the used random variables.

Using (1.2)-(1.4), we construct a new approximation by transferring a part of the remainder to the approximation. In this connection we expand the term

$$I_{000}(L_1^2 a, t, h) = \int_t^{t+h} \left(\int_t^\theta \left(\int_t^{\theta_1} L_1^2 a(\theta_2, X(\theta_2)) \, d\theta_2 \right) d\theta_1 \right) d\theta$$
$$= L_1^2 a(t, X(t)) h^3/6 + \rho',$$

where
$$E\rho' = O(h^4 + \varepsilon^2 h^3),$$
$$E[(\rho')^2]^{1/2} = O(h^4 + \varepsilon h^{7/2} + \varepsilon^2 h^3) = O(h^4 + \varepsilon^2 h^3).$$

The new approximation has the form
$$\tilde{X}(t+h) = \bar{X}(t+h) + L_1^2 a(t, X(t)) h^3/6, \tag{1.6}$$

where $\bar{X}(t+h)$ is taken from (1.3). The remainder $\tilde{\rho}$ of the approximation (1.6) can be obtained from ρ_1 if we substitute ρ' instead of $h^3 I_{000}(L_1^2 a, t, h)$. It is clear that

$$E\tilde{\rho} = O(h^4 + \varepsilon^2 h^2), \quad E(\tilde{\rho}^2)^{1/2} = O(h^4 + \varepsilon^2 h).$$

Of course, the approximation (1.6) can be derived by another way, for instance, from an approximation with local order three but the suggested way is the simplest.

In this subsection we have demonstrated the basic idea of the chapter. In contrast to the general case smallness of terms of an approximation for a system with small noise and of its remainder depends not only on time increment h but also on small parameter ε. This circumstance, as shown above, allows us to construct new numerical methods by excluding complicated terms, for instance, multiple Ito integrals, from a method and including them in its remainder. New methods are efficient as to simulation of the used random variables and have small mean-square errors in the sense of product $\varepsilon^i h^j$. Moreover, the methods contain fewer terms with operators than the corresponding schemes for a general system.

3.1.2 Theorem on mean-square global estimate

Usually after reducing, estimates of remainder of a concrete one-step approximation are sufficiently simple and often contain only two terms (for instance, see (1.5)). However, it may not be the case. For instance, the sum $h^3 + \varepsilon h^{3/2} + \varepsilon^2 h$ cannot be reduced. Detailed analysis of possible errors gives the form of estimates which is exactly used in Theorem 1.1 stated below, i.e., the conditions of Theorem 1.1 are natural.

Introduce the notation: ρ is a local error of a method, R is a global error (we also call it as method error, mean-square error or error if it does not lead to misunderstanding), r_0 is a fixed natural number, S_1 is either an empty set or a subset of positive integers p which are less than r_0, S_2 is either empty set or a subset of positive integers and semi-integers q which are less than r_0, i.e.,

$$S_1 \subset \{p : 0 < p < r_0, \ p \text{ is an integer}\},$$

$$S_2 \subset \{q : 0 < q < r_0, \ q \text{ is either an integer or semi-integer}\}.$$

Below in Theorem 1.1 the sum $\sum_{p \in S_1}$ ($\sum_{q \in S_2}$) must be replaced by zero if S_1 (S_2) is an empty set.

Theorem 1.1. *Let $\bar{X}_{t,x}(t+h)$ be an approximation of the exact solution $X_{t,x}(t+h)$ of the system (0.2) with initial condition $X(t) = \bar{X}(t) = x$ and $J_1(p)$ and $J_2(q)$ be decreasing functions with natural values. If the inequalities*

$$|E\rho| = |E\left(X_{t,x}(t+h) - \bar{X}_{t,x}(t+h)\right)|$$

$$\leq K\left(1 + |x|^2\right)^{1/2} \left(h^{r_c} + \sum_{p \in S_1} h^p \varepsilon^{2J_1(p)} \right), \quad (1.7)$$

$$(E\rho^2)^{1/2} = (E|X_{t,x}(t+h) - \bar{X}_{t,x}(t+h)|^2)^{1/2}$$

$$\leq K\left(1 + |x|^2\right)^{1/2} \left(h^{r_o} + \sum_{q \in S_2} h^q \varepsilon^{J_2(q)} \right) \quad (1.8)$$

hold, then

$$(E|X_{t_0,X_0}(t_k) - \bar{X}_{t_0,X_0}(t_k)|^2)^{1/2} \leq K\left(1 + E|X_0|^2\right)^{1/2}$$

$$\times \left(h^{r_0-1} + \sum_{p \in S_1} h^{p-1} \varepsilon^{2J_1(p)} - \sum_{q \in S_2} h^{q-1/2} \varepsilon^{J_2(q)} \right), \quad (1.9)$$

where the constant K does not depend on discretization step h, parameter ε, $0 \leq \varepsilon \leq \varepsilon_0$, and $k = 1, \ldots, N$.

The proof of Theorem 1.1 is similar to the proof of Theorem 1.1.1, and here it is omitted.

Applying Theorem 1.1, for example, to the method based on the one-step approximation (1.3), we obtain that its mean-square global error is estimated by $O(h^2 + \varepsilon^2 h + \varepsilon^2 h^{1/2}) = O(h^2 + \varepsilon^2 h^{1/2})$. This error is sufficiently small because of the small factor ε^2 at $h^{1/2}$.

It follows from Theorem 1.1 that if $r_0 > 1$ and the set S_1 is either empty or every number p of S_1 is greater than one and the set S_2 is either empty or every number q of S_2 is greater than $1/2$, then the corresponding method converges (cf. Theorem 1.1.1). However, the primary meaning of Theorem 1.1 is not that it gives convergence order of a method but is that it gives a method error in terms of h and ε.

3.1.3 Selection of time increment h depending on parameter ε

In practice the parameter ε is small but fixed, and we can usually choose only the step h. Nevertheless, asymptotic behavior of method error under $\varepsilon \to 0$, when h is chosen depending on ε, is interesting in many respects. And the inequality (1.9) makes such an analysis possible.

Let us choose time increment h so that $h = C\varepsilon^\alpha$, $\alpha > 0$. Then the error of a method can be estimated in powers of small parameter ε

$$(E|X_{t_0,X_0}(t_k) - \bar{X}_{t_0,X_0}(t_k)|^2)^{1/2} = O(\varepsilon^\beta),$$

where $\beta = \min\{\alpha(r_0 - 1), \min_{p \in S_1}(\alpha(p-1) + 2J_1(p)), \min_{q \in S_2}(\alpha(q - 1/2) + J_2(q))\}$. The parameter α and a method may be so that a certain term of this method is smaller than $O(\varepsilon^\beta)$. Such a term may be omitted and, in spite of this, the order of the method error does not change with respect to ε.

Let us analyze the method based on the one-step approximation (1.3). If $h = C\varepsilon^\alpha$, the mean-square global error of the method (1.3) is estimated by $O(\varepsilon^{2\alpha} + \varepsilon^{2+\alpha/2})$. Let us choose α be equal to one. In this case the method error is estimated by $O(\varepsilon^2)$, the order of the terms $\varepsilon L_1 \sigma_r I_{0r}$ and $\varepsilon \Lambda_r a I_{r0}$ is equal to $O(\varepsilon^{5/2})$, and their omission gives $O(\varepsilon^2)$ to the mean-square error. So, in the case of $\alpha = 1$ these terms may be omitted, and that does not lead to substantial increase of the error. Thus, we obtain the new method

$$X_{k+1} = X_k + \varepsilon \sum_{r=1}^{q}(\sigma_r I_r)_k + a_k h + L_1 a_k h^2/2, \qquad (1.10)$$

$$(ER^2)^{1/2} = O(h^2 + \varepsilon h + \varepsilon^2 h^{1/2}),$$

where $\sigma_{r_k} = \sigma_r(t_k, X_k)$, $a_k = a(t_k, X_k)$, $(I_r)_k = I_r(t_k, h)$. It is clear that if $h = C\varepsilon^\alpha$, where $\alpha \leq 1$ or $\alpha \geq 2$, the errors of the methods (1.3) and (1.10) have the same order with respect to ε. But if $h = C\varepsilon^\alpha$, $1 < \alpha < 2$, the method (1.10) has the lower order with respect to ε than the method (1.3).

3.1.4 (h, ε)-approach versus (ε, h)-approach

In this chapter we construct numerical methods by (h, ε)-approach, for instance, see the methods (1.3) and (1.10). According to (h, ε)-approach, we expand the exact solution $X(t)$ of the system (0.2) in powers of time increment h and obtain an expansion which is similar to the stochastic Taylor-type expansion (see Sect. 1.2.2). Then we regroup terms of the expansion with respect to their $h^i \varepsilon^j$ factors and decide which terms must be included in a method. Such a decision depends on the desired mean-square error of the method and on computational complexity of an expansion term, especially on complexity of simulation of the used random variables.

(ε, h)-approach is based on another idea. First, the exact solution of the system (0.2) is expanded in powers of small parameter ε, for instance,

$$\bar{X}(t) = X^0(t) + \varepsilon X^1(t),\qquad(1.11)$$

$$R = X(t) - \bar{X}(t) = O(\varepsilon^2),$$

where $X^0(t)$ and $X^1(t)$ are found as the solutions of the original system under $\varepsilon = 0$ and its system of the first approximation:

$$dX^0 = a(t, X^0)\,dt,\ X^0(0) = X_0,\qquad(1.12)$$

$$dX^1 = a'_x(t, X^0)X^1\,dt + \sum_{r=1}^{q}\sigma_r(t, X^0)\,dw_r,\ X^1(0) = 0.\qquad(1.13)$$

The system (1.12) is the system of deterministic differential equations for which, as is generally known, efficient high-order numerical methods exist, for example,

$$X^0_{k+1} = X^0_k + a_k h + (a\,a'_x + a'_t)_k h^2/2,\qquad(1.14)$$

$$X^0_0 = X_0,\ R_0 = O(h^2),$$

where $a_k = a(t_k, X^0_k)$, the $d \times d$-matrix $(a'_x)_k$ is equal to $\partial a(t_k, X^0_k)/\partial x$, d-vector $(a'_t)_k$ is equal to $\partial a(t_k, X^0_k)/\partial t$, R_0 is the error of the method. The system (1.13) is the system of SDEs with additive noise. The Euler method for the system (1.13) has the form

$$X^1_{k+1} = X^1_k + \sum_{r=1}^{q}(\sigma_r I_r)_k + (a'_x X^1)_k h,\qquad(1.15)$$

$$X^1_0 = 0,\ (E(R_1)^2)^{1/2} = O(h),$$

where $\sigma_{r_k} = \sigma_r(t_k, X^0_k)$, $(a'_x)_k = \partial a(t_k, X^0_k)/\partial x$, R_1 is the error of the method. So, we obtain the method (1.11), (1.14), (1.15) for numerical solution of the system (0.2) with the error $O(h^2 + \varepsilon^2)$.

One can see that (h, ε)-approach and (ε, h)-approach are essentially different. If time increment h tends to zero, a method, constructed by (ε, h)-approach, does not converge to the exact solution and converges to $X^0(t) + \varepsilon X^1(t)$. In contrast to (ε, h)-approach, (h, ε)-approach gives a method which always converges to the exact solution of the system (0.2) when $h \to 0$. Our aim is to derive numerical methods for solution of the system (0.2) with small but fixed parameter $\varepsilon > 0$. That is why (h, ε)-approach is more preferable than (ε, h)-approach.

3.2 Some concrete mean-square methods for systems with small noise

Our aim is to construct methods with small mean-square errors (provided that ε is a small parameter) and with simply simulated random variables.

3.2 Some concrete mean-square methods for systems with small noise

Herein we restrict ourselves to the methods which contain the following Ito integrals

$$I_r = h^{1/2}\xi_r, \quad I_{r0} = h^{3/2}(\eta_r/\sqrt{3}+\xi_r)/2, \quad I_{0r} = hI_r - I_{r0},$$

$$J_r = \int_0^h \vartheta w_r(\vartheta)\,d\vartheta = h^{5/2}(\xi_r/3 + \eta_r/(4\sqrt{3}) + \zeta_r/(12\sqrt{5})),$$

$$I_{r00} = hI_{r0} - J_r, \quad I_{0r0} = 2J_r - hI_{r0}, \quad I_{00r} = h^2 I_r/2 - J_r, \tag{2.1}$$

where ξ_r, η_r, ζ_r are independent normally distributed $N(0,1)$ random variables with zero mean and unit standard deviation. The used random variables (Ito integrals) of all the proposed methods are simulated at each step according to the formulas (2.1).

In this section we restrict ourselves to the set of most common and, in our opinion, useful methods and illustrate the proposed approach to numerical solution of a stochastic system with small noise. A lot of other methods can be derived. First, by adding or omitting some terms one can obtain methods that are similar to the ones given below but have other mean-square errors, for instance, $O(h^5 + \cdots)$, $O(h^6 + \cdots)$. Second, it is possible to derive other types of methods, for instance, implicit Runge–Kutta methods. The presented methods are obtained using the arguments of Sect. 3.1 and further expansion of $X(t+h)$. Their detailed derivation is omitted.

3.2.1 Taylor-type numerical methods

Method $O(h + \cdots)$. The simplest numerical method is the Euler one:

$$X_{k+1} = X_k + \varepsilon \sum_{r=1}^q (\sigma_r\, I_r)_k + a_k h, \tag{2.2}$$

$$E\rho = O(h^2), \quad (E\rho^2)^{1/2} = O(h^2 + \varepsilon^2 h),$$
$$(ER^2)^{1/2} = O(h + \varepsilon^2 h^{1/2}).$$

Methods $O(h^2 + \cdots)$ and $O(h^3 + \cdots)$. These methods are based on the one-step approximations which have been derived in Sect. 3.1. The mean-square error of the method (1.3) is equal to $O(h^2 + \varepsilon^2 h^{1/2})$. The mean-square error of the method (1.10) is estimated by $O(h^2 + \varepsilon h + \varepsilon^2 h^{1/2})$. The method based on the one-step approximation (1.6) has the error $(ER^2)^{1/2} = O(h^3 + \varepsilon^2 h^{1/2})$.

Methods $O(h^4 + \cdots)$. The following method is obtained:

$$X_{k+1} = X_k + \varepsilon \sum_{r=1}^{q}(\sigma_r I_r)_k + a_k h + \varepsilon \sum_{r=1}^{q}(L_1 \sigma_r I_{0r})_k + \varepsilon \sum_{r=1}^{q}(\Lambda_r a I_{r0})_k$$

$$+ L_1 a_k h^2/2 + \varepsilon \sum_{r=1}^{q}(L_1^2 \sigma_r I_{00r})_k + \varepsilon \sum_{r=1}^{q}(L_1 \Lambda_r a I_{0r0})_k$$

$$+ \varepsilon \sum_{r=1}^{q}(\Lambda_r L_1 a I_{r00})_k + L_1^2 a_k h^3/6 + L_1^3 a_k h^4/24, \qquad (2.3)$$

$$E\rho = O(h^5 + \varepsilon^2 h^2), \quad (E\rho^2)^{1/2} = O(h^5 + \varepsilon^2 h),$$
$$(ER^2)^{1/2} = O(h^4 + \varepsilon^2 h^{1/2}).$$

In some cases the derived methods may be improved due to special properties of a concrete system. For instance, let us consider the commutative case, i.e., when $\Lambda_i \sigma_r = \Lambda_r \sigma_i$, or a system with one noise. For these systems we obtain

$$X_{k+1} = A(t_k, X_k, h; (\xi, \eta, \zeta)_k) + \varepsilon^2 \sum_{i=1}^{q-1} \sum_{r=i+1}^{q}(\Lambda_i \sigma_r I_i I_r)_k$$

$$+ \varepsilon^2 \sum_{i=1}^{q}(\Lambda_i \sigma_i (I_i^2 - h)/2)_k + \varepsilon^2 L_2 a_k h^2/2, \qquad (2.4)$$

$$E\rho = O(h^5 + \varepsilon^2 h^3), \quad (E\rho^2)^{1/2} = O(h^5 + \varepsilon^2 h^2 + \varepsilon^3 h^{3/2}),$$
$$(ER^2)^{1/2} = O(h^4 + \varepsilon^2 h^{3/2} + \varepsilon^3 h),$$

where $A(t_k, X_k, h; (\xi, \eta, \zeta)_k)$ is equal to the right-hand side of (2.3).

Note that for the system with one noise the term $\varepsilon^2 \sum_{i=1}^{q-1}\sum_{r=i+1}^{q} (\Lambda_i \sigma_r I_i I_r)_k$ is neglected. One can see that the error of the method (2.4) is smaller than the error of the scheme (2.3). Moreover, the mean-square order of the method (2.4) is equal to one, while the mean-square order of the method (2.3) is equal to one-half.

3.2.2 Runge–Kutta methods

To reduce calculations of derivatives in the methods of Sect. 3.2.1, we propose Runge–Kutta schemes (in fact, they are Runge–Kutta-type methods because they need calculation of some derivatives).

3.2 Some concrete mean-square methods for systems with small noise   181

Method $O(h^2 + \cdots)$. The following method is obtained:

$$X_{k+1} = X_k + \varepsilon \sum_{r=1}^{q} (\sigma_r I_r)_k$$

$$+ \left[a(t_k + h, X_k + \varepsilon \sum_{r=1}^{q} (\sigma_r I_r)_k + a_k h) + a_k \right] h/2$$

$$+ \varepsilon \sum_{r=1}^{q} (L_1 \sigma_r I_{0r})_k + \varepsilon \sum_{r=1}^{q} [\Lambda_r a(I_{r0} - I_r h/2)]_k, \qquad (2.5)$$

$$E\rho = O(h^3), \quad (E\rho^2)^{1/2} = O(h^3 + \varepsilon^2 h),$$
$$(ER^2)^{1/2} = O(h^2 + \varepsilon^2 h^{1/2}).$$

Method $O(h^3 + \cdots)$. The following method is obtained:

$$X_{k+1} = X_k + (k_1 + 4k_2 + k_3)/6 + \varepsilon \sum_{r=1}^{q} (\sigma_r I_r)_k$$

$$+ \varepsilon \sum_{r=1}^{q} (L_1 \sigma_r I_{0r})_k + \varepsilon \sum_{r=1}^{q} (\Lambda_r a I_{r0})_k, \qquad (2.6)$$

$$(ER^2)^{1/2} = O(h^3 + \varepsilon^2 h^{1/2}),$$

where

$$k_1 = ha(t_k, X_k), \quad k_2 = ha(t_k + h/2, X_k + k_1/2),$$
$$k_3 = ha(t_{k+1}, X_k - k_1 + 2k_2).$$

Methods $O(h^4 + \cdots)$. The following method is obtained:

$$X_{k+1} = X_k + (k_1 + 2k_2 + 2k_3 + k_4)/6 + \varepsilon \sum_{r=1}^{q} (\sigma_r I_r)_k$$

$$+ \varepsilon \sum_{r=1}^{q} (L_1 \sigma_r I_{0r})_k + \varepsilon \sum_{r=1}^{q} (\Lambda_r a I_{r0})_k + \varepsilon \sum_{r=1}^{q} (L_1^2 \sigma_r I_{00r})_k$$

$$+ \varepsilon \sum_{r=1}^{q} (L_1 \Lambda_r a I_{0r0})_k + \varepsilon \sum_{r=1}^{q} (\Lambda_r L_1 a I_{r00})_k, \qquad (2.7)$$

$$(ER^2)^{1/2} = O(h^4 + \varepsilon^2 h^{1/2}),$$

where

$$k_1 = ha(t_k, X_k), \quad k_2 = ha(t_k + h/2, X_k + k_1/2),$$
$$k_3 = ha(t_k + h/2, X_k + k_2/2), \quad k_4 = ha(t_{k+1}, X_k + k_3).$$

In the commutative case the method (2.7) can be improved as in Sect. 3.2.1. The simpler method

$$X_{k+1} = X_k + (k_1 + 2k_2 + 2k_3 + k_4)/6 + \varepsilon \sum_{r=1}^{q} (\sigma_r I_r)_k, \qquad (2.8)$$

where k_i are calculated as in (2.7), has the larger error in comparison with (2.7):

$$(ER^2)^{1/2} = O(h^4 + \varepsilon h + \varepsilon^2 h^{1/2}).$$

3.2.3 Implicit methods

Methods $O(h + \cdots)$. The one-parameter family of implicit Euler schemes has the form

$$X_{k+1} = X_k + \varepsilon \sum_{r=1}^{q} (\sigma_r I_r)_k + \alpha h a_k + (1-\alpha) h a_{k+1}, \qquad (2.9)$$

$$0 \leq \alpha \leq 1, \quad (ER^2)^{1/2} = O(h + \varepsilon^2 h^{1/2}).$$

Methods $O(h^2 + \cdots)$. The two-parameter family of implicit schemes (1.3.7) applied to (0.2) has the form

$$X_{k+1} = X_k + \varepsilon \sum_{r=1}^{q} (\sigma_r I_r)_k + \alpha h a_k + (1-\alpha) h a_{k+1}$$

$$+ \varepsilon \sum_{r=1}^{q} (\Lambda_r a(I_{r0} - (1-\alpha) I_r h))_k + \varepsilon \sum_{r=1}^{q} (L_1 \sigma_r I_{0r})_k$$

$$+ \beta(2\alpha - 1) L_1 a_k h^2/2 + (1-\beta)(2\alpha - 1) L_1 a_{k+1} h^2/2, \qquad (2.10)$$

$$0 \leq \alpha \leq 1, \ 0 \leq \beta \leq 1,$$

$$E\rho = O(h^3 + \varepsilon^2 h^2), \quad (E\rho^2)^{1/2} = O(h^3 + \varepsilon^2 h),$$

$$(ER^2)^{1/2} = O(h^2 + \varepsilon^2 h^{1/2}).$$

If $\alpha = 1/2$, we obtain the trapezoidal method which is the simplest of the family (2.10):

$$X_{k+1} = X_k + \varepsilon \sum_{r=1}^{q} (\sigma_r I_r)_k + h(a_k + a_{k+1})/2$$

$$+ \varepsilon \sum_{r=1}^{q} (L_1 \sigma_r I_{0r})_k + \varepsilon \sum_{r=1}^{q} (\Lambda_r a(I_{r0} - I_r h/2))_k, \qquad (2.11)$$

$$(ER^2)^{1/2} = O(h^2 + \varepsilon^2 h^{1/2}).$$

In the commutative case or in the case of one noise the methods (2.10)-(2.11) can be improved as the method (2.3) in Sect. 3.2.1.

3.2.4 Stratonovich SDEs with small noise

For some physical applications Stratonovich interpretation of a stochastic system is preferable. The stochastic system in Stratonovich sense

$$dX = a(t, X)\, dt + \varepsilon \sum_{r=1}^{q} \sigma_r(t, X) \circ dw_r\,, \quad X(t_0) = X_0\,, \tag{2.12}$$

is equivalent to the following system of the Ito SDEs

$$dX = [a(t, X) + \frac{\varepsilon^2}{2} \sum_{r=1}^{q} \frac{\partial \sigma_r}{\partial x}(t, X)\, \sigma_r(t, X)]\, dt + \varepsilon \sum_{r=1}^{q} \sigma_r(t, X)\, dw_r\,, \tag{2.13}$$

$$X(t_0) = X_0\,.$$

In the general case ($\varepsilon = 1$) numerical methods constructed for the Ito system are easily rewritten for the Stratonovich system by adding the term $\frac{1}{2}\sum_{r=1}^{q}(\partial \sigma_r/\partial x)\, \sigma_r$ to the drift (see Sect. 1.1.3). However, in the case of small noise the additional term is multiplied by small factor ε^2. So, the Stratonovich system with small noise (2.12) is distinguished from the Ito system $dX = a(t,X)dt + \varepsilon \sum_{r=1}^{q} \sigma_r(t,X)dw_r$ by the small component in the drift, and constructing a numerical method for the system (2.13), one must take the magnitude of the additional term into account.

Most of the methods for the system (2.12) are obtained from methods for the Ito system (0.2) by adding the term $\frac{\varepsilon^2}{2}\sum_{r=1}^{q}(\partial \sigma_r/\partial x)\, \sigma_r\, h$. Namely, for the Stratonovich system (2.12) the methods (1.3), (1.6), (1.10), (2.2), (2.3), (2.5), (2.6), (2.7), (2.8) acquire the form

$$X_{k+1} = A(t_k, X_k, h; (\xi, \eta, \zeta)_k) + \frac{\varepsilon^2}{2} \sum_{r=1}^{q} (\frac{\partial \sigma_r}{\partial x}\sigma_r)_k\, h\,, \tag{2.14}$$

where the expressions $A(t_k, X_k, h; (\xi, \eta, \zeta)_k)$ are calculated according to the same rules as the right-hand sides of the corresponding methods for Ito systems, and the corresponding errors have the same order of smallness.

In the commutative case we obtain

$$X_{k+1} = A(t_k, X_k, h; (\xi, \eta, \zeta)_k) + \frac{\varepsilon^2}{2} \sum_{r=1}^{q} (\frac{\partial \sigma_r}{\partial x}\sigma_r)_k\, h$$

$$+ \varepsilon^2 \sum_{i=1}^{q-1} \sum_{r=i+1}^{q} (\Lambda_i \sigma_r I_i I_r)_k + \varepsilon^2 \sum_{i=1}^{q} \left[\Lambda_i \sigma_i (I_i^2 - h)/2\right]_k$$

$$+ \varepsilon^2 L_1 \left[\sum_{r=1}^{q} \frac{\partial \sigma_r}{\partial x}\sigma_r\right]_k h^2/4 + \varepsilon^2 (\tilde{L}_2 a)_k\, h^2/2\,, \tag{2.15}$$

$$(ER^2)^{1/2} = O(h^4 + \varepsilon^2 h^{3/2} + \varepsilon^3 h),$$

where $A(t_k, X_k, h; (\xi, \eta, \zeta)_k)$ is the right-hand side of (2.3) and

$$\tilde{L}_2 = \frac{1}{2} \sum_{r=1}^{q} \left(\sigma_r, \frac{\partial}{\partial x}\right)^2 + \frac{1}{2} \sum_{r=1}^{q} \left(\frac{\partial \sigma_r}{\partial x} \sigma_r, \frac{\partial}{\partial x}\right).$$

Analogously, the method for the Stratonovich system, which corresponds to the method (2.7) for the Ito system, can be improved in the commutative case.

The one-parameter family of implicit methods for the Stratonovich system has the form

$$X_{k+1} = A(t_k, X_k, h; (\xi)_k)$$
$$+ \frac{\varepsilon^2}{2} \sum_{r=1}^{q} \left[\alpha \left(\frac{\partial \sigma_r}{\partial x} \sigma_r\right)_k + (1-\alpha)\left(\frac{\partial \sigma_r}{\partial x} \sigma_r\right)_{k+1}\right] h, \quad (2.16)$$

$$(ER^2)^{1/2} = O(h + \varepsilon^2 h^{1/2}),$$

where $A(t_k, X_k, h; (\xi)_k)$ is the right-hand side of (2.9).

The two-parameter family of implicit methods for the Stratonovich system has the form

$$X_{k+1} = A(t_k, X_k, h; (\xi, \eta)_k)$$
$$+ \frac{\varepsilon^2}{2} \sum_{r=1}^{q} \left[\alpha \left(\frac{\partial \sigma_r}{\partial x} \sigma_r\right)_k + (1-\alpha)\left(\frac{\partial \sigma_r}{\partial x} \sigma_r\right)_{k+1}\right] h, \quad (2.17)$$

$$(ER^2)^{1/2} = O(h^2 + \varepsilon^2 h^{1/2}),$$

where $A(t_k, X_k, h; (\xi, \eta)_k)$ is the right-hand side of (2.10).

3.2.5 Mean-square methods for systems with small additive noise

One of the important particular cases of the system (0.2) is the system with additive noise:

$$dX = a(t, X) dt + \varepsilon \sum_{r=1}^{q} \sigma_r(t) dw_r. \quad (2.18)$$

Note that in this case the Stratonovich system coincides with the Ito system.

For the system (2.18) we obtain the Taylor-type and Runge–Kutta methods with the errors $O(h^2 + \varepsilon h)$, $O(h^2 + \varepsilon^2 h)$, $O(h^2 + \varepsilon^2 h^{3/2})$, $O(h^3 + \varepsilon^2 h^{3/2})$, $O(h^4 + \varepsilon^2 h^{3/2})$ which are similar to the methods of Sects. 3.2.1-3.2.3. We also obtain the implicit schemes which follow from the schemes of Sect. 3.2.3. Herein we restrict ourselves to the methods with errors like $O(h^4 + \cdots)$ which, from our point of view, are the most interesting.

The Taylor-type method with the error $O(h^4 + \varepsilon^2 h^{3/2})$ is

$$X_{k+1} = X_k + \varepsilon \sum_{r=1}^{q} (\sigma_r I_r)_k + a_k h + \varepsilon \sum_{r=1}^{q} \left(\frac{d\sigma_r}{dt} I_{0r}\right)_k$$

$$+\varepsilon \sum_{r=1}^{q} (\Lambda_r a I_{r0})_k + (L_1 + \varepsilon^2 L_2) a_k h^2 / 2 + \varepsilon \sum_{r=1}^{q} \left(\frac{d^2\sigma_r}{dt^2} I_{00r}\right)_k$$

$$+\varepsilon \sum_{r=1}^{q} (L_1 \Lambda_r a I_{0r0})_k + \varepsilon \sum_{r=1}^{q} (\Lambda_r L_1 a I_{r00})_k$$

$$+L_1^2 a_k h^3 / 6 + L_1^3 a_k h^4 / 24, \tag{2.19}$$

$$E\rho = O(h^5 + \varepsilon^2 h^3), \quad (E\rho^2)^{1/2} = O(h^5 + \varepsilon^2 h^2),$$
$$(ER^2)^{1/2} = O(h^4 + \varepsilon^2 h^{3/2}).$$

The Runge–Kutta method with the error $O(h^4 + \varepsilon^2 h^{3/2})$ is

$$X_{k+1} = X_k + (k_1 + 2k_2 + 2k_3 + k_4)/6 + \varepsilon \sum_{r=1}^{q} (\sigma_r I_r)_k + \varepsilon \sum_{r=1}^{q} \left(\frac{d\sigma_r}{dt} I_{0r}\right)_k$$

$$+\varepsilon \sum_{r=1}^{q} (\Lambda_r a I_{r0})_k + \varepsilon^2 L_2 a_k h^2 / 2 + \varepsilon \sum_{r=1}^{q} \left(\frac{d^2\sigma_r}{dt^2} I_{00r}\right)_k$$

$$+\varepsilon \sum_{r=1}^{q} (L_1 \Lambda_r a I_{0r0})_k + \varepsilon \sum_{r=1}^{q} (\Lambda_r L_1 a I_{r00})_k, \tag{2.20}$$

$$(ER^2)^{1/2} = O(h^4 + \varepsilon^2 h^{3/2}),$$

where

$$k_1 = ha(t_k, X_k), \quad k_2 = ha(t_k + h/2, X_k + k_1/2),$$
$$k_3 = ha(t_k + h/2, X_k + k_2/2), \quad k_4 = ha(t_{k+1}, X_k + k_3).$$

It is possible to obtain the simpler method which coincides with the scheme (2.8) but its mean-square error in the case of additive noise is equal to $(ER^2)^{1/2} = O(h^4 + \varepsilon h)$.

3.3 Numerical tests of mean-square methods

3.3.1 Simulation of Lyapunov exponent of a linear system with small noise

It is known [6, 124] that one can investigate stability of a dynamical stochastic system by Lyapunov exponents. The negativeness of upper Lyapunov exponents is an indication of system stability. It is usually impossible to derive analytical expressions for Lyapunov exponents. In this case numerical approaches are useful. For the first time an algorithm of numerical computation of Lyapunov exponents was proposed in [284]. The algorithm is based on weak schemes (see also Sect. 3.8).

Here we calculate Lyapunov exponent as a convenient example to illustrate the effectiveness (in comparison with ordinary mean-square schemes) of the proposed methods. Although the weak schemes are usually more efficient than mean-square ones, our approach is interesting in itself because we find the exponent together with the real trajectory.

Let us consider the following two-dimensional linear Ito stochastic system

$$dX = AXdt + \varepsilon \sum_{r=1}^{q} B_r X \, dw_r, \tag{3.1}$$

where X is a two-dimensional vector, A and B_r are constant 2×2-matrices, $w_r(t)$ are independent standard Wiener processes, $\varepsilon > 0$ is a small parameter. In ergodic case the unique Lyapunov exponent λ of the system (3.1) exists [124] and

$$\lambda = \lim_{t \to \infty} \frac{1}{t} E(ln|X(t)|) = \lim_{t \to \infty} \frac{1}{t} ln|X(t)|, \tag{3.2}$$

where $X(t)$, $t \geq 0$, is a non-trivial solution of the system (3.1). The last equality of (3.2) holds with probability one. A non-trivial solution of the system (3.1) is asymptotically stable with probability one if and only if the Lyapunov exponent λ is negative [124].

In [6,8] the expansion of Lyapunov exponent of the system (3.1) in powers of small parameter ε was obtained. In the case of

$$A = \begin{bmatrix} a & c \\ -c & a \end{bmatrix}, \quad B_r = \begin{bmatrix} b_r & d_r \\ -d_r & b_r \end{bmatrix}, \quad r = 1, \ldots, q, \tag{3.3}$$

the Lyapunov exponent of the system (3.1) is exactly equal to [8]

$$\lambda = a + \frac{\varepsilon^2}{2} \sum_{r=1}^{q} [(d_r)^2 - (b_r)^2]. \tag{3.4}$$

To test the mean-square methods proposed in this chapter, we choose the case (3.3) of the system (3.1) with two independent noises. We calculate the function $\bar{\lambda}(t)$:

$$\bar{\lambda}(t) = \frac{1}{t} \ln|\bar{X}(t)| \approx \frac{1}{t} \ln|X(t)| \tag{3.5}$$

which in the limit of large time ($t \to \infty$) tends to an approximation of the Lyapunov exponent λ. The approximation $\bar{X}(t)$ of the exact solution $X(t)$ of the system (3.1) is simulated by three mean-square schemes: (i) the first-order method (1.0.5) with the error $O(h)$ which in our case is efficient as to

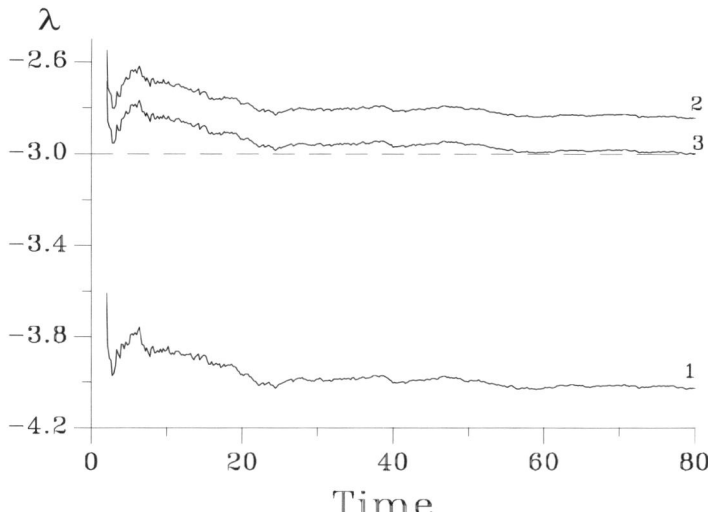

Fig. 3.1. Lyapunov exponent. Time dependence of the function $\bar{\lambda}(t)$, $t \geq 2$, for $a = -3$, $c = 1$, $b_1 = b_2 = 1$, $d_1 = 1$, $d_2 = -1$, $\varepsilon = 0.1$, $X^1(0) = 0$, $X^2(0) = 1$ and time step $h = 0.3$. The solution of the system (3.1)-(3.3) is approximated by (1) the method with the error $O(h)$, (2) the Runge–Kutta method with the error $O(h^2 + \varepsilon h + \varepsilon^2 h^{1/2})$, and (3) the Runge–Kutta method with the error $O(h^4 + \varepsilon h + \varepsilon^2 h^{1/2})$. Dashed line is the exact value of the Lyapunov exponent λ ($\lambda = -3$).

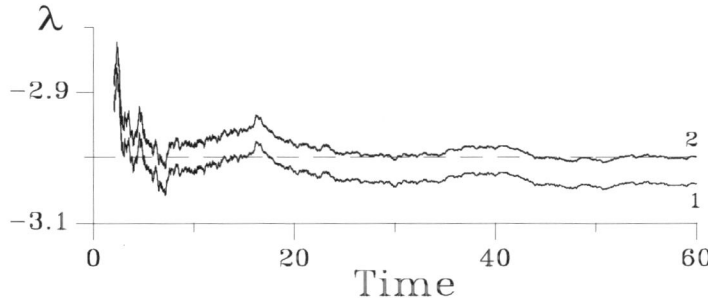

Fig. 3.2. Lyapunov exponent. Time dependence of the function $\bar{\lambda}(t)$, $t \geq 2$, for $h = 0.01$, the other parameters are the same as in Fig. 3.1. The solution of the system (3.1)-(3.3) is approximated by (1) the method with the error $O(h)$ and (2) the Runge–Kutta methods with the errors $O(h^2 + \varepsilon h + \varepsilon^2 h^{1/2})$ and $O(h^4 + \varepsilon h + \varepsilon^2 h^{1/2})$. Dashed line is the exact value of the Lyapunov exponent λ ($\lambda = -3$).

simulation of the used random variables due to commutativity of the matrices B_r, $r = 1, \ldots, q$, (ii) the simplified version of the Runge–Kutta scheme (2.5) with the error $O(h^2 + \varepsilon h + \varepsilon^2 h^{1/2})$, and (iii) the Runge–Kutta scheme (2.8) with the error $O(h^4 + \varepsilon h + \varepsilon^2 h^{1/2})$ (see Figs. 3.1 and 3.2).

3.3.2 Stochastic model of a laser

Our second example is devoted to a stochastic model of laser [88, 248] which can be written as the following Stratonovich system

$$dX^1 = (\alpha_0 X^1 - \beta_0 X^2 - (AX^1 - BX^2)XX^*) dt$$
$$+\varepsilon \left[\sum_{i=1}^{2} (\alpha_i X^1 - \beta_i X^2) \circ dw_i + \sigma\, dw_3 \right]$$
$$dX^2 = (\beta_0 X^1 + \alpha_0 X^2 - (BX^1 + AX^2)XX^*) dt$$
$$+\varepsilon \left[\sum_{i=1}^{2} (\beta_i X^1 + \alpha_i X^2) \circ dw_i + \sigma\, dw_4 \right], \qquad (3.6)$$

where

$$X = X^1 + i X^2, \quad X^* = X^1 - i X^2.$$

For $\varepsilon = 0$ the system (3.6) becomes deterministic. In the case of $\alpha_0/A > 0$ it has asymptotically stable limit cycle $(X^1)^2 + (X^2)^2 = \alpha_0/A$. The radius $\rho = |X|$ under $\varepsilon = 0$ satisfies the equation

$$d\rho/dt = \rho(\alpha_0 - A\rho^2)$$

and does not depend on the detuning parameters β_0 and B. The value ρ^2 for $\varepsilon \neq 0$ satisfies the Stratonovich equation

$$d\rho^2 = 2\rho^2(\alpha_0 - A\rho^2)dt + 2\varepsilon\rho^2(\alpha_1 \circ dw_1 + \alpha_2 \circ dw_2)$$
$$+2\varepsilon\sigma(X_1 \circ dw_3 + X_2 \circ dw_4)$$

and also does not depend on β_0 and B. But the difference equations, which are the result of applying numerical methods to the system (3.6), essentially depend not only on the choice of a scheme and time step but also on the detuning parameters, and growing of $|\beta_0 - B|$ leads to vanishing of stable cycle. Therefore, to solve the system (3.6), one must use high-order schemes or choose sufficiently small time step. Since the system (3.6) contains multiplicative noise and does not belong to the class of systems with commutative noise, the Euler method is the highest order scheme among known mean-square methods with easily simulated random variables (see Chap. 1). The Euler method has the mean-square error $O(h + \varepsilon^2 h^{1/2})$ and in the case of large $|\beta_0 - B|$ too small step h is required. On the other hand, for instance, the method with the mean-square error $O(h^4 + \varepsilon h + \varepsilon^2 h^{1/2})$ allows us to obtain sufficiently accurate approximations of solutions of the system (3.6) and, particularly, to simulate phase trajectories.

The radius $\rho = |X_k|$ of a typical trajectory is plotted in Figs. 3.3 and 3.4. In Fig. 3.3 the radius ρ is calculated with the time step $h = 0.005$ by the Euler scheme and by the Runge–Kutta scheme with the error $O(h^4 + \varepsilon h + \varepsilon^2 h^{1/2})$ which corresponds to the method (2.8) for the Ito system. In this case both

3.3 Numerical tests of mean-square methods 189

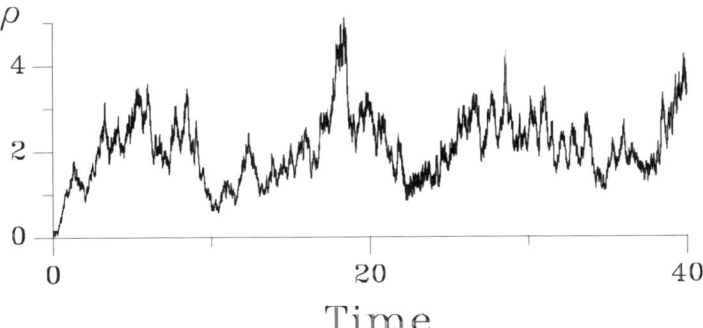

Fig. 3.3. Stochastic model of a laser. Time dependence of the radius $\rho = |X_k|$ for $\alpha_0 = 0.5$, $\beta_0 = 1$, $A = 0.1$, $B = 0.4$, $\varepsilon = 0.3$, $\alpha_i = \beta_i = \sigma = 1$, $i = 1, 2$, $X^1(0) = X^2(0) = 0$, and time step $h = 0.005$. The solution X_k of the system (3.6) is approximated by the Euler method and by the Runge–Kutta method with the error $O(h^4 + \varepsilon h + \varepsilon^2 h^{1/2})$.

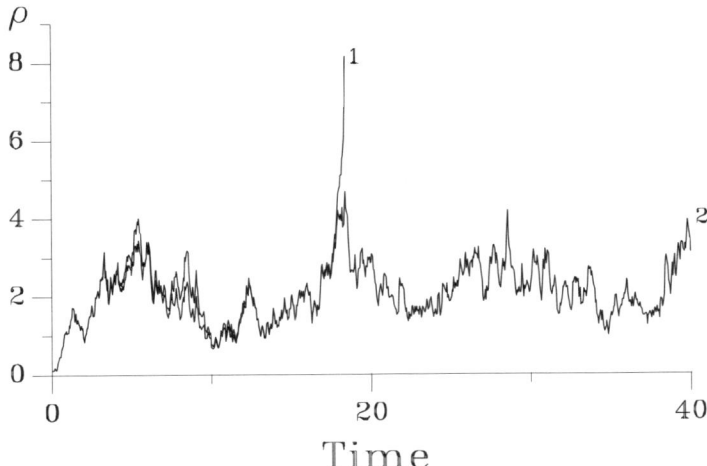

Fig. 3.4. Stochastic model of a laser. Time dependence of the radius $\rho = |X_k|$ for time step $h = 0.05$, the other parameters are the same as in Fig. 3.3. The solution X_k of the system (3.6) is approximated by (1) the Euler method and (2) the Runge–Kutta method with the error $O(h^4 + \varepsilon h + \varepsilon^2 h^{1/2})$.

methods give the same results, and the trajectories plotted in Fig. 3.3 can be considered as exact. As seen in Fig. 3.4, if one chooses the larger time step ($h = 0.05$), the Runge–Kutta scheme gives quite well results (compare with Fig. 3.3) but the Euler method becomes unstable. In all cases we use the same sample paths for the Wiener processes.

Note that the Runge–Kutta method (2.8) and the corresponding method for the Stratonovich system can be improved up to $(ER^2)^{1/2} = O(h^4 + \varepsilon^2 h^{1/2})$ (see the method (2.7)).

3.4 The main theorem on error estimation and general approach to construction of weak methods

We construct weak schemes for (0.1) on the basis of the proposed mean-square methods. See Chap. 2 for definitions and main notions (in particular, Definition 2.1.1 of the class **F**) related to numerical integration of SDEs in the weak sense.

Theorem 4.1. *Assume that the following conditions hold.*

(1) *The coefficients of the system* (0.1) *are continuous and satisfy a Lipschitz condition with respect to* $x \in \mathbf{R}^d$, *they and their partial derivatives up to a sufficiently high order belong to the class* **F**;

(2) *The error of a one-step approximation* $\bar{X}_{t,x}(t+h)$ *of the exact solution* $X_{t,x}(t+h)$ *of the system* (0.1) *with initial condition* $X(t) = \bar{X}(t) = x$ *is estimated by*

$$|Ef(X_{t,x}(t+h)) - Ef(\bar{X}_{t,x}(t+h))|$$
$$\leq K(x)[h^{p+1} + \sum_{l \in S} h^{l+1} \varepsilon^{J(l)}], \quad K(x) \in \mathbf{F}, \quad (4.1)$$

where the function $f(x)$ *and its partial derivatives up to a sufficiently high order belong to the class* **F**, S *is a subset of the positive integers* $\{1, 2, \ldots, p\}$, J *is a decreasing function on* S *with natural values;*

(3) *For a sufficiently large integer* m *the moments* $E|\bar{X}_k|^m$ *exist and are uniformly bounded with respect to* N, $k = 0, 1, \ldots, N$, *and* $0 \leq \varepsilon \leq \varepsilon_0$.

Then for any N *and* $k = 0, 1, \ldots, N$

$$|Ef(X_{t_0, X_0}(t_k)) - Ef(\bar{X}_{t_0, X_0}(t_k))| \leq K[h^p + \sum_{l \in S} h^l \varepsilon^{J(l)}], \quad (4.2)$$

where the constant K *does not depend on* h, ε, *and* k.

The proof of Theorem 4.1 differs only little from the proof of Theorem 2.2.1 and is therefore omitted.

According to Theorem 4.1, to estimate the global error of a method we need properties of the corresponding one-step approximation, i.e., to prove an error for a weak method we need the estimate (4.1). By using the Taylor expansion of the function f, it is possible to obtain the estimate (4.1), provided the inequalities (cf. Theorem 2.2.1):

3.4 The main theorem on error estimation and general approach

$$|E\prod_{j=1}^{s}\Delta^{i_j} - E\prod_{j=1}^{s}\bar{\Delta}^{i_j}| \leq K(x)(h^{p+1} + \sum_{l\in S}h^{l+1}\varepsilon^{J(l)}), \quad s = 1,\ldots,\bar{s}-1,$$

$$\Delta^{i_j} = X^{i_j}(t+h) - x^{i_j}, \quad \bar{\Delta}^{i_j} = \bar{X}^{i_j}(t+h) - x^{i_j},$$
$$X(t) = \bar{X}(t) = x, \quad i_j = 1,\ldots,d, \tag{4.3}$$

$$E\prod_{j=1}^{\bar{s}}|\bar{\Delta}^{i_j}| \leq K(x)(h^{p+1} + \sum_{l\in S}h^{l+1}\varepsilon^{J(l)}), \quad i_j = 1,\ldots,d, \tag{4.4}$$

$$E\prod_{j=1}^{\bar{s}}|\Delta^{i_j}| \leq K(x)(h^{p+1} + \sum_{l\in S}h^{l+1}\varepsilon^{J(l)}), \quad i_j = 1,\ldots,d. \tag{4.5}$$

hold. The number $\bar{s}$ in (4.3)-(4.5) must be such that $h^k(\varepsilon h^{1/2})^{\bar{s}-k} = O(h^{p+1} + \sum_{l\in S}h^{l+1}\varepsilon^{J(l)})$ for even $\bar{s} - k$ where $0 \leq k \leq \bar{s}$. The last assertion follows from an analysis of the Taylor expansion of $Ef(X_{t,x}(t+h))$ in powers of h and ε. We underline that $\bar{s}$ is greater than or equal to $(2+2\min\{l \in S\})$ and, as a rule, less than $2p + 2$.

In the case of the system (0.1) the one-step approximation (2.1.10) can be written in the form

$$\tilde{X}(t+h) = x + \varepsilon\sum_{r=1}^{q}\sigma_r I_r + h(a + \varepsilon^2 b) + \varepsilon^2\sum_{i,r=1}^{q}\Lambda_i\sigma_r I_{ir}$$
$$+ \varepsilon\sum_{r=1}^{q}(L_1 + \varepsilon^2 L_2)\sigma_r I_{0r} + \varepsilon\sum_{r=1}^{q}\Lambda_r(a + \varepsilon^2 b)I_{r0}$$
$$+ h^2(L_1 + \varepsilon^2 L_2)(a + \varepsilon^2 b)/2, \tag{4.6}$$

$$X(t+h) = \tilde{X}(t+h) + \tilde{\rho}. \tag{4.7}$$

The coefficients σ_r, a, b, $\Lambda_i\sigma_r$, etc. in (4.6) are calculated at the point (t,x), and $\tilde{\rho}$ in (4.7) is the remainder.

The weak method based on (4.6) has global error $O(h^2)$ (cf. (2.2.18) and Theorem 2.2.3), and for the system (0.1) it is written as

$$X_{k+1} = X_k + \varepsilon h^{1/2}\sum_{r=1}^{q}(\sigma_r\xi_r)_k + h(a + \varepsilon^2 b)_k + \varepsilon^2 h\sum_{i,r=1}^{q}(\Lambda_i\sigma_r\xi_{ir})_k$$
$$+ \varepsilon h^{3/2}\sum_{r=1}^{q}\left[(L_1 + \varepsilon^2 L_2)\sigma_r\xi_r\right]_k/2 + \varepsilon h^{3/2}\sum_{r=1}^{q}\left[\Lambda_r(a + \varepsilon^2 b)\xi_r\right]_k/2$$
$$+ h^2\left[(L_1 + \varepsilon^2 L_2)(a + \varepsilon^2 b)\right]_k/2, \tag{4.8}$$

where the random variables are simulated according to

$$\xi_{ir} = (\xi_i\xi_r - \gamma_{ir}\zeta_i\zeta_r)/2, \quad \gamma_{ir} = \begin{cases} -1, & i < j \\ 1, & i \geq j \end{cases},$$
$$P(\xi = 0) = 2/3, \quad P(\xi = \pm\sqrt{3}) = 1/6, \quad P(\zeta = \pm 1) = 1/2, \tag{4.9}$$

ξ_r and ζ_i are mutually independent.

192   3 Numerical methods for SDEs with small noise

A higher-order method based on a one-step approximation of weak order 4 would be too complicated. However, as will be shown below, a small modification of the one-step approximation (4.6) leads to an efficient method with the local error $O(h^4 + \varepsilon^2 h^3)$.

The remainder $\tilde{\rho}$ in (4.7) contains terms with factors $\varepsilon^3 h^{3/2}$, $\varepsilon^2 h^2$, $\varepsilon^4 h^2$, $\varepsilon h^{5/2}$, $\varepsilon^3 h^{5/2}$, $\varepsilon^5 h^{5/2}$. Contribution of these terms to the local error of the one-step weak approximation is not worse than $O(\varepsilon^2 h^3)$, which follows from properties of Ito integrals (see Lemmas 1.2.1, 2.1.3 and 2.1.5). Additionally, $\tilde{\rho}$ contains the term $h^3 L_1^2 a/6$, which yields a contribution to the local error of the one-step weak approximation equal to $O(h^3)$, and terms, for example, $\varepsilon^2 h^3$, h^4, etc., which contribute to the local error not more than $O(\varepsilon^2 h^3)$. It is clear that moving the term $h^3 L_1^2 a/6$ from $\tilde{\rho}$ to $\tilde{X}$ leads to a new one-step approximation which is not essentially more complicated, but is at the same time considerably more accurate than the previous one. Of course, this reasoning requires a strict proof which can be found in [202].

On this way we get the weak method of order $O(h^3 + \varepsilon^2 h^2)$:

$$X_{k+1} = X_k + \varepsilon h^{1/2} \sum_{r=1}^{q} (\sigma_r \xi_r)_k + h(a + \varepsilon^2 b)_k + \varepsilon^2 h \sum_{i,r=1}^{q} (\Lambda_i \sigma_r \xi_{ir})_k$$

$$+ \varepsilon h^{3/2} \sum_{r=1}^{q} \left[(L_1 + \varepsilon^2 L_2)\sigma_r \xi_r\right]_k /2 + \varepsilon h^{3/2} \sum_{r=1}^{q} \left[\Lambda_r(a + \varepsilon^2 b)\xi_r\right]_k /2$$

$$+ h^2 \left[(L_1 + \varepsilon^2 L_2)(a + \varepsilon^2 b)\right]_k /2 + h^3 (L_1^2 a)_k /6 \qquad (4.10)$$

with the random variables $(\xi_r)_k$ and $(\xi_{ir})_k$ simulated at each step according to (4.9). This method is of second order in h just like the standard second-order method (4.8), which provides an error $O(h^2)$. But the term h^2 in the error of the method (4.10) is multiplied by ε^2. That is why the new method has smaller error than the standard scheme (4.8).

Above we shifted a term, which is sufficiently simply to simulate, from the remainder to the method, thereby reducing the error. However, we can also shift some more complicated terms, multiplied by ε^α, from the method to the corresponding remainder. Such a procedure reduces the computational costs (which, of course, is important for applications) while it does not lead to a substantial increase of the error.

For instance, by shifting the complicated (from the computational point of view) terms $\varepsilon^3 h^{3/2} L_2 \sigma_r \xi_r /2$ and $\varepsilon^4 h^2 L_2 b/2$ to the remainder, we obtain a further method to solve the system (0.1). It can be seen that such a method has a global error $O(h^3 + \varepsilon^4 h)$. Moreover, if we additionally transfer the terms $\varepsilon^2 h \Lambda_i \sigma_r \xi_{ir}$ and $\varepsilon^3 h^{3/2} \Lambda_r b \xi_r /2$ to the remainder, it can be proved that we do not loose the accuracy of the corresponding method, with respect to both h and ε. We finally arrive at the method

$$X_{k+1} = X_k + \varepsilon h^{1/2} \sum_{r=1}^{q} (\sigma_r \xi_r)_k + h(a + \varepsilon^2 b)_k$$

$$+ \varepsilon h^{3/2} \sum_{r=1}^{q} (L_1 \sigma_r \xi_r)_k / 2 + \varepsilon h^{3/2} \sum_{r=1}^{q} (\Lambda_r a \xi_r)_k / 2$$

$$+ h^2 \left[(L_1 + \varepsilon^2 L_2) a \right]_k / 2 + \varepsilon^2 h^2 (L_1 b)_k / 2 + h^3 (L_1^2 a)_k / 6, \quad (4.11)$$

the global error of which is $O(h^3 + \varepsilon^4 h)$. It is sufficient for its realization to simulate only q independent random variables ξ_r according to the law $P(\xi = \pm 1) = 1/2$. The time-step order of the method (4.11) is equal to one, i.e., it is lower than the time-step order of the method (4.10) and of the standard method (4.8). Nevertheless, for small ε the error behavior of the method (4.11) is acceptable. If we, for instance, choose a time-step h with $h = C \varepsilon^\alpha$, $0 < \alpha < 4$, the method (4.11) even beats the standard method (4.8) as far as the degree of smallness with respect to ε is concerned. Furthermore, if we choose a time-step $h = C \varepsilon^\alpha$, $0 < \alpha \le 2$, then the method (4.11) is not worse than method (4.10) in the same sense. We want to emphasize that, additionally, the method (4.11) requires fewer calculations of both the number of simulated random variables and the number of arithmetic operations.

Thus, we have briefly explained how to construct weak methods for a system with small noise. Let us stress that to put these new methods on a sound basis one must thoroughly analyze the remainder and apply Theorem 4.1.

3.5 Some concrete weak methods

We aim at constructing weak methods which have small errors and are sufficiently effective with respect to their computational costs. Below we present the methods without detailed derivation like we did in Sect. 3.2.

3.5.1 Taylor-type methods

For the system (0.1) we obtain Taylor-type weak methods with errors $O(h^2 + \varepsilon^2 h)$, $O(h^2 + \varepsilon^4 h)$, $O(h^3 + \varepsilon^2 h)$, $O(h^3 + \varepsilon^4 h)$, $O(h^3 + \varepsilon^2 h^2)$, $O(h^3 + \varepsilon^4 h^2)$, $O(h^4 + \varepsilon^2 h)$, $O(h^4 + \varepsilon^2 h^2 + \varepsilon^4 h)$, $O(h^4 + \varepsilon^4 h)$, $O(h^4 + \varepsilon^2 h^2)$, $O(h^4 + \varepsilon^4 h^2)$. In Sect. 3.4 we have derived weak methods with errors $O(h^3 + \varepsilon^4 h)$ and $O(h^3 + \varepsilon^2 h^2)$. More methods can be derived in the same manner. In this subsection we state several methods with errors $O(h^4 + \cdots)$. Others can be found in the preprint [200]. There one also can find some implicit methods.

By the approach stated above it is possible to derive methods with errors $O(h^5 + \cdots)$, $O(h^6 + \cdots)$, etc. But we do not write them because most popular deterministic schemes have orders not higher than 4. Note that it is also possible to derive methods with errors $O(h^3 + \varepsilon^6 h^2)$, $O(h^4 + \varepsilon^6 h^2)$, $O(h^4 + \varepsilon^\alpha h^3)$, $\alpha = 2, \ldots, 8$, however they require huge computational efforts.

The method with error $R = O(h^4 + \varepsilon^2 h)$ has the form

$$X_{k+1} = X_k + \varepsilon h^{1/2} \sum_{r=1}^{q} (\sigma_r \xi_r)_k + h(a + \varepsilon^2 b)_k$$
$$+ h^2 (L_1 a)_k / 2 + h^3 (L_1^2 a)_k / 6 + h^4 (L_1^3 a)_k / 24, \qquad (5.1)$$

where the random variables ξ_r are distributed as

$$P(\xi = \pm 1) = 1/2. \qquad (5.2)$$

The method with error $R = O(h^4 + \varepsilon^4 h)$ has the form

$$X_{k+1} = X_k + \varepsilon h^{1/2} \sum_{r=1}^{q} (\sigma_r \xi_r)_k + h(a + \varepsilon^2 b)_k$$
$$+ \varepsilon h^{3/2} \sum_{r=1}^{q} [L_1 \sigma_r (\xi_r/2 - \eta_r)]_k + \varepsilon h^{3/2} \sum_{r=1}^{q} [\Lambda_r a(\xi_r/2 + \eta_r)]_k$$
$$+ h^2 \left[L_1 (a + \varepsilon^2 b)\right]_k / 2 + \varepsilon^2 h^2 (L_2 a)_k / 2$$
$$+ \varepsilon h^{5/2} \sum_{r=1}^{q} \left[(L_1^2 \sigma_r + L_1 \Lambda_r a + \Lambda_r L_1 a) \xi_r\right]_k / 6 + h^3 \left[L_1^2 (a + \varepsilon^2 b)\right]_k / 6$$
$$+ \varepsilon^2 h^3 \left[(L_1 L_2 + L_2 L_1) a\right]_k / 6 + h^4 (L_1^3 a)_k / 24, \qquad (5.3)$$

where the random variables ξ_r and η_r are distributed as

$$P(\xi = \pm 1) = 1/2, \; P(\eta = \pm 1/\sqrt{12}) = 1/2. \qquad (5.4)$$

The method with error $R = O(h^4 + \varepsilon^2 h^2)$ has the form

$$X_{k+1} = X_k + \varepsilon h^{1/2} \sum_{r=1}^{q} (\sigma_r \xi_r)_k + h(a + \varepsilon^2 b)_k + \varepsilon^2 h \sum_{i,r=1}^{q} (\Lambda_i \sigma_r \xi_{ir})_k$$
$$+ \varepsilon h^{3/2} \sum_{r=1}^{q} \left[(L_1 + \varepsilon^2 L_2) \sigma_r \xi_r\right]_k / 2 + \varepsilon h^{3/2} \sum_{r=1}^{q} \left[\Lambda_r (a + \varepsilon^2 b) \xi_r\right]_k / 2$$
$$+ h^2 \left[(L_1 + \varepsilon^2 L_2)(a + \varepsilon^2 b)\right]_k / 2 - h^3 (L_1^2 a)_k / 6 + h^4 (L_1^3 a)_k / 24, \; (5.5)$$

where the random variables ξ_r and ξ_{ir} are simulated according to

$$P(\xi = 0) = 2/3, \; P(\xi = \pm \sqrt{3}) = 1/6, \; \xi_{ir} = (\xi_i \xi_r - \gamma_{ir} \zeta_i \zeta_r)/2,$$
$$\gamma_{ir} = \begin{cases} -1, & i < r \\ 1, & i \geq r \end{cases},$$
$$P(\zeta = \pm 1) = 1/2, \qquad (5.6)$$

or

$$P(\xi = 0) = 2/3, \ P(\xi = \pm\sqrt{3}) = 1/6,$$
$$\xi_{ir} = (\xi_i \xi_r - \zeta_{ir})/2, \ \zeta_{ii} = 1, \ \zeta_{ir} = -\zeta_{ri}, \ i \neq r,$$
$$P(\zeta_{ir} = \pm 1) = 1/2, \ i < r. \tag{5.7}$$

The method with error $R = O(h^4 + \varepsilon^4 h^2)$ has the form

$$X_{k+1} = X_k + \varepsilon h^{1/2} \sum_{r=1}^{q} (\sigma_r \xi_r)_k + h(a + \varepsilon^2 b)_k + \varepsilon^2 h \sum_{i,r=1}^{q} (\Lambda_i \sigma_r \xi_{ir})_k$$
$$+ \varepsilon h^{3/2} \sum_{r=1}^{q} \left[(L_1 + \varepsilon^2 L_2) \sigma_r (\xi_r - \mu_r) \right]_k + \varepsilon h^{3/2} \sum_{r=1}^{q} \left[\Lambda_r (a + \varepsilon^2 b) \mu_r \right]_k$$
$$+ h^2 \left[(L_1 + \varepsilon^2 L_2)(a + \varepsilon^2 b) \right]_k / 2$$
$$+ \varepsilon h^{5/2} \sum_{r=1}^{q} \left[(L_1^2 \sigma_r + L_1 \Lambda_r a + \Lambda_r L_1 a) \xi_r \right]_k / 6$$
$$+ h^3 \left[L_1^2 (a + \varepsilon^2 b) \right]_k / 6 + \varepsilon^2 h^3 \left[(L_1 L_2 + L_2 L_1) a \right]_k / 6$$
$$+ h^4 (L_1^3 a)_k / 24, \tag{5.8}$$

where ξ_r, ξ_{ir}, and μ_r are simulated, for example, according to

$$\xi_{ir} = (\xi_i \xi_r - \gamma_{ir} \zeta_i \zeta_r)/2, \ \gamma_{ir} = \begin{cases} -1, \ i < r \\ 1, \ i \geq r \end{cases}, \ P(\zeta = \pm 1) = 1/2,$$
$$P(\xi = 0) = 2/3, \ P(\xi = \pm\sqrt{3}) = 1/6, \ \mu_r = \xi_r/2 + \zeta_r/\sqrt{12}. \tag{5.9}$$

Remark 5.1. Let us discuss how to choose the increment h given ε, i.e., the interdependence of the time increment h and the parameter ε in the methods of this section. We first choose the time increment h to be $h = C\varepsilon^\alpha$. Then the global error of a method can be estimated in powers of the small parameter ε by
$$R = O(\varepsilon^\beta),$$
where
$$\beta = \min \left\{ \alpha p, \min_{l \in S} (\alpha l + J(l)) \right\}.$$

If $h = C\varepsilon^\alpha$, the method (5.8) has $R = O(\varepsilon^{4\alpha} + \varepsilon^{2\alpha+4})$, while the method (5.1) yields $R = O(\varepsilon^{4\alpha} + \varepsilon^{\alpha+2})$. In the case of $0 < \alpha \leq 2/3$, both errors are bounded by $O(\varepsilon^{4\alpha})$, and so both methods have the same order with respect to ε. However, if $\alpha > 2/3$, the method (5.8) has higher order with respect to ε than (5.1) (for instance, if $\alpha = 2$, we have $O(\varepsilon^8)$ for (5.8) and $O(\varepsilon^4)$ for (5.1)). Thus, in the case of a comparatively large time increment h compared to ε (this is of interest mainly if ε is sufficiently small, i.e., when the error estimated by ε^β is not large), complicated methods like (5.8) and sufficiently simple methods like (5.1) have the same order in ε. In such a situation simple methods are usually preferable because of their considerably lower computational costs. But if one wants to reach an error of high order with respect to ε, complicated methods are preferable.

3.5.2 Runge–Kutta methods

Below we consider (i) full (derivative free) Runge–Kutta schemes and (ii) Runge–Kutta schemes without derivatives of the coefficients $a(t,x)$ and $b(t,x)$ but with derivatives of the diffusion coefficients $\sigma_r(t,x)$ (semi-Runge–Kutta schemes) which may be useful in the case of simple functions σ_r.

It is known (see Chap. 2) that in the case of a general system with $\varepsilon \equiv 1$ there are no entirely constructive higher-order Runge–Kutta schemes. For systems with small noise we obtain full Runge–Kutta methods with errors $O(h^2+\varepsilon^2 h)$, $O(h^2+\varepsilon^4 h)$, $O(h^3+\varepsilon^2 h)$, $O(h^3+\varepsilon^4 h)$, $O(h^4+\varepsilon^2 h)$, $O(h^4+\varepsilon^2 h^2 + \varepsilon^4 h)$, and $O(h^4+\varepsilon^4 h)$. For higher orders we have succeeded in constructing semi-Runge–Kutta schemes with errors $O(h^3 + \varepsilon^2 h^2)$, $O(h^3 + \varepsilon^4 h^2)$, $O(h^4 + \varepsilon^2 h^2)$ and $O(h^4 + \varepsilon^4 h^2)$.

In this subsection we state several methods with errors $O(h^2 + \cdots)$ and $O(h^4 + \cdots)$. Other Runge–Kutta methods can be found in the preprint [200].

To construct Runge–Kutta methods for system (0.1), we use deterministic Runge–Kutta methods as a subsidiary tool. To this end, we select specific deterministic schemes which from our point of view are most appropriate. Obviously, it is possible to derive families of stochastic Runge–Kutta methods which are similar to the proposed ones but use different deterministic Runge–Kutta schemes.

Methods $O(h^2 + \cdots)$. The method with error $R = O(h^2 + \varepsilon^2 h)$ has the form

$$X_{k+1} = X_k + \varepsilon h^{1/2} \sum_{r=1}^{q} (\sigma_r \xi_r)_k + \varepsilon^2 h b_k + h(a_k + a(t_{k+1}, X_k + h a_k))/2, \quad (5.10)$$

where ξ_r are as in (5.2).

The method with error $R = O(h^2 + \varepsilon^4 h)$ has the form

$$X_{k+1} = X_k + \varepsilon h^{1/2} \sum_{r=1}^{q} (\sigma_r(t_k, X_k) + \sigma_r(t_{k+1}, X_k + h a_k)) \xi_{r_k}/2$$

$$+ h \left[a_k + a(t_{k+1}, X_k + \varepsilon h^{1/2} \sum_{r=1}^{q} (\sigma_r \xi_r)_k + h(a + \varepsilon^2 b)_k) \right]/2$$

$$+ \varepsilon^2 h(b_k + b(t_{k+1}, X_k + h a_k))/2, \quad (5.11)$$

where ξ_r are as in (5.2).

Methods $O(h^4 + \cdots)$. The method with error $R = O(h^4 + \varepsilon^2 h)$ has the form

$$X_{k+1} = X_k + \varepsilon h^{1/2} \sum_{r=1}^{q} (\sigma_r \xi_r)_k + \varepsilon^2 h b_k + (k_1 + 2k_2 + 2k_3 + k_4)/6, \quad (5.12)$$

where
$$k_1 = ha_k, \quad k_2 = ha(t_{k+1/2}, X_k + k_1/2), \quad k_3 = ha(t_{k+1/2}, X_k + k_2/2),$$
$$k_4 = ha(t_{k+1}, X_k + k_3), \tag{5.13}$$

and ξ_r are as in (5.2).

The method with error $R = O(h^4 + \varepsilon^2 h^2 + \varepsilon^4 h)$ has the form

$$X_{k+1} = X_k + \varepsilon h^{1/2} \sum_{r=1}^{q} [\sigma_r(t_k, X_k) + \sigma_r(t_{k+1}, X_k + ha_k)] \xi_{r_k}/2$$
$$+ (k_1 + 2k_2 + 2k_3 + k_4)/6 + \varepsilon^2 h \left[b_k + b(t_{k+1}, X_k + ha_k)\right]/2, \tag{5.14}$$

where
$$k_1 = ha_k, \quad k_2 = ha(t_{k+1/2}, X_k + k_1/2),$$
$$k_3 = ha(t_{k+1/2}, X_k + \varepsilon h^{1/2} \sum_{r=1}^{q} (\sigma_r \xi_r)_k + k_2/2),$$
$$k_4 = ha(t_{k+1}, X_k + \varepsilon h^{1/2} \sum_{r=1}^{q} (\sigma_r \xi_r)_k + k_3 + 3\varepsilon^2 h b_k), \tag{5.15}$$

and ξ_r are as in (5.2).

The method with error $R = O(h^4 + \varepsilon^4 h)$ has the form

$$X_{k+1} = X_k + \varepsilon h^{1/2} \sum_{r=1}^{q} [\sigma_r(t_k, X_k)(\xi_r + 6\eta_r)_k$$
$$+ 4\sigma_r(t_{k+1/2}, X_k + k_2/2)\xi_{r_k} + \sigma_r(t_{k+1}, X_k + k_1)(\xi_r - 6\eta_r)_k]/6$$
$$+ h[a(t_k, X_k + \varepsilon h^{1/2} \sum_{r=1}^{q} (\sigma_r \eta_r)_k) - a(t_k, X_k - \varepsilon h^{1/2} \sum_{r=1}^{q} (\sigma_r \eta_r)_k)]/2$$
$$+ (k_1 + 2k_2 + 2k_3 + k_4)/6 + \varepsilon^2 (l_1 + 3l_2)/4, \tag{5.16}$$

where
$$k_1 = ha_k, \quad k_2 = ha(t_{k+1/2}, X_k + k_1/2),$$
$$k_3 = ha(t_{k+1/2}, X_k + \varepsilon h^{1/2} \sum_{r=1}^{q} (\sigma_r \xi_r)_k + k_2/2 + \varepsilon^2 l_1/4 + 3\varepsilon^2 l_2/4),$$
$$k_4 = ha(t_{k+1}, X_k + \varepsilon h^{1/2} \sum_{r=1}^{q} \sigma_r(t_{k+1}, X_k + k_1)\xi_{r_k} + k_3 + \varepsilon^2 l_1),$$
$$l_1 = hb_k, \quad l_2 = hb(t_k + 2h/3, X_k + 2k_1/9 + 4k_2/9), \tag{5.17}$$

and ξ_r, η_r are simulated as in (5.4). This full Runge–Kutta method requires six recalculations of the function $a(t, x)$, three recalculations of the functions $\sigma_r(t, x)$, and two recalculations of the function $b(t, x)$.

The method with error $R = O(h^4 + \varepsilon^2 h^2)$ has the form

$$X_{k+1} = X_k + \varepsilon h^{1/2} \sum_{r=1}^{q} [\sigma_r(t_k, X_k) + \sigma_r(t_{k+1}, X_k)] \xi_{r_k}/2$$

$$+\varepsilon^2 h \sum_{i,r=1}^{q} (\Lambda_i \sigma_r \xi_{ir})_k + \varepsilon h^{3/2} \sum_{r=1}^{q} \sum_{i=1}^{n} \left(a^i \frac{\partial \sigma_r}{\partial x^i} \xi_r\right)_k /2$$

$$+\varepsilon^3 h^{3/2} \sum_{r=1}^{q} (L_2 \sigma_r \xi_r)_k/2 + (k_1 + 2k_2 + 2k_3 + k_4)/6$$

$$+\varepsilon^2 h [b_k + b(t_{k+1}, X_k + \varepsilon h^{1/2} \sum_{r=1}^{q}(\sigma_r \xi_r)_k + h(a + \varepsilon^2 b)_k)]/2, \quad (5.18)$$

where k_i, $i = 1, \ldots, 4$, are from (5.15) and the used random variables ξ_r, ξ_{ir} are simulated as in the method (5.5). The method (5.18) contains first and second derivatives of the functions σ_r with respect to x.

Note that in the case of a single noise ($q = 1$) we succeeded in constructing a full Runge–Kutta method with error $O(h^4 + \varepsilon^2 h^2)$ [200].

The method with error $R = O(h^4 + \varepsilon^4 h^2)$ has the form

$$X_{k+1} = X_k + \varepsilon h^{1/2} \sum_{r=1}^{q} [\sigma_r(t_k, X_k)(\xi_r + 6\eta_r)_k$$

$$+ 4\sigma_r(t_{k+1/2}, X_k + k_2/2)\xi_{r_k} + \sigma_r(t_{k+1}, X_k + k_1)(\xi_r - 6\eta_r)_k]/6$$

$$+ h[a(t_k, X_k + \varepsilon h^{1/2} \sum_{r=1}^{q}(\sigma_r \eta_r)_k) - a(t_k, X_k - \varepsilon h^{1/2} \sum_{r=1}^{q}(\sigma_r \eta_r)_k)]/2$$

$$+\varepsilon^2 h \sum_{i,r=1}^{q} (\Lambda_i \sigma_r \xi_{ir})_k + \varepsilon^3 h^{3/2} \sum_{r=1}^{q}(L_2 \sigma_r \xi_r)_k/2$$

$$+(k_1 + 2k_2 + 2k_3 + k_4)/6 + \varepsilon^2(l_1 + 3l_2)/4, \quad (5.19)$$

where

$$k_1 = ha_k, \quad k_2 = ha(t_{k+1/2}, X_k + k_1/2),$$

$$k_3 = ha(t_{k+1/2}, X_k + \varepsilon h^{1/2} \sum_{r=1}^{q}(\sigma_r \xi_r)_k + k_2/2 + \varepsilon^2 l_1/4 + 3\varepsilon^2 l_2/4),$$

$$k_4 = ha(t_{k+1}, X_k + \varepsilon h^{1/2} \sum_{r=1}^{q} \sigma_r(t_{k+1}, X_k + k_1)\xi_{r_k} + k_3 + \varepsilon^2 l_1),$$

$$l_1 = hb(t_k, X_k + \varepsilon h^{1/2}(1+\sqrt{3})\sum_{r=1}^{q}(\sigma_r \xi_r)_k/2),$$

$$l_2 = hb(t_k + 2h/3, X_k + 2\varepsilon^2 l_1/3 + 2k_1/9 + 4k_2/9$$
$$+ \varepsilon h^{1/2}(3-\sqrt{3})\sum_{r=1}^{q}(\sigma_r \xi_r)_k/6), \quad (5.20)$$

and the used random variables are simulated using

$$P(\xi = 0) = 2/3, \ P(\xi = \pm\sqrt{3}) = 1/6, \ P(\zeta = \pm 1) = 1/2,$$

$$\xi_{ir} = (\xi_i \xi_r - \gamma_{ir} \zeta_i \zeta_r)/2, \ \gamma_{ir} = \begin{cases} -1, & i < r \\ 1, & i \geq r \end{cases},$$

$$\eta_r = \zeta_r/\sqrt{12}. \quad (5.21)$$

Remark 5.2. The stochastic system in the Stratonovich sense

$$dX = a(t,X)dt + \varepsilon^2 c(t,X)dt + \varepsilon \sum_{r=1}^{q} \sigma_r(t,X) \circ dw_r, \ X(t_0) = X_0, \quad (5.22)$$

is equivalent to the system in the Ito sense

$$dX = a(t,X)dt + \varepsilon^2 b(t,X)dt + \varepsilon \sum_{r=1}^{q} \sigma_r(t,X) \, dw_r, \quad (5.23)$$

where

$$b(t,x) = c(t,x) + \frac{1}{2}\sum_{r=1}^{q} \frac{\partial \sigma_r}{\partial x}(t,x) \, \sigma_r(t,x). \quad (5.24)$$

In Sects. 3.5.1 and 3.5.2 we have proposed weak methods for the Ito system having the form of (5.23). Thus, the methods of Sects. 3.5.1 and 3.5.2 are also appropriate for the Stratonovich system (5.22). Note that the full Runge–Kutta methods of Sect. 3.5.2 are no longer full when applied to system (5.22), since $b(t,x)$ in (5.24) contains derivatives $\partial \sigma_r/\partial x$. However, if the diffusion coefficients σ_r are simple functions, the methods of Sect. 3.5.2 may be efficient and useful for the Stratonovich system (5.22). Nevertheless, in some cases we obtain the full Runge–Kutta schemes for (5.22) [200].

3.5.3 Weak methods for systems with small additive noise

Consider the system with small additive noise

$$dX = a(t,X)dt + \varepsilon \sum_{r=1}^{q} \sigma_r(t) \, dw_r, \ X(t_0) = X_0. \quad (5.25)$$

For the system (5.25) we obtain methods with the errors estimated by $O(h^3 + \varepsilon^6 h^2)$, $O(h^3)$, $O(h^4 + \varepsilon^2 h^3 + \varepsilon^6 h^2)$, $O(h^4 + \varepsilon^6 h^2)$, $O(h^4 + \varepsilon^2 h^3)$, $O(h^4 + \varepsilon^4 h^3)$ and also with the same orders as in Sects. 3.5.1 and 3.5.2. Methods $O(h^4 + \varepsilon^6 h^3)$ and $O(h^4 + \varepsilon^8 h^3)$ are too complicated, and therefore we do not write them. Note that in [200] we also give a few full Runge–Kutta methods for the system with small colored noise, for instance, a scheme with error $O(h^4 + \varepsilon^2 h^3)$.

Methods for the system (5.25) with the same orders as in Sects. 3.5.1 and 3.5.2 follow from the corresponding methods for a general system with small noise taking into account that for the system (5.25) we have

$$\Lambda_r \sigma_i = 0, \ L_2 \sigma_i = 0, \ L_1 \sigma_i = \frac{d\sigma_i}{dt}, \ b = 0.$$

The Runge–Kutta methods $O(h^2 + \cdots)$ and $O(h^4 + \cdots)$ easily follow from the corresponding methods of Sect. 3.5.2. Fortunately, the methods (5.18) and (5.19) for the system with additive noise become fully (derivative free) Runge–Kutta schemes. Let us give a more detailed exposition of Taylor-type methods.

Methods $O(h^2 + \cdots)$ easily follow from the corresponding methods of Sect. 3.5.1, and here we do not write them.

The method $O(h^3 + \varepsilon^6 h^2)$ is written as

$$X_{k+1} = X_k + \varepsilon h^{1/2} \sum_{r=1}^{q} (\sigma_r \xi_r)_k + h a_k + \varepsilon h^{3/2} \sum_{r=1}^{q} \left(\frac{d\sigma_r}{dt} (\xi_r/2 - \eta_r) \right)_k$$

$$+ \varepsilon h^{3/2} \sum_{r=1}^{q} \Lambda_r a (\xi_r/2 + \eta_r)_k + h^2 (L_1 + \varepsilon^2 L_2) a_k / 2$$

$$+ \varepsilon^2 h^2 \sum_{r=1}^{q} \sum_{i=1}^{q} [\Lambda_i \Lambda_r a (\xi_i \xi_r - \zeta_i \zeta_r)]_k / 6$$

$$+ \varepsilon h^{5/2} \sum_{r=1}^{q} [(\frac{d^2 \sigma_r}{dt^2} + (L_1 + \varepsilon^2 L_2) \Lambda_r a + \Lambda_r (L_1 + \varepsilon^2 L_2) a) \xi_r]_k / 6$$

$$+ h^3 (L_1 + \varepsilon^2 L_2)^2 a_k / 6, \tag{5.26}$$

where the random variables ξ_r, η_r and ζ_r are simulated as

$$P(\xi = 0) = 2/3, \ P(\xi = \pm\sqrt{3}) = 1/6, \ P(\eta = \pm 1/\sqrt{12}) = 1/2,$$
$$P(\zeta = \pm 1) = 1/2. \tag{5.27}$$

The method $O(h^3)$ has the same form (5.27) but requires simulation of the needed random variables by the laws

$$P(\xi = 0) = 1/3, \ P(\xi = \pm 1) = 3/10, \ P(\xi = \pm\sqrt{6}) = 1/30,$$
$$P(\eta = \pm 1/\sqrt{12}) = 1/2, \ P(\zeta = \pm 1) = 1/2. \tag{5.28}$$

This method coincides with the third-order weak method (2.3.34) for a general system with additive noise ($\varepsilon = 1$).

Methods $O(h^4 + \cdots)$ for the system (5.25), except the methods $O(h^4 + \varepsilon^2 h^3 + \varepsilon^6 h^2)$, $O(h^4 + \varepsilon^6 h^2)$, $O(h^4 + \varepsilon^2 h^3)$, $O(h^4 + \varepsilon^4 h^3)$, are obtained from the corresponding methods of Sect. 3.5.1.

The method $O(h^4 + \varepsilon^2 h^3 + \varepsilon^6 h^2)$ is written as

$$X_{k+1} = X_k + \varepsilon h^{1/2} \sum_{r=1}^{q} (\sigma_r \xi_r)_k + h a_k + \varepsilon h^{3/2} \sum_{r=1}^{q} \left(\frac{d\sigma_r}{dt} (\xi_r/2 - \eta_r) \right)_k$$

$$+ \varepsilon h^{3/2} \sum_{r=1}^{q} [\Lambda_r a(\xi_r/2 + \eta_r)]_k + h^2 (L_1 + \varepsilon^2 L_2) a_k / 2$$

$$+ \varepsilon^2 h^2 \sum_{r=1}^{q} \sum_{i=1}^{q} (\Lambda_i \Lambda_r a(\xi_i \xi_r - \zeta_i \zeta_r))_k / 6$$

$$+ \varepsilon h^{5/2} \sum_{r=1}^{q} \left[\left(\frac{d^2 \sigma_r}{dt^2} + (L_1 + \varepsilon^2 L_2) \Lambda_r a + \Lambda_r (L_1 + \varepsilon^2 L_2) a \right) \xi_r \right]_k / 6$$

$$+ h^3 (L_1 + \varepsilon^2 L_2)^2 a_k / 6 + h^4 L_1^3 a_k / 24 \,, \tag{5.29}$$

where the random variables are as in (5.27).

The method $O(h^4 + \varepsilon^6 h^2)$ has the form

$$X_{k+1} = X_k + \varepsilon h^{1/2} \sum_{r=1}^{q} (\sigma_r \xi_r)_k + h a_k + \varepsilon h^{3/2} \sum_{r=1}^{q} \left(\frac{d\sigma_r}{dt} (\xi_r/2 - \eta_r) \right)_k$$

$$+ \varepsilon h^{3/2} \sum_{r=1}^{q} [\Lambda_r a(\xi_r/2 + \eta_r)]_k + h^2 (L_1 + \varepsilon^2 L_2) a_k / 2$$

$$+ \varepsilon^2 h^2 \sum_{r=1}^{q} \sum_{i=1}^{q} [\Lambda_i \Lambda_r a(\xi_i \xi_r - \zeta_i \zeta_r)]_k / 6$$

$$+ \varepsilon h^{5/2} \sum_{r=1}^{q} [(L_1 + \varepsilon^2 L_2) \Lambda_r a \xi_r]_k / 6$$

$$+ \varepsilon h^{5/2} \sum_{r=1}^{q} [\Lambda_r (L_1 + \varepsilon^2 L_2) a (\xi_r/6 + \eta_r/2)]_k$$

$$+ \varepsilon h^{5/2} \sum_{r=1}^{q} \left(\frac{d^2 \sigma_r}{dt^2} (\xi_r/6 - \eta_r/2) \right)_k + h^3 (L_1 + \varepsilon^2 L_2)^2 a_k / 6$$

$$+ \varepsilon h^{7/2} \sum_{r=1}^{q} \left[\left(\Lambda_r L_1^2 a + L_1 \Lambda_r L_1 a + L_1^2 \Lambda_r a + \frac{d^3 \sigma_r}{dt^3} \right) \xi_r \right]_k / 24$$

$$+ h^4 L_1^3 a_k / 24 + \varepsilon^2 h^4 (L_2 L_1^2 a + L_1^2 L_2 a + L_1 L_2 L_1 a)_k / 24 \,, \tag{5.30}$$

where the random variables are simulated as in (5.27).

The method $O(h^4+\varepsilon^2 h^3)$ has the form (5.29) but the random variables are simulated as in (5.28). Note that this method distinguishes from the method $O(h^3)$ by the additional term $h^4 L_1^3 a_k/24$ only.

The method $O(h^4+\varepsilon^4 h^3)$ has the same form (5.30) as the method $O(h^4+\varepsilon^6 h^2)$ but the needed random variables are from (5.28).

3.6 Expansion of the global error in powers of h and ε

It was shown in Sect. 2.2.3 that it is possible to expand the global errors of methods for stochastic systems in powers of time increment h. Below we expand the global error not only in powers of the time increment h but also in powers of the small parameter ε. Therefore, we cannot directly apply Theorem 2.2.5 here.

Theorem 6.1. *The global error of the method*

$$X_{k+1} = X_k + \varepsilon h^{1/2} \sum_{r=1}^{q}(\sigma_r \xi_r)_k + h(a+\varepsilon^2 b)_k + h^2(L_1 a)_k/2, \qquad (6.1)$$

$$P(\xi = \pm 1) = 1/2,$$

is

$$R = O(h^2+\varepsilon^2 h) = C_1(\varepsilon)h^2 + \varepsilon^2 C_2(\varepsilon)h + O(h^3+\varepsilon^2 h^2), \qquad (6.2)$$

where the functions $C_i(\varepsilon)$, $i=1,2$, do not depend on h and are equal to $C_i(\varepsilon) = C_i^0 + O(\varepsilon^2)$, and the constants C_i^0 do not depend on both h and ε.

The proof of this theorem is analogous to the proof of Theorem 2.2.5 and here it is omitted (see details in [202]). The same proof shows that the expansions of the global error for other methods can be obtained in the same way as the expansion (6.2) for the method (6.1). For instance, for the method (5.3) with error $O(h^4+\varepsilon^4 h)$, we have

$$R = C_1(\varepsilon)h^4 + \varepsilon^2 C_2(\varepsilon)h^3 + \varepsilon^4 C_3(\varepsilon)h^2 + \varepsilon^4 C_4(\varepsilon)h + O(h^5+\varepsilon^6 h^2).$$

An expansion like (6.2) can be used for derivation of extrapolation schemes as follows. Simulate $u^\varepsilon(t_0,X_0) = Ef(X_{t_0,X_0}^\varepsilon(T))$ twice using the method (6.1) for given ε with the time steps $h_1 = h$ and $h_2 = \alpha h$, $\alpha > 0$, $\alpha \neq 1$. We obtain $\bar{u}^{\varepsilon,h_1}(t_0,X_0) = Ef(\bar{X}_{t_0,X_0}^{\varepsilon,h_1}(T))$ and $\bar{u}^{\varepsilon,h_2}(t_0,X_0) = Ef(\bar{X}_{t_0,X_0}^{\varepsilon,h_2}(T))$, respectively. We can expand

$$u^\varepsilon = \bar{u}^{\varepsilon,h_1} + C_1(\varepsilon)h_1^2 + \varepsilon^2 C_2(\varepsilon)h_1 + O(h^3+\varepsilon^2 h^2)$$

and

$$u^\varepsilon = \bar{u}^{\varepsilon,h_2} + C_1(\varepsilon)h_2^2 + \varepsilon^2 C_2(\varepsilon)h_2 + O(h^3+\varepsilon^2 h^2).$$

This yields

$$\varepsilon^2 C_2(\varepsilon) = \varepsilon^2 \bar{C}_2(\varepsilon) - C_1^0 \times (h_1 + h_2) + O(h^2 + \varepsilon^2 h), \quad (6.3)$$

where $\varepsilon^2 \bar{C}_2(\varepsilon)$ is given by

$$\varepsilon^2 \bar{C}_2(\varepsilon) = (\bar{u}^{\varepsilon,h_1} - \bar{u}^{\varepsilon,h_2})/(h_2 - h_1).$$

On the other hand, setting $\varepsilon = 0$ and using the method (6.1) with the time steps h_1 and h_2, we obtain $\bar{u}^{0,h_1}(t_0, X_0) = f(\bar{X}^{0,h_1}_{t_0,X_0}(T))$ and $\bar{u}^{0,h_2}(t_0, X_0) = f(\bar{X}^{0,h_2}_{t_0,X_0}(T))$, where $\bar{X}^{0,h_i}_{t_0,X_0}(t)$ is the corresponding approximation of the solution $X^0_{t_0,X_0}(t)$ of the deterministic system. Then the Runge extrapolation method yields

$$C_1(0) = C_1^0 = \bar{C}_1^0 + O(h), \quad (6.4)$$

where $\bar{C}_1^0$ can be calculated by

$$\bar{C}_1^0 = \left(\bar{u}^{0,h_1} - \bar{u}^{0,h_2}\right)/(h_2^2 - h_1^2),$$

By (6.3) and (6.4) we obtain an improved value $\bar{u}^{\varepsilon}_{imp}$ with error $O(h^3 + \varepsilon^2 h^2)$ by letting

$$\bar{u}^{\varepsilon}_{imp} = \bar{u}^{\varepsilon,h_1} + \varepsilon^2 \bar{C}_2(\varepsilon)h_1 - \bar{C}_1^0 h_1 h_2. \quad (6.5)$$

In the same spirit, using three recalculations of $u^{\varepsilon}(t_0, X_0) = Ef(X^{\varepsilon}_{t_0,X_0}(T))$ by the method (6.1) for given ε and with not equal time steps, one can also find $C_1(\varepsilon)$ and $C_2(\varepsilon)$ from (6.2) and obtain yet another improved value.

We conclude that according to our approach to the construction of weak methods for a system with small noise, we can shift some terms, which contribute to the error proportionally to $h^i \varepsilon^j$, from the method to its remainder and vice versa. By calculating the constants $C_i(\varepsilon)$ it is possible to estimate the proper weights of the terms in the sums above and select the most appropriate scheme for solving a given system with small noise, both keeping computational costs low and accuracy high.

3.7 Reduction of the Monte Carlo error

Here we apply the method of important sampling from Sect. 2.4.1 in the case of systems with small noise.

Together with the system (0.1), consider the following (cf. (2.4.3)):

$$dX = a(t, X)dt + \varepsilon^2 b(t, X)dt - \varepsilon \sum_{r=1}^{q} \mu_r(t, X)\sigma_r(t, X)dt + \varepsilon \sum_{r=1}^{q} \sigma_r(t, X)dw_r,$$

$$dY = \sum_{r=1}^{q} \mu_r(t, X) Y\, dw_r, \quad (7.1)$$

where μ_r and Y are scalars.

According to the Girsanov theorem, we have for any μ_r:

$$yEf(X_{s,x}(T))|_{(0.1)} = E\left(Y_{s,x,y}(T)f(X_{s,x}(T))\right)|_{(7.1)}. \qquad (7.2)$$

The function $u(s,x) = Ef(X_{s,x}(T))|_{(0.1)}$ satisfies the equation

$$Lu \equiv \frac{\partial u}{\partial s} + \sum_{i=1}^{d} a^i \frac{\partial u}{\partial x^i} + \varepsilon^2 \sum_{i=1}^{d} b^i \frac{\partial u}{\partial x^i} + \frac{\varepsilon^2}{2} \sum_{r=1}^{q} \sum_{i=1}^{d} \sum_{j=1}^{d} \sigma_r^i \sigma_r^j \frac{\partial^2}{\partial x^i \partial x^j} = 0 \qquad (7.3)$$

subject to the following condition at the instant T:

$$u(T,x) = f(x). \qquad (7.4)$$

Under sufficiently mild conditions on the coefficients and on the function f, the solution $u(s,x) = u^\varepsilon(s,x)$ of the problem (7.3)-(7.4) has the form [72, Chap. 2]:

$$u^\varepsilon(s,x) = u^0(s,x) + \varepsilon^2 u^1(s,x;\varepsilon). \qquad (7.5)$$

The function u^0 satisfies the first-order partial differential equation

$$\frac{\partial u}{\partial s} + \sum_{i=1}^{d} a^i \frac{\partial u}{\partial x^i} = 0 \qquad (7.6)$$

under the condition (7.4). Obviously, the solution of (7.6) has the form

$$u^0(s,x) = f(X^0_{s,x}(T)), \qquad (7.7)$$

where $X^0_{s,x}$ is the solution of the Cauchy problem for the deterministic system of differential equations

$$\frac{dX}{dt} = a(t,X), \quad X(s) = x. \qquad (7.8)$$

Applying the Ito formula along the solution of the system (7.1), we get the following expression (note that here $Lu = 0$):

$$d\left[u(t, X_{s,x}(t))Y_{s,x,y}(t)\right] = LuY\, dt - \varepsilon \sum_{r=1}^{q} \mu_r(\sigma_r, \frac{\partial u}{\partial x}) Y\, dt$$

$$+ \varepsilon \sum_{r=1}^{q} (\sigma_r, \frac{\partial u}{\partial x}) Y\, dw_r(t) + u \sum_{r=1}^{q} \mu_r Y\, dw_r(t)$$

$$+ \varepsilon \sum_{r=1}^{q} (\sigma_r, \frac{\partial u}{\partial x}) \mu_r Y\, dt$$

$$= \sum_{r=1}^{q} \left(\varepsilon(\sigma_r, \frac{\partial u}{\partial x}) + \mu_r u\right) Y\, dw_r(t).$$

Then

$$u(t, X_{s,x}(t))Y_{s,x,y}(t) = u(s,x)y + \int_s^t \sum_{r=1}^q \left(\varepsilon(\sigma_r, \frac{\partial u}{\partial x}) + \mu_r u\right) Y dw_r(t). \quad (7.9)$$

If we suppose that $t = T$, $y = 1$, $\mu_r \equiv 0$, we obtain

$$f(X_{s,x}(T)) = u(s,x) + \int_s^T \varepsilon \sum_{r=1}^q (\sigma_r, \frac{\partial u}{\partial x}) dw_r(t).$$

Therefore

$$Df(X_{s,x}(T)) = \varepsilon^2 \int_s^T E \left[\sum_{r=1}^q (\sigma_r, \frac{\partial u}{\partial x})\right]^2 dt \quad (7.10)$$

because $u(s,x) = Ef(X_{s,x}(T))|_{(0.1)}$.

Thus, if we calculate $Ef(X(T))$ by the Monte Carlo technique using a weak method for solving the system (0.1), then the Monte Carlo error, evaluated by $c[Df(\bar{X}(T))/N]^{1/2}$ and close to $c[Df(X(T))/N]^{1/2}$, contains a small factor equal to ε.

As can be seen from (7.2), the mean value

$$EZ = E\left(Y_{s,x,y}(T)f(X_{s,x}(T))\right)|_{(7.1)}$$

does not depend on μ_r, whereas $D\left(Y_{s,x,y}(T)f(X_{s,x}(T))\right)|_{(7.1)}$ does depend on μ_r. So below we will select functions μ_r, $r = 1, \ldots, q$, such that the variance DZ becomes less than the variance (7.10).

Assume that $f > 0$. Then $u^0 > 0$. Note that if the function f is not positive but there are constants K and C such that $Kf + C > 0$, we can take the function $g = Kf + C$ instead. Then we can simulate Eg and finally obtain Ef.

Setting $t = T$, $y = 1$ in (7.9), and

$$\mu_r = -\frac{\varepsilon}{u^0}\left(\sigma_r, \frac{\partial u^0}{\partial x}\right), \quad r = 1, \ldots, q, \quad (7.11)$$

we obtain

$$f(X_{s,x}(T))Y = u(s,x) + \int_s^T \varepsilon^3 \sum_{r=1}^q \left[(\sigma_r, \frac{\partial u^1}{\partial x}) - (\sigma_r, \frac{\partial u^0}{\partial x})\frac{u^1}{u^0}\right] dw_r(t).$$

Therefore

$$D\left[f(X_{s,x}(T))Y\right] = \varepsilon^6 \int_s^T E \left(\sum_{r=1}^q \left[(\sigma_r, \frac{\partial u^1}{\partial x}) - (\sigma_r, \frac{\partial u^0}{\partial x})\frac{u^1}{u^0}\right]\right)^2 dt.$$

Hence, the Monte Carlo error for the system (7.1) with μ_r from (7.11) inherits a small factor equal to ε^3.

The system (7.1) with μ_r from (7.11) is again a system with small noise, and all the methods proposed above are suitable for finding its solution. We observe that, even if the number M of simulations is small, the Monte Carlo error for this system will be reasonably small. Of course, in order to apply the approach outlined above, we must know the function $u^0(s,x)$.

3.8 Simulation of the Lyapunov exponent of a linear system with small noise by weak methods

To test weak methods proposed in this chapter, we use the two-dimensional linear Ito stochastic system (3.1) again:

$$dX = AX\,dt + \varepsilon \sum_{r=1}^{q} B_r X\,dw_r, \qquad (8.1)$$

where X is a two-dimensional vector, A and B_r are constant 2×2-matrices, w_r are independent standard Wiener processes, $\varepsilon > 0$ is a small parameter.

D. Talay [284] proposed a numerical approach to calculating Lyapunov exponents based on ergodic property. Using weak methods, Lyapunov exponents are calculated by simulating a single trajectory. This procedure is appealing as it is intuitive and computationally cheap. However, it is difficult to analyze the errors arising from this approach. Below we calculate Lyapunov exponent of (8.1) as a convenient example to illustrate efficiency of the proposed methods. We also pay attention to analysis of the errors.

In the ergodic case there exist a unique Lyapunov exponent λ of system (8.1) (cf. [124]), with

$$\lambda = \lim_{t \to \infty} \frac{1}{t} E\rho(t) = \lim_{t \to \infty} \frac{1}{t}\rho(t) \quad a.s.,$$

where $\rho(t) = \ln|X(t)|$, and $X(t),\ t \geq 0$, is a non-trivial solution of the system (8.1).

If $D(\rho(t)) \to \infty$ for $t \to \infty$ then [124]

$$E\left(\frac{\rho(t)}{t} - \lambda\right)^2 = D\left(\frac{\rho(t)}{t}\right)(1 + \varphi^2(t)), \qquad (8.2)$$

where $\varphi(t) \to 0$ for $t \to \infty$. It is not difficult to show that $D(\rho(t)/t) \to 0$ for $t \to \infty$. From (8.2) and the equality

$$D\left(\frac{\rho(t)}{t}\right) = E\left(\frac{\rho(t)}{t} - \lambda\right)^2 - \left[E\left(\frac{\rho(t)}{t}\right) - \lambda\right]^2,$$

3.8 Simulation of the Lyapunov exponent by weak methods

we have

$$\left| E\left(\frac{\rho(t)}{t}\right) - \lambda \right| = \varphi(t) \left[D\left(\frac{\rho(t)}{t}\right) \right]^{1/2}. \tag{8.3}$$

Herein we consider system (8.1) with the matrices A and B_r of the form

$$A = \begin{pmatrix} a & c \\ -c & a \end{pmatrix}, \quad B_r = \begin{pmatrix} b_r & d_r \\ -d_r & b_r \end{pmatrix}, \quad r = 1, 2 \tag{8.4}$$

In this case the Lyapunov exponent is given by (3.4) (cf. [8]).

By the Monte Carlo technique we numerically calculate the function

$$\lambda(T) = \frac{1}{T} E\rho(T) \approx \bar{\lambda}(T) = \frac{1}{T} E\bar{\rho}(T), \quad \bar{\rho}(T) = \ln |\bar{X}(T)|. \tag{8.5}$$

The function $\lambda(t)$ in the limit of large time ($t \to \infty$) tends to the Lyapunov exponent λ. In this case three errors arise: (a) the method error, i.e., $|E\rho(T)/T - E\bar{\rho}(T)/T|$, (b) the Monte Carlo error which is bounded by $c[D(\bar{\rho}(T)/T)]^{1/2}/\sqrt{M}$, and (c) the error with respect to the choice of integration time T (see (8.3)).

As can be seen from our computational results, the third error, i.e., $|\lambda(T) - \lambda| = |E(\rho(T)/T) - \lambda|$, is negligibly small, at any rate for $T \geq 2$, as compared to both the method error and the Monte Carlo error.

In our case the function $[D(\bar{\rho}(T)/T)]^{1/2}$ tends to zero with rate $1/\sqrt{T}$. So the Monte Carlo error is proportional to $1/\sqrt{TM}$. Therefore, to reduce the Monte Carlo error we can increase either M or T. As far as the computational costs are concerned, it does not matter whether we increase M or T. In our case Talay's approach requires the same computational costs as the simulation of Lyapunov exponents by the Monte Carlo technique. But using Monte Carlo simulations we find both $E\bar{\rho}(T)/T$ and $D(\bar{\rho}(T)/T)$, which is useful for estimating errors.

We simulate the system (8.1) by four different weak schemes: (i) the method (6.1) with error $O(h^2 + \varepsilon^2 h)$, which is the simplest method among the weak schemes proposed in this chapter, (ii) the method with error $O(h^2 + \varepsilon^4 h)$:

Table 8.1. Simulation of $\bar{\lambda}(T)$ for $a = -2$, $c = 1$, $b_1 = b_2 = 2$, $d_1 = 1$, $d_2 = -1$, $\varepsilon = 0.2$, $X^1(0) = 0$, $X^2(0) = 1$, $T = 10$, and for various steps h averaged over M realizations, where $M = 4 \cdot 10^4$ for the methods $O(h^2 + \cdots)$ and $M = 1 \cdot 10^6$ for the method $O(h^4 + \varepsilon^4 h^2)$. The exact solution is $\lambda = -2.12$.

h	$O(h^2 + \varepsilon^2 h)$	$O(h^2 + \varepsilon^4 h)$	$O(h^2)$	$O(h^4 + \varepsilon^4 h^2)$
0.3	-2.461 ± 0.004	-2.067 ± 0.002	-2.067 ± 0.002	-2.1228 ± 0.0004
0.2	-2.290 ± 0.003	-2.106 ± 0.002	-2.097 ± 0.002	-2.1195 ± 0.0004
0.1	-2.186 ± 0.002	-2.1198 ± 0.0018	-2.1140 ± 0.0017	-2.1192 ± 0.0004
0.05	-2.150 ± 0.002	-2.1219 ± 0.0018	-2.1186 ± 0.0018	-2.1197 ± 0.0004

$$X_{k+1} = X_k + \varepsilon h^{1/2} \sum_{r=1}^{q} (\sigma_r \xi_r)_k + h(a + \varepsilon^2 b)_k + \varepsilon h^{3/2} \sum_{r=1}^{q} (L_1 \sigma_r \xi_r)_k / 2$$

$$+ \varepsilon h^{3/2} \sum_{r=1}^{q} (\Lambda_r a \xi_r)_k / 2 + h^2 \left[L_1(a + \varepsilon^2 b) \right]_k / 2 + \varepsilon^2 h^2 (L_2 a)_k / 2, \quad (8.6)$$

where the random variables ξ_r are distributed as $P(\xi = \pm 1) = 1/2$, (iii) the standard method (4.8) with error $O(h^2)$, (iv) the semi-Runge–Kutta scheme (5.19) with error $O(h^4 + \varepsilon^4 h^2)$, which is the most accurate scheme among the weak methods proposed in this chapter for general systems with small noise.

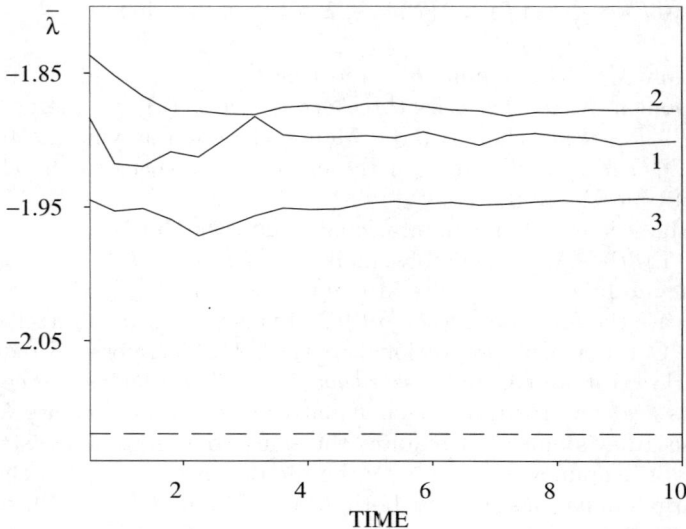

Fig. 8.1. Time dependence of the function $\bar{\lambda}(T) = E\bar{\rho}(T)/T$ for time step $h = 0.45$. The other parameters are the same as in Table 8.1. The solution of the system (8.1), (8.4) is approximated by (1) the method (6.1), (2) the method (8.6), and (3) the standard method (4.8). The dashed line shows the exact value of the Lyapunov exponent λ ($\lambda = -2.12$). The number of realizations is $M = 400$ which ensures that the Monte Carlo errors at $T \geq 7$ are not greater than 0.04 for curve 1 and not greater than 0.02 for curves 2, 3 and they are less than the method errors.

The results in Table 8.1 are approximations of $\lambda(T)$ calculated as

$$\bar{\lambda}(T) \simeq \frac{1}{M} \sum_{m=1}^{M} \bar{\rho}^{(m)}(T)/T$$

$$\pm \frac{2}{\sqrt{M}} \left(\frac{1}{M} \sum_{m=1}^{M} \left[\bar{\rho}^{(m)}(T)/T\right]^2 - \left[\frac{1}{M} \sum_{m=1}^{M} \bar{\rho}^{(m)}(T)/T\right]^2 \right)^{1/2}.$$

3.8 Simulation of the Lyapunov exponent by weak methods

We can infer from Table 8.1 and Fig. 8.1 that the proposed methods for systems with small noise require less computational effort than the standard ones.

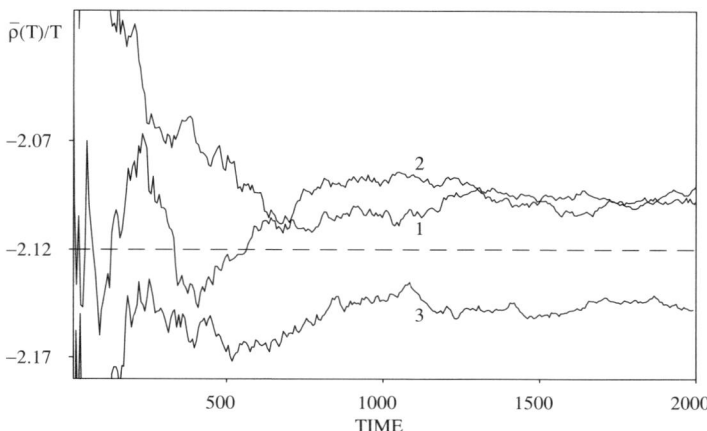

Fig. 8.2. Time dependence of the function $\bar{\rho}(T)/T$ computed along a single trajectory using (1) the method (8.6) with $h = 0.1$, (2) the standard method (4.8) with $h = 0.1$, and (3) the method (5.19) with $h = 0.3$. The other parameters are the same as in Table 8.1. The dashed line shows the exact value of the Lyapunov exponent λ ($\lambda = -2.12$).

The data of Table 8.1 show that the methods $O(h^2 + \varepsilon^2 h)$, $O(h^2 + \varepsilon^4 h)$, and $O(h^2)$ can be improved by using the expansion from Sect. 3.6. For $\varepsilon = 0.2$, for instance, one can calculate $C_1(\varepsilon)$ and $C_2(\varepsilon)$ from the expansion of the global error of the method (6.1) (see Theorem 6.1) to obtain $C_1(\varepsilon) \approx 2.1$ and $C_2(\varepsilon) \approx 10.2$. Let us emphasize that if some constants in the error expansion have opposite signs then the error will become a non-monotonous function of the time step h and thus may increase while h is decreasing. Such behavior is demonstrated in Table 8.1 (see the methods $O(h^2 + \varepsilon^4 h)$ and $O(h^4 + \varepsilon^4 h^2)$).

In Fig. 8.2 we show the time dependence of the function $\bar{\rho}(T)/T$ when taking Talay's approach to calculate Lyapunov exponents, i.e., along a single trajectory of a weak scheme. One can see that in this case our methods give accurate results and allow us to reduce computational costs. By the Monte Carlo simulations using the method (5.19) with $h = 0.2$, $T = 10$, $M = 10^6$ we achieve an accuracy of $\approx 0.5 \times 10^{-3}$ (see Table 8.1).

Remark 8.1. Note that the function $\ln |x|$ does not belong to the class **F**. Thus, if $\lambda > 0$, to deduce strict conclusions one can consider the function $\ln(1 + |x|)$ instead of $\ln |x|$. The function $\ln(1 + |x|)$ already belongs to the class **F** and $\lim_{t \to \infty} \ln(1 + |X(t)|)/t = \lim_{t \to \infty} \ln(|X(t)|)/t$. As can be seen by carrying out numerical tests, simulations of the function $\ln(1 + |X(t)|)$ yield the same results as simulations of $\ln|X(t)|$. For $\lambda < 0$ one can either switch to the function $\ln(1 + 1/|x|)$ or to the system

$$dX = (\gamma I + AX)dt + \varepsilon \sum_{r=1}^{q} B_r X\, dw_r \qquad (8.7)$$

instead of the system (8.1). The Lyapunov exponent of the system (8.7) is equal to $\gamma + \lambda$, and if we choose γ such that $\gamma + \lambda > 0$, we can use the function $\ln(1 + |x|)$ again.

4 Stochastic Hamiltonian systems and Langevin-type equations

In this chapter we construct specific methods for two important classes of stochastic systems which often occur in physical applications. First we consider *stochastic Hamiltonian systems* which can be written in the form of the Stratonovich SDEs:

$$dP = f(t, P, Q)dt + \sum_{r=1}^{m} \sigma_r(t, P, Q) \circ dw_r(t), \ P(t_0) = p,$$

$$dQ = g(t, P, Q)dt + \sum_{r=1}^{m} \gamma_r(t, P, Q) \circ dw_r(t), \ Q(t_0) = q, \qquad (0.1)$$

where P, Q, f, g, σ_r, γ_r are n-dimensional column-vectors with the components P^i, Q^i, f^i, g^i, σ_r^i, γ_r^i, $i = 1, \ldots, n$, and $w_r(t)$, $r = 1, \ldots, m$, are independent standard Wiener processes.

We denote by $X(t; t_0, x) = (P^\top(t; t_0, p, q), Q^\top(t; t_0, p, q))^\top$, $t_0 \leq t \leq T$, the solution of the problem (0.1). A more detailed notation is $X(t; t_0, x; \omega)$, where ω is an elementary event. It is known that $X(t; t_0, x; \omega)$ is a phase flow (diffeomorphism) for almost every ω. See its properties in, e.g. [21, 55, 110].

If there are functions $H_r(t, p, q)$, $r = 0, \ldots, m$, such that (see [21] and Sect. 4.1 below)

$$\begin{aligned} f^i(t, p, q) &= -\partial H_0/\partial q^i, \quad g^i(t, p, q) = \partial H_0/\partial p^i, \\ \sigma_r^i(t, p, q) &= -\partial H_r/\partial q^i, \quad \gamma_r^i(t, p, q) = \partial H_r/\partial p^i, \\ r &= 1, \ldots, m, \quad i = 1, \ldots, n, \end{aligned} \qquad (0.2)$$

then the phase flow of (0.1) (like the phase flow of a deterministic Hamiltonian system) preserves symplectic structure:

$$dP \wedge dQ = dp \wedge dq, \qquad (0.3)$$

i.e., the sum of the oriented areas of projections onto the coordinate planes $(p^1, q^1), \ldots, (p^n, q^n)$ is an integral invariant [7]. To avoid confusion, we note that the differentials in (0.1) and (0.3) have different meaning. In (0.1) P, Q are treated as functions of time and p, q are fixed parameters, while differentiation in (0.3) is made with respect to the initial data p, q.

212   4 Stochastic Hamiltonian systems and Langevin-type equations

Let $P_k, Q_k, k = 0, \ldots, N, t_{k+1} - t_k = h_{k+1}, t_N = T$, be a method for (0.1) based on the one-step approximation $\bar{P} = \bar{P}(t+h; t, p, q), \bar{Q} = \bar{Q}(t+h; t, p, q)$. We say that the method preserves symplectic structure if

$$d\bar{P} \wedge d\bar{Q} = dp \wedge dq. \tag{0.4}$$

A lot of attention in deterministic numerical analysis has been paid to symplectic integration of Hamiltonian systems (see, e.g. [96, 97, 261] and references therein). This interest is motivated by the fact that symplectic integrators in comparison with usual numerical schemes allow us to simulate Hamiltonian systems on very long time intervals with high accuracy. As it will be shown in this chapter, symplectic methods for stochastic Hamiltonian systems proposed in Sects. 4.2-4.6 have significant advantages over standard schemes for SDEs.

Sections 4.2-4.4 deal with mean-square symplectic methods [194, 195] while Section 4.6 is devoted to weak symplectic integration [213]. We construct symplectic methods for general stochastic Hamiltonian systems (0.1)-(0.2) as well as higher-order symplectic schemes for Hamiltonian systems with separable Hamiltonians and Hamiltonian systems with additive noise. Symplectic integrators for other specific stochastic Hamiltonian systems such as Hamiltonian systems with colored noise, Hamiltonian systems with small noise, etc. can be found in [193–195, 213]. We also propose volume-preserving methods for stochastic Liouvillian systems (see Sect. 4.5).

It is natural to expect that making use of numerical methods, which are close, in a sense, to symplectic ones, also has some advantages when applying to stochastic systems close to Hamiltonian ones. An important and fairly large class of such systems is *Langevin-type equations* which can be written as the following system of Ito SDEs

$$dP = f(t, Q)dt - \nu \tilde{f}(t, P, Q)dt + \sum_{r=1}^{m} \sigma_r(t, Q)dw_r(t), \ P(t_0) = p,$$

$$dQ = g(P)dt, \ Q(t_0) = q, \tag{0.5}$$

where $P, Q, f, \tilde{f}, g, \sigma_r$ are n-dimensional column-vectors, ν is a parameter, and $w_r(t), r = 1, \ldots, m$, are independent standard Wiener processes. It is not difficult to verify that this system has the same form in the sense of Stratonovich (see also Example 1.2.8).

The Langevin-type equations (0.5) have the widespread occurrence in models from physics, chemistry, and biology. They are used in dissipative particle dynamics (see, e.g., [247] and references therein), in molecular simulations (see, e.g., [114, 266] and references therein), for studying lattice dynamics in strongly anharmonic crystals [86], descriptions of noise-induced transport in stochastic ratchets [144], investigations of the dispersion of passive tracers in turbulent flows (see [262, 294] and references therein), etc. In the second part of this chapter (Sects. 4.7-4.8) we construct special numerical methods (we call them as quasi-symplectic) which preserve some specific properties of the Langevin-type equations [213]. The proposed methods are such that they degenerate to symplectic methods when the system degenerates to a Hamiltonian one and their law of phase volume contractivity is close to the exact one.

4.1 Preservation of symplectic structure

Consider the system (0.1). Our urgent aim is to indicate a class of stochastic systems, which preserve symplectic structure, i.e., satisfy the condition (0.3).

Using the formula of change of variables in differential forms, we obtain

$$dP \wedge dQ = dP^1 \wedge dQ^1 + \cdots + dP^n \wedge dQ^n$$
$$= \sum_{k=1}^{n} \sum_{l=k+1}^{n} \sum_{i=1}^{n} \left(\frac{\partial P^i}{\partial p^k} \frac{\partial Q^i}{\partial p^l} - \frac{\partial P^i}{\partial p^l} \frac{\partial Q^i}{\partial p^k} \right) dp^k \wedge dp^l$$
$$+ \sum_{k=1}^{n} \sum_{l=k+1}^{n} \sum_{i=1}^{n} \left(\frac{\partial P^i}{\partial q^k} \frac{\partial Q^i}{\partial q^l} - \frac{\partial P^i}{\partial q^l} \frac{\partial Q^i}{\partial q^k} \right) dq^k \wedge dq^l$$
$$+ \sum_{k=1}^{n} \sum_{l=1}^{n} \sum_{i=1}^{n} \left(\frac{\partial P^i}{\partial p^k} \frac{\partial Q^i}{\partial q^l} - \frac{\partial P^i}{\partial q^l} \frac{\partial Q^i}{\partial p^k} \right) dp^k \wedge dq^l \, .$$

Hence the phase flow of (0.1) preserves symplectic structure if and only if

$$\sum_{i=1}^{n} \frac{D(P^i, Q^i)}{D(p^k, p^l)} = 0, \quad k \neq l, \tag{1.1}$$

$$\sum_{i=1}^{n} \frac{D(P^i, Q^i)}{D(q^k, q^l)} = 0, \quad k \neq l, \tag{1.2}$$

and

$$\sum_{i=1}^{n} \frac{D(P^i, Q^i)}{D(p^k, q^l)} = \delta_{kl}, \quad k, l = 1, \ldots, n \, . \tag{1.3}$$

Introduce the notation

$$P_p^{ik} = \frac{\partial P^i}{\partial p^k}, \quad P_q^{ik} = \frac{\partial P^i}{\partial q^k}, \quad Q_p^{ik} = \frac{\partial Q^i}{\partial p^k}, \quad Q_q^{ik} = \frac{\partial Q^i}{\partial q^k} \, .$$

For a fixed k, we obtain that P_p^{ik}, Q_p^{ik}, $i = 1, \ldots, n$, obey the following system of SDEs

$$dP_p^{ik} = \sum_{\alpha=1}^{n} \left(\frac{\partial f^i}{\partial p^\alpha} P_p^{\alpha k} + \frac{\partial f^i}{\partial q^\alpha} Q_p^{\alpha k} \right) dt + \sum_{r=1}^{m} \sum_{\alpha=1}^{n} \left(\frac{\partial \sigma_r^i}{\partial p^\alpha} P_p^{\alpha k} + \frac{\partial \sigma_r^i}{\partial q^\alpha} Q_p^{\alpha k} \right) \circ dw_r,$$
$$P_p^{ik}(t_0) = \delta_{ik} \, ,$$

$$dQ_p^{ik} = \sum_{\alpha=1}^{n}\left(\frac{\partial g^i}{\partial p^\alpha}P_p^{\alpha k} + \frac{\partial g^i}{\partial q^\alpha}Q_p^{\alpha k}\right)dt + \sum_{r=1}^{m}\sum_{\alpha=1}^{n}\left(\frac{\partial \gamma_r^i}{\partial p^\alpha}P_p^{\alpha k} + \frac{\partial \gamma_r^i}{\partial q^\alpha}Q_p^{\alpha k}\right)\circ dw_r,$$
$$Q_p^{ik}(t_0) = 0. \tag{1.4}$$

Analogously, for a fixed k, P_q^{ik}, Q_q^{ik}, $i=1,\ldots,n$, satisfy the system

$$dP_q^{ik} = \sum_{\alpha=1}^{n}\left(\frac{\partial f^i}{\partial p^\alpha}P_q^{\alpha k} + \frac{\partial f^i}{\partial q^\alpha}Q_q^{\alpha k}\right)dt + \sum_{r=1}^{m}\sum_{\alpha=1}^{n}\left(\frac{\partial \sigma_r^i}{\partial p^\alpha}P_q^{\alpha k} + \frac{\partial \sigma_r^i}{\partial q^\alpha}Q_q^{\alpha k}\right)\circ dw_r,$$
$$P_q^{ik}(t_0) = 0,$$

$$dQ_q^{ik} = \sum_{\alpha=1}^{n}\left(\frac{\partial g^i}{\partial p^\alpha}P_q^{\alpha k} + \frac{\partial g^i}{\partial q^\alpha}Q_q^{\alpha k}\right)dt + \sum_{r=1}^{m}\sum_{\alpha=1}^{n}\left(\frac{\partial \gamma_r^i}{\partial p^\alpha}P_q^{\alpha k} + \frac{\partial \gamma_r^i}{\partial q^\alpha}Q_q^{\alpha k}\right)\circ dw_r,$$
$$Q_q^{ik}(t_0) = \delta_{ik}. \tag{1.5}$$

The coefficients in (1.4) and (1.5) are calculated at (t,P,Q) with $P = P(t) = [P^1(t;t_0,p,q),\ldots,P^n(t;t_0,p,q)]^\top$, $Q = Q(t) = [Q^1(t;t_0,p,q),\ldots,Q^n(t;t_0,p,q)]^\top$ being a solution to (0.1).

Consider the condition (1.1). Clearly,

$$\frac{D(P^i(t_0),Q^i(t_0))}{D(p^k,p^l)} = \frac{D(p^i,q^i)}{D(p^k,p^l)} = 0.$$

Therefore, (1.1) is fulfilled if and only if

$$\sum_{i=1}^{n} d\frac{D(P^i(t),Q^i(t))}{D(p^k,p^l)} = 0. \tag{1.6}$$

Due to (1.4), we get

$$d\frac{\partial P^i}{\partial p^k}\frac{\partial Q^i}{\partial p^l} = dP_p^{ik}(t)Q_p^{il}(t)$$
$$= \sum_{\alpha=1}^{n}\left[\left(\frac{\partial f^i}{\partial p^\alpha}P_p^{\alpha k} + \frac{\partial f^i}{\partial q^\alpha}Q_p^{\alpha k}\right)Q_p^{il}\right.$$
$$\left.+ \left(\frac{\partial g^i}{\partial p^\alpha}P_p^{\alpha l} + \frac{\partial g^i}{\partial q^\alpha}Q_p^{\alpha l}\right)P_p^{ik}\right]dt$$
$$+ \sum_{r=1}^{m}\sum_{\alpha=1}^{n}\left[\left(\frac{\partial \sigma_r^i}{\partial p^\alpha}P_p^{\alpha k} + \frac{\partial \sigma_r^i}{\partial q^\alpha}Q_p^{\alpha k}\right)Q_p^{il}\right.$$
$$\left.+ \left(\frac{\partial \gamma_r^i}{\partial p^\alpha}P_p^{\alpha l} + \frac{\partial \gamma_r^i}{\partial q^\alpha}Q_p^{\alpha l}\right)P_p^{ik}\right]\circ dw_r.$$

4.1 Preservation of symplectic structure

Then (1.6) holds if and only if the following equalities take place:

$$\sum_{i=1}^{n}\sum_{\alpha=1}^{n}\left(\frac{\partial f^i}{\partial p^\alpha}P_p^{\alpha k}Q_p^{il} + \frac{\partial f^i}{\partial q^\alpha}Q_p^{\alpha k}Q_p^{il} + \frac{\partial g^i}{\partial p^\alpha}P_p^{\alpha l}P_p^{ik} + \frac{\partial g^i}{\partial q^\alpha}Q_p^{\alpha l}P_p^{ik}\right.$$
$$\left. - \frac{\partial f^i}{\partial p^\alpha}P_p^{\alpha l}Q_p^{ik} - \frac{\partial f^i}{\partial q^\alpha}Q_p^{\alpha l}Q_p^{ik} - \frac{\partial g^i}{\partial p^\alpha}P_p^{\alpha k}P_p^{il} - \frac{\partial g^i}{\partial q^\alpha}Q_p^{\alpha k}P_p^{il}\right) = 0, \quad (1.7)$$

$$\sum_{i=1}^{n}\sum_{\alpha=1}^{n}\left(\frac{\partial \sigma_r^i}{\partial p^\alpha}P_p^{\alpha k}Q_p^{il} + \frac{\partial \sigma_r^i}{\partial q^\alpha}Q_p^{\alpha k}Q_p^{il} + \frac{\partial \gamma_r^i}{\partial p^\alpha}P_p^{\alpha l}P_p^{ik} + \frac{\partial \gamma_r^i}{\partial q^\alpha}Q_p^{\alpha l}P_p^{ik}\right.$$
$$\left. - \frac{\partial \sigma_r^i}{\partial p^\alpha}P_p^{\alpha l}Q_p^{ik} - \frac{\partial \sigma_r^i}{\partial q^\alpha}Q_p^{\alpha l}Q_p^{ik} - \frac{\partial \gamma_r^i}{\partial p^\alpha}P_p^{\alpha k}P_p^{il} - \frac{\partial \gamma_r^i}{\partial q^\alpha}Q_p^{\alpha k}P_p^{il}\right) = 0, \quad (1.8)$$
$$r = 1,\ldots,m.$$

It is not difficult to check that if the functions $f^i(t,p,q)$, $g^i(t,p,q)$ are such that

$$\frac{\partial f^i}{\partial p^\alpha} + \frac{\partial g^\alpha}{\partial q^i} = 0, \quad \frac{\partial f^i}{\partial q^\alpha} = \frac{\partial f^\alpha}{\partial q^i}, \quad \frac{\partial g^i}{\partial p^\alpha} = \frac{\partial g^\alpha}{\partial p^i}, \quad i,\alpha = 1,\ldots,n, \quad (1.9)$$

then (1.7) holds, and if the functions $\sigma_r^i(t,p,q)$, $\gamma_r^i(t,p,q)$, $r = 1,\ldots,m$, are such that

$$\frac{\partial \sigma_r^i}{\partial p^\alpha} + \frac{\partial \gamma_r^\alpha}{\partial q^i} = 0, \quad \frac{\partial \sigma_r^i}{\partial q^\alpha} = \frac{\partial \sigma_r^\alpha}{\partial q^i}, \quad \frac{\partial \gamma_r^i}{\partial p^\alpha} = \frac{\partial \gamma_r^\alpha}{\partial p^i}, \quad i,\alpha = 1,\ldots,n, \quad (1.10)$$

then (1.8) holds. Thus, if the relations (1.9)-(1.10) take place, the condition (1.1) is fulfilled.

The condition (1.2) also holds when (1.9)-(1.10) are true. This can be proved analogously by using (1.5) instead of (1.4).

Now consider the condition (1.3). Clearly,

$$\sum_{i=1}^{n}\frac{D(P^i(t_0),Q^i(t_0))}{D(p^k,q^l)} = \sum_{i=1}^{n}\frac{D(p^i,q^i)}{D(p^k,q^l)} = \delta_{kl}.$$

Then the condition (1.3) is fulfilled if and only if

$$\sum_{i=1}^{n}d\frac{D(P^i(t),Q^i(t))}{D(p^k,q^l)} = 0.$$

Using the same arguments again, we prove that the relations (1.9)-(1.10) ensure this condition as well.

Finally, noting that the relations (0.2) imply (1.9)-(1.10), we obtain the following proposition.

Theorem 1.1. *The phase flow of the system of SDEs*
$$dP^i = -\frac{\partial H}{\partial q^i}(t, P, Q)dt - \sum_{r=1}^{m}\frac{\partial H_r}{\partial q^i}(t, P, Q) \circ dw_r(t)$$
$$dQ^i = \frac{\partial H}{\partial p^i}(t, P, Q)dt + \sum_{r=1}^{m}\frac{\partial H_r}{\partial p^i}(t, P, Q) \circ dw_r(t), \quad i = 1, \ldots, n,$$

with Hamiltonians $H(t,p,q)$, $H_r(t,p,q)$, $r = 1, \ldots, m$, *preserves symplectic structure.*

Corollary 1.2. *The phase flow of a Hamiltonian system with additive noise preserves symplectic structure.*

4.2 Mean-square symplectic methods for stochastic Hamiltonian systems

4.2.1 General stochastic Hamiltonian systems

As is known [261], in the case of deterministic general Hamiltonian systems symplectic Runge–Kutta (RK) methods are all implicit. Hence it is natural to expect that to construct symplectic methods for general stochastic Hamiltonian systems, fully implicit methods are needed. Here, using the results on fully implicit methods from Sect. 1.3, we construct symplectic methods for the general Hamiltonian system (0.1), (0.2). Its Ito form reads

$$dP = fdt + \frac{1}{2}\sum_{r=1}^{m}\sum_{j=1}^{n}\frac{\partial \sigma_r}{\partial p^j}\sigma_r^j dt + \frac{1}{2}\sum_{r=1}^{m}\sum_{j=1}^{n}\frac{\partial \sigma_r}{\partial q^j}\gamma_r^j dt + \sum_{r=1}^{m}\sigma_r dw_r(t)$$

$$dQ = gdt + \frac{1}{2}\sum_{r=1}^{m}\sum_{j=1}^{n}\frac{\partial \gamma_r}{\partial p^j}\sigma_r^j dt + \frac{1}{2}\sum_{r=1}^{m}\sum_{j=1}^{n}\frac{\partial \gamma_r}{\partial q^j}\gamma_r^j dt + \sum_{r=1}^{m}\gamma_r dw_r(t). \quad (2.1)$$

Introduce the following implicit method:

$$P_{k+1} = P_k + fh - \frac{1}{2}\sum_{r=1}^{m}\sum_{j=1}^{n}(\frac{\partial \sigma_r}{\partial p^j}\sigma_r^j - \frac{\partial \sigma_r}{\partial q^j}\gamma_r^j)h + \sum_{r=1}^{m}\sigma_r (\zeta_{rh})_k \sqrt{h}$$

$$Q_{k+1} = Q_k + gh - \frac{1}{2}\sum_{r=1}^{m}\sum_{j=1}^{n}(\frac{\partial \gamma_r}{\partial p^j}\sigma_r^j - \frac{\partial \gamma_r}{\partial q^j}\gamma_r^j)h + \sum_{r=1}^{m}\gamma_r (\zeta_{rh})_k \sqrt{h}, \quad (2.2)$$

where all the functions have t, P_{k+1}, Q_k as their arguments.

We recall (cf. (1.3.37)) that the random variables ζ_{rh} here are such that

$$\zeta_{rh} = \begin{cases} \xi_r, & |\xi_r| \leq A_h, \\ A_h, & \xi_r > A_h, \\ -A_h, & \xi_r < -A_h, \end{cases} \quad (2.3)$$

where ξ_r are independent $\mathcal{N}(0,1)$-distributed random variables and $A_h = \sqrt{2c|\ln h|}$, $c \geq 1$.

4.2 Mean-square symplectic methods for stochastic Hamiltonian systems

Theorem 2.1. *The implicit method (2.2) for the system (2.1), (0.2) (or for the system (0.1), (0.2)) is symplectic and of mean-square order* $1/2$.

Proof. The method (2.2) belongs to the family (1.3.57) and, consequently, the assertion about its order of convergence follows from Sect. 1.3.6. Let us prove symplecticness of the method. It is convenient to write the one-step approximation corresponding to (2.2) in the form

$$\bar{P}^i = p^i - \frac{\partial H_0}{\partial q^i} h - \frac{1}{2} \sum_{r=1}^{m} \sum_{j=1}^{n} \frac{\partial^2 H_r}{\partial q^i \partial p^j} \frac{\partial H_r}{\partial q^j} h$$

$$- \frac{1}{2} \sum_{r=1}^{m} \sum_{j=1}^{n} \frac{\partial^2 H_r}{\partial q^i \partial q^j} \frac{\partial H_r}{\partial p^j} h - \sum_{r=1}^{m} \frac{\partial H_r}{\partial q^i} \zeta_{rh} \sqrt{h}$$

$$\bar{Q}^i = q^i + \frac{\partial H_0}{\partial p^i} h + \frac{1}{2} \sum_{r=1}^{m} \sum_{j=1}^{n} \frac{\partial^2 H_r}{\partial p^i \partial p^j} \frac{\partial H_r}{\partial q^j} h$$

$$+ \frac{1}{2} \sum_{r=1}^{m} \sum_{j=1}^{n} \frac{\partial^2 H_r}{\partial p^i \partial q^j} \frac{\partial H_r}{\partial p^j} h + \sum_{r=1}^{m} \frac{\partial H_r}{\partial p^i} \zeta_{rh} \sqrt{h}, \qquad (2.4)$$

where $i = 1, \ldots, n$ and all the functions have $t, \bar{P}, q$ as their arguments. Introduce the function $F(t, p, q)$ (h, ζ_{rh} are fixed here):

$$F(t, p, q) = H_0(t, p, q)h + \frac{1}{2} \sum_{r=1}^{m} \sum_{j=1}^{n} \frac{\partial H_r}{\partial q^j}(t, p, q) \frac{\partial H_r}{\partial p^j}(t, p, q) h$$

$$+ \sum_{r=1}^{m} H_r(t, p, q) \zeta_{rh} \sqrt{h}.$$

Then (2.4) can be written as

$$\bar{P}^i = p^i - \frac{\partial F}{\partial q^i}(t, \bar{P}, q)$$

$$\bar{Q}^i = q^i + \frac{\partial F}{\partial p^i}(t, \bar{P}, q). \qquad (2.5)$$

We have

$$\sum_{i=1}^{n} d\bar{P}^i \wedge d\bar{Q}^i = \sum_{i=1}^{n} d\bar{P}^i \wedge (dq^i + \sum_{j=1}^{n} F''_{p^i p^j} d\bar{P}^j + \sum_{j=1}^{n} F''_{p^i q^j} dq^j)$$

$$= \sum_{i=1}^{n} d\bar{P}^i \wedge dq^i + \sum_{i=1}^{n} \sum_{j=1}^{n} F''_{p^i p^j} d\bar{P}^i \wedge d\bar{P}^j + \sum_{i=1}^{n} \sum_{j=1}^{n} F''_{p^i q^j} d\bar{P}^i \wedge dq^j.$$

Since $d\bar{P}^i \wedge d\bar{P}^j = -d\bar{P}^j \wedge d\bar{P}^i$, we get

218   4 Stochastic Hamiltonian systems and Langevin-type equations

$$\sum_{i=1}^{n} d\bar{P}^i \wedge d\bar{Q}^i = \sum_{i=1}^{n} d\bar{P}^i \wedge dq^i + \sum_{i=1}^{n}\sum_{j=1}^{n} F''_{p^i q^j} d\bar{P}^i \wedge dq^j$$

$$= \sum_{i=1}^{n} d\bar{P}^i \wedge dq^i + \sum_{i=1}^{n}\sum_{j=1}^{n} F''_{q^i p^j} d\bar{P}^j \wedge dq^i. \qquad (2.6)$$

Further

$$d\bar{P}^i = dp^i - \sum_{j=1}^{n} F''_{q^i p^j} d\bar{P}^j - \sum_{j=1}^{n} F''_{q^i q^j} dq^j.$$

Substituting $\sum_{j=1}^{n} F''_{q^i p^j} d\bar{P}^j$ from here in (2.6), we obtain

$$\sum_{i=1}^{n} d\bar{P}^i \wedge d\bar{Q}^i = \sum_{i=1}^{n} d\bar{P}^i \wedge dq^i + \sum_{i=1}^{n}(dp^i - d\bar{P}^i - \sum_{j=1}^{n} F''_{q^i q^j} dq^j) \wedge dq^i$$

$$= \sum_{i=1}^{n} dp^i \wedge dq^i - \sum_{i=1}^{n}\sum_{j=1}^{n} F''_{q^i q^j} dq^j \wedge dq^i = \sum_{i=1}^{n} dp^i \wedge dq^i.$$

□

A more general symplectic method for the Hamiltonian system (0.1), (0.2) has the form

$$P_{k+1} = P_k + f(t_k + \beta h, \alpha P_{k+1} + (1-\alpha)P_k, (1-\alpha)Q_{k+1} + \alpha Q_k)h$$
$$+ (\frac{1}{2} - \alpha)\sum_{r=1}^{m}\sum_{j=1}^{n}(\frac{\partial \sigma_r}{\partial p^j}\sigma_r^j - \frac{\partial \sigma_r}{\partial q^j}\gamma_r^j)h + \sum_{r=1}^{m} \sigma_r\,(\zeta_{rh})_k\,\sqrt{h}$$

$$Q_{k+1} = Q_k + g(t_k + \beta h, \alpha P_{k+1} + (1-\alpha)P_k, (1-\alpha)Q_{k+1} + \alpha Q_k)h$$
$$+ (\frac{1}{2} - \alpha)\sum_{r=1}^{m}\sum_{j=1}^{n}(\frac{\partial \gamma_r}{\partial p^j}\sigma_r^j - \frac{\partial \gamma_r}{\partial q^j}\gamma_r^j)h + \sum_{r=1}^{m} \gamma_r\,(\zeta_{rh})_k\,\sqrt{h}, \qquad (2.7)$$

where $\sigma_r, \gamma_r, r = 1, \ldots, m$, and their derivatives are calculated at $(t_k, \alpha P_{k+1} + (1-\alpha)P_k, (1-\alpha)Q_{k+1} + \alpha Q_k)$, and $\alpha, \beta \in [0,1]$ are parameters.

Using arguments similar to ones in the proof of Theorem 2.1, we obtain the theorem.

Theorem 2.2. *The implicit method (2.7) for the system (0.1), (0.2) (or for the system (2.1), (0.2)) is symplectic and of mean-square order* $1/2$.

The method (2.2) is a particular case of (2.7) when $\alpha = 1$, $\beta = 0$. If $\alpha = \beta = 1/2$ the method (2.7) becomes the midpoint method (cf. (1.3.61)):

4.2 Mean-square symplectic methods for stochastic Hamiltonian systems

$$P_{k+1} = P_k + f(t_k + \frac{h}{2}, \frac{P_k + P_{k+1}}{2}, \frac{Q_k + Q_{k+1}}{2})h$$

$$+ \sum_{r=1}^{m} \sigma_r(t_k, \frac{P_k + P_{k+1}}{2}, \frac{Q_k + Q_{k+1}}{2})(\zeta_{rh})_k \sqrt{h}$$

$$Q_{k+1} = Q_k + g(t_k + \frac{h}{2}, \frac{P_k + P_{k+1}}{2}, \frac{Q_k + Q_{k+1}}{2})h$$

$$+ \sum_{r=1}^{m} \gamma_r(t_k, \frac{P_k + P_{k+1}}{2}, \frac{Q_k + Q_{k+1}}{2})(\zeta_{rh})_k \sqrt{h}. \qquad (2.8)$$

Remark 2.3. In the commutative case, i.e., when $\Lambda_i b_r = \Lambda_r b_i$ or in the case of a system with one noise (i.e., $m = 1$) the symplectic method (2.8) for (0.1), (0.2) has the first mean-square order of convergence.

Remark 2.4. In the case of Hamiltonians that are separable in the noise part, i.e., when $H_r(t, p, q) = U_r(t, q) + V_r(t, p)$, $r = 1, \ldots, m$, we can obtain symplectic methods for (0.1), (0.2) which are explicit in stochastic terms and do not need truncated random variables. For instance, (2.2) acquires the form

$$P_{k+1} = P_k + f(t_k, P_{k+1}, Q_k)h$$

$$+ \frac{h}{2} \sum_{r=1}^{m} \sum_{j=1}^{n} \frac{\partial \sigma_r}{\partial q^j}(t_k, Q_k) \cdot \gamma_r^j(P_{k+1}) + \sum_{r=1}^{m} \sigma_r(t_k, Q_k) \Delta_k w_r,$$

$$Q_{k+1} = Q_k + g(t_k, P_{k+1}, Q_k)h$$

$$- \frac{h}{2} \sum_{r=1}^{m} \sum_{j=1}^{n} \frac{\partial \gamma_r}{\partial p^j}(P_{k+1}) \cdot \sigma_r^j(t_k, Q_k) + \sum_{r=1}^{m} \gamma_r(t_k, P_{k+1}) \Delta_k w_r. \qquad (2.9)$$

Note that the method (2.9) is implicit in the deterministic terms. See fully explicit symplectic methods for some systems with separable Hamiltonians in the next remark and subsection.

Of course, if it is necessary, fully implicit methods which require truncated random variables can be used in the case of separable Hamiltonians as well.

Remark 2.5. It is possible to construct fully explicit symplectic methods for the following partitioned system:

$$dP = f(t, Q)dt + \sum_{r=1}^{m} \sigma_r(t, Q) \circ dw_r(t), \ P(t_0) = p,$$

$$dQ = g(P)dt + \sum_{r=1}^{m} \gamma_r(t)dw_r(t), \ Q(t_0) = q, \qquad (2.10)$$

with $f^i = -\partial U_0/\partial q^i$, $g^i = \partial V_0/\partial p^i$, $\sigma_r^i = -\partial U_r/\partial q^i$, $r = 1, \ldots, m$, $i = 1, \ldots, n$.

For instance, the explicit partitioned Runge–Kutta (PRK) method (cf. (2.17)-(2.18))

$$\mathcal{Q}_1 = Q_k + \alpha h g(P_k),$$

$$\mathcal{P}_1 = P_k + h f(t_k + \alpha h, \mathcal{Q}_1) + \frac{h}{2} \sum_{r=1}^{m} \sum_{j=1}^{n} \frac{\partial \sigma_r}{\partial q^j}(t_k, \mathcal{Q}_1) \cdot \gamma_r^j(t_k),$$

$$\mathcal{Q}_2 = \mathcal{Q}_1 + (1-\alpha) h g(\mathcal{P}_1), \qquad (2.11)$$

$$P_{k+1} = \mathcal{P}_1 + \sum_{r=1}^{m} \sigma_r(t_k, \mathcal{Q}_2) \Delta_k w_r,$$

$$Q_{k+1} = \mathcal{Q}_2 + \sum_{r=1}^{m} \gamma_r(t_k) \Delta_k w_r, \ k = 0, \ldots, N-1, \qquad (2.12)$$

with the parameter $0 \le \alpha \le 1$ is symplectic and of the mean-square order $1/2$.

A particular case of the system (2.10) is considered in the next subsection, where explicit symplectic methods of a higher order are proposed.

4.2.2 Explicit methods in the case of separable Hamiltonians

Consider a special case of the Hamiltonian system (0.1), (0.2) such that

$$H_0(t,p,q) = V_0(p) + U_0(t,q), \ H_r(t,p,q) = U_r(t,q), \ r=1,\ldots,m. \qquad (2.13)$$

In this case we get the following system in the sense of Stratonovich

$$dP = f(t,Q)dt + \sum_{r=1}^{m} \sigma_r(t,Q) \circ dw_r(t), \ P(t_0) = p,$$

$$dQ = g(P)dt, \ Q(t_0) = q, \qquad (2.14)$$

with

$$f^i = -\partial U_0/\partial q^i, \ g^i = \partial V_0/\partial p^i, \ \sigma_r^i = -\partial U_r/\partial q^i, \qquad (2.15)$$
$$r = 1, \ldots, m, \ i = 1, \ldots, n.$$

We note that it is not difficult to consider a slightly more general separable Hamiltonian $H_0(t,p,q) = V_0(t,p) + U_0(t,q)$ but we restrict ourselves to H_0 from (2.13). It is obvious that the system (2.14) has the same form in the sense of Ito.

For $V_0(p) = \frac{1}{2}(M^{-1}p, p)$ with M a constant, symmetric, invertible matrix, the system (2.14) takes the form

$$dP = f(t,Q)dt + \sum_{r=1}^{m} \sigma_r(t,Q) dw_r(t), \ P(t_0) = p,$$

$$dQ = M^{-1}Pdt, \ Q(t_0) = q. \qquad (2.16)$$

4.2 Mean-square symplectic methods for stochastic Hamiltonian systems

This system can be written as a second-order differential equation with multiplicative noise.

Due to specific features of the system (2.14), (2.15), we have succeeded in construction of explicit partitioned Runge–Kutta (PRK) methods of a higher order.

First-order methods. A PRK method for (2.14) has the form (cf. (2.11)-(2.12)):

$$\mathcal{Q}_1 = Q_k + \alpha h g(P_k), \quad \mathcal{P}_1 = P_k + h f(t_k + \alpha h, \mathcal{Q}_1),$$
$$\mathcal{Q}_2 = \mathcal{Q}_1 + (1-\alpha) h g(\mathcal{P}_1), \qquad (2.17)$$

$$P_{k+1} = \mathcal{P}_1 + \sum_{r=1}^{m} \sigma_r(t_k, \mathcal{Q}_2) \Delta_k w_r, \quad Q_{k+1} = \mathcal{Q}_2, \quad k = 0, \ldots, N-1, \quad (2.18)$$

where $0 \leq \alpha \leq 1$ is a parameter.

Theorem 2.6. *The explicit method (2.17)-(2.18) for the system (2.14) with (2.15) is symplectic and of first mean-square order.*

Proof. In the case of the system (2.14) the operators Λ_r take the form $\Lambda_r = (\sigma_r, \partial/\partial p)$. Since σ_r do not depend on p, we get $\Lambda_i \sigma_j = 0$. It is known (see Sect. 1.2.3) that in such a case the Euler method has the first mean-square order of accuracy. Comparing the method (2.17)-(2.18) with the Euler method, it is not difficult to get that the method (2.17)-(2.18) is of the first mean-square order as well.

Due to (2.15), $\partial \sigma_r^i / \partial q^j = \partial \sigma_r^j / \partial q^i$. Using this, we obtain $dP_{k+1} \wedge dQ_{k+1} = d\mathcal{P}_1 \wedge d\mathcal{Q}_2$. It is easy to prove that $d\mathcal{P}_1 \wedge d\mathcal{Q}_2 = d\mathcal{P}_1 \wedge d\mathcal{Q}_1 = dP_k \wedge dQ_k$. Therefore the method (2.17)-(2.18) is symplectic. □

Remark 2.7. By swapping the roles of p and q, we can propose the following symplectic method of the first mean-square order for the system (2.14)-(2.15):

$$\mathcal{P} = P_k + \alpha h f(t_k, Q_k), \quad \mathcal{Q} = Q_k + h g(\mathcal{P}) \qquad (2.19)$$

$$P_{k+1} = \mathcal{P} + (1-\alpha) h f(t_{k+1}, \mathcal{Q}) + \sum_{r=1}^{m} \sigma_r(t_k, \mathcal{Q}) \Delta_k w_r, \quad Q_{k+1} = \mathcal{Q}. \quad (2.20)$$

Methods of order 3/2. Consider the relations

$$\mathcal{P}_i = p + h \sum_{j=1}^{s} \alpha_{ij} f(t + c_j h, \mathcal{Q}_j) + \sum_{j=1}^{s} \sum_{r=1}^{m} \sigma_r(t + d_j h, \mathcal{Q}_j) \left(\lambda_{ij} \varphi_r + \mu_{ij} \psi_r \right),$$

$$\mathcal{Q}_i = q + h \sum_{j=1}^{s} \hat{\alpha}_{ij} g(\mathcal{P}_j), \quad i = 1, \ldots, s, \qquad (2.21)$$

$$\bar{P} = p + h\sum_{i=1}^{s}\beta_i f(t+c_i h, \mathcal{Q}_i) + \sum_{i=1}^{s}\sum_{r=1}^{m}\sigma_r(t+d_i h, \mathcal{Q}_i)(\nu_i\varphi_r + \varkappa_i\psi_r),$$

$$\bar{Q} = q + h\sum_{i=1}^{s}\hat{\beta}_i g(\mathcal{P}_i), \qquad (2.22)$$

where φ_r, ψ_r do not depend on p and q, the parameters α_{ij}, $\hat{\alpha}_{ij}$, β_i, $\hat{\beta}_i$, λ_{ij}, μ_{ij}, ν_i, $\varkappa_i$ satisfy the conditions

$$\beta_i\hat{\alpha}_{ij} + \hat{\beta}_j\alpha_{ji} - \beta_i\hat{\beta}_j = 0, \quad \nu_i\hat{\alpha}_{ij} + \hat{\beta}_j\lambda_{ji} - \nu_i\hat{\beta}_j = 0,$$
$$\varkappa_i\hat{\alpha}_{ij} + \hat{\beta}_j\mu_{ji} - \varkappa_i\hat{\beta}_j = 0, \quad i,j = 1,\ldots,s, \qquad (2.23)$$

and c_i, d_i are arbitrary parameters.

If $\sigma_r \equiv 0$, the relations (2.21)-(2.22) coincide with a general form of s-stage PRK methods for deterministic differential equations (see, e.g., [261, p. 34]). It is known [261, 277] that the symplectic condition holds for $\bar{P}$, $\bar{Q}$ from (2.21)-(2.22) with (2.23) in the case of $\sigma_r \equiv 0$. By a generalization of the proof of Theorem 6.2 from [261], we prove the following lemma.

Lemma 2.8. *The relations* (2.21)-(2.22) *with conditions* (2.23) *preserve symplectic structure, i.e.,* $d\bar{P} \wedge d\bar{Q} = dp \wedge dq$.

Proof. Denote for a while:

$$f_i = f(t+c_i h, \mathcal{Q}_i), \; g_i = g(\mathcal{P}_i), \; \sigma_{ri} = \sigma_r(t+d_i h, \mathcal{Q}_i).$$

We get

$$d\bar{P} \wedge d\bar{Q} = dp \wedge dq + h\sum_{j=1}^{s}\hat{\beta}_j dp \wedge dg_j + h\sum_{i=1}^{s}\beta_i df_i \wedge dq$$
$$+ h^2\sum_{i=1}^{s}\sum_{j=1}^{s}\beta_i\hat{\beta}_j df_i \wedge dg_j + \sum_{i=1}^{s}\sum_{r=1}^{m}(\nu_i\varphi_r + \varkappa_i\psi_r)\, d\sigma_{ri} \wedge dq$$
$$+ h\sum_{i=1}^{s}\sum_{j=1}^{s}\sum_{r=1}^{m}(\nu_i\varphi_r + \varkappa_i\psi_r)\hat{\beta}_j d\sigma_{ri} \wedge dg_j. \qquad (2.24)$$

Then we express $dp \wedge dg_j$ from

$$d\mathcal{P}_j \wedge dg_j = dp \wedge dg_j + h\sum_{i=1}^{s}\alpha_{ji}df_i \wedge dg_j + \sum_{i=1}^{s}\sum_{r=1}^{m}\left(\lambda_{ji}\varphi_r + \mu_{ji}\psi_r\right)d\sigma_{ri} \wedge dg_j$$

and substitute it in (2.24). Analogously, we act with $df_i \wedge dq$ and $d\sigma_{ri} \wedge dq$ finding them from the expressions for $df_i \wedge d\mathcal{Q}_i$ and $d\sigma_{ri} \wedge d\mathcal{Q}_i$. As a result, using (2.23), we obtain

4.2 Mean-square symplectic methods for stochastic Hamiltonian systems

$$d\bar{P} \wedge d\bar{Q} = dp \wedge dq + h \sum_{i=1}^{s} \hat{\beta}_i dP_i \wedge dg_i + h \sum_{i=1}^{s} \beta_i df_i \wedge dQ_i$$

$$+ \sum_{i=1}^{s} \sum_{r=1}^{m} (\nu_i \varphi_r + \varkappa_i \psi_r) \, d\sigma_{ri} \wedge dQ_i.$$

Taking into account that the wedge product is skew-symmetric, the vector-functions f, g, σ_r are gradients, f, σ_r do not depend on p, and g does not depend on q, it is not difficult to see that each of the terms $dP_i \wedge dg_i$, $df_i \wedge dQ_i$, $d\sigma_{ri} \wedge dQ_i$ vanishes. Therefore $d\bar{P} \wedge d\bar{Q} = dp \wedge dq$. □

Introduce the 2-stage explicit PRK method for the system (2.14), (2.15):

$$\mathcal{Q}_1 = Q_k, \quad \mathcal{P}_1 = P_k + \frac{h}{4} f(t_k, \mathcal{Q}_1) + \frac{1}{2} \sum_{r=1}^{m} \sigma_r(t_k, \mathcal{Q}_1) \left(3(J_{r0})_k - \Delta_k w_r\right),$$

$$\mathcal{Q}_2 = \mathcal{Q}_1 + \frac{2}{3} h g(\mathcal{P}_1),$$

$$\mathcal{P}_2 = \mathcal{P}_1 + \frac{3}{4} h f(t_k + \frac{2}{3} h, \mathcal{Q}_2) + \frac{3}{2} \sum_{r=1}^{m} \sigma_r(t_k + \frac{2}{3} h, \mathcal{Q}_2) \left(-(J_{r0})_k + \Delta_k w_r\right),$$

(2.25)

$$P_{k+1} = \mathcal{P}_2, \quad Q_{k+1} = \mathcal{Q}_2 + \frac{h}{3} g(\mathcal{P}_2), \quad k = 0, \ldots, N-1, \tag{2.26}$$

where

$$J_{r0} := \frac{1}{h} \int_{t}^{t+h} (w_r(\vartheta) - w_r(t)) \, d\vartheta. \tag{2.27}$$

We recall (see 1.5.7) that the random variables $\Delta_k w_r(h)$, $(J_{r0})_k$ have a Gaussian joint distribution, and they can be simulated at each step by $2m$ independent $\mathcal{N}(0,1)$-distributed random variables ξ_{rk} and η_{rk}, $r = 0, \ldots, m$:

$$\Delta_k w_r(h) = \xi_{rk} \sqrt{h}, \quad (J_{r0})_k = \left(\xi_{rk}/2 + \eta_{rk}/\sqrt{12}\right) \sqrt{h}.$$

As a result, the method (2.25)-(2.26) takes the constructive form.

Theorem 2.9. *The explicit PRK method (2.25)-(2.26) for system (2.14), (2.15) preserves symplectic structure and has the mean-square order 3/2.*

Proof. The method (2.25)-(2.26) has the form of (2.21)-(2.22) and its parameters satisfy the conditions (2.23). Then, Lemma 2.8 implies that this method preserves symplectic structure.

The mean-square order of convergence of (2.25)-(2.26) is proved using the general theory of numerical integration of SDEs of Chap. 1 (see details in [195]). □

Remark 2.10. In the case of $\sigma_r = 0$, $r = 1, \ldots, m$, the method (2.25)-(2.26) coincides with the well-known deterministic symplectic PRK method of the second order. Adapting other explicit deterministic second-order PRK methods from [261, 277], it is possible to construct other explicit symplectic methods of the order 3/2 for the system (2.14), (2.15).

Remark 2.11. In the case of a more general system than (2.14) methods of the order 3/2 require simulation of repeated Ito integrals which is a laborious problem from the computational point of view.

Lemma 2.8 can be generalized for the general separable case, i.e., for the system (0.1), (0.2) with $H_r = V_r(p) + U_r(t, q)$, $r = 0, 1, \ldots, m$, and it can also be generalized for the general stochastic Hamiltonian system (0.1), (0.2). In the case of systems with one noise repeated Ito integrals can effectively be simulated and generalizations of Lemma 2.8 can be used for constructing high-order symplectic methods for Hamiltonian systems with one noise (i.e., when $m = 1$).

4.3 Mean-square symplectic methods for Hamiltonian systems with additive noise

4.3.1 The case of a general Hamiltonian

In this subsection we consider the general Hamiltonian system with additive noise

$$dP = f(t, P, Q)dt + \sum_{r=1}^{m} \sigma_r(t)dw(t), \ P(t_0) = p,$$

$$dQ = g(t, P, Q)dt + \sum_{r=1}^{m} \gamma_r(t)dw(t), \ Q(t_0) = q, \quad (3.1)$$

$$f^i = -\partial H/\partial q^i, \qquad g^i = \partial H/\partial p^i, \ i = 1, \ldots, n, \quad (3.2)$$

where P, Q, f, g, σ_r, γ_r are n-dimensional column-vectors, $w_r(t)$, $r = 1, \ldots, m$, are independent standard Wiener processes, and $H(t, p, q)$ is a Hamiltonian.

First-order methods. Consider the two-parameter family of implicit methods

$$\mathcal{P} = P_k + hf(t_k + \beta h, \alpha \mathcal{P} + (1-\alpha)P_k, (1-\alpha)\mathcal{Q} + \alpha Q_k),$$
$$\mathcal{Q} = Q_k + hg(t_k + \beta h, \alpha \mathcal{P} + (1-\alpha)P_k, (1-\alpha)\mathcal{Q} + \alpha Q_k), \quad (3.3)$$

4.3 Mean-square methods for Hamiltonian systems with additive noise

$$P_{k+1} = \mathcal{P} + \sum_{r=1}^{m} \sigma_r(t_k)\Delta_k w_r, \quad Q_{k+1} = \mathcal{Q} + \sum_{r=1}^{m} \gamma_r(t_k)\Delta_k w_r, \quad (3.4)$$

$$k = 0, \ldots, N-1,$$

where $\Delta_k w_r(h) := w_r(t_k + h) - w_r(t_k)$ and the parameters $\alpha, \beta \in [0, 1]$.

When $\sigma_r = 0$, $\gamma_r = 0$, $r = 1, \ldots, m$, this family coincides with the known family of symplectic methods for deterministic Hamiltonian systems (see [277]).

The following lemma guarantees the unique solvability of (3.3) with respect to $\mathcal{P}$, $\mathcal{Q}$ for any P_k, Q_k and sufficiently small h.

Lemma 3.1. *Let $F(x; c, s)$ be a continuous d-dimensional vector-function depending on $x \in \mathbf{R}^d$, $c \in \mathbf{R}^d$, and $s \in S$, where S is a set from an R^l. Suppose F has the first partial derivatives $\partial F^i/\partial x^j$, $i, j = 1, \ldots, d$, which are uniformly bounded in $\mathbf{R}^d \times \mathbf{R}^d \times S$. Then there is an $h_0 > 0$ such that the equation*

$$x = c + hF(x; c, s) + \nu \quad (3.5)$$

is uniquely solvable with respect to x for $0 < h \leq h_0$ and any $c \in \mathbf{R}^d$, $\nu \in \mathbf{R}^d$, $s \in S$. The solution of equation (3.5) can be found by the method of simple iteration with an arbitrary initial approximation.

The proof of this lemma is not difficult and it is omitted. The next lemma is true for system (3.1) with arbitrary f and g (i.e., f and g may not obey the condition (3.2)).

Lemma 3.2. *The mean-square order of the methods (3.3)-(3.4) for the system (3.1) is equal to 1.*

The proof is based on comparison of the one-step approximation of the method (3.3)-(3.4) with the one-step approximation of the Euler method.

The one-step approximation $\tilde{P}$, $\tilde{Q}$ of the method (3.3)-(3.4) is such that $d\tilde{P} = d\mathcal{P}$, $d\tilde{Q} = d\mathcal{Q}$. Hence $d\tilde{P} \wedge d\tilde{Q} = d\mathcal{P} \wedge d\mathcal{Q}$. The relations for $\mathcal{P}$, $\mathcal{Q}$ coincide with ones for the one-step approximation corresponding to the deterministic symplectic method [277]. Therefore, the method (3.3)-(3.4) is symplectic as well. From here and Lemma 3.2, we get the theorem.

Theorem 3.3. *The method (3.3)-(3.4) for the system (3.1)-(3.2) preserves symplectic structure and has the first mean-square order of convergence.*

Now consider another family of symplectic methods for system (3.1):

$$P_{k+1} = P_k + hf(t_k + \beta h, \alpha P_{k+1} + (1-\alpha)P_k, (1-\alpha)Q_{k+1} + \alpha Q_k)$$
$$+ \sum_{r=1}^{m} \sigma_r(t_k)\Delta_k w_r,$$
$$Q_{k+1} = Q_k + hg(t_k + \beta h, \alpha P_{k+1} + (1-\alpha)P_k, (1-\alpha)Q_{k+1} + \alpha Q_k)$$
$$+ \sum_{r=1}^{m} \gamma_r(t_k)\Delta_k w_r, \quad k = 0, \ldots, N-1, \quad (3.6)$$

with the parameters $\alpha, \beta \in [0, 1]$.

For sufficiently small h, the equations (3.6) are uniquely solvable with respect to P_{k+1}, Q_{k+1} according to Lemma 3.1.

Theorem 3.4. *The method (3.6) for the system (3.1)-(3.2) preserves symplectic structure and has the first mean-square order of convergence.*

Proof. Comparing the one-step approximation of the method (3.6) with the one-step approximation of the Euler method, one can establish that the mean-square order of the method (3.6) is equal to 1.

Now we check symplecticness of the method. Let $\tilde{P}$, $\tilde{Q}$ be the one-step approximation corresponding to the method (3.6). Introduce

$$\hat{p} = p + \alpha \sum_{r=1}^{m} \sigma_r(t)\Delta w_r, \quad \hat{q} = q + (1-\alpha)\sum_{r=1}^{m} \gamma_r(t)\Delta w_r,$$

$$\hat{P} = \tilde{P} - (1-\alpha)\sum_{r=1}^{m} \sigma_r(t)\Delta w_r, \quad \hat{Q} = \tilde{Q} - \alpha \sum_{r=1}^{m} \gamma_r(t)\Delta w_r.$$

We have

$$\hat{P} = \hat{p} + hf(t+\beta h, \alpha\hat{P} + (1-\alpha)\hat{p}, (1-\alpha)\hat{Q} + \alpha\hat{q}),$$

$$\hat{Q} = \hat{q} + hg(t+\beta h, \alpha\hat{P} + (1-\alpha)\hat{p}, (1-\alpha)\hat{Q} + \alpha\hat{q}).$$

The relations for $\hat{P}$, $\hat{Q}$ coincide with the one-step approximation corresponding to the symplectic deterministic method. Therefore, $d\hat{P} \wedge d\hat{Q} = d\hat{p} \wedge d\hat{q}$. Further, it is obvious that $d\hat{P} \wedge d\hat{Q} = d\tilde{P} \wedge d\tilde{Q}$ and $d\hat{p} \wedge d\hat{q} = dp \wedge dq$. Consequently, $d\tilde{P} \wedge d\tilde{Q} = dp \wedge dq$, i.e., the method (3.6) is symplectic. □

Methods of order 3/2. For $i = 1, \ldots, s$, consider the relations

$$\mathcal{P}_i = p + h\sum_{j=1}^{s} \alpha_{ij} f(t + c_j h, \mathcal{P}_j, \mathcal{Q}_j) + \varphi_i,$$

$$\mathcal{Q}_i = q + h\sum_{j=1}^{s} \alpha_{ij} g(t + c_j h, \mathcal{P}_j, \mathcal{Q}_j) + \psi_i, \qquad (3.7)$$

$$\bar{P} = p + h\sum_{i=1}^{s} \beta_i f(t + c_i h, \mathcal{P}_i, \mathcal{Q}_i) + \eta,$$

$$\bar{Q} = q + h\sum_{i=1}^{s} \beta_i g(t + c_i h, \mathcal{P}_i, \mathcal{Q}_i) + \zeta, \qquad (3.8)$$

where φ_i, ψ_i, η, ζ do not depend on p and q, the parameters α_{ij} and β_i satisfy the conditions

4.3 Mean-square methods for Hamiltonian systems with additive noise

$$\beta_i\alpha_{ij} + \beta_j\alpha_{ji} - \beta_i\beta_j = 0, \quad i,j = 1,\ldots,s, \quad (3.9)$$

and c_i are arbitrary parameters.

The equations (3.7) are uniquely solvable with respect to $\mathcal{P}_i$, $\mathcal{Q}_i$, $i = 1,\ldots,s$, for any p, q, φ_i, ψ_i, η, ζ and sufficiently small h according to Lemma 3.1.

If $\varphi_i = \psi_i = \eta = \zeta = 0$, the relations (3.7)-(3.8) coincide with a general form of s-stage Runge–Kutta (RK) methods for deterministic differential equations. It is known (see, e.g., Theorem 6.1 in [261]) that the symplectic condition $d\bar{P} \wedge d\bar{Q} = dp \wedge dq$ holds for $\bar{P}$, $\bar{Q}$ from (3.7)-(3.8) with (3.9) and $\varphi_i = \psi_i = \eta = \zeta = 0$. Generalizing this result for arbitrary $\varphi_i, \psi_i, \eta, \zeta$, we obtain the following lemma.

Lemma 3.5. *The relations* (3.7)-(3.8) *with condition* (3.9) *preserve symplectic structure, i.e.,* $d\bar{P} \wedge d\bar{Q} = dp \wedge dq$.

The lemma is proved in [194] by a generalization of the proof of Theorem 6.1 from [261] (see also Lemma 2.8 in the previous section).

The next lemma will be used in Theorem 3.7 for the Hamiltonian system (3.1)-(3.2). Consider the general (not necessarily Hamiltonian) system with additive noise

$$dX = a(t,X)dt + \sum_{r=1}^{m} b_r(t)dw_r(t), \quad X(t_0) = X_0, \quad (3.10)$$

and introduce the parametric family of one-step approximations for (3.10):

$$\mathsf{X}_1 = x + \frac{\alpha}{2}ha(t + \frac{\alpha}{2}h, \mathsf{X}_1) + \sum_{r=1}^{m} b_r(t)\left(\lambda_1 J_{r0} + \mu_1 \Delta w_r\right),$$

$$\mathsf{X}_2 = x + \alpha ha(t + \frac{\alpha}{2}h, \mathsf{X}_1) + \frac{1-\alpha}{2}ha(t + \frac{1+\alpha}{2}h, \mathsf{X}_2)$$

$$+ \sum_{r=1}^{m} b_r(t)\left(\lambda_2 J_{r0} + \mu_2 \Delta w_r\right),$$

$$\bar{X} = x + h\left[\alpha a(t + \frac{\alpha}{2}h, \mathsf{X}_1) + (1-\alpha)a(t + \frac{1+\alpha}{2}h, \mathsf{X}_2)\right]$$

$$+ \sum_{r=1}^{m} b_r(t)\Delta w_r + \sum_{r=1}^{m} b'_r(t)I_{0r}, \quad (3.11)$$

where

$$\Delta w_r := w_r(t+h) - w_r(t), \quad I_{0r} := \int_t^{t+h}(\vartheta - t)\,dw_r(\vartheta),$$

$$J_{r0} := \frac{1}{h}\int_t^{t+h}(w_r(\vartheta) - w_r(t))\,d\vartheta, \quad (3.12)$$

and the parameters α, λ_1, λ_2, μ_1, μ_2 are such that

$$\alpha\lambda_1 + (1-\alpha)\lambda_2 = 1, \quad \alpha\mu_1 + (1-\alpha)\mu_2 = 0, \tag{3.13}$$

$$\alpha\left(\frac{\lambda_1^2}{3} + \lambda_1\mu_1 + \mu_1^2\right) + (1-\alpha)\left(\frac{\lambda_2^2}{3} + \lambda_2\mu_2 + \mu_2^2\right) = \frac{1}{2}. \tag{3.14}$$

For example, the following set of parameters satisfies (3.13)-(3.14):

$$\alpha = \frac{1}{2}, \ \lambda_1 = \lambda_2 = 1, \ \mu_1 = -\mu_2 = \frac{1}{\sqrt{6}}. \tag{3.15}$$

Note that the random variables Δw_r and J_{r0} are of the same mean-square order $O(h^{1/2})$.

Lemma 3.6. *The method based on the one-step approximation (3.11) with conditions (3.13)-(3.14) is of the mean-square order 3/2.*

Proof. Due to properties of the Wiener process and Ito integrals, we get

$$E\Delta w_i = 0, \ E\Delta w_i \Delta w_j = \delta_{ij}h, \ E\Delta w_i \Delta w_j \Delta w_k = 0, \ E(\Delta w_i)^4 = 3h^2,$$

$$EJ_{i0} = 0, \ EJ_{i0}J_{j0} = \delta_{ij}\frac{h}{3}, \ EJ_{i0}J_{j0}J_{k0} = 0, \ E(J_{i0})^4 = \frac{h^2}{3},$$

$$E\Delta w_i J_{j0} = \delta_{ij}\frac{h}{2}, \ E\Delta w_i \Delta w_j J_{k0} = 0, \ E\Delta w_i J_{j0} J_{k0} = 0. \tag{3.16}$$

Let $\Delta X_i := X_i - x$, $i = 1, 2$. We have

$$|E\Delta X_i| = O(h), \ E(\Delta X_i)^{2l} = O(h^l), \ l = 1, 2, 3, 4, \ i = 1, 2,$$

$$\left|E(\Delta X_i)^3\right| = O(h^2). \tag{3.17}$$

Expand (3.11):

$$\Delta X_1 = \frac{\alpha}{2}ha(t,x) + \sum_{r=1}^{m} b_r(t)(\lambda_1 J_{r0} + \mu_1 \Delta w_r) + \rho_1, \tag{3.18}$$

$$\Delta X_2 = \frac{1+\alpha}{2}ha(t,x) + \sum_{r=1}^{m} b_r(t)(\lambda_2 J_{r0} + \mu_2 \Delta w_r) + \rho_2, \tag{3.19}$$

$$\bar{X} = x + \sum_{r=1}^{m} b_r(t)\Delta w_r + \sum_{r=1}^{m} b'_r(t) I_{0r} + ha(t,x)$$

$$+ h\sum_{i=1}^{d} \frac{\partial a}{\partial x^i}(t,x)\left(\alpha\Delta X_1^i + (1-\alpha)\Delta X_2^i\right) + \frac{h^2}{2}\frac{\partial a}{\partial t}(t,x)$$

$$+ \frac{h}{2}\sum_{i,j=1}^{d} \frac{\partial^2 a}{\partial x^i \partial x^j}(t,x)\left(\alpha\Delta X_1^i \Delta X_1^j + (1-\alpha)\Delta X_2^i \Delta X_2^j\right) + \bar{\rho}. \tag{3.20}$$

4.3 Mean-square methods for Hamiltonian systems with additive noise

Using (3.16)-(3.17), one can obtain

$$|E\rho_i| = O(h^2), \ |E\rho_i^l \Delta X_i^k| = O(h^2), \ E\rho_i^2 = O(h^3), \quad (3.21)$$

$$|E\bar{\rho}| = O(h^3), \ E\bar{\rho}^2 = O(h^5). \quad (3.22)$$

Substituting (3.18)-(3.19) in (3.20) and using (3.13), we get

$$\bar{X} = x + \sum_{r=1}^{m} b_r \Delta w_r + \sum_{r=1}^{m} b'_r I_{0r} + ha + \frac{h^2}{2}\frac{\partial a}{\partial t} + \frac{h^2}{2}\sum_{i=1}^{d}\frac{\partial a}{\partial x^i}a^i$$

$$+ h\sum_{r=1}^{m}\sum_{i=1}^{d} b_r^i \frac{\partial a}{\partial x^i} J_{r0} + \frac{h^2}{4}\sum_{r=1}^{m}\sum_{i,j=1}^{d}\frac{\partial^2 a}{\partial x^i \partial x^j}b_r^i b_r^j + R, \quad (3.23)$$

$$R = \frac{h}{2}\sum_{r,l=1}^{m}\sum_{i,j=1}^{d}\frac{\partial^2 a}{\partial x^i \partial x^j}b_r^i b_l^j \left[\alpha\left(\lambda_1 J_{r0} + \mu_1 \Delta w_r\right)\left(\lambda_1 J_{l0} + \mu_1 \Delta w_l\right)\right.$$

$$\left.+(1-\alpha)\left(\lambda_2 J_{r0} + \mu_2 \Delta w_r\right)\left(\lambda_2 J_{l0} + \mu_2 \Delta w_l\right)\right] - \frac{h^2}{4}\sum_{r=1}^{m}\sum_{i,j=1}^{d}\frac{\partial^2 a}{\partial x^i \partial x^j}b_r^i b_r^j + \rho,$$

where the coefficients and their derivatives are calculated at (t,x) and ρ satisfies the same relations as $\bar{\rho}$ (see (3.22)).

The relations (3.16) and (3.14) imply

$$E[\alpha\left(\lambda_1 J_{r0} + \mu_1 \Delta w_r\right)\left(\lambda_1 J_{l0} + \mu_1 \Delta w_l\right)$$

$$+(1-\alpha)\left(\lambda_2 J_{r0} + \mu_2 \Delta w_r\right)\left(\lambda_2 J_{l0} + \mu_2 \Delta w_l\right)] = \frac{h}{2}\delta_{rl}. \quad (3.24)$$

Using (3.16), (3.21)-(3.22), and (3.24), it is not difficult to get that

$$|ER| = O(h^3), \ \left(ER^2\right)^{1/2} = O(h^2). \quad (3.25)$$

Comparing (3.23) with the one-step approximation of the standard method of mean-square order $3/2$ for systems with additive noise (1.5.8), we obtain that the method (3.11) is of mean-square order $3/2$. □

Now we return to the Hamiltonian system with additive noise (3.1). Consider the parametric family of methods:

$$\mathcal{P}_1 = P_k + \frac{\alpha}{2}hf(t_k + \frac{\alpha}{2}h, \mathcal{P}_1, \mathcal{Q}_1) + \sum_{r=1}^{m}\sigma_r(t_k)\left(\lambda_1\left(J_{r0}\right)_k + \mu_1 \Delta_k w_r\right),$$

$$\mathcal{Q}_1 = Q_k + \frac{\alpha}{2}hg(t_k + \frac{\alpha}{2}h, \mathcal{P}_1, \mathcal{Q}_1) + \sum_{r=1}^{m}\gamma_r(t_k)\left(\lambda_1\left(J_{r0}\right)_k + \mu_1 \Delta_k w_r\right),$$

$$\mathcal{P}_2 = P_k + \alpha h f(t_k + \frac{\alpha}{2}h, \mathcal{P}_1, \mathcal{Q}_1) + \frac{1-\alpha}{2} h f(t_k + \frac{1+\alpha}{2}h, \mathcal{P}_2, \mathcal{Q}_2)$$
$$+ \sum_{r=1}^{m} \sigma_r(t_k) \left(\lambda_2 (J_{r0})_k + \mu_2 \Delta_k w_r\right),$$
$$\mathcal{Q}_2 = Q_k + \alpha h g(t_k + \frac{\alpha}{2}h, \mathcal{P}_1, \mathcal{Q}_1) + \frac{1-\alpha}{2} h g(t_k + \frac{1+\alpha}{2}h, \mathcal{P}_2, \mathcal{Q}_2)$$
$$+ \sum_{r=1}^{m} \gamma_r(t_k) \left(\lambda_2 (J_{r0})_k + \mu_2 \Delta_k w_r\right),$$
$$P_{k+1} = P_k + h \left[\alpha f(t_k + \frac{\alpha}{2}h, \mathcal{P}_1, \mathcal{Q}_1) + (1-\alpha) f(t_k + \frac{1+\alpha}{2}h, \mathcal{P}_2, \mathcal{Q}_2)\right]$$
$$+ \sum_{r=1}^{m} \sigma_r(t_k) \Delta_k w_r + \sum_{r=1}^{m} \sigma'_r(t_k) (I_{0r})_k,$$
$$Q_{k+1} = Q_k + h \left[\alpha g(t_k + \frac{\alpha}{2}h, \mathcal{P}_1, \mathcal{Q}_1) + (1-\alpha) g(t_k + \frac{1+\alpha}{2}h, \mathcal{P}_2, \mathcal{Q}_2)\right]$$
$$+ \sum_{r=1}^{m} \gamma_r(t_k) \Delta_k w_r + \sum_{r=1}^{m} \gamma'_r(t_k) (I_{0r})_k, \qquad (3.26)$$

where the parameters α, λ_1, λ_2, μ_1, μ_2 satisfy (3.13)-(3.14). The formula (3.26) contains the random variables $\Delta_k w_r(h)$, $(J_{r0})_k$, $(I_{0r})_k$ whose joint distribution is Gaussian. As usual, they can be simulated at each step by $2m$ independent $N(0,1)$-distributed random variables ξ_{rk} and η_{rk}, $r = 0, \ldots, m$:

$$\Delta_k w_r(h) = \sqrt{h} \xi_{rk}, \quad (J_{r0})_k = \sqrt{h} \left(\xi_{rk}/2 + \eta_{rk}/\sqrt{12}\right),$$
$$(I_{0r})_k = h^{3/2} \left(\xi_{rk}/2 - \eta_{rk}/\sqrt{12}\right). \qquad (3.27)$$

For $\sigma_r \equiv 0$, $\gamma_r \equiv 0$, $r = 1, \ldots, m$, the method (3.26) is reduced to the well-known second-order symplectic Runge–Kutta method for deterministic Hamiltonian systems (see, e.g., [261, p. 101]). Let us note that using this deterministic method with $\alpha = 0$ (the midpoint rule), another implicit 3/2-order method for Hamiltonian systems with noise was proposed in [298], however without preserving symplectic structure.

The one-step approximation corresponding to method (3.26) is of the form (3.11). Therefore, due to Lemma 3.6, the method (3.26) is of the mean-square order 3/2. Moreover, this one-step approximation is of the form (3.7) with $s = 2$ and

$$\varphi_1 = \sum_{r=1}^{m} \sigma_r \left(\lambda_1 J_{r0} + \mu_1 \Delta w_r\right), \quad \varphi_2 = \sum_{r=1}^{m} \sigma_r \left(\lambda_2 J_{r0} + \mu_2 \Delta w_r\right),$$
$$\psi_1 = \sum_{r=1}^{m} \gamma_r \left(\lambda_1 J_{r0} + \mu_1 \Delta w_r\right), \quad \psi_2 = \sum_{r=1}^{m} \gamma_r \left(\lambda_2 J_{r0} + \mu_2 \Delta w_r\right),$$

4.3 Mean-square methods for Hamiltonian systems with additive noise

$$\eta = \sum_{r=1}^{m} \sigma_r \Delta w_r + \sum_{r=1}^{m} \sigma'_r I_{0r}, \quad \zeta = \sum_{r=1}^{m} \gamma_r \Delta w_r + \sum_{r=1}^{m} \gamma'_r I_{0r},$$

$$\alpha_{11} = \frac{\alpha}{2}, \ \alpha_{12} = 0, \ \alpha_{21} = \alpha, \ \alpha_{22} = \frac{1-\alpha}{2}, \ \beta_1 = \alpha,$$

$$\beta_2 = 1 - \alpha, \ c_1 = \frac{\alpha}{2}, \ c_2 = \frac{1+\alpha}{2}.$$

This set of parameters α_{ij}, β_i, $i,j = 1,2$, satisfies the conditions (3.9). Then due to Lemma 3.5, the method (3.26) is symplectic. Thus, we have obtained the following theorem.

Theorem 3.7. *Under conditions (3.13)-(3.14) on the parameters, the method (3.26) for the system (3.1)-(3.2) preserves symplectic structure and has the mean-square order 3/2.*

4.3.2 The case of separable Hamiltonians

Here we consider the Hamiltonian system with additive noise (3.1), which Hamiltonian has the special structure

$$H(t, p, q) = V(p) + U(t, q). \tag{3.28}$$

We note that it is not difficult to consider a slightly more general Hamiltonian $H(t,p,q) = V(t,p) + U(t,q)$ but we restrict ourselves here to (3.28). In the case of separable Hamiltonian (3.28) the system (3.1) takes the partitioned form

$$dP = f(t, Q)dt + \sum_{r=1}^{m} \sigma_r(t) dw_r(t), \ P(t_0) = p,$$

$$dQ = g(P)dt + \sum_{r=1}^{m} \gamma_r(t) dw_r(t), \ Q(t_0) = q, \tag{3.29}$$

where $f^i = -\partial U/\partial q^i$, $g^i = \partial V/\partial p^i$, $i = 1, \ldots, n$.

Obviously, the implicit symplectic methods from the previous subsection can be applied to the partitioned system (3.29), and they take a more simple form in this case (we do not write them down here). We recall that there are no explicit symplectic RK methods for the general system (3.1)-(3.2). However, for the partitioned system (3.29) it is possible to construct explicit symplectic methods just as in the deterministic case.

Explicit first-order methods. On the basis of the known family of deterministic PRK methods [260, 261, 277], we construct the family of explicit partitioned methods for stochastic system (3.29):

$$\mathcal{Q} = Q_k + \alpha h g(P_k), \quad \mathcal{P} = P_k + hf(t_k + \alpha h, \mathcal{Q}),$$

$$Q_{k+1} = \mathcal{Q} + (1-\alpha)hg(\mathcal{P}) - \sum_{r=1}^{m} \gamma_r(t_k)\Delta_k w_r,$$

$$P_{k+1} = \mathcal{P} + \sum_{r=1}^{m} \sigma_r(t_k)\Delta_k w_r, \quad k = 0, \ldots, N-1. \qquad (3.30)$$

Since the expressions for dP_{k+1}, dQ_{k+1} coincide with the ones corresponding to the deterministic symplectic method, the method (3.30) is symplectic. Further, it is not difficult to show that the method (3.30) has the first mean-square order of accuracy. As a result, we obtain the following theorem.

Theorem 3.8. *The explicit partitioned method (3.30) for the system (3.29) preserves symplectic structure and has the first mean-square order of convergence.*

Remark 3.9. In the special cases of $\alpha = 0$ and $\alpha = 1$ the method (3.30) takes a more simple form. In these cases it requires evaluation of each of the coefficients f, g once per step only.

Remark 3.10. It is possible to propose other symplectic first-order methods for (3.29) on the basis of the same deterministic PRK methods as above. For instance, the method

$$\mathcal{Q} = Q_k + \alpha h g(P_k) + \sum_{r=1}^{m} \gamma_r(t_k)\Delta_k w_r,$$

$$\mathcal{P} = P_k + hf(t_k + \alpha h, \mathcal{Q}) + \sum_{r=1}^{m} \sigma_r(t_k)\Delta_k w_r,$$

$$Q_{k+1} = \mathcal{Q} + (1-\alpha)hg(\mathcal{P}), \quad P_{k+1} = \mathcal{P}, \quad k = 0, \ldots, N-1, \qquad (3.31)$$

is of the first mean-square order and symplectic.

Explicit methods of order 3/2. Here using specificity of the system (3.29), we construct a 3/2-order symplectic *explicit Runge–Kutta* method (other symplectic methods for (3.29) are given in [193]).

Introduce the relations (cf. (3.7)-(3.8)):

$$\mathcal{P}_i = p + h\sum_{j=1}^{s} \alpha_{ij} f(t + c_j h, \mathcal{Q}_j) + \varphi_i,$$

$$\mathcal{Q}_i = q + h\sum_{j=1}^{s} \hat{\alpha}_{ij} g(\mathcal{P}_j) + \psi_i, \quad i = 1, \ldots, s, \qquad (3.32)$$

$$\bar{P} = p + h\sum_{i=1}^{s} \beta_i f(t + c_i h, \mathcal{Q}_i) + \eta, \quad \bar{Q} = q + h\sum_{i=1}^{s} \hat{\beta}_i g(\mathcal{P}_i) + \zeta, \qquad (3.33)$$

4.3 Mean-square methods for Hamiltonian systems with additive noise

where φ_i, ψ_i, η, ζ do not depend on p and q, the parameters α_{ij}, $\hat{\alpha}_{ij}$, β_i and $\hat{\beta}_i$ satisfy the conditions

$$\beta_i \hat{\alpha}_{ij} + \hat{\beta}_j \alpha_{ji} - \beta_i \hat{\beta}_j = 0, \quad i,j = 1,\ldots,s, \tag{3.34}$$

and c_i are arbitrary parameters.

If $\varphi_i = \psi_i = \eta = \zeta = 0$, the relations (3.32)-(3.33) coincide with a general form of s-stage PRK methods for deterministic differential equations. By a generalization of the proof of Theorem 6.2 from [261] (see also Lemma 3.5 here), it is not difficult to prove the following lemma.

Lemma 3.11. *The relations* (3.32)-(3.33) *with condition* (3.34) *preserve symplectic structure, i.e.,* $d\bar{P} \wedge d\bar{Q} = dp \wedge dq$.

Introduce the parametric family of 2-stage explicit PRK methods for the system (3.29):

$$\mathcal{Q}_1 = Q_k + \sum_{r=1}^m \gamma_r(t_k)\left(\hat{\lambda}_1(J_{r0})_k + \hat{\mu}_1 \Delta_k w_r\right),$$

$$\mathcal{P}_1 = P_k + h\beta_1 f(t_k + c_1 h, \mathcal{Q}_1) + \sum_{r=1}^m \sigma_r(t_k)\left(\lambda_1(J_{r0})_k + \mu_1 \Delta_k w_r\right),$$

$$\mathcal{Q}_2 = Q_k + h\hat{\beta}_1 g(\mathcal{P}_1) + \sum_{r=1}^m \gamma_r(t_k)\left(\hat{\lambda}_2(J_{r0})_k + \hat{\mu}_2 \Delta_k w_r\right),$$

$$\mathcal{P}_2 = P_k + h\sum_{i=1}^2 \beta_i f(t_k + c_i h, \mathcal{Q}_i) + \sum_{r=1}^m \sigma_r(t_k)\left(\lambda_2(J_{r0})_k + \mu_2 \Delta_k w_r\right),$$

$$\tag{3.35}$$

$$P_{k+1} = P_k + \sum_{r=1}^m \sigma_r(t_k)\Delta_k w_r + \sum_{r=1}^m \sigma'_r(t_k)(I_{0r})_k + h\sum_{i=1}^2 \beta_i f(t_k + c_i h, \mathcal{Q}_i),$$

$$Q_{k+1} = Q_k + \sum_{r=1}^m \gamma_r(t_k)\Delta_k w_r + \sum_{r=1}^m \gamma'_r(t_k)(I_{0r})_k + h\sum_{i=1}^2 \hat{\beta}_i g(\mathcal{P}_i), \tag{3.36}$$

where the parameters β_i, $\hat{\beta}_i$, c_i, λ_i, $\hat{\lambda}_i$, μ_i, $\hat{\mu}_i$, $i=1,2$, satisfy the conditions

$$\beta_1 + \beta_2 = 1, \ \hat{\beta}_1 + \hat{\beta}_2 = 1, \ \hat{\beta}_2\beta_1 = 1/2, \ c_1 = 0, \ c_2 = \hat{\beta}_1, \tag{3.37}$$

$$\beta_1 \hat{\mu}_1 + \beta_2 \hat{\mu}_2 = 0, \ \hat{\beta}_1 \mu_1 + \hat{\beta}_2 \mu_2 = 0,$$

$$\beta_1 \hat{\lambda}_1 + \beta_2 \hat{\lambda}_2 = 1, \ \hat{\beta}_1 \lambda_1 + \hat{\beta}_2 \lambda_2 = 1,$$

$$\beta_1\left(\frac{\hat{\lambda}_1^2}{3} + \hat{\lambda}_1\hat{\mu}_1 + \hat{\mu}_1^2\right) + \beta_2\left(\frac{\hat{\lambda}_2^2}{3} + \hat{\lambda}_2\hat{\mu}_2 + \hat{\mu}_2^2\right) = \frac{1}{2},$$

$$\hat{\beta}_1\left(\frac{\lambda_1^2}{3} + \lambda_1\mu_1 + \mu_1^2\right) + \hat{\beta}_2\left(\frac{\lambda_2^2}{3} + \lambda_2\mu_2 + \mu_2^2\right) = \frac{1}{2}, \tag{3.38}$$

and Δw_r, I_{0r}, J_{r0} are defined in (3.12).

For example, the following set of parameters satisfies (3.37)-(3.38):

$$\beta_1 = \frac{1}{4}, \ \beta_2 = \frac{3}{4}, \ \hat{\beta}_1 = \frac{2}{3}, \ \hat{\beta}_2 = \frac{1}{3}, \ \lambda_1 = \lambda_2 = \hat{\lambda}_1 = \hat{\lambda}_2 = 1,$$

$$\mu_1 = \frac{1}{2\sqrt{3}}, \ \mu_2 = -\frac{1}{\sqrt{3}}, \ \hat{\mu}_1 = \frac{1}{\sqrt{2}}, \ \hat{\mu}_2 = -\frac{1}{3\sqrt{2}}. \qquad (3.39)$$

It is not difficult to see that the method (3.35)-(3.36) has the form of (3.32)-(3.33) and its parameters satisfy the conditions (3.34). Then, Lemma 3.11 implies that this method preserves symplectic structure. Using ideas of the proof of Lemma 3.6, we establish that the method (3.35)-(3.36) with (3.37)-(3.38) is of mean-square order 3/2. Thus we have proved the following theorem.

Theorem 3.12. *Under conditions (3.37)-(3.38), the explicit PRK method (3.35)-(3.36) for system (3.29) preserves symplectic structure and has the mean-square order 3/2.*

4.3.3 The case of Hamiltonian $H(t,p,q) = \frac{1}{2}p^\top M^{-1}p + U(t,q)$

Now we propose symplectic methods for the Hamiltonian system (3.29), when $\gamma_r(t) = 0$ and the separable Hamiltonian has the special form

$$H(t,p,q) = \frac{1}{2}p^\top M^{-1}p + U(t,q), \qquad (3.40)$$

with M a constant, symmetric, invertible matrix (i.e., the kinetic energy $V(p)$ in (3.28) is equal to $\frac{1}{2}p^\top M^{-1}p$). In this case the system (3.29) reads

$$dP = f(t,Q)dt + \sum_{r=1}^{m} \sigma_r(t)dw_r(t), \ P(t_0) = p,$$

$$dQ = M^{-1}Pdt, \ Q(t_0) = q, \qquad (3.41)$$

$$f^i = -\partial U/\partial q^i, \ i = 1,\ldots,n. \qquad (3.42)$$

This system can be written as a second-order differential equation with additive noise

$$\frac{d^2Q}{dt^2} = M^{-1}f(t,Q) + M^{-1}\sum_{r=1}^{m}\sigma_r(t)\dot{w}_r(t). \qquad (3.43)$$

Clearly, the symplectic methods from the previous Sects. 4.3.1 and 4.3.2 can be applied to (3.41)-(3.42). Due to specific features of this system, these methods have a more simple form here. In this subsection we restrict ourselves to explicit methods of orders 2 and 3.

4.3 Mean-square methods for Hamiltonian systems with additive noise   235

Explicit methods of order two. One can prove that the method (3.35)-(3.36) in application to (3.41)-(3.42) is of the mean-square order 2. Further, on the basis of the Störmer–Verlet method (the deterministic second-order symplectic method), we construct the method for the system (3.41)-(3.42):

$$\mathcal{Q} = Q_k + \frac{h}{2}M^{-1}P_k\,,$$

$$P_{k+1} = P_k + \sum_{r=1}^{m}\sigma_r(t_k)\Delta_k w_r + hf(t_k + \frac{h}{2}, \mathcal{Q}) + \sum_{r=1}^{m}\sigma'_r(t_k)(I_{0r})_k$$

$$Q_{k+1} = Q_k + hM^{-1}P_k + \sum_{r=1}^{m}M^{-1}\sigma_r(t_k)(I_{r0})_k + \frac{h^2}{2}M^{-1}f(t_k + \frac{h}{2}, \mathcal{Q})\,,$$

(3.44)

$$k = 0, \ldots, N-1\,.$$

Theorem 3.13. *The explicit method (3.44) for the system (3.41)-(3.42) is symplectic and of the mean-square order 2.*

Other methods of order two are given in [193]. In [265] a symplectic method of mean-square order one for (3.41)-(3.42) is proposed on the basis of the Störmer–Verlet method.

Explicit methods of order three. Introduce the integrals

$$(I_{0r})_k = \int_{t_k}^{t_{k+1}} (\vartheta - t_k)\,dw_r(\vartheta)\,, \quad (I_{r0})_k = \int_{t_k}^{t_{k+1}}(w_r(\vartheta) - w_r(t_k))\,d\vartheta\,,$$

$$(I_{00r})_k := \frac{1}{2}\int_{t_k}^{t_{k+1}}(\vartheta - t_k)^2\,dw_r(\vartheta)\,, \quad (I_{0r0})_k := \int_{t_k}^{t_{k+1}}\int_{t_k}^{\vartheta_1}(\vartheta_2 - t_k)\,dw_r(\vartheta_2)d\vartheta_1\,,$$

$$(I_{r00})_k := \int_{t_k}^{t_{k+1}}\int_{t_k}^{\vartheta_1}(w_r(\vartheta_2) - w_r(t_k))\,d\vartheta_2 d\vartheta_1\,,$$

$$(J_r)_k = \int_{t_k}^{t_{k+1}}(\vartheta - t_k)(w_r(\vartheta) - w_r(t_k))\,d\vartheta\,. \quad (3.45)$$

Joint distribution of the random variables $\Delta_k w_r(h)$, $(I_{0r})_k$, $(I_{r0})_k$, $(I_{0r0})_k$, $(I_{r00})_k$, $(I_{00r})_k$ is Gaussian. They can be simulated at each step by $3m$ independent $N(0,1)$-distributed random variables ξ_{rk}, η_{rk}, and ζ_{rk}, $r = 1, \ldots, m$:

$$\Delta_k w_r = h^{1/2}\xi_{rk}\,, \quad (I_{r0})_k = h^{3/2}(\eta_{rk}/\sqrt{3} + \xi_{rk})/2\,,$$

$$(I_{0r})_k = h\Delta_k w_r - (I_{r0})_k\,, \quad (J_r)_k = h^{5/2}(\xi_{rk}/3 + \eta_{rk}/(4\sqrt{3}) + \zeta_{rk}/(12\sqrt{5}))\,,$$

$$(I_{r00})_k = h(I_{r0})_k - (J_r)_k\,, \quad (I_{0r0})_k = 2(J_r)_k - h(I_{r0})_k\,,$$

$$(I_{00r})_k = h^2\Delta_k w_r/2 - (J_r)_k\,. \quad (3.46)$$

Clearly, for $\sigma_r = 0$, $r = 1, \ldots, m$, the stochastic system (3.41) is reduced to the deterministic system

$$\frac{dp}{dt} = f(t, q), \quad \frac{dq}{dt} = M^{-1} p. \qquad (3.47)$$

The following lemma is true for system (3.41) with an arbitrary f (i.e., f may not obey the condition (3.42)). Its proof is available in [193].

Lemma 3.14. *Let $\bar{q} = q + G(t+h; t, p, q)$, $\bar{p} = p + F(t+h; t, p, q)$ be a one-step approximation of the third-order explicit method for the deterministic system (3.47). Suppose an n-dimensional (deterministic) variable $\mathcal{Q} = \mathcal{Q}(t+h; t, p, q)$ is such that*

$$|\mathcal{Q} - q| = O(h).$$

Then, the method

$$P_{k+1} = P_k + F(t+h; t, P_k, \mathcal{Q}_k) + \sum_{r=1}^{m} \sigma_r(t_k) \Delta_k w_r + \sum_{r=1}^{m} \sigma'_r(t_k)(I_{0r})_k$$

$$+ \sum_{r=1}^{m} \sigma''_r(t_k)(I_{00r})_k + \sum_{r=1}^{m} \sum_{i=1}^{n} (M^{-1} \sigma_r(t_k))^i \frac{\partial f}{\partial q^i}(t_k, \mathcal{Q}_k)(I_{r00})_k,$$

$$Q_{k+1} = Q_k + G(t+h; t, P_k, \mathcal{Q}_k) + \sum_{r=1}^{m} M^{-1} \sigma_r(t_k)(I_{r0})_k$$

$$+ \sum_{r=1}^{m} M^{-1} \sigma'_r(t_k)(I_{0r0})_k \qquad (3.48)$$

is of mean-square order 3 for the system (3.41) with an arbitrary f.

Using the known deterministic third-order symplectic method (see [261, 275, 277]), we obtain the following method for system (3.41)-(3.42):

$$\mathcal{Q}_1 = Q_k + \frac{7}{24} h M^{-1} P_k, \quad \mathcal{P}_1 = P_k + \frac{2}{3} h f(t_k + \frac{7h}{24}, \mathcal{Q}_1),$$

$$\mathcal{Q}_2 = \mathcal{Q}_1 + \frac{3}{4} h M^{-1} \mathcal{P}_1, \quad \mathcal{P}_2 = \mathcal{P}_1 - \frac{2}{3} h f(t_k + \frac{25h}{24}, \mathcal{Q}_2),$$

$$\mathcal{Q}_3 = \mathcal{Q}_2 - \frac{1}{24} h M^{-1} \mathcal{P}_2, \quad \mathcal{P}_3 = \mathcal{P}_2 + h f(t_k + h, \mathcal{Q}_3), \qquad (3.49)$$

$$P_{k+1} = \mathcal{P}_3 + \sum_{r=1}^{m} \sigma_r(t_k) \Delta_k w_r + \sum_{r=1}^{m} \sigma'_r(t_k)(I_{0r})_k$$

$$+ \sum_{r=1}^{m} \sigma''_r(t_k)(I_{00r})_k + \sum_{r=1}^{m} \sum_{i=1}^{n} (M^{-1} \sigma_r(t_k))^i \frac{\partial f}{\partial q^i}(t_k, \mathcal{Q}_3)(I_{r00})_k,$$

$$Q_{k+1} = \mathcal{Q}_3 + \sum_{r=1}^{m} M^{-1} \sigma_r(t_k)(I_{r0})_k + \sum_{r=1}^{m} M^{-1} \sigma'_r(t_k)(I_{0r0})_k, \qquad (3.50)$$

$$k = 0, \ldots, N-1.$$

Theorem 3.15. *The explicit method* (3.49)-(3.50) *for the system* (3.41)-(3.42) *is symplectic and of mean-square order* 3.

Proof. It is not difficult to check that $dP_{k+1} \wedge dQ_{k+1} = dP_3 \wedge dQ_3$. The expression for $dP_3 \wedge dQ_3$ coincides with the one corresponding to the deterministic third-order symplectic method. This implies that the method (3.49)-(3.50) is symplectic. By Lemma 3.14 we get that the method has the mean-square order 3. □

4.4 Numerical tests of mean-square symplectic methods

4.4.1 Kubo oscillator

The system of SDEs in the sense of Stratonovich (Kubo oscillator)

$$dX^1 = -aX^2 dt - \sigma X^2 \circ dw(t), \quad X^1(0) = x^1,$$
$$dX^2 = aX^1 dt + \sigma X^1 \circ dw(t), \quad X^2(0) = x^2, \quad (4.1)$$

is often used for testing numerical methods. Here a and σ are constants and $w(t)$ is a one-dimensional standard Wiener process.

The phase flow of this system preserves symplectic structure. Moreover, the quantity $\mathcal{H}(x^1, x^2) = (x^1)^2 + (x^2)^2$ is conservative for this system, i.e.,

$$\mathcal{H}(X^1(t), X^2(t)) = \mathcal{H}(x^1, x^2) \text{ for } t \geq 0.$$

This means that a phase trajectory of (4.1) belongs to the circle with center at the origin and of radius $\sqrt{\mathcal{H}(x^1, x^2)}$.

We test three methods here. In application to (4.1) the symplectic PRK method (2.9) takes the form:

$$X_{k+1}^1 = X_k^1 - aX_k^2 h - \frac{\sigma^2}{2} X_{k+1}^1 h - \sigma X_k^2 \Delta_k w,$$
$$X_{k+1}^2 = X_k^2 + aX_{k+1}^1 h + \frac{\sigma^2}{2} X_k^2 h + \sigma X_{k+1}^1 \Delta_k w. \quad (4.2)$$

This method is implicit in the deterministic part only.

The midpoint method (2.8) applied to the system with one noise (4.1) reads

$$X_{k+1}^1 = X_k^1 - a \frac{X_k^2 + X_{k+1}^2}{2} h - \sigma \frac{X_k^2 + X_{k+1}^2}{2} (\zeta_h)_k \sqrt{h},$$
$$X_{k+1}^2 = X_k^2 + a \frac{X_k^1 + X_{k+1}^1}{2} h + \sigma \frac{X_k^1 + X_{k+1}^1}{2} (\zeta_h)_k \sqrt{h}. \quad (4.3)$$

This is a fully implicit method. Note that due to specific features of the system (4.1), the formula (4.3) is valid (solvable) not only in the case of the truncated random variable ζ_h but also if we put $\Delta_k w$ instead of $(\zeta_h)_k \sqrt{h}$.

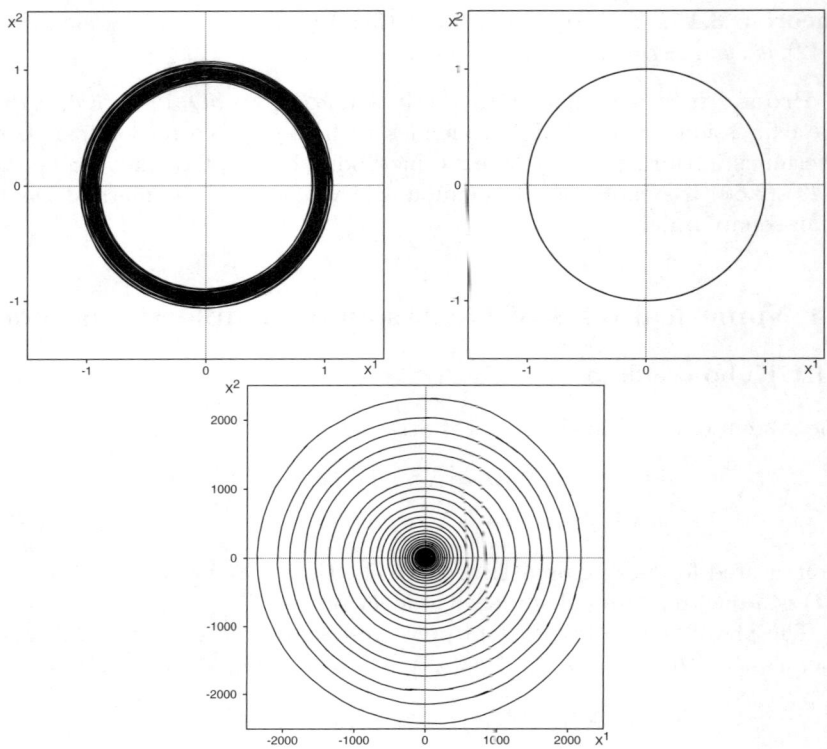

Fig. 4.1. A sample phase trajectory of (4.1) with $X^1(0) = 1$, $X^2(0) = 0$ obtained by the symplectic method (4.2) (*top left*), the midpoint method (4.3) (*top right*), and by the Euler method (4.4) (*bottom*) for $a = 2$, $\sigma = 0.3$, $h = 0.02$ on the time interval $t \leq 200$.

The method (4.3) is of first mean-square order. The method (4.2) is of mean-square order $1/2$ as well as the Euler method:

$$X^1_{k+1} = X^1_k - aX^2_k h - \frac{\sigma^2}{2} X^1_k h - \sigma X^2_k \Delta_k w,$$
$$X^2_{k+1} = X^2_k + aX^1_k h - \frac{\sigma^2}{2} X^2_k h + \sigma X^1_k \Delta_k w, \qquad (4.4)$$

which, of course, is not symplectic.

Figure 4.1 gives approximations of a sample phase trajectory of (4.1) simulated by the symplectic methods (4.2) and (4.3) and by the Euler method (4.4). The initial condition is $x^1 = 1$, $x^2 = 0$. The corresponding exact phase trajectory belongs to the circle with center at the origin and with the unit radius.

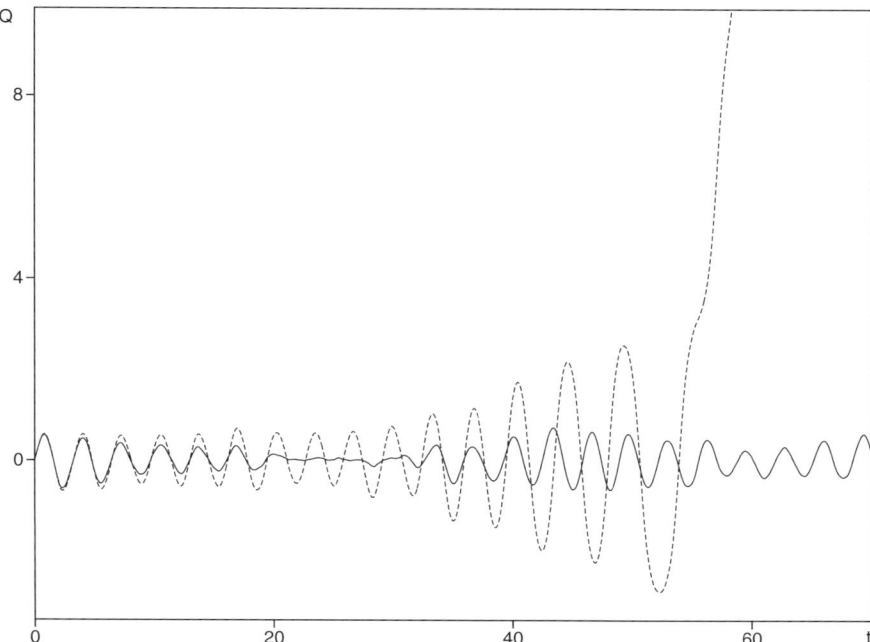

Fig. 4.2. A sample trajectory of (4.5) for $\omega = 2$, $\sigma_1 = 0.2$, $\sigma_2 = 0.1$, $h = 0.02$. *Solid line* – the symplectic method (4.7), *dashed line* – the Euler method (4.6).

We see that the Euler method is not appropriate for simulation of the oscillator (4.1) on long time intervals while the symplectic methods preserve conservative properties of the Kubo oscillator.

These experiments also demonstrate that the midpoint method is much more accurate than the other methods applied. It is not difficult to check that $\mathcal{H}(x^1, x^2)$ is conserved by the midpoint method (4.3) but it is not conserved by the symplectic PRK method (4.2). This is similar to the deterministic case. Indeed, it is known [260, 261] that symplectic deterministic RK methods (e.g., the midpoint scheme) conserve all quadratic functions that are conserved by the Hamiltonian system being integrated, while deterministic PRK methods do not possess this property.

4.4.2 A model for synchrotron oscillations of particles in storage rings

In [265] a model describing synchrotron oscillations of particles in storage rings under the influence of external fluctuating electromagnetic fields was considered. This model can be written in the following form

$$dP = -\omega^2 \sin(Q)dt - \sigma_1 \cos(Q)dw_1 - \sigma_2 \sin(Q)dw_2,$$
$$dQ = Pdt. \qquad (4.5)$$

P and Q are scalars here. The system (4.5) is of the form (2.14) and therefore its phase flow preserves symplectic structure.

The Euler method for (4.5) takes the form

$$P_{k+1} = P_k - h\omega^2 \sin(Q_k) - h^{1/2}(\sigma_1 \cos(Q_k)\Delta_k w_1 + \sigma_2 \sin(Q_k)\Delta_k w_2),$$
$$Q_{k+1} = Q_k + hP_k. \qquad (4.6)$$

In application to (4.5) the explicit symplectic method (2.17)-(2.18) with $\alpha = 1$ is written as

$$\mathcal{Q} = Q_k + hP_k,$$
$$P_{k+1} = P_k - h\omega^2 \sin(\mathcal{Q}) - h^{1/2}(\sigma_1 \cos(\mathcal{Q})\Delta_k w_1 + \sigma_2 \sin(\mathcal{Q})\Delta_k w_2),$$
$$Q_{k+1} = \mathcal{Q}. \qquad (4.7)$$

Both methods are of first mean-square order.

Approximations of a sample trajectory of (4.5) simulated by the symplectic method (4.7) and the Euler method (4.6) are plotted on Fig. 4.2. The trajectory obtained by the symplectic method with $h = 0.02$ (solid line) visually coincides with the one obtained for a smaller step, e.g. for $h = 0.002$, using the same sample paths for the Wiener processes, i.e., this trajectory visually coincides with the exact solution of (4.5). This figure clearly demonstrates that the Euler method (dashed line) is unacceptable for simulation of the solution to (4.5) on a long time interval while the symplectic method (4.7) produces quite accurate results despite both methods have the same mean-square order of accuracy.

In Appendix A.1 an implementation of the Euler method (4.6) and the explicit symplectic method (4.7) is considered and the program is given (see Listing A.1), by which sample trajectories plotted on Fig. 4.2 were obtained.

4.4.3 Linear oscillator with additive noise

In this section we consider the following Hamiltonian system with additive noise:

$$dX^1 = X^2 dt + \sigma dw_1(t), \quad X^1(0) = X_0^1,$$
$$dX^2 = -X^1 dt + \gamma dw_2(t), \quad X^2(0) = X_0^2. \qquad (4.8)$$

We have for the solution $X = (X^1, X^2)^\top$ of (4.8):

$$X(t_{k+1}) = FX(t_k) + u_k, \quad X(0) = X_0, \quad k = 0, 1, \ldots, N-1, \qquad (4.9)$$

where

4.4 Numerical tests of mean-square symplectic methods   241

$$F = \begin{bmatrix} \cos h & \sin h \\ -\sin h & \cos h \end{bmatrix},$$

$$u_k = \begin{bmatrix} \sigma \int_{t_k}^{t_{k+1}} \cos(t_{k+1}-s)dw_1(s) + \gamma \int_{t_k}^{t_{k+1}} \sin(t_{k+1}-s)dw_2(s) \\ -\sigma \int_{t_k}^{t_{k+1}} \sin(t_{k+1}-s)dw_1(s) + \gamma \int_{t_k}^{t_{k+1}} \cos(t_{k+1}-s)dw_2(s) \end{bmatrix}.$$

In application to (4.8) the explicit symplectic method (3.31) with $\alpha = 1$ takes the form

$$X_{k+1}^2 = X_k^2 - hX_k^1 + \gamma \Delta_k w_2, \quad X_{k+1}^1 = X_k^1 + hX_{k+1}^2 + \sigma \Delta_k w_1. \quad (4.10)$$

The method (4.10) can be written as

$$X_{k+1} = HX_k + v_k, \quad k = 0, 1, \ldots, N-1, \quad (4.11)$$

where $X_k = (X_k^1, X_k^2)^\top$,

$$H = \begin{bmatrix} 1-h^2 & h \\ -h & 1 \end{bmatrix}, \quad v_k = \begin{bmatrix} \sigma \Delta_k w_1 + \gamma h \Delta_k w_2 \\ \gamma \Delta_k w_2 \end{bmatrix}.$$

Our nearest aim is to analyze propagation of the error $r_k := X_k - X(t_k)$. We get

$$X(t_k) = F^k X_0 + F^{k-1} u_0 + F^{k-2} u_1 + \cdots + u_{k-1}, \quad (4.12)$$

$$X_k = H^k X_0 + H^{k-1} v_0 + H^{k-2} v_1 + \cdots + v_{k-1}. \quad (4.13)$$

Proposition 4.1. *Suppose T and h are such that Th^2 is sufficiently small. Then for $k = 0, 1, \ldots, N$, $T = Nh$, the following inequality holds*

$$\|H^k - F^k\| \leq \frac{h}{2} + \frac{kh^3}{24} + O(h^2 + Th^3) \leq \frac{h}{2} + \frac{Th^2}{24} + O(h^2 + Th^3). \quad (4.14)$$

Proof. Clearly

$$F^k = \begin{bmatrix} \cos kh & \sin kh \\ -\sin kh & \cos kh \end{bmatrix}.$$

Let us represent H as $H = G\Lambda G^{-1}$ with Λ and G such that $\Lambda = \mathrm{diag}(\lambda_1, \lambda_2)$, $\lambda_{1,2} = 1 - \frac{h^2}{2} \pm ih\sqrt{1 - \frac{h^2}{4}}$, and the columns of the matrix G are eigenvectors of H corresponding to the eigenvalues λ_1, λ_2. We write the matrices Λ and G in the form

$$\Lambda = \begin{bmatrix} e^{i\varphi} & 0 \\ 0 & e^{-i\varphi} \end{bmatrix}, \quad G = \begin{bmatrix} 1 & 1 \\ e^{i\psi} & e^{-i\psi} \end{bmatrix},$$

where $0 < \varphi, \psi < \frac{\pi}{2}$, $\cos\varphi = 1 - \frac{h^2}{2}$, $\cos\psi = \frac{h}{2}$. We obtain: $H^k = G\Lambda^k G^{-1}$,

$$H^k - F^k = G(\Lambda^k - G^{-1} F^k G)G^{-1}, \quad (4.15)$$

$$\Lambda^k - G^{-1}F^k G$$
$$= \begin{bmatrix} e^{ki\varphi} - e^{kih} - i\sin kh \dfrac{1-\sin\psi}{\sin\psi} & -\dfrac{i\sin kh \times e^{-i\psi}}{\sin\psi}\cos\psi \\ \dfrac{i\sin kh \times e^{i\psi}}{\sin\psi}\cos\psi & e^{-ki\varphi} - e^{-kih} + i\sin kh \dfrac{1-\sin\psi}{\sin\psi} \end{bmatrix}.$$
(4.16)

Let us represent this matrix $\Lambda^k - G^{-1}F^k G$ as the sum $D_1 + D_2$, where $D_2 = diag(e^{ki\varphi} - e^{kih}, e^{-ki\varphi} - e^{-kih})$. It is not difficult to show that (the norms of matrices are Euclidean)

$$||G|| = \sqrt{2}(1 + O(h)), \ ||G^{-1}|| = \dfrac{\sqrt{2}}{2}(1 + O(h)),$$
$$||D_1|| \leq \dfrac{h}{2}(1 + O(h)), \ ||D_2|| = 2|\sin\dfrac{k\varphi - kh}{2}|. \quad (4.17)$$

Taking into account that $\varphi = \arcsin(h\sqrt{1 - \dfrac{h^2}{4}}) = h + \dfrac{h^3}{24} + O(h^5)$, $kh \leq T$, $k = 0, 1, \ldots, N$, and the assumption on smallness of Th^2, we get

$$||D_2|| \leq \dfrac{kh^3}{24} + O(h^2) \leq \dfrac{Th^2}{24} + O(h^2), \ k = 0, 1, \ldots, N. \quad (4.18)$$

The inequality (4.14) follows from (4.15)-(4.18). □

Using Proposition 4.1, we prove the following assertion.

Proposition 4.2. *Let T and h be such that Th^2 is sufficiently small. Suppose $E|X_0|^2 \leq C$. Then the mean-square error is estimated as*

$$(E|r_k|^2)^{1/2} \leq K \times (T^{1/2}h + T^{3/2}h^2), \ k = 0, 1, \ldots, N. \quad (4.19)$$

In application to (4.8) the Euler method can be written in the form

$$\bar{X}_{k+1} = \bar{H}\bar{X}_k + \bar{v}_k = \begin{bmatrix} 1 & h \\ -h & 1 \end{bmatrix}\bar{X}_k + \begin{bmatrix} \sigma\Delta_k w_1 \\ \gamma\Delta_k w_2 \end{bmatrix}. \quad (4.20)$$

Analogously to (4.15)-(4.16), we get $\bar{H}^k - F^k = \bar{G}(\bar{\Lambda}^k - \bar{G}^{-1}F^k\bar{G})\bar{G}^{-1} := \bar{G}\bar{D}\bar{G}^{-1}$ with

$$\bar{\Lambda} = \begin{bmatrix} 1+ih & 0 \\ 0 & 1-ih \end{bmatrix}, \ \bar{G} = \begin{bmatrix} 1 & 1 \\ i & -i \end{bmatrix},$$
$$\bar{D} = \begin{bmatrix} (1+ih)^k - e^{ihk} & 0 \\ 0 & (1-ih)^k - e^{-ihk} \end{bmatrix}.$$

Further, $||\bar{G}|| = \sqrt{2}$, $||\bar{G}^{-1}|| = \sqrt{2}/2$, and

$$||\bar{D}|| = [((1+h^2)^{k/2} - 1)^2 + 4(1+h^2)^{k/2}\sin^2\dfrac{k(\varphi - h)}{2}]^{1/2}$$
$$\leq [(e^{Th/2} - 1)^2 + 4e^{Th/2}\sin^2\dfrac{k(\varphi - h)}{2}]^{1/2},$$

where $\varphi = \arcsin \dfrac{h}{\sqrt{1+h^2}} \simeq \dfrac{h}{\sqrt{1+h^2}} + \dfrac{1}{6}\dfrac{h^3}{(1+h^2)^{3/2}}$, $\varphi - h \simeq -\dfrac{h^3}{3}$. Hence if Th is small then

$$\|\bar{D}\| \leq [(e^{Th/2} - 1)^2 + 4e^{Th/2} \sin^2 \dfrac{k(\varphi - h)}{2}]^{1/2} \simeq e^{Th/2} - 1 \simeq Th/2,$$

and it is not difficult to show that the mean-square error of the Euler method is estimated as $O(T^{3/2}h)$.

Consequently, the Euler method can be used on the interval $[0, T_E]$ if $T_E^{3/2}h$ is sufficiently small. Due to Proposition 4.2, the error of the symplectic method (4.10) on $[0, T_S]$ with $T_S = T_E^2$ is equal to $O(T_E h + T_E^3 h^2)$, i.e., the symplectic method is applicable on essentially longer time intervals than the Euler method. Of course, the Euler method possesses the worse properties than the symplectic method since the absolute values of the eigenvalues of $\bar{H}$ are greater than 1.

Finally, consider the optimal method from [180, p. 61] (the method also uses only the increments $\Delta_k w$ as the information regarding $w(t)$ but it uses this information optimally):

$$\hat{X}_{k+1} = \hat{H}\hat{X}_k + \hat{v}_k$$
$$= \begin{bmatrix} \cos h & \sin h \\ -\sin h & \cos h \end{bmatrix} \hat{X}_k$$
$$+ \dfrac{1}{h} \begin{bmatrix} \sigma \sin h \times \Delta_k w_1 + 2\gamma \sin^2 \dfrac{h}{2} \times \Delta_k w_2 \\ -2\sigma \sin^2 \dfrac{h}{2} \times \Delta_k w_1 + \gamma \sin h \times \Delta_k w_2 \end{bmatrix}. \quad (4.21)$$

Evidently, this method is symplectic. And, as $\hat{H} = F$, it has no error in the absence of noise. We get for its error:

$$E|\hat{r}_N|^2 = \sum_{m=0}^{N-1} E|\hat{v}_m - u_m|^2 = N(\sigma^2 + \gamma^2)\dfrac{h^3}{12} + N \times O(h^5) \simeq \dfrac{\sigma^2 + \gamma^2}{12} Th^2.$$

Consequently, the error of the optimal method is estimated as $O(T^{1/2}h)$. This implies that the method (4.21) is applicable on the longer time interval $[0, T_O] = [0, T_E^3]$ than the symplectic method (4.10).

To guarantee the same sample paths for the Wiener processes in realization of the exact, symplectic, and Euler methods, we simulate six independent $\mathcal{N}(0,1)$-distributed random variables $\xi_{1,k+1}, \eta_{1,k+1}, \zeta_{1,k+1}, \xi_{2,k+1}, \eta_{2,k+1}, \zeta_{2,k+1}$ at every step $k+1 = 1, \ldots, N-1$. It is not difficult to show that the needed random variables can be evaluated as

$$\Delta_k w_i = \sqrt{h}\xi_{i,k+1}, \quad \int_{t_k}^{t_{k+1}} \cos(t_{k+1} - s) dw_i(s) = \dfrac{1}{\sqrt{h}} \sin h \times \xi_{i,k+1} + c_1 \eta_{i,k+1},$$

$$\int_{t_k}^{t_{k+1}} \sin(t_{k+1} - s) dw_i(s) = \dfrac{2}{\sqrt{h}} \sin^2 \dfrac{h}{2} \times \xi_{i,k+1} + c_2 \eta_{i,k+1} + c_3 \zeta_{i,k+1}, \quad i = 1, 2,$$

where

$$c_1 = (\frac{1}{2}h + \frac{1}{4}\sin 2h - \frac{\sin^2 h}{h})^{1/2}, \quad c_2 = \frac{1}{c_1}(\frac{1}{2}\sin^2 h - \frac{2}{h}\sin^2\frac{h}{2}\sin h),$$

$$c_3 = (\frac{1}{2}h - \frac{1}{4}\sin 2h - \frac{4}{h}\sin^4\frac{h}{2} - c_2^2)^{1/2}.$$

In the numerical tests we simulate the system (4.8) by (i) the exact formula (4.9), (ii) the symplectic method (4.10), and (iii) the Euler method (4.20). Figure 4.3 corresponds to the time interval $[0, 128]$ which approximately contains 20 oscillations of (4.8) (note that the period of free oscillations of (4.8) is equal to 2π).

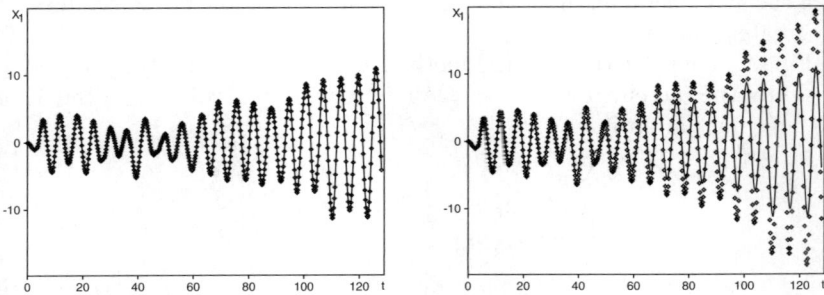

Fig. 4.3. A sample trajectory of the solution to (4.8) for $\sigma = 0$, $\gamma = 1$, $X_1(0) = X_2(0) = 0$ obtained by the exact formulae (4.9) (*solid line*), the symplectic method (4.10) with $h = 0.02$ (*points* on the *left* figure), and the Euler method (4.20) with $h = 0.02$ (*points* on the *right* figure). The points of the symplectic and Euler methods are plotted once per 10 steps, i.e., once per each interval 0.2.

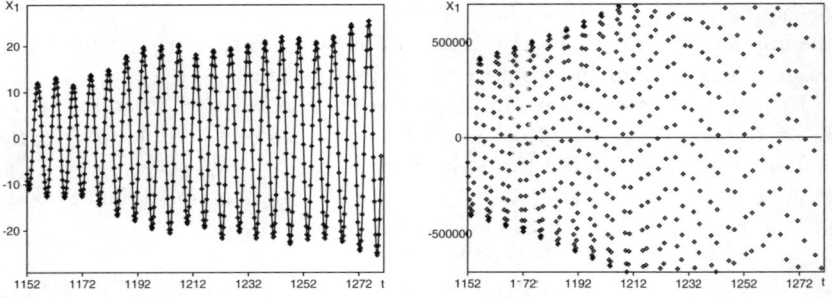

Fig. 4.4. Another part of the same sample trajectory as in Fig. 4.3. *Solid line* - the exact solution, *points* - the symplectic method (*left*) and the Euler method (*right*).

4.4 Numerical tests of mean-square symplectic methods   245

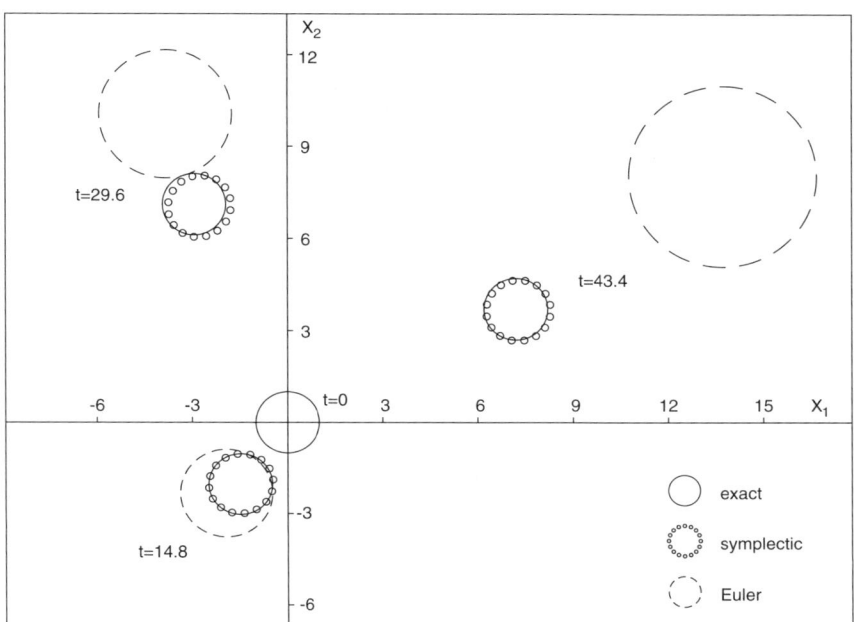

Fig. 4.5. The evolution of domains in the phase plane of system (4.8) for $\sigma = 0$, $\gamma = 1$. Images of the initial unit circle are obtained at three time moments by the exact mapping, by the mapping in the case of the symplectic method (4.10) with $h = 0.05$, and by the mapping in the case of the Euler method (4.20) with $h = 0.05$.

The results clearly demonstrate that the Euler method is unacceptable for simulation of the Hamiltonian system (4.8) on a long time interval. After 10 oscillations (Fig. 4.3) the norm of its error is already half of the norm of the solution, and after 200 oscillations (see Fig. 4.4) the amplitude of oscillations simulated by the Euler method is greater than the exact amplitude in 50 000 times.

In contrast to the Euler method, the symplectic method reproduces oscillations of the system (4.8) quite accurately. After 10 oscillations (Fig. 4.3) the norm of its error is approximately 2% of the norm of the solution. But it is more astonishing that after 200 oscillations (see Fig. 4.4) the relative error remains the same. The error of the amplitude of oscillations on the considered time interval is also about 2%. As is known, a symplectic method in application to a deterministic oscillator preserves conservative properties of solutions, in particular their boundedness on infinite time interval. One can say that the symplectic method generates a discrete conservative system ("discrete linear oscillator"). It turns out that behavior of this system affected by noise (which is also discrete) is qualitatively identical to the behavior of the continuous Hamiltonian system with noise. For instance, the approximate solution adequately reproduces an increase of the amplitude of the oscillations.

Figure 4.5 presents evolution of domains in the phase plane of system (4.8). The initial domain is the circle with center at the origin and with the unit radius. We plot images of this circle, which are obtained at three time moments by the exact mapping, by the mapping in the case of the symplectic method (4.10), and by the mapping in the case of the Euler method (4.20). For the considered system (4.8), exact images of the unit circle are circles of the unit radius shifted from the origin due to the action of noise. In the case of the Euler method these images are also circles but with increasing radius. In the case of symplectic method (4.10) the images of the initial circle are ellipses. In spite of the fact that the symplectic method (4.10) and the Euler method (4.20) have the same mean-square order of accuracy, these ellipses approximate the exact images essentially better than the circles obtained by the Euler method.

4.5 Liouvillian methods for stochastic systems preserving phase volume

In the previous sections we considered some Hamiltonian methods for stochastic Hamiltonian systems. These systems (as well as the methods) preserve the symplectic structure and, consequently, preserve the phase volume. In this section we deal with a more general class of systems which preserve the phase volume but may not preserve the symplectic structure.

Let us start with the deterministic d-dimensional system

$$\frac{dX}{dt} = a(t, X), \quad X(t_0) = x, \tag{5.1}$$

the phase flow $X(t; t_0, x)$ of which preserves the phase volume. Note that the dimension d may be odd.

Let $D_0 \in \mathbf{R}^d$ be a domain with finite volume. The transformation $X(t; t_0, x)$ maps D_0 into the domain D_t. The volume V_t of the domain D_t is equal to

$$V_t = \int_{D_t} dX^1 \ldots dX^d = \int_{D_0} \left| \frac{D(X^1, \ldots, X^d)}{D(x^1, \ldots, x^d)} \right| dx^1 \ldots dx^d.$$

Then, the volume-preserving condition consists in the equality

$$\left| \frac{D(X^1(t), \ldots, X^d(t))}{D(x^1, \ldots, x^d)} \right| = 1 \tag{5.2}$$

or, equivalently, it consists in preservation of the d-form $dX^1 \wedge dX^2 \wedge \cdots \wedge dX^d$.

4.5 Liouvillian methods for stochastic systems preserving phase volume

According to the Liouville theorem, the phase flow of (5.1) preserves phase volume if and only if

$$\frac{\partial a^1(t,x)}{\partial x^1} + \cdots + \frac{\partial a^d(t,x)}{\partial x^d} = div\, a = 0. \tag{5.3}$$

Numerical methods preserving the phase volume are called *Liouvillian* [275, 279]. Due to our best knowledge, there are no constructive Liouvillian methods for the deterministic system (5.1), (5.3) of a general form (see [60, 242, 275, 279] and references therein). Some constructive Liouvillian methods for particular cases of (5.1), (5.3) can be found in [60, 242, 275, 279]. It was shown in [279] that certain methods known to be symplectic are also phase volume preserving. However, it was also demonstrated that in general the relation between these two properties is rather delicate: neither of them implies the other.

Consider the Cauchy problem for the d-dimensional system of SDEs in the sense of Ito:

$$dX = a(t,X)dt + \sum_{r=1}^{m} b_r(t,X)dw_r(t), \ X(t_0) = x, \tag{5.4}$$

the phase flow $X(t; t_0, x; \omega)$ of which preserves phase volume, i.e., for which the condition (5.2) holds.

It is known (see [4, 140] and also [193]) that the phase flow of (5.4) preserves phase volume if and only if

$$div\,(a - \frac{1}{2}\sum_{r=1}^{m} \frac{\partial b_r}{\partial x} b_r) = 0, \ \ div\, b_r = 0, \ r = 1, \ldots, m. \tag{5.5}$$

Let X_k, $k = 0, \ldots, N$, $t_{k+1} - t_k = h_{k+1}$, $t_N = t_0 + T$:

$$X_0 = X(t_0), \ \ X_{k+1} = \bar{X}_{t_k, X_k}(t_{k+1}),$$

be a mean-square method for (5.4) based on the one-step approximation $\bar{X}_{t,x}(t+h) = \bar{X}(t+h; t, x)$. It is clear that a method preserves phase volume if its one-step approximation satisfies the equality

$$\left|\frac{D(\bar{X}^1, \ldots, \bar{X}^d)}{D(x^1, \ldots, x^d)}\right| = 1 \tag{5.6}$$

or equivalently

$$d\bar{X}^1 \wedge \cdots \wedge d\bar{X}^d = dx^1 \wedge \cdots \wedge dx^d. \tag{5.7}$$

Taking into account that there are no constructive Liouvillian methods for a general deterministic Liouvillian system, we restrict ourselves here to some particular cases of the stochastic system (5.4), (5.5).

4.5.1 Liouvillian methods for partitioned systems with multiplicative noise

Consider the particular case of (5.4):

$$dX = f(t,Y)dt + \sum_{r=1}^{m} \sigma_r(t,Y)\, dw_r(t)\,, \quad X(t_0) = x\,,$$

$$dY = g(t,X)dt + \sum_{r=1}^{m} \gamma_r(t)\, dw_r(t)\,, \quad Y(t_0) = y\,, \qquad (5.8)$$

where X, f, σ_r are l-dimensional column vectors and Y, g, γ_r are n-dimensional column vectors.

It is not difficult to check that the coefficients of (5.8) satisfy (5.5), i.e., the phase flow of system (5.8) preserves phase volume. Note that if $l = n$ and there are U_r, $r = 0, \ldots, m$, and V_0 such that $f^i = -\partial U_0/\partial y^i$, $g^i = \partial V_0/\partial x^i$, and $\sigma_r = -\partial U_r/\partial y^i$, $r = 1, \ldots m$, $i = 1, \ldots, l$, then the system (5.8) possesses the symplectic property (cf. (2.10), we pay attention that the system (2.10) is in the sense of Stratonovich).

Introduce the PRK method for (5.8) (cf. (2.11)-(2.12)):

$$\begin{aligned}
\mathsf{Y}_1 &= Y_k + \alpha h g(t_k, X_k),\\
\mathsf{X}_1 &= X_k + h f(t_k + \alpha h, \mathsf{Y}_1),\\
\mathsf{Y}_2 &= \mathsf{Y}_1 + (1-\alpha) h g(t_{k+1}, \mathsf{X}_1),
\end{aligned} \qquad (5.9)$$

$$X_{k+1} = \mathsf{X}_1 + \sum_{r=1}^{m} \sigma_r(t_k, \mathsf{Y}_2)\Delta_k w_r,$$

$$Y_{k+1} = \mathsf{Y}_2 + \sum_{r=1}^{m} \gamma_r(t_k)\Delta_k w_r, \quad k = 0, \ldots, N-1, \qquad (5.10)$$

with the parameter $0 \leq \alpha \leq 1$.

If $\sigma_r = \gamma_r = 0, r = 1, \ldots, m$, this method coincides with the deterministic Liouvillian method [242, 275, 279].

Theorem 5.1. *The method (5.9)-(5.10) for the system (5.8) is Liouvillian and of mean-square order $1/2$.*

Proof. Let us check that the one-step approximation $\bar{X}, \bar{Y}$ corresponding to (5.9)-(5.10) satisfies (5.7). Using properties of exterior products, we obtain

4.5 Liouvillian methods for stochastic systems preserving phase volume   249

$$d\bar{X}^1 \wedge \cdots \wedge d\bar{X}^l \wedge d\bar{Y}^1 \cdots \wedge d\bar{Y}^n = (dX_1^1 + \sum_{r=1}^{m}\sum_{j=1}^{n}\frac{\partial \sigma_r^1}{\partial y^j}dY_2^j) \wedge \cdots$$

$$\wedge (dX_1^{l-1} + \sum_{r=1}^{m}\sum_{j=1}^{n}\frac{\partial \sigma_r^{l-1}}{\partial y^j}dY_2^j) \wedge (dX_1^l + \sum_{r=1}^{m}\sum_{j=1}^{n}\frac{\partial \sigma_r^l}{\partial y^j}dY_2^j) \wedge dY_2^1 \wedge \cdots \wedge dY_2^n$$

$$= (dX_1^1 + \sum_{r=1}^{m}\sum_{j=1}^{n}\frac{\partial \sigma_r^1}{\partial y^j}dY_2^j) \wedge \cdots \wedge (dX_1^{l-1} + \sum_{r=1}^{m}\sum_{j=1}^{n}\frac{\partial \sigma_r^{l-1}}{\partial y^j}dY_2^j)$$

$$\wedge (dX_1^l \wedge dY_2^1 \wedge \cdots \wedge dY_2^n + \sum_{r=1}^{m}\sum_{j=1}^{n}\frac{\partial \sigma_r^l}{\partial y^j}dY_2^j \wedge dY_2^1 \wedge \cdots \wedge dY_2^n)$$

$$= (dX_1^1 + \sum_{r=1}^{m}\sum_{j=1}^{n}\frac{\partial \sigma_r^1}{\partial y^j}dY_2^j) \wedge \cdots \wedge (dX_1^{l-1} + \sum_{r=1}^{m}\sum_{j=1}^{n}\frac{\partial \sigma_r^{l-1}}{\partial y^j}dY_2^j)$$

$$\wedge dX_1^l \wedge dY_2^1 \wedge \cdots \wedge dY_2^n$$

$$= \cdots = dX_1^1 \wedge \cdots \wedge dX_1^l \wedge dY_2^1 \wedge dY_2^2 \wedge \cdots \wedge dY_2^n. \quad (5.11)$$

Since (5.9) corresponds to the deterministic Liouvillian method, it follows from (5.11) that the method (5.9)-(5.10) is Liouvillian.

To prove the mean-square order of (5.9)-(5.10), we compare it with the Euler method as usual. □

Now put $\gamma_r = 0$, $r = 1, \ldots, m$, in (5.8) (cf. (2.14)):

$$dX = f(t, Y)dt + \sum_{r=1}^{m}\sigma_r(t, Y)dw_r(t), \ X(t_0) = x,$$

$$dY = g(t, X)dt, \ Y(t_0) = y. \quad (5.12)$$

The Liouvillian method (5.9)-(5.10) in application to (5.12) is of first mean-square order (cf. Theorem 2.6).

Introduce the PRK method for (5.12):

$$Y_1 = Y_k, \ X_1 = X_k + \frac{h}{4}f(t_k, Y_1) + \frac{1}{2}\sum_{r=1}^{m}\sigma_r(t_k, Y_1)\left(3(J_{r0})_k - \Delta_k w_r\right),$$

$$Y_2 = Y_1 + \frac{2}{3}hg(t_k + \frac{h}{4}, X_1),$$

$$X_2 = X_1 + \frac{3}{4}hf(t_k + \frac{2}{3}h, Y_2) + \frac{3}{2}\sum_{r=1}^{m}\sigma_r(t_k + \frac{2}{3}h, Y_2)$$

$$\times (-(J_{r0})_k + \Delta_k w_r), \quad (5.13)$$

$$X_{k+1} = X_2, \ Y_{k+1} = Y_2 + \frac{h}{3}g(t_{k+1}, X_2), \ k = 0, \ldots, N-1. \quad (5.14)$$

This method applied to (2.14) gives the symplectic method (2.25)-(2.26).

Theorem 5.2. *The method (5.13)-(5.14) for the system (5.12) is Liouvillian and of mean-square order* $3/2$.

Proof. By the arguments similar to ones used to obtain (5.11) in Theorem 5.1, we prove that the one-step approximation corresponding to (5.13)-(5.14) satisfies the volume-preserving condition (5.7). For a proof of the mean-square order see Theorem 2.9. □

4.5.2 Liouvillian methods for a volume-preserving system with additive noise

The d-dimensional system with additive noise

$$dX = a(t, X)dt + \sum_{r=1}^{m} b_r(t)dw_r(t), \ X(t_0) = x, \tag{5.15}$$

possesses the volume-preserving property if and only if the condition (5.3) holds.

Theorem 5.3. *Let* $\bar{X} = X + A(t, X, \bar{X}; h)$ *be a one-step approximation corresponding to the first-order Liouvillian method for the deterministic system* (5.1), (5.3). *Then the method for the stochastic system* (5.15), (5.3):

$$X_{k+1} = X_k + A(t_k, X_k, X_{k+1}; h) + \sum_{r=1}^{m} b_r(t_k)\Delta_k w_r \tag{5.16}$$

is Liouvillian and of first mean-square order.

Proof. We have for the one-step approximation $\bar{X}$ corresponding to (5.16): $d\bar{X}^i = dx^i + dA^i$, $i = 1, \ldots, d$. Since these expressions coincide with the ones for the deterministic Liouvillian method, the approximation $\bar{X}$ satisfies (5.7) and the method is Liouvillian. The mean-square order of (5.16) easily follows from the general theory of Chap. 1. □

Due to this theorem, construction of first-order Liouvillian methods for Liouvillian systems with additive noise reduces to construction of such methods for deterministic Liouvillian systems. For instance, consider the following Liouvillian system

$$dX^i = a^i(t, X^1, \ldots, X^{i-1}, X^{i+1}, \ldots, X^d)dt + \sum_{r=1}^{m} b_r^i(t)dw_r(t), \tag{5.17}$$

$$X(t_0) = x, \ i = 1, \ldots, d.$$

In [242] an explicit first-order Liouvillian method for the deterministic system (5.1) with $a(t, x)$ as in (5.17) was proposed. Using it, we obtain

$$X_{k+1}^i = X_k^i + ha^i(t_k, X_{k+1}^1, \ldots, X_{k+1}^{i-1}, X_k^{i+1}, \ldots, X_k^d) + \sum_{r=1}^m b_r^i(t_k)\Delta_k w_r,$$

(5.18)

$$i = 1, \ldots, d, \ k = 0, \ldots, N-1.$$

Corollary 5.4. *The method* (5.18) *for* (5.17) *is Liouvillian and of first mean-square order.*

Note that the Liouvillian method (5.9)-(5.10) for the system (5.8) with $\sigma_r(t,y) = \sigma_r(t)$, $r = 1, \ldots, m$, (the partitioned system with additive noise) is of first mean-square order. Further, for the partitioned system (5.8) with $\sigma_r(t,y) = \sigma_r(t)$, $r = 1, \ldots, m$, a parametric family of two-stage explicit Liouvillian PRK methods of mean-square 3/2 is derived. The form of these methods coincide with the symplectic method (3.35)-(3.39). Let us also note that for the particular case of system (5.12) with $\sigma_r(t,y) = \sigma_r(t)$ and $g(t,x) = M^{-1}x$ where M is a constant, symmetric, invertible matrix, we construct a Liouvillian method of the third mean-square order. The form of this method coincides with the third-order symplectic method (3.49)-(3.50).

4.6 Weak symplectic methods for stochastic Hamiltonian systems

4.6.1 Hamiltonian systems with multiplicative noise

In this subsection weak symplectic methods for Hamiltonian systems with multiplicative noise are constructed. First we consider the general case and then treat the case of separable Hamiltonians.

Implicit first-order methods for general stochastic Hamiltonian systems. Here all the methods are fully implicit (i.e., implicit in both deterministic and stochastic components). Let us recall that in the case of deterministic general Hamiltonian systems symplectic RK methods are all implicit [261].

On the basis of the symplectic method of mean-square order 1/2 (2.7), we propose the weak method:

$$P_{k+1} = P_k + hf(t_k + \beta h, \alpha P_{k+1} + (1-\alpha)P_k, (1-\alpha)Q_{k+1} + \alpha Q_k)$$
$$+ h(\frac{1}{2} - \alpha)\sum_{r=1}^m \sum_{j=1}^n (\frac{\partial \sigma_r}{\partial p^j}\sigma_r^j - \frac{\partial \sigma_r}{\partial q^j}\gamma_r^j) + h^{1/2}\sum_{r=1}^m \sigma_r \xi_{rk},$$

$$Q_{k+1} = Q_k + hg(t_k + \beta h, \alpha P_{k+1} + (1-\alpha)P_k, (1-\alpha)Q_{k+1} + \alpha Q_k)$$
$$+ h(\frac{1}{2} - \alpha)\sum_{r=1}^m \sum_{j=1}^n (\frac{\partial \gamma_r}{\partial p^j}\sigma_r^j - \frac{\partial \gamma_r}{\partial q^j}\gamma_r^j) + h^{1/2}\sum_{r=1}^m \gamma_r \xi_{rk}, \quad (6.1)$$

where $\sigma_r, \gamma_r, r = 1, \ldots, m$, and their derivatives are calculated at $(t_k, \alpha P_{k+1} + (1-\alpha)P_k, (1-\alpha)Q_{k+1} + \alpha Q_k)$, the parameters $\alpha, \beta \in [0,1]$, and ξ_{rk} are i.i.d. random variables with the law

$$P(\xi = \pm 1) = 1/2. \tag{6.2}$$

Note that if $\alpha = \beta = 1/2$ the method (6.1) becomes the derivative-free (midpoint) method. The method requires solution of a nonlinear equation at each step (its solvability is proved within the next theorem).

Theorem 6.1. *The implicit method* (6.1) *for the system* (0.1), (0.2) *is symplectic and of first weak order.*

Proof. The symplecticness is proved as in Theorem 2.2. Let us prove convergence of the method. Denote by $\bar{X} = \bar{X}(t+h;t,x) = (\bar{P}^\mathsf{T}, \bar{Q}^\mathsf{T})^\mathsf{T}$ the one-step approximation corresponding to the method (6.1):

$$\bar{P} = p + hf(t+\beta h, \alpha \bar{P} + (1-\alpha)p, (1-\alpha)\bar{Q} + \alpha q)$$
$$+ h(\frac{1}{2} - \alpha)\sum_{r=1}^{m}\sum_{j=1}^{n}(\frac{\partial \sigma_r}{\partial p^j}\sigma_r^j - \frac{\partial \sigma_r}{\partial q^j}\gamma_r^j) + h^{1/2}\sum_{r=1}^{m}\sigma_r \xi_r,$$

$$\bar{Q} = q + hg(t+\beta h, \alpha \bar{P} + (1-\alpha)p, (1-\alpha)\bar{Q} + \alpha q)$$
$$+ h(\frac{1}{2} - \alpha)\sum_{r=1}^{m}\sum_{j=1}^{n}(\frac{\partial \gamma_r}{\partial p^j}\sigma_r^j - \frac{\partial \gamma_r}{\partial q^j}\gamma_r^j) + h^{1/2}\sum_{r=1}^{m}\gamma_r \xi_r, \tag{6.3}$$

where $\sigma_r, \gamma_r, r = 1, \ldots, m$, and their derivatives are calculated at $(t, \alpha \bar{P} + (1-\alpha)p, (1-\alpha)\bar{Q} + \alpha q)$.

Using a Lipschitz condition on the coefficients of the system (0.1), we prove (cf. Lemma 1.3.7) that there are constants $K > 0$ and $h_0 > 0$ such that for any $h \leq h_0$, $t_0 \leq t \leq t_0 + T$, $x = (p^\mathsf{T}, q^\mathsf{T})^\mathsf{T} \in \mathbf{R}^d$, $d = 2n$, the equation (6.3) has a unique solution $\bar{X}$ which satisfies the inequality

$$|\bar{X} - x| \leq K(1 + |x|)\sqrt{h}, \tag{6.4}$$

and this solution can be found by the method of simple iteration with $x = (p^\mathsf{T}, q^\mathsf{T})^\mathsf{T}$ as the initial approximation.

The condition (2.2.4) with $p = 1$ of Theorem 2.2.1 holds for the approximation (6.3) due to (6.4) (to avoid a confusion, we note that in Theorem 2.2.1 we denote by p the order of a method while in this chapter p means the initial condition for $P(t)$). Let us check fulfillment of the condition (2.2.3) with $p = 1$. To this end, introduce the weak Euler approximation $\hat{X} = (\hat{P}^\mathsf{T}, \hat{Q}^\mathsf{T})^\mathsf{T}$ for the Stratonovich system (0.1), (0.2):

4.6 Weak symplectic methods for stochastic Hamiltonian systems

$$\hat{P} = p + hf + \frac{h}{2}\sum_{r=1}^{m}\sum_{j=1}^{n}\left(\frac{\partial \sigma_r}{\partial p^j}\sigma_r^j + \frac{\partial \sigma_r}{\partial q^j}\gamma_r^j\right) + h^{1/2}\sum_{r=1}^{m}\sigma_r \xi_r,$$

$$\hat{Q} = q + hg + \frac{h}{2}\sum_{r=1}^{m}\sum_{j=1}^{n}\left(\frac{\partial \gamma_r}{\partial p^j}\sigma_r^j + \frac{\partial \gamma_r}{\partial q^j}\gamma_r^j\right) + h^{1/2}\sum_{r=1}^{m}\gamma_r \xi_r, \quad (6.5)$$

where f, g and σ_r, γ_r, $r = 1, \ldots, m$, and their derivatives are calculated at (t, p, q).

Expanding the terms in the right-hand side of (6.3) around (t, p, q) and using (6.4) and the corresponding conditions on smoothness and boundedness of the coefficients, it is not difficult to obtain that

$$\left|E\left(\prod_{j=1}^{s}\hat{\Delta}^{i_j} - \prod_{j=1}^{s}\bar{\Delta}^{i_j}\right)\right| \leq K(x)h^2, \ s = 1, 2, 3, \ i_j = 1, \ldots, 2n, \ K(x) \in \mathbf{F}, \quad (6.6)$$

where $\bar{\Delta}^i := \bar{X}^i - x^i$, $\hat{\Delta}^i := \hat{X}^i - x^i$.

Taking into account (6.6) and the fact that the Euler approximation (6.5) satisfies (2.2.3) with $p = 1$, we get that the approximation (6.3) satisfies (2.2.3) with $p = 1$ as well.

Finally, to check the fourth condition of Theorem 2.2.1, we use Lemma 2.2.2 which ensures existence and uniform boundedness of the moments $E|\bar{X}_k|^{\bar{m}}$ under the conditions: (i) $|E\bar{\Delta}| \leq K(1+|x|)h$ and (ii) $|\bar{\Delta}| \leq M(\xi)(1+|x|)\sqrt{h}$ with $M(\xi)$ having moments of all orders. The inequalities (6.6) and $|E\hat{\Delta}| \leq K(1+|x|)h$ imply fulfillment of the condition (i), while the condition (ii) holds here due to (6.4). □

Remark 6.2. In the case of separable Hamiltonians at noise, i.e., when $H_r(t, p, q) = U_r(t, q) + V_r(t, p)$, $r = 1, \ldots, m$, the method (6.1) with $\alpha = 1$, $\beta = 0$ acquires the form

$$P_{k+1} = P_k + f(t_k, P_{k+1}, Q_k)h$$
$$+ \frac{h}{2}\sum_{r=1}^{m}\sum_{j=1}^{n}\frac{\partial \sigma_r}{\partial q^j}(t_k, Q_k) \cdot \gamma_r^j(t_k, P_{k+1}) + h^{1/2}\sum_{r=1}^{m}\sigma_r(t_k, Q_k)\xi_{rk},$$

$$Q_{k+1} = Q_k + g(t_k, P_{k+1}, Q_k)h$$
$$- \frac{h}{2}\sum_{r=1}^{m}\sum_{j=1}^{n}\frac{\partial \gamma_r}{\partial p^j}(t_k, P_{k+1}) \cdot \sigma_r^j(t_k, Q_k) + h^{1/2}\sum_{r=1}^{m}\gamma_r(t_k, P_{k+1})\xi_{rk} \quad (6.7)$$

with not too complicated implicitness. Besides, when the Hamiltonians are such that $H_0(t, p, q) = V_0(t, p) + U_0(t, q)$ and $H_r(t, p, q) = \Gamma_r^\top(t)p + U_r(t, q)$, $r = 1, \ldots, m$, $\Gamma_r(t)$ are n-dimensional vectors, one obtains full explicit symplectic methods.

Explicit first-order methods in the case of separable Hamiltonians.
Now we consider a special case of the Hamiltonian system (0.1), (0.2) such that

$$H_0(t,p,q) = V_0(p) + U_0(t,q), \ H_r(t,p,q) = U_r(t,q), \ r = 1, \ldots, m. \quad (6.8)$$

In this case we get the following system

$$dP = f(t,Q)dt + \sum_{r=1}^{m} \sigma_r(t,Q)dw_r(t), \ P(t_0) = p,$$
$$dQ = g(P)dt, \ Q(t_0) = q, \quad (6.9)$$

with

$$f^i = -\partial U_0/\partial q^i, \ g^i = \partial V_0/\partial p^i, \ \sigma_r^i = -\partial U_r/\partial q^i, \ r = 1,\ldots,m, \ i = 1,\ldots,n. \quad (6.10)$$

Recall that the system (6.9) has the same form in the sense of Stratonovich. Due to specific features of the system (6.9), (6.10) we have succeeded in construction of explicit PRK methods of a higher order.

On the basis of the mean-square PRK method (2.17)-(2.18) we obtain the weak PRK method for (6.9):

$$\mathcal{Q}_1 = Q_k + \alpha h g(P_k), \ \mathcal{P}_1 = P_k - hf(t_k + \alpha h, \mathcal{Q}_1),$$
$$\mathcal{Q}_2 = \mathcal{Q}_1 + (1-\alpha)hg(\mathcal{P}_1), \quad (6.11)$$

$$P_{k+1} = \mathcal{P}_1 + h^{1/2}\sum_{r=1}^{m}\sigma_r(t_k,\mathcal{Q}_2)\xi_{rk}, \ Q_{k+1} = \mathcal{Q}_2, \ k = 0,\ldots,N-1, \quad (6.12)$$

where $0 \le \alpha \le 1$ is a parameter and ξ_{rk} are i.i.d. random variables with the law (6.2).

Theorem 6.3. *The explicit method* (6.11)-(6.12) *for the system* (6.9), (6.10) *is symplectic and of first weak order.*

Proof. Due to (6.10), $\partial \sigma_r^i/\partial q^j = \partial \sigma_r^j/\partial q^i$. Using this, we obtain $dP_{k+1} \wedge dQ_{k+1} = d\mathcal{P}_1 \wedge d\mathcal{Q}_2$. It is easy to prove that $d\mathcal{P}_1 \wedge d\mathcal{Q}_2 = d\mathcal{P}_1 \wedge d\mathcal{Q}_1 = dP_k \wedge dQ_k$. Therefore the method (6.11)-(6.12) is symplectic. The order of convergence is proved as in Theorem 6.1 (even simpler). □

Remark 6.4. By swapping the roles of p and q we can propose another symplectic method of first weak order for the system (6.9), (6.10). Namely, instead of (6.11)-(6.12) one can propose

$$\mathcal{P}_1 = P_k + \alpha h f(t_k, Q_k), \ \mathcal{Q}_1 = Q_k + hg(\mathcal{P}_1),$$
$$\mathcal{P}_2 = \mathcal{P}_1 + (1-\alpha)hf(t_k + h, \mathcal{Q}_1), \quad (6.13)$$

$$P_{k+1} = \mathcal{P}_2 + h^{1/2}\sum_{r=1}^{m}\sigma_r(t_k,\mathcal{Q}_1)\xi_{rk}, \ Q_{k+1} = \mathcal{Q}_1, \ k = 0,\ldots,N-1. \quad (6.14)$$

4.6 Weak symplectic methods for stochastic Hamiltonian systems

Explicit second-order method in the case of separable Hamiltonians. Introduce the explicit PRK method for the system (6.9), (6.10):

$$\mathcal{Q}_1 = Q_k + \frac{h}{2}g(P_k), \quad \mathcal{P}_1 = P_k + hf(t_k + \frac{h}{2}, \mathcal{Q}_1)$$

$$+ h^{1/2}\sum_{r=1}^{m}\sigma_r(t_k + \frac{h}{2}, \mathcal{Q}_1)\xi_{rk},$$

$$P_{k+1} = \mathcal{P}_1, \quad Q_{k+1} = \mathcal{Q}_1 + \frac{h}{2}g(\mathcal{P}_1), \quad k = 0, \ldots, N-1, \qquad (6.15)$$

where ξ_{rk} are i.i.d. random variables with the law

$$P(\xi = 0) = 2/3, \quad P(\xi = \pm\sqrt{3}) = 1/6. \qquad (6.16)$$

It follows from Lemma 3.11 that this method is symplectic. Comparing (6.15) with the standard Taylor-type second-order weak method (2.1.31) applied to (6.9), we prove that the method (6.15) is of weak order 2.

Theorem 6.5. *The explicit method* (6.15) *for the system* (6.9), (6.10) *is symplectic and of second weak order.*

4.6.2 Hamiltonian systems with additive noise

Consider Hamiltonian systems with additive noise

$$dP = f(t, P, Q)dt + \sum_{r=1}^{m}\sigma_r(t)dw_r(t), \quad P(t_0) = p,$$

$$dQ = g(t, P, Q)dt + \sum_{r=1}^{m}\gamma_r(t)dw_r(t), \quad Q(t_0) = q, \qquad (6.17)$$

where f and g satisfy (0.2).

The first-order method for (6.17) follows from the method (6.1).

Implicit second-order methods in the case of general Hamiltonian system. On the basis of a mean-square symplectic method of order 3/2 (see (3.26)), we construct the weak method:

$$\mathcal{P}_1 = P_k + \frac{\alpha}{2}hf(t_k + \frac{\alpha}{2}h, \mathcal{P}_1, \mathcal{Q}_1) + \lambda_1 h^{1/2}\sum_{r=1}^{m}\sigma_r(t_k + \frac{h}{2})\xi_{rk},$$

$$\mathcal{Q}_1 = Q_k + \frac{\alpha}{2}hg(t_k + \frac{\alpha}{2}h, \mathcal{P}_1, \mathcal{Q}_1) + \lambda_1 h^{1/2}\sum_{r=1}^{m}\gamma_r(t_k + \frac{h}{2})\xi_{rk},$$

256   4 Stochastic Hamiltonian systems and Langevin-type equations

$$\mathcal{P}_2 = P_k + \alpha h f(t_k + \frac{\alpha}{2}h, \mathcal{P}_1, \mathcal{Q}_1) + \frac{1-\alpha}{2}hf(t_k + \frac{1+\alpha}{2}h, \mathcal{P}_2, \mathcal{Q}_2)$$

$$+ \lambda_2 h^{1/2} \sum_{r=1}^{m} \sigma_r(t_k + \frac{h}{2})\xi_{rk},$$

$$\mathcal{Q}_2 = Q_k + \alpha h g(t_k + \frac{\alpha}{2}h, \mathcal{P}_1, \mathcal{Q}_1) + \frac{1-\alpha}{2}hg(t_k + \frac{1+\alpha}{2}h, \mathcal{P}_2, \mathcal{Q}_2)$$

$$+ \lambda_2 h^{1/2} \sum_{r=1}^{m} \gamma_r(t_k + \frac{h}{2})\xi_{rk},$$

$$P_{k+1} = P_k + h\left[\alpha f(t_k + \frac{\alpha}{2}h, \mathcal{P}_1, \mathcal{Q}_1) + (1-\alpha)f(t_k + \frac{1+\alpha}{2}h, \mathcal{P}_2, \mathcal{Q}_2)\right]$$

$$+ h^{1/2}\sum_{r=1}^{m}\sigma_r(t_k + \frac{h}{2})\xi_{rk},$$

$$Q_{k+1} = Q_k + h\left[\alpha g(t_k + \frac{\alpha}{2}h, \mathcal{P}_1, \mathcal{Q}_1) + (1-\alpha)g(t_k + \frac{1+\alpha}{2}h, \mathcal{P}_2, \mathcal{Q}_2)\right]$$

$$+ h^{1/2}\sum_{r=1}^{m}\gamma_r(t_k + \frac{h}{2})\xi_{rk}, \qquad (6.18)$$

where the parameters α, λ_1, λ_2 are such that

$$\alpha\lambda_1 + (1-\alpha)\lambda_2 = \frac{1}{2}, \quad \alpha\lambda_1^2 + (1-\alpha)\lambda_2^2 = \frac{1}{2}, \qquad (6.19)$$

and ξ_{rk} are i.i.d. random variables with the law (6.16).

For example, the following set of parameters satisfies (6.19):

$$\alpha = \frac{1}{2}, \ \lambda_1 = 0, \ \lambda_2 = 1. \qquad (6.20)$$

The symplecticness follows from Lemma 3.5. The order of convergence is proved similarly to the proof of Theorem 6.1 comparing (6.18) with the standard Taylor-type second-order weak method (2.1.31) applied to (6.17).

Theorem 6.6. *The implicit method* (6.18), (6.19) *for the system* (6.17) *is symplectic and of second weak order.*

A third-order method in a particular case of Hamiltonian system. Now we propose a symplectic weak method of order three for the system with additive noise:

$$dP = f(t,Q)dt + \sum_{r=1}^{m}\sigma_r(t)dw_r(t), \ f^i(t,Q) = -\frac{\partial U_0}{\partial q^i}, \ P(t_0) = p,$$

$$dQ = M^{-1}Pdt, \ Q(t_0) = q. \qquad (6.21)$$

4.6 Weak symplectic methods for stochastic Hamiltonian systems   257

On the basis of a symplectic mean-square method of order 3 (see (3.49)-(3.50)), we construct the weak method:

$$\mathcal{Q}_1 = Q_k + \frac{7}{24}hM^{-1}\mathcal{P}_k, \quad \mathcal{P}_1 = P_k + \frac{2}{3}hf(t_k + \frac{7h}{24}, \mathcal{Q}_1),$$

$$\mathcal{Q}_2 = \mathcal{Q}_1 + \frac{3}{4}hM^{-1}\mathcal{P}_1, \quad \mathcal{P}_2 = \mathcal{P}_1 - \frac{2}{3}hf(t_k + \frac{25h}{24}, \mathcal{Q}_2),$$

$$\mathcal{Q}_3 = \mathcal{Q}_2 - \frac{1}{24}hM^{-1}\mathcal{P}_2, \quad \mathcal{P}_3 = \mathcal{P}_2 + hf(t_k + h, \mathcal{Q}_3), \qquad (6.22)$$

$$P_{k+1} = \mathcal{P}_3 + h^{1/2}\sum_{r=1}^{m}\sigma_r(t_k)\xi_{rk} + h^{3/2}\sum_{r=1}^{m}\sigma'_r(t_k)(\xi_r/2 - \eta_r)_k$$

$$+h^{5/2}\sum_{r=1}^{m}\sigma''_r(t_k)\xi_{rk}/6 + h^{5/2}\sum_{r=1}^{m}\sum_{i=1}^{n}(M^{-1}\sigma_r(t_k))^i\frac{\partial f}{\partial q^i}(t_k,\mathcal{Q}_3)\xi_{rk}/6,$$

$$Q_{k+1} = \mathcal{Q}_3 + h^{3/2}\sum_{r=1}^{m}M^{-1}\sigma_r(t_k)(\xi_r/2 + \eta_r)_k + h^{5/2}\sum_{r=1}^{m}M^{-1}\sigma'_r(t_k)\xi_{rk}/6,$$

$$(6.23)$$

$$k = 0, \ldots, N-1,$$

where ξ_{rk}, η_{rk} are mutually independent random variables distributed by the laws

$$P(\xi = 0) = \frac{1}{3}, \quad P(\xi = \pm 1) = \frac{3}{10}, \quad P(\xi = \pm\sqrt{6}) = \frac{1}{30},$$

$$P(\eta = \pm 1/\sqrt{12}) = \frac{1}{2}. \qquad (6.24)$$

The symplecticness of this method follows from Theorem 3.15. The order of convergence can be proved by standard arguments using the fact that the corresponding mean-square method (3.49)-(3.50) has the third order of convergence or by comparing the method (6.22)-(6.23) with the weak method of order 3 (see (2.3.34)) applied to (6.21).

Theorem 6.7. *The explicit method (6.22)-(6.23) for the system (6.21) is symplectic and of third weak order.*

4.6.3 Numerical tests

Kubo oscillator. Consider the Kubo oscillator (cf. (4.1))

$$dX^1 = -aX^2 dt - \sigma X^2 \circ dw(t), \quad X^1(0) = x^1,$$

$$dX^2 = aX^1 dt + \sigma X^1 \circ dw(t), \quad X^2(0) = x^2, \qquad (6.25)$$

where a and σ are constants and $w(t)$ is a one-dimensional standard Wiener process.

The quantity $\mathcal{H}(x^1, x^2) = (x^1)^2 + (x^2)^2$ is conservative for this system:

$$\mathcal{H}(X^1(t), X^2(t)) = \mathcal{H}(x^1, x^2) \text{ for } t \geq 0.$$

Here we test three specific methods of weak order one. The weak Euler method in application to (6.25) takes the form:

$$X^1_{k+1} = X^1_k - haX^2_k - h\frac{\sigma^2}{2}X^1_k - h^{1/2}\sigma X^2_k \xi_k,$$

$$X^2_{k+1} = X^2_k + haX^1_k - h\frac{\sigma^2}{2}X^2_k + h^{1/2}\sigma X^1_k \xi_k, \qquad (6.26)$$

where ξ_k are i.i.d random variables with the law (6.2).

The weak midpoint method (the symplectic method (6.1) with $\alpha = 1/2$) is written for the autonomous system (6.25) as

$$X^1_{k+1} = X^1_k - ha\frac{X^2_k + X^2_{k+1}}{2} - h^{1/2}\sigma\frac{X^2_k + X^2_{k+1}}{2}\xi_k,$$

$$X^2_{k+1} = X^2_k + ha\frac{X^1_k + X^1_{k+1}}{2} + h^{1/2}\sigma\frac{X^1_k + X^1_{k+1}}{2}\xi_k. \qquad (6.27)$$

This is an implicit method in both deterministic and stochastic terms.

When applied to (6.25), the PRK method (6.7) has the form:

$$X^1_{k+1} = X^1_k - haX^2_k - h\frac{\sigma^2}{2}X^1_{k+1} - h^{1/2}\sigma X^2_k \xi_k,$$

$$X^2_{k+1} = X^2_k + haX^1_{k+1} + h\frac{\sigma^2}{2}X^2_k + h^{1/2}\sigma X^1_{k+1}\xi_k. \qquad (6.28)$$

This method is symplectic and of first weak order. It is implicit in the deterministic part only.

Let us analyze how accurately these methods approximate $E\mathcal{H}(X^1(t), X^2(t))$. In the case of the Euler method we obtain

$$E\mathcal{H}(X^1_k, X^2_k) = (1 + h^2(a^2 + \frac{\sigma^4}{4}))^k \times \mathcal{H}(x^1, x^2)$$

$$\geq \exp(\frac{1}{2}(a^2 + \frac{\sigma^4}{4})ht_k) \times \mathcal{H}(x^1, x^2), \qquad (6.29)$$

i.e., the quantity grows exponentially fast as t increases.

It is not difficult to check that $\mathcal{H}(x^1, x^2)$ is conserved by the midpoint method (6.27). But the PRK method (6.28) does not preserve the quantity $\mathcal{H}(x^1, x^2)$. See a similar discussion concerning the mean-square midpoint method in Sect. 4.4.1.

4.6 Weak symplectic methods for stochastic Hamiltonian systems

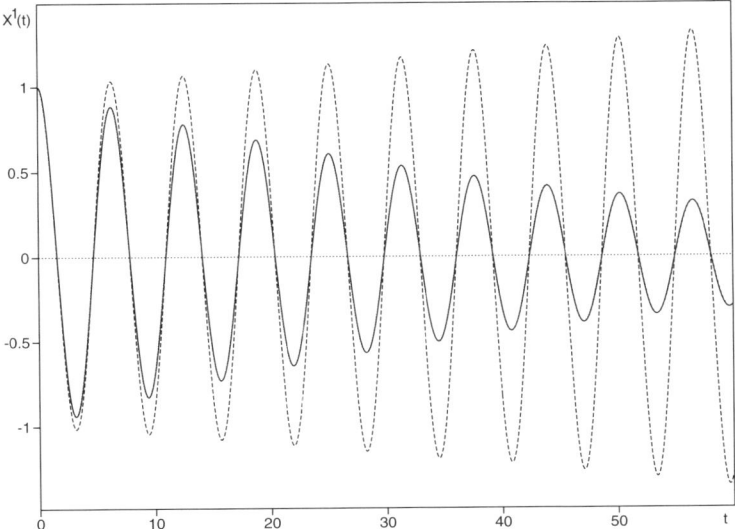

Fig. 6.1. The Kubo oscillator (6.25). Simulation of $EX^1(t)$ with $X^1(0) = 1$, $X^2(0) = 0$, $a = 1$, $\sigma = 0.2$, $h = 0.05$ on the time interval $t \leq 60$. The results obtained by the Euler method (6.26) – *dashed line*. The results obtained by the midpoint method (6.27) and the symplectic method (6.28) visually coincide with the exact solution (6.30) (*solid line*). The Monte Carlo error is not greater than 0.03 with probability 0.95.

Further, it is not difficult to find the following exact expressions for the Kubo oscillator (6.25):

$$EX^1_{0,x}(t) = e^{-\sigma^2 t/2}(x^1 \cos at - x^2 \sin at),$$
$$EX^2_{0,x}(t) = e^{-\sigma^2 t/2}(x^2 \cos at + x^1 \sin at). \qquad (6.30)$$

Figure 6.1 gives results of Monte Carlo simulation of $EX^1_{0,x}(t)$ by the methods (6.26), (6.27), and (6.28). We see that the Euler method is not appropriate for simulation of the oscillator (6.25) on long time intervals while the symplectic methods produce quite accurate results.

A model for synchrotron oscillations of particles in storage rings. Consider a model describing synchrotron oscillations of particles in storage rings under the influence of external fluctuating electromagnetic fields from [265] (cf. (4.5)):

$$dP = -\omega^2 \sin(Q)dt - \sigma_1 \cos(Q)dw_1 - \sigma_2 \sin(Q)dw_2,$$
$$dQ = Pdt, \qquad (6.31)$$

where P and Q are scalars.

Here we test four weak methods: two first-order methods (the Euler method, which is not symplectic, and the symplectic method (6.11)-(6.12)) and two second-order methods (the standard second-order weak method (2.2.18) and the symplectic method (6.15)).

The weak Euler method for (6.31) takes the form

$$P_{k+1} = P_k - h\omega^2 \sin(Q_k) - h^{1/2}(\sigma_1 \cos(Q_k)\xi_{1k} + \sigma_2 \sin(Q_k)\xi_{2k}),$$
$$Q_{k+1} = Q_k + hP_k, \qquad (6.32)$$

where ξ_{1k}, ξ_{2k} are i.i.d random variables with the law (6.2).

In application to (6.31) the first-order symplectic method (6.11)-(6.12) with $\alpha = 1$ is written as

$$\mathcal{Q} = Q_k + hP_k,$$
$$P_{k+1} = P_k - h\omega^2 \sin(\mathcal{Q}) - h^{1/2}(\sigma_1 \cos(\mathcal{Q})\xi_{1k} + \sigma_2 \sin(\mathcal{Q})\xi_{2k}),$$
$$Q_{k+1} = \mathcal{Q}, \qquad (6.33)$$

where ξ_{1k}, ξ_{2k} are i.i.d random variables with the law (6.2).

The standard second-order method (2.2.18) applied to (6.31) has the form

$$P_{k+1} = P_k - h^{1/2}(\sigma_1 \cos(Q_k)\xi_{1k} + \sigma_2 \sin(Q_k)\xi_{2k}) - h\omega^2 \sin(Q_k)$$
$$+ \frac{h^{3/2}}{2}(\sigma_1 \sin(Q_k)\xi_{1k} - \sigma_2 \cos(Q_k)\xi_{2k})P_k - \frac{h^2}{2}\omega^2 \cos(Q_k)P_k,$$
$$Q_{k+1} = Q_k + hP_k - \frac{h^{3/2}}{2}(\sigma_1 \cos(Q_k)\xi_{1k} + \sigma_2 \sin(Q_k)\xi_{2k})$$
$$- \frac{h^2}{2}\omega^2 \sin(Q_k), \qquad (6.34)$$

where ξ_{1k}, ξ_{2k} are i.i.d. random variables with the law (6.16).

The second-order symplectic method (6.15) is written for the system (6.31) as

$$\mathcal{Q}_1 = Q_k + \frac{h}{2}P_k, \quad \mathcal{P}_1 = P_k - h\omega^2 \sin(\mathcal{Q}_1) - h^{1/2}(\sigma_1 \cos(\mathcal{Q}_1)\xi_{1k}$$
$$+ \sigma_2 \sin(\mathcal{Q}_1)\xi_{2k}),$$
$$P_{k+1} = \mathcal{P}_1, \quad Q_{k+1} = \mathcal{Q}_1 + \frac{h}{2}\mathcal{P}_1, \qquad (6.35)$$

where ξ_{1k}, ξ_{2k} are i.i.d. random variables with the law (6.16).

Consider the quantity

$$\mathcal{E}(p, q) = \frac{p^2}{2} - \omega^2 \cos(q).$$

Its mean value $E\mathcal{E}(P(t), Q(t))$ is treated in physical literature (see, e.g., [265] and references therein) as a mean energy of the system. Under the assumption $\sigma_1 = \sigma_2 = \sigma$ one can obtain that

4.7 Quasi-symplectic mean-square methods for Langevin-type equations   261

$$E\mathcal{E}(P_{0,p,q}(t), Q_{0,p,q}(t)) = \mathcal{E}(p,q) + \frac{\sigma^2}{2}t. \qquad (6.36)$$

In Table 6.1 we compare results produced by the four methods given above. We have two types of errors in numerical simulations here: error of a weak method used and a Monte Carlo error. The results in the table are approximations of $E\mathcal{E}(\bar{P}(t), \bar{Q}(t))$ calculated as

$$E\mathcal{E}(\bar{P}(t), \bar{Q}(t)) \doteq \frac{1}{M}\sum_{m=1}^{M} \mathcal{E}(\bar{P}^{(m)}(t), \bar{Q}^{(m)}(t)) \pm 2\sqrt{\frac{\bar{D}_M}{M}}, \qquad (6.37)$$

where

$$\bar{D}_M = \frac{1}{M}\sum_{m=1}^{M}[\mathcal{E}(\bar{P}^{(m)}(t), \bar{Q}^{(m)}(t))]^2 - \left[\frac{1}{M}\sum_{m=1}^{M}\mathcal{E}(\bar{P}^{(m)}(t), \bar{Q}^{(m)}(t))\right]^2,$$

i.e., $E\mathcal{E}(\bar{P}(t), \bar{Q}(t))$ belongs to the interval defined in this formula with probability 0.95 (we recall that for sufficiently small h the sampling variance is sufficiently close to the variance of $\mathcal{E}(\bar{P}(t), \bar{Q}(t))$). Note that the "$\pm$" reflects the Monte Carlo error only, it does not reflect the error of a method.

The above experiments with the model (6.31) demonstrate once again superiority of symplectic methods in comparison with nonsymplectic ones. We note that the authors of [265] are interested in systems with small noise. Effective symplectic methods in the weak sense for Hamiltonian systems with small noise can be obtained using ideas from Chap. 3.

4.7 Quasi-symplectic mean-square methods for Langevin-type equations

In Sect. 4.7.1, we construct mean-square quasi-symplectic methods for Langevin equations which are an important particular case of (0.5) when $f(t,q) =$

Table 6.1. The model (6.31). Simulation of $E\mathcal{E}(P(t), Q(t))$ with $P(0) = 1$, $Q(0) = 0$, $\omega = 4$, $\sigma_1 = \sigma_2 = 0.3$, $t = 200$ for various time steps h by the Euler method (6.32), the first-order symplectic method (6.33), the standard second-order method (6.34), and the second-order symplectic method (6.35). The exact solution is -6.5. M is a number of independent realizations in the Monte Carlo simulation. Note that the "$\pm$" reflects the Monte Carlo error only (cf. (6.37)), it does not reflect the error of a method.

h	M	(6.32)	(6.33)	(6.34)	(6.35)
0.1	10^5	493.3 ± 0.3	-6.268 ± 0.059	462.2 ± 0.6	-6.316 ± 0.059
0.05	10^5	966.1 ± 0.7	-6.397 ± 0.059	0.896 ± 0.094	-6.421 ± 0.058
0.01	$4 \cdot 10^6$	234.5 ± 0.06	-6.503 ± 0.009	-6.456 ± 0.009	-6.502 ± 0.009

$f(q)$, $\tilde{f}(t,p,q) = \Gamma p$, Γ is an $n \times n$-dimensional constant matrix, $g(p) = M^{-1}p$, M is a positive definite matrix, and $\sigma_r(t,q) = \sigma_r$, $r = 1, \ldots, m$, are constant vectors. The proposed methods are such that they degenerate to symplectic methods when the system degenerates to a Hamiltonian one and their law of phase volume contractivity is close to the law of the considered system. To construct numerical methods, we use the splitting technique (see, e.g. [261, 271, 311]) and some ideas of [278], where methods for deterministic second-order differential equations with similar properties were obtained. In Sect. 4.7.2, we generalize mean-square methods of Sect. 4.7.1 to the Langevin-type equations (0.5) and also to more general systems.

4.7.1 Langevin equation: Linear damping and additive noise

Consider the Langevin equation

$$dP = f(Q)dt - \nu\Gamma P dt + \sum_{r=1}^{m} \sigma_r dw_r(t), \quad P(t_0) = p,$$
$$dQ = M^{-1}P dt, \quad Q(t_0) = q, \tag{7.1}$$

where P, Q, f are n-dimensional column-vectors, σ_r, $r = 1, \ldots, m$, are n-dimensional constant column-vectors, Γ is an $n \times n$-dimensional constant matrix, $\nu \geq 0$ is a parameter, M is a positive definite matrix, and $w_r(t)$, $r = 1, \ldots, m$, are independent standard Wiener processes. If there is a scalar function $U_0(q)$ such that

$$f^i(q) = -\frac{\partial U_0}{\partial q^i}, \quad i = 1, \ldots, n, \tag{7.2}$$

and if $\nu = 0$, then the system (7.1) is a Hamiltonian system with additive noise, i.e., its phase flow preserves symplectic structure.

The system (7.1) can be written as the second-order differential equation with additive noise:

$$M\ddot{Q} = f(Q) - \nu\Gamma M\dot{Q} + \sum_{r=1}^{m} \sigma_r \dot{w}_r.$$

Let $D_0 \in \mathbf{R}^d$, $d = 2n$, be a domain with finite volume. This domain may be random. We suppose that $D_0 = D_0(\omega)$ is independent of the Wiener processes $w_r(t)$, $t \in [t_0, t_0 + T]$. The transformation $(p,q) \mapsto (P,Q)$ maps D_0 into the domain D_t. The volume V_t of the domain D_t is equal to

$$V_t = \int_{D_t} dP^1 \ldots dP^n dQ^1 \ldots dQ^n$$
$$= \int_{D_0} \left| \frac{D(P^1, \ldots, P^n, Q^1, \ldots, Q^n)}{D(p^1, \ldots, p^n, q^1, \ldots, q^n)} \right| dp^1 \ldots dp^n dq^1 \ldots dq^n. \tag{7.3}$$

4.7 Quasi-symplectic mean-square methods for Langevin-type equations

In the case of the system (7.1) the Jacobian J is equal to

$$J = \frac{D(P^1, \ldots, P^n, Q^1, \ldots, Q^n)}{D(p^1, \ldots, p^n, q^1, \ldots, q^n)} = \exp\left(-\nu\, tr\Gamma \cdot (t - t_0)\right). \tag{7.4}$$

That is, the system (7.1) preserves phase volume for $\nu = 0$. If $\nu > 0$ and $tr\Gamma > 0$ then phase-volume contractivity takes place.

Our aim is to propose mean-square methods based on the one-step approximations

$$\bar{P} = \bar{P}(t+h; t, p, q), \bar{Q} = \bar{Q}(t+h; t, p, q)$$

such that

RL1. *The method applied to (7.1)-(7.2) degenerates to a symplectic method when $\nu = 0$, i.e., $d\bar{P} \wedge d\bar{Q} = dp \wedge dq$ for $\nu = 0$ and f from (7.2);*

RL2. *The Jacobian*

$$\bar{J} = \frac{D(\bar{P}, \bar{Q})}{D(p, q)}$$

does not depend on p, q.

Methods for (7.1) satisfying the conditions RL1 and RL2 are called *quasi-symplectic*.

As it is understood, a method is convergent and, consequently, $\bar{J}$ is close to J at any rate. The requirement RL2 is natural since the Jacobian J of the original system (7.1) does not depend on p, q. RL2 reflects the structural properties of the system which are connected with the law of phase volume contractivity. It is often possible to reach a stronger property consisting in the equality $\bar{J} = J$. However, such an requirement is too restrictive in general. In the context of deterministic equations the requirement RL2 was introduced in [278].

To construct methods satisfying RL1-RL2, we use ideas of splitting technique (see, e.g. [261, 271]). In connection with (7.1), introduce the systems

$$dP_I = f(Q_I)dt + \sum_{r=1}^{m} \sigma_r dw_r(t), \quad P_I(t_0) = p,$$

$$dQ_I = M^{-1} P_I dt, \quad Q_I(t_0) = q, \tag{7.5}$$

$$\frac{dP_{II}}{dt} = -\nu \Gamma P_{II}, \quad P_{II}(0) = p, \tag{7.6}$$

and denote their solutions as $P_I(t; t_0, p, q)$, $Q_I(t; t_0, p, q)$ and $P_{II}(t; p)$, respectively. The system (7.5) with $f(q)$ from (7.2) is a Hamiltonian system with additive noise. The system (7.6) is a deterministic linear system with constant coefficients, and its solution $P_{II}(t; p)$ can be found explicitly.

First-order methods. Let $\bar{P}_I = \bar{P}_I(t_0+h; t_0, p, q)$, $\bar{Q}_I = \bar{Q}_I(t_0+h; t_0, p, q)$ be a one-step approximation of a symplectic first-order mean-square method for (7.5), (7.6) (any explicit or implicit method from Sect. 4.3 can be used). Its Jacobian is equal to one, i.e.,

$$\frac{D(\bar{P}_I(t_0+h; t_0, p, q), \bar{Q}_I(t_0+h; t_0, p, q))}{D(p,q)} = 1.$$

We construct the one-step approximation $\bar{P}$, $\bar{Q}$ for the solution of (7.1)-(7.2) as follows

$$\begin{aligned}\bar{P} &= \bar{P}(t_0+h; t_0, p, q) := P_{II}(h; \bar{P}_I(t_0+h; t_0, p, q)), \\ \bar{Q} &= \bar{Q}(t_0+h; t_0, p, q) := \bar{Q}_I(t_0+h; t_0, p, q).\end{aligned} \quad (7.7)$$

We have

$$\bar{J} = \frac{D(\bar{P}, \bar{Q})}{D(p,q)} = \frac{D(P_{II}, \bar{Q}_I)}{D(\bar{P}_I, \bar{Q}_I)} \frac{D(\bar{P}_I, \bar{Q}_I)}{D(p,q)} = J. \quad (7.8)$$

Further, if $\nu = 0$, then $\bar{P} = \bar{P}_I$, $\bar{Q} = \bar{Q}_I$, i.e., the approximation (7.7) degenerates to the symplectic method for (7.1)-(7.2) with $\nu = 0$. Thus, the approximation $\bar{P}$, $\bar{Q}$ satisfies both requirements RL1 and RL2.

It is not difficult to prove the following lemma.

Lemma 7.1. *Let $\bar{P}_I$, $\bar{Q}_I$ be a one-step approximation corresponding to any first-order mean-square method for the system (7.5). Then $\bar{P}$, $\bar{Q}$ defined in (7.7) is a one-step approximation of the first-order mean-square method for the system (7.1).*

Thus, due to (7.8), we obtain the theorem.

Theorem 7.2. *Let $\bar{P}_I$, $\bar{Q}_I$ be a one-step approximation corresponding to a symplectic first-order mean-square method for the system (7.5), (7.2). Then $\bar{P}$, $\bar{Q}$ defined in (7.7) is a one-step approximation of the first-order mean-square method for the system (7.1) such that (i) it is symplectic when applied to (7.1)-(7.2) with $\nu = 0$, (ii) its phase volume changes according to the same law as the phase volume of (7.1) does, i.e., the Jacobians $\bar{J} = D(\bar{P}, \bar{Q})/D(p,q)$ and $J = D(P, Q)/D(p,q)$ are equal.*

Remark 7.3. Theorem 7.2 also holds for the method based on the following one-step approximation:

$$\begin{aligned}\bar{P} &= \bar{P}(t_0+h; t_0, p, q) := \bar{P}_I(t_0+h; t_0, P_{II}(h;p), q), \\ \bar{Q} &= \bar{Q}(t_0+h; t_0, p, q) := \bar{Q}_I(t_0+h; t_0, P_{II}(h;p), q).\end{aligned} \quad (7.9)$$

Remark 7.4. In practice, it can be more convenient to use an approximation $\bar{P}_{II}$ of the solution to (7.6) instead of the exact solution P_{II} in (7.7) (or (7.9)). Since (7.6) is a deterministic equation, we can exploit a high-order

4.7 Quasi-symplectic mean-square methods for Langevin-type equations

deterministic scheme in order to obtain $\bar{P}_{II}$. In this case the Jacobian $\bar{J}$ approximates the original Jacobian J with the accuracy of the deterministic scheme. Due to the linearity of (7.6), this $\bar{J}$ does not depend on the initial data p, q (it depends on $\nu\Gamma$ and h only).

There is another possibility to propose methods for (7.1) satisfying RL1-RL2. It consists in direct application of symplectic methods. For instance, the parametric first-order family of implicit methods (3.3)-(3.4) in application to (7.1) takes the form

$$\bar{P} = p + hf((1-\alpha)\bar{Q} + \alpha q) - h\nu\Gamma \cdot (\alpha\bar{P} + (1-\alpha)p) + \sum_{r=1}^{m} \sigma_r \Delta w_r ,$$

$$\bar{Q} = q + hM^{-1}(\alpha\bar{P} + (1-\alpha)p) . \qquad (7.10)$$

However, it satisfies the requirement RL2 for $\alpha = 0$ and $\alpha = 1$ only. Moreover, due to their specific structure, not all the symplectic methods (see, for example, the explicit method (3.30)) can be directly applied to the Langevin equation (7.1) itself. Thus, on the way of the direct application of symplectic methods to (7.1) we have rather restrictive opportunities. Nevertheless, we can obtain on this way some new methods.

Second-order methods. In order to construct second-order methods for the Langevin equation (7.1) with the properties RL1 and RL2, we use ideas of the method of fractional steps [261,271,311]. In the deterministic case (i.e., when $\sigma_r = 0$, $r = 1, \ldots, m$) a second-order method satisfying RL1 and RL2 can be based on the following one-step approximation

$$\bar{P} = \bar{P}(t_0 + h; t_0, p, q) := P_{II}(\frac{h}{2}; \bar{P}_I(t_0 + h; t_0, P_{II}(\frac{h}{2}; p), q)),$$

$$\bar{Q} = \bar{Q}(t_0 + h; t_0, p, q) := \bar{Q}_I(t_0 + h; t_0, P_{II}(\frac{h}{2}; p), q), \qquad (7.11)$$

where $\bar{P}_I$, $\bar{Q}_I$ corresponds to a one-step approximation of a symplectic method for (7.5), (7.2) with $\sigma_r = 0$.

In the stochastic case the interconnection between terms in (7.1) is more complicated and a correction to (7.11) is needed to obtain second-order accuracy. Consider the following approximation for solution of (7.1):

$$\bar{P} = \bar{P}(t_0 + h; t_0, p, q) := P_{II}(\frac{h}{2}; \bar{P}_I(t_0 + h; t_0, P_{II}(\frac{h}{2}; p), q))$$

$$- \nu \sum_{r=1}^{m} \Gamma \sigma_r (I_{r0} - \frac{h}{2}\Delta w_r),$$

$$\bar{Q} = \bar{Q}(t_0 + h; t_0, p, q) := \bar{Q}_I(t_0 + h; t_0, P_{II}(\frac{h}{2}; p), q), \qquad (7.12)$$

where $\bar{P}_I$, $\bar{Q}_I$ is a one-step approximation corresponding to a symplectic (explicit or implicit) second-order mean-square method for (7.5), (7.2) (such methods are available in Sect. 4.3),

$$I_{r0} = \int_{t_0}^{t} (w_r(s) - w_r(t_0))\, ds\,.$$

Lemma 7.5. *Let $\bar{P}_I$, $\bar{Q}_I$ be a one-step approximation corresponding to any second-order mean-square method for the system (7.5). Then $\bar{P}, \bar{Q}$ defined in (7.12) is a one-step approximation of the second-order mean-square method for the system (7.1).*

Proof. Due to the assumption, we can write

$$\bar{P}_I(t_0+h;t_0,p,q) = p + \sum_{r=1}^{m} \sigma_r \Delta w_r + h f(q) + \frac{h^2}{2}\sum_{i=1}^{n} (M^{-1}p)^i \frac{\partial f}{\partial q^i} + r_1,$$

$$\bar{Q}_I(t_0+h;t_0,p,q) = q + h M^{-1}p + \sum_{r=1}^{m} M^{-1}\sigma_r I_{r0} + \frac{h^2}{2} M^{-1}f(q) + r_2, \quad (7.13)$$

where the remainders r_1 and r_2 are such that

$$|Er_i| = O(h^3),\quad Er_i^2 = O(h^5)\,,\quad i = 1,2\,.$$

We also have

$$P_{II}(h;p) = p - h\nu\Gamma p + \frac{h^2}{2}\nu^2\Gamma^2 p + \rho,\quad \rho = O(h^3)\,. \quad (7.14)$$

We obtain from (7.12)-(7.14) that

$$\bar{P} = p + \sum_{r=1}^{m} \sigma_r \Delta w_r + h\left(f(q) - \nu\Gamma p\right) - \nu \sum_{r=1}^{m}\Gamma\sigma_r I_{r0}$$
$$+ \frac{h^2}{2}\left[\sum_{i=1}^{n}(M^{-1}p)^i \frac{\partial f}{\partial q^i} + \nu^2\Gamma^2 p - \nu\Gamma f(q)\right] + R_1\,,$$

$$\bar{Q} = q + h M^{-1}p + \sum_{r=1}^{m} M^{-1}\sigma_r I_{r0} + \frac{h^2}{2} M^{-1}\left[f(q) - \nu\Gamma p\right] + R_2\,,\quad (7.15)$$

where R_1 and R_2 are such that

$$|ER_i| = O(h^3),\quad ER_i^2 = O(h^5)\,,\quad i = 1,2\,.$$

It is not difficult to show that the standard Taylor-type mean-square method of order 3/2 for systems with additive noise (see (1.5.8)) has the second order of accuracy when it is applied to (7.1). Comparing the one-step approximation of this standard method with (7.15), we obtain that the method based on (7.12) is of mean-square order 2. □

One can easily check that the approximation (7.12) satisfies our requirements RL1 and RL2. The following theorem summarizes the result.

4.7 Quasi-symplectic mean-square methods for Langevin-type equations

Theorem 7.6. *Let $\bar{\mathcal{P}}_I$, $\bar{\mathcal{Q}}_I$ be a one-step approximation corresponding to a symplectic second-order mean-square method for the system (7.5), (7.2). Then $\bar{\mathcal{P}}$, $\bar{\mathcal{Q}}$ defined in (7.12) is a one-step approximation of the second-order mean-square method for the system (7.1)-(7.2) such that (i) it is symplectic when applied to (7.1)-(7.2) with $\nu = 0$, (ii) its phase volume changes according to the same law as the phase volume of (7.1)-(7.2) does.*

Let us give a concrete example of a method based on (7.12):

$$\mathcal{P}_1 = P_{II}(\frac{h}{2}; P_k), \quad \mathcal{Q}_1 = Q_k + \frac{h}{2}M^{-1}\mathcal{P}_1,$$

$$\mathcal{P}_2 = \mathcal{P}_1 + \sum_{r=1}^{m} \sigma_r \Delta_k w_r + h f(\mathcal{Q}_1),$$

$$\mathcal{Q}_2 = Q_k + hM^{-1}P_k + \sum_{r=1}^{m} M^{-1}\sigma_r (I_{r0})_k + \frac{h^2}{2}M^{-1}f(\mathcal{Q}_1),$$

$$P_{k+1} = P_{II}(\frac{h}{2}; \mathcal{P}_2) - \nu \sum_{r=1}^{m} \Gamma \sigma_r (I_{r0} - \frac{h}{2}\Delta w_r), \quad Q_{k+1} = \mathcal{Q}_2, \quad (7.16)$$

$$k = 0, \ldots, N-1.$$

To obtain (7.16), we use the explicit symplectic second-order PRK method (3.44), which is a generalization of the Störmer–Verlet method.

The random variables $\Delta_k w_r$, $(I_{r0})_k$ have a Gaussian joint distribution, and they can be simulated at each step by $2m$ mutually independent $\mathcal{N}(0,1)$-distributed random variables ξ_{rk} and η_{rk}, $r = 0, \ldots, m$:

$$\Delta_k w_r = h^{1/2}\xi_{rk}, \quad (I_{r0})_k = h^{3/2}(\xi_{rk} + \eta_{rk}/\sqrt{3})/2. \quad (7.17)$$

Note that Remark 7.4 is applicable here if one will approximate $P_{II}(t)$ using a deterministic method of one-step order not less than 3.

In molecular dynamics several methods based on the deterministic Störmer–Verlet method are used for simulation of the Langevin equation (7.1) with diagonal matrix Γ (see [114, 266] and references therein). Effective numerical methods for this type of Langevin equations can be constructed by, for instance, the following splitting

$$dP_I = -\nu \Gamma P_I\, dt + \sum_{r=1}^{m} \sigma_r dw_r(t), \quad dQ_I = M^{-1}P_I\, dt, \quad dP_{II} = f(q)dt.$$

Since P_I, Q_I satisfy the linear system with additive noise, they can be simulated exactly. A number of concrete schemes satisfying our requirements RL1-RL2 can be derived using the exact P_I, Q_I and a deterministic symplectic method. Such a second-order method based on the Störmer–Verlet scheme coincides with the method proposed in [266]. In the case of the unit matrix Γ it has the form:

$$\mathcal{P}_1 = P_k + \frac{h}{2}f(Q_k),$$

$$\mathcal{P}_2 = e^{-\nu h}\mathcal{P}_1 + \sum_{r=1}^{m}\sigma_r(\Delta_k w_r - \tilde{I}_{rk}),$$

$$\mathcal{Q}_2 = Q_k + M^{-1}\frac{1-e^{-\nu h}}{\nu}\mathcal{P}_1 + \frac{M^{-1}}{\nu}\sum_{r=1}^{m}\sigma_r\tilde{I}_{rk},$$

$$P_{k+1} = \mathcal{P}_2 + \frac{h}{2}f(\mathcal{Q}_2), \quad Q_{k+1} = \mathcal{Q}_2,$$

where

$$\tilde{I}_{rk} := \int_{t_k}^{t_{k+1}} \left(1 - e^{-\nu(t_{k+1}-s)}\right) dw_r(s).$$

The random variables $\Delta_k w_r$, $\tilde{I}_{rk}$ have a Gaussian joint distribution. They can be simulated at each step by $2m$ independent $\mathcal{N}(0,1)$-distributed random variables ξ_{rk} and η_{rk}, $r = 0, \ldots, m$. As a result, the above method can be written in the constructive form.

Third-order methods. Using ideas of the method of fractional steps, as we did above, it is possible to construct a third-order method for (7.1) which satisfies the requirements RL1 and RL2. But such a method contains two fractional steps at which we have to approximate the Hamiltonian system (7.5), (7.2) using a third-order symplectic method. This makes a method too complicated, and we will use another approach. In [278] a similar problem for deterministic second-order differential equations was solved by a modification of symplectic Runge–Kutta–Nyström (RKN) methods from [276]. Here we modify the symplectic RKN method (3.49)-(3.50) from Sect. 4.3.3 using some ideas of [278].

As a result, we obtain the method

$$\mathcal{Q}_1 = Q_k + \frac{7}{24}hM^{-1}P_k, \quad \mathcal{P}_1 = P_k + \frac{7}{24}h\left[f(\mathcal{Q}_1) - \nu\Gamma\mathcal{P}_1\right],$$

$$\mathcal{Q}_2 = Q_k + \frac{25}{24}hM^{-1}P_k + \frac{h^2}{2}M^{-1}\left[f(\mathcal{Q}_1) - \nu\Gamma\mathcal{P}_1\right],$$

$$\mathcal{P}_2 = P_k + \frac{2}{3}h\left[f(\mathcal{Q}_1) - \nu\Gamma\mathcal{P}_1\right] + \frac{3}{8}h\left[f(\mathcal{Q}_2) - \nu\Gamma\mathcal{P}_2\right]$$

$$\mathcal{Q}_3 = Q_k + hM^{-1}P_k + \frac{17}{36}h^2M^{-1}\left[f(\mathcal{Q}_1) - \nu\Gamma\mathcal{P}_1\right]$$

$$+ \frac{1}{36}h^2M^{-1}\left[f(\mathcal{Q}_2) - \nu\Gamma\mathcal{P}_2\right],$$

$$\mathcal{P}_3 = P_k + \frac{2}{3}h\left[f(\mathcal{Q}_1) - \nu\Gamma\mathcal{P}_1\right] - \frac{2}{3}h\left[f(\mathcal{Q}_2) - \nu\Gamma\mathcal{P}_2\right]$$

$$+ h\left[f(\mathcal{Q}_3) - \nu\Gamma\mathcal{P}_3\right], \tag{7.18}$$

4.7 Quasi-symplectic mean-square methods for Langevin-type equations   269

$$P_{k+1} = \mathcal{P}_3 + \sum_{r=1}^{m} \sigma_r \Delta_k w_r - \nu \sum_{r=1}^{m} \Gamma \sigma_r \cdot (I_{r0})_k$$
$$+ \sum_{r=1}^{m} \left[\sum_{i=1}^{n} (M^{-1}\sigma_r)^i \frac{\partial f}{\partial q^i}(\mathcal{Q}_3) + \nu^2 \Gamma^2 \sigma_r \right] (I_{r00})_k,$$
$$Q_{k+1} = \mathcal{Q}_3 + \sum_{r=1}^{m} M^{-1}\sigma_r \cdot (I_{r0})_k - \nu \sum_{r=1}^{m} M^{-1}\Gamma \sigma_r (I_{r00})_k,$$
$$k = 0, \ldots, N-1, \tag{7.19}$$

where

$$(I_{r00})_k := \int_{t_k}^{t_k+h} \int_{t_k}^{\vartheta_1} (w_r(\vartheta_2) - w_r(t_k)) \, d\vartheta_2 d\vartheta_1 .$$

Joint distribution of the random variables $\Delta_k w_r$, $(I_{r0})_k$, $(I_{r00})_k$ is Gaussian. They can be simulated at each step by $3m$ independent $\mathcal{N}(0,1)$-distributed random variables ξ_{rk}, η_{rk}, and ζ_{rk}, $r = 0, \ldots, m$:

$$\Delta_k w_r = h^{1/2} \xi_{rk}, \quad (I_{r0})_k = h^{3/2}(\xi_{rk} + \eta_{rk}/\sqrt{3})/2,$$
$$(I_{r00})_k = h^{5/2}(\xi_{rk} + \sqrt{3}\eta_{rk}/2 - \zeta_{rk}/(2\sqrt{5}))/6. \tag{7.20}$$

The method (7.18)-(7.19) is implicit in the components $\mathcal{P}_1, \mathcal{P}_2, \mathcal{P}_3$ and can easily be resolved at each step since the dependence on $\mathcal{P}$ is linear.

For $\nu = 0$ the method (7.18)-(7.19) coincides with the third-order symplectic method (3.49)-(3.50) and so it satisfies the requirement RL1. For $\sigma_r = 0$, $r = 1, \ldots, m$, (deterministic case), the RKN method (7.18)-(7.19) satisfies conditions set up in [278]. These conditions ensure that the Jacobian of the deterministic RKN method depends on $\nu\Gamma$ and h only, more precisely:

$$\bar{J}_0 = \bar{J}_0(h, \nu\Gamma) := \frac{D(\mathcal{P}_3, \mathcal{Q}_3)}{D(P_k, Q_k)}$$
$$= \frac{\det(I - \frac{3}{8} h\nu\Gamma) \det(I + \frac{25}{24} h\nu\Gamma)}{\det(I + \frac{7}{24} h\nu\Gamma) \det(I + \frac{3}{8} h\nu\Gamma) \det(I + h\nu\Gamma)},$$

where I is the $n \times n$ unit matrix.

We have

$$\bar{J} := \frac{D(P_{k+1}, Q_{k+1})}{D(P_k, Q_k)} = \frac{D(P_{k+1}, Q_{k+1})}{D(\mathcal{P}_3, \mathcal{Q}_3)} \frac{D(\mathcal{P}_3, \mathcal{Q}_3)}{D(P_k, Q_k)} = \bar{J}_0 ,$$

i.e., the Jacobian $\bar{J}$ does not depend on the initial data P_k, Q_k. Further, it is possible to adopt the proof of Lemma 3.14 and prove that the method (7.18)-(7.19) is of mean-square order 3. Thus, we obtain the theorem.

Theorem 7.7. *The method (7.18)-(7.19) for the system (7.1) is of mean-square order 3 and it is such that (i) it is symplectic when applied to (7.1)-(7.2) with $\nu = 0$, (ii) the Jacobian $D(P_{k+1}, Q_{k+1})/D(P_k, Q_k)$ (i.e., the change of phase volume per step) does not depend on P_k, Q_k.*

4.7.2 Langevin-type equation: Nonlinear damping and multiplicative noise

Here we generalize methods of Sect. 4.7.1 to the Langevin-type system (cf. (0.5)):

$$dP = f(t,Q)dt - \nu \tilde{f}(t,P,Q)dt + \sum_{r=1}^{m} \sigma_r(t,Q)dw_r(t), \quad P(t_0) = p,$$
$$dQ = g(P)dt, \quad Q(t_0) = q, \qquad (7.21)$$

where P, Q, f, $\tilde{f}$, g, σ_r are n-dimensional column-vectors, ν is a parameter, and $w_r(t)$, $r = 1, \ldots, m$, are independent standard Wiener processes. Note that the system (7.21) has the same form in the sense of Stratonovich.

If there are Hamiltonians $H_0(t,p,q) = V_0(p) + U_0(t,q)$ and $H_r(t,q)$, $r = 1, \ldots, m$, such that

$$f^i = -\partial H_0/\partial q^i, \quad g^i = \partial H_0/\partial p^i, \quad \sigma_r^i = -\partial H_r/\partial q^i, \quad i = 1, \ldots, n, \qquad (7.22)$$

and if $\nu = 0$, then (7.21) is a Hamiltonian system with multiplicative noise (cf. (2.14)-(2.15)).

Our aim is to construct methods for (7.21) such that they inherit the properties RL1-RL2 of the quasi-symplectic methods for the Langevin equation (7.1), more precisely we require

RLT1. *The methods become symplectic when the system degenerate to a Hamiltonian one;*

RLT2. *The methods degenerate to those satisfying the requirement* RL2 *from Sect. 4.7.1 when the system degenerates to the Langevin equation (7.1).*

We recall that the Euler method for general systems with multiplicative noise is of order $1/2$. But due to specific features of system (7.21), the Euler method (and other usual methods of order $1/2$) applied to (7.21) is of order 1. Therefore, we start with methods of order 1.

First-order methods based on splitting. In connection with (7.21) introduce the systems (cf. (7.5)-(7.6)):

$$dP_I = f(t,Q_I)dt + \sum_{r=1}^{m} \sigma_r(t,Q_I)dw_r(t), \quad P_I(t_0) = p,$$
$$dQ_I = g(P_I)dt, \quad Q_I(t_0) = q, \qquad (7.23)$$

$$\frac{dP_{II}}{dt} = -\nu \tilde{f}(t,P_{II},q), \quad P_{II}(t_0) = p, \qquad (7.24)$$

and denote their solutions as $P_I(t;t_0,p,q)$, $Q_I(t;t_0,p,q)$ and $P_{II}(t;t_0,p,q)$, respectively.

4.7 Quasi-symplectic mean-square methods for Langevin-type equations

The system (7.23), (7.22) is a Hamiltonian system with separable Hamiltonians. Symplectic integrators for such systems are proposed in Sect. 4.2.2. The system (7.24) is deterministic.

Let $\bar{P}_I$, $\bar{Q}_I$ be a one-step approximation corresponding to a symplectic method for (7.23), (7.22) and $\bar{P}_{II}$ be a one-step approximation of a deterministic method for (7.24). Introduce the approximation for (7.21) as follows

$$\bar{P} = \bar{P}(t_0 + h; t_0, p, q)$$
$$:= \bar{P}_{II}(t_0 + h; t_0, \bar{P}_I(t_0 + h; t_0, p, q), \bar{Q}_I(t_0 + h; t_0, p, q)),$$
$$\bar{Q} = \bar{Q}(t_0 + h; t_0, p, q) := \bar{Q}_I(t_0 + h; t_0, p, q). \qquad (7.25)$$

Clearly, the approximation (7.25) satisfies the requirements RLT1 and RLT2. Further, using arguments similar to those in the proof of Lemma 7.1, we prove the following theorem.

Theorem 7.8. *Let $\bar{P}_I$, $\bar{Q}_I$ be a one-step approximation corresponding to a symplectic first-order mean-square method for the system (7.23), (7.22) and $\bar{P}_{II}$ be a one-step approximation corresponding to a first-order deterministic method for the system (7.24). Then $\bar{P}$, $\bar{Q}$ defined in (7.25) is a one-step approximation of the first-order mean-square method for the system (7.21) such that (i) it is symplectic when applied to (7.21)-(7.22) with $\nu = 0$, (ii) it satisfies the requirement RL2 from Sect. 4.7.1 when (7.21) degenerates to the Langevin equation (7.1).*

Let us give a concrete example of a first-order splitting method (to this end we use the PRK method (2.17)-(2.18) from Sect. 4.2.2):

$$\mathcal{Q}_1 = Q_k + \alpha h g(P_k), \quad \mathcal{P}_1 = P_k + h f(t_k + \alpha h, \mathcal{Q}_1),$$
$$\mathcal{Q}_2 = \mathcal{Q}_1 + (1-\alpha)h g(\mathcal{P}_1), \quad \mathcal{P}_2 = P_k + h f(t_k + \alpha h, \mathcal{Q}_1)$$
$$+ \sum_{r=1}^{m} \sigma_r(t_k, \mathcal{Q}_2) \Delta_k w_r,$$
$$Q_{k+1} = \mathcal{Q}_2, \quad P_{k+1} = \mathcal{P}_2 - h\nu \tilde{f}(t_k, \mathcal{P}_2, \mathcal{Q}_2). \qquad (7.26)$$

The particular case of system (7.21), when $\tilde{f}(t, p, q) = \Gamma(q)p$, Γ is an $m \times m$-dimensional matrix, is of a special interest, in particular due to its application in dissipative particle dynamics (see, e.g. [247] and references therein). In this case the system (7.24) becomes deterministic linear system with constant coefficients, which can be solved exactly. If in addition to $f_\nu(t, p, q) = \Gamma(q)p$ the system (7.21) is with additive noise (i.e., $\sigma_r(t, q) = \sigma_r(t)$, $r = 1, \ldots, q$) and $g(p) = M^{-1}p$, then the method (7.28) (see below) becomes of mean-square order 2. An important example of such systems is the Van der Pol oscillator under external excitations

$$\ddot{Q} = -\omega^2 Q + \varepsilon^2(1-Q^2)\dot{Q} + \sigma \dot{w}.$$

Further, our approach can easily be applied to a more general system of Stratonovich SDEs

$$dP = \left(f(t,P,Q) - \nu\tilde{f}(t,P,Q)\right)dt + \sum_{r=1}^{m}\sigma_r(t,P,Q)\circ dw_r(t),\ P(t_0)=p,$$

$$dQ = (g(t,P,Q) - \nu\tilde{g}(t,P,Q))\,dt + \sum_{r=1}^{m}\gamma_r(t,P,Q)\circ dw_r(t),\ Q(t_0)=q,$$

(7.27)

where $\nu \geq 0$ is a parameter, P, Q and all the coefficients are n-dimensional column-vectors, and f, g, σ_r, γ_r satisfy (0.2). For $\nu = 0$ it coincides with the general Hamiltonian system (0.1). As usual, we can split (7.27) in two parts: in the Hamiltonian system (0.1) and the deterministic system, and then use a relation like (7.25) to approximate (7.27). In such an approximation we have $\bar{P}_I$, $\bar{Q}_I$ corresponding to a full implicit symplectic method from Sect. 4.2.1. As a result, we obtain the approximation $\bar{P}$, $\bar{Q}$ for (7.27) which satisfies the requirements RLT1-RLT2. Such a method for (7.27) based on an approximation of this kind has the mean-square order $1/2$.

Methods of order 3/2. Using the fractional step method, we propose the following approximation for (7.21):

$$\bar{P}(t_0+h;t_0,p,q)$$

$$:= \bar{P}_{II}(t_0+\frac{h}{2};t_0,\bar{P}_I(t_0+h;t_0,\bar{P}_{II}(t_0+\frac{h}{2};t_0,p,q),q),$$

$$\bar{Q}_I(t_0+h;t_0,\bar{P}_{II}(t_0+\frac{h}{2};t_0,p,q),q))$$

$$-\nu\sum_{r=1}^{m}\sum_{i=1}^{n}\sigma_r^i\frac{\partial\tilde{f}}{\partial p^i}(t_0,p,q)\left[I_{r0}-\frac{h}{2}\Delta w_r\right] - \frac{h^2}{4}\nu\frac{\partial\tilde{f}}{\partial t}(t_0,p,q),$$

$$\bar{Q}(t_0+h;t_0,p,q) := \bar{Q}_I(t_0+h;t_0,\bar{P}_{II}(t_0+\frac{h}{2};t_0,p,q),q), \qquad (7.28)$$

where $\bar{P}_I$, $\bar{Q}_I$ is a one-step approximation corresponding to a symplectic method of order $3/2$ for (7.23), (7.22) (such methods are available in Sect. 4.2.2) and $\bar{P}_{II}$ is a one-step approximation of a second-order deterministic method for (7.24).

By argument similar to those exploited in previous sections, we prove the following theorem.

Theorem 7.9. *Let $\bar{P}_I$, $\bar{Q}_I$ be a one-step approximation corresponding to a symplectic mean-square method of order $3/2$ for the system (7.23), (7.22), and $\bar{P}_{II}$ be a one-step approximation corresponding to a second-order deterministic method for the system (7.24). Then $\bar{P}$, $\bar{Q}$ defined in (7.28) is the one-step approximation of mean-square method of order $3/2$ for the system (7.21) which satisfies the requirements RLT1-RLT2.*

As it is also noted before, if $\tilde{f}(t,p,q) = \Gamma(q)p$ then $P_{II}(t)$ can be found explicitly.

4.8 Quasi-symplectic weak methods for Langevin-type equations

Symplectic methods in the weak sense proposed in Sect. 4.6 together with the ideas of Sect. 4.7 allow us to derive efficient weak methods for Langevin-type equations.

4.8.1 Langevin equation: Linear damping and additive noise

In this subsection we propose weak methods for the Langevin equation (7.1), which satisfy the requirements RL1-RL2 from Sect. 4.7.1.

Using the splitting ideas presented in Sect. 4.7.1, we obtain the first-order method.

Theorem 8.1. *Let $\bar{P}_I$, $\bar{Q}_I$ be a one-step approximation corresponding to a symplectic method of first weak order for the system (7.1), (7.2). Then $\bar{P}$, $\bar{Q}$ defined in (7.7) or in (7.9) is a one-step approximation of the method of first weak order for the system (7.1) which satisfies the requirements RL1-RL2.*

As for $\bar{P}_I$, $\bar{Q}_I$ appearing in the above theorem, one can take the approximation corresponding to the symplectic implicit method (6.1) or to the explicit one (6.11)-(6.12).

Remark 8.2. The implicit method (6.1) can directly be applied to the Langevin equation (7.1). Of course, it satisfies the requirement $RL1$. The method (6.1) satisfies the requirement RL2 for $\alpha = 0$ and $\alpha = 1$ only (see also the discussion after (7.10) in Sect. 4.7.1).

Now we construct a method of weak order 2. To this end, consider the following approximation for (7.1) (cf. (7.12)):

$$\bar{P} = \bar{P}(t_0+h; t_0, p, q) := P_{II}(\frac{h}{2}; \bar{P}_I(t_0+h; t_0, P_{II}(\frac{h}{2}; p), q)),$$
$$\bar{Q} = \bar{Q}(t_0+h; t_0, p, q) := \bar{Q}_I(t_0+h; t_0, P_{II}(\frac{h}{2}; p), q), \qquad (8.1)$$

where $\bar{P}_I$, $\bar{Q}_I$ is a one-step approximation corresponding to any symplectic weak second-order method for (7.5), (7.2) (e.g., one can use the implicit method (6.18) or the explicit method (6.15)), and $P_{II}(t)$ is the exact solution of (7.6).

Theorem 8.3. *Let $\bar{P}_I$, $\bar{Q}_I$ be a one-step approximation corresponding to a symplectic method of second weak order for the system (7.5), (7.6). Then $\bar{P}$, $\bar{Q}$ defined in (8.1) is a one-step approximation of the method of second weak order for the system (7.1) which satisfies the requirements RL1-RL2.*

Note that Remark 7.4 is applicable for both first and second-order methods.

To get a method of weak order three for (7.1), we modify the symplectic RKN method (6.22)-(6.23) as we did in Sect. 4.7.1 in the case of mean-square methods. On this way we obtain the following method

$$\mathcal{Q}_1 = Q_k + \frac{7}{24}hM^{-1}P_k, \quad \mathcal{P}_1 = P_k + \frac{7}{24}h\left[f(\mathcal{Q}_1) - \nu \Gamma \mathcal{P}_1\right],$$

$$\mathcal{Q}_2 = Q_k + \frac{25}{24}hM^{-1}P_k + \frac{h^2}{2}M^{-1}\left[f(\mathcal{Q}_1) - \nu \Gamma \mathcal{P}_1\right],$$

$$\mathcal{P}_2 = P_k + \frac{2}{3}h\left[f(\mathcal{Q}_1) - \nu \Gamma \mathcal{P}_1\right] + \frac{3}{8}h\left[f(\mathcal{Q}_2) - \nu \Gamma \mathcal{P}_2\right],$$

$$\mathcal{Q}_3 = Q_k + hM^{-1}P_k + \frac{17}{36}h^2 M^{-1}\left[f(\mathcal{Q}_1) - \nu \Gamma \mathcal{P}_1\right]$$

$$+ \frac{1}{36}h^2 M^{-1}\left[f(\mathcal{Q}_2) - \nu \Gamma \mathcal{P}_2\right],$$

$$\mathcal{P}_3 = P_k + \frac{2}{3}h\left[f(\mathcal{Q}_1) - \nu \Gamma \mathcal{P}_1\right] - \frac{2}{3}h\left[f(\mathcal{Q}_2) - \nu \Gamma \mathcal{P}_2\right]$$

$$+ h\left[f(\mathcal{Q}_3) - \nu \Gamma \mathcal{P}_3\right], \tag{8.2}$$

$$P_{k+1} = \mathcal{P}_3 + h^{1/2}\sum_{r=1}^{m}\sigma_r \xi_{rk} - \nu h^{3/2}\sum_{r=1}^{m} \Gamma \sigma_r \cdot (\xi_r/2 + \eta_r)_k$$

$$+ h^{5/2}\sum_{r=1}^{m}\left[\sum_{i=1}^{n}(M^{-1}\sigma_r)^i \frac{\partial f}{\partial q^i}(\mathcal{Q}_3) + \nu^2 \Gamma^2 \sigma_r\right]\xi_{rk}/6,$$

$$Q_{k+1} = \mathcal{Q}_3 + h^{3/2}\sum_{r=1}^{m}M^{-1}\sigma_r \cdot (\xi_r/2 + \eta_r)_k - \nu h^{5/2}\sum_{r=1}^{m}M^{-1}\Gamma \sigma_r \xi_{rk}/6,$$

$$k = 0, \ldots, N-1, \tag{8.3}$$

where ξ_{rk}, η_{rk} are mutually independent random variables distributed by the laws (6.24).

The weak order of this method can be proved by standard arguments from Chap. 2 and its phase-volume contractivity properties are proved by the same arguments as those before Theorem 7.7.

Theorem 8.4. *The method (8.2)-(8.3) for the system (7.1) has third weak order and satisfies the requirements RL1-RL2.*

4.8.2 Langevin-type equation: Nonlinear damping and multiplicative noise

In this subsection we propose weak methods for the Langevin-type equation (7.21) which satisfy the requirements RLT1-RLT2 from Sect. 4.7.2. As for first-order methods, we have the theorem.

Theorem 8.5. *Let $\bar{P}_I$, $\bar{Q}_I$ be a one-step approximation corresponding to a symplectic method of first weak order for the system (7.23), (7.22), and $\bar{P}_{II}$ be a one-step approximation corresponding to a first-order deterministic method for the system (7.24). Then $\bar{P}$, $\bar{Q}$ defined in (7.25) is a one-step approximation of the method of first weak order for the system (7.21) which satisfies the requirements RLT1-RLT2.*

A concrete method based on $\bar{P}$, $\bar{Q}$ from the above theorem can be written using the implicit symplectic method (6.1) or the explicit one (6.11)-(6.12) for $\bar{P}_I$, $\bar{Q}_I$. Further, as in the case of mean-square methods, the proposed approach can be generalized to a more general system of the form (7.27) (see the comment after (7.27) in Sect. 4.7.2).

By the method of fractional steps (as in Sect. 4.7) we construct the second-order weak method for (7.21) on the basis of the symplectic method (6.15). The method has the form

$$\mathcal{P}_1 = \bar{P}_{II}(t_k + \frac{h}{2}; t_k, P_k, Q_k), \quad \mathcal{Q}_1 = Q_k + \frac{h}{2}g(\mathcal{P}_1),$$

$$\mathcal{P}_2 = \mathcal{P}_1 + hf(t_k + \frac{h}{2}, \mathcal{Q}_1) + h^{1/2}\sum_{r=1}^{m}\sigma_r(t_k + \frac{h}{2}, \mathcal{Q}_1)\xi_{rk},$$

$$\mathcal{Q}_2 = \mathcal{Q}_1 + \frac{h}{2}g(\mathcal{P}_2),$$

$$P_{k+1} = \bar{P}_{II}(t_k + \frac{h}{2}; t_k, \mathcal{P}_2, \mathcal{Q}_2) - \frac{h^2}{4}\nu\frac{\partial \tilde{f}}{\partial t}(t_k, P_k, Q_k), \quad Q_{k+1} = \mathcal{Q}_2, \quad (8.4)$$

$$k = 0, \ldots, N-1,$$

where ξ_{rk} are i.i.d. random variables with the law (6.16) and $\bar{P}_{II}$ is a one-step approximation of any second-order deterministic method for system (7.24).

Using a specific approximation instead of $\bar{P}_{II}$, it is possible to modify the method (8.4) in such a way that it will become a derivative-free method (i.e., the correction with the derivative $\partial \tilde{f}/\partial t$ can be incorporated in $\bar{P}_{II}$) but we do not consider this here.

The following theorem holds for the method (8.4).

Theorem 8.6. *The method (8.4) for the system (7.21) has the second weak order and satisfies the requirements RLT1-RLT2.*

We note that for $\tilde{f}(t, p, q) = \Gamma(q)p$, Γ – $m \times m$ dimensional matrix, $P_{II}(t)$ can be found explicitly. Consequently, we can put P_{II} instead of $\bar{P}_{II}$ in (8.4).

4.8.3 Numerical examples

Linear oscillator with linear damping under external random excitation. Let us consider the linear oscillator with linear damping term and additive noise

$$dX^1 = \omega X^2 dt$$
$$dX^2 = (-\omega X^1 - \nu X^2)dt + \frac{\sigma}{\omega}dw(t), \qquad (8.5)$$

where $w(t)$ is a standard Wiener process, ω, ν, σ are positive constants. The system (8.5) is dissipative, its invariant measure μ is Gaussian $\mathcal{N}(0, R)$ with the density

$$\rho(x) = (2\pi)^{-1}(\det R)^{-1/2}\exp\left\{-\frac{1}{2}(R^{-1}x, x)\right\}, \qquad (8.6)$$

where $R = (\sigma^2/2\nu\omega^2)I$ is the covariance matrix for the two-dimensional process $X = (X^1, X^2)^\top$, I denotes the identity matrix.

The discrete system obtained by the explicit Euler scheme has the form

$$\bar{X}^1_{k+1} = \bar{X}^1_k + \omega \bar{X}^2_k h$$
$$\bar{X}^2_{k+1} = \bar{X}^2_k - (\omega \bar{X}^1_k + \nu \bar{X}^2_k)h + \frac{\sigma}{\omega}\Delta_k w. \qquad (8.7)$$

The eigenvalues of the homogeneous part of (8.7) are

$$\lambda_{1,2} = 1 - \frac{\nu h}{2} \pm h\sqrt{\frac{\nu^2}{4} - \omega^2}. \qquad (8.8)$$

We consider the case when the damping term is small, and that is why we suppose that

$$\frac{\nu}{2} < \omega. \qquad (8.9)$$

If (8.9) is fulfilled, then $|\lambda_{1,2}|^2 = 1 - \nu h + \omega^2 h^2$, and consequently (8.7) is asymptotically stable if and only if

$$h < \frac{\nu}{\omega^2}. \qquad (8.10)$$

In this case, the system (8.7) possesses a unique invariant measure $\mu_h(x)$ with a Gaussian density $\rho_h(x)$ corresponding to the normal law $\mathcal{N}(0, R_h)$ with zero mean and the covariance matrix

$$R_h = \frac{\sigma^2}{\omega^2 \varkappa}\begin{bmatrix} 1 - \nu h/2 + \omega^2 h^2/2 & -\omega h/2 \\ -\omega h/2 & 1 \end{bmatrix},$$

where

$$\varkappa := 2\nu - 2\omega^2 h - \nu^2 h + \frac{3\nu\omega^2 h^2}{2} - \frac{\omega^4 h^3}{2}.$$

4.8 Quasi-symplectic weak methods for Langevin-type equations

Due to (8.9) and (8.10), it is possible to prove that $\varkappa > 0$. The elements of R_h can be represented as

$$R_h^{jj} = \frac{\sigma^2}{2\nu\omega^2}\left(1 + \frac{\omega^2 h}{\nu} + O(h\nu) + O\left(\frac{h^2}{\nu^2}\right)\right), \ j = 1, 2,$$

$$R_h^{ij} = \frac{\sigma^2}{2\nu\omega^2}\left(-\frac{\omega h}{2} - \frac{\omega^3 h^2}{2\nu} + O(h^2\nu) + O\left(\frac{h^3}{\nu^2}\right)\right), \ i \neq j,$$

where, for instance, $O\left(\frac{h^2}{\nu^2}\right)$ satisfies the inequality $\left|O\left(\frac{h^2}{\nu^2}\right)\right| \leq C\frac{h^2}{\nu^2}$ for all $\nu > 0$, $h > 0$ such that the ratio h/ν is sufficiently small, C is a positive number.

Therefore, if one would like to approximate $\mu(x)$ by $\mu_h(x)$ quite accurately, then the step h must be essentially less than ν/ω^2, i.e., just the fulfillment of the stability condition (8.10) is not enough. Suppose our aim is to evaluate

$$\int |x|^2 d\mu(x) = \int |x|^2 \rho(x) dx = \lim_{T \to \infty} E|X_x(T)|^2,$$

where $X_x(t)$ is the solution of (8.5) with $X_x(0) = x$.

We can approximate the limit by $E|X_x(T)|^2$ under a sufficiently large T. To evaluate $E|X_x(T)|^2$ by the explicit Euler method, we need to perform $N = T/h$ steps of (8.7). If the damping factor ν is small then the time T is rather large and the step h of the Euler method should be very small to satisfy the above condition $h \ll \nu/\omega^2$. Consequently, the number N is huge, and the Euler method is not appropriate for numerical solution of this problem under small ν.

Let us apply the implicit Euler method to system (8.5):

$$\bar{X}^1_{k+1} = \bar{X}^1_k + \omega \bar{X}^2_{k+1} h$$
$$\bar{X}^2_{k+1} = \bar{X}^2_k - (\omega \bar{X}^1_{k+1} + \nu \bar{X}^2_{k+1})h + \frac{\sigma}{\omega}\Delta_k w. \quad (8.11)$$

The eigenvalues of the homogeneous part of (8.11) are

$$\lambda_{1,2} = 1 - \frac{\nu h + 2\omega^2 h^2}{2(1 + \nu h + \omega^2 h^2)} \pm \frac{\sqrt{\nu^2 h^2 - 4\omega^2 h^2}}{2(1 + \nu h + \omega^2 h^2)}.$$

Under (8.9), the eigenvalues are again complex numbers and

$$|\lambda_{1,2}|^2 = 1 - \frac{\nu h + \omega^2 h^2}{1 + \nu h + \omega^2 h^2}.$$

Therefore, in contrast to the explicit Euler method, we need not any restriction on h for asymptotic stability. This can give rise to the illusion about a possibility to choose a comparatively big step h in the implicit Euler scheme. However, the coming evaluations show that such an illusion is very dangerous. Indeed, the system (8.11) possesses a unique invariant measure $\mu_h(x)$ corresponding to the normal law $\mathcal{N}(0, R_h)$ with zero mean and the covariance matrix R_h with the elements

$$R_h^{jj} = \frac{\sigma^2}{2\nu\omega^2}(1 - \frac{\omega^2 h}{\nu} + O(h\nu) + O(\frac{h^2}{\nu^2})), \ j = 1, 2,$$

$$R_h^{ij} = \frac{\sigma^2}{2\nu\omega^2}(\frac{\omega h}{2} - \frac{\omega^3 h^2}{2\nu} + O(h^2) + O(\frac{h^3}{\nu^2})), \ i \neq j,$$

and we are again forced to take a very small h to reach a satisfactory accuracy.

Now let us use the quasi-symplectic method based on the one-step approximation (7.7) with $\bar{P}_I, \bar{Q}_I$ from (3.30) with $\alpha = 0$. For simplicity we take $\bar{P}_{II} = p - h\nu p$ instead of the exact P_{II} (see Remark 7.4). As a result, we get

$$\bar{X}_{k+1}^1 = \bar{X}_k^1 + \omega h(\bar{X}_k^2 - \omega h \bar{X}_k^1)$$

$$\bar{X}_{k+1}^2 = (\bar{X}_k^2 - \omega h \bar{X}_k^1 + \frac{\sigma}{\omega}\Delta_k w)(1 - \nu h). \quad (8.12)$$

In this case, if

$$\frac{\nu}{2} < \omega - \frac{\omega^2 h}{2},$$

the eigenvalues $\lambda_{1,2}$ are complex and

$$|\lambda_{1,2}|^2 = 1 - \nu h.$$

For all not too large h the system (8.12) is asymptotically stable and possesses a unique invariant measure with a Gaussian density. The corresponding normal law has zero mean and the covariance matrix with the elements

$$R_h^{11} = \frac{\sigma^2}{2\nu\omega^2}(1 - 2\nu h + O(h^2)), \quad R_h^{22} = \frac{\sigma^2}{2\nu\omega^2}(1 - \frac{3}{2}\nu h + O(h^2)),$$

$$R_h^{ij} = \frac{\sigma^2}{2\nu\omega^2}(\frac{\omega h}{2} - \frac{5}{4}\omega\nu h^2 + O(h^3)), \ i \neq j.$$

We see that the implicit Euler method has advantages in comparison with the explicit Euler method due to its better stability properties. But both of them require too small step to reach a sufficient accuracy, in particular, if ν is small. At the same time, the quasi-symplectic method (8.12) gives very good results for very big steps. This is important, for instance, for the problem of computing a mean due to an invariant law which needs numerical integration on very long time intervals.

As an example, we evaluate $E\left(X^1(T)\right)^2$ for a large T by weak analogues of the implicit Euler method (8.11) and the quasi-symplectic method (8.12) (i.e., we replace $\Delta_k w$ in these methods by $h^{1/2}\xi_k$, ξ_k are i.i.d. random variables with the law (6.2)). Notice that the moments $E\left(X^i(t)X^j(t)\right)$, $i, j = 1, 2$, satisfy a system of linear differential equations and $E\left(X^1(T)\right)^2$ can be found exactly. The results of simulation are presented on Fig. 8.1. We see that even

4.8 Quasi-symplectic weak methods for Langevin-type equations

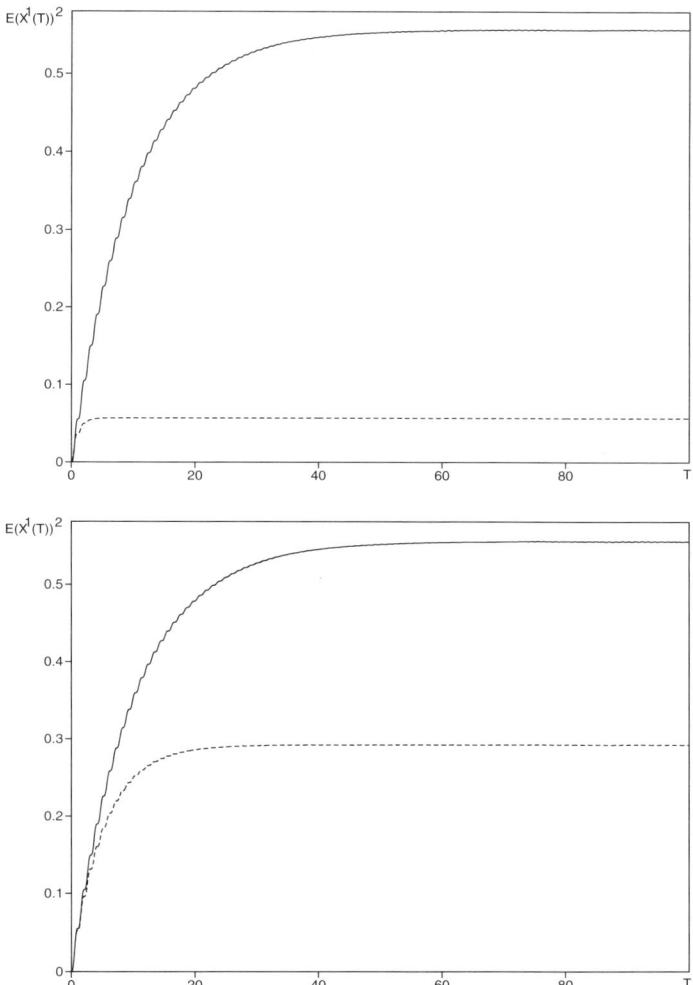

Fig. 8.1. The linear oscillator with linear damping (8.5). Behavior of $E(X^1(T))^2$ with $X^1(0) = 0$, $X^2(0) = 0$, $\omega = 3$, $\nu = 0.1$, $\sigma = 1$, $h = 0.1$ (*top*) and $h = 0.01$ (*bottom*) on the time interval $t \leq 100$ in the case of the weak implicit Euler method (*dashed line*) and the weak quasi-symplectic method (*solid line*, which visually coincides with the exact dependence $E(X^1(T))^2$). The Monte Carlo error is not greater than 0.00005 (*top*) and 0.0003 (*bottom*) for the Euler method and 0.0005 for the quasi-symplectic method with probability 0.95.

for the small step $h = 0.01$ the implicit Euler method tends to a wrong limit with increasing T while the quasi-symplectic method gives quite accurate results, e.g., for $h = 0.1$. The explicit Euler method is unstable for $h = 0.1$ (see (8.10)).

An oscillator with cubic restoring force under external random excitation. Consider the oscillator with cubic restoring force and additive noise

$$\ddot{Q} = Q - Q^3 - \nu \dot{Q} + \sigma \dot{w}, \tag{8.13}$$

i.e., (7.1) with $U_0(q) = \dfrac{1}{4}q^4 - \dfrac{1}{2}q^2$. The dynamical system (8.13) is ergodic (see, e.g., [171]) and its invariant measure has the density

$$\rho(p,q) = C \exp\left(-\dfrac{\nu}{\sigma^2}\left(p^2 + \dfrac{1}{2}q^4 - q^2\right)\right), \tag{8.14}$$

where C is defined by the normalization condition.

Here we compare an implicit quasi-symplectic method and the implicit Euler scheme. We use the implicit quasi-symplectic method based on the one-step approximation (7.7) and on the weak implicit symplectic method (6.1) with $\alpha = 1/2$. For simplicity we take $\bar{P}_{II} = p - h\nu p$ instead of the exact P_{II} (see Remark 7.4). As a result, we get for (8.13):

$$\bar{P}_I = P_k + h\left(\dfrac{\bar{Q}_I + Q_k}{2} - \dfrac{(\bar{Q}_I + Q_k)^3}{8}\right) + h^{1/2}\sigma\xi_k,$$
$$\bar{Q}_I = Q_k + h(\bar{P}_I + P_k)/2,$$
$$P_{k+1} = (1 - \nu h)\bar{P}_I, \quad Q_{k+1} = \bar{Q}_I, \tag{8.15}$$

where ξ_k are i.i.d. random variables with the law (6.2).

In application to (8.13) the weak implicit Euler scheme has the form

$$P_{k+1} = P_k + h\left(Q_{k+1} - Q_{k+1}^3 - \nu P_{k+1}\right) + h^{1/2}\sigma\xi_k$$
$$Q_{k+1} = Q_k + hP_{k+1}, \tag{8.16}$$

where ξ_k are i.i.d. random variables with the law (6.2).

Figure 8.2 gives results of evaluation of $E\left(Q(T)\right)^2$ for a large T by these two methods. We see that even for such a small step as $h = 0.01$ the implicit Euler method tends to a wrong limit with increasing T, while the quasi-symplectic method gives quite accurate results, e.g., for $h = 0.25$.

Now consider the *explicit* quasi-symplectic method based on the one-step approximation (7.7) and on the weak explicit symplectic method (6.11)-(6.12) with $\alpha = 0$. We take $\bar{P}_{II} = p - h\nu p$ instead of the exact P_{II} again. This method for (8.13) is written as

4.8 Quasi-symplectic weak methods for Langevin-type equations   281

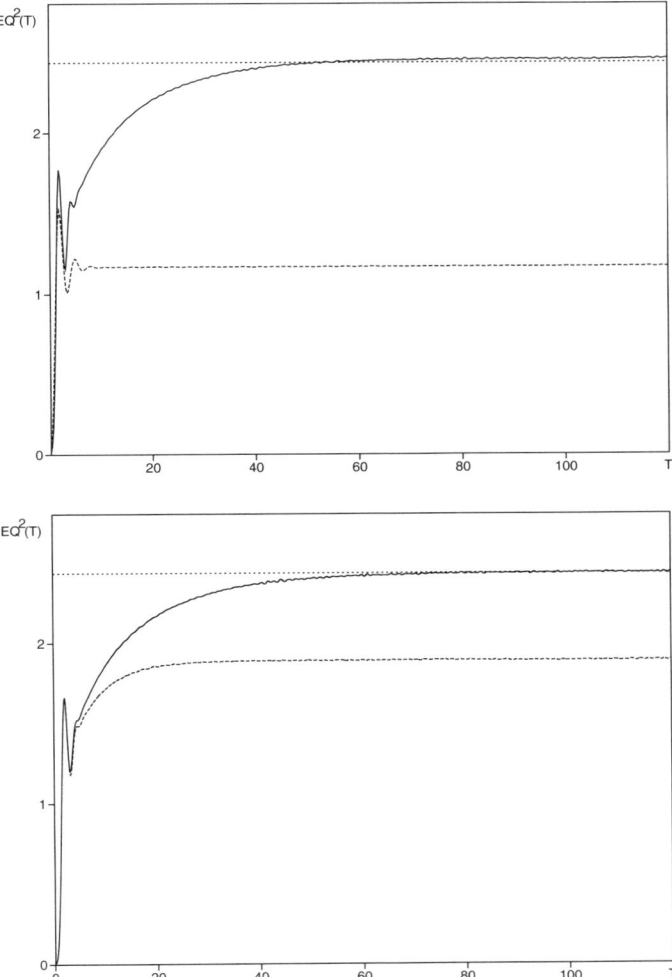

Fig. 8.2. The oscillator with cubic restoring force (8.13). Behavior of $E(Q(T))^2$ with $P(0) = 0$, $Q(0) = 0$, $\nu = 0.05$, $\sigma = 1$, $h = 0.25$ (*top*) and $h = 0.01$ (*bottom*) on the time interval $t \leq 120$ in the case of the weak implicit Euler method (8.16) (*dashed line*) and the weak quasi-symplectic method (8.15) (*solid line*). The Monte Carlo error is not greater than 0.005 with probability 0.95. The dotted line presents the limit value of $E(Q(T))^2$ as $T \to \infty$ evaluated due to $\int_{-\infty}^{\infty} \int_{-\infty}^{\infty} q^2 \rho(p,q) dp\, dq$ with the invariant measure $\rho(p,q)$ from (8.14). This value is equal to 2.435.

$$P_{k+1} = (1 - \nu h)\left(P_k + h\left(Q_k - Q_k^3\right) + h^{1/2}\sigma\xi_k\right)$$
$$Q_{k+1} = Q_k + h\left(P_k + h\left(Q_k - Q_k^3\right)\right). \qquad (8.17)$$

Since this quasi-symplectic method is explicit, it is much simpler than (8.15). However, for comparatively large h the difference system (8.17) has unstable behavior (e.g., for ν, σ as in Fig. 8.2 and $h = 0.2$). Most likely, for all sufficiently small h the system (8.17) acquires stable behavior (of course, this assertion requires further investigation). For instance, $E\bar{Q}^2(T)$ obtained by (8.17) for ν, σ as in Fig. 8.2 and $h = 0.1$ visually coincides with the results obtained by the implicit quasi-symplectic method (8.15). Thus, even an explicit quasi-symplectic method can effectively be used for solution of Langevin equations on long time intervals in contrast to the implicit Euler method which is more complicated than (8.17).

In Appendix A.2 an implementation of the weak implicit quasi-symplectic method (8.15) and the weak implicit Euler method (8.16) is considered and the program is given (see Listing A.2), by which the results presented on Fig. 8.2 were obtained.

5 Simulation of space
and space-time bounded diffusions

"Ordinary" mean-square methods from Chap. 1, intended to solve SDEs on a finite time interval, are based on a time discretization (sampling). The space-time point, corresponding to an "ordinary" one-step approximation constructed at a time point t_k, lies on the d-dimensional plane $t = t_k$, which belongs to the $(d+1)$-dimensional semi-space $[T_0, \infty) \times \mathbf{R}^d$. The "ordinary" mean-square methods give both time and phase components of the approximate trajectory. They ensure smallness of time increments at each step, but space increments can take arbitrary large values with some probability. To approximate SDEs in a bounded domain, we have to control space increments at each step in a way such that the constructed approximation belongs to the bounded domain. Of course, this cannot be achieved by "ordinary" mean-square methods, and thus special methods are required.

In the first part of this chapter (Sects. 5.1-5.2) we propose a mean-square approximation of an autonomous diffusion process in a space bounded domain [184,187], while in the second part (Sects. 5.3-5.5) we construct algorithms for space-time diffusions in space-time bounded domains [204]. In the last section (Sect. 5.6), we give a brief survey on simulation of reflected diffusions.

In Sect. 5.1 we consider the autonomous system of SDEs without drift

$$dX = \chi_{\tau_x > t} \sigma(X) dw(t), \ X(0) = x, \tag{0.1}$$

in a bounded domain $G \subset \mathbf{R}^d$ with boundary ∂G. Here $w(t) = (w^1(t), \ldots, w^d(t))^\top$, $t \geq 0$, is a standard $\mathcal{F}_t$-measurable Wiener process of dimension d defined on a probability space $(\Omega, \mathcal{F}, P)$, where $\mathcal{F}_t$ is a nondecreasing family of σ-subalgebras of $\mathcal{F}$; $X = (X^1, \ldots, X^d)^\top$ is a vector of dimension d, $\sigma(x) = \{\sigma^{ij}(x)\}$ is a matrix of dimension $d \times d$, τ_x is a random time at which the path $X_x(t)$ leaves the region G.

In addition to (0.1), we introduce the system with the coefficients frozen at x:

$$d\bar{X} = \sigma(x) dw(t), \ \bar{X}(0) = x. \tag{0.2}$$

Let $U_r \subset \mathbf{R}^d$ be an open sphere of radius r with center at the origin and with the boundary ∂U_r. Denote by $\bar{\theta}$ the first time at which the process $w(t)$ leaves the sphere U_r. Clearly, $w(\bar{\theta})$ has the uniform distribution on ∂U_r. Let $U_r^\sigma(x)$ be an open ellipsoid with the boundary $\partial U_r^\sigma(x)$ obtained from the sphere U_r by the linear transformation $\sigma(x)$ and the shift x. It is assumed that r is small enough to satisfy the inclusion: $U_r^\sigma(x) \subset G$. The solution $\bar{X}_x(t)$ of the problem (0.2) at the time $\bar\theta$ is equal to

$$\bar{X}_x(\bar\theta) = x + \sigma(x)w(\bar\theta),\ \bar{X}_x(\bar\theta) \in \partial U_r^\sigma(x), \qquad (0.3)$$

and $\bar\theta$ is the first exit time of the trajectory $\bar{X}_x(t)$ from $U_r^\sigma(x)$. It turns out that $\bar{X}_x(\bar\theta)$ is close to $X_x(\bar\theta)$ in the mean-square sense (of course, if $\tau_x \le \bar\theta$ then $X_x(\bar\theta) = X_x(\tau_x)$). So, the point $\bar{X}_x(\bar\theta)$ is an approximation of a point which belongs to the phase trajectory starting at x.

The distribution function of the random instant of time $\bar\theta$ at which the process $w(t)$ reaches the boundary of the sphere U_r can be found from the formula

$$Ee^{-\lambda\bar\theta} = \int_0^\infty e^{-\lambda s} p(s)\, ds,$$

where $p(s)$ is the distribution density of $\bar\theta$, $\lambda > 0$. The quantity $Ee^{-\lambda\bar\theta}$ can be obtained explicitly:

$$Ee^{-\lambda\bar\theta} = \frac{1}{\Gamma(d/2)} \left[\sum_{k=0}^\infty \frac{1}{k!\Gamma(k+d/2)} \left(\sqrt{\lambda/2}\, r\right)^{2k} \right]^{-1},$$

and the inverse Laplace transformation can be used to find $p(s)$. In practice, this is a quite laborious way of modeling $\bar\theta$.

However, the construction of the point $(\bar\theta, \bar{X}_x(\bar\theta))$ amounts to modeling $\bar\theta$ and $\bar{X}_x(\bar\theta)$ separately because of their independence. It is important to underline that if we are interested in phase trajectories only, it is possible to simulate them without modeling $\bar\theta$. To simulate $\bar{X}(\bar\theta)$, we need only $w(\bar\theta)$ which has the uniform distribution on ∂U_r, i.e., modeling the point $\bar{X}_x(\bar\theta) \in \partial U_r^\sigma(x)$ is a fairly simple problem.

Let $\bar{X}_0 = x$, $\bar{X}_1 = \bar{X}_x(\bar\theta)$. We find the point $\bar{X}_2$ on the boundary $\partial U_r^\sigma(\bar{X}_1)$ in the same way as we found $\bar{X}_1$ coming from $\bar{X}_0 = x$. Then we construct $\bar{X}_3$ and so on until a point $\bar{X}_{\bar\nu}$ with a random subscript $\bar\nu$ (see Algorithm 1.4). As a result, the sequence $\bar{X}_0, \ldots, \bar{X}_{\bar\nu}$ is obtained which can be considered as a mean-square approximation of the phase trajectory of the solution $X_x(t)$. If the point $\bar{X}_{\bar\nu}$ is sufficiently close to the boundary ∂G, it is possible to simulate the exit point $X_x(\tau_x)$. In Sect. 5.2, we construct an approximation for an autonomous system with *drift* in a space bounded domain.

Thus, the algorithm of Sects. 5.1-5.2 is based on a space discretization (quantization) using *a random walk over small spheres*. It gives the points which are close in the mean-square sense to the points of the real phase trajectory for SDEs in the space bounded domain. To realize the algorithm, the exit point of the Wiener process from a d-dimensional ball has to be constructed at each step. The algorithm gives only the phase component of the approximate trajectory without modelling the corresponding time component like the algorithm over touching spheres [218]. The space-time point lies on the d-dimensional lateral surface of a semi-cylinder with sphere base

5 Simulation of space and space-time bounded diffusions   285

in the $(d+1)$-dimensional semi-space $[T_0, \infty) \times \mathbf{R}^d$. The algorithm ensures smallness of the phase increments at each step, but the non-simulated time increments can take arbitrary large values with some probability.

In Sects. 5.3-5.5, the mean-square approximations are considered which control boundedness of both space increments and time increments at each step. In addition they give approximate values for both phase and time components of the space-time diffusion in the space-time bounded domain. The space-time point lies on a bounded d-dimensional manifold. It is possible to solve this problem in a constructive manner by the implementation of a space-time discretization by a random walk over boundaries of small space-time parallelepipeds.

In Sect. 5.4, we consider the system of SDEs

$$dX = \chi_{\tau_{t,x}>s}b(s,X)ds + \chi_{\tau_{t,x}>s}\sigma(s,X)dw(s), \quad X(t) = X_{t,x}(t) = x, \quad (0.4)$$

in a space-time bounded domain $Q = [T_0, T_1) \times G \subset \mathbf{R}^{d+1}$. Here X and b are d-dimensional vectors, σ is a $d \times d$-matrix, $(w(s), \mathcal{F}_s)$, $s \geq T_0$, is a d-dimensional standard Wiener process defined on a probability space $(\Omega, \mathcal{F}, P)$, G is a bounded open domain in $\mathbf{R}^d$, and the Markov moment $\tau_{t,x}$ is the first-passage time of the process $(s, X_{t,x}(s))$, $s \geq t$, to $\Gamma = \overline{Q} \setminus Q$. The set Γ is a part of the boundary ∂Q consisting of the lateral surface and the upper base of the cylinder $\overline{Q}$. We put $X_{t,x}(s) = X_{t,x}(\tau_{t,x})$ under $s \geq \tau_{t,x}$, and thus, the process $(s, X_{t,x}(s))$ is defined for all $t \leq s < T_1$. The coefficients $b^i(s, x)$ and $\sigma^{ij}(s, x)$, $(s, x) \in \overline{Q}$, and the boundary ∂G are assumed to be sufficiently smooth, while the strict ellipticity condition is imposed on the matrix $a(s, x) := \sigma(s, x)\sigma^\top(s, x)$.

The mean-square approximations for (0.4) are based on a space-time discretization by *a random walk over boundaries of small space-time parallelepipeds*. It turns out that the first exit point $(\bar{\theta}, w(\bar{\theta}))$ of the space-time Brownian motion $(s, w(s))$, $s > 0$, from the space-time parallelepiped $\Pi_{r,l} = [0, lr^2) \times C_r$, where $C_r \subset \mathbf{R}^d$ is a cube with center at the origin and edge length equal to $2r$, can be simulated in a sufficiently easy way (some aspects of the space-time Brownian motion under $d = 1$ are considered in [113]). To construct a one-step approximation, we introduce the system with frozen coefficients (both t, x fixed)

$$d\bar{X} = b(t, x)ds + \sigma(t, x)dw(s), \quad \bar{X}(t) = x. \quad (0.5)$$

As an approximation of the point $(t + \bar{\theta}, X_{t,x}(t + \bar{\theta}))$ of the space-time diffusion $(s, X_{t,x}(s))$, $s \geq t$, we take the point $(t + \bar{\theta}, \bar{X}_{t,x}(t + \bar{\theta}))$, where $\bar{X}_{t,x}(t + \bar{\theta})$ is a solution of (0.5):

$$\bar{X}_{t,x}(t + \bar{\theta}) = x + b(t, x)\bar{\theta} + \sigma(t, x)(w(t + \bar{\theta}) - w(t)), \quad (0.6)$$

and $(\bar{\theta}, w(t + \bar{\theta}) - w(t))$ is the exit point of the space-time Brownian motion $(s - t, w(s) - w(t))$, $s > t$, from the space-time parallelepiped $\Pi_{r,l}$.

The point $(t+\bar{\theta}, \bar{X}_{t,x}(t+\bar{\theta}))$ lies on the lateral surface or on the upper base of a certain parallelepiped obtained from $\Pi_{r,l}$ by a linear transformation, i.e., it is constructed on a bounded d-dimensional manifold in contrast to the "ordinary" mean-square approximations of Chap. 1 and to the approximations of Sects. 5.1-5.2, which are constructed on the d-dimensional unbounded manifolds.

On the basis of the one-step approximation (0.6), we form a Markov chain $(\bar{\vartheta}_k, \bar{X}_k)$ which belongs to Q at each step and approximates the points $(\bar{\vartheta}_k, X(\bar{\vartheta}_k))$ of the trajectory $(s, X_{t,x}(s))$, $s \geq t$, in the mean-square sense (see Algorithm 4.3). Section 5.3 is devoted to simulation of space-time Brownian motions which is the basis for our algorithms. A global algorithm is proposed and convergence theorems are proved in Sect. 5.4. An approximation for the space-time exit point is also constructed in Sect. 5.4. Numerical examples are given in Sect. 5.5.

5.1 Mean-square approximation for autonomous SDEs without drift in a space bounded domain

In this section we consider the autonomous system of SDEs (see (0.1))

$$dX = \chi_{\tau_x > t} \sigma(X) dw(t), \ X(0) = x, \tag{1.1}$$

in a bounded domain $G \subset \mathbf{R}^d$ with a boundary ∂G.

The following conditions are assumed to be satisfied:

(i) G is a convex open bounded set with the twice continuously differentiable boundary ∂G;
(ii) the coefficients $\sigma^{ij}(x)$ belong to the class $C^{(2)}(\bar{G})$;
(iii) the matrix

$$a(x) = \sigma(x)\sigma^\top(x), \ a(x) = \{a^{ij}(x)\},$$

satisfies the strict ellipticity condition, i.e.,

$$\lambda_1^2 = \min_{x \in \bar{G}} \min_{1 \leq i \leq d} \lambda_i^2(x) > 0,$$

where $\lambda_1^2(x) \leq \lambda_2^2(x) \leq \cdots \leq \lambda_d^2(x)$ are eigenvalues of the matrix $a(x)$.

Also introduce $\lambda_d^2 = \max_{x \in \bar{G}} \lambda_d^2(x)$. Then for any $x \in \bar{G}$, $y \in \mathbf{R}^d$ the following inequality

$$\lambda_1^2 \sum_{i=1}^d (y^i)^2 \leq \sum_{i,j=1}^d a^{ij}(x) y^i y^j \leq \lambda_d^2 \sum_{i=1}^d (y^i)^2 \tag{1.2}$$

holds.

5.1 Mean-square approximation for SDEs in a space bounded domain

Due to (1.2), the first exit time τ_x, at which the path $X_x(t)$ leaves the region G, is finite with probability one. We shall consider the process $X_x(t)$ defined on $0 \leq t < \infty$ regarding it as the stopped one after τ_x.

A local approximation theorem is given in Sect. 5.1.1. In Sect. 5.1.2 we prove two convergence theorems. The first one is devoted to approximation properties of the sequence $\bar{X}_0, \ldots, \bar{X}_{\bar{\nu}}$ till leaving an open domain $D \subset G$ with $\rho(\partial D, \partial G) > 0$ which does not depend on r. In the second convergence theorem, the point $\bar{X}_{\bar{\nu}}$ belongs to a boundary layer which decreases in a definite way with decreasing r, i.e., $\bar{X}_{\bar{\nu}}$ becomes sufficiently close to ∂G with decreasing r (more precisely, $\rho(\bar{X}_{\bar{\nu}}, \partial G) = O(r^{1-\varepsilon})$ with a sufficiently small $\varepsilon > 0$). In both situations the mean-square order of accuracy is equal to $O(r)$. The second theorem is important for approximation of the exit point $X_x(\tau_x)$. It is shown (Sect. 5.1.3) that this point can be approximated by $\bar{X}_{\bar{\nu}}$ with a mean-square order which is close to $O(\sqrt{r})$.

5.1.1 Local approximation of diffusion in a space bounded domain

In Sects. 5.1.1-5.1.3, $X_x(t)$ is the solution of the problem (1.1), $X_{t_0,x}(t)$, $t \geq t_0$, is the solution of the system from (1.1) with initial data $X(t_0) = x$, and $\bar{X}_x(t)$ is found from (0.2).

Let Γ_δ be the interior of a δ-neighborhood of the boundary ∂G belonging to G. Obviously, if $x \in G \setminus \Gamma_{2\lambda_d r}$, then the inclusion $U_r^\sigma(x) \subset U_{2r}^\sigma(x) \subset G$ holds for all sufficiently small r.

Theorem 1.1. *For every natural number n there exists a constant $K > 0$ such that for any sufficiently small $r > 0$ and for any $x \in G \setminus \Gamma_{2\lambda_d r}$ the following inequality*

$$E|X_x(\bar{\theta}) - \bar{X}_x(\bar{\theta})|^{2n} \leq K r^{4n} \tag{1.3}$$

holds.

Proof. Let the Markov moment $\bar{\theta}$ be the first time at which the process $X_x(t)$ leaves the ellipsoid $U_{2r}^\sigma(x)$. First we prove the theorem for $n = 1$. We have

$$E|X_x(\bar{\theta}) - \bar{X}_x(\bar{\theta})|^2 = E\left|\int_0^{\bar{\theta}} (\chi_{\tau_x > s}\sigma(X_x(s)) - \sigma(x))dw(s)\right|^2$$

$$= E \int_0^{\bar{\theta}} |\chi_{\tau_x > s}\sigma(X_x(s)) - \sigma(x)|^2 ds$$

$$= E \int_0^{\bar{\theta} \wedge \theta} |\sigma(X_x(s)) - \sigma(x)|^2 ds$$

$$+ E \int_{\bar{\theta} \wedge \theta}^{\bar{\theta}} |\chi_{\tau_x > s}\sigma(X_x(s)) - \sigma(x)|^2 ds$$

$$\leq E \int_0^{\bar{\theta} \wedge \theta} |\sigma(X_x(s)) - \sigma(x)|^2 ds + K \times E(\bar{\theta} - \bar{\theta} \wedge \theta). \tag{1.4}$$

Here the notation $|x|$ means the Euclidean norm of a vector x and $|\sigma|$ means $(\operatorname{tr}\sigma\sigma^\top)^{1/2}$ of a matrix σ.

Since $E\bar\theta = r^2/d$, then $E(\bar\theta \wedge \theta) \leq r^2/d$. Further, $X_x(s) \in U^\sigma_{2r}(x)$ for $s \in (0, \bar\theta \wedge \theta)$. Therefore

$$E|X_x(\bar\theta \wedge \theta) - \bar X_x(\bar\theta \wedge \theta)|^2 = E\int_0^{\bar\theta \wedge \theta} |\sigma(X_x(s)) - \sigma(x)|^2 ds$$
$$\leq Kr^2 E(\bar\theta \wedge \theta) \leq Kr^4. \qquad (1.5)$$

Using (1.2), it is easy to show that if $\xi \in \overline U^\sigma_r(x)$, $\eta \in \partial U^\sigma_{2r}(x)$, then $|\xi - \eta| \geq \lambda_1 r$. Because $\bar X_x(\bar\theta \wedge \theta) \in \overline U^\sigma_r(x)$, $X_x(\theta) \in \partial U^\sigma_{2r}(x)$, we have for every $m > 0$

$$E(\chi_{\theta<\bar\theta}|X_x(\bar\theta \wedge \theta) - \bar X_x(\bar\theta \wedge \theta)|^m) = E(\chi_{\theta<\bar\theta}|X_x(\theta) - \bar X_x(\bar\theta \wedge \theta)|^m)$$
$$\geq P(\theta < \bar\theta)\lambda_1^m r^m. \qquad (1.6)$$

On the other hand

$$E(\chi_{\theta<\bar\theta}|X_x(\bar\theta \wedge \theta) - \bar X_x(\bar\theta \wedge \theta)|^m)$$
$$\leq (P(\theta < \bar\theta))^{1/2} \times (E|X_x(\bar\theta \wedge \theta) - \bar X_x(\bar\theta \wedge \theta)|^{2m})^{1/2}$$
$$= (P(\theta < \bar\theta))^{1/2} \times (E|\int_0^{\bar\theta \wedge \theta}(\sigma(X_x(s)) - \sigma(x))dw(s)|^{2m})^{1/2}. \qquad (1.7)$$

Let i be one of the indices $1,\ldots,d$. Introduce the variable

$$Z(t) = X_x^i(\bar\theta \wedge \theta \wedge t) - \bar X_x^i(\bar\theta \wedge \theta \wedge t)$$
$$= \int_0^{\bar\theta \wedge \theta \wedge t} \sum_{j=1}^d (\sigma^{ij}(X_x(s)) - \sigma^{ij}(x))dw^j(s)$$
$$= \int_0^t \chi_{\bar\theta \wedge \theta \geq s}\varphi(s)dw(s),$$

where $\varphi(s)$ is the i-th row vector of the matrix $\sigma(X_x(s)) - \sigma(x)$. We do not write the index i at Z and φ because this does not lead to any misunderstanding. Clearly, $Z(t)$, $t \geq 0$, is a uniformly bounded scalar, and

$$|\varphi(s)| \leq |\sigma(X_x(s)) - \sigma(x)| \leq Kr, \ 0 \leq s \leq \bar\theta \wedge \theta.$$

We have for every natural $m \geq 1$:

$$dZ^{2m}(t) = 2mZ^{2m-1}(t)\chi_{\bar\theta \wedge \theta \geq t}\varphi(t)dw(t) + m(2m-1)Z^{2m-2}(t)\chi_{\bar\theta \wedge \theta \geq t}|\varphi(t)|^2 dt.$$

Hence

$$EZ^{2m}(t) = m(2m-1)E\int_0^t Z^{2m-2}(s)\chi_{\bar\theta \wedge \theta \geq s}|\varphi(s)|^2 ds$$
$$\leq Km(2m-1)r^2 \times E(\bar\theta \wedge \theta \times \max_{0 \leq s \leq t}|Z(s)|^{2m-2}).$$

5.1 Mean-square approximation for SDEs in a space bounded domain

Applying the Hölder inequality with $p = 2m/(2m-2)$ (see such a reception, e.g., in [82]) and taking into account that (see (6.4.29))

$$E(\bar{\theta} \wedge \theta)^m \leq E\bar{\theta}^m \leq \frac{m!}{d^m} r^{2m},$$

we get

$$\begin{aligned}
E|Z(t)|^{2m} &\leq Km(2m-1)r^2 \times (E \max_{0 \leq s \leq t} |Z(s)|^{2m})^{(2m-2)/2m} \\
&\quad \times (E(\bar{\theta} \wedge \theta)^m)^{1/m} \\
&\leq Km(2m-1)r^4 \times (E \max_{0 \leq s \leq t} |Z(s)|^{2m})^{(2m-2)/2m}. \quad (1.8)
\end{aligned}$$

As $Z(t)$ is a martingale, we can use the Doob inequality

$$E \max_{0 \leq s \leq t} |Z(s)|^{2m} \leq \left(\frac{2m}{2m-1}\right)^{2m} E|Z(t)|^{2m}.$$

Now we obtain from (1.8):

$$E|Z(t)|^{2m} \leq K r^{4m},$$

where K does not depend on t (of course, K depends on m).
Hence

$$E \left| \int_0^{\bar{\theta} \wedge \theta} (\sigma(X_x(s)) - \sigma(x)) dw(s) \right|^{2m} \leq K r^{4m}. \quad (1.9)$$

The inequalities (1.6), (1.7), and (1.9) imply

$$P(\theta < \bar{\theta}) \times \lambda_1^m r^m \leq K \times (P(\theta < \bar{\theta}))^{1/2} \times r^{2m}.$$

Therefore, for every positive m (recall K depends on m)

$$P(\theta < \bar{\theta}) \leq K r^{2m}. \quad (1.10)$$

Further,

$$\begin{aligned}
E(\bar{\theta} - \theta \wedge \theta) &= E\chi_{\theta < \bar{\theta}}(\bar{\theta} - \theta \wedge \theta) \leq (P(\theta < \bar{\theta}))^{1/2} (E(\bar{\theta} - \theta \wedge \theta)^2)^{1/2} \\
&\leq (P(\theta < \bar{\theta}))^{1/2} (E\bar{\theta}^2)^{1/2} \leq K(P(\theta < \bar{\theta}))^{1/2} r^2,
\end{aligned}$$

whence

$$E(\bar{\theta} - \theta \wedge \theta) \leq K r^{m+2}. \quad (1.11)$$

Using this inequality for $m = 2$ together with (1.4) and (1.5), we arrive at (1.3) for $n = 1$. Thus, the theorem is proved for $n = 1$.

For an arbitrary positive integer n, we get

$$E|X_x(\bar\theta) - \bar X_x(\bar\theta)|^{2n}$$
$$= E|\int_0^{\bar\theta\wedge\theta}(\sigma(X_x(s))-\sigma(x))dw(s) + \int_{\bar\theta\wedge\theta}^{\bar\theta}(\chi_{\tau_x>s}\sigma(X_x(s))-\sigma(x))dw(s)|^{2n}$$
$$\leq KE|\int_0^{\bar\theta\wedge\theta}(\sigma(X_x(s))-\sigma(x))dw(s)|^{2n}$$
$$+KE|\int_{\bar\theta\wedge\theta}^{\bar\theta}(\chi_{\tau_x>s}\sigma(X_x(s))-\sigma(x))dw(s)|^{2n}, \tag{1.12}$$

where the constant K depends on n only. The first term on the right-hand side is bounded by Kr^{4n} due to (1.9). The second term can be bounded as follows (see (1.4) and (1.11) for $m = 4n - 2$):

$$E|\int_{\bar\theta\wedge\theta}^{\bar\theta}[\chi_{\tau_x>s}\sigma(X_x(s))-\sigma(x)]dw(s)|^{2n}$$
$$= E(|\int_{\bar\theta\wedge\theta}^{\bar\theta}[\chi_{\tau_x>s}\sigma(X_x(s))-\sigma(x)]dw(s)|^2$$
$$\times|X_x(\bar\theta)-X_x(\bar\theta\wedge\theta)-\bar X_x(\bar\theta)+\bar X_x(\bar\theta\wedge\theta)|^{2n-2})$$
$$\leq KE|\int_{\bar\theta\wedge\theta}^{\bar\theta}[\chi_{\tau_x>s}\sigma(X_x(s))-\sigma(x)]dw(s)|^2 \leq KE(\bar\theta-\bar\theta\wedge\theta) \leq Kr^{4n}.$$

Now (1.12) implies (1.3). Theorem 1.1 is proved. □

Remark 1.2. Clearly, the inequality (1.10) remains true if θ is the first time at which the process $X_x(t)$ leaves the ellipsoid $U^\sigma_{(1+\alpha)r}(x)$ for any $\alpha > 0$. Therefore, the condition $x \in G\backslash\Gamma_{2\lambda_d r}$ in Theorem 1.1 may be replaced by $x \in G\backslash\Gamma_{(1+\alpha)\lambda_d r}$, $\alpha > 0$. Moreover, it is not difficult to show that the theorem remains true under the condition $x \in G\backslash\Gamma_{(1+r^\beta)\lambda_d r}$ if only $0 \leq \beta < 2$. But for definiteness we take here and in what follows the layer $\Gamma_{2\lambda_d r}$.

Remark 1.3. For convenience let us adduce formulas for simulation of the uniform distribution on the surface of the sphere in d-dimensional space with center at the origin and with radius r. The point has this distribution if coordinates of the point are

$$x_1 = r\cos\varphi_1$$
$$x_2 = r\sin\varphi_1\cos\varphi_2$$
$$\ldots\ldots\ldots\ldots$$
$$x_{d-1} = r\sin\varphi_1\sin\varphi_2\times\cdots\times\sin\varphi_{d-2}\cos\varphi_{d-1}$$
$$x_d = r\sin\varphi_1\sin\varphi_2\times\cdots\times\sin\varphi_{d-2}\sin\varphi_{d-1},$$
$$0 \leq \varphi_1 \leq \pi,\ 0 \leq \varphi_2 \leq \pi,\ldots,\ 0 \leq \varphi_{d-2} \leq \pi,\ 0 \leq \varphi_{d-1} \leq 2\pi,$$

$$I_{d-2}(\varphi_1) = I_{d-2}(\pi)\gamma_1$$
$$\ldots\ldots\ldots\ldots$$
$$I_1(\varphi_{d-2}) = I_1(\pi)\gamma_{d-2}$$

5.1 Mean-square approximation for SDEs in a space bounded domain 291

$$\varphi_{d-1} = 2\pi\gamma_{d-1},$$

where $\gamma_1, \ldots, \gamma_{d-1}$ are independent random variables uniformly distributed on $[0,1]$ and

$$I_k(\varphi) = \int_0^\varphi \sin^k \vartheta \, d\vartheta, \ k = 1, \ldots, d-2.$$

5.1.2 Global algorithm for diffusion in a space bounded domain

Algorithm 1.4 Let $\bar{\theta}_1$ be the first time at which the Wiener process $w(t)$ leaves the sphere U_r, $\bar{\theta}_1 + \bar{\theta}_2$ be the first time at which the process $w(t) - w(\bar{\theta}_1)$, $t \geq \bar{\theta}_1$, leaves the same sphere U_r and so on. Let $x \in G \backslash \Gamma_{2\lambda_d r}$. We construct a recurrence sequence of random vectors $\bar{X}_k$, $k = 0, 1, \ldots, \bar{\nu}$:

$$\bar{X}_0 = x$$
$$\bar{X}_1 = \bar{X}_0 + \sigma(\bar{X}_0)w(\bar{\theta}_1)$$
$$\ldots\ldots\ldots\ldots\ldots\ldots$$
$$\bar{X}_{k+1} = \bar{X}_k + \sigma(\bar{X}_k)(w(\bar{\theta}_1 + \cdots + \bar{\theta}_{k+1}) - w(\bar{\theta}_1 + \cdots + \bar{\theta}_k)),$$
$$\ldots\ldots\ldots\ldots\ldots\ldots$$

where $\bar{\nu} = \bar{\nu}_x$ is the first number for which $\bar{X}_k \in \Gamma_{2\lambda_d r}$. See Remark 1.3 for modelling $w(\bar{\theta}_1 + \cdots + \bar{\theta}_{k+1}) - w(\bar{\theta}_1 + \cdots + \bar{\theta}_k)$.

Of course, the random moment $\bar{\nu}$ also depends on the domain $G \backslash \Gamma_{2\lambda_d r}$ which is left by $\bar{X}_{\bar{\nu}}$. Therefore, the more detailed notation for $\bar{\nu} = \bar{\nu}_x$ is $\bar{\nu} = \bar{\nu}_x(G \backslash \Gamma_{2\lambda_d r})$. Let us set $\bar{\theta}_k = 0$ and $\bar{X}_k = \bar{X}_{\bar{\nu}}$ for $k > \bar{\nu}$.

We have obtained the random walk

$$\bar{X}_0, \ldots, \bar{X}_k, \ldots,$$

which stops at a random step $\bar{\nu}$. It is a Markov chain. We start consideration of properties of this Markov chain by obtaining some average characteristics of $\bar{\nu} = \bar{\nu}_x$.

In connection with the homogeneous Markov chain $\bar{X}_k$, we introduce the one-step transition function

$$P(x, B) = P\left(\bar{X}_1 \in B \mid \bar{X}_0 = x\right)$$

where B is a Borel set belonging to $\bar{G}$.

Define an operation P acting on functions $v(x)$, $x \in \bar{G}$, by the formula

$$Pv(x) = \int_{\bar{G}} P(x, dy) v(y) = Ev\left(\bar{X}_1\right), \ \bar{X}_0 = x,$$

and an operator

$$Av(x) = Pv(x) - v(x)$$

which is called the generator of the chain. The generator gives an average increment of the function v on the trajectory of the considered chain per step.

Consider the boundary value problem in $\bar{G}$:

$$Pv(x) - v(x) = -g(x), \quad x \in G \setminus \Gamma_{2\lambda_d r}, \qquad (1.13)$$

$$v(x) = 0, \quad x \in \Gamma_{2\lambda_d r}, \qquad (1.14)$$

which is connected with the chain $\bar{X}_k$.

The solution of the problem is the following function (see [301]):

$$v(x) = E \sum_{k=0}^{\bar{\nu}_x - 1} g(\bar{X}_k), \quad \bar{X}_0 = x. \qquad (1.15)$$

If $g \equiv 1$ then

$$v(x) = E\bar{\nu}_x .$$

Further, if $v(x)$ is the solution of the boundary value problem (1.13)-(1.14) with the function $g(x)$ satisfying the inequality

$$g(x) \geq 1$$

in $G \setminus \Gamma_{2\lambda_d r}$, then, thanks to (1.15), we obtain

$$E\bar{\nu}_x \leq v(x). \qquad (1.16)$$

Lemma 1.5. *There exists a constant $C > 0$ depending only on a diameter of the domain G such that the inequality*

$$E\bar{\nu}_x \leq \frac{C}{\lambda_1^2 r^2} \qquad (1.17)$$

holds.

Proof. Introduce the function

$$V(x) = \begin{cases} A^2 - x^2, & x \in G \setminus \Gamma_{2\lambda_d r}, \\ 0, & x \in \Gamma_{2\lambda_d r}, \end{cases}$$

where constant A^2 is such that for all $x \in \bar{G}$ we have

$$A^2 - x^2 \geq 0, \; x \in \bar{G}.$$

This function satisfies the boundary condition (1.14).

Let a point x be such that $U_r^\sigma(x) \subset G \setminus \Gamma_{2\lambda_d r}$. Now we evaluate $PV(x) - V(x)$. The measure $P(x, B)$ concentrates on $\partial U_r^\sigma(x)$, and, due to the inclusion $U_r^\sigma(x) \subset G \setminus \Gamma_{2\lambda_d r}$, the function $V(y)$ on $\partial U_r^\sigma(x)$ is equal to

5.1 Mean-square approximation for SDEs in a space bounded domain

$A^2 - y^2$. Let dS be an area element of the surface ∂U_r and S be the area of this surface. We have

$$PV(x) = EV(\bar{X}_1) = EV(x + \sigma(x)w(\bar{\theta}))$$
$$= \frac{1}{S}\int_{\partial U_r}\left(A^2 - (x + \sigma(x)z)^2\right)dS$$
$$= A^2 - x^2 - \frac{2}{S}\int_{\partial U_r}(x, \sigma(x)z)\,dS - \frac{1}{S}\int_{\partial U_r}(\sigma(x)z)^2\,dS. \quad (1.18)$$

Clearly,
$$\int_{\partial U_r}(x, \sigma(x)z)\,dS = 0,$$

and, due to the strict ellipticity condition

$$(\sigma(x)z)^2 \geq \lambda_1^2 \sum_{i=1}^n (z^i)^2 = \lambda_1^2 r^2,$$

the equality (1.18) implies

$$PV(x) - V(x) \leq -\lambda_1^2 r^2. \quad (1.19)$$

Now let $x \in G \setminus \Gamma_{2\lambda_d r}$ but the part of $U_r^\sigma(x)$ can belong to $\Gamma_{2\lambda_d r}$. We temporarily introduce the function $\bar{V}(y)$ which is equal to $A^2 - y^2$ on the entire surface $\partial U_r^\sigma(x)$. Therefore, as in (1.18) and (1.19), we obtain

$$P\bar{V}(x) = A^2 - x^2 - \frac{1}{S}\int_{\partial U_r}(\sigma(x)z)^2\,dS \leq A^2 - x^2 - \lambda_1^2 r^2.$$

Since $V(y) \leq \bar{V}(y)$ on $\partial U_r^\sigma(x)$, we have $PV(x) \leq P\bar{V}(x)$ and, consequently, the inequality (1.19) is proved for all $x \in G \setminus \Gamma_{2\lambda_d r}$.

It obviously follows from (1.19) that the function

$$v(x) = \frac{V(x)}{\lambda_1^2 r^2}$$

satisfies (1.13)-(1.14) with $g \geq 1$. Hence, in view of (1.16), we obtain (1.17) with $C = \max_{x \in \bar{G}} V(x)$. $\square$

Remark 1.6. It is clear that the proof of Lemma 1.5 remains the same if we use the function $V(x)$ of the form:

$$V(x) = \begin{cases} A^2 + (a, x) - x^2, & x \in G \setminus \Gamma_{2\lambda_d r}, \\ 0, & x \in \Gamma_{2\lambda_d r}, \end{cases}$$

where the constant A^2 and the vector a are such that for all $x \in \bar{G}$ the inequality

holds.

On account of the choice of a and A^2, the bound (1.17) can be strengthened. For example, let $x^* \in G$ be a point such that
$$rad\, G = \max_{x \in \partial G} (x - x^*)^2 = \min_{y \in \bar{G}} \max_{x \in \partial G} (x - y)^2.$$

If we take V as
$$V(x) = \begin{cases} rad\, G - (x - x^*)^2, & x \in G \setminus \Gamma_{2\lambda_d r}, \\ 0, & x \in \Gamma_{2\lambda_d r}, \end{cases}$$

we obtain the inequality
$$E\nu_x \leq \frac{rad\, G - (x - x^*)^2}{\lambda_1^2 r^2}, \quad x \in G \setminus \Gamma_{2\lambda_d r}.$$

Below we will need the result from [117, p. 297], which we present here in the form convenient for our purposes.

Lemma 1.7. *Let*
$$AV_1(x) \leq -\left(1 - e^{-\alpha}\right) V_1(x), \quad x \in G \setminus \Gamma_{2\lambda_d r},$$
where $V_1(x) \geq 1$, $x \in \bar{G}$. Then the inequality
$$Ee^{\alpha \bar{\nu}_x} \leq V_1(x)$$

holds.

Lemma 1.8. *For all sufficiently small r the inequality*
$$E\left(1 + \frac{\lambda_1^2}{1 + C} r^2\right)^{\bar{\nu}_x} \leq 1 + C, \quad C = \max_{x \in \bar{G}} V(x), \qquad (1.20)$$

is valid.

Proof. We take $V_1 = V + 1$. From (1.19) we get
$$AV_1(x) = PV_1(x) - V_1(x) \leq -\lambda_1^2 r^2, \quad x \in G \setminus \Gamma_{2\lambda_d r}.$$

For all sufficiently small r we obtain
$$AV_1(x) \leq -\lambda_1^2 r^2 \frac{V_1(x)}{1 + C} = -\left(1 - e^{\ln(1 - \lambda_1^2 r^2/(1+C))}\right) V_1(x), \quad x \in G \setminus \Gamma_{2\lambda_d r}.$$

Then it follows from Lemma 1.7 that
$$E \exp\left(-\bar{\nu}_x \ln\left(1 - \frac{\lambda_1^2 r^2}{1 + C}\right)\right) = E\left(\frac{1}{1 - \lambda_1^2 r^2/(1+C)}\right)^{\bar{\nu}_x} \leq V_1(x) \leq 1 + C$$

which implies (1.20). □

5.1 Mean-square approximation for SDEs in a space bounded domain

Corollary 1.9. *The probability $P(\bar{\nu}_x \geq L/r^2)$ decreases exponentially as L increases. More precisely, for every $L > 0$*

$$P(\bar{\nu}_x \geq L/r^2) \leq (1+C)e^{-\alpha_r \lambda_1^2 L/(1+C)} \qquad (1.21)$$

with $\alpha_r \to 1$ as $r \to 0$. The constant C in (1.21) is the same as in (1.20).

The corollary easily follows from Lemma 1.8 by Chebyshev's inequality. To prove convergence theorems, we need the following lemma.

Lemma 1.10. *For every natural number n there exists a constant $K > 0$ such that for any sufficiently small $r > 0$ and for any $x, y \in G \setminus \Gamma_{2\lambda_d r}$ the inequality*

$$E | \int_0^{\bar{\theta}} (\chi_{\tau_x > s} \sigma(X_x(s)) - \chi_{\tau_y > s} \sigma(X_y(s))) dw(s) |^{2n}$$
$$\leq K|x-y|^{2n} r^{2n} + K r^{4n} \qquad (1.22)$$

holds.

Proof. We have

$$\int_0^{\bar{\theta}} (\chi_{\tau_x > s} \sigma(X_x(s)) - \chi_{\tau_y > s} \sigma(X_y(s))) dw(s)$$
$$= \int_0^{\bar{\theta}} (\chi_{\tau_x > s} \sigma(X_x(s)) - \sigma(x)) dw(s) - \int_0^{\bar{\theta}} (\chi_{\tau_y > s} \sigma(X_y(s)) - \sigma(y)) dw(s)$$
$$+ \int_0^{\bar{\theta}} (\sigma(x) - \sigma(y)) dw(s)$$
$$= (X_x(\bar{\theta}) - \bar{X}_x(\bar{\theta})) - (X_y(\bar{\theta}) - \bar{X}_y(\bar{\theta})) + (\sigma(x) - \sigma(y)) w(\bar{\theta}).$$

Hence

$$| \int_0^{\bar{\theta}} (\chi_{\tau_x > s} \sigma(X_x(s)) - \chi_{\tau_y > s} \sigma(X_y(s))) dw(s) |^{2n}$$
$$= |(X_x(\bar{\theta}) - \bar{X}_x(\bar{\theta})) - (X_y(\bar{\theta}) - \bar{X}_y(\bar{\theta})) + (\sigma(x) - \sigma(y)) \times w(\bar{\theta})|^{2n}$$
$$\leq K|X_x(\bar{\theta}) - \bar{X}_x(\bar{\theta})|^{2n} + K|X_y(\bar{\theta}) - \bar{X}_y(\bar{\theta})|^{2n} + K|\sigma(x) - \sigma(y)|^{2n} |w(\bar{\theta})|^{2n},$$

where the constant K depends on n only.

Now Theorem 1.1 and the relations

$$|\sigma(x) - \sigma(y)| \leq K|x-y|, \quad |w(\bar{\theta})|^{2n} = r^{2n}$$

imply (1.22). □

Let D be an open domain such that $\bar{D} \subset G$ and $\Delta := \rho(\partial D, \partial G)$. We consider $r < \Delta$ so that $D \subset G \setminus \Gamma_{2\lambda_d r}$. Let $x \in D$ and $\bar{\nu} = \bar{\nu}_x = \bar{\nu}_x(D)$

be the first moment at which $\bar{X}_{\bar{\nu}} \in G \backslash D$. For brevity, we preserve the old notation $\bar{\nu}$ for the new Markov moment $\bar{\nu}_x(D)$ as this does not cause any confusion. As before, we set $\bar{\theta}_k = 0$ and $\bar{X}_k = \bar{X}_{\bar{\nu}}$ for $k > \bar{\nu}$, i.e., we stop the above constructed trajectory $\bar{X}_k$ at the moment $\bar{\nu} = \bar{\nu}_x(D) < \bar{\nu}_x(G \backslash \Gamma_{2\lambda_d r})$. Therefore, the inequality (1.17) is fulfilled for the moment $\bar{\nu} = \bar{\nu}_x(D)$ as well.

Now consider the sequence

$$X_0 = x$$
$$X_1 = X_x(\bar{\theta}_1)$$
$$\dots\dots\dots\dots\dots\dots\dots$$
$$X_{k+1} = X_x(\bar{\theta}_1 + \cdots + \bar{\theta}_{k+1}) = X_{\bar{\theta}_1 + \cdots + \bar{\theta}_k, X_k}(\bar{\theta}_1 + \cdots + \bar{\theta}_{k+1})$$
$$\dots\dots\dots\dots\dots\dots\dots$$

which is connected with the solution of the system (0.1).

If $\bar{\theta}_1 + \cdots + \bar{\theta}_k \geq \tau_x$ then, of course, $X_k = X_x(\tau_x)$, and if $k > \bar{\nu} = \bar{\nu}_x(D)$ then $X_k = X_{\bar{\nu}}$ as $\bar{\theta}_{\bar{\nu}+1} = \cdots = \bar{\theta}_k = 0$. Thus, X_k stops at a random step $\bar{\nu} \wedge \kappa$, where $\kappa = \min\{k : \bar{\theta}_1 + \cdots + \bar{\theta}_k > \tau_x\}$ if $\tau_x < \bar{\theta}_1 + \cdots + \bar{\theta}_{\bar{\nu}}$ and $\kappa = \bar{\nu}$ otherwise. The sequence X_k, just as $\bar{X}_k$, is a Markov chain. Furthermore, both $\bar{X}_k$ and X_k are martingales over σ-algebras $\mathcal{F}_0 = \{\emptyset, \Omega\}, \mathcal{F}_k = \mathcal{F}_{\bar{\theta}_1 + \cdots + \bar{\theta}_k}$, $k = 1, 2, \ldots$.

Consider the sequences $\bar{X}_k$, X_k for $N = L/r^2$ steps. The closeness of $\bar{X}_k$ to X_k for N steps is established in the following theorem.

Theorem 1.11. *Let $\bar{\nu} = \bar{\nu}_x(D)$ be the first exit time of the approximate trajectory $\bar{X}_k$ from the domain D. There exist constants $K > 0$ and $\gamma > 0$ (which do not depend on x, r, L, and Δ) such that for any $x \in D$ and for any sufficiently small $r > 0$ the inequality*

$$(E \max_{1 \leq k \leq \bar{\nu} \wedge N} |X_k - \bar{X}_k|^2)^{1/2} = (E \max_{1 \leq k \leq N} |X_k - \bar{X}_k|^2)^{1/2} \leq \frac{K}{\Delta} e^{\gamma L} r \quad (1.23)$$

holds.

Proof. Let ν be the first number at which $X_\nu \in \Gamma_{2\lambda_d r}$. More precisely,

$$\nu = \begin{cases} \min\{k : X_k \in \Gamma_{2\lambda_d r}, \ k \leq \bar{\nu}\}, \\ \infty, \ X_k \notin \Gamma_{2\lambda_d r}, \ k = 1, \ldots, \bar{\nu}. \end{cases} \quad (1.24)$$

Clearly, for a sufficiently small r (if only $D \subset G \backslash \Gamma_{2\lambda_d r}$ and $3\lambda_d r \leq \Delta/2$)

$$|X_\nu - \bar{X}_\nu| \geq \frac{\Delta}{2}, \text{ if } \nu \leq \bar{\nu}. \quad (1.25)$$

Introduce the sequences $\bar{X}_{\nu \wedge m}$, $X_{\nu \wedge m}$ stopped at ν and the differences

$$d_m = X_{\nu \wedge m} - \bar{X}_{\nu \wedge m}, \ m = 0, 1, \ldots.$$

As ν is a Markov moment with respect to the system of σ-algebras $(\mathcal{F}_m)$, the stopped sequences $(\bar{X}_{\nu \wedge m}, \mathcal{F}_m)$, $(X_{\nu \wedge m}, \mathcal{F}_m)$ and $(d_m, \mathcal{F}_m)$ are martingales.

5.1 Mean-square approximation for SDEs in a space bounded domain

The sequence $\bar{X}_{\nu \wedge m}$ $(X_{\nu \wedge m})$ is the Markov chain $\bar{X}_m$ (X_m) stopped at the moment ν. This is equivalent to the fact that $\bar{\theta}_m = 0$ not only for $m > \bar{\nu}$ but also for $m > \nu$, i.e., we may consider $\bar{\theta}_m = 0$ for $m > \bar{\nu} \wedge \nu$. Consequently, if $\bar{\nu} \wedge \nu = k$ then $d_k = d_{k+1} = \cdots = d_N$. This implies $d_k^2 = d_{k+1}^2 = \cdots = d_N^2$.

We have

$$d_m = d_1 \chi_{\bar{\nu} \wedge \nu = 1} + \cdots + d_{m-1} \chi_{\bar{\nu} \wedge \nu = m-1} + d_m \chi_{\bar{\nu} \wedge \nu \geq m}$$

and

$$d_{m-1} = d_1 \chi_{\bar{\nu} \wedge \nu = 1} + \cdots + d_{m-2} \chi_{\bar{\nu} \wedge \nu = m-2} + d_{m-1} \chi_{\bar{\nu} \wedge \nu = m-1} + d_{m-1} \chi_{\bar{\nu} \wedge \nu \geq m} \ .$$

Therefore

$$d_m = d_{m-1} + (d_m - d_{m-1}) \chi_{\bar{\nu} \wedge \nu \geq m} \ . \tag{1.26}$$

Analogously,

$$d_m^2 = d_{m-1}^2 + (d_m^2 - d_{m-1}^2) \chi_{\bar{\nu} \wedge \nu \geq m} \ .$$

We get

$$\begin{aligned} d_m &= X_m - \bar{X}_m = X_x(\bar{\theta}_1 + \cdots + \bar{\theta}_m) - \bar{X}_m \\ &= X_{\bar{\theta}_1 + \cdots + \bar{\theta}_{m-1}, X_{m-1}}(\bar{\theta}_1 + \cdots + \bar{\theta}_m) - \bar{X}_m \\ &= X_{\bar{\theta}_1 + \cdots + \bar{\theta}_{m-1}, X_{m-1}}(\bar{\theta}_1 + \cdots + \bar{\theta}_m) - X_{\bar{\theta}_1 + \cdots + \bar{\theta}_{m-1}, \bar{X}_{m-1}}(\bar{\theta}_1 + \cdots + \bar{\theta}_m) \\ &\quad + X_{\bar{\theta}_1 + \cdots + \bar{\theta}_{m-1}, \bar{X}_{m-1}}(\bar{\theta}_1 + \cdots + \bar{\theta}_m) - \bar{X}_m \ . \end{aligned} \tag{1.27}$$

The first difference at the right-hand side of (1.27) is the error of the solution due to the error in the initial data at the time $(\bar{\theta}_1 + \cdots + \bar{\theta}_{m-1})$ accumulated to the $(m-1)$-st step. The second difference is the one-step error at the m-th step.

For $m \leq \bar{\nu} \wedge \nu$ the vectors $\bar{X}_{m-1}$ and X_{m-1} belong to $G \setminus \Gamma_{2\lambda_d r}$, and we obtain from the equality (1.27):

$$\begin{aligned} &\chi_{\bar{\nu} \wedge \nu \geq m} d_m \\ &= \chi_{\bar{\nu} \wedge \nu \geq m}(X_{m-1} + \int_{\bar{\theta}_1 + \cdots + \bar{\theta}_{m-1}}^{\bar{\theta}_1 + \cdots + \bar{\theta}_m} \chi(s) \times \sigma(X_{\bar{\theta}_1 + \cdots + \bar{\theta}_{m-1}, X_{m-1}}(s)) dw(s)) \\ &\quad - \chi_{\bar{\nu} \wedge \nu \geq m}(\bar{X}_{m-1} + \int_{\bar{\theta}_1 + \cdots + \bar{\theta}_{m-1}}^{\bar{\theta}_1 + \cdots + \bar{\theta}_m} \bar{\chi}(s) \times \sigma(X_{\bar{\theta}_1 + \cdots + \bar{\theta}_{m-1}, \bar{X}_{m-1}}(s)) dw(s)) \\ &\quad + \chi_{\bar{\nu} \wedge \nu \geq m}(X_{\bar{\theta}_1 + \cdots + \bar{\theta}_{m-1}, \bar{X}_{m-1}}(\bar{\theta}_1 + \cdots + \bar{\theta}_m) - \bar{X}_m) \ . \end{aligned} \tag{1.28}$$

Here

$$\chi(s) := \chi_{\tau(\bar{\theta}_1 + \cdots + \bar{\theta}_{m-1}, X_{m-1}) > s}, \quad \bar{\chi}(s) := \chi_{\tau(\bar{\theta}_1 + \cdots + \bar{\theta}_{m-1}, \bar{X}_{m-1}) > s},$$

where $\tau(\bar{\theta}_1 + \cdots + \bar{\theta}_{m-1}, x)$ is a random time at which the path $X_{\bar{\theta}_1 + \cdots + \bar{\theta}_{m-1}, x}(t)$ leaves the region G.

For brevity, we also introduce the following notation

$$\sigma(s) := \sigma(X_{\bar{\theta}_1+\cdots+\bar{\theta}_{m-1}, X_{m-1}}(s)), \quad \bar{\sigma}(s) := \sigma(X_{\bar{\theta}_1+\cdots+\bar{\theta}_{m-1}, \bar{X}_{m-1}}(s)).$$

From (1.28) and (1.26) we obtain

$$d_m - d_{m-1} = (d_m - d_{m-1})\chi_{\bar{\nu}\wedge\nu\geq m}$$
$$= \chi_{\bar{\nu}\wedge\nu\geq m}\int_{\bar{\theta}_1+\cdots+\bar{\theta}_{m-1}}^{\bar{\theta}_1+\cdots+\bar{\theta}_m} (\chi(s)\times\sigma(s) - \bar{\chi}(s)\times\bar{\sigma}(s))dw(s)$$
$$+\chi_{\bar{\nu}\wedge\nu\geq m}(X_{\bar{\theta}_1+\cdots+\bar{\theta}_{m-1},\bar{X}_{m-1}}(\bar{\theta}_1+\cdots+\bar{\theta}_m) - \bar{X}_m). \quad (1.29)$$

Due to $\mathcal{F}_{m-1}$-measurability of the random variable $\chi_{\bar{\nu}\wedge\nu\geq m}$, the equality (1.29) implies

$$E(d_m - d_{m-1})^2$$
$$\leq 2E\chi_{\bar{\nu}\wedge\nu\geq m}E(|\int_{\bar{\theta}_1+\cdots+\bar{\theta}_{m-1}}^{\bar{\theta}_1+\cdots+\bar{\theta}_m}(\chi(s)\times\sigma(s) - \bar{\chi}(s)\times\bar{\sigma}(s))dw(s)|^2 \mid \mathcal{F}_{m-1})$$
$$+2E\chi_{\bar{\nu}\wedge\nu\geq m}E(|X_{\bar{\theta}_1+\cdots+\bar{\theta}_{m-1},\bar{X}_{m-1}}(\bar{\theta}_1+\cdots+\bar{\theta}_m) - \bar{X}_m|^2 \mid \mathcal{F}_{m-1}).$$

By the conditional versions of Lemma 1.10 and Theorem 1.1 with $n = 1$, we obtain

$$E(d_m - d_{m-1})^2 \leq Kr^2 E(\chi_{\bar{\nu}\wedge\nu\geq m}d_{m-1}^2) + Kr^4 \leq Kr^2 Ed_{m-1}^2 + Kr^4, \quad (1.30)$$

where the constant K does not depend on x, r, L, and Δ.

Because $(d_m, \mathcal{F}_m)$ is a martingale, we have

$$Ed_m^2 = Ed_{m-1}^2 + E(d_m - d_{m-1})^2. \quad (1.31)$$

The relations (1.30) and (1.31) imply

$$Ed_m^2 \leq Ed_{m-1}^2 + Kr^2 Ed_{m-1}^2 + Kr^4, \quad d_0 = 0.$$

From here we get for $N = L/r^2$:

$$Ed_N^2 = E|X_{\nu\wedge N} - \bar{X}_{\nu\wedge N}|^2 \leq [(1+Kr^2)^{L/r^2} - 1]\times Kr^2 \leq Ke^{2\gamma L}r^2, \quad (1.32)$$

where the constant $\gamma > 0$ does not depend on x, r, L, and Δ.

Further, it is not difficult to obtain that $X_{\bar{\nu}\wedge\nu\wedge N} = X_{\nu\wedge N}$, $\bar{X}_{\bar{\nu}\wedge\nu\wedge N} = \bar{X}_{\nu\wedge N}$. Indeed, this is evident for $\bar{\nu} \geq \nu \wedge N$. And this is valid for $\bar{\nu} < \nu \wedge N$ because both X and $\bar{X}$ stop after the moment $\bar{\nu}$. Hence,

$$E|X_{\bar{\nu}\wedge\nu\wedge N} - \bar{X}_{\bar{\nu}\wedge\nu\wedge N}|^2 \leq Ke^{2\gamma L}r^2. \quad (1.33)$$

Let us prove now that

5.1 Mean-square approximation for SDEs in a space bounded domain   299

$$P(\nu \leq \bar{\nu} \wedge N) \leq K \frac{e^{2\gamma L}}{\Delta^2} r^2. \tag{1.34}$$

In fact, due to (1.25), we have

$$E\chi_{\nu \leq \bar{\nu} \wedge N}|X_{\bar{\nu} \wedge \nu \wedge N} - \bar{X}_{\bar{\nu} \wedge \nu \wedge N}| = E\chi_{\nu \leq \bar{\nu} \wedge N}|X_\nu - \bar{X}_\nu|$$
$$\geq P(\nu \leq \bar{\nu} \wedge N) \times \frac{\Delta}{2}. \tag{1.35}$$

On the other hand, using (1.32), we get

$$E\chi_{\nu \leq \bar{\nu} \wedge N}|X_{\bar{\nu} \wedge \nu \wedge N} - \bar{X}_{\bar{\nu} \wedge \nu \wedge N}|$$
$$\leq (P(\nu \leq \bar{\nu} \wedge N))^{1/2} \times (E|X_{\bar{\nu} \wedge \nu \wedge N} - \bar{X}_{\bar{\nu} \wedge \nu \wedge N}|^2)^{1/2}$$
$$\leq K(P(\nu \leq \bar{\nu} \wedge N))^{1/2} \times e^{\gamma L} r. \tag{1.36}$$

The relations (1.35) and (1.36) imply (1.34).

Since $X_{\bar{\nu} \wedge N} = X_N$, $\bar{X}_{\bar{\nu} \wedge N} = \bar{X}_N$, we obtain from (1.33) and (1.34):

$$E|X_N - \bar{X}_N|^2 = E|X_{\bar{\nu} \wedge N} - \bar{X}_{\bar{\nu} \wedge N}|^2$$
$$= E\chi_{\nu \geq \bar{\nu} \wedge N}|X_{\bar{\nu} \wedge N} - \bar{X}_{\bar{\nu} \wedge N}|^2 + E\chi_{\nu < \bar{\nu} \wedge N}|X_{\bar{\nu} \wedge N} - \bar{X}_{\bar{\nu} \wedge N}|^2$$
$$= E\chi_{\nu \geq \bar{\nu} \wedge N}|X_{\bar{\nu} \wedge \nu \wedge N} - \bar{X}_{\bar{\nu} \wedge \nu \wedge N}|^2$$
$$\quad + E\chi_{\nu < \bar{\nu} \wedge N}|X_{\bar{\nu} \wedge N} - \bar{X}_{\bar{\nu} \wedge N}|^2$$
$$\leq E|X_{\bar{\nu} \wedge \nu \wedge N} - \bar{X}_{\bar{\nu} \wedge \nu \wedge N}|^2 + KP(\nu \leq \bar{\nu} \wedge N)$$
$$\leq K\frac{e^{2\gamma L}}{\Delta^2} r^2. \tag{1.37}$$

Using Doob's inequality for the martingale $(X_m - \bar{X}_m, \mathcal{F}_m)$, we arrive at (1.23). Theorem 1.11 is proved. □

Remark 1.12. It follows from Theorem 1.14 that it is possible to avoid the multiplier $1/\Delta$ in (1.23), i.e., the following inequality

$$(E \max_{1 \leq k \leq \bar{\nu} \wedge N}|X_k - \bar{X}_k|^2)^{1/2} = (E \max_{1 \leq k \leq N}|X_k - \bar{X}_k|^2)^{1/2} \leq Ke^{\gamma L} r \tag{1.38}$$

is valid.

Theorem 1.13. *Let* $\bar{\nu} = \bar{\nu}_x(D)$. *The inequality*

$$(E \max_{1 \leq k \leq \bar{\nu}}|X_k - \bar{X}_k|^2)^{1/2} \leq K(e^{\gamma L} r/\Delta + e^{-\alpha_r \lambda_1^2 L/2(1+C)}) \tag{1.39}$$

holds.

Proof. Introduce two sets: $\mathcal{C} = \{\bar{\nu} \leq L/r^2\}$ and $\Omega \backslash \mathcal{C} = \{\bar{\nu} > L/r^2\}$. Due to (1.21) and (1.23), we have (below l is the diameter of G):

$$E\left|X_{\bar{\nu}} - \bar{X}_{\bar{\nu}}\right|^2 = E(\left|X_{\bar{\nu}} - \bar{X}_{\bar{\nu}}\right|^2; \mathcal{C}) + E(\left|X_{\bar{\nu}} - \bar{X}_{\bar{\nu}}\right|^2; \Omega \setminus \mathcal{C})$$
$$= E(\left|X_{\bar{\nu} \wedge N} - \bar{X}_{\bar{\nu} \wedge N}\right|^2; \mathcal{C}) + E(\left|X_{\bar{\nu}} - \bar{X}_{\bar{\nu}}\right|^2; \Omega \setminus \mathcal{C})$$
$$\leq E(\left|X_{\bar{\nu} \wedge N} - \bar{X}_{\bar{\nu} \wedge N}\right|^2) + l^2 P(\Omega \setminus \mathcal{C})$$
$$\leq K \frac{e^{2\gamma L}}{\Delta^2} r^2 + l^2 (1 + C) e^{-\alpha_r \lambda_1^2 L/(1+C)}, \qquad (1.40)$$

whence (1.39) follows. □

The domain D in Theorems 1.11 and 1.13 is not changed with decreasing r. Now consider the domain $G \setminus \Gamma_{cr^{1-1/n}}$, where $c > 0$ is a certain number and $n \geq 2$ is a natural number. Let $x \in G$ and let r be sufficiently small such that $\Gamma_{cr^{1-1/n}} \supset \Gamma_{2\lambda_d r}$ and $x \in G \setminus \Gamma_{cr^{1-1/n}}$. We construct the approximate phase trajectory $\bar{X}_k$ till its exit into the layer $\Gamma_{cr^{1-1/n}}$, i.e., we stop the approximate trajectory, which was constructed at the beginning of this subsection, at the moment $\bar{\nu} = \bar{\nu}_x(G \setminus \Gamma_{cr^{1-1/n}})$. This stopping moment satisfies the inequality

$$\bar{\nu}_x(G \setminus \Gamma_{cr^{1-1/n}}) < \bar{\nu}_x(G \setminus \Gamma_{2\lambda_d r}).$$

As before, we preserve the same notation both for $\bar{X}_k$ with the new stopping moment and for the very stopping moment $\bar{\nu} = \bar{\nu}_x(G \setminus \Gamma_{cr^{1-1/n}})$ as there is no risk of ambiguity. And as before $N = L/r^2$. Theorems 1.14-1.16 are proved in [187].

Theorem 1.14. *Let $\bar{\nu} = \bar{\nu}_x(G \setminus \Gamma_{cr^{1-1/n}})$ be the first exit time of the approximate trajectory $\bar{X}_k$ from the domain $G \setminus \Gamma_{cr^{1-1/n}}$. There exist constants $K > 0$ and $\gamma > 0$ (which do not depend on x, r, and L) such that for any sufficiently small $r > 0$ the inequality*

$$(E \max_{1 \leq k \leq \bar{\nu} \wedge N} \left|X_k - \bar{X}_k\right|^2)^{1/2} = (E \max_{1 \leq k \leq N} \left|X_k - \bar{X}_k\right|^2)^{1/2} \leq K e^{\gamma L} r \quad (1.41)$$

holds.

Theorem 1.15. *Let $\bar{\nu} = \bar{\nu}_x(G \setminus \Gamma_{cr^{1-1/n}})$. The inequality*

$$(E \max_{1 \leq k \leq \bar{\nu}} \left|X_k - \bar{X}_k\right|^2)^{1/2} \leq K(e^{\gamma L} r + e^{-\alpha_r \lambda_1^2 L/2(1+C)})$$

is valid.

Theorem 1.16. *Let $n > 1$, $l \geq 1$ be some natural numbers and $\bar{\nu} = \bar{\nu}_x(G \setminus \Gamma_{cr^{1-1/n}})$ be the first exit moment of the approximate trajectory $\bar{X}_k$ from the domain $G \setminus \Gamma_{cr^{1-1/n}}$. There exist constants $K > 0$ and $\gamma > 0$ (which do not depend on x, r, L) such that for any sufficiently small $r > 0$ the inequality*

$$(E \max_{1 \leq k \leq \bar{\nu} \wedge N} \left|X_k - \bar{X}_k\right|^{2l})^{1/2l} = (E \max_{1 \leq k \leq N} \left|X_k - \bar{X}_k\right|^{2l})^{1/2l} \leq K e^{\gamma L} r \quad (1.42)$$

holds.

5.1.3 Simulation of exit point $X_x(\tau_x)$

In the previous subsection we constructed the point $\bar{X}_N = \bar{X}_{\bar{\nu} \wedge N}$, where $N = L/r^2$, $\bar{\nu} = \bar{\nu}_x(G \backslash \Gamma_{cr^{1-1/n}})$. What distance is between $\bar{X}_N$ and the exit point $X_x(\tau_x)$? What point on ∂G can be taken as an approximation of $X_x(\tau_x)$?

On the set $\mathcal{C} = \{\bar{\nu} \leq L/r^2\}$ we have $\bar{X}_N = \bar{X}_{\bar{\nu}} \in \Gamma_{cr^{1-1/n}}$. Let $\xi_x(\omega)$, $\omega \in \mathcal{C}$, be a point on ∂G such that

$$|\bar{X}_N - \xi_x| \leq cr^{1-1/n}, \ \omega \in \mathcal{C}. \tag{1.43}$$

It is natural to take this point as an approximation of the exit point $X_x(\tau_x)$ if $\bar{X}_N \in \Gamma_{cr^{1-1/n}}$. Due to Theorem 1.14 and (1.43), we obtain

$$E(|\bar{X}_N - \xi_x|^2 \,; \mathcal{C}) \leq K(c^2 + e^{2\gamma L}) \times r^{2-2/n}. \tag{1.44}$$

Lemma 1.17. *There exists a constant K such that for any $x \in \bar{G}$, $y \in \partial G$ the inequality*

$$E(X_x(\tau_x) - y)^2 \leq K|x - y|$$

holds.

Proof. Consider the Dirichlet problem

$$\frac{1}{2} \sum_{i,j=1}^{d} a^{ij}(x) \frac{\partial^2 u}{\partial x^i \partial x^j} = 0, \ x \in G,$$

$$u|_{\partial G} = (x - y)^2.$$

The solution of this problem is

$$u_y(x) = E(X_x(\tau_x) - y)^2.$$

Due to the conditions $(i) - (iii)$ from the beginning of Sect. 5.1, $u_y \in C^{(4)}(\bar{G})$ (see [216]). Since $u_y(y) = 0$, we obtain

$$u_y(x) = u_y(x) - u_y(y) \leq K|x - y|.$$

□

We have defined the variable $\xi_x(\omega)$ on $\mathcal{C}$ only. Now we complete the definition by letting $\xi_x(\omega)$ for $\omega \in \Omega \backslash \mathcal{C}$ be, e.g., the point on ∂G nearest to $\bar{X}_N$.

According to Lemma 1.17, we have

$$E((X_x(\tau_x) - \xi_x)^2 \mid \mathcal{F}_N) = E((X_{\bar{X}_N}(\tau_{\bar{X}_N}) - \xi_x)^2 \mid \mathcal{F}_N) \leq K|\bar{X}_N - \xi_x|.$$

Since $\mathcal{C} \in \mathcal{F}_N$, the above inequality and (1.44) imply

$$E((X_x(\tau_x) - \xi_x)^2; \mathcal{C}) \le KE(|X_N - \xi_x|; \mathcal{C})$$
$$\le K(E(|X_N - \xi_x|^2; \mathcal{C}))^{1/2} \le K\left(c + e^{\gamma L}\right) \times r^{1-1/n}.$$

We can also evaluate the expectation $E(X_x(\tau_x) - \xi_x)^2$ analogously to (1.40). As a result, we obtain the following theorem.

Theorem 1.18. *Let $\xi_x(\omega) \in \partial G$ be the nearest point to $\bar{X}_N$. Then (for clearness we reduce some non-essential constants):*

$$(E([X_x(\tau_x) - \xi_x]^2; \mathcal{C}))^{1/2} \le Ke^{\gamma L/2} \times r^{1/2 - 1/2n},$$
$$[E(X_x(\tau_x) - \xi_x)^2]^{1/2} \le Ke^{\gamma L/2} \times r^{1/2 - 1/2n} + Ke^{-\alpha_r \lambda_1^2 L/2(1+C)}.$$

5.2 Systems with drift in a space bounded domain

In this section we construct a local approximation for the autonomous system with drift

$$dX = \chi_{\tau_x > t} b(X) dt + \chi_{\tau_x > t} \sigma(X) dw(t), \quad X(0) = x_0. \tag{2.1}$$

Freezing the coefficients at the point x_0, we obtain the system

$$d\bar{X} = b(x_0) dt + \sigma(x_0) dw(t), \quad \bar{X}(0) = x_0. \tag{2.2}$$

Owing to the drift $b(x_0)$, the symmetry is broken down. The distribution of the process $\bar{X}(\bar{\theta})$ on the surface of some ellipsoid, where $\bar{\theta}$ is the first exit time of $\bar{X}$ from the ellipsoid, is already not simple. To seek another surface with a simple distribution of $\bar{X}_{x_0}(\bar{\theta})$ on it is not an easy problem as well.

Instead of (2.2), let us take the process satisfying another equation:

$$d\bar{X} = \bar{b}(\bar{X}) dt + \bar{\sigma}(\bar{X}) dw(t), \quad \bar{X}(0) = x_0, \tag{2.3}$$

where

$$\bar{b}(x_0) = b(x_0), \quad \bar{\sigma}(x_0) = \sigma(x_0).$$

It is clear that the solutions of (2.1) and (2.3) are close in a small neighborhood of the point x_0. If we are able to find $\bar{b}, \bar{\sigma}$, and some surface such that it is easy to construct the exit point $\bar{X}_{x_0}(\bar{\theta})$ then the problem of a local approximation of phase trajectories for systems with drift will be solved.

To this end, let us consider a sufficiently smooth one-to-one transformation $g: U_r^\sigma(x_0) \longrightarrow V_r^\sigma(x_0)$ with the inverse transformation $f: V_r^\sigma(x_0) \longrightarrow U_r^\sigma(x_0)$, where $V_r^\sigma(x_0) \subset G$ is a set. We have $f(g(x)) = x$ for $x \in U_r^\sigma(x_0)$ and $g(f(x)) = x$ for $x \in V_r^\sigma(x_0)$.

Theorem 2.1. *Let a transformation g be such that the relations*

$$g(x_0) = x_0 = f(x_0), \tag{2.4}$$

5.2 Systems with drift in a space bounded domain

$$\frac{1}{2}\sum_{m,k=1}^{d}\frac{\partial^{2}g}{\partial x^{m}\partial x^{k}}\left(x_{0}\right)\sum_{j=1}^{d}\sigma^{mj}\left(x_{0}\right)\sigma^{kj}\left(x_{0}\right)=b\left(x_{0}\right), \qquad (2.5)$$

and

$$\left\{\frac{\partial g^{i}}{\partial x^{m}}\left(x_{0}\right)\right\}=I \qquad (2.6)$$

are fulfilled (I is the identity matrix).

Then the system (2.3) with

$$\bar{b}\left(x\right)=\frac{1}{2}\sum_{m,k=1}^{d}\frac{\partial^{2}g}{\partial x^{m}\partial x^{k}}\left(f\left(x\right)\right)\sum_{j=1}^{d}\sigma^{mj}\left(x_{0}\right)\sigma^{kj}\left(x_{0}\right) \qquad (2.7)$$

and

$$\bar{\sigma}\left(x\right)=\left\{\sum_{m=1}^{d}\frac{\partial g^{i}}{\partial x^{m}}\left(f\left(x\right)\right)\sigma^{mj}\left(x_{0}\right)\right\} \qquad (2.8)$$

has the solution

$$\bar{X}\left(t\right)=g\left(x_{0}+\sigma\left(x_{0}\right)w\left(t\right)\right),\ t\in\left[0,\bar{\theta}\right),$$

which belongs to $V_{r}^{\sigma}\left(x_{0}\right)$. For this solution, $\bar{\theta}$ is the first time at which $\bar{X}\left(t\right)$ leaves the domain $V_{r}^{\sigma}\left(x_{0}\right)$.

Proof. It is clear that for $0\leq t\leq\bar{\theta}$

$$\bar{X}\left(t\right)=g\left(x_{0}+\sigma\left(x_{0}\right)w\left(t\right)\right)\in V_{r}^{\sigma}\left(x_{0}\right),\ \bar{X}\left(0\right)=g\left(x_{0}\right)=x_{0},$$
$$\bar{X}\left(\bar{\theta}\right)=g\left(x_{0}+\sigma\left(x_{0}\right)w\left(\bar{\theta}\right)\right)\in\partial V_{r}^{\sigma}\left(x_{0}\right),$$

and

$$x_{0}+\sigma\left(x_{0}\right)w\left(t\right)=f\left(\bar{X}\left(t\right)\right),\ 0\leq t\leq\bar{\theta}. \qquad (2.9)$$

Using Ito's formula and (2.9), we obtain

$$d\bar{X}=dg\left(x_{0}+\sigma\left(x_{0}\right)w\left(t\right)\right)$$
$$=\sum_{m=1}^{d}\frac{\partial g}{\partial x^{m}}\left(x_{0}+\sigma\left(x_{0}\right)w\left(t\right)\right)\sum_{j=1}^{d}\sigma^{mj}\left(x_{0}\right)dw^{j}\left(t\right)$$
$$+\frac{1}{2}\sum_{m,k=1}^{d}\frac{\partial^{2}g}{\partial x^{m}\partial x^{k}}\left(x_{0}+\sigma\left(x_{0}\right)w\left(t\right)\right)\sum_{j=1}^{d}\sigma^{mj}\left(x_{0}\right)\sigma^{kj}\left(x_{0}\right)dt$$
$$=\frac{1}{2}\sum_{m,k=1}^{d}\frac{\partial^{2}g}{\partial x^{m}\partial x^{k}}\left(f\left(\bar{X}\left(t\right)\right)\right)\sum_{j=1}^{d}\sigma^{mj}\left(x_{0}\right)\sigma^{kj}\left(x_{0}\right)dt$$
$$+\sum_{m=1}^{d}\frac{\partial g}{\partial x^{m}}\left(f\left(\bar{X}\left(t\right)\right)\right)\sum_{j=1}^{d}\sigma^{mj}\left(x_{0}\right)dw^{j}\left(t\right).$$

Due to (2.4)-(2.6), it is obvious now that $\bar{X}(t)$ is the solution of the system (2.3) with the coefficients given by (2.7)-(2.8). □

To obtain some constructive methods, we form the transformation

$$g(x) = \left(g^1\left(x^1,\ldots,x^d\right),\ldots,g^d\left(x^1,\ldots,x^d\right)\right)^\top$$

as

$$g(x) = \left(g^1\left(x^1\right),\ldots,g^d\left(x^d\right)\right)^\top,$$

where each component is a function of a single argument. Then the inversion of g is reduced to the inversion of functions of a single argument. Let $f^i(x^i)$ be the inverse function to $g^i(x^i)$. Then, the system (2.3) with (2.7)-(2.8) takes the form

$$d\bar{X}^i = \frac{1}{2}(g^i)''\left(f^i(\bar{X}^i)\right)\sum_{j=1}^d\left(\sigma^{ij}(x_0)\right)^2 dt$$

$$+(g^i)'\left(f^i(\bar{X}^i)\right)\sum_{j=1}^d \sigma^{ij}(x_0)\, dw^j(t), \quad i=1,\ldots,d. \quad (2.10)$$

The restrictions (2.5) and (2.6) are written now as

$$\frac{1}{2}(g^i)''(x_0^i) = \frac{b^i(x_0)}{\sum_{k=1}^d\left(\sigma^{ik}(x_0)\right)^2} = \beta_0^i, \quad i=1,\ldots,d,$$

and

$$(g^i)'(x_0^i) = 1, \quad i=1,\ldots,d.$$

Let us give some concrete methods of constructing functions g^i.

The first method is based on the function

$$g^i(x^i) = x^i + \beta_0^i\left(x^i - x_0^i\right)^2, \quad \beta_0^i = \frac{b^i(x_0)}{\sum_{k=1}^d[\sigma^{ik}(x_0)]^2}.$$

The corresponding inverse function $f^i(x^i)$ is equal to

$$f^i(x^i) = \begin{cases} x_i^0 + \left(-1 + \sqrt{1+4\beta_0^i\left(x^i - x_0^i\right)}\right)/2\beta_0^i, & \beta_0^i \neq 0, \\ x^i, & \beta_0^i = 0. \end{cases}$$

The system (2.10) acquires the form

$$d\bar{X}^i = b^i(x_0)\, dt + \sqrt{1+4\beta_0^i\left(\bar{X}^i - x_0^i\right)}\sum_{j=1}^d \sigma^{ij}(x_0)\, dw^j(t),$$

$$\bar{X}^i(0) = x_0^i,$$

5.2 Systems with drift in a space bounded domain

and on the interval $[0, \bar{\theta}]$ it has the solution

$$\bar{X}^i(t) = x_0^i + \sum_{j=1}^{d} \sigma^{ij}(x_0) w^j(t) + \beta_0^i \left(\sum_{j=1}^{d} \sigma^{ij}(x_0) w^j(t) \right)^2,$$

which at the moment $\bar{\theta}$ belongs to $\partial V_r^\sigma(x_0)$: $\bar{X}(\bar{\theta}) \in \partial V_r^\sigma(x_0)$. The equation of the surface $\partial V_r^\sigma(x_0)$ has the form

$$(\sigma^{-1}(x_0)\lambda, \sigma^{-1}(x_0)\lambda) = r^2, \qquad (2.11)$$

where the vector λ has components λ^i, $i = 1, \ldots, d$, of the form

$$\lambda^i = \begin{cases} \left(-1 + \sqrt{1 + 4\beta_0^i (x^i - x_0^i)} \right) / 2\beta_0^i, & \beta_0^i \neq 0, \\ x^i - x_0^i, & \beta_0^i = 0. \end{cases}$$

The second method is based on the functions

$$g^i(x^i) = x_0^i + \frac{x^i - x_0^i}{1 - \beta_0^i (x^i - x_0^i)}.$$

The function f^i is equal to

$$f^i(x^i) = x_0^i + \frac{x^i - x_0^i}{1 + \beta_0^i (x^i - x_0^i)}.$$

The system (2.10) acquires the form

$$d\bar{X}^i = \left(1 + \beta_0^i (\bar{X}^i - x_0^i)\right)^3 b^i(x_0) dt$$

$$+ \left(1 + \beta_0^i (\bar{X}^i - x_0^i)\right)^2 \sum_{j=1}^{d} \sigma^{ij}(x_0) dw^j(t), \quad \bar{X}^i(0) = x_0^i,$$

and on the interval $[0, \bar{\theta}]$ it has the solution

$$\bar{X}^i(t) = x_0^i + \frac{\sum_{j=1}^{d} \sigma^{ij}(x_0) w^j(t)}{1 - \beta_0^i \sum_{j=1}^{d} \sigma^{ij}(x_0) w^j(t)}.$$

The equation of the surface $\partial V_r^\sigma(x_0)$ has the form (2.11) with

$$\lambda^i = \frac{x^i - x_0^i}{1 + \beta_0^i (x^i - x_0^i)}, \quad i = 1, \ldots, d.$$

This surface is a slightly deformed ellipsoid.

5.3 Space-time Brownian motion

In this section we propose an algorithm for simulation of the exit point of the space-time Brownian motion from a space-time parallelepiped. This algorithm is the basis for the one-step approximation of the solution to (0.4) considered in Sect. 5.4. To make the exposition self-contained, we start with some auxiliary knowledge and distributions of a one-dimensional Wiener process.

5.3.1 Auxiliary knowledge

Let G be a bounded domain in $\mathbf{R}^d$, $Q = [T_0, T_1) \times G$ be a cylinder in $\mathbf{R}^{d+1}$, $\Gamma = \overline{Q} \backslash Q$. The set Γ is a part of the boundary of the cylinder Q consisting of the upper base and the lateral surface.

Consider the first boundary value problem for the equation of parabolic type

$$\frac{\partial u}{\partial t} + \frac{1}{2} \sum_{i,j=1}^{d} a^{ij}(t,x) \frac{\partial^2 u}{\partial x^i \partial x^j} + \sum_{i=1}^{d} b^i(t,x) \frac{\partial u}{\partial x^i} + c(t,x) u + e(t,x) = 0, \ (t,x) \in Q, \tag{3.1}$$

with the initial condition on the upper base

$$u(T_1, x) = f(x), \ x \in \overline{G}, \tag{3.2}$$

and the boundary condition on the lateral surface

$$u(t, x) = g(t, x), \ T_0 \le t \le T_1, \ x \in \partial G. \tag{3.3}$$

Introduce the function φ defined on Γ which is equal to $f(x)$ on the upper base and is equal to $g(t, x)$ on the lateral surface. Then the conditions (3.2)-(3.3) may be rewritten shortly as

$$u \mid \Gamma = \varphi. \tag{3.4}$$

The coefficients $a^{ij} = a^{ji}$ are assumed to satisfy the property of strict ellipticity in $\overline{Q}$, i.e.,

$$\lambda_1^2 = \min_{(t,x) \in \overline{Q}} \min_{1 \le i \le d} \lambda_i^2(t, x) > 0,$$

where $\lambda_1^2(t, x) \le \lambda_2^2(t, x) \le \cdots \le \lambda_d^2(t, x)$ are eigenvalues of the matrix $a(t,x) = \{a^{ij}(t,x)\}$.

Let $\lambda_d^2 = \max_{(t,x) \in \overline{Q}} \lambda_d^2(t, x)$. Then, for any $(t, x) \in \overline{Q}$ and $y \in \mathbf{R}^d$ the inequality

$$\lambda_1^2 \sum_{i=1}^{d} (y^i)^2 \le \sum_{i,j=1}^{d} a^{ij}(t,x) y^i y^j \le \lambda_d^2 \sum_{i=1}^{d} (y^i)^2 \tag{3.5}$$

holds.

The solution to the problem (3.1), (3.4) has the following probabilistic representation [48], [82, p. 299]

$$u(t,x) = E\left[\varphi(\tau, X_{t,x}(\tau))Y_{t,x,1}(\tau) + Z_{t,x,1,0}(\tau)\right], \quad (3.6)$$

where $X_{t,x}(s)$, $Y_{t,x,y}(s)$, $Z_{t,x,y,z}(s)$, $s \geq t$, is the solution of the Cauchy problem to the following system of stochastic differential equations

$$\begin{aligned} dX &= b(s,X)ds + \sigma(s,X)dw(s), \quad X(t) = x, \\ dY &= c(s,X)Y ds, \quad Y(t) = y, \\ dZ &= e(s,X)Y ds, \quad Z(t) = z. \end{aligned} \quad (3.7)$$

Here the point (t,x) belongs to Q, $\tau = \tau_{t,x}$ is the first-passage time of the trajectory $(s, X_{t,x}(s))$ to the boundary Γ. In the system (3.7), Y and Z are scalars, $w(s) = (w^1(s), \ldots, w^d(s))^\top$ is a d-dimensional standard Wiener process, $b(s,x)$ is a column-vector of dimension d compounded from the coefficients $b^i(s,x)$, $\sigma(s,x)$ is a matrix of dimension $d \times d$ which is obtained from the equation

$$\sigma(s,x)\sigma^\top(s,x) = a(s,x), \quad a(s,x) = \{a^{ij}(s,x)\}. \quad (3.8)$$

Setting in (3.1), (3.7)

$$c = 0, \ e = 0, \ f = 0, \ g = \chi_{(\partial G)_0}(x), \quad (3.9)$$

where $(\partial G)_0 \subseteq \partial G$, we get the formula:

$$u(t,x) = P(\tau_{t,x} < T_1, \ X_{t,x}(\tau_{t,x}) \in (\partial G)_0), \ T_0 \leq t < T_1, \quad (3.10)$$

where the time $\tau_{t,x}$ is the first-passage time of the trajectory $X_{t,x}(s)$ to the boundary ∂G.

In particular, if

$$c = 0, \ e = 0, \ f = 0, \ g = 1, \quad (3.11)$$

then

$$u(t,x) = P(\tau_{t,x} < T_1), \ T_0 \leq t < T_1. \quad (3.12)$$

Setting in (3.1), (3.7)

$$c = 0, \ e = 0, \ f = \chi_{G_0}(x), \ g = 0, \quad (3.13)$$

where $G_0 \subset G$, we get the formula:

$$u(t,x) = P(\tau_{t,x} \geq T_1, \ X_{t,x}(T_1) \in G_0). \quad (3.14)$$

In autonomous case (i.e., a^{ij}, b^i, c, e, g do not depend on t) we shall consider the first boundary value problem for parabolic equations in the following form:

$$\frac{\partial u}{\partial t} = \frac{1}{2}\sum_{i,j=1}^{d} a^{ij}(x)\frac{\partial^2 u}{\partial x^i \partial x^j} + \sum_{i=1}^{d} b^i(x)\frac{\partial u}{\partial x^i}$$
$$+c(x)u + e(x), \quad t > 0, \ x \in G, \tag{3.15}$$

$$u(0,x) = f(x), \quad x \in \overline{G}, \tag{3.16}$$

$$u(t,x) = g(x), \quad t > 0, \ x \in \partial G. \tag{3.17}$$

Using (3.9)-(3.10) and (3.13)-(3.14), it is not difficult to obtain that the function
$$u(t,x) = P(\tau_{0,x} < t, \ X_{0,x}(\tau_{0,x}) \in (\partial G)_0), \ t > 0, \tag{3.18}$$
is the solution of the problem (3.15)-(3.17) under (3.9); the function
$$u(t,x) = P(\tau_{0,x} < t), \ t > 0, \tag{3.19}$$
is the solution of the problem (3.15)-(3.17) under (3.11); the function
$$u(t,x) = P(\tau_{0,x} \geq t, \ X_{0,x}(t) \in G_0) \tag{3.20}$$
is the solution of the problem (3.15)-(3.17) under (3.13).

Here $X_{0,x}(s)$ is the solution to the Cauchy problem
$$dX = b(X)ds + \sigma(X)dw(s), \ X(0) = x, \tag{3.21}$$

and $\tau_{0,x}$ is the first-passage time of the trajectory $X_{0,x}(s)$ to the boundary ∂G.

5.3.2 Some distributions for one-dimensional Wiener process

A part of distributions for the Wiener process, which we give in this section (see Sects. 5.3.2 and 5.3.3), may be found in the literature. For instance, in [23, 49, 113] some distributions for the one-dimensional Wiener process are written down in a certain form. But we do not know whether all the distributions needed for our goals are available in the literature. Moreover, we need various analytical forms of one and the same distribution due to computational aspects. That is why, for completeness of the exposition, we derive all the distributions here and give them in the forms, which are suitable for practical realization.

Introduce the first-passage time $\tau_x := \tau_{0,x}$ of the one-dimensional Wiener process $x+W(t)$, $-1 \leq x \leq 1$, $t > 0$, to the boundary of the interval $[-1,1]$. Derive the formulas for
$$u(t,x) = P(\tau_x < t).$$
From (3.15)-(3.17) under (3.11), we obtain that the function (see (3.19))
$$v(t,x) = u(t,x) - 1 = P(\tau_x < t) - 1$$

satisfies the following boundary value problem

$$\frac{\partial v}{\partial t} = \frac{1}{2}\frac{\partial^2 v}{\partial x^2}, \ t>0, \ -1<x<1, \tag{3.22}$$

$$v(0,x) = -1, \ v(t,-1) = v(t,1) = 0. \tag{3.23}$$

By the method of separation of variables, we get the formula

$$P(\tau_x < t) = 1 - \frac{4}{\pi}\sum_{k=0}^{\infty}\frac{(-1)^k}{2k+1}\cos\frac{\pi(2k+1)x}{2}\exp(-\frac{1}{8}\pi^2(2k+1)^2 t). \tag{3.24}$$

Further, extending the initial data in (3.22)-(3.23) by the odd way on the whole axis and solving the obtained Cauchy problem, we get another form for the same distribution

$$P(\tau_x < t) = 1 - \int_{-1}^{1} G(t,x,y)\,dy, \tag{3.25}$$

where

$$G(t,x,y) = \frac{1}{\sqrt{2\pi t}}\sum_{k=-\infty}^{\infty}[\exp(-\frac{1}{2t}(x-4k-y)^2)$$

$$-\exp(-\frac{1}{2t}(x-(4k+2)+y)^2)]. \tag{3.26}$$

We shall use the formulas (3.24) and (3.25) under $x=0$. Denote $\tau = \tau_0$,

$$\mathcal{P}(t) := P(\tau < t),$$

and introduce the density $\mathcal{P}'(t)$. From (3.24) and (3.25) one can obtain the following lemma.

Lemma 3.1. *Let τ be the first-passage time of the one-dimensional standard Wiener process $W(t)$ to the boundary of the interval $[-1,1]$. Then the following formulas for its distribution and density take place*

$$\mathcal{P}(t) = 1 - \frac{4}{\pi}\sum_{k=0}^{\infty}\frac{(-1)^k}{2k+1}\exp(-\frac{1}{8}\pi^2(2k+1)^2 t), \ t>0, \tag{3.27}$$

and

$$\mathcal{P}(t) = 2\sum_{k=0}^{\infty}(-1)^k \operatorname{erfc}\frac{2k+1}{\sqrt{2t}}, \ t>0, \tag{3.28}$$

$$\mathcal{P}'(t) = \frac{\pi}{2}\sum_{k=0}^{\infty}(-1)^k(2k+1)\exp(-\frac{1}{8}\pi^2(2k+1)^2 t), \ t>0, \tag{3.29}$$

and

$$\mathcal{P}'(t) = \frac{2}{\sqrt{2\pi t^3}}\sum_{k=0}^{\infty}(-1)^k(2k+1)\exp(-\frac{1}{2t}(2k+1)^2), \ t>0. \tag{3.30}$$

Recall
$$\text{erfc } x = \frac{2}{\sqrt{\pi}} \int_x^\infty \exp(-s^2)\, ds, \quad \text{erfc } 0 = 1. \tag{3.31}$$

The formulas (3.27) and (3.29) are suitable for calculations under large t, and the formulas (3.28) and (3.30) are suitable under small t. The remainders of the series (3.29) and (3.30) are evaluated by the quantities

$$r_k(t) = \frac{\pi}{2}(2k+3)\exp\left(-\frac{1}{8}\pi^2(2k+3)^2 t\right)$$

and

$$\rho_k(t) = \frac{2}{\sqrt{2\pi t^3}}(2k+3)\exp\left(-\frac{1}{2t}(2k+3)^2\right),$$

respectively.

These quantities coincide for $t = \dfrac{2}{\pi}$ and

$$r_k(t) < r_k\left(\frac{2}{\pi}\right), \quad t > \frac{2}{\pi},$$

$$\rho_k(t) < r_k\left(\frac{2}{\pi}\right), \quad t < \frac{2}{\pi}.$$

If we take k, for example, equal to 2, then

$$r_2\left(\frac{2}{\pi}\right) = \frac{7\pi}{2} e^{-49\pi/4} < 2.13 \times 10^{-16},$$

and consequently,

$$\bar{\mathcal{P}}'(t) = \begin{cases} \dfrac{2}{\sqrt{2\pi t^3}}(e^{-1/2t} - 3e^{-9/2t} + 5e^{-25/2t}), & 0 < t < \dfrac{2}{\pi}, \\ \dfrac{\pi}{2}(e^{-\pi^2 t/8} - 3e^{-9\pi^2 t/8} + 5e^{-25\pi^2 t/8}), & t > \dfrac{2}{\pi}, \end{cases} \tag{3.32}$$

differs from $\mathcal{P}'(t)$ by a quantity of 2.13×10^{-16} on the whole interval $[0, \infty)$.

It is not difficult to evaluate that

$$\bar{\mathcal{P}}(t) = \int_0^t \bar{\mathcal{P}}'(s)\, ds \tag{3.33}$$

differs from $\mathcal{P}(t)$ on the whole interval $[0, \infty)$ by $(8/7\pi)e^{-49\pi/4} < 7.04 \times 10^{-18}$. Such an exactness is quite sufficient for practical calculations. See the curves of the distribution $\mathcal{P}(t)$ and its density $\mathcal{P}'(t)$ in Fig. 3.1.

Denote the inverse function to $\mathcal{P}$ by $\mathcal{P}^{-1}$, and let γ be a random variable uniformly distributed on $[0, 1]$. Then the random variable

$$\tau = \mathcal{P}^{-1}(\gamma)$$

is distributed by the law $\mathcal{P}(t)$.

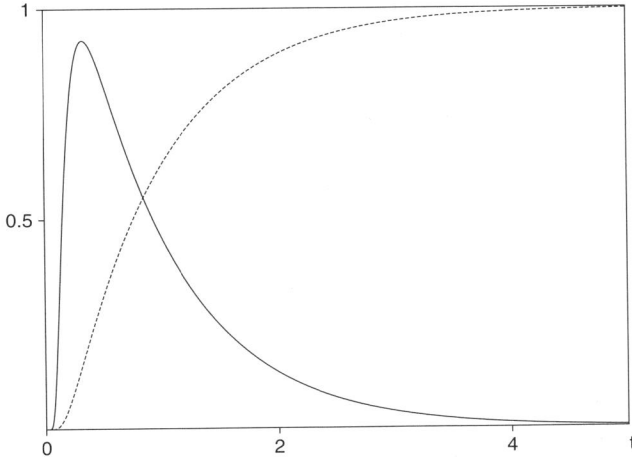

Fig. 3.1. The distribution function $\mathcal{P}(t)$ and the density $\mathcal{P}'(t)$.

To simulate this law in practice, we have to solve the following equation

$$\bar{\mathcal{P}}(t) = \gamma. \tag{3.34}$$

Let us note that because of the analytical simplicity of the function $\bar{\mathcal{P}}(t)$ it is natural to use the Newton method for solving (3.34).

Lemma 3.2. *For the conditional probability*

$$\mathcal{Q}(\beta;t) := P(W(t) < \beta \,/\, |W(s)| < 1,\ 0 < s < t),$$

where $-1 < \beta \le 1$, *the following equalities hold:*

$$\begin{aligned}
\mathcal{Q}(\beta;t) &= \frac{P(W(t) < \beta,\ \tau \ge t)}{P(\tau \ge t)} \\
&= \frac{1}{1-\mathcal{P}(t)} \frac{2}{\pi} \sum_{k=0}^{\infty} \frac{1}{2k+1} \left((-1)^k + \sin\frac{\pi(2k+1)\beta}{2} \right) \\
&\quad \times \exp(-\frac{1}{8}\pi^2(2k+1)^2 t),
\end{aligned} \tag{3.35}$$

and

$$\mathcal{Q}(\beta;t) = \frac{1}{1-\mathcal{P}(t)} \times$$
$$\times \sum_{k=0}^{\infty} \frac{(-1)^k}{2} \left(\operatorname{erfc} \frac{2k-1}{\sqrt{2t}} - \operatorname{erfc} \frac{2k+\beta}{\sqrt{2t}} \right.$$
$$\left. - \operatorname{erfc} \frac{2k+2-\beta}{\sqrt{2t}} + \operatorname{erfc} \frac{2k+3}{\sqrt{2t}} \right). \qquad (3.36)$$

Proof. The first equality in (3.35) flows out of equivalence of the events $(|W(s)| < 1,\ 0 < s < t)$ and $(\tau \geq t)$. Let us prove the second one. To this end consider the probability
$$u(t,x) = P(\tau_x \geq t,\ \alpha \leq x + W(t) < \beta),$$
where $\alpha \geq -1$.

Due to (3.15)-(3.17), (3.20) under (3.13), this probability is the solution of the boundary value problem:
$$\frac{\partial u}{\partial t} = \frac{1}{2}\frac{\partial^2 u}{\partial x^2},\ t > 0,\ -1 < x < 1, \qquad (3.37)$$
$$u(0,x) = \chi_{[\alpha,\beta]}(x),\ u(t,-1) = u(t,1) = 0,\ t > 0. \qquad (3.38)$$

Solving this problem, we get
$$u(t,x) = \frac{2}{\pi} \sum_{k=1}^{\infty} \frac{1}{k} \sin \frac{\pi k(\alpha+\beta)}{2} \sin \frac{\pi k(\beta-\alpha)}{2} \sin \pi k x \exp(-\frac{1}{2}\pi^2 k^2 t)$$
$$+ \frac{4}{\pi} \sum_{k=0}^{\infty} \frac{1}{2k+1} \sin \frac{\pi(2k+1)(\beta-\alpha)}{4} \cos \frac{\pi(2k+1)(\beta+\alpha)}{4}$$
$$\times \cos \frac{\pi(2k+1)x}{2} \exp(-\frac{1}{8}\pi^2(2k+1)^2 t).$$

As $P(W(t) < \beta,\ \tau \geq t) = u(t,0)$ under $\alpha = -1,\ x = 0$, we arrive at (3.35) from here. The equality (3.36) follows from
$$u(t,x) = \frac{1}{\sqrt{2\pi t}} \int_\alpha^\beta G(t,x,y)dy$$
obtained analogously to (3.25). □

Let us note that the series (3.35) and (3.36) are of the Leibniz type, (3.35) is convenient for calculations under large t, and (3.36) is convenient under small t. We draw our attention to the denominator $(1 - \mathcal{P}(t))$ in (3.35), which is close to zero for $t \gg 1$. But it is not difficult to transform (3.35) to the form proper for calculations. See the curves of the distribution $\mathcal{Q}(\beta;t)$ for some values of t in Fig. 3.2.

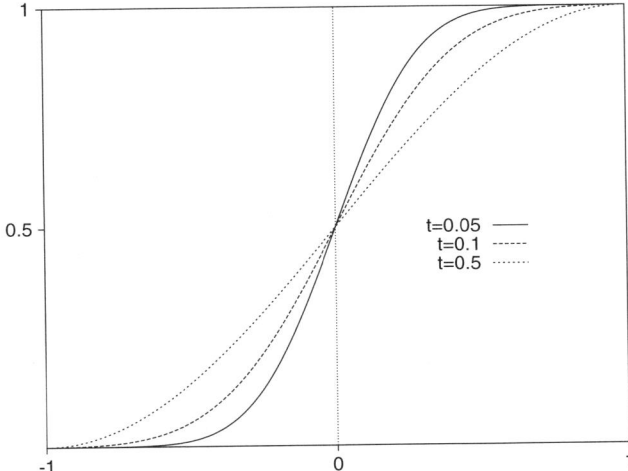

Fig. 3.2. The distribution function $Q(\beta;\cdot)$; under $t \geq 0.5$ the curves coincide visually.

Let the function $\mathcal{Q}^{-1}(\,\cdot\,;t)$ be the inverse function to $\mathcal{Q}(\,\cdot\,;t)$ for every fixed t and γ be a random variable uniformly distributed on $[0,1]$. Then the random variable

$$\xi = \mathcal{Q}^{-1}(\gamma;t)$$

has $\mathcal{Q}(\beta;t)$ as its distribution function.

5.3.3 Simulation of exit time and exit point of Wiener process from a cube

Let $C \subset \mathbf{R}^d$ be a d-dimensional cube with center at the origin and with edge length equal to 2. We suppose all the edges of the cube to be parallel to the coordinate axes, i.e., $C = \{x = (x^1,\ldots,x^d) : |x^i| < 1,\ i = 1,\ldots,d\}$. Let $W(s) = (W^1(s),\ldots,W^d(s))^\top$ be a d-dimensional standard Wiener process, τ be the first-passage time of $W(s)$ to the boundary ∂C of the cube C.

Let us give the following evident result in the form of a lemma.

Lemma 3.3. *The distribution function $\mathcal{P}_d(t)$ for τ is equal to*

$$\mathcal{P}_d(t) = P(\tau < t) = 1 - (1 - \mathcal{P}(t))^d \tag{3.39}$$

and the random variable

$$\tau = \mathcal{P}^{-1}(1 - \gamma^{1/d}) \tag{3.40}$$

is distributed by the law $\mathcal{P}_d(t)$.

Our nearest goal is to construct an algorithm for simulation of the point $(\tau, W(\tau))$. To this end, we obtain some distributions connected with the d-dimensional Wiener process.

Lemma 3.4. *Let τ^j be the first-passage time of the component $W^j(t)$ to the boundary of the interval $[-1, 1]$. Then*

$$P(\bigcap_{i \neq j}(W^i(\tau^j) < \beta^i, \ |W^i(s)| < 1, \ 0 < s < \tau^j)/\tau^j)$$

$$= (1 - \mathcal{P}(\tau^j))^{d-1} \prod_{i \neq j} \mathcal{Q}(\beta^i; \tau^j). \qquad (3.41)$$

Proof. We shall use an assertion of the following kind: if $\zeta \geq 0$ is $\tilde{\mathcal{F}}$-measurable (where $\tilde{\mathcal{F}}$ is a σ-subalgebra of a general σ-algebra $\mathcal{F}$), a random variable $\varphi(t, \omega)$ for every $t \geq 0$ does not depend on $\tilde{\mathcal{F}}$ and $E\varphi(t, \omega) = h(t)$, then $E(\varphi(\zeta, \omega)/\tilde{\mathcal{F}}) = h(\zeta)$ (see [82, p. 67], [138, p. 158]).

Due to Lemma 3.2 and independence of the processes $W^i(s)$, we get for any $t \geq 0$,

$$P\left(\bigcap_{i \neq j}(W^i(t) < \beta^i, \ |W^i(s)| < 1, \ 0 < s < t)\right) = (1 - \mathcal{P}(t))^{d-1} \prod_{i \neq j} \mathcal{Q}(\beta^i; t).$$

This equality implies (3.41) in accordance with the above-mentioned assertion because the processes $W^i(s)$, $i \neq j$, do not depend on the process $W^j(s)$. □

Introduce the random variable $\varkappa$ which takes the value j for $\omega \in \{\omega : W^j(\tau) = \pm 1\}$. This variable is defined uniquely with probability 1, and $P(\varkappa = j) = 1/d$. Let $\nu := W^\varkappa(\tau)$. Clearly, the distribution law for ν is given by $P(\nu = -1) = P(\nu = 1) = 1/2$.

Lemma 3.5. *The following equality takes place:*

$$P(\varkappa = j, \ \tau < \theta, \ \bigcap_{i \neq j}(W^i(\tau) < \beta^i))$$

$$= \int_0^\theta (1 - \mathcal{P}(\vartheta))^{d-1} \prod_{i \neq j} \mathcal{Q}(\beta^i; \vartheta) \mathcal{P}'(\vartheta) d\vartheta. \qquad (3.42)$$

Proof. We have

$$P(\varkappa = j, \ \tau < \theta, \ \bigcap_{i \neq j}(W^i(\tau) < \beta^i))$$

$$= P(\bigcap_{i \neq j}(W^i(\tau^j) < \beta^i, \ |W^i(s)| < 1, \ 0 < s < \tau^j), \ \tau^j < \theta)$$

$$= \int_0^\theta P(\bigcap_{i \neq j}(W^i(\tau^j) < \beta^i, \ |W^i(s)| < 1, \ 0 < s < \tau^j)/\ \tau^j = \vartheta) d\mathcal{P}_{\tau^j}(\vartheta),$$

where $\mathcal{P}_{\tau^j}(\vartheta)$ is the distribution function for τ^j. Clearly $\mathcal{P}_{\tau^j}(\vartheta) = \mathcal{P}(\vartheta)$. Now the assertion (3.42) arises from Lemma 3.4. □

Lemma 3.6. *The following equality holds:*

$$P(\bigcap_{i \neq j}(W^i(\tau) < \beta^i)/\varkappa = j, \ \tau = \theta) = \prod_{i \neq j} \mathcal{Q}(\beta^i; \theta) \ . \tag{3.43}$$

Proof. The random variables $\varkappa$ and τ are independent. Indeed, $P(\varkappa = 1, \ \tau < \theta) = \cdots = P(\varkappa = d, \ \tau < \theta)$ on the strength of symmetry. Hence $P(\varkappa = i, \ \tau < \theta) = (1/d)P(\tau < \theta) = P(\varkappa = i)P(\tau < \theta)$. Further (see (3.39))

$$dP(\varkappa = j, \ \tau < \theta) = \frac{1}{d} d\mathcal{P}_d(\theta) = (1 - \mathcal{P}(\theta))^{d-1} \mathcal{P}'(\theta) d\theta \ .$$

From here we get

$$P(\varkappa = j, \ \tau < \theta, \ \bigcap_{i \neq j}(W^i(\tau) < \beta^i))$$

$$= \int_0^\theta P(\bigcap_{i \neq j}(W^i(\tau) < \beta^i)/\varkappa = j, \ \tau = \vartheta)(1 - \mathcal{P}(\vartheta))^{d-1}\mathcal{P}'(\vartheta)d\vartheta \ . \tag{3.44}$$

Comparing (3.42) with (3.44), we obtain (3.43). □

Let us note that the point $(\tau, W(\tau)) \in [0, \infty) \times \partial C$, i.e., this point belongs to the lateral surface of the unbounded semi-cylinder $[0, \infty) \times C$ with cubic base in $(d+1)$-dimensional space of variables $(t, x^1, \ldots, x^d)$.

Theorem 3.7. (*Algorithm for simulating exit point to lateral surface of cylinder with cubic base*). *Let $\varkappa, \ \nu, \ \gamma, \gamma^1, \ldots, \gamma^{d-1}$ be independent random variables. Let $\varkappa$ and ν be simulated by the laws $P(\varkappa = j) = 1/d$, $j = 1, \ldots, d$; $P(\nu = \pm 1) = 1/2$, and let $\gamma, \gamma^1, \ldots, \gamma^{d-1}$ be uniformly distributed on $[0, 1]$.*
Then the point $(\tau, \xi) = (\tau, \xi^1, \ldots, \xi^d)$ with

$$\tau = \mathcal{P}^{-1}(1 - \gamma^{1/d}), \ \xi^1 = \mathcal{Q}^{-1}(\gamma^1; \tau), \ldots, \ \xi^{\varkappa - 1} = \mathcal{Q}^{-1}(\gamma^{\varkappa - 1}; \tau),$$
$$\xi^\varkappa = \nu, \ \xi^{\varkappa + 1} = \mathcal{Q}^{-1}(\gamma^\varkappa; \tau), \ldots, \ \xi^d = \mathcal{Q}^{-1}(\gamma^{d-1}; \tau) \tag{3.45}$$

has the same distribution as $(\tau, W(\tau))$.

This theorem is a simple consequence of Lemmas 3.3 and 3.6.

Corollary 3.8. *Let $C_r = \{x = (x^1, \ldots, x^d) : |x^i| < r, \ i = 1, \ldots, d\} \subset \mathbf{R}^d$ be a d-dimensional cube with center at the origin and with edge length equal to $2r$. Let $\bar{\theta}$ be the first-passage time of the d-dimensional standard Wiener process $w(s)$ to the boundary ∂C_r of the cube C_r. Then the point*

$$(\bar{\theta}, \bar{w}) = (r^2 \tau, r\xi),$$

where (τ, ξ) is simulated by the algorithm for simulating exit point to lateral surface of cylinder with the cubic base C, has the same distribution as $(\bar{\theta}, w(\bar{\theta}))$.

The proof easily follows from the fact that if $W(t)$ is a Wiener process, then $w(t) = rW(t/r^2)$ is a Wiener process as well.

5.3.4 Simulation of exit point of the space-time Brownian motion from a space-time parallelepiped with cubic base

Now let us consider the space-time parallelepiped $\Pi = [0, l) \times C \subset \mathbf{R}^{d+1}$, where the cube $C \subset \mathbf{R}^d$ is defined as above. We will construct an algorithm for simulating the exit point $(\tau(l), W(\tau(l)))$ from the parallelepiped Π. The random variable $\tau(l)$ is found as $\min(\tau, l)$, where τ is the first-passage time of $W(s)$ to the boundary ∂C as above. The distribution function of $\tau(l)$ is equal to

$$P(\tau(l) < t) = \begin{cases} 1 - (1 - \mathcal{P}(t))^d, & t \leq l, \\ 1, & t > l. \end{cases} \quad (3.46)$$

Theorem 3.9. *(Algorithm for simulating exit point from a space-time parallelepiped with cubic base).* Let $\iota, \varkappa, \nu, \gamma, \gamma^1, \ldots, \gamma^{d-1}$ be independent random variables. Let ι be simulated by the law

$$P(\iota = -1) = 1 - (1 - \mathcal{P}(l))^d, \quad P(\iota = 1) = (1 - \mathcal{P}(l))^d,$$

and the random variables $\varkappa, \nu, \gamma, \gamma^1, \ldots, \gamma^{d-1}$ be simulated as in Theorem 3.7.

Then a random point $(\tau(l), \xi)$, distributed as the exit point $(\tau(l), W(\tau(l)))$, is simulated by the following algorithm.

If the simulated value of ι is equal to -1, then the point $(\tau(l), \xi)$ belongs to the lateral surface of Π, and

$$\tau(l) = \mathcal{P}^{-1}(1 - [1 - \gamma(1 - (1 - \mathcal{P}(l))^d]^{1/d}),$$
$$\xi^1 = \mathcal{Q}^{-1}(\gamma^1; \tau(l)), \ldots, \xi^{\varkappa-1} = \mathcal{Q}^{-1}(\gamma^{\varkappa-1}; \tau(l)), \xi^{\varkappa} = \nu,$$
$$\xi^{\varkappa+1} = \mathcal{Q}^{-1}(\gamma^{\varkappa}; \tau(l)), \ldots, \xi^d = \mathcal{Q}^{-1}(\gamma^{d-1}; \tau(l));$$

otherwise, when $\iota = 1$, the point $(\tau(l), \xi)$ belongs to the upper base of Π, and

$$\tau(l) = l,$$
$$\xi^1 = \mathcal{Q}^{-1}(\gamma; l), \xi^2 = \mathcal{Q}^{-1}(\gamma^1; l), \ldots, \xi^d = \mathcal{Q}^{-1}(\gamma^{d-1}; l).$$

Proof. Using Lemma 3.3, we have

$$P(\tau(l) < l) = P(\tau < l) = 1 - (1 - \mathcal{P}(l))^d,$$
$$P(\tau(l) = l) = P(\tau \geq l) = (1 - \mathcal{P}(l))^d. \quad (3.47)$$

The conditional probability $P(\tau(l) < t / \tau(l) < l)$ is equal to

$$P(\tau(l) < t / \tau(l) < l) = \frac{P((\tau(l) < t) \cap (\tau(l) < l))}{P(\tau(l) < l)}$$
$$= \chi_{[l,\infty)}(t) + \chi_{[0,l)}(t) \frac{P(\tau < t)}{P(\tau < l)},$$

and the random variable $\mathcal{P}^{-1}(1-[1-\gamma(1-(1-\mathcal{P}(l))^d]^{1/d})$ is distributed by the law $P(\tau(l)<t/\tau(l)<l)$.

Carrying out reasoning similar to Lemmas 3.4, 3.5, and 3.6, we obtain

$$P(\bigcap_{i\neq j}(W^i(\tau(l))<\beta^i)/\varkappa=j,\tau(l)=\theta<l)=\chi_{[0,l)}(\theta)\prod_{i\neq j}\mathcal{Q}(\beta^i;\theta). \quad (3.48)$$

Further, the equality

$$P(\bigcap_{i=1}^d(W^i(\tau(l))<\beta^i)/\tau(l)=l)=P(\bigcap_{i=1}^d(W^i(l)<\beta^i)/\tau\geq l)$$

$$=\frac{1}{P(\tau\geq l)}P(\bigcap_{i=1}^d(W^i(l)<\beta^i,\ |W^i(s)|<1,\ 0<s<l))$$

$$=\frac{1}{\prod_{i=1}^d P(\tau^i\geq l)}P(\bigcap_{i=1}^d(W^i(l)<\beta^i,\tau^i\geq l))$$

$$=\prod_{i=1}^d \mathcal{Q}(\beta^i;l) \quad (3.49)$$

holds due to the mutual independence of the components W^i, $i=1,\ldots,d$, and Lemma 3.2. Now the statement of the theorem easily follows from (3.47)-(3.49). □

The following corollary has the same proof as Corollary 3.8.

Corollary 3.10. *Let $\Pi_{r,l}=[0,lr^2)\times C_r=\{(t,x)=(t,x^1,\ldots,x^d):\ 0\leq t<lr^2,\ |x^i|<r,\ i=1,\ldots,d\}\subset\mathbf{R}^{d+1}$ be a space-time parallelepiped. Let $\bar\theta$ be the first-passage time of the process $(s,w(s))$, $s>0$, to the boundary $\partial\Pi_{r,l}$. Then the point*

$$(\bar\theta,\bar w)=(r^2\tau(l),r\xi),$$

where $(\tau(l),\xi)$ is simulated by the algorithm for simulating the exit point from the space-time parallelepiped Π, has the same distribution as $(\bar\theta,w(\bar\theta))$.

5.4 Approximations for SDEs in a space-time bounded domain

The simulation of the space-time Brownian motion considered in the previous section is the basis for our algorithms. Local approximation theorem is proved in Sect. 5.4.1. A global algorithm is proposed and convergence theorems are proved in Sect. 5.4.2. An approximation for the space-time exit point is constructed in Sect. 5.4.3. Section 5.4.4 briefly describes an algorithm for (0.4) which is based on simulation of exit points for the Brownian motion with drift. Numerical examples are given in Sect. 5.5.

5.4.1 Local mean-square approximation in a space-time bounded domain

Let us consider the system of SDEs (see (0.4))

$$dX = \chi_{\tau_{t,x}>s} b(s,X)ds + \chi_{\tau_{t,x}>s} \sigma(s,X)dw(s), \quad X(t) = X_{t,x}(t) = x, \quad (4.1)$$

in a space-time bounded domain $Q = [T_0, T_1] \times G \subset \mathbf{R}^{d+1}$. Here and in the next subsections we assume that the coefficients of (4.1) belong to the class $C^{1,2}([T_0,T_1] \times \overline{G})$, the boundary ∂G of the domain G is twice continuously differentiable, and the strict ellipticity condition is imposed on the matrix $a(s,x) = \sigma(s,x)\sigma^\top(s,x)$ (see (3.5)).

Introduce the space-time parallelepiped $\Pi_{r,l}^{\sigma(t,x)}(x)$:

$$\Pi_{r,l}^{\sigma(t,x)}(x) = \bigcup_{0 \leq s < lr^2} \{t+s\} \times C_r^{\sigma(t,x)}(x + b(t,x)s),$$

where $(t,x) \in Q$ and $C_r^{\sigma(t,x)}(x+b(t,x)s)$ is the space parallelepiped in $\mathbf{R}^d$ obtained from the open cube C_r by the linear transformation $\sigma(t,x)$ and the shift $x + b(t,x)s$, and as in the previous section, C_r is the cube with center at the origin and with edges of length $2r$ which are parallel to the coordinate axes. The time size of $\Pi_{r,l}^{\sigma(t,x)}(x)$ is taken lr^2 since the characteristic exit time of a diffusion process from a space cube of linear size r is proportional to r^2.

Let Γ_δ be an intersection of a δ-neighborhood of the set Γ with the domain Q. Recall that the set Γ is a part of the boundary ∂Q consisting of the lateral surface and the upper base of the cylinder $\overline{Q}$. The size δ of the layer Γ_δ may depend on r. The condition of strict ellipticity ensures for any $\beta > 0$ the existence of a constant $\alpha > 0$ such that under all sufficiently small r for every point $(t,x) \in Q \backslash \Gamma_{\alpha r}$ the following relations take place:

$$\Pi_{r,l}^{\sigma(t,x)}(x) \subset Q, \quad \min_{0 \leq s \leq lr^2} \rho(\partial C_r^{\sigma(t,x)}(x+b(t,x)s), \partial G) \geq \beta r. \quad (4.2)$$

Indeed, due to the property of strict ellipticity, we get

$$\max_{0 \leq s \leq lr^2} \rho(x, \partial C_r^{\sigma(t,x)}(x+b(t,x)s)) \leq lr^2 \max_{(s,y)\in\overline{Q}} |b(s,y)| + 2r\sqrt{d}\lambda_d.$$

It is easy to see that if we take

$$\alpha = lr \max_{(s,y)\in\overline{Q}} |b(s,y)| + 2\sqrt{d}\lambda_d + \beta,$$

then for a sufficiently small r the relations (4.2) are fulfilled. The values β, α, and r used below are assumed to ensure (4.2).

To construct a one-step approximation for the system (4.1), we consider the system with frozen coefficients (see (0.5)):

5.4 Approximations for SDEs in a space-time bounded domain

$$d\bar{X} = b(t,x)ds + \sigma(t,x)dw(s), \quad \bar{X}(t) = x, \ (t,x) \in Q\backslash\Gamma_{\alpha r}. \quad (4.3)$$

Let $\bar{\theta}$ be the first-passage time of the process $(s - t, w(s) - w(t))$, $s > t$, to the boundary $\partial\Pi_{r,l}$ of the space-time parallelepiped $\Pi_{r,l} = [0, lr^2] \times C_r \subset \mathbf{R}^{d+1}$. Clearly, $\bar{\theta} \leq lr^2$. The point $(\bar{\theta}, w(t+\bar{\theta}) - w(t))$ is simulated in accordance with Corollary 3.10.

Let us take the point $(t + \bar{\theta}, \bar{X}_{t,x}(t+\bar{\theta}))$ with $\bar{X}_{t,x}(t+\bar{\theta})$ calculated by

$$\bar{X}_{t,x}(t+\bar{\theta}) = x + b(t,x)\bar{\theta} + \sigma(t,x)(w(t+\bar{\theta}) - w(t)) \quad (4.4)$$

as an approximation of the point $(t+\bar{\theta}, X_{t,x}(t+\bar{\theta}))$, $(t,x) \in Q\backslash\Gamma_{\alpha r}$, where $X_{t,x}(s)$ is a solution of the system (4.1). Recall that if $t + \bar{\theta} \geq \tau_{t,x}$, then $X_{t,x}(t+\bar{\theta}) = X_{t,x}(\tau_{t,x})$.

The point $(t+\bar{\theta}, \bar{X}_{t,x}(t+\bar{\theta}))$ belongs to the lateral surface or to the upper base of the space-time parallelepiped $\Pi_{r,l}^{\sigma(t,x)}(x) \subset Q$. It follows from (4.2) that

$$\rho(\bar{X}_{t,x}(t+s), \partial G) \geq \beta r, \ 0 \leq s \leq lr^2. \quad (4.5)$$

Theorem 4.1. *For every natural m there exists a constant $K > 0$ such that for any sufficiently small r and for any point $(t,x) \in Q\backslash\Gamma_{\alpha r}$ the inequality*

$$E\left|X_{t,x}(t+\bar{\theta}) - \bar{X}_{t,x}(t+\bar{\theta})\right|^{2m} \leq K r^{4m} \quad (4.6)$$

holds.

Proof. We have (see (4.1)) that $\tau_{t,x} \leq T_1$, $X_{t,x}(s) \in G$ for $s \in [t, \tau_{t,x})$, and $X_{t,x}(s) = X_{t,x}(\tau_{t,x})$ for $s \geq \tau_{t,x}$.

Let us rewrite the local error in the form

$$E\left|X_{t,x}(t+\bar{\theta}) - \bar{X}_{t,x}(t+\bar{\theta})\right|^{2m}$$

$$= E\left|\int_t^{t+\bar{\theta}} \left(\chi_{\tau_{t,x}>s} b(s, X_{t,x}(s)) - b(t,x)\right) ds\right.$$

$$\left. + \int_t^{t+\bar{\theta}} \left(\chi_{\tau_{t,x}>s} \sigma(s, X_{t,x}(s)) - \sigma(t,x)\right) dw(s)\right|^{2m}$$

$$\leq K E\left|\int_t^{t+\bar{\theta}} \left(\chi_{\tau_{t,x}>s} b(s, X_{t,x}(s)) - b(t,x)\right) ds\right|^{2m}$$

$$+ K E\left|\int_t^{(t+\bar{\theta})\wedge\tau_{t,x}} (\sigma(s, X_{t,x}(s)) - \sigma(t,x)) dw(s)\right|^{2m}$$

$$+ K E\left|\int_{(t+\bar{\theta})\wedge\tau_{t,x}}^{t+\bar{\theta}} \sigma(t,x) dw(s)\right|^{2m}. \quad (4.7)$$

We obtain for the first term in (4.7):

$$K E\left|\int_t^{t+\bar{\theta}} \left(\chi_{\tau_{t,x}>s} b(s, X_{t,x}(s)) - b(t,x)\right) ds\right|^{2m} \leq K E\bar{\theta}^{2m} \leq K r^{4m} \quad (4.8)$$

because of boundedness of $b(s,x)$, $(s,x) \in \bar{Q}$, and $\bar{\theta} \leq lr^2$.

Below we need the following inequality for Ito integrals in the case of the scalar Wiener process (see, e.g., [82, p.26]):

$$E(\int_t^{t+T} \varphi(s)dw(s))^{2m} \leq (m(2m-1))^{m-1}T^{m-1}\int_t^{t+T} E\varphi^{2m}(s)ds, \quad (4.9)$$
$$m = 1, 2, \ldots.$$

Clearly, in the case of the d-dimensional Wiener process the inequality (4.9) implies

$$E|\int_t^{t+T} \varphi(s)dw(s)|^{2m} \leq KT^{m-1}\int_t^{t+T} E\sum_{i,j=1}^d (\varphi^{ij}(s))^{2m}ds, \quad m = 1, 2, \ldots,$$
$$(4.10)$$

where the constant K depends on m. If φ is bounded, we also have

$$E|\int_t^{t+T} \varphi(s)dw(s)|^{2m} \leq KT^m, \quad m = 1, 2, \ldots. \quad (4.11)$$

Due to the inequality (4.10), smoothness of $\sigma(s,x)$ for $(s,x) \in \overline{Q}$, and $(t+\bar{\theta}) \wedge \tau_{t,x} \leq t + lr^2$, we obtain for the second term of (4.7):

$$KE|\int_t^{(t+\bar{\theta})\wedge\tau_{t,x}} (\sigma(s,X(s)) - \sigma(t,x))\,dw(s)|^{2m}$$

$$= KE|\int_t^{t+lr^2} \chi_{(t+\bar{\theta})\wedge\tau_{t,x}>s} (\sigma(s,X_{t,x}(s)) - \sigma(t,x))\,dw(s)|^{2m}$$

$$\leq Kr^{2m-2}\int_t^{t+lr^2} E(\chi_{(t+\bar{\theta})\wedge\tau_{t,x}>s}\sum_{i,j=1}^d |\sigma^{ij}(s,X_{t,x}(s)) - \sigma^{ij}(t,x)|^{2m})\,ds$$

$$\leq Kr^{2m-2}\int_t^{t+lr^2} \left(E\chi_{\tau_{t,x}>s}|X_{t,x}(s) - x|^{2m} + (s-t)^{2m}\right)\,ds$$

$$\leq Kr^{2m-2}\int_t^{t+lr^2} E\chi_{\tau_{t,x}>s}|X_{t,x}(s) - x|^{2m}\,ds + Kr^{6m}. \quad (4.12)$$

Further,

$$E\chi_{\tau_{t,x}>s}|X_{t,x}(s) - x|^{2m} = E\chi_{\tau_{t,x}>s}|\int_t^s b(s',X_{t,x}(s'))\,ds'$$
$$+ \int_t^s \sigma(s',X_{t,x}(s'))dw(s')|^{2m},$$

whence, due to (4.11),

$$E\chi_{\tau_{t,x}>s}|X_{t,x}(s) - x|^{2m} \leq K(s-t)^{2m} + K(s-t)^m.$$

5.4 Approximations for SDEs in a space-time bounded domain

Substituting this inequality in (4.12), we obtain

$$KE \left| \int_t^{(t+\bar{\theta}) \wedge \tau_{t,x}} (\sigma(s, X_{t,x}(s)) - \sigma(t,x)) \, dw(s) \right|^{2m} \leq K r^{4m}. \quad (4.13)$$

It follows from the inequalities (4.8) and (4.13) that

$$E \left| X_{t,x}(\tau_{t,x} \wedge (t+\bar{\theta})) - \bar{X}_{t,x}(\tau_{t,x} \wedge (t+\bar{\theta})) \right|^{2m} \leq K r^{4m}. \quad (4.14)$$

Now let us estimate the third term in (4.7). Due to (4.10), we have

$$KE \left| \int_{(t+\bar{\theta}) \wedge \tau_{t,x}}^{t+\bar{\theta}} \sigma(t,x) \, dw(s) \right|^{2m}$$

$$= KE \left| \int_t^{t+lr^2} \left(\chi_{(t+\bar{\theta})>s} - \chi_{(t+\bar{\theta}) \wedge \tau_{t,x}>s} \right) \sigma(t,x) dw(s) \right|^{2m}$$

$$\leq K r^{2m-2} \int_t^{t+lr^2} E \left(\chi_{(t+\bar{\theta})>s} - \chi_{(t+\bar{\theta}) \wedge \tau_{t,x}>s} \right) ds$$

$$= K r^{2m-2} E \chi_{\tau_{t,x}<(t+\bar{\theta})} \left((t+\bar{\theta}) - (t+\bar{\theta}) \wedge \tau_{t,x} \right)$$

$$\leq K r^{2m} P(\tau_{t,x} < t + \bar{\theta}). \quad (4.15)$$

Evaluate the probability $P(\tau_{t,x} < t + \bar{\theta})$ using the reception from Sect. 5.1.1. If $\tau_{t,x} < t + \bar{\theta}$, then $\tau_{t,x} < T_1$ and, consequently, $X_{t,x}(\tau_{t,x}) \in \partial G$. At the same time, according to (4.5),

$$\rho(\bar{X}_{t,x}(\tau_{t,x} \wedge (t+\bar{\theta})), \partial G) \geq \beta r.$$

Therefore,

$$E \left(\chi_{\tau_{t,x}<t+\bar{\theta}} \left| X_{t,x}(\tau_{t,x} \wedge (t+\bar{\theta})) - \bar{X}_{t,x}(\tau_{t,x} \wedge (t+\bar{\theta})) \right|^m \right)$$
$$\geq P(\tau_{t,x} < t + \bar{\theta}) (\beta r)^m, \ m = 1, 2, \dots .$$

On the other hand, due to (4.14), we have

$$P(\tau_{t,x} < t + \bar{\theta}) (\beta r)^m$$
$$\leq E \left(\chi_{\tau_{t,x}<t+\bar{\theta}} \left| X_{t,x}(\tau_{t,x} \wedge (t+\bar{\theta})) - \bar{X}_{t,x}(\tau_{t,x} \wedge (t+\bar{\theta})) \right|^m \right)$$
$$\leq \sqrt{P(\tau_{t,x} < t + \bar{\theta})} \left[E \left| X_{t,x}(\tau_{t,x} \wedge (t+\bar{\theta})) - \bar{X}_{t,x}(\tau_{t,x} \wedge (t+\bar{\theta})) \right|^{2m} \right]^{1/2}$$
$$\leq K r^{2m} \sqrt{P(\tau_{t,x} < t + \bar{\theta})}.$$

Consequently,

$$P(\tau_{t,x} < t + \bar{\theta}) \leq K r^{2m}, \ m = 1, 2, \dots . \quad (4.16)$$

Now the inequality (4.7) together with (4.8), (4.13), and (4.15) gives (4.6). Theorem 4.1 is proved. □

Remark 4.2. In the case of additive noise, i.e., when $\sigma(t,x) \equiv \sigma(t)$, we prove under the assumptions of Theorem 4.1 that

$$E\left|X_{t,x}(t+\bar{\theta}) - \bar{X}_{t,x}(t+\bar{\theta})\right|^{2m} \leq K\, r^{6m}.$$

5.4.2 Global algorithm in a space-time bounded domain

In this subsection we construct a random walk over small space-time parallelepipeds based on the one-step approximation (4.4).

Algorithm 4.3 Let $(\bar{\theta}_1, w(t+\bar{\theta}_1) - w(t))$ be the first exit point of the process $(s-t, w(s)-w(t))$, $s > t$, from the parallelepiped $\Pi_{r,l}$ simulated in accordance with Corollary 3.10, $(\bar{\theta}_2, w(t+\bar{\theta}_1+\bar{\theta}_2) - w(t+\bar{\theta}_1))$ be the exit point of the process $(s-t-\bar{\theta}_1, w(s) - w(t+\bar{\theta}_1))$, $s > t+\bar{\theta}_1$, from the parallelepiped $\Pi_{r,l}$, and so on.

Suppose that $(t,x) \in Q\backslash \Gamma_{\alpha r}$. Then, we construct the recurrence sequence $(\bar{\vartheta}_k, \bar{X}_k)$, $k = 0, 1, \ldots, \bar{\nu}$:

$$\bar{\vartheta}_0 = t, \ \bar{X}_0 = x,$$
$$\bar{\vartheta}_k = \bar{\vartheta}_{k-1} + \bar{\theta}_k,$$
$$\bar{X}_k = \bar{X}_{k-1} + b(\bar{\vartheta}_{k-1}, \bar{X}_{k-1})\bar{\theta}_k + \sigma(\bar{\vartheta}_{k-1}, \bar{X}_{k-1})(w(\bar{\vartheta}_k) - w(\bar{\vartheta}_{k-1})),$$

$$k = 1, \ldots, \bar{\nu},$$

where the number $\bar{\nu} = \bar{\nu}_{t,x}$ is the first one for which $(\bar{\vartheta}_k, \bar{X}_k) \in \Gamma_{\alpha r}$. If $(t,x) \in \Gamma_{\alpha r}$, we put $\bar{\nu} = 0$.

Let $(\bar{\vartheta}_k, \bar{X}_k) = (\bar{\vartheta}_{\bar{\nu}}, \bar{X}_{\bar{\nu}})$ for $k > \bar{\nu}$. The obtained sequence $(\bar{\vartheta}_k, \bar{X}_k)$, $k = 0, 1, \ldots$, is a Markov chain stopping at the Markov moment $\bar{\nu}$. It is clear that the random number of steps $\bar{\nu}$ depends on the domain $Q\backslash\Gamma_{\alpha r}$. That is why, the more rigorous notation for $\bar{\nu}$ is $\bar{\nu}_{t,x}(Q\backslash\Gamma_{\alpha r})$.

Using the technique described in Sect. 5.1.2, we prove the theorems on average characteristics of $\bar{\nu}_{t,x} = \bar{\nu}_{t,x}(Q\backslash\Gamma_{\alpha r})$.

Theorem 4.4. *The mean number of steps $\bar{\nu}_{t,x}(Q\backslash\Gamma_{\alpha r})$ is estimated as*

$$E\bar{\nu}_{t,x}(Q\backslash\Gamma_{\alpha r}) \leq \frac{K}{r^2}, \tag{4.17}$$

where the positive constant K does not depend on r.

Theorem 4.5. *For every $L > 0$, the inequality*

$$P\left\{\bar{\nu}_{t,x}(Q\backslash\Gamma_{\alpha r}) \geq \frac{L}{r^2}\right\} \leq (1 + T_1 - T_0)\exp(-c_r \frac{\gamma}{1+T_1-T_0}L), \tag{4.18}$$

$$c_r \to 1 \text{ as } r \to 0,$$

is valid.

5.4 Approximations for SDEs in a space-time bounded domain

Further, we need two auxiliary lemmas. Their proofs are available in [204].

Lemma 4.6. *There exists a constant K such that for all r small enough and all $(t, x) \in Q \backslash \Gamma_{\alpha r}$ the inequality*

$$\left| E(X_{t,x}(t + \bar{\theta}_1) - \bar{X}_{t,x}(t + \bar{\theta}_1)) \right| \leq K r^4 \tag{4.19}$$

is valid.

Lemma 4.7. *Let the random variable Z be defined by the relation*

$$X_{t,x}(t + \bar{\theta}_1) - X_{t,y}(t + \bar{\theta}_1) = x - y + Z \,.$$

Then for every natural m there exists a positive constant K such that for any r small enough and all $(t, x), (t, y) \in Q \backslash \Gamma_{\alpha r}$ the inequalities

$$E|Z|^m \leq K r^m \left(|x - y|^m + r^m \right), \tag{4.20}$$

$$|EZ| \leq K r^2 \left(|x - y| + r^2 \right) \tag{4.21}$$

hold.

Remark 4.8. If $\sigma(t, x) \equiv \sigma(t)$, we obtain under the assumptions of Lemma 4.7 (cf. Remark 4.2):

$$E|Z|^m \leq K r^m \left(|x - y|^m + r^{2m} \right) \,.$$

For every $\varepsilon \in (0, 1]$ and any $\beta > 0$ it is possible to introduce the layer $\Gamma_{\alpha r^{1-\varepsilon}}$ with a constant α such that under a sufficiently small r and for every $(t, x) \in Q \backslash \Gamma_{\alpha r^{1-\varepsilon}}$ the following relations together with the relations (4.2) take place:

$$\Pi_{r,l}^{\sigma(t,x)}(x) \subset Q, \quad \min_{0 \leq s \leq lr^2} \rho(\partial C_r^{\sigma(t,x)}(x + b(t,x)s), \partial G) \geq \beta r^{1-\varepsilon} \,.$$

Clearly, $\Gamma_{\alpha r} \subset \Gamma_{\alpha r^{1-\varepsilon}}$.

The Markov moment $\bar{\nu}_{t,x}(Q \backslash \Gamma_{\alpha r^{1-\varepsilon}})$, when the chain $(\bar{\vartheta}_k, \bar{X}_k)$ leaves the domain $Q \backslash \Gamma_{\alpha r^{1-\varepsilon}}$, satisfies the inequality

$$\bar{\nu}_{t,x}(Q \backslash \Gamma_{\alpha r^{1-\varepsilon}}) \leq \bar{\nu}_{t,x}(Q \backslash \Gamma_{\alpha r}) \,.$$

We shall use the old notation $(\bar{\vartheta}_k, \bar{X}_k)$ for the new Markov chain, which is constructed by the same rules as above but stops in the layer $\Gamma_{\alpha r^{1-\varepsilon}}$ at the new Markov moment $\bar{\nu} = \bar{\nu}_{t,x}(Q \backslash \Gamma_{\alpha r^{1-\varepsilon}})$. We believe that the use of the same notation $(\bar{\vartheta}_k, \bar{X}_k)$ for various Markov chains and $\bar{\nu}$ for various stopping moments will cause no confusion below.

Consider the sequence $(\bar{\vartheta}_k, X_k)$, $k = 0, 1, \ldots$, with X_k:

$$X_0 = x,$$
$$X_1 = X_{t,x}(\bar{\vartheta}_1)$$
$$\cdots \cdots \cdots$$
$$X_k = X_{t,x}(\bar{\vartheta}_k) = X_{\bar{\vartheta}_{k-1}, X_{k-1}}(\bar{\vartheta}_k)$$
$$\cdots \cdots \cdots$$

connected with the system (4.1).

The sequence $(\bar{\vartheta}_k, X_k)$ is a Markov chain, which stops at the random moment $\bar{\nu}$ due to $\bar{\vartheta}_k = \bar{\vartheta}_{\bar{\nu}}$ under $k > \bar{\nu}$.

The following theorem states the closeness of X_k and $\bar{X}_k$ for $N = L/r^2$ steps (see its proof in [204]).

Theorem 4.9. *Let $\bar{\nu} = \bar{\nu}_{t,x}(Q \setminus \Gamma_{\alpha r^{1-\varepsilon}})$, $0 < \varepsilon \leq 1$, be the first exit moment of the Markov chain $(\bar{\vartheta}_i, \bar{X}_i)$, $i = 1, 2, \ldots$, from the domain $Q \setminus \Gamma_{\alpha r^{1-\varepsilon}}$. Then, there exist constants $K > 0$ and $\gamma > 0$ such that for all r small enough the inequality*

$$\left(E \left|X_{N \wedge \bar{\nu}} - \bar{X}_{N \wedge \bar{\nu}}\right|^2\right)^{1/2} = \left(E \left|X_N - \bar{X}_N\right|^2\right)^{1/2} \leq K e^{\gamma L} r$$

holds.

Remark 4.10. In the case of additive noise (see Remarks 4.2 and 4.8) we have faster convergence:

$$\left(E \left|X_{N \wedge \bar{\nu}} - \bar{X}_{N \wedge \bar{\nu}}\right|^2\right)^{1/2} = \left(E \left|X_N - \bar{X}_N\right|^2\right)^{1/2} \leq K e^{\gamma L} r^2$$

under the assumptions of Theorem 4.9.

Theorem 4.11. *Let $\bar{\nu} = \bar{\nu}_{t,x}(Q \setminus \Gamma_{\alpha r^{1-\varepsilon}})$, $0 < \varepsilon \leq 1$, be the first exit moment of the Markov chain $(\bar{\vartheta}_i, \bar{X}_i)$, $i = 1, 2, \ldots$, from the domain $Q \setminus \Gamma_{\alpha r^{1-\varepsilon}}$. Then, there exist constants $K > 0$ and $\gamma > 0$ such that for all r small enough the inequality*

$$\left(E \left|X_{\bar{\nu}} - \bar{X}_{\bar{\nu}}\right|^2\right)^{1/2} \leq K \left(e^{\gamma L} r + e^{-c_r \gamma L/2}\right)$$

holds.

Proof. Introduce two sets $\mathcal{C} = \{\bar{\nu} \leq L/r^2\}$ and $\Omega \setminus \mathcal{C} = \{\bar{\nu} > L/r^2\}$. Let l be a diameter of G. Using Theorems 4.5 and 4.9, we obtain

$$E \left|X_{\bar{\nu}} - \bar{X}_{\bar{\nu}}\right|^2 = E \left(\left|X_{\bar{\nu}} - \bar{X}_{\bar{\nu}}\right|^2; \mathcal{C}\right) + E \left(\left|X_{\bar{\nu}} - \bar{X}_{\bar{\nu}}\right|^2; \Omega \setminus \mathcal{C}\right)$$
$$= E \left(\left|X_{N \wedge \bar{\nu}} - \bar{X}_{N \wedge \bar{\nu}}\right|^2; \mathcal{C}\right) + E \left(\left|X_{\bar{\nu}} - \bar{X}_{\bar{\nu}}\right|^2; \Omega \setminus \mathcal{C}\right)$$
$$\leq E \left|X_N - \bar{X}_N\right|^2 + l^2 P(\Omega \setminus \mathcal{C}) \leq K e^{2\gamma L} r^2 + K e^{-c_r \gamma L}.$$

□

5.4 Approximations for SDEs in a space-time bounded domain   325

Remark 4.12. Let $\bar{\nu} = \bar{\nu}_{t,x}(Q\backslash\Gamma_{\alpha r^{1-\varepsilon}})$, $0 < \varepsilon \leq 1$, be the first exit moment of the Markov chain $(\bar{\vartheta}_i, \bar{X}_i)$, $i = 1, 2, \ldots$, from the domain $Q\backslash\Gamma_{\alpha r^{1-\varepsilon}}$. Then, for every natural m there exist constants $K > 0$ and $\gamma > 0$ such that for all r small enough the following inequalities:

$$E\left|X_N - \bar{X}_N\right|^{2m} \leq K e^{2\gamma L} r^{2m}, \tag{4.22}$$

$$E\left|X_{\bar{\nu}} - \bar{X}_{\bar{\nu}}\right|^{2m} \leq K(e^{2\gamma L} r^{2m} + e^{-c_r \gamma L}), \tag{4.23}$$

and

$$P(\tau_{t,x} < \bar{\vartheta}_N) \leq K e^{2\gamma L} r^{2n}, \quad n = 1, 2, \ldots, \tag{4.24}$$

hold [204].

5.4.3 Approximation of exit point $(\tau, X(\tau))$

In this subsection we are interested in an approximation of the exit point $(\tau_{t,x}, X_{t,x}(\tau_{t,x}))$ of the space-time diffusion $(s, X_{t,x}(s))$, $s \geq t$, from the space-time domain Q. For the sake of simplicity in proofs we restrict ourselves to the case of the convex domain G.

We have $(\bar{\vartheta}_N, \bar{X}_N) = (\bar{\vartheta}_{\bar{\nu}}, \bar{X}_{\bar{\nu}}) \in \Gamma_{\alpha r^{1-\varepsilon}}$ on the set $\mathcal{C} = \{\bar{\nu} \leq L/r^2\}$. Let $(\bar{\tau}_{t,x}, \xi_{t,x})(\omega)$, $\omega \in \mathcal{C}$, be a point on Γ defined as follows. If $\bar{\vartheta}_{\bar{\nu}} \geq T_1 - \alpha r^{1-\varepsilon}$ then $\bar{\tau}_{t,x} = T_1$ and $\xi_{t,x} = \bar{X}_{\bar{\nu}} \in G$, otherwise (i.e., when $\rho(\bar{X}_{\bar{\nu}}, \partial G) \leq \alpha r^{1-\varepsilon}$) $\bar{\tau}_{t,x} = \bar{\vartheta}_{\bar{\nu}}$ and a point $\xi_{t,x} \in \partial G$ is such that

$$\left|\bar{X}_{\bar{\nu}} - \xi_{t,x}\right| \leq \alpha r^{1-\varepsilon}, \quad \omega \in \mathcal{C}. \tag{4.25}$$

To complete the definition of $(\bar{\tau}_{t,x}, \xi_{t,x})(\omega)$ on the set $\Omega\backslash\mathcal{C}$, we put $\bar{\tau}_{t,x}$ be equal to $\bar{\vartheta}_N$ and $\xi_{t,x}$ be a point on ∂G nearest to $\bar{X}_N$. It is natural to take the point $(\bar{\tau}_{t,x}, \xi_{t,x})$ as an approximate one to the exit point $(\tau_{t,x}, X_{t,x}(\tau_{t,x}))$.

Below we need the following lemma [204] (it is analogous to Lemma 1.17).

Lemma 4.13. *There exists a constant $K > 0$ such that for all $(t, x) \in \overline{Q}$ and $y \in \partial G$ the inequalities*

$$E(\tau_{t,x} - t) \leq K|x - y|, \tag{4.26}$$

$$E\left(X_{t,x}(\tau_{t,x}) - y\right)^2 \leq K|x - y| \tag{4.27}$$

are valid.

Theorem 4.14. *Let $\bar{\nu} = \bar{\nu}_{t,x}(Q\backslash\Gamma_{\alpha r^{1-\varepsilon}})$, $0 < \varepsilon \leq 1$, be the first exit moment of the Markov chain $(\bar{\vartheta}_i, \bar{X}_i)$, $i = 1, 2, \ldots$, from the domain $Q\backslash\Gamma_{\alpha r^{1-\varepsilon}}$. Then, there exist positive constants K and γ such that for all r small enough the inequalities*

$$\left[E\left(\left|X_{t,x}(\tau_{t,x}) - \xi_{t,x}\right|^2; \mathcal{C}\right)\right]^{1/2} \leq K r^{(1-\varepsilon)/2} \tag{4.28}$$

and
$$\left[E\left|X_{t,x}(\tau_{t,x}) - \xi_{t,x}\right|^2\right]^{1/2} \leq K(r^{(1-\varepsilon)/2} + e^{-c_r \gamma L/2}) \qquad (4.29)$$
hold.

Proof. Consider the distance between $X_{t,x}(\tau_{t,x})$ and $\xi_{t,x}$ on $\mathcal{C}$:
$$E\left(\left|X_{t,x}(\tau_{t,x}) - \xi_{t,x}\right|^2; \mathcal{C}\right)$$
$$= E\left(\chi_{\bar{\vartheta}_N \geq T_1 - \alpha r^{1-\varepsilon}} \left|X_{t,x}(\tau_{t,x}) - \xi_{t,x}\right|^2; \mathcal{C}\right)$$
$$+ E\left(\chi_{\bar{\vartheta}_N < T_1 - \alpha r^{1-\varepsilon}} \left|X_{t,x}(\tau_{t,x}) - \xi_{t,x}\right|^2; \mathcal{C}\right). \qquad (4.30)$$

We get for the first term of (4.30):
$$E\left(\chi_{\bar{\vartheta}_N \geq T_1 - \alpha r^{1-\varepsilon}} \left|X_{t,x}(\tau_{t,x}) - \xi_{t,x}\right|^2; \mathcal{C}\right)$$
$$= E\chi_{\bar{\vartheta}_N \geq T_1 - \alpha r^{1-\varepsilon}} \left|X_{t,x}(\tau_{t,x}) - \bar{X}_N\right|^2$$
$$\leq 2E\chi_{\bar{\vartheta}_N \geq T_1 - \alpha r^{1-\varepsilon}} \left|X_{t,x}(\tau_{t,x}) - X_N\right|^2 + 2E\left|X_N - \bar{X}_N\right|^2. \qquad (4.31)$$

Due to Theorem 4.9, the second term of (4.31) is estimated by $Ke^{2\gamma L} r^2$, and we have for the first term of (4.31):
$$E\chi_{\bar{\vartheta}_N \geq T_1 - \alpha r^{1-\varepsilon}} \left|X_{t,x}(\tau_{t,x}) - X_N\right|^2$$
$$= E\left|\chi_{\bar{\vartheta}_N \geq T_1 - \alpha r^{1-\varepsilon}}(X_{t,x}(\tau_{t,x}) - X_N)\right|^2$$
$$\leq 2E\left|\int_{\bar{\vartheta}_N \wedge \tau_{t,x}}^{\tau_{t,x}} \chi_{\bar{\vartheta}_N \geq T_1 - \alpha r^{1-\varepsilon}} b(s, X_{t,x}(s)) ds\right|^2$$
$$+ 2E\left|\int_{\bar{\vartheta}_N \wedge \tau_{t,x}}^{\tau_{t,x}} \chi_{\bar{\vartheta}_N \geq T_1 - \alpha r^{1-\varepsilon}} \sigma(s, X_{t,x}(s)) dw(s)\right|^2$$
$$\leq KE\chi_{\bar{\vartheta}_N \geq T_1 - \alpha r^{1-\varepsilon}} \left(\tau_{t,x} - \tau_{t,x} \wedge \bar{\vartheta}_N\right)^2$$
$$+ KE\chi_{\bar{\vartheta}_N \geq T_1 - \alpha r^{1-\varepsilon}} \left(\tau_{t,x} - \tau_{t,x} \wedge \bar{\vartheta}_N\right)$$
$$\leq KE\chi_{\bar{\vartheta}_N \geq T_1 - \alpha r^{1-\varepsilon}}(T_1 - \bar{\vartheta}_N)^2 + KE\chi_{\bar{\vartheta}_N \geq T_1 - \alpha r^{1-\varepsilon}}(T_1 - \bar{\vartheta}_N) \leq K r^{1-\varepsilon},$$

whence it follows that
$$E\left(\chi_{\bar{\vartheta}_N \geq T_1 - \alpha r^{1-\varepsilon}} \left|X_{t,x}(\tau_{t,x}) - \xi_{t,x}\right|^2; \mathcal{C}\right) \leq K r^{1-\varepsilon}. \qquad (4.32)$$

Consider the second term of (4.30). Due to its definition, the point $\xi_{t,x}(\omega)$, $\omega \in \mathcal{C}$, belongs to ∂G if $\bar{\vartheta}_N < T_1 - \alpha r^{1-\varepsilon}$. Then by the conditional version of Lemma 4.13, we get (note that $\xi_{t,x}$ is measurable with respect to $\mathcal{F}_N$)

$$E\left(\chi_{\bar{\vartheta}_N<T_1-\alpha r^{1-\varepsilon}}\left|X_{t,x}(\tau_{t,x})-\xi_{t,x}\right|^2;\mathcal{C}\right)$$
$$=E\left(\chi_{\bar{\vartheta}_N<T_1-\alpha r^{1-\varepsilon}}E\left(\left|X_{\bar{\vartheta}_N,X_N}(\tau_{\bar{\vartheta}_N,X_N})-\xi_{t,x}\right|^2/\mathcal{F}_N\right);\mathcal{C}\right)$$
$$\leq KE\left(\chi_{\bar{\vartheta}_N<T_1-\alpha r^{1-\varepsilon}}\left|X_N-\xi_{t,x}\right|;\mathcal{C}\right).$$

Theorem 4.9 and the inequality (4.25) imply

$$E\left(\chi_{\bar{\vartheta}_N<T_1-\alpha r^{1-\varepsilon}}\left|X_N-\xi_{t,x}\right|;\mathcal{C}\right)$$
$$\leq\left[E\left(\chi_{\bar{\vartheta}_N<T_1-\alpha r^{1-\varepsilon}}\left|X_N-\xi_{t,x}\right|^2;\mathcal{C}\right)\right]^{1/2}$$
$$\leq\left[2E\left|X_N-\bar{X}_N\right|^2+2\left(E\chi_{\bar{\vartheta}_N<T_1-\alpha r^{1-\varepsilon}}\left|\bar{X}_N-\xi_{t,x}\right|^2;\mathcal{C}\right)\right]^{1/2}$$
$$\leq Ke^{\gamma L}r+2\alpha r^{1-\varepsilon}\leq Kr^{1-\varepsilon}. \tag{4.33}$$

Thus,

$$E\left(\chi_{\bar{\vartheta}_N<T_1-\alpha r^{1-\varepsilon}}\left|X_{t,x}(\tau_{t,x})-\xi_{t,x}\right|^2;\mathcal{C}\right)\leq K\,r^{1-\varepsilon}.$$

Substituting this inequality and the inequality (4.32) in (4.30), we get (4.28).
The inequality (4.29) is obtained by Theorem 4.5 analogously to the proof of Theorem 4.11. Theorem 4.14 is proved. □

Theorem 4.15. *Under the assumptions of Theorem 4.14, the inequalities*

$$E\left(\left|\tau_{t,x}-\bar{\tau}_{t,x}\right|;\mathcal{C}\right)\leq K\,r^{1-\varepsilon}, \tag{4.34}$$

$$E\left|\tau_{t,x}-\bar{\tau}_{t,x}\right|\leq K(r^{1-\varepsilon}+e^{-\alpha_r\gamma L}) \tag{4.35}$$

hold.

Proof. Recall that $\tau_{t,x}\leq T_1$, $\bar{\vartheta}_N\leq T_1$. Further, $\bar{\tau}_{t,x}=T_1$ under $\bar{\vartheta}_N\geq T_1-\alpha r^{1-\varepsilon}$ and $\bar{\tau}_{t,x}=\bar{\vartheta}_N$ otherwise. Consequently, $\bar{\tau}_{t,x}\geq\bar{\vartheta}_N$. Let below $\tau:=\tau_{t,x}$, $\bar{\tau}:=\bar{\tau}_{t,x}$.

Consider the difference $|\tau-\bar{\tau}|$ on the set $\mathcal{C}$. We have

$$E\left(|\tau-\bar{\tau}|;\mathcal{C}\right)=E\left((\bar{\tau}-\tau\wedge\bar{\tau});\mathcal{C}\right)+E\left((\tau-\tau\wedge\bar{\tau});\mathcal{C}\right). \tag{4.36}$$

We get for the first term:

$$E\left((\bar{\tau}-\tau\wedge\bar{\tau});\mathcal{C}\right)\leq E(\bar{\tau}-\tau\wedge\bar{\tau})=E\chi_{\tau<\bar{\tau}}(\bar{\tau}-\tau\wedge\bar{\tau})$$
$$=E\chi_{\tau<\bar{\vartheta}_N}(\bar{\tau}-\tau\wedge\bar{\tau})+E\chi_{\bar{\vartheta}_N\leq\tau<\bar{\tau}}(\bar{\tau}-\tau\wedge\bar{\tau})$$
$$\leq(T_1-T_0)P(\tau<\bar{\vartheta}_N)+E\chi_{T_1-\alpha r^{1-\varepsilon}\leq\tau<T_1}(T_1-\tau).$$

Then using (4.24) under $n=1$, we obtain

$$E\left((\bar{\tau}-\tau\wedge\bar{\tau});\mathcal{C}\right)\leq Ke^{2\gamma L}\cdot r^2+\alpha r^{1-\varepsilon}\leq Kr^{1-\varepsilon}. \tag{4.37}$$

Consider the second term of (4.36). Due to $\xi_{t,x} \in \partial G$ under $\bar{\vartheta}_N < T_1 - \alpha r^{1-\varepsilon}$, Lemma 4.13, and the inequality (4.33), we get

$$E\left((\tau - \tau \wedge \bar{\tau}); \mathcal{C}\right) = E\left(\chi_{\bar{\tau} < \tau}(\tau - \tau \wedge \bar{\tau}); \mathcal{C}\right)$$
$$= E\left(\chi_{\bar{\vartheta}_N < \tau} \chi_{\bar{\vartheta}_N < T_1 - \alpha r^{1-\varepsilon}}(\tau - \tau \wedge \bar{\vartheta}_N); \mathcal{C}\right)$$
$$= E\left(\chi_{\bar{\vartheta}_N < T_1 - \alpha r^{1-\varepsilon}}(\tau_{\bar{\vartheta}_N, X_N} - \bar{\vartheta}_N); \mathcal{C}\right)$$
$$= E\left(\chi_{\bar{\vartheta}_N < T_1 - \alpha r^{1-\varepsilon}} E(\tau_{\bar{\vartheta}_N, X_N} - \bar{\vartheta}_N / \mathcal{F}_N); \mathcal{C}\right)$$
$$\leq KE\left(\chi_{\bar{\vartheta}_N < T_1 - \alpha r^{1-\varepsilon}} |X_N - \xi_{t,x}|; \mathcal{C}\right) \leq K r^{1-\varepsilon}.$$

Substituting this inequality and the inequality (4.37) in (4.36), we get (4.34).

The inequality (4.35) is obtained by Theorem 4.5 analogously to the proof of Theorem 4.11. Theorem 4.15 is proved. □

5.4.4 Simulation of space-time Brownian motion with drift

In this section we have dealt with the one-step approximation $(t+\bar{\theta}, \bar{X}_{t,x}(t+\bar{\theta}))$, $(t,x) \in Q \backslash \Gamma_{\alpha r}$ (see (4.4)), which is based on the simulation of the exit point $(\bar{\theta}, w(t+\bar{\theta}) - w(t))$ of the process $(s-t, w(s) - w(t))$, $s > t$, from the space-time parallelepiped $\Pi_{r,l} = [0, lr^2] \times C_r$ with the cubic base C_r.

It is possible to derive other constructive one-step approximations. In this subsection we briefly consider a one-step approximation based on a simulation of exit points for the Brownian motion with drift $W_\mu(s)$:

$$W_\mu(s) = \mu s + W(s), \ W_\mu(0) = 0,$$

where μ is a d-dimensional fixed vector and $W(s)$ is a d-dimensional standard Wiener process.

If $(\bar{\theta}, w_\mu(t+\bar{\theta}) - w_\mu(t))$ is the first exit point of the process $(s-t, w_\mu(s) - w_\mu(t))$, $s > t$, under $\mu = \sigma^{-1}(t,x)b(t,x)$, $(t,x) \in Q \backslash \Gamma_{\alpha r}$, from the space-time parallelepiped $[0,l] \times C_r$, $l \leq T_1 - t$, then it is easy to see that the approximation

$$\bar{X}_{t,x}(t+\bar{\theta}) = x + \sigma(t,x)(w_\mu(t+\bar{\theta}) - w_\mu(t)) \tag{4.38}$$

belongs to the space parallelepiped $\overline{C}_r^\sigma(x)$ even under not small l.

Then we are able to ensure belonging of $\bar{X}_{t,x}(t+\bar{\theta})$ to $\bar{G}$, and, consequently, $(t+\bar{\theta}, \bar{X}_{t,x}(t+\bar{\theta}))$ to $\bar{Q}$, but the smallness of time size of the space-time parallelepiped $[0,l] \times C_r$ is already not required in contrast to the approximation (4.4).

The approximation (4.38) is more universal than the approximation (4.4). However, the approximation (4.4) is simpler in a computational sense than (4.38) and is quite appropriate for the majority of problems.

Algorithms on simulating exit points for the Brownian motion with drift $W_\mu(s)$ are available in [204].

5.5 Numerical examples

The numerical methods proposed in this chapter are widely applicable. These methods are the first ones which can constructively approximate trajectories of a diffusion process in bounded domains. They can also be applied to solving boundary value problems through a Monte Carlo technique on a level with the weak methods. Let us underline that the methods from this chapter give an estimator for a solution to the Dirichlet problem for parabolic and elliptic equations with constant coefficients which do not contain the error of numerical integration in comparison with the weak methods (see the next chapter).

Here we give three numerical examples. The first and the second examples deal with solving boundary value problems. In the second example an elliptic problem is considered, nevertheless, we need the simulation of the space-time exit points. The third example concerns the stability analysis of linear autonomous system of SDEs and uses simulation of space-time trajectories essentially.

Example 5.1. Let us consider an application of random walks over touching space-time parallelepipeds to the Dirichlet problem for parabolic equation (3.1)-(3.3) in the case when the coefficients are constant. This problem has the probabilistic representation (3.6)-(3.7), which we use for the Monte Carlo procedure here.

Let $(\bar{\vartheta}_k, \bar{X}_k)$ be a Markov chain which is formed analogously to the one of Sect. 5.4.2 but wandering is realized over touching space-time parallelepipeds (instead of small space-time parallelepipeds in Sect. 5.4.2) and is finished in the layer Γ_δ at a random step $\bar{\nu}$, where $\delta > 0$ is a sufficiently small constant. The equation with frozen coefficients (4.4), which we are able to simulate exactly, coincides with (3.7) when its coefficients are constant. Consequently, the chain $(\bar{\vartheta}_k, \bar{X}_k)$ coincides with the chain $(\bar{\vartheta}_k, X_k)$. In the considered case, the solution $u(t,x)$ to the Dirichlet problem (3.1)-(3.3) under $c = 0$ and $e = 0$ is simulated as (see (3.6))

$$u(t,x) \doteq \bar{u}(t,x) = \frac{1}{M} \sum_{m=1}^{M} \varphi(\bar{X}_{\bar{\nu}}^{(m)}) \pm 2[\bar{D}/M]^{1/2},$$

where

$$\varphi(\bar{X}_{\bar{\nu}}^{(m)}) = \begin{cases} f(\bar{X}_{\bar{\nu}}^{(m)}), & \bar{\vartheta}_{\bar{\nu}}^{(m)} \in (T_1 - \delta, T_1], \\ g(\bar{X}_{\bar{\nu}}^{(m)}), & \bar{\vartheta}_{\bar{\nu}}^{(m)} \notin (T_1 - \delta, T_1], \end{cases}$$

$$\bar{D} = \frac{1}{M} \sum_{m=1}^{M} \left[\varphi(\bar{X}_{\bar{\nu}}^{(m)}))\right]^2 - \left[\frac{1}{M} \sum_{m=1}^{M} \varphi(\bar{X}_{\bar{\nu}}^{(m)})\right]^2,$$

and M is a number of independent Markov chains $\left(\bar{\vartheta}_k^{(m)}, \bar{X}_k^{(m)}\right)$, $m = 1, \ldots, M$.

Table 5.1. Test results for the boundary value problem (5.1)-(5.3). The exact solution $u(1, 0.7, 0.4) = 0.4796$ ($\delta = 0.00001$).

M	$\bar{u}(1, 0.7, 0.4) \pm 2[\bar{D}/M]^{1/2}$	$E\bar{\nu}$
1000	0.4460 ± 0.0527	3.142
4000	0.4780 ± 0.0270	3.257
10^5	0.4782 ± 0.0054	3.272

Because the simulated values $(\bar{\vartheta}_k, \bar{X}_k)$ coincide with the points of exact solution (ϑ_k, X_k) here, the estimator $\bar{u}(t, x)$ does not contain the error of numerical integration (naturally, there are Monte Carlo error depending on M and the error due to approximation of the boundary conditions depending on δ).

The mean number of steps of the random walk over touching spheres up to the boundary of space domain G is estimated by $C \ln(l/2\delta)$ (see, e.g., [58, 257] and also Theorem 6.4.12 in Sect. 6.4.3), if G is convex and l is its diameter. In our case the value of $\bar{\nu}$ is also estimated by $C \ln(l/2\delta)$.

Another Monte Carlo approach, whereby a random walk is made on a maximum square and the differential Laplace operator is approximated by a difference one, was proposed in [107].

As an illustration, we take the following heat equation in the domain $Q = [0, T_1) \times G$, $G = \{x = (x_1, x_2) : |x_1| < 2, |x_2| < 1\}$ (this example is similar to one in [107]):

$$\frac{\partial u}{\partial t} = \frac{1}{2} \Delta u, \ t > 0, \ |x_1| < 2, \ |x_2| < 1, \tag{5.1}$$

with the initial and boundary conditions

$$u(0, x) = 2, \tag{5.2}$$

$$u(t, x)|_{\partial G} = 0, \ t > 0. \tag{5.3}$$

By changing time $t = T_1 - s$ in (5.1)-(5.3), we obtain the corresponding boundary value problem (like (3.1)-(3.3)) with the initial condition on the upper base. The results of numerical test are presented in Table 5.1.

Example 5.2. Consider the boundary value problem for biharmonic equation

$$L^2 u + c_1(x) L u + c_2(x) u = f(x), \ x \in G \subset \mathbf{R}^d, \tag{5.4}$$

$$u|_{\partial G} = \varphi(x), \ L u|_{\partial G} = \psi(x), \tag{5.5}$$

where L is an operator of elliptic type:

$$L = \frac{1}{2} \sum_{i,j=1}^{d} a^{ij}(x) \frac{\partial^2}{\partial x^i \partial x^j} + \sum_{i=1}^{d} b^i(x) \frac{\partial}{\partial x^i},$$

and $c_1(x)$, $c_2(x)$, $f(x)$, $\varphi(x)$, and $\psi(x)$ are some known functions.

Introducing the function $v = Lu$, we obtain the system of elliptic equations

$$Lu - v = 0, \ x \in G, \ u\,|_{\partial G} = \varphi(x), \tag{5.6}$$

$$Lv + c_1(x)v + c_2(x)u = f(x), \ x \in G, \ v\,|_{\partial G} = \psi(x). \tag{5.7}$$

Let us give a probabilistic representation of the solution to the problem (5.6)-(5.7) (the first probabilistic representation for the problem (5.6)-(5.7) in the case of constant c_1 and c_2 is obtained in [123]). To this end introduce the system of SDEs

$$dX = b(X)\,ds + \sigma(X)\,dw(s), \tag{5.8}$$

$$\frac{dY_1}{ds} = c_2(X)Y_2$$
$$\frac{dY_2}{ds} = -Y_1 + c_1(X)Y_2, \tag{5.9}$$

where $w(s)$ is a standard d-dimensional Wiener process, $b(x)$ is the d-dimensional vector with the components $b^i(x)$, Y_1 and Y_2 are scalars, and $\sigma(x)$ is a matrix that is obtained from the equality

$$a(x) = \sigma(x)\sigma^\mathsf{T}(x), \ a(x) = \{a^{ij}(x)\}.$$

Under some conditions on the coefficients of the problem (5.6)-(5.7), its solution $(u(x), v(x))$ has the following form (see [177]):

$$u(x) = E\left[\varphi(X_x(\tau))Y_1^{(1)}(\tau) + \psi(X_x(\tau))Y_2^{(1)}(\tau)\right]$$
$$- E\int_0^\tau f(X_x(s))Y_2^{(1)}(s)\,ds,$$
$$v(x) = E\left[\varphi(X_x(\tau))Y_1^{(2)}(\tau) + \psi(X_x(\tau))Y_2^{(2)}(\tau)\right]$$
$$- E\int_0^\tau f(X_x(s))Y_2^{(2)}(s)\,ds, \tag{5.10}$$

where τ is the first exit time of the process $X_x(s)$, $X(0) = x$, from the domain G, and $(Y_1^{(1)}, Y_2^{(1)})$ is the solution of the system (5.9) with initial data $Y_1^{(1)}(0) = 1$, $Y_2^{(1)}(0) = 0$, and $(Y_1^{(2)}, Y_2^{(2)})$ has the initial data $Y_1^{(2)}(0) = 0$, $Y_2^{(2)}(0) = 1$.

The probabilistic representation (5.8)-(5.10) for the boundary value problem (5.4)-(5.5) can be used for solving the problem (5.4)-(5.5) by implementation of the random walk over small space-time parallelepipeds through the Monte Carlo technique. If the coefficients of the elliptic operator L and the scalars c_1, c_2, f are constant, we can use the random walk over touching space parallelepipeds that gives an estimator which is free from the error of numerical integration. Note that in this case a sufficient condition, under which the representation (5.10) is valid, consists in $c_1 \leq 0$, $c_2 \geq 0$.

As an illustration, consider the following two-dimensional problem in the square $G = \{x = (x_1, x_2) : |x_1| < 1, |x_2| < 1\}$:

$$\frac{1}{4}\Delta^2 u = 1, \ x \in G, \tag{5.11}$$

$$u \mid_{\partial G} = \varphi(x), \ \varphi(x_1, \pm 1) = \frac{1 + x_1^4}{12}, \ \varphi(\pm 1, x_2) = \frac{1 + x_2^4}{12},$$

$$\frac{1}{2}\Delta u \mid_{\partial G} = \psi(x), \ \psi(x_1, \pm 1) = \frac{1 + x_1^2}{2}, \ \psi(\pm 1, x_2) = \frac{1 + x_2^2}{2}. \tag{5.12}$$

Introducing the function $v = \frac{1}{2}\Delta u$ as above, we obtain the system of elliptic equations

$$\frac{1}{2}\Delta u - v = 0, \ x \in G, \ u \mid_{\partial G} = \varphi(x) \tag{5.13}$$

$$\frac{1}{2}\Delta v = 1, \ x \in G, \ v \mid_{\partial G} = \psi(x). \tag{5.14}$$

Its exact solution is

$$u(x) = \frac{x_1^4 + x_2^4}{12}, \ v(x) = \frac{x_1^2 + x_2^2}{2}.$$

Of course, one can solve the problem (5.13)-(5.14) sequentially: first find the function v from the problem (5.14) and then u from (5.13). But such an approach requires the knowledge of the function v in the whole domain G even if one needs the solution (u, v) only at individual points of the domain G. In the last case, the Monte Carlo approach is more preferable.

For the system (5.13)-(5.14), the formulas (5.8)-(5.10) acquire the form

$$u(x) = E\varphi(x + w(\tau)) - E[\tau\psi(x + w(\tau))] + \frac{1}{2}E\tau^2,$$

$$v(x) = E\psi(x + w(\tau)) - E\tau,$$

where τ is the first exit time of the process $x + w(s)$ from the domain G.

To simulate the point $(\tau, x + w(\tau))$, we use the random walk over touching space squares, which is finished in a δ-neighborhood of the boundary ∂G belonging to G. Recall that we are able to exactly simulate both the exit point and the exit time of the Wiener process from a square in accordance with Theorem 3.7. Then for the same reasons as in Example 5.1, the corresponding estimator $(\bar{u}, \bar{v})$ does not contain the error of numerical integration. The notice on the mean number of steps $E\bar{\nu}$ from Example 5.1 is also valid here. Let us underline that the method of random walk over touching spheres in the space domain G cannot be applied to this problem, because we essentially

Table 5.2. Test results for the boundary value problem (5.11)-(5.12) ($\delta = 0.00001$).

M	x_1	x_2	$u(x_1, x_2)$	$\bar{u}(x_1, x_2)$	$v(x_1, x_2)$	$\bar{v}(x_1, x_2)$	$E\bar{\nu}$
10^4	0.3	0.5	0.00588	0.0051 ± 0.0037	0.17000	0.1656 ± 0.0082	4
10^5				0.0059 ± 0.0012		0.1698 ± 0.0026	4
10^6				0.0058 ± 0.0004		0.1700 ± 0.0008	4
10^4	0.7	0.8	0.05414	0.0539 ± 0.0022	0.56500	0.5598 ± 0.0064	4
10^5				0.0541 ± 0.0006		0.5638 ± 0.0020	4
10^6				0.0542 ± 0.0002		0.5646 ± 0.0006	4
10^4	0.9	0.9	0.10935	0.1090 ± 0.0011	0.81000	0.8067 ± 0.0039	3
10^5				0.1092 ± 0.0003		0.8088 ± 0.0012	3
10^6				0.1093 ± 0.0001		0.8097 ± 0.0004	3

use the simulation of both the exit point $x + w(\tau)$ and the exit time τ. The results of numerical tests are given in Table 5.2.

See Appendix A.3 for an implementation of the random walk over touching space squares. By the program from Listing A.9 the results presented in Table 5.2 were obtained.

Example 5.3. Consider the second-order linear autonomous Ito system of SDEs

$$dX = AX\, dt + \sum_{i=1}^{2} B_i X\, dw_i(t), \qquad (5.15)$$

where X is a two-dimensional vector, A and B_i, $i = 1, 2$, are constant 2×2-matrices, $w_i(t)$, $i = 1, 2$, are independent standard Wiener processes.

Various characteristics describing asymptotic behavior of solutions of the system (5.15), such as the Lyapunov exponent, the moment Lyapunov exponents, the stability index, and some others, are considered in [5, 6, 124] (see also references therein). The Lyapunov exponent λ^* of system (5.15) (cf. [124]) is defined as

$$\lambda^* := \lim_{t \to \infty} \frac{1}{t} E \ln |X_x(t)| = \lim_{t \to \infty} \frac{1}{t} \ln |X_x(t)| \text{ a.s.}, \qquad (5.16)$$

and the moment Lyapunov exponent $g(p)$ is defined as

$$g(p) := \lim_{t \to \infty} \frac{1}{t} E \ln |X_x(t)|^p, \ p \in \mathbf{R}, \qquad (5.17)$$

where $X_x(t)$, $t \geq 0$, is a nontrivial solution to system (5.15).

The limits λ^* and $g(p)$ exist, and they are independent of x, $x \neq 0$, in the ergodic case. The limit $g(p)$ is a convex analytic function of $p \in \mathbf{R}$, $g(0) = 0$, $g(p)/p$ increases with growing p, and

$$g'(0) = \lambda^*. \tag{5.18}$$

If $\lambda^* < 0$ then the trivial solution to system (5.15) is a.s. asymptotically stable. It is well known and follows from (5.18) that in this case $g(p)$ is negative for all sufficiently small $p > 0$, i.e., the solution $X = 0$ of (5.15) is p-stable for such p. If $g(p) \to +\infty$ as $p \to +\infty$, then the equation

$$g(p) = 0 \tag{5.19}$$

has the unique root $\gamma^* > 0$, which is known as the stability index.

It is clear that the solution $X = 0$ of (5.15) is p-stable for $0 < p < \gamma^*$ and p-unstable for $p > \gamma^*$. The stability index γ^* is connected with the asymptotic behavior of the probability $V_\delta(x)$ of the exit of $X_x(t)$ from the ball $|x| < \delta$ (see [15]): $V_\delta(x) := P\{\sup_{t\geq 0} |X_x(t)| > \delta\}$, $|x|/\delta \to 0$. It turns out that there exists a constant $K > 0$ such that for all $\delta > 0$ and $|x| < \delta$ the following inequality takes place:

$$\frac{1}{K}(|x|/\delta)^{\gamma^*} \leq V_\delta(x) \leq K\,(|x|/\delta)^{\gamma^*}. \tag{5.20}$$

The unstable case, when the equation (5.19) has a negative root γ^*, is considered analogously [15].

The stability properties of the system (5.15) can also be characterized by the exit time τ of $X_x(t)$ from a certain neighborhood of the origin. In [141] the value of $Ee^{-\mu\tau}$, $\mu > 0$, is simulated. By the algorithms proposed in this chapter, we are able to evaluate the distribution function $P(\tau < t)$, which may be a good characteristic for description of transient behavior related to the system (5.15). Naturally, we are also able to evaluate functionals on τ, e.g., $Ee^{-\mu\tau}$.

We take the following particular case of the two-dimensional system (5.15) for our numerical tests:

$$dX_1 = (aX_1 + cX_2)\,ds + b_1 X_1\,dw_1(s) + b_2 X_2\,dw_2(s)$$
$$dX_2 = (-cX_1 + aX_2)\,ds + b_1 X_2\,dw_1(s) - b_2 X_1\,dw_2(s),$$
$$X(0) = X_x(0) = x. \tag{5.21}$$

The function $g(p)$, the Lyapunov exponent λ^*, and the stability index γ^* for this system are equal to [185]:

$$g(p) = p \times (a + \frac{1}{2}(b_2^2 - b_1^2)) + \frac{1}{2}p^2 b_1^2,$$
$$\lambda^* = g'(0) = a + \frac{1}{2}(b_2^2 - b_1^2),$$
$$\gamma^* = -\frac{2a + (b_2^2 - b_1^2)}{b_1^2}. \tag{5.22}$$

Here we evaluate the distribution function $P(\tau < t)$, where τ is the first exit time of $X_x(s)$ under $X(0) = (1,1)^\intercal$ from the square $G = \{(x_1, x_2):$

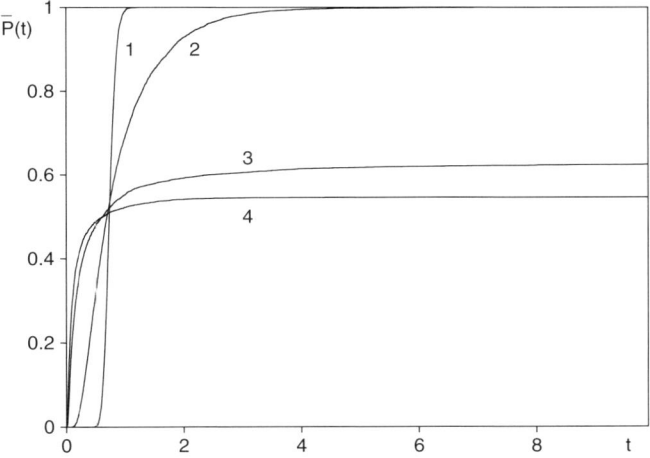

Fig. 5.1. The distribution function $\bar{P}(t)$ for $a = -1$, $c = 1$, $b_2 = 2$, $X(0) = (1,1)^\mathsf{T}$, $r = 0.02$, $M = 5000$, and for various b_1: (1) $b_1 = 0.1$ ($\lambda^* = 0.995$, $\gamma^* = -199$), (2) $b_1 = 0.6$ ($\lambda^* = 0.82$, $\gamma^* = -4.556$), (3) $b_1 = \sqrt{5}$ ($\lambda^* = -1.5$, $\gamma^* = 0.6$), and (4) $b_1 = 3$ ($\lambda^* = -3.5$, $\gamma^* = 0.778$).

$|x_i| < 3$, $i = 1, 2\}$. To simulate the system (5.21), we use the random walk over boundaries of small space-time parallelepipeds constructed in Sect. 5.4.2. The algorithm allows finding $\bar{\tau}$ (see Sect. 5.4.3), which is close to τ. The sampling distribution function $\bar{P}_M(t)$ is calculated as

$$\bar{P}_M(t) = \begin{cases} 0, & t \leq \bar{\tau}_1^{(M)}, \\ m/M, & \bar{\tau}_m^{(M)} < t \leq \bar{\tau}_{m+1}^{(M)}, \\ 1, & t > \bar{\tau}_M^{(M)}, \end{cases}$$

where $\{\bar{\tau}_1^{(M)}, \ldots, \bar{\tau}_M^{(M)}\}$ is a sample point of size M sorting in the ascending order, it corresponds to the random variable $\bar{\tau}$.

The sampling function $\bar{P}_M(t)$ is close to the distribution function $\bar{P}(t) = P(\bar{\tau} < t)$ under a sufficiently big M, and $\bar{P}(t)$ is close to $P(\tau < t)$ under a sufficiently small r. We control the accuracy of our simulations by increasing M and decreasing r. We select M and r such that the curves $\bar{P}_M(t)$ are visually almost identical under larger values of M and smaller values of r. The parameter l of space-time parallelepipeds $\Pi_{r,l}$ used in the simulations is taken equal to 1.

Figure 5.1 presents the behavior $\bar{P}(t) \doteq \bar{P}_M(t)$ under fixed a, c, b_2, and various b_1. Increase of b_1 leads to stabilization (see formulas (5.22)). It is interesting to note (see Fig. 5.1) that the probability of the exit of $X_x(s)$ from G at small times t under $\lambda^* > 0$ (unstable case) is lower than the

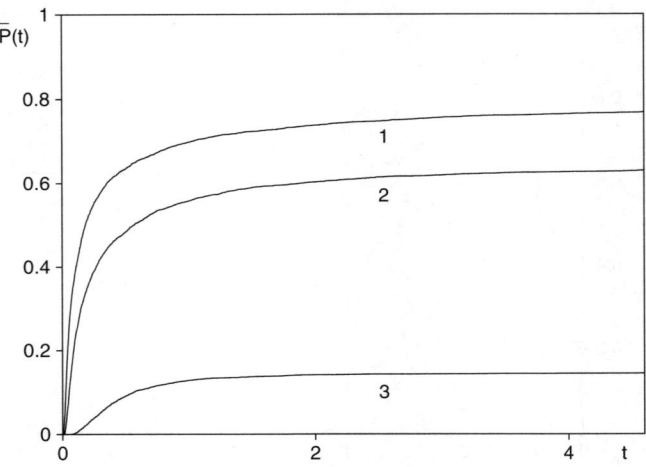

Fig. 5.2. The distribution function $\bar{P}(t)$ for $a = -1$, $c = 1$, $X(0) = (1,1)^\mathsf{T}$, $\lambda^* = -1.5$, $M = 5000$, and for various γ^*: (1) $\gamma^* = 1/3$ ($b_1 = 3$, $b_2 = 2.828$, $r = 0.02$), (2) $\gamma^* = 0.6$ ($b_1 = \sqrt{5}$, $b_2 = 2$, $r = 0.02$), and (3) $\gamma^* = 2.479$ ($b_1 = 1.1$, $b_2 = 0.4683$, $r = 0.05$).

corresponding probability under $\lambda^* < 0$ (stable case). It may be explained in the following way. The radius $\rho(s) = \sqrt{X_1^2(s) + X_2^2(s)}$ satisfies the equation

$$d\rho = \left(a + \frac{b_2^2}{2}\right)\rho\, ds + b_1 \rho\, dw_1(s). \qquad (5.23)$$

Due to the selection of the parameters, the Lyapunov exponent λ^* is positive (unstable case) under relatively small b_1 and large b_2. In this case, the first term of (5.23) plays the main role and the influence of noise is relatively small. So there is a lag time before the trajectory $X_x(s)$ leaves the domain G. In the stable case our parameters are such that b_1 is large and the second term of (5.23) plays an essential role. Then the trajectory $X_x(s)$ can leave the domain G during a small time interval with a rather large probability.

Figure 5.2 illustrates the behavior of $\bar{P}(t)$ under fixed a, c, and $\lambda^* = a + (b_2^2 - b_1^2)/2$ for various values of the stability index γ^* (see (5.22)). One can see that the probability of the exit of the trajectory $X_x(s)$ from G decreases with increasing of γ^* in accordance with (5.20).

Figures 5.1 and 5.2 also demonstrate that in the unstable case the trajectory leaves the neighborhood of the origin during a finite time interval with the probability equal to 1 (see the curves 1 and 2 on Fig. 5.1). However, in the stable case, the probability $P(\tau < \infty)$ of leaving the neighborhood of the origin by the trajectory is less than 1. This probability decreases with decrease of the Lyapunov exponent λ^* (see the curves 3 and 4 on Fig. 5.1) and with increase of the stability index γ^* (see Fig. 5.2).

5.6 Mean-square approximation of diffusion with reflection

Consider an autonomous system of SDEs with normal reflection at the boundary

$$dX = b(X)\,I_G(X)\,ds + \sigma(X)\,I_G(X)\,dw(s) - I_{\partial G}(X)\,d\mu(s), \quad X(0) = x, \quad (6.1)$$

in a convex domain G in $\mathbf{R}^d$ with nonempty interior. Here $w(t)$, $t \geq 0$, is a standard d-dimensional Wiener process, $b(x)$ is a d-dimensional vector, $\sigma(x)$ is a $d \times d$-dimensional matrix, $\mu(t)$, $t \geq 0$, is the local time of the process X on the boundary ∂G of the domain G, and $I_A(x)$ is the indicator function of a set A. The domain G may be unbounded. For instance, G can be a convex polyhedron, i.e., $G = \cap_{i=1}^n G_i$, where G_i are half-spaces; in particular, G can be a half-space.

We recall that the local time is a continuous, nondecreasing random process which increases on the set $\{t \geq 0,\ X(t) \in \partial G\}$ only. The Lebesgue measure of this set is zero. A tutorial on processes with reflection is available in, e.g., [70, Section 1.6].

We briefly observe two types of mean-square methods for (6.1): the projection scheme [35, 154, 234, 235, 267, 268] and the penalization scheme [157, 235, 268].

Introduce the projection map onto $\bar{G}$:

$$\Pi(x) = \arg\min_{y \in \bar{G}} |x - y|, \quad x \in \mathbf{R}^d.$$

We note that if $x \in \bar{G}$ then $\Pi(x) = x$.

The projection scheme can be written in the form

$$X_{k+1} = \Pi\left(X_k + h\,b(X_k) + \sigma(X_k)\Delta w_k\right), \quad (6.2)$$

where, as usual, h is a time step and $\Delta w_k := w(t_{k+1}) - w(t_k)$. It is clear that according to this scheme we have the standard mean-square Euler approximations (see Sect. 1.1.5) while a point $\hat{X}_{k+1} = X_k + h\,b(X_k) + \sigma(X_k)\Delta w_k$ is inside the domain $\bar{G}$. If $\hat{X}_{k+1} \notin \bar{G}$ then the next value X_{k+1} is taken as the projection of $\hat{X}_{k+1}$ on the boundary ∂G.

To introduce the penalization scheme, we need an additional notation:

$$\beta(x) := x - \Pi(x), \quad \beta_\lambda(x) := \beta(x)/\lambda, \quad x \in \mathbf{R}^d.$$

We note that $\beta(x) = 0$ for $x \in \bar{G}$. Consider the system of SDEs
$$dX_\lambda = b(X_\lambda)ds + \sigma(X_\lambda)\,dw(s) - \beta_\lambda(X_\lambda)\,ds, \quad X(0) = x. \tag{6.3}$$
It is known [172] that for any $0 < \alpha < 1$ the estimate
$$E \sup_{t \in [0,T]} |X(t) - X_\lambda(t)|^2 \leq C\lambda^{1-\alpha} \tag{6.4}$$
holds under some conditions on the coefficients of (6.1).

Approximating the solution of (6.3) by the standard mean-square Euler scheme and choosing $\lambda = h$, one obtains the penalization scheme for (6.1):
$$\begin{aligned}X_{k+1} &= X_k + h\,b(X_k) - h\,b_\lambda(X_k) + \sigma(X_k)\,\Delta w_k \\ &= \Pi(X_k) + h\,b(X_k) + \sigma(X_k)\,\Delta w_k.\end{aligned} \tag{6.5}$$

The choice $\lambda = h$ in (6.5) can be explained in the following way (further details are available in [235]). Suppose we have applied the mean-square Euler method to (6.3). Then the penalization scheme contains the penalty term $h\beta_\lambda(X_k)$. If $\lambda < h$, the penalty term may push X_{k+1} inwards too much for G because when this term is nonzero it is generically of order $h^{3/2}/\lambda$ which, if $\lambda = h^{1+\varepsilon}$, $\varepsilon > 0$, is too large compared with the increments of X which are of order $h^{1/2}$. For small $h \leq \lambda$ the mean-square error of the penalization scheme is estimated as $O([h\ln(1/h)]^{1/4} + \lambda^{1/2-\alpha})$ for any $\alpha > 0$ [235] (cf. (6.4) and Theorem 6.1). For $\lambda = h^{1/2}$ the error of the method is $O(h^{1/4-\alpha})$ for any $\alpha > 0$ while for $\lambda = h$ the estimates given in Theorem 6.1 are valid.

The following convergence theorem was proved in [268] (see also [234, 235]).

Theorem 6.1. *Assume that the coefficients of (6.1) satisfy a Lipschitz condition. Let $\bar{X}(t) := X_k$ for $t \in [t_k, t_{k+1})$, where X_k are obtained by either the projection scheme (6.2) or by the penalization scheme (6.5). Then for every $T > 0$ and natural number p the following estimates hold:*
(i) in the case of a general convex domain G:
$$\left(E \sup_{t \in [0,T]} |X(t) - \bar{X}(t)|^{2p}\right)^{1/2p} = O([h\ln(1/h)]^{1/4}),$$
(ii) in the case of a convex polyhedron G:
$$\left(E \sup_{t \in [0,T]} |X(t) - \bar{X}(t)|^{2p}\right)^{1/2p} = O([h\ln(1/h)]^{1/2}).$$

In the case of a convex polyhedron it was possible to obtain the rate of convergence $O([h\ln(1/h)]^{1/2})$ (see [234, 235, 267, 268]) due to the fact from [47] that an estimate of $\sup_{t \in [0,T]} |X(t) - \bar{X}(t)|^{2p}$ does not contain a term with local time. It is probably impossible to obtain the rate of convergence $O([h\ln(1/h)]^{1/2})$ for other convex domains in $\mathbf{R}^d$ (see further details in [234, 267, 268]).

6 Random walks for linear boundary value problems

Chapter 2 is devoted to weak numerical methods for SDEs which are suitable for solving the Cauchy problem for linear parabolic equations. The previous chapter deals with mean-square approximations of SDEs in bounded domains, and its results can be applied for solving boundary value problems. However, since solutions of boundary value problems for parabolic and elliptic equations can be represented as expectations of solutions of the corresponding systems of SDEs in bounded domains, one can apply far more simple weak approximations which certainly should be subject to limitations related to nonexit from the bounded domains. Such weak methods are considered in this chapter.

In Sects. 6.1 and 6.2, we present various weak methods for the Dirichlet problem for parabolic equations obtained in [181, 182, 211]. Section 6.3 deals with random walks for the Dirichlet problem for elliptic equations [186, 196, 211]. In Sects. 6.4-6.5, specific weak methods for the Dirichlet problem for elliptic equations are constructed [186]. Estimates of errors of these specific methods contain only low-order derivatives of the solution to the Dirichlet problem. This can be important, in particular, for problems with a boundary layer (i.e., for problems with small parameter at higher derivatives). Section 6.6 is devoted to solution of the Neumann problem for linear parabolic equations [181, 183]. Some other random walks for linear boundary value problems are proposed in [41]. This chapter and the monograph [141] have a common object of investigation, but our approach is essentially different from [141]. See other approaches to the Monte Carlo methods for solving partial differential equations in [58, 257] and references therein.

6.1 Algorithms for solving the Dirichlet problem based on time-step control

Let G be a bounded domain in $\mathbf{R}^d$ and $Q = [T_0, T_1) \times G$ be a cylinder in $\mathbf{R}^{d+1}$, $\Gamma = \bar{Q} \backslash Q$ be the part of the cylinder's boundary consisting of the upper base and lateral surface. Consider the Dirichlet problem for the parabolic equation:

$$\frac{\partial u}{\partial t} + \frac{1}{2}\sum_{i,j=1}^{d} a^{ij}(t,x)\frac{\partial^2 u}{\partial x^i \partial x^j} + \sum_{i=1}^{d} b^i(t,x)\frac{\partial u}{\partial x^i} + c(t,x)u$$

$$+ g(t,x) = 0, \quad (t,x) \in Q, \tag{1.1}$$

$$u\mid_\Gamma = \varphi(t,x). \tag{1.2}$$

The form of equation (1.1) is convenient for a probabilistic approach: the "initial" condition is prescribed at the final time moment $t = T_1$, and the equation is considered for $t < T_1$.

We assume that the coefficients $a^{ij} = a^{ji}$ satisfy the strict ellipticity condition in $\bar{Q}$ and that conditions hold which guarantee existence of the sufficiently smooth in $\bar{Q}$ solution of (1.1)-(1.2). We recall [73] that these conditions consist in some smoothness requirements imposed on the functions $a^{ij}(t,x)$, $b^i(t,x)$, $c(t,x)$, $g(t,x)$, on the boundary ∂G of the domain G, and on the function φ which is finally assumed to satisfy (1.1) on the boundary of the upper base of $\bar{Q}$.

The solution of the problem (1.1)-(1.2) has the following probabilistic representation:

$$u(t,x) = E\left[\varphi(\tau, X_{t,x}(\tau))Y_{t,x,1}(\tau) + Z_{t,x,1,0}(\tau)\right], \tag{1.3}$$

where $X_{t,x}(s)$, $Y_{t,x,y}(s)$, $Z_{t,x,y,z}(s)$, $s \geq t$, is the solution of the Cauchy problem for the system of SDEs:

$$dX = (b(s,X) - \sigma(s,X)\mu(s,X))\,ds + \sigma(s,X)\,dw(s), \quad X(t) = x, \tag{1.4}$$
$$dY = c(s,X)Y\,ds + \mu^\mathsf{T}(s,X)Y\,dw(s), \quad Y(t) = y, \tag{1.5}$$
$$dZ = g(s,X)Y\,ds + F^\mathsf{T}(s,X)Y\,dw(s), \quad Z(t) = z, \tag{1.6}$$

$(t,x) \in Q$, and $\tau = \tau_{t,x}$ is the first exit time of the trajectory $(s, X_{t,x}(s))$ to the boundary Γ. In (1.4)-(1.6), $w(s) = (w^1(s), \ldots, w^d(s))^\mathsf{T}$ is a standard Wiener process, $b(s,x)$ is a d-dimensional column-vector composed of the coefficients $b^i(s,x)$, the $d \times d$ matrix $\sigma(s,x)$ is obtained from the formula $\sigma(s,x)\sigma^\mathsf{T}(s,x) = a(s,x)$, $a(s,x) = \{a^{ij}(s,x)\}$, $i,j = 1,\ldots,d$, $\mu(s,x)$ and $F(s,x)$ are arbitrary d-dimensional column-vectors sufficiently smooth in $\bar{Q}$, Y and Z are scalars.

For $\mu(s,x) = 0$ and $F(s,x) = 0$, the formula (1.3) gives the standard probabilistic representation [48, 82]. For $\mu(s,x) \neq 0$ and $F(s,x) = 0$, (1.3) follows from Girsanov's theorem. It is also clear that the term $F(s,x) \neq 0$ does not affect the validity of this formula. Thus, the mean of the random variable appearing under the symbol of expectation in (1.3) does not depend on μ and F. At the same time, other properties of this random variable can essentially depend on μ and F. In particular, choosing μ or/and F, it is possible to reach zero variance of this random variable (see Sect. 2.4). This property is of great importance since we use the Monte Carlo technique for solving linear boundary value problems by probabilistic methods. This is a

6.1 Algorithms for solving the Dirichlet problem based on time-step control

reason for us to construct random walks taking into account dependence on μ and F.

To realize the representation (1.3), we need an approximation of the trajectory $(s, X(s))$, and this approximation should satisfy some restrictions related to nonexit from the domain $\bar{Q}$. Such approximations are considered in Sects. 6.1 and 6.2.

Approximations of this section (originally proposed in [181,182]) are based on controlling a time step of numerical integration of the system (1.4). The step is chosen so that (of course, aside of reaching a required accuracy) the next state of a Markov chain approximating in the weak sense the solution of (1.4) remains in the domain $\bar{Q}$ with probability one. This leads to a decrease of the time step when the chain is close to the boundary Γ of the domain Q. The chain is stopped in a narrow zone near the boundary so that values of the solution $u(t, x)$ in this zone can be approximated quite accurately by the known values of the function φ on the boundary. Another type of approximations is considered in Sect. 6.2. There the step of numerical integration of the system (1.4) is constant for points belonging to a certain time layer $t = t_k$. But when a point is close to the boundary, we make an intermediate (auxiliary) step of the random walk, which preserves the point in the time layer $t = t_k$.

In Sect. 6.1.1, some one-step weak approximations for (1.4)-(1.6) are given together with their error analysis while convergence theorems are proved in Sect. 6.1.2. In this section we restrict ourselves to the case of $\mu = 0$ and $F = 0$ in (1.4)-(1.6).

In Chap. 9 we investigate regular oscillations in systems with stochastic resonance and noise-induced unidirectional transport. To find domains of parameters for which these phenomena are observed, we need to solve some boundary value problems for linear parabolic equations. It turns out that the methods proposed in this section are most relevant for solving the problems from Chap. 9.

6.1.1 Theorems on one-step approximation

Let $(t_0, x_0) \in Q$, ξ be a point with uniform distribution on the surface of an open unit ball $U_1 \subset \mathbf{R}^d$ with center at the origin, and $r > 0$ be a number. Introduce an open ellipsoid $U_r^{\sigma(t_0,x_0)}(x_0)$ obtained from the ball U_1 by the linear transformation $r\sigma(t_0, x_0)$ and the shift $x_0 + b(t_0, x_0)r^2/d$. For any point $(t_0, x_0) \in Q$, there is an $r > 0$ such that $x_0 \in U_r^{\sigma(t_0,x_0)}(x_0)$ and the cylinder $\Pi_r^{\sigma(t_0,x_0)}(t_0, x_0) = [t_0, t_0 + r^2/d) \times U_r^{\sigma(t_0,x_0)}(x_0) \subset Q$. Below r is always assumed to be sufficiently small. Obviously, the point (t_1, X_1) with

$$t_1 = t_0 + \frac{r^2}{d}, \quad X_1 = x_0 + b(t_0, x_0)\frac{r^2}{d} + r\sigma(t_0, x_0)\xi, \qquad (1.7)$$

belongs to the boundary of the upper base of $\bar{\Pi}_r^{\sigma(t_0,x_0)}(t_0, x_0)$.

We also introduce
$$Y_1 = y_0 + c(t_0, x_0)y_0\frac{r^2}{d}, \quad Z_1 = z_0 + g(t_0, x_0)y_0\frac{r^2}{d}. \tag{1.8}$$

Theorem 1.1. *Let the function f be the restriction to $\bar{\Pi}_r^{\sigma(t_0,x_0)}(t_0, x_0)$ of the solution of the problem (1.1)-(1.2) having in $\bar{Q}$ continuous derivatives $D_x^m D_t^k u$, $0 \le m + 2k \le 4$, $k = 0, 1$. Let $X_{t_0,x_0}(s)$, $Y_{t_0,x_0,y_0}(s)$, $Z_{t_0,x_0,y_0,z_0}(s)$ be the solution of the system (1.4)-(1.6) and ϑ be the exit time of the process $(s, X_{t_0,x_0}(s))$ from $\Pi_r^{\sigma(t_0,x_0)}(t_0, x_0)$ (i.e., the point $(\vartheta, X_{t_0,x_0}(\vartheta))$ belongs either to the upper base or to the lateral surface of the cylinder $\bar{\Pi}_r^{\sigma(t_0,x_0)}(t_0, x_0)$). Then*

$$E\left[f(t_1, X_1)Y_1 + Z_1 - f(\vartheta, X_{t_0,x_0}(\vartheta))Y_{t_0,x_0,y_0}(\vartheta) - Z_{t_0,x_0,y_0,z_0}(\vartheta)\right] = O(r^4). \tag{1.9}$$

Proof. We have
$$E\left[f(\vartheta, X_{t_0,x_0}(\vartheta))Y_{t_0,x_0,y_0}(\vartheta) + Z_{t_0,x_0,y_0,z_0}(\vartheta)\right] = u(t_0, x_0)y_0 + z_0.$$

Since $f(t_1, X_1) = u(t_1, X_1)$, the difference under the expectation symbol in (1.9) is equal to

$$u(t_1, X_1)Y_1 + Z_1 - u(t_0, x_0)y_0 - z_0$$
$$= u(t_0 + \frac{r^2}{d}, x_0 + b(t_0, x_0)\frac{r^2}{d} + r\sigma(t_0, x_0)\xi) \times (y_0 + c(t_0, x_0)y_0\frac{r^2}{d})$$
$$+ z_0 + g(t_0, x_0)y_0\frac{r^2}{d} - u(t_0, x_0)y_0 - z_0$$
$$= [u(t_0, x_0) + \frac{\partial u}{\partial t}(t_0, x_0)\frac{r^2}{d}$$
$$+ \sum_{i=1}^d \frac{\partial u}{\partial x^i}(t_0, x_0)\{b(t_0, x_0)\frac{r^2}{d} + r\sigma(t_0, x_0)\xi\}^i$$
$$+ \frac{1}{2}\sum_{i,j=1}^d \frac{\partial^2 u}{\partial x^i \partial x^j}(t_0, x_0)\{b(t_0, x_0)\frac{r^2}{d} + r\sigma(t_0, x_0)\xi\}^i$$
$$\times \{b(t_0, x_0)\frac{r^2}{d} + r\sigma(t_0, x_0)\xi\}^j]$$
$$\times (1 + c(t_0, x_0)\frac{r^2}{d})y_0 + \rho + g(t_0, x_0)y_0\frac{r^2}{d} - u(t_0, x_0)y_0. \tag{1.10}$$

The remainder ρ contains the terms of the form
$$\frac{\partial^2 u}{\partial t \partial x^i}(t_0, x_0)\frac{r^2}{d}r\{\sigma(t_0, x_0)\xi\}^i,$$
$$\frac{\partial^3 u}{\partial x^i \partial x^j \partial x^k}(t_0, x_0)r^3\{\sigma(t_0, x_0)\xi\}^i\{\sigma(t_0, x_0)\xi\}^j\{\sigma(t_0, x_0)\xi\}^k,$$

6.1 Algorithms for solving the Dirichlet problem based on time-step control

having zero expectations, and it also contains the terms of order $O(r^4)$ (uniform in $(t_0, x_0) \in Q$). This is due to the fact that $E\xi^i = 0$, $E\xi^i\xi^j\xi^k = 0$, $i, j, k = 1, \ldots, d$, and due to the assumption that $u(t, x)$ has continuous in $\bar{Q}$ derivatives:
$$\frac{\partial^2 u}{\partial t^2}, \ \frac{\partial^3 u}{\partial x^i \partial x^j \partial x^k}, \ \frac{\partial^4 u}{\partial x^i \partial x^j \partial x^k \partial x^l}.$$

Using the equalities $E\xi^i\xi^j = \delta_{ij}/d$, $i, j = 1, \ldots, d$, we obtain from (1.10):

$$E[u(t_1, X_1)Y_1 + Z_1 - u(t_0, x_0)y_0 - z_0]$$
$$= \left(\frac{\partial u}{\partial t} + \frac{1}{2} \sum_{i,j=1}^{d} a^{ij} \frac{\partial^2 u}{\partial x^i \partial x^j} + \sum_{i=1}^{d} b^i \frac{\partial u}{\partial x^i} + cu + g \right) \frac{r^2}{d} y_0 + O(r^4) = O(r^4),$$

i.e., (1.9) is proved.

For a completeness of the exposition, we derive the relations:

$$E\xi^i = 0, \quad E\xi^i\xi^j = \delta_{ij}/d, \quad E\xi^i\xi^j\xi^k = 0, \quad i, j, k = 1, \ldots, d. \tag{1.11}$$

The first and the third groups of equalities in (1.11) follow from the symmetry. The equalities $E\xi^i\xi^i = 1/d$, $i = 1, \ldots, d$, follow from the relation $(\xi^1)^2 + \cdots + (\xi^d)^2 = 1$. To prove $E\xi^i\xi^j = 0$, $i \neq j$, we note that ξ has the same distribution as $w(\theta)$, where θ is the exit time of the d-dimensional Wiener process from the unit ball U_1. Further, the function $v = x^i x^j$, $i \neq j$, is the solution of the problem

$$\frac{1}{2} \Delta v = 0, \quad v|_{\partial U_1} = x^i x^j.$$

From here $0 = v(0) = Ew^i(\theta)w^j(\theta) = E\xi^i\xi^j$, and we have obtained the relations (1.11). Theorem 1.1 is proved. □

Now let $C_1 \subset \mathbf{R}^d$ be the cube with center at the origin and with the coordinates of the vertices equal to ± 1, i.e., its edges are parallel to the coordinate axes and their length is equal to 2. Introduce the parallelepiped $C_r^{\sigma(t_0, x_0)}(x_0)$ obtained from the cube C_1 by the linear transformation $r\sigma(t_0, x_0)$ and the shift $x_0 + b(t_0, x_0)r^2/d$ and also the cylinder $\Pi_r^{\sigma(t_0,x_0)}(t_0, x_0) = [t_0, t_0 + r^2/d] \times C_r^{\sigma(t_0,x_0)}(x_0) \subset Q$. Let ν be the random vector whose coordinates ν^i, $i = 1, \ldots, d$, are mutually independent random variables taking the values ± 1 with probability $1/2$. It is clear that the point (t_1, X_1) with

$$t_1 = t_0 + r^2, \quad X_1 = x_0 + b(t_0, x_0)r^2 + \sigma(t_0, x_0)r\nu, \tag{1.12}$$

finds itself on the vertices of the upper base of $\bar{\Pi}_r^{\sigma(t_0,x_0)}(t_0, x_0)$.

We also introduce

$$Y_1 = y_0 + c(t_0, x_0)y_0 r^2, \quad Z_1 = z_0 + g(t_0, x_0)y_0 r^2. \tag{1.13}$$

Theorem 1.2. *Let the function f be the restriction to $\bar{\Pi}_r^{\sigma(t_0,x_0)}(t_0,x_0)$ of the solution of the problem (1.1)-(1.2) having in $\bar{Q}$ continuous derivatives $D_x^m D_t^k u$, $0 \le m + 2k \le 4$, $k = 0, 1$. Let $X_{t_0,x_0}(s)$, $Y_{t_0,x_0,y_0}(s)$, $Z_{t_0,x_0,y_0,z_0}(s)$ be the solution of the system (1.4)-(1.6) and ϑ be the exit time of the process $(s, X_{t_0,x_0}(s))$ from $\Pi_r^{\sigma(t_0,x_0)}(t_0,x_0)$. Then (1.9) holds.*

Proof. Consider the difference

$$d_1 = u(t_1, X_1)Y_1 + Z_1 - u(t_0, x_0)y_0 - z_0. \tag{1.14}$$

Doing transformations of d_1 as in the proof of Theorem 1.1 and then taking the expectation of d_1, we obtain (1.9). □

Remark 1.3. In Theorems 1.1 and 1.2 we obtain that one-step orders of accuracy of the approximations (1.7)-(1.8) and (1.12)-(1.13) are equal to $O(r^4)$. The time step is proportional to r^2 and the proposed methods have one-step order two with respect to the time step as the weak Euler method (2.0.21).

Remark 1.4. "Ordinary" weak methods (see Chap. 2) usually involve an analysis of the difference

$$E\left[f(\bar{X}_{t_0,x_0}(t_0+h)) - f(X_{t_0,x_0}(t_0+h))\right],$$

i.e., the d-dimensional manifold in $(d+1)$-dimensional space of variables $(t, x^1, \ldots, x^d)$, on which the function f is defined, has a special form $t = t_1$, and, in particular, it is unbounded. In Theorems 1.1 and 1.2 this manifold is already bounded, it is a part of the cylinder's boundary. Then, associating t_1 with ϑ, it is natural to have t_1 in more general methods as a random variable.

Introduce the one-step approximation:

$$t_1 = t_0 + \Delta t(t_0, x_0, \xi; r), \quad X_1 = x_0 + \Delta x(t_0, x_0, \xi; r),$$
$$Y_1 = y_0 + \Delta y(t_0, x_0, \xi; r) y_0, \quad Z_1 = z_0 + \Delta z(t_0, x_0, \xi; r) y_0, \tag{1.15}$$

where ξ is a random vector and r is a small parameter, possibly multidimensional, connected with the size of steps in $t, x^1, \ldots, x^d$. Here we restrict ourselves to a one-dimensional r. We assume that for all possible values of ξ increments Δt and Δx are such that the point $(t_1, X_1) \in \bar{Q}$.

Our nearest aim is to obtain sufficient conditions which ensure the order of accuracy $O(r^4)$ for the approximation (1.15). To this end, we substitute (1.15) in (1.14) and expand the difference d_1 in the powers of Δt, Δx^i, Δy, Δz. Assuming that

$$E\Delta t^2 = O(r^4), \quad E\Delta t \Delta x^i = O(r^4), \quad E\Delta x^i \Delta x^j \Delta x^k = O(r^4),$$
$$E\Delta t \Delta y = O(r^4), \quad E\Delta x^i \Delta y = O(r^4), \quad i,j,k = 1,\ldots,d, \tag{1.16}$$

we get

6.1 Algorithms for solving the Dirichlet problem based on time-step control

$$d_1 = u(t_0, x_0) + \frac{\partial u}{\partial t}(t_0, x_0) y_0 \Delta t + \sum_{i=1}^{d} \frac{\partial u}{\partial x^i}(t_0, x_0) y_0 \Delta x^i$$

$$+ \frac{1}{2} \sum_{i,j=1}^{d} \frac{\partial^2 u}{\partial x^i \partial x^j}(t_0, x_0) y_0 \Delta x^i \Delta x^j + \Delta z + \rho,$$

where $E\rho = O(r^4)$.

The relation $Ed_1 = O(r^4)$ holds provided that in addition to (1.16) the random increments Δt, Δx^i, Δy, Δz satisfy the conditions:

$$E\Delta t = r^2 + O(r^4), \quad E\Delta x^i = b^i(t_0, x_0) r^2 + O(r^4),$$
$$E\Delta x^i \Delta x^j = a^{ij}(t_0, x_0) r^2 + O(r^4), \quad E\Delta x^i \Delta y = c(t_0, x_0) y_0 r^2 + O(r^4),$$
$$E\Delta z = g(t_0, x_0) y_0 r^2 + O(r^4), \quad i, j = 1, \ldots, d. \tag{1.17}$$

Thus, we have proved the following theorem.

Theorem 1.5. *Assume that the solution of* (1.1)-(1.2) *has continuous derivatives* $D_x^m D_t^k u$, $0 \leq m + 2k \leq 4$, $k = 0, 1$, *in* $\bar{Q}$. *Let the approximation* (1.15) *give a point* (t_1, X_1) *such that for all sufficiently small* r *the line segment connecting the points* (t_0, x_0) *and* (t_1, X_1) *entirely belongs to* $\bar{Q}$. *Then, under the conditions* (1.16)-(1.17) *the approximation* (1.15) *has a one-step order of accuracy* $O(r^4)$, *i.e.*,

$$E[u(t_1, X_1) Y_1 + Z_1 - u(t_0, x_0) y_0 - z_0] = O(r^4). \tag{1.18}$$

It is obvious that the conditions (1.16)-(1.17) are fulfilled both in Theorem 1.1 (with r^2 replaced by r^2/d) and in Theorem 1.2.

Example 1.6. Let Δt be a random variable taking two values 0 and $2r^2$ with probability $1/2$. Consider the approximation

$$t_1 = t_0 + \Delta t,$$
$$X_1 = x_0 + [b(t_0, x_0) r^2 + \frac{r}{\sqrt{2}} \sigma(t_0, x_0) \nu] \frac{2r^2 - \Delta t}{r^2},$$
$$Y_1 = y_0 + c(t_0, x_0) y_0 r^2, \quad Z_1 = z_0 + g(t_0, x_0) y_0 r^2, \tag{1.19}$$

where $\nu = (\nu^1, \ldots, \nu^d)^\mathsf{T}$ is a random vector with mutually independent components taking the values ± 1 with probability $1/2$. In addition, Δt and ν are also independent. The relations (1.16)-(1.17) are easily verified. Thus, the approximation (1.19) has one-step order of accuracy $O(r^4)$. Note that in the case of the classical heat equation $\partial u / \partial t + (1/2) \partial^2 u / \partial x^2 = 0$ (considered with the reverse time, of course) the configuration of knots in the scheme (1.19) is the same as in the simplest implicit finite-difference scheme.

Example 1.7. Let Δt be a random variable taking two values 0 and r with probabilities $1-r$ and r, respectively. Consider the approximation

$$t_1 = t_0 + \Delta t, \quad X_1 = x_0 + b(t_0, x_0)\, r^2 + r\sigma(t_0, x_0)\, \nu,$$
$$Y_1 = y_0 + c(t_0, x_0)\, y_0 r^2, \quad Z_1 = z_0 + g(t_0, x_0)\, y_0 r^2. \tag{1.20}$$

Here the relations (1.17) are fulfilled, but we have $E\Delta t^2 = O(r^3)$ only instead of $E\Delta t^2 = O(r^4)$ in (1.16). Therefore, the one-step order of accuracy of (1.20) is equal to $O(r^3)$. In this example the point (t, X) jumps from one time layer to another with a larger step (r instead of r^2) but such jumps occur comparatively rarely.

Now we construct methods of a higher order. Here we restrict ourselves to the known weak approximations of solutions of SDEs (see Chap. 2).

Suppose the coefficients of the system (1.4)-(1.6) are defined in the strip $[T_0, T_1] \times \mathbf{R}^d$. Consider a weak approximation $\bar{X}_{t_0, x_0}(t_0 + h)$, $\bar{Y}_{t_0, x_0, y_0}(t_0 + h)$, $\bar{Z}_{t_0, x_0, y_0, z_0}(t_0 + h)$ of the solution of (1.4)-(1.6) with one-step error $O(h^{p+1})$. We recall (see Chap. 2) that this means the fulfillment of the relation

$$Ef(\bar{X}_{t_0,x_0}(t_0+h), \bar{Y}_{t_0,x_0,y_0}(t_0+h), \bar{Z}_{t_0,x_0,y_0}(t_0+h))$$
$$-Ef(X_{t_0,x_0}(t_0+h), Y_{t_0,x_0,y_0}(t_0+h), Z_{t_0,x_0,y_0}(t_0+h))$$
$$= O(h^{p+1}) \tag{1.21}$$

for a sufficiently wide class of functions $f(x, y, z)$. In particular, (1.21) holds for a function of the form $f(x, y, z) = f(x)y + z$.

Theorem 1.8. *Assume that the coefficients and the solution of the problem (1.1)-(1.2) are sufficiently smooth functions in $\bar{Q}$. Let an approximation $\bar{X}_{t_0,x_0}(t_0+h)$, $\bar{Y}_{t_0,x_0,y_0}(t_0+h)$, $\bar{Z}_{t_0,x_0,y_0,z_0}(t_0+h)$ have one-step weak order of accuracy $O(h^{p+1})$. Let for all sufficiently small h the point $(t_0+h, \bar{X}_{t_0,x_0}(t_0+h)) \in \bar{Q}$ provided $(t_0, x_0) \in Q$. Then*

$$E[u(t_0+h, \bar{X}_{t_0,x_0}(t_0+h))\bar{Y}_{t_0,x_0,y_0}(t_0+h)$$
$$+\bar{Z}_{t_0,x_0,y_0,z_0}(t_0+h) - u(t_0, x_0)\, y_0 - z_0]$$
$$= O(h^{p+1}), \tag{1.22}$$

i.e., the input approximation has the same order of accuracy in the bounded domain.

Proof. We extend the coefficients $a^{ij}(t, x)$, $b^i(t, x)$, $c(t, x)$ and the solution $u(t, x)$ of the problem (1.1)-(1.2) onto the strip $[T_0, T_1] \times \mathbf{R}^d$ preserving their properties of smoothness and boundedness [174]. This extension can be done in such a way that the property of strict ellipticity is also preserved (with a new constant). As an extension of $g(t, x)$, we take the evaluated value:

6.1 Algorithms for solving the Dirichlet problem based on time-step control

$$g(t,x) = -\frac{\partial u}{\partial t}(t,x) - \frac{1}{2}\sum_{i,j=1}^{d} a^{ij}(t,x)\frac{\partial^2 u}{\partial x^i \partial x^j}(t,x) - \sum_{i=1}^{d} b^i(t,x)\frac{\partial u}{\partial x^i}(t,x)$$
$$-c(t,x)u(t,x), \quad (t,x) \in [T_0, T_1] \times \mathbf{R}^d.$$

As a result, we obtain the extension of the boundary value problem (1.1)-(1.2) to the Cauchy problem, the characteristic system of which is the system (1.4)-(1.6) with the extended coefficients. Then

$$u(t_0, x_0) y_0 + z_0 = E[u(t_0 + h, X_{t_0,x_0}(t_0 + h))Y_{t_0,x_0,y_0}(t_0 + h)$$
$$+ Z_{t_0,x_0,y_0,z_0}(t_0 + h)].$$

Substituting this expression in (1.22), we obtain an expression of the form (1.21), in which the function $f(x, y, z)$ is equal to $u(t_0+h, x)y+z$ (here t_0+h plays the role of a parameter). Theorem 1.8 is proved. □

In Chap. 2, there are weak methods with one-step error $O(h^3)$ for general systems of SDEs and with one-step error $O(h^4)$ for systems with additive noise. Since we can choose bounded random variables in these methods, we can use them for solving the boundary value problem (1.1)-(1.2).

Example 1.9. Let in (1.1):

$$\frac{1}{2}\sum_{i,j=1}^{d} a^{ij}(t,x)\frac{\partial^2}{\partial x^i \partial x^j} = \frac{1}{2}\Delta,$$

where Δ is the Laplace operator. Then, the system (1.4)-(1.6) takes the form (recall that in this section we restrict ourselves to the case $\mu = 0$, $F = 0$):

$$dX = b(s, X)ds + dw(s)$$
$$dY = c(s, X)Y \, ds$$
$$dZ = g(s, X)Y \, ds.$$

As an example, we write the method with one-step error $O(r^6)$ (cf. (2.1.31)):

$$t_1 = t_0 + r^2,$$

$$X_1 = x_0 + r\xi + r^2 b(t_0, x_0) + \frac{1}{2}r^3 \sum_{i=1}^{d} \frac{\partial b}{\partial x^i}(t_0, x_0)\xi^i$$

$$+\frac{1}{2}r^4 \left(\frac{\partial b}{\partial t} + \sum_{i=1}^{d} b^i \frac{\partial b}{\partial x^i} + \frac{1}{2}\sum_{i=1}^{d} \frac{\partial^2 b}{(\partial x^i)^2} \right)(t_0, x_0),$$

$$Y_1 = y_0 + r^2 c(t_0, x_0) y_0 + \frac{1}{2}r^3 y_0 \sum_{i=1}^{d} \frac{\partial c}{\partial x^i}(t_0, x_0)\xi^i$$

$$+\frac{1}{2}r^4 y_0 \left(\frac{\partial c}{\partial t} + \sum_{i=1}^{d} b^i \frac{\partial c}{\partial x^i} + c^2 + \frac{1}{2}\sum_{i=1}^{d} \frac{\partial^2 c}{(\partial x^i)^2} \right)(t_0, x_0),$$

$$Z_1 = z_0 + r^2 g(t_0, x_0) y_0 + \frac{1}{2} r^3 y_0 \sum_{i=1}^{d} \frac{\partial g}{\partial x^i}(t_0, x_0) \xi^i$$

$$+ \frac{1}{2} y_0 r^4 \left(\frac{\partial g}{\partial t} + \sum_{i=1}^{d} b^i \frac{\partial g}{\partial x^i} + cg + \frac{1}{2} \sum_{i=1}^{d} \frac{\partial^2 g}{(\partial x^i)^2} \right)(t_0, x_0), \quad (1.23)$$

where $\xi = (\xi^1, \ldots, \xi^d)^\top$ is a random vector with mutually independent components distributed as $P(\xi^i = 0) = 2/3$, $P(\xi^i = \pm\sqrt{3}) = 1/6$. Note that X_1 takes a finite number of values.

6.1.2 Numerical algorithms and convergence theorems

Let $r > 0$ be sufficiently small. Denote by Γ_{r^2} the intersection of the r^2-neighborhood of the boundary Γ with the domain Q. We construct the algorithm based on the one-step approximation (1.7)-(1.8).

Algorithm 1.10 *Let*

$$t_{k+1} = t_k + \frac{r_{k+1}^2}{d},$$

$$X_{k+1} = X_k + b(t_k, X_k) \frac{r_{k+1}^2}{d} + r_{k+1} \sigma(t_k, X_k) \xi_{k+1}, \quad X_0 = x_0,$$

$$Y_{k+1} = Y_k + c(t_k, X_k) Y_k \frac{r_{k+1}^2}{d}, \quad Y_0 = 1,$$

$$Z_{k+1} = Z_k + g(t_k, X_k) Y_k \frac{r_{k+1}^2}{d}, \quad Z_0 = 0, \quad k = 0, 1, \ldots, \qquad (1.24)$$

where $\xi_1, \ldots, \xi_k, \ldots$ *are mutually independent random variables each of which has the same distribution as ξ in* (1.7).

The sequence r_k determining the algorithm is constructed as follows. Define the function $\rho_r(t, x)$, $(t, x) \in Q \backslash \Gamma_{r^2}$: if $\bar{\Pi}_r^{\sigma(t,x)}(t, x) \in Q$, we set $\rho_r(t, x) = r$; otherwise we find $\rho_r(t, x) < r$ such that $\bar{\Pi}_{\rho_r(t,x)}^{\sigma(t,x)}(t, x)$ touches the boundary Γ. In this way the function $\rho_r(t, x)$ is defined everywhere in $Q \backslash \Gamma_{r^2}$.

Let $(t_k, X_k) \in Q \backslash \Gamma_{r^2}$. Set $r_{k+1} = \rho_r(t_k, X_k)$ and then obtain the next point $(t_{k+1}, X_{k+1}, Y_{k+1}, Z_{k+1})$ according to (1.24). This walk terminates at a random step $\varkappa$ as soon as $(t_\varkappa, X_\varkappa) \in \Gamma_{r^2}$. Then we find the point $(\bar{t}_\varkappa, \bar{X}_\varkappa)$ on Γ which is the nearest to $(t_\varkappa, X_\varkappa)$.

It is proved (see Theorem 1.12 below) that $(\bar{t}_\varkappa, \bar{X}_\varkappa)$ weakly approximates $(\tau, X_{t_0, x_0}(\tau))$ with the accuracy $O(r^2)$, i.e.,

$$E[\varphi(\bar{t}_\varkappa, \bar{X}_\varkappa) Y_\varkappa + Z_\varkappa - u(t_0, x_0)]$$
$$= E[\varphi(\bar{t}_\varkappa, \bar{X}_\varkappa) Y_\varkappa + Z_\varkappa - \varphi(\tau, X_{t_0, x_0}(\tau)) Y_{t_0, x_0, 1}(\tau) - Z_{t_0, x_0, 1, 0}(\tau)]$$
$$= O(r^2). \qquad (1.25)$$

6.1 Algorithms for solving the Dirichlet problem based on time-step control

On the basis of the one-step approximation (1.12)-(1.13), we construct the following algorithm.

Algorithm 1.11 *Let*

$$
\begin{aligned}
t_{k+1} &= t_k + r_{k+1}^2, \\
X_{k+1} &= X_k + b(t_k, X_k)\, r_{k+1}^2 + \sigma(t_k, X_k)\, r_{k+1}\nu_{k+1}, \quad X_0 = x_0, \\
Y_{k+1} &= Y_k + c(t_k, X_k)\, Y_k r_{k+1}^2, \quad Y_0 = 1, \\
Z_{k+1} &= Z_k + g(t_k, X_k)\, Y_k r_{k+1}^2, \quad Z_0 = 0, \quad k = 0, 1, \ldots,
\end{aligned} \quad (1.26)
$$

where $\nu_1, \ldots, \nu_k, \ldots$ are mutually independent random variables distributed as ν in (1.12). Define the function $\rho_r(t,x)$, $(t,x) \in Q \backslash \Gamma_{r^2}$: if $\Pi_r^{\sigma(t,x)}(t,x) \in Q$, then $\rho_r(t,x) = r$, otherwise we find $\rho_r(t,x) < r$ such that $\bar{\Pi}_{\rho_r(t,x)}^{\sigma(t,x)}(t,x)$ touches the boundary Γ. Then we recursively define the sequence $r_{k+1} = \rho_r(t_k, X_k)$ and the random walk in accordance with (1.26) up to the moment $\varkappa$ when the point $(t_\varkappa, X_\varkappa)$ reaches Γ_{r^2}.

Note that in the case of convex G the point $X_\varkappa \in \partial G$ (of course, if $t_\varkappa < \tau$), and hence $(t_\varkappa, X_\varkappa) = (\bar{t}_\varkappa, \bar{X}_\varkappa) \in \Gamma$. The relation (1.25) also holds here.

The sequences (t_k, X_k, Y_k, Z_k) defined in (1.24) and (1.26) are Markov chains. The pair (t_k, X_k), which is of primary interest in what follows, is also a Markov chain.

The more general one-step approximation (1.15) determines a Markov chain if we take r as a control parameter with a corresponding synthesizing function $r = \rho_r(t,x)$. Let an approximation of the form (1.15) have one-step order of accuracy $O(r^{2s+2})$, $s \geq 1$. Then it is natural to introduce a narrower neighborhood $\Gamma_{r^{2s}}$ of the boundary Γ instead of Γ_{r^2}. According to (1.15), we construct the Markov chain:

$$
\begin{aligned}
t_{k+1} &= t_k + \Delta t(t_k, X_k, \xi_{k+1}; \rho_r(t_k, X_k)), \\
X_{k+1} &= X_k + \Delta x(t_k, X_k, \xi_{k+1}; \rho_r(t_k, X_k)), \quad X_0 = x_0, \\
Y_{k+1} &= Y_k + \Delta y(t_k, X_k, \xi_{k+1}; \rho_r(t_k, X_k))\, Y_k, \quad Y_0 = 1, \\
Z_{k+1} &= Z_k + \Delta z(t_k, X_k, \xi_{k+1}; \rho_r(t_k, X_k))\, Y_k, \quad Z_0 = 0, \quad k = 0, 1, \ldots.
\end{aligned} \quad (1.27)
$$

Assume the function $\rho_r(t,x)$ to be such that for $(t_k, X_k) \in Q \backslash \Gamma_{r^{2s}}$ the point $(t_{k+1}, X_{k+1}) \in \bar{Q}$. The random walk stops at the moment $\varkappa$ when $(t_\varkappa, X_\varkappa)$ reaches $\Gamma_{r^{2s}}$. As before, $(\bar{t}_\varkappa, \bar{X}_\varkappa)$ is the nearest to $(t_\varkappa, X_\varkappa)$ point on the boundary Γ.

Theorem 1.12. *Suppose $E\varkappa = O(1/r^2)$ and the components Y_k, Z_k of the chain (1.27) are uniformly bounded. Let the approximation (1.15) have one-step order of accuracy $O(r^{2s+2})$, $s \geq 1$, uniformly in $(t_0, x_0) \in Q$ for uniformly bounded y_0, z_0. Then $(\bar{t}_\varkappa, \bar{X}_\varkappa)$ approximates $(\tau, X_{t_0,x_0}(\tau))$ with the weak order of accuracy $O(r^{2s})$, i.e.,*

$$|E[\varphi(\bar{t}_\varkappa, \bar{X}_\varkappa)Y_\varkappa + Z_\varkappa - \varphi(\tau, X_{t_0,x_0}(\tau))Y_{t_0,x_0,1}(\tau) - Z_{t_0,x_0,1,0}(\tau)]|$$
$$= |E[\varphi(\bar{t}_\varkappa, \bar{X}_\varkappa)Y_\varkappa + Z_\varkappa - u(t_0, x_0)]| \leq Kr^{2s}. \tag{1.28}$$

Proof. Since $\varkappa$ is finite, we have

$$u(\bar{t}_\varkappa, \bar{X}_\varkappa)Y_\varkappa + Z_\varkappa - u(t_0, x_0)$$
$$= \left(u(\bar{t}_\varkappa, \bar{X}_\varkappa)Y_\varkappa + Z_\varkappa\right) - (u(t_\varkappa, X_\varkappa)Y_\varkappa + Z_\varkappa)$$
$$+ \sum_{k=0}^{\varkappa-1} [u(t_{k+1}, X_{k+1})Y_{k+1} + Z_{k+1} - (u(t_k, X_k)Y_k + Z_k)]$$
$$= u(\bar{t}_\varkappa, \bar{X}_\varkappa)Y_\varkappa - u(t_\varkappa, X_\varkappa)Y_\varkappa$$
$$+ \sum_{k=0}^{\infty} [u(t_{k+1}, X_{k+1})Y_{k+1} + Z_{k+1} - (u(t_k, X_k)Y_k + Z_k)], \tag{1.29}$$

where $(X_k, Y_k, Z_k) = (X_\varkappa, Y_\varkappa, Z_\varkappa)$ for $k \geq \varkappa$. We also put $r_{k+1} = 0$ for $k \geq \varkappa$.

According to the assumption of the theorem on the order of the approximation (1.15), we have for $(t_k, X_k) \in Q \setminus \Gamma_{r^{2s}}$:

$$|E\left(u(t_{k+1}, X_{k+1})Y_{k+1} + Z_{k+1} - u(t_k, X_k)Y_k - Z_k | t_k, X_k\right)|$$
$$\leq C\rho_r^{2s+2}(t_k, X_k) = Cr_{k+1}^{2s+2}. \tag{1.30}$$

Since the distance between the points $(t_\varkappa, X_\varkappa)$ and $(\bar{t}_\varkappa, \bar{X}_\varkappa)$ is not larger than r^{2s} and $Y_\varkappa$ is bounded, the absolute value of the first term in (1.29) is of order $O(r^{2s})$. Further, due to the condition $E\varkappa < \infty$ and the uniform boundedness of the terms of series in (1.29), expectation of the series is equal to the series of expectations of the terms. Using (1.29) and (1.30), we obtain

$$|E[\varphi(\bar{t}_\varkappa, \bar{X}_\varkappa)Y_\varkappa + Z_\varkappa - u(t_0, x_0)]|$$
$$\leq Cr^{2s} + \sum_{k=0}^{\infty} |EE\left(u(t_{k+1}, X_{k+1})Y_{k+1} + Z_{k+1} - u(t_k, X_k)Y_k - Z_k | t_k, X_k\right)|$$
$$\leq Cr^{2s} + Cr^{2s} E \sum_{k=0}^{\varkappa-1} r_{k+1}^2 \leq Cr^{2s}\left(1 + r^2 E\varkappa\right) \leq Kr^{2s}.$$

Theorem 1.12 is proved. □

The conditions of Theorem 1.12 are quite natural. Indeed, Y and Z satisfy the system (1.4)-(1.6) where the coefficients $c(s, X(s))$ and $g(s, X(s))$ are bounded due to the boundedness of $X(s)$ and finiteness of the integration interval. This implies (recall that $\mu = 0$ and $F = 0$) boundedness of $Y(s)$ and $Z(s)$, and the assumption on uniform boundedness of Y and Z becomes justified. Further, the assumption on uniformity in $(t, x) \in Q$ of the accuracy order of the approximation is justified by the corresponding smoothness of

6.1 Algorithms for solving the Dirichlet problem based on time-step control

solution u of the problem (1.1)-(1.2) in $\bar{Q}$. The condition $E\varkappa = O(1/r^2)$ is connected with the fact that for a lot of methods the discrete process (t_k, X_k) mainly walks with a large step and the step decreases only when the walk is close to the boundary. Near the boundary this walk is similar to a random walk over touching spheres which terminates rapidly. For instance, the mean number of steps of the random walk over touching spheres before reaching an ε-neighborhood of the boundary is estimated as $|\ln \varepsilon|$ (in our case this value is estimated as $|\ln r^{2s}| \sim |\ln r|$). Of course, trajectories of the Markov chains (t_k, X_k) can be complicated, and a point can spend some time near the boundary and then jump inside the domain. Nevertheless, the majority of steps is spent on approaching the boundary and it is estimated as $O(1/r^2)$. A rigorous justification of the relation $E\varkappa = O(1/r^2)$ involves consideration of a boundary value problem for the corresponding Markov chain.

Now we present some results for such a problem in connection with the chain (t_k, X_k). If we take t, x instead of t_0, x_0 in the first two formulas of (1.27), then the obtained relations determine the transition of the chain per step. Let $v(t, x)$ be a certain function defined in $\bar{Q}$. Introduce the one-step transition function:

$$Pv(t, x) = Ev(t_1, X_1)$$

and consider the boundary value problem in $\bar{Q}$:

$$Pv - v = -1, \quad (t, x) \in Q \setminus \Gamma_{r^{2s}}, \tag{1.31}$$
$$v(t, x) = 0, \quad (t, x) \in \Gamma_{r^{2s}}. \tag{1.32}$$

As is known [301], the solution of this boundary value problem is $v(t,x) = E\varkappa$, where $\varkappa$ corresponds to the chain (t_k, X_k) starting from the point (t, x).

Further, if we find a function $v(t, x) \geq 0$ in $Q \setminus \Gamma_{r^{2s}}$ and $v(t, x) = 0$ in $\Gamma_{r^{2s}}$ such that

$$Pv - v \leq -c < 0, \tag{1.33}$$

then

$$E\varkappa \leq \frac{v(t, x)}{c}. \tag{1.34}$$

As an example, let us consider the method (1.26) in detail. Let $r > 0$ be a sufficiently small number, $\rho_r(t, x)$ be the function defined above. The formulas

$$t_1 = \begin{cases} t + \Delta t(t, x, \xi; \rho_r(t, x)) = t + \rho_r^2, & (t, x) \in Q \setminus \Gamma_{r^2}, \\ t, & (t, x) \in \Gamma_{r^2}, \end{cases}$$

$$X_1 = \begin{cases} x + \Delta x(t, x, \xi; \rho_r(t, x)) = x + b(t, x)\rho_r^2 + \sigma(t, x)\rho_r \nu, & (t, x) \in Q \setminus \Gamma_{r^2}, \\ x, & (t, x) \in \Gamma_{r^2}, \end{cases} \tag{1.35}$$

define the transition of the chain per step.

Theorem 1.13. *The method (1.26) converges with the weak order of accuracy $O(r^2)$. Mean number of steps is equal to $E\varkappa = O(1/r^2)$.*

Proof. To prove the theorem, it is enough to obtain an estimate of $E\varkappa$ because the other conditions of Theorem 1.12 obviously hold. To estimate $E\varkappa$, we use (1.33)-(1.34) with v of the form:

$$v(t,x) = \begin{cases} T_1 - t + v_0, & (t,x) \in Q \setminus \Gamma_{r^2}, \\ 0, & (t,x) \in \Gamma_{r^2}, \end{cases} \qquad (1.36)$$

where $v_0 > 0$ is a constant. If a point (t,x) is such that $\rho_r(t,x) = r$, then the operator P applied to the function $v = T_1 - t$ gives $T_1 - t - r^2$ when $T_1 - t - r^2 \geq r^2$ and 0 when $T_1 - t - r^2 < r^2$. The operator P applied to the constant v_0 gives a value which is not greater than this constant (speaking here about functions, we consider them in $Q \setminus \Gamma_{r^2}$ only; for instance, the constant v_0 is the function of (t,x) which is equal to v_0 for $(t,x) \in Q \setminus \Gamma_{r^2}$ and equal to 0 for $(t,x) \in \Gamma_{r^2}$). As a result, for such points (t,x) we have

$$Pv - v \leq -r^2.$$

If a point (t,x) is such that $\rho_r(t,x) < r$, then the boundary of the upper base of the corresponding cylinder (right parallelepiped) either entirely belongs to the upper base of the cylinder $\bar{Q}$ (i.e., when $t_1 = T_1$) or touches the lateral surface of the cylinder $\bar{Q}$. In the first case the operator P gives zero for both $T_1 - t$ and v_0, and thus

$$Pv - v \leq -v_0.$$

In the second case, for a sufficiently small r the point (t_1, X_1) get into Γ_{r^2} with a probability $p \geq p_0$, where p_0 is independent of r. Hence the operator P applied to the constant v_0 is less than v_0 at least by $p_0 v_0$. Thus

$$Pv - v \leq -p_0 v_0.$$

It follows from the above estimates that we can take r^2 as c in (1.33)-(1.34) and $v(t,x) \leq T_1 - t + v_0$. Theorem 1.13 is proved. □

Remark 1.14. The same function v allows us to prove convergence theorems for the methods from Examples 1.6, 1.7, and 1.9. We emphasize that the order of accuracy of the method from Example 1.9 is $O(r^4)$, but the mean number of steps is still $O(1/r^2)$. The common property of all these methods, due to which the function satisfying the conditions (1.33)-(1.34) is so easily constructed, is as follows: for random walks over touching cylinders the probability that the point gets into r^{2s}-neighborhood of the boundary is bounded from below by a certain number p_0 which is independent of r. In fact, it would be enough if this probability were bounded from below by Kr^2, $K > 0$. Then, we have the inequality $Pv - v \leq -Kr^2 v_0$ instead of $Pv - v \leq -p_0 v_0$, and, due to (1.33)-(1.34), the relation $E\varkappa = O(1/r^2)$ remains valid. We note in connection with the method (1.24) that if the radius of the d-dimensional sphere is not larger that r, then the area of its segment of the height r^2 is larger than $Kr^{(d-1)/2}$, where $K > 0$ is a constant. Taking this into account, it

6.2 The simplest walk for the Dirichlet problem for parabolic equations   353

is possible to prove a convergence theorem like Theorem 1.13 for the method (1.24) in the case of $d = 1, \ldots, 5$ using the function (1.36). If $d > 5$ then the function of a more complicated form is required to prove such a theorem.

Remark 1.15. The majority of the results of this section can be carried over to elliptic equations (see also Sect. 6.3).

6.2 The simplest random walk for the Dirichlet problem for parabolic equations

In Sect. 6.1, approximations are based on controlling a time step of numerical integration of the system (1.4). In this section the step of numerical integration of the system (1.4) is constant for points belonging to a certain time layer $t = t_k$. But when a point is close to the boundary, we make an intermediate (auxiliary) step of the random walk, which preserves the point in the time layer $t = t_k$. The result of this auxiliary step is such that the point, which is close to the boundary, is replaced by two points with some probabilities using an interpolation. One of these new points belongs to the boundary and if it is realized, the walk terminates. The other point is inside the domain so that, starting from it, we can make a new step of numerical integration without leaving the domain $\bar{Q}$. The approach of Sect. 6.1 is probably more universal. However, methods of Sect. 6.2 are of independent interest from both theoretical and applied points of view due to their simplicity. The methods of Sect. 6.2 are the simplest among methods of order $O(h)$ because inside the domain we use the Euler weak approximations and near the boundary we exploit linear interpolation. In addition, these methods have a layer structure and, in particular, the only step of numerical integration can be used. Due to this fact it becomes possible to apply extrapolation methods.

The main algorithm of random walk is proposed in Sect. 6.2.1. In Sect. 6.2.2, its convergence with weak order of accuracy $O(h)$ is proved. Some similar algorithms of random walks are given in Sect. 6.2.3. One of this algorithms is distinguished by being the most simple but its order of convergence is $O(\sqrt{h})$. Finally, some numerical tests of the proposed methods are presented in Sect. 6.2.4. We note that the simplest random walk is used for evaluating the price and hedging strategy of American options in [192]. Numerical tests are presented there as well.

6.2.1 The algorithm of the simplest random walk

We apply the weak explicit Euler approximation with the simplest simulation of noise to the system (1.4)-(1.6) (cf. (2.0.21)):

$$X_{t,x}(t+h) \approx X = x + h\left(b(t,x) - \sigma(t,x)\,\mu(t,x)\right) + h^{1/2}\sigma(t,x)\,\xi, \quad (2.1)$$
$$Y_{t,x,y}(t+h) \approx Y = y + hc(t,x)\,y + h^{1/2}\mu^{\mathsf{T}}(t,x)\,y\,\xi, \quad (2.2)$$
$$Z_{t,x,y,z}(t+h) \approx Z = z + hg(t,x)\,y + h^{1/2}F^{\mathsf{T}}(t,x)\,y\,\xi, \quad (2.3)$$

where $h > 0$ is a step of integration (a sufficiently small number), $\xi = (\xi^1, \ldots, \xi^d)^\mathsf{T}$, ξ^i, $i = 1, \ldots, d$, are mutually independent random variables taking the values ± 1 with probability $1/2$. Clearly, the random vector X takes 2^d different values.

Introduce the set of points close to the boundary (a boundary zone) $S_{t,h} \subset \bar{G}$ on the layer t: we say that $x \in S_{t,h}$ if at least one of the 2^d values of the vector X is outside $\bar{G}$. It is not difficult to see that due to compactness of $\bar{Q}$ there is a constant $\lambda > 0$ such that if the distance from $x \in G$ to the boundary ∂G is equal to or greater than $\lambda\sqrt{h}$ then x is outside the boundary zone and, therefore, for such x all the realizations of the random variable X belong to $\bar{G}$.

Since restrictions connected with nonexit from the domain $\bar{G}$ should be imposed on an approximation of the system (1.4), the formulas (2.1)-(2.3) can be used only for the points $x \in \bar{G}\backslash S_{t,h}$ on the layer t, and a special construction is required for points from the boundary zone.

Let $x \in S_{t,h}$. Denote by $x^\pi \in \partial G$ the projection of the point x on the boundary of the domain G (the projection is unique because h is sufficiently small and ∂G is smooth) and by $n(x^\pi)$ the unit vector of internal normal to ∂G at x^π. Introduce the random vector $X^\pi_{x,h}$ taking two values x^π and $x + h^{1/2}\lambda n(x^\pi)$ with probabilities $p = p_{x,h}$ and $q = q_{x,h} = 1 - p_{x,h}$, respectively, where

$$p_{x,h} = \frac{h^{1/2}\lambda}{|x + h^{1/2}\lambda n(x^\pi) - x^\pi|}.$$

If $v(x)$ is a twice continuously differentiable function with the domain of definition $\bar{G}$, then an approximation of $v(x)$ by the expectation $Ev(X^\pi_{x,h})$ corresponds to linear interpolation and

$$v(x) = Ev(X^\pi_{x,h}) + O(h) = pv(x^\pi) + qv(x + h^{1/2}\lambda n(x^\pi)) + O(h). \quad (2.4)$$

We emphasize that the second value $x + h^{1/2}\lambda n(x^\pi)$ does not belong to the boundary zone. We also note that p is always greater than $1/2$ (since the distance from x to ∂G is less than $h^{1/2}\lambda$) and that if $x \in \partial G$ then $p = 1$ (since in this case $x^\pi = x$).

Let a point $(t_0, x_0) \in Q$. We would like to find the value $u(t_0, x_0)$. Introduce a discretization of the interval $[t_0, T]$, for definiteness the equidistant one:

$$t_0 < t_1 < \cdots < t_N = T, \quad h := (T - t_0)/N.$$

To approximate the solution of the system (1.4), we construct a Markov chain (t_k, X_k) which stops when it reaches the boundary Γ at a random step $\varkappa$. The number k takes nonnegative integer values not greater than N.

We set $X'_0 = x_0$. If $X'_0 \notin S_{t_0,h}$ then we take $X_0 = X'_0$. If $X'_0 \in S_{t_0,h}$ then the random variable X_0 takes two values: either $X'^\pi_0 \in \partial G$ with probability $p_{X'_0,h}$ or $X'_0 + h^{1/2}\lambda n(X'^\pi_0) \notin S_{t_0,h}$ with probability $q_{X'_0,h}$. If $X_0 = X'^\pi_0$ (i.e., $(t_0, X_0) \in \Gamma$), then we put $\varkappa = 0$, $X_\varkappa = X'^\pi_0$, and the random walk is finished.

6.2 The simplest walk for the Dirichlet problem for parabolic equations   355

Let X_k, $k < N$, be constructed and either $X_k \in \partial G$ (i.e., the chain is stopped at one of the instances $t_0, \ldots, t_k$ and, consequently, $\varkappa \leq k$) or $X_k \notin S_{t_k,h}$. Now our aim is to construct X_{k+1} with the same properties. Suppose that the chain does not stop until t_k inclusively, i.e., $\varkappa > k$. Introduce X'_{k+1} due to (2.1) with $t = t_k$, $x = X_k$, $\xi = \xi_k$ (we suppose that all ξ_k are mutually independent and distributed as ξ):

$$X'_{k+1} = X_k + h\left(b(t_k, X_k) - \sigma(t_k, X_k)\,\mu(t_k, X_k)\right) + h^{1/2}\sigma(t_k, X_k)\,\xi_k. \quad (2.5)$$

If $k + 1 = N$ then we set $X_{k+1} = X'_{k+1}$. If $k + 1 < N$, we obtain X_{k+1} using X'_{k+1} as we got X_0 using X'_0. More precisely, we use the following rule. If $X'_{k+1} \notin S_{t_{k+1},h}$ then we take $X_{k+1} = X'_{k+1}$. If $X'_{k+1} \in S_{t_{k+1},h}$ then the random variable X_{k+1} takes two values: either $X'^{\pi}_{k+1} \in \partial G$ with probability $p_{X'_{k+1},h}$ or $X'_{k+1} + h^{1/2}\lambda n(X'^{\pi}_{k+1}) \notin S_{t_{k+1},h}$ with probability $q_{X'_{k+1},h}$. If $X_{k+1} = X'^{\pi}_{k+1}$ (i.e., $(t_{k+1}, X_{k+1}) \in \Gamma$), then we put $\varkappa = k + 1$, $X_\varkappa = X'^{\pi}_{k+1}$, and the random walk is finished.

So, the main steps in constructing the chain (t_k, X_k) are as follows. The state X_0 is formed by the rule: if $X'_0 := x_0$ does not belong to the boundary zone $S_{t_0,h}$ ($X'_0 \notin S_{t_0,h}$), we put $X_0 = X'_0$; if $X'_0 \in S_{t_0,h}$, the state X_0 is defined as the random variable which takes two values: either X'^{π}_0 on the boundary ∂G or the other value in the internal part of the domain G. It can be said that the first value concentrates a "stopping" behavior of the original process X while the second value concentrates a "walking" behavior. Further, if the stopping moment $\varkappa$ is larger than zero, i.e., if the realized value of X_0 does not belong to the boundary, X'_1 is obtained by the step according to the usual formula. Then we get X_1 using X'_1 just as X_0 has been got by X'_0, and so on.

We also introduce an extended chain (t_k, X_k, Y_k, Z_k). We put $Y_0 = 1$ and $Z_0 = 0$. Let $\varkappa > k$ and X_k, Y_k, Z_k be known. Then Y_{k+1} and Z_{k+1} are evaluated in accordance with the system (2.2)-(2.3) for $t = t_k$, $x = X_k$, $y = Y_k$, $z = Z_k$, $\xi = \xi_k$. The extended chain stops at the same step $\varkappa$.

Below we write the constructed algorithm formally.

Algorithm 2.1

STEP 0. $X'_0 = x_0$, $Y_0 = 1$, $Z_0 = 0$, $k = 0$.

STEP 1.   If $X'_k \notin S_{t_k,h}$ then $X_k = X'_k$ and go to STEP 3.
If $X'_k \in S_{t_k,h}$ then either $X_k = X'^{\pi}_k$ with probability $p_{X'_k,h}$ or $X_k = X'_k + h^{1/2}\lambda n(X'^{\pi}_k)$ with probability $q_{X'_k,h}$.

STEP 2.   If $X_k = X'^{\pi}_k$ then STOP and $\varkappa = k$,
$X_\varkappa = X'^{\pi}_k$, $Y_\varkappa = Y_k$, $Z_\varkappa = Z_k$.

STEP 3.   Simulate ξ_k and find X'_{k+1}, Y_{k+1}, Z_{k+1} according to (2.5), (2.2)-(2.3) for $t = t_k$, $x = X_k$, $y = Y_k$, $z = Z_k$,
$\xi = \xi_k$.

STEP 4.   If $k + 1 = N$, STOP and $\varkappa = N$, $X_\varkappa = X'_N$, $Y_\varkappa = Y_N$,
$Z_\varkappa = Z_N$, otherwise $k := k + 1$ and return to STEP 1.

It may happen so that it is more rational to choose both h and λ depending on the chain's state: h_k and λ_k. Then, in Theorem 2.2 (see below) one should put $h = \max_{0 \leq k < N} h_k$. In practice, one can take $\lambda_k = |\sigma(t_k, X_k)|$, possibly with small corrections.

6.2.2 Convergence theorem

Denote by ν_{t_0,x_0} the number of those t_k at which X'_k gets into the set $S_{t_k,h}$. The event $\{\nu_{t_0,x_0} > n\}$ implies an event such that first n trials in a certain trial scheme are unsuccessful (Γ is not attained by (t_k, X_k)). Moreover, the probability of each unsuccess is less than $1/2$. Therefore, the following estimate takes place:

$$P\{\nu_{t_0,x_0} > n\} \leq \frac{1}{2^n}. \tag{2.6}$$

But these arguments are not completely rigorous since the probability of attaining the boundary by X_k depends on X'_k. Along with the chain X_k, let us consider a chain $\tilde{X}_k$ which differs from the original one in the following way only: when X'_k gets into the boundary zone $S_{t_k,h}$, each time a coin is thrown so that the new chain hits the boundary at X'^{π}_k with probability $1/2$ and it hits the same state as the original chain with probability $1/2$, i.e., the state $X'_k + h^{1/2}\lambda n(X'^{\pi}_k) \notin S_{t_k,h}$. Since the new chain terminates with a smaller probability than the original chain, we have $\nu_{t_0,x_0} \leq \tilde{\nu}_{t_0,x_0}$. Therefore, $P\{\nu_{t_0,x_0} > n\} \leq P\{\tilde{\nu}_{t_0,x_0} > n\}$. But the arguments before (2.6) are rigorous for $\tilde{\nu}_{t_0,x_0}$, i.e., $P\{\tilde{\nu}_{t_0,x_0} > n\} \leq 1/2^n$. The last two inequalities imply (2.6). As a conclusion, we state the following lemma.

Lemma 2.2. *The inequality (2.6) holds together with the inequalities*

$$P\{\nu_{t_0,x_0} = n\} \leq \frac{1}{2^{n-1}}, \tag{2.7}$$

$$E\nu_{t_0,x_0} \leq C, \quad E\nu^2_{t_0,x_0} \leq C, \tag{2.8}$$

where C does not depend on t_0, x_0, h.

We extend the definition of the constructed chain for all k by the rule: if $k > \varkappa$, then $(t_k, X_k, Y_k, Z_k) = (t_\varkappa, X_\varkappa, Y_\varkappa, Z_\varkappa)$.

We assume that the functions $a^{ij}(t,x)$, $b^i(t,x)$, $c(t,x)$, $g(t,x)$ together with their first partial derivatives in t, x^m and second partial derivatives in x^m are continuous in $\bar{Q}$, the domain G has twice continuously differentiable boundary ∂G, the function $\varphi(t,x)$, $(t,x) \in \Gamma$, is at least of the same smoothness as the solution $u(t,x)$, and, finally, the function $\varphi(t,x)$ satisfies (1.1) on the boundary of the upper base of the cylinder $\bar{Q}$. Then, the classical solution $u(t,x)$ of the problem (1.1)-(1.2) has continuous in $\bar{Q}$ derivatives (see, e.g., [73]):

6.2 The simplest walk for the Dirichlet problem for parabolic equations

$$\frac{\partial^2 u}{\partial t^2}, \ \frac{\partial^3 u}{\partial t \partial x^i \partial x^j}, \ \frac{\partial^3 u}{\partial x^i \partial x^j \partial x^m}, \ \frac{\partial^4 u}{\partial x^i \partial x^j \partial x^m \partial x^l}.$$

Now we prove a lemma on one-step error. Introduce

$$d_k = u(t_k, X_k) Y_k + Z_k - u(t_k, X'_k) Y_k - Z_k,$$
$$d'_k = u(t_{k+1}, X_{k+1}) Y_{k+1} + Z_{k+1} - u(t_k, X_k) Y_k - Z_k,$$
$$k = 0, \ldots, N-1.$$

Clearly, X'_k belongs to the layer $t = t_k$; the variable d_k can be nonzero in the case of $X'_k \in S_{t_k,h}$ only; if $\varkappa > k$, then $X_k \notin S_{t_k,h}$ and all the 2^d realizations of the random variable X'_{k+1} belong to $\bar{G}$; if $\varkappa \leq k$, then $t_k = t_{k+1} = t_\varkappa$, $X'_{k+1} = X_k = X_\varkappa$, $Y_{k+1} = Y_k = Y_\varkappa$, $Z_{k+1} = Z_k = Z_\varkappa$ and, consequently, $d'_k = 0$. We also note that it is not difficult to show that for sufficiently small h the component Y_k is positive.

Lemma 2.3. *The following inequalities hold:*

$$|E(d_k \mid X'_k, Y_k, Z_k)| \leq Ch Y_k I_{S_{t_k,h}}(X'_k) \chi_{\varkappa \geq k}, \tag{2.9}$$

$$|E(d'_k \mid X_k, Y_k, Z_k)| \leq Ch^2 Y_k \chi_{\varkappa > k}. \tag{2.10}$$

Proof. The inequality (2.9) follows from the interpolation relation (2.4) and from the reminder before the lemma. The inequality (2.10) is a consequence of the fact that the one-step accuracy order of the Euler method (2.1)-(2.3) in the weak sense is $O(h^2)$. □

Theorem 2.4. *Algorithm 2.1 has weak order of accuracy $O(h)$, i.e., the inequality*

$$|E(\varphi(t_\varkappa, X_\varkappa) Y_\varkappa + Z_\varkappa) - u(t_0, x_0)| \leq Ch \tag{2.11}$$

holds with C independent of t_0, x_0, h.

Proof. We have

$$R := E(\varphi(t_\varkappa, X_\varkappa) Y_\varkappa + Z_\varkappa) - u(t_0, x_0)$$
$$= E(u(t_\varkappa, X_\varkappa) Y_\varkappa + Z_\varkappa) - u(t_0, x_0)$$
$$= E(u(t_0, X_0) Y_0 + Z_0) - u(t_0, x_0)$$
$$+ E \sum_{k=0}^{\varkappa-1} (u(t_{k+1}, X_{k+1}) Y_{k+1} + Z_{k+1} - u(t_k, X_k) Y_k - Z_k)$$
$$= E \sum_{k=0}^{N-1} (d_k + d'_k) = \sum_{k=0}^{N-1} EE(d'_k \mid X_k, Y_k, Z_k) + \sum_{k=0}^{N-1} EE(d_k \mid X'_k, Y_k, Z_k).$$

Due to (2.10), the absolute value of the first sum is estimated by $Ch^2 \times \sum_{k=0}^{N-1} EY_k$. It is not difficult to show that $EY_k \leq C$, $k = 0, \ldots, N-1$. Thus, the first sum is $O(h)$ uniformly in t_0, x_0, h. Using (2.9) and Lemma 2.2, we estimate the second sum:

$$\left| \sum_{k=0}^{N-1} EE(d_k \mid X_k', Y_k, Z_k) \right| \leq Ch E \sum_{k=0}^{N-1} Y_k I_{S_{t_k,h}}(X_k') \chi_{\varkappa \geq k}$$

$$\leq Ch E \left(\max_{0 \leq k \leq N-1} Y_k \times \sum_{k=0}^{N-1} I_{S_{t_k,h}}(X_k') \right)$$

$$= Ch E \left(\max_{0 \leq k \leq N-1} Y_k \times \nu_{t_0,x_0} \right)$$

$$\leq Ch \left(E \max_{0 \leq k \leq N-1} Y_k^2 \right)^{1/2} \left(E \nu_{t_0,x_0}^2 \right)^{1/2}$$

$$\leq Ch \left(E \max_{0 \leq k \leq N-1} Y_k^2 \right)^{1/2}. \qquad (2.12)$$

Let $c(t,x) \leq \bar{c}$, $(t,x) \in \bar{Q}$, where $\bar{c}$ is a constant. Introduce the sequence

$$\tilde{Y}_0 = 1, \ \tilde{Y}_k = \tilde{Y}_{k-1}(1 + h\bar{c} + h^{1/2}\mu(t_{k-1}, X_{k-1})\xi_{k-1}), \ k \leq \varkappa;$$
$$\tilde{Y}_k = \tilde{Y}_\varkappa, \ k > \varkappa.$$

It is evident that $Y_k \leq \tilde{Y}_k$. It is not difficult to see that the sequence $V_k = (1 + h\bar{c})^{-k}\tilde{Y}_k$ is a martingale. Hence,

$$E \max_{0 \leq k \leq N-1} V_k^2 \leq 4EV_{N-1}^2 \leq C$$

(as before, here C is independent of h), and this together with (2.12) implies (2.11). Theorem 2.4 is proved. $\square$

We note that despite some steps of Algorithm 2.1 are very rough (their one-step errors are $O(h)$), this algorithm converges, and, moreover, its global order of convergence is $O(h)$. We have this rate of convergence due to the fact that the number of rough steps (on average) is bounded from above by a constant which does not grow as $h \to 0$.

Remark 2.5. If we suppose that the global error R of Algorithm 2.1 can be expand in powers of time step h:

$$R = C_0 h + O(h^2), \qquad (2.13)$$

then it becomes possible to use the extrapolation method and to obtain a method of order two applying two times Algorithm 2.1 with different time steps. Namely, let $\bar{u}^{h_1}$ and $\bar{u}^{h_2}$ be approximations of $u(t_0, x_0)$ calculated according to Algorithm 2.1. Then (see Sect. 2.2.3)

$$\bar{u}_{imp} = \bar{u}^{h_1} \frac{h_2}{h_2 - h_1} - \bar{u}^{h_2} \frac{h_1}{h_2 - h_1}, \ u(t_0, x_0) = \bar{u}_{imp} + O(h^2). \qquad (2.14)$$

6.2 The simplest walk for the Dirichlet problem for parabolic equations

We do not give here a proof of (2.13) (this is apparently more difficult in comparison with the proof of Theorem 2.2.5) however our experiments confirm the natural rule expressed by formula (2.14).

6.2.3 Other random walks

The next algorithm is obtained by a simplification of Algorithm 2.1. Indeed, as soon as X_k gets into the boundary domain $S_{t_k,h}$, the random walk terminates, i.e., $\varkappa = k$, and $\bar{X}_\varkappa = X_k^\pi$, $Y_\varkappa = Y_k$, $Z_\varkappa = Z_k$ is taken as the final state of the Markov chain. Let us write this algorithm formally.

Algorithm 2.6
  STEP 0. $X_0 = x_0$, $Y_0 = 1$, $Z_0 = 0$, $k = 0$.

  STEP 1. If $X_k \notin S_{t_k,h}$ then go to STEP 2.
  If $X_k \in S_{t_k,h}$ then STOP and $\varkappa = k$, $\bar{X}_\varkappa = X_k^\pi$,
  $Y_\varkappa = Y_k$, $Z_\varkappa = Z_k$.

  STEP 2. Simulate ξ_k and find X_{k+1}, Y_{k+1}, Z_{k+1} according to
  (2.1)-(2.3) for $t = t_k$, $x = X_k$, $y = Y_k$,
  $z = Z_k$, $\xi = \xi_k$.

  STEP 3. If $k + 1 = N$, STOP and $\varkappa = N$, $\bar{X}_\varkappa = X_N$, $Y_\varkappa = Y_N$,
  $Z_\varkappa = Z_N$, otherwise $k := k + 1$ and return to STEP 1.

In this algorithm one-step errors of all steps except the last one is $O(h^2)$. The error on the last step (i.e., on the step $\varkappa$) is estimated as $O(\sqrt{h})$. It is easy to prove the following theorem (we note in passing that the conditions on smoothness of the solution and parameters mentioned before Lemma 2.3 can be weakened for this theorem).

Theorem 2.7. *Algorithm 2.6 has weak order of accuracy* $O(\sqrt{h})$:

$$|E(\varphi(t_\varkappa, \bar{X}_\varkappa) Y_\varkappa + Z_\varkappa) - u(t_0, x_0)| \leq C\sqrt{h},$$

where C is independent of t_0, x_0, h.

A similar random walk was proposed in [41]. In contrast to Algorithm 2.6, which terminates when the chain gets into the boundary zone $S_{t_k,h} \subset \bar{G}$, the method from [41] terminates when the chain exits from $\bar{G}$ and the projection of the point, having exited from the domain, on the boundary ∂G is taken as the final state of the chain.

Remark 2.8. If we suppose that the global error R of Algorithm 2.6 can be expanded as
$$R = C_0 h^{1/2} + O(h),$$
then analogously to the rule (2.14) we obtain

$$\bar{u}_{imp} = \bar{u}^{h_1} \frac{h_2^{1/2}}{h_2^{1/2} - h_1^{1/2}} - \bar{u}^{h_2} \frac{h_1^{1/2}}{h_2^{1/2} - h_1^{1/2}}, \ u(t_0, x_0) = \bar{u}_{imp} + O(h). \quad (2.15)$$

Now we construct a random walk with a more accurate one-step approximation for $x \in S_{t,h}$ than in Algorithm 2.1. Let $x \in S_{t,h}$. We denote by $\xi^{(i)} = \left(\xi^{(i),1}, \ldots, \xi^{(i),d}\right)^\mathsf{T}$, $i = 1, \ldots, 2^d$, the values of the vector ξ from (2.1), and we assign indices to these values so that $\xi^{(2^{d-1}+i)} = -\xi^{(i)}, i = 1, \ldots, 2^{d-1}$. We denote by $X^{(i)}$ the value of the vector X from (2.1) corresponding to $\xi^{(i)}$. At least one of the points $X^{(i)} \notin \bar{G}$ since $x \in S_{t,h}$. We connect the point x with those $X^{(i)}$ which are outside $\bar{G}$ by the curves $\eta^{(i)}(\theta)$:

$$\eta^{(i)}(\theta) = (x + \theta h \left(b(t, x) - \sigma(t, x)\mu(t, x)\right) + \sqrt{\theta h}\, \sigma(t, x)\, \xi^{(i)}), \ \theta \in [0, 1].$$

There is a value $\theta = \theta^{(i)}$, $0 < \theta^{(i)} < 1$, (it is unique because h is sufficiently small and ∂G is smooth) such that the point $\eta^{(i)} := \eta^{(i)}(\theta^{(i)})$ belongs to the boundary ∂G. We put $\theta^{(i)} = 1$ and $\eta^{(i)} = X^{(i)}$ for those points $X^{(i)}$ which belong to $\bar{G}$.

Introduce the pair:

$$(\vartheta^{(i)}, \eta^{(i)}) := (t + \theta^{(i)}h, \eta^{(i)}). \quad (2.16)$$

Let α be the random variable taking values $\{1, \ldots, 2^d\}$ and distributed by the law

$$p_i = P\{\alpha = i\} = \frac{\gamma}{\sqrt{\theta^{(i)}} \left(\sqrt{\theta^{(i)}} + \sqrt{\theta^{(2^{d-1}+i)}}\right)}, \ i = 1, \ldots, 2^{d-1},$$

$$p_i = P\{\alpha = i\} = \frac{\gamma}{\sqrt{\theta^{(i)}} \left(\sqrt{\theta^{(i)}} + \sqrt{\theta^{(i-2^{d-1})}}\right)}, \ i = 2^{d-1}+1, \ldots, 2^d, \quad (2.17)$$

where γ is uniquely found from the condition $\sum_{i=1}^{2^d} P\{\alpha = i\} = 1$.

Now we construct a Markov chain (ϑ_k, X_k) which stops reaching the boundary Γ at a random step $\varkappa$. The index k takes nonnegative integer values not greater than N, and $\vartheta_k = t_k$, at least for $k < \varkappa$. We put $\vartheta_0 = t_0$, $X_0 = x_0$. Let $(\vartheta_k, X_k) \in Q$, i.e., $\varkappa > k$, $k \leq N-1$, $\vartheta_k = t_k$. Let us define $(\vartheta_{k+1}, X_{k+1})$. If $X_k \notin S_{t_k, h}$, then $\vartheta_{k+1} = t_{k+1}$ and X_{k+1} is found in accordance with (2.1) for $t = t_k$, $x = X_k$, $\xi = \xi_k$. If $X_k \in S_{t_k, h}$, then

$$(\vartheta_{k+1}, X_{k+1}) = (\vartheta_k^{(\alpha_k)}, \eta_k^{(\alpha_k)}),$$

where $(\vartheta_k^{(\alpha_k)}, \eta_k^{(\alpha_k)})$ is found as $(\vartheta^{(i)}, \eta^{(i)})$ from (2.16) for $t = t_k$, $x = X_k$, $i = \alpha_k$ (both ξ_k and α_k are independent of previous history; α_k is distributed as α; it is also clear that at each step we simulate either ξ_k or α_k). In this case, due to the construction of the pair $(\vartheta_k^{(\alpha_k)}, X_k^{(\alpha_k)})$, the point X_{k+1} either

6.2 The simplest walk for the Dirichlet problem for parabolic equations

belongs to ∂G with probability $p \geq 1/2^d$ and the random walk terminates (in this case $\varkappa = k+1$, $(\vartheta_\varkappa, X_\varkappa) = (\vartheta_{k+1}, X_{k+1})$) or $X_{k+1} \notin \partial G$ (this means that the realized α_k is such that $\theta_k^{(\alpha_k)} = 1$ and, therefore, $\vartheta_{k+1} = t_{k+1}$) and for $k+1 < N$ the random walk should be continued. For $k+1 = N$ the point $(\vartheta_{k+1}, X_{k+1}) \in \Gamma$ and the random walk terminates with $\varkappa = N$, $(\vartheta_\varkappa, X_\varkappa) = (T_1, X_N)$. Thus, the chain (ϑ_k, X_k) has been constructed.

Note that (ϑ_k, X_k) remains in the domain $\bar{Q}$ with probability 1; $\varkappa$ takes nonnegative integer values not greater than N.

Now we introduce an extended Markov chain $(\vartheta_k, X_k, Y_k, Z_k)$. We put $Y_0 = 1$, $Z_0 = 0$. Let $\varkappa > k$ and Y_k, Z_k be known. If $X_k \notin S_{t_k, h}$, then Y_{k+1}, Z_{k+1} are evaluated in accordance with (2.1)-(2.3) for $t = t_k$, $x = X_k$, $y = Y_k$, $z = Z_k$, $\xi = \xi_k$. If $X_k \in S_{t_k, h}$, then

$$Y_{k+1} = Y_k + \theta_k^{(\alpha_k)} h c(\vartheta_k, X_k) Y_k + \sqrt{\theta_k^{(\alpha_k)} h} \, \mu^\mathsf{T}(\vartheta_k, X_k) Y_k \xi^{(\alpha_k)},$$

$$Z_{k+1} = Z_k + \theta_k^{(\alpha_k)} h g(\vartheta_k, X_k) Y_k + \sqrt{\theta_k^{(\alpha_k)} h} \, F^\mathsf{T}(\vartheta_k, X_k) Y_k \xi^{(\alpha_k)}, \quad (2.18)$$

where $\theta_k^{(i)}$, $i = 1, \ldots, 2^d$, and α_k are the same as in evaluation of $(\vartheta_{k+1}, X_{k+1})$.

The constructed algorithm is written in the form.

Algorithm 2.9

STEP 0. $X_0 = x_0$, $Y_0 = 1$, $Z_0 = 0$, $k = 0$.

STEP 1. If $X_k \notin S_{t_k, h}$ then go to STEP 3.
If $X_k \in S_{t_k, h}$ then simulate α_k according to the distribution (2.17), find $(\vartheta_{k+1}, X_{k+1}) = (\vartheta_k^{(\alpha_k)}, \eta_k^{(\alpha_k)})$ according to (2.16) for $i = \alpha_k$, $t = \vartheta_k$, $x = X_k$ and find Y_{k+1}, Z_{k+1} according to (2.18).

STEP 2. If $X_{k+1} \in \partial G$ then STOP and put $(\vartheta_\varkappa, X_\varkappa, Y_\varkappa, Z_\varkappa)$
$= (\vartheta_{k+1}, X_{k+1}, Y_{k+1}, Z_{k+1})$; otherwise go to STEP 4.

STEP 3. Put $\vartheta_{k+1} = t_{k+1}$, simulate ξ_k, and find X_{k+1}, Y_{k+1}, Z_{k+1} according to (2.1)-(2.3) for $t = \vartheta_k$, $x = X_k$, $y = Y_k$, $z = Z_k$, $\xi = \xi_k$.

STEP 4. If $k+1 = N$, STOP and put $(\vartheta_\varkappa, X_\varkappa, Y_\varkappa, Z_\varkappa) = (\vartheta_N, X_N, Y_N, Z_N)$; otherwise put $k := k+1$ and return to STEP 1.

We prove the following lemma on one-step error of Algorithm 2.9.

Lemma 2.10. *The following inequality is valid:*

$$|E(u(\vartheta_{k+1}, X_{k+1}) Y_{k+1} + Z_{k+1} - u(\vartheta_k, X_k) Y_k - Z_k \mid X_k, Y_k, Z_k)|$$
$$\leq Ch^2 Y_k I_{G \setminus S_{t_k, h}}(X_k) \chi_{\varkappa > k} + Ch^{3/2} Y_k I_{S_{t_k, h}}(X_k) \chi_{\varkappa > k}. \quad (2.19)$$

362   6 Random walks for linear boundary value problems

Proof. For $X_k \in G \setminus S_{t_k,h}$ the inequality (2.19) is a consequence of the fact that the one-step accuracy order of the Euler method (2.1)-(2.3) in the weak sense is $O(h^2)$.

Let $X_k \in S_{t_k,h}$. Then we have

$$\rho := E(u(\vartheta_{k+1}, X_{k+1}) Y_{k+1} + Z_{k+1} - u(\vartheta_k, X_k) Y_k - Z_k \mid X_k, Y_k, Z_k)$$

$$= \sum_{i=1}^{2d} p_i \left(u(\vartheta_{k+1}^{(i)}, X_{k+1}^{(i)}) Y_{k+1}^{(i)} + Z_{k+1}^{(i)} \right) - u(\vartheta_k, X_k) Y_k - Z_k, \qquad (2.20)$$

where p_α is from (2.17).

Substituting the expressions for $\vartheta_{k+1}^{(i)}$, $X_{k+1}^{(i)}$, $Y_{k+1}^{(i)}$, $Z_{k+1}^{(i)}$ in (2.20) and expanding $u(\vartheta_{k+1}^{(i)}, X_{k+1}^{(i)})$ in a series in powers of h around the point (ϑ_k, X_k), we obtain

$$\rho = \sum_{i=1}^{2d} p_i Y_k \left\{ \left[u(\vartheta_k, X_k) + \frac{\partial u}{\partial t} \theta_k^{(i)} h + \sum_{j=1}^{d} \frac{\partial u}{\partial x^j} \left(\theta_k^{(i)} h (b^j - (\sigma\mu)^j) \right. \right. \right.$$

$$\left. \left. + \sqrt{\theta_k^{(i)} h} (\sigma\xi^{(i)})^j \right) + \frac{1}{2} \sum_{j,l=1}^{d} \frac{\partial^2 u}{\partial x^j \partial x^l} \theta_k^{(i)} h (\sigma\xi^{(i)})^j (\sigma\xi^{(i)})^l + O(h^{3/2}) \right]$$

$$\left. \times \left(1 + \theta_k^{(i)} h c + \sqrt{\theta_k^{(i)} h} \mu^\mathsf{T} \xi^{(i)} \right) + \left(\theta_k^{(i)} h g + \sqrt{\theta_k^{(i)} h} F^\mathsf{T} \xi^{(i)} \right) \right\}$$

$$+ \sum_{i=1}^{2d} p_i Z_k - u(\vartheta_k, X_k) Y_k - Z_k, \qquad (2.21)$$

where the derivatives of u and the coefficients b, σ, μ, c, g, F are evaluated at the point (ϑ_k, X_k).

Rearranging the terms in (2.21), we get

$$\rho = Y_k h \sum_{i=1}^{2d} p_i \theta_k^{(i)} \left[\frac{\partial u}{\partial t} + \frac{1}{2} \sum_{j,l=1}^{d} \frac{\partial^2 u}{\partial x^j \partial x^l} \sum_{m=1}^{d} \sigma_{jm} \sigma_{lm} + \sum_{j=1}^{d} \frac{\partial u}{\partial x^j} b^j + cu + g \right]$$

$$+ Y_k \sqrt{h} \sum_{i=1}^{2d} p_i \sqrt{\theta_k^{(i)}} \left[\sum_{j=1}^{d} \frac{\partial u}{\partial x^j} (\sigma\xi^{(i)})^j + u\mu^\mathsf{T} \xi^{(i)} + F^\mathsf{T} \xi^{(i)} \right]$$

$$- Y_k h \sum_{j=1}^{d} \frac{\partial u}{\partial x^j} (\sigma\mu)^j \sum_{i=1}^{2d} p_i \theta_k^{(i)} + Y_k h \sum_{j=1}^{d} \frac{\partial u}{\partial x^j} \sum_{i=1}^{2d} p_i \theta_k^{(i)} \mu^\mathsf{T} \xi^{(i)} (\sigma\xi^{(i)})^j$$

$$+ \frac{1}{2} Y_k h \sum_{j,l=1}^{d} \frac{\partial^2 u}{\partial x^j \partial x^l} \sum_{n,m=1,n\neq m}^{d} \sum_{i=1}^{2d} p_i \theta_k^{(i)} \sigma_{jn} \xi^{(i),n} \sigma_{lm} \xi^{(i),m} + O(h^{3/2}).$$

$$(2.22)$$

6.2 The simplest walk for the Dirichlet problem for parabolic equations

The first sum in (2.22) is equal to zero since $u(t,x)$ satisfies the equation (1.1). We have from (2.17): $p_i\sqrt{\theta_k^{(i)}} = p_{2^{d-1}+i}\sqrt{\theta_k^{(2^{d-1}+i)}}$, $i = 1,\ldots,2^{d-1}$, and, therefore, for any $r = 1,\ldots,d$ we obtain

$$\sum_{i=1}^{2^d} p_i\sqrt{\theta_k^{(i)}}\,\xi^{(i),r} = \sum_{i=1}^{2^{d-1}} \xi^{(i),r}\left(p_i\sqrt{\theta_k^{(i)}} - p_{2^{d-1}+i}\sqrt{\theta_k^{(2^{d-1}+i)}}\right) = 0,$$

whence the second sum in (2.22) is equal to zero. It follows from (2.17) that $p_i\theta_k^{(i)} + p_{2^{d-1}+i}\theta_k^{(2^{d-1}+i)} = \gamma$. Then it is not difficult to check that for $n \neq m$:

$$\sum_{i=1}^{2^d} p_i\theta_k^{(i)}\,\xi^{(i),n}\xi^{(i),m} = \gamma\sum_{i=1}^{2^{d-1}} \xi^{(i),n}\xi^{(i),m} = \frac{\gamma}{2}\sum_{i=1}^{2^d} \xi^{(i),n}\xi^{(i),m} = 0\,.$$

This implies that the last sum in (2.22) is equal to zero and the sum before the last one is equal to $Y_k h \sum_{j=1}^{d} \frac{\partial u}{\partial x^j}(\sigma\mu)^j \sum_{i=1}^{2^d} p_i\theta_k^{(i)}$. Thus, $\rho = O(h^{3/2})$. Lemma 2.10 is proved. $\square$

The next theorem is proved similarly to the proof of Theorem 2.4.

Theorem 2.11. *Algorithm 2.9 has weak order of accuracy $O(h)$:*

$$|E(\varphi(\vartheta_\varkappa, X_\varkappa)Y_\varkappa + Z_\varkappa) - u(t_0, x_0)| \leq Ch\,,$$

where C is independent of t_0, x_0, h.

We note that the chain $(\vartheta_k, X_k, Y_k, Z_k)$ is more expensive from computational point of view, but due to smaller errors in the boundary domain, it allows us to obtain more accurate results than the chain from Sect. 6.2.1. Further, if instead of the Euler approximation X, Y, Z from (2.1)-(2.3) we use an approximation X, Y, Z with one-step error $O(h^{5/2})$ or higher, then the corresponding modification of Algorithm 2.9 has weak order $O(h^{3/2})$. For instance, we can exploit the second-order weak method (2.2.18) from Sect. 2.2.1 for this purpose.

Remark 2.12. Algorithm 2.9 is used in Sect. 8.1.3 to construct a layer (deterministic) method for solving the Dirichlet problem for semilinear parabolic equations.

Remark 2.13. We note that the algorithms from Sects. 6.2.1 and 6.2.3 can be applied in the case when the domain Q has the form

$$Q = \{(t,x) : T_0 \leq t < T,\ x \in G_t\},$$

i.e., when G depends on t (see [192]).

6.2.4 Numerical tests

Consider the Dirichlet problem for the *heat* equation:

$$\frac{\partial u}{\partial t} = \frac{1}{2}\left(1.21 - x_2^2 - x_3^2\right)\frac{\partial^2 u}{\partial x_1^2} + \frac{1}{2}\frac{\partial^2 u}{\partial x_2^2} + \frac{1}{2}\frac{\partial^2 u}{\partial x_3^2} + 6(1 - 0.5e^{-t})$$
$$\times (x_1^2(1.21 - x_2^2 - x_3^2) + x_2^2) + 0.5e^{-t}(1.21 - x_1^4 - x_2^4),$$
$$t \in (0, T], \ x \in U_1, \qquad (2.23)$$

$$u(0, x) = \frac{1}{2}(1.21 - x_1^4 - x_2^4), \quad x \in \bar{U}_1,$$
$$u(t, x) = (1.21 - x_1^4 - x_2^4)(1 - 0.5e^{-t}), \ t \in [0, T], \ x \in \partial U_1, \quad (2.24)$$

where $U_1 \subset \mathbf{R}^3$ is a unit ball with center at the origin. This problem has the solution:

$$u(t, x) = (1.21 - x_1^4 - x_2^4)(1 - 0.5e^{-t}).$$

By changing time $t = T - s$ the problem (2.23)-(2.24) is rewritten in the form (1.1)-(1.2) which is suitable for the probabilistic approach.

The results of simulation of $u(0.6, 0, 0, 0)$ and $u(15, 0, 0, 0)$ by Algorithms 2.1 and 2.6 are given in Table 2.1. The values in Table 2.1 are approximations of $\bar{u} = E(\varphi(t_\varkappa, X_\varkappa)Y_\varkappa + Z_\varkappa)$ evaluated as

$$\bar{u} \doteq \frac{1}{M}\sum_{m=1}^{M}\left(\varphi\left(t_\varkappa^{(m)}, X_\varkappa^{(m)}\right)Y_\varkappa^{(m)} + Z_\varkappa^{(m)}\right) \pm 2\sqrt{\frac{\bar{D}_M}{M}}, \qquad (2.25)$$

where

$$\bar{D}_M = \frac{1}{M}\sum_{m=1}^{M}\left[\varphi\left(t_\varkappa^{(m)}, X_\varkappa^{(m)}\right)Y_\varkappa^{(m)} + Z_\varkappa^{(m)}\right]^2$$
$$- \left[\frac{1}{M}\sum_{m=1}^{M}\left(\varphi\left(t_\varkappa^{(m)}, X_\varkappa^{(m)}\right)Y_\varkappa^{(m)} + Z_\varkappa^{(m)}\right)\right]^2.$$

Thus, assuming that the sampling variance is sufficiently close to $\bar{D}_M$, $\bar{u}$ belongs to the interval defined in (2.25) with probability 0.95. The Monte Carlo error for values in Table 2.1 is not greater than errors of numerical integration. As it follows from this table, results of the experiment are in quite good agreement with theoretical results. Numerical tests of Algorithm 2.9 also gave results corresponding with the theory. We have checked also the rules given by Remarks 2.5 and 2.8. For example, using the results of lower Table 2.1 for $h_1 = 0.0001$, $h_2 = 0.0016$, we get due to formula (2.15): $\bar{u}(0.6, 0, 0, 0) = 0.8780 \pm 0.0004$, $\bar{u}(15, 0, 0, 0) = 1.2098 \pm 0.0008$.

In Appendix A.4 an implementation of Algorithm 2.1 is considered and the program is given (see Listing A.10), by which the results of upper Table 2.1 were obtained.

Table 2.1. Parabolic problem. The results of approximate solution of the problem (2.23)-(2.24) according to Algorithm 2.1 (*the upper table*) and Algorithm 2.6 (*the lower table*) for $\mu \equiv 0$, $F \equiv 0$. The exact solution $u(0.6,0,0,0) \doteq 0.87797$, $u(15,0,0,0) \doteq 1.21000$.

h	M	$\bar{u}(0.6,0,0,0)$	$\approx E\varkappa$	$\bar{u}(15,0,0,0)$	$\approx E\varkappa$
0.04	$1 \cdot 10^3$	0.7805 ± 0.0127	7	1.0340 ± 0.0189	7
0.01	$4 \cdot 10^4$	0.8449 ± 0.0022	30	1.1520 ± 0.0035	32
0.0016	$1 \cdot 10^6$	0.8716 ± 0.0005	193	1.1997 ± 0.0007	207
0.0001	$4 \cdot 10^6$	0.8776 ± 0.0002	3111	1.2095 ± 0.0004	3326

h	M	$\bar{u}(0.6,0,0,0)$	$\approx E\varkappa$	$\bar{u}(15,0,0,0)$	$\approx E\varkappa$
0.04	$1 \cdot 10^3$	0.7127 ± 0.0115	6	0.9560 ± 0.0167	6
0.01	$4 \cdot 10^4$	0.7941 ± 0.0020	27	1.0719 ± 0.0030	28
0.0016	$1 \cdot 10^6$	0.8440 ± 0.0004	183	1.1526 ± 0.0007	193
0.0001	$4 \cdot 10^6$	0.8695 ± 0.0002	3064	1.1955 ± 0.0004	3262

6.3 Random walks for the elliptic Dirichlet problem

Consider the Dirichlet problem for elliptic equation

$$\frac{1}{2}\sum_{i,j=1}^{d} a^{ij}(x) \frac{\partial^2 u}{\partial x^i \partial x^j} + \sum_{i=1}^{d} b^i(x) \frac{\partial u}{\partial x^i} + c(x)\,u + g(x) = 0, \ x \in G, \quad (3.1)$$

$$u\,|_{\partial G} = \varphi(x). \quad (3.2)$$

We assume that the coefficients $a^{ij} = a^{ji}$ satisfy the strict ellipticity condition in $\bar{G}$, i.e.,

$$\lambda_1^2 = \min_{x \in \bar{G}} \min_{1 \leq i \leq d} \lambda_i^2(x) > 0,$$

where $\lambda_1^2(x) \leq \lambda_2^2(x) \leq \cdots \leq \lambda_d^2(x)$ are eigenvalues of the matrix $a(x) = \{a^{ij}(x)\}$. Let $\lambda_d^2 = \max_{x \in \bar{G}} \lambda_d^2(x)$. Then, for any $x \in \bar{G}$ and $y \in \mathbf{R}^d$ the following inequality takes place:

$$\lambda_1^2 \sum_{i=1}^{d}(y^i)^2 \leq \sum_{i,j=1}^{d} a^{ij}(x)\,y^i y^j \leq \lambda_d^2 \sum_{i=1}^{d}(y^i)^2. \quad (3.3)$$

We also suppose that conditions hold which guarantee existence of the unique solution $u(x)$ of the problem (3.1)-(3.2) from the class $C^4(\bar{G})$. We recall [216] that it is sufficient for the above to require that the functions $a^{ij}(x)$, $b^i(x)$, $c(x)$, $g(x)$ are from the class $C^2(\bar{G})$, G is an open domain with twice continuously differentiable boundary ∂G, $\varphi(x) \in C^4(\bar{G})$, and $c(x) \leq 0$, $x \in \bar{G}$.

The solution of the problem (3.1)-(3.2) has the probabilistic representation:
$$u(x) = E\left[\varphi(X_x(\tau))Y_{x,1}(\tau) + Z_{x,1,0}(\tau)\right], \tag{3.4}$$
where $X_x(s)$, $Y_{x,y}(s)$, $Z_{x,y,z}(s)$, $s \geq 0$, is the solution of the Cauchy problem for the system of SDEs:

$$dX = (b(X) - \sigma(X)\mu(X))\,ds + \sigma(X)\,dw(s),\quad X(0) = x, \tag{3.5}$$
$$dY = c(X)Y\,ds + \mu^\mathsf{T}(X)Y\,dw(s),\quad Y(0) = y, \tag{3.6}$$
$$dZ = g(X)Y\,ds + F^\mathsf{T}(X)Y\,dw(s),\quad Z(0) = z, \tag{3.7}$$

$x \in G$, and $\tau = \tau_x$ is the first exit time of the trajectory $X_x(s)$ to the boundary ∂G.

The notation here and in what follows is similar to the notation in Sects. 6.1-6.2.

6.3.1 The simplest random walk for elliptic equations

In this subsection, the results of Sect. 6.2 are carried over to the elliptic problem (3.1)-(3.2). Here we restrict ourselves to the case $\mu \equiv 0$.

First we construct an algorithm of random walk, which is similar to Algorithm 2.1. We apply the weak explicit Euler approximation with the simplest simulation of noise to the system (3.5)-(3.7):

$$X_{t,x}(t+h) \approx X = x + hb(x) + h^{1/2}\sigma(x)\xi, \tag{3.8}$$
$$Y_{t,x,y}(t+h) \approx Y = y + hc(x)\,y, \tag{3.9}$$
$$Z_{t,x,y,z}(t+h) \approx Z = z + hg(x)\,y + h^{1/2}F^\mathsf{T}(x)\,y\,\xi. \tag{3.10}$$

Introduce the boundary zone $S_h \subset \bar{G}$: $x \in S_h$ if at least one of the 2^d values of the vector X is outside $\bar{G}$. Let a constant $\lambda > 0$ be such that if the distance from $x \in G$ to the boundary ∂G is equal to or greater than $\lambda\sqrt{h}$, then x is outside the boundary zone and, therefore, for such x all the realizations of the random variable X belong to $\bar{G}$.

Let $x \in S_h$. Introduce the random vector $X^\pi_{x,h}$ taking two values x^π and $x + h^{1/2}\lambda n(x^\pi)$ with probabilities $p = p_{x,h}$ and $q = q_{x,h} = 1 - p_{x,h}$, respectively, where

$$p_{x,h} = \frac{h^{1/2}\lambda}{|x + h^{1/2}\lambda n(x^\pi) - x^\pi|},$$

$x^\pi \in \partial G$ is the projection of the point x on the boundary ∂G, and $n(x^\pi)$ is the unit vector of internal normal to ∂G at x^π.

To approximate the solution of the system (3.5), we construct a Markov chain X_k which stops when it reaches the boundary ∂G at a random step $\varkappa$.

We set $X'_0 = x_0$. If $X'_0 \notin S_h$ then we take $X_0 = X'_0$. If $X'_0 \in S_h$ then the random variable X_0 takes two values: either $X'^\pi_0 \in \partial G$ with probability

6.3 Random walks for the Dirichlet problem for elliptic equations   367

$p_{X'_0,h}$ or $X'_0 + h^{1/2}\lambda n(X_0^{'\pi}) \notin S_h$ with probability $q_{X'_0,h}$. If $X_0 = X_0^{'\pi}$, we put $\varkappa = 0$, $X_\varkappa = X_0^{'\pi}$, and the random walk is finished. Let X_k be constructed and either $X_k \in \partial G$ (i.e., the chain is stopped at one of the previous steps and, consequently, $\varkappa \leq k$) or $X_k \notin S_h$. Now our aim is to construct X_{k+1} with the same properties. Suppose that the chain does not stop until the step k inclusively, i.e., $\varkappa > k$. Introduce X'_{k+1} due to (3.8) with $t = t_k$, $x = X_k$, $\xi = \xi_k$:

$$X'_{k+1} = X_k + hb(X_k) + h^{1/2}\sigma(X_k)\,\xi_k\,. \tag{3.11}$$

Now we obtain X_{k+1} using X'_{k+1} as we got X_0 using X'_0. More precisely, we use the following rule. If $X'_{k+1} \notin S_h$ then we take $X_{k+1} = X'_{k+1}$. If $X'_{k+1} \in S_h$ then the random variable X_{k+1} takes two values: either $X_{k+1}^{'\pi} \in \partial G$ with probability $p_{X'_{k+1},h}$ or $X'_{k+1} + h^{1/2}\lambda n(X_{k+1}^{'\pi}) \notin S_h$ with probability $q_{X'_{k+1},h}$. If $X_{k+1} = X_{k+1}^{'\pi}$, we put $\varkappa = k+1$, $X_\varkappa = X_{k+1}^{'\pi}$, and the random walk is finished. So, the random walk X_k has been constructed.

Clearly, X_k remains in the domain $\bar{G}$ with probability 1.

We also introduce an extended chain (X_k, Y_k, Z_k). We put $Y_0 = 1$ and $Z_0 = 0$. Let $\varkappa \geq k+1$ and Y_k, Z_k be known. Then the values Y_{k+1}, Z_{k+1} are evaluated in accordance with the system (3.9)-(3.10) for $x = X_k$, $y = Y_k$, $z = Z_k$, $\xi = \xi_k$.

The constructed algorithm can be written as follows.

Algorithm 3.1
STEP 0. $X'_0 = x_0$, $Y_0 = 1$, $Z_0 = 0$, $k = 0$.

STEP 1.   If $X'_k \notin S_{t_k,h}$ then $X_k = X'_k$ and go to STEP 3.
If $X'_k \in S_{t_k,h}$ then either $X_k = X_k^{'\pi}$ with probability $p_{X'_k,h}$ or $X_k = X'_k + h^{1/2}\lambda n(X_k^{'\pi})$ with probability $q_{X'_k,h}$.

STEP 2.   If $X_k = X_k^{'\pi}$ then STOP and $\varkappa = k$,
$X_\varkappa = X_k^{'\pi}$, $Y_\varkappa = Y_k$, $Z_\varkappa = Z_k$.

STEP 3.   Simulate ξ_k and find X'_{k+1}, Y_{k+1}, Z_{k+1} according to (3.11), (3.9)-(3.10) for $x = X_k$, $y = Y_k$, $z = Z_k$, $\xi = \xi_k$.

STEP 4.   Put $k := k+1$ and return to STEP 1.

This algorithm is similar to Algorithm 2.1 for parabolic equation. At the same time, we emphasize that here $\varkappa$ can take arbitrary large values in contrast to Algorithm 2.1.

Denote by ν_{x_0} the number of those k at which X'_k gets into the set S_h.

Lemma 3.2. *The following inequalities hold:*

$$P\{\nu_{x_0} = n\} \leq \frac{1}{2^{n-1}}\,, \quad E\nu_{x_0} \leq C, \tag{3.12}$$

$$E\varkappa \leq \frac{C}{h}, \qquad (3.13)$$

where C does not depend on t_0 and h.

Proof. The inequalities (3.12) are proved analogously to Lemma 2.2. Let us prove the inequality (3.13). Denote by μ_x the number of steps which the chain X'_k starting from $x \in G\backslash S_h$ spends in the domain $G\backslash S_h$ before X'_k gets into S_h. In connection with the chain X'_k we consider the boundary value problem (cf. (1.31)-(1.32)):

$$PV - V = -f(x), \quad x \in G\backslash S_h, \qquad (3.14)$$
$$V(x) = 0, \quad x \in S_h, \qquad (3.15)$$

where P is the one-step transition operator: $PV(x) = EV(X'_1)$, $X'_0 = x$.

It is known [301] that the solution of this problem is the function:

$$V(x) = E \sum_{k=0}^{\mu_x - 1} f(X'_k). \qquad (3.16)$$

If we find the solution $V(x)$ of (3.14)-(3.15) with a function $f(x)$ which everywhere in $G\backslash S_h$ satisfies the condition

$$f(x) \geq I_{G\backslash S_h}(x), \qquad (3.17)$$

then, due to (3.16),

$$E\mu_x \leq V(x). \qquad (3.18)$$

We take $V(x)$ of the form [124]

$$V(x) = \begin{cases} A^2 - |x + B|^{2n}, & x \in G\backslash S_h, \\ 0, & x \in S_h, \end{cases} \qquad (3.19)$$

where B is a d-dimensional vector such that

$$\min_{x \in \bar{G}} |x + B| \geq C > 0, \qquad (3.20)$$

n is a sufficiently large natural number (how to choose it is shown below), $A^2 = \max_{x \in \bar{G}} |x + B|^{2n}$. The function $V(x)$ satisfies the boundary condition (3.15).

Let $x \in G\backslash S_h$, then $X'_1 = x + hb(x) + h^{1/2}\sigma(x)\xi$. It is not difficult to obtain (recall that $a = \sigma\sigma^\mathsf{T}$):

$$PV(x) - V(x) = -hn|x + B|^{2n-4}$$
$$\times \left[2|x + B|^2(x + B, b(x)) + |x + B|^2 \sum_{i=1}^{d} a^{ii} \right.$$
$$\left. + 2(n-1) \sum_{i,j=1}^{d} a^{ij}(x)(x^i + B^i)(x^j + B^j) \right] + O(h^2). \qquad (3.21)$$

6.3 Random walks for the Dirichlet problem for elliptic equations 369

The relations (3.21) and (3.3) imply

$$PV - V \leq -hn|x + B|^{2n-2}\left[2(x + B, b(x)) + (2n - 2 + d)\lambda_1^2\right] + O(h^2).$$

We select n so that for all $x \in \bar{G}$:

$$2(x + B, b(x)) + (2n - 2 + d)\lambda_1^2 \geq C > 0$$

that is always possible. Then, for a sufficiently small h we obtain $PV - V \leq -\gamma h$, where $\gamma > 0$ is independent of h and x.

Obviously, the function $v(x) = V(x)/(\gamma h)$ is the solution of the problem (3.14)-(3.15) with $f(x) \geq 1$. Therefore, (see (3.17)-(3.19)) $E\mu_x \leq A^2/(\gamma h)$. From here and the inequalities (3.12), it is not difficult to obtain the estimate (3.13). Lemma 3.2 is proved. □

We extend the definition of the constructed chain for all k by the rule: if $k > \varkappa$, then $(X_k, Y_k, Z_k) = (X_\varkappa, Y_\varkappa, Z_\varkappa)$. Introduce

$$d_k = u(X_k)\,Y_k + Z_k - u(X_k')\,Y_k - Z_k,$$
$$d_k' = u(X_{k+1}')\,Y_{k+1} + Z_{k+1} - u(X_k)\,Y_k - Z_k, \quad k = 0, 1, \ldots.$$

The lemma on one-step errors is proved analogously to Lemma 2.3.

Lemma 3.3. *The following inequalities hold:*

$$|E(d_k \mid X_k', Y_k, Z_k)| \leq ChY_k I_{S_h}(X_k')\chi_{\varkappa \geq k}, \tag{3.22}$$

$$|E(d_k' \mid X_k, Y_k, Z_k)| \leq Ch^2 Y_k \chi_{\varkappa > k}. \tag{3.23}$$

Theorem 3.4. *Algorithm 3.1 has weak order of accuracy $O(h)$:*

$$|E(\varphi(X_\varkappa)\,Y_\varkappa + Z_\varkappa) - u(x_0)| \leq Ch, \tag{3.24}$$

where C is independent of x_0, h.

Proof. We have

$$R := E(\varphi(X_\varkappa)\,Y_\varkappa + Z_\varkappa) - u(x_0) = E(u(X_\varkappa)\,Y_\varkappa + Z_\varkappa) - u(x_0)$$
$$= E(u(X_0)\,Y_0 + Z_0) - u(x_0)$$
$$+ E\sum_{k=0}^{\varkappa-1}(u(X_{k+1})\,Y_{k+1} + Z_{k+1} - u(X_k)\,Y_k - Z_k)$$
$$= E\sum_{k=0}^{\infty}(d_k + d_k') = \sum_{k=0}^{\infty}EE(d_k' \mid X_k, Y_k, Z_k) + \sum_{k=0}^{\infty}EE(d_k \mid X_k', Y_k, Z_k).$$

It is obvious that $Y_k > 0$ and (since $c(x) \leq 0$) $Y_k \leq 1$, $k = 0, 1, \ldots$. Using Lemmas 3.2 and 3.3, we obtain

$$|R| \le Ch^2 \sum_{k=0}^{\infty} E(Y_k \chi_{\varkappa > k}) + Ch \sum_{k=0}^{\infty} E(Y_k I_{S_h}(X_k) \chi_{\varkappa \ge k})$$
$$\le Ch^2 E\varkappa + Ch E\nu_{x_0} \le Ch,$$

where C is independent of x_0, h. Theorem 3.4 is proved. $\square$

Remark 3.5. It is possible to construct algorithms of random walks analogous to Algorithms 2.6 and 2.9 for the elliptic problem (3.1)-(3.2).

6.3.2 Other methods for elliptic problems

In this subsection we construct two methods for the elliptic problem (3.1)-(3.2) which are similar to the methods for the parabolic problem from Sect. 6.1. Here we use the probabilistic representation (3.4)-(3.7) with

$$\mu(x) = \sigma^{-1}(x)b(x), \qquad (3.25)$$

and we restrict ourselves to the case $F(x) = 0$. Then, the system (3.5)-(3.7) has the form

$$dX = \sigma(X)dw(t), \quad X(0) = x, \qquad (3.26)$$
$$dY = c(X)Y dt + (\sigma^{-1}(X)b(X))^\top Y dw(t), \quad Y(0) = y, \qquad (3.27)$$
$$dZ = g(X)Y dt, \quad Z(0) = z. \qquad (3.28)$$

Denote by $U_r \subset \mathbf{R}^d$ a sphere of radius r with center at the origin. Let $x \in G$. Consider the one-step approximation of the solution to the system (3.26)-(3.28):

$$X_1 = x + \sigma(x)w(\vartheta) \qquad (3.29)$$
$$Y_1 = y + yc(x)\frac{r^2}{d} + y\mu^\top(x)w(\vartheta) \qquad (3.30)$$
$$Z_1 = z + yg(x)\frac{r^2}{d}, \qquad (3.31)$$

where $w(\vartheta)$ has the uniform distribution on the sphere ∂U_r and r is such that the ellipsoid $(\sigma^{-1}(x)(X-x), \sigma^{-1}(x)(X-x)) = r^2$ belongs to $\bar{G}$.

It is not difficult to prove the following lemma (see details in [186]).

Lemma 3.6. *The one-step order of the approximation (3.29)-(3.31) with respect to r is equal to 4:*

$$|E[u(X_1)Y_1 + Z_1] - [u(x)y + z]| \le Cyr^4,$$

where C depends on derivatives of $u(x)$ up to fourth order.

6.3 Random walks for the Dirichlet problem for elliptic equations 371

Denote by Γ_δ the interior of a δ-neighborhood of the boundary ∂G belonging to $\bar{G}$. On the basis of the one-step approximation (3.29)-(3.31) we construct the numerical algorithm.

Algorithm 3.7 *Let $X_0 = x \in G$, $Y_0 = 1$, $Z_0 = 0$, $k = 0$ and r be a small positive number. Let ξ be a point uniformly distributed on a unit sphere with center at the origin and $\xi_1, \xi_2, \ldots$ be independent random points distributed as ξ (see Remark 5.1.3). If $X_k \in G\backslash\Gamma_{\lambda_d r}$ then we set $r_{k+1} = r$, and if $X_k \in \Gamma_{\lambda_d r}\backslash\Gamma_{r^2}$ then we find a number r_{k+1} such that the ellipsoid $(\sigma^{-1}(X_k)(X - X_k), \sigma^{-1}(X_k)(X - X_k)) = r_{k+1}^2$ touches the boundary ∂G. In both cases we evaluate $X_{k+1}, Y_{k+1}, Z_{k+1}$ in accordance with the system*

$$X_{k+1} = X_k + \sigma(X_k) r_{k+1} \xi_{k+1} \tag{3.32}$$

$$Y_{k+1} = Y_k + Y_k c(X_k) \frac{r_{k+1}^2}{d} + Y_k \mu^\top(X_k) r_{k+1} \xi_{k+1} \tag{3.33}$$

$$Z_{k+1} = Z_k + Y_k g(X_k) \frac{r_{k+1}^2}{d}, \tag{3.34}$$

where $\mu(x)$ is from (3.25).

If $X_{k+1} \in \Gamma_{r^2}$ then the algorithm is stopped and we take $(\bar{X}_\varkappa, Y_\varkappa, Z_\varkappa)$ with $\varkappa = k + 1$, $\bar{X}_\varkappa = X_\varkappa^\pi$ as the final state of the Markov chain (recall that $x^\pi \in \partial G$ is the projection of the point x on the boundary of the domain G). If $X_{k+1} \in G\backslash\Gamma_{r^2}$ then we put $k := k+1$ and continue the algorithm.

It is possible to prove (see [186] and also Sect. 6.4.3) that the average number of steps for this algorithm is equal to $O(1/r^2)$ and if $c(x) \leq -c_0 < 0$ then this algorithm has the second order of convergence with respect to r.

Theorem 3.8. *Let $c(x) \leq -c_0 < 0$. Then Algorithm 3.7 has the second order of convergence with respect to r, i.e., for all sufficiently small r*

$$|E(\varphi(\bar{X}_\varkappa) Y_\varkappa - Z_\varkappa) - u(x)| \leq K r^2.$$

Now let us propose another algorithm.

Algorithm 3.9 *Let $X_0 = x \in G$, $Y_0 = 1$, $Z_0 = 0$, $k = 0$ and r be a small positive number. Introduce the random vector η which coordinates η^i, $i = 1, \ldots, d$, are mutually independent random variables taking values $\pm 1/\sqrt{d}$ with probability $1/2$. Let $\eta_1, \eta_2, \ldots$ be independent random points distributed as η. If $X_k \in G\backslash\Gamma_{\lambda_d r}$ then we set $r_{k+1} = r$, and if $X_k \in \Gamma_{\lambda_d r}\backslash\partial G$ then we find a minimal number r_{k+1} such that one of the points from the set $\{X : X = X_k + \sigma(X_k) r_{k+1} \eta_{k+1}\}$ belongs to ∂G. In both cases we evaluate $X_{k+1}, Y_{k+1}, Z_{k+1}$ in accordance with the system*

$$X_{k+1} = X_k + \sigma(X_k) r_{k+1} \eta_{k+1} \tag{3.35}$$

$$Y_{k+1} = Y_k + Y_k c(X_k)\frac{r_{k+1}^2}{d} + Y_k \mu^\top(X_k) r_{k+1} \eta_{k+1} \qquad (3.36)$$

$$Z_{k+1} = Z_k + Y_k g(X_k)\frac{r_{k+1}^2}{d}, \qquad (3.37)$$

where $\mu(x)$ is from (3.25).

In the second case the point X_{k+1} with probability $1/2^d$ falls on ∂G. Let $\varkappa = \varkappa_x$ be the first number at which $X_\varkappa \in \partial G$. The random walk is stopped at this random step $\varkappa$ and $(X_\varkappa, Y_\varkappa, Z_\varkappa)$ is taken as the final state of the Markov chain.

Note that the random walk X_k obtained in Algorithm 3.9 gets a finite number of values at each step (it is equal to 2^d) in contrast to the random walk in Algorithm 3.7 and it does not require any neighborhood Γ_δ of the boundary ∂G.

It is not difficult to prove that the average number of steps for Algorithm 3.9 is equal to $O(1/r^2)$ and if $c(x) \leq -c_0 < 0$ then this algorithm has the second order of convergence with respect to r [186].

Theorem 3.10. *Let* $c(x) \leq -c_0 < 0$. *Then Algorithm 3.9 has the second order of convergence with respect to* r, *i.e., for all sufficiently small* r

$$|E(\varphi(X_\varkappa)Y_\varkappa - Z_\varkappa) - u(x)| \leq K r^2.$$

6.3.3 Numerical tests

Consider the Dirichlet problem for the elliptic equation (cf. (2.23)-(2.24)):

$$\frac{1}{2}(1.21 - x_2^2 - x_3^2)\frac{\partial^2 u}{\partial x_1^2} + \frac{1}{2}\frac{\partial^2 u}{\partial x_2^2} + \frac{1}{2}\frac{\partial^2 u}{\partial x_3^2}$$
$$+ 6x_1^2(1.21 - x_2^2 - x_3^2) + 6x_2^2 = 0, \; x \in U_1, \qquad (3.38)$$
$$u\,|_{\partial U_1} = 1.21 - x_1^4 - x_2^4. \qquad (3.39)$$

The solution of this problem is

$$u(x) = 1.21 - x_1^4 - x_2^4.$$

The results of simulation of $u(0,0,0)$ and $u(0.5,0,0)$ by Algorithms 3.1 and 3.7 are given in Tables 3.1 and 3.2, respectively. The values in Tables 3.1 (Table 3.2) are approximations of $\bar{u} = E(\varphi(X_\varkappa)Y_\varkappa + Z_\varkappa)$ ($\bar{u} = E(\varphi(\bar{X}_\varkappa)Y_\varkappa + Z_\varkappa)$) and evaluated by a formula analogous to (2.25). The Monte Carlo error for values in the tables is not greater than errors of numerical integration. Analyzing the results presented in Table 3.1, we can conclude that the error of numerical integration by Algorithm 3.1 is proportional to Ch and the mean number of steps $E\varkappa$ is proportional to K/h. We see from Table 3.2 that the error of Algorithm 3.7 is proportional to Cr^2 and the mean number

Table 3.1. Elliptic problem. The results of approximate solution of the problem (3.38)-(3.39) according to Algorithm 3.1 for $F \equiv 0$. The exact solution $u(0,0,0) = 1.21$, $u(0.5,0,0) = 1.1475$.

h	M	$\bar{u}(0,0,0)$	$\approx E\varkappa$	$\bar{u}(0.5,0,0)$	$\approx E\varkappa$
0.04	10^3	1.0337 ± 0.0180	7	0.9361 ± 0.0239	5.1
0.01	4×10^4	1.1504 ± 0.0035	32	1.0810 ± 0.0042	23.6
0.0016	10^6	1.1999 ± 0.0007	207	1.1359 ± 0.0009	155
0.0001	4×10^6	1.2093 ± 0.0004	3325	1.1467 ± 0.0004	2477

Table 3.2. Elliptic problem. The results of approximate solution of the problem (3.38)-(3.39) according to Algorithm 3.7. The exact solution $u(0,0,0) = 1.21$, $u(0.5,0,0) = 1.1475$.

r	M	$\bar{u}(0,0,0)$	$\approx E\varkappa$	$\bar{u}(0.5,0,0)$	$\approx E\varkappa$
0.4	10^4	1.0346 ± 0.0062	5.8	0.9382 ± 0.0075	4.0
0.2	10^4	1.1578 ± 0.0071	23.8	1.0912 ± 0.0082	17.8
0.1	10^4	1.1910 ± 0.0073	99.0	1.1289 ± 0.0084	74.0
0.08	10^4	1.1983 ± 0.0074	153.5	1.1354 ± 0.0085	113.6

of steps $E\varkappa$ is proportional to K/r^2. To avoid a confusion, we recall that h is step in time and r is step in space and $h \sim r^2$. Comparing the results of Tables 3.1 and 3.2, we obtain that the constant C at h in the error of numerical integration is larger for Algorithm 3.1 while the constant K at $1/h$ in the estimate of the mean number of steps $E\varkappa$ is larger for Algorithm 3.7.

We can reduce the Monte Carlo error in simulations by Algorithm 3.1. As it was mentioned in Sect. 6.1, it is possible to select the function F so that the variance $\bar{D}$ of the random variable $\bar{\eta} = \varphi(X_\varkappa) Y_\varkappa + Z_\varkappa$ related to the discrete system will decrease. Such a selection allows us to reduce computational costs. For instance, if we take

$$F^\intercal = \left(4x_1^3 \sqrt{1.21 - x_2^2 - x_3^2},\ 4x_2^3,\ 0 \right),$$

then it is not difficult to show that the variance D of the random variable $\eta = \varphi(X_x(\tau)) Y_{x,1}(\tau) + Z_{x,1,0}(\tau)$ related to the system of SDEs is equal to zero (see Theorem 2.4.5). Since the accuracy order of the method is $O(h)$, the variance $\bar{D}$ satisfies the inequality $\bar{D} \leq Ch$, where C is independent of x, h. Therefore, the Monte Carlo error is bounded from above by $C\sqrt{h/M}$. In numerical experiments, making use of the function F selected as above leads to a significant decrease of computational costs in comparison with simulations when $F \equiv 0$.

6.4 Specific random walks for elliptic equations and boundary layer

In this section, we construct some effective methods for the Dirichlet problem for the elliptic equation (cf. (3.1)-(3.2)):

$$\frac{a^2}{2}\Delta u + \sum_{i=1}^{d} b^i(x)\frac{\partial u}{\partial x^i} + c(x)u + g(x) = 0, \quad x \in G, \quad u\,|_{\partial G} = \varphi(x), \quad (4.1)$$

We assume that

(i) G is an open bounded set with twice continuously differentiable boundary ∂G;

(ii) the coefficients $a^{ij}(x)$, $b^i(x)$, $c(x)$, $g(x)$ belong to the class $C^2(\bar{G})$, $c(x) \leq 0$, $\varphi \in C^4(\partial G)$.

These conditions ensure the existence of the unique solution $u(x)$ of the problem (4.1) belonging to the class $C^4(\bar{G})$ [216].

The solution of the problem (4.1) has the probabilistic representation (cf. (3.4), (3.26)-(3.28)):

$$u(x) = E(\varphi(X_x(\tau))Y_{x,1}(\tau) + Z_{x,1,0}(\tau)), \quad (4.2)$$

where $X_x(t)$, $Y_{x,y}(t)$, $Z_{x,y,z}(t)$, $t \geq 0$, is the solution of the system

$$dX = a\,dw(t), \quad X(0) = x, \quad (4.3)$$

$$dY = c(X)Y\,dt + \frac{1}{a}b^\mathsf{T}(X)Y\,dw(t), \quad Y(0) = y, \quad (4.4)$$

$$dZ = g(X)Y\,dt, \quad Z(0) = z. \quad (4.5)$$

In (4.3)-(4.5), $b(x) = (b^1(x), \ldots, b^d(x))^\mathsf{T}$, $w(t) = (w^1(t), \ldots, w^d(t))^\mathsf{T}$ is a standard Wiener process which is defined on a probabilistic space $(\Omega, \mathcal{F}, P)$ and measurable with respect to the flow $\mathcal{F}_t$, $t \geq 0$, and τ is the first passage time of the trajectory $X_x(t)$ to the boundary ∂G. For definiteness, we always set $y > 0$. The simplicity of the equation (4.3) allows us to simulate its solutions exactly.

Let Γ_δ (Γ_{ar}) be the interior of a δ-neighborhood (of an ar-neighborhood) of the boundary ∂G belonging to $\bar{G}$. Usually, r is taken sufficiently small and δ is taken smaller than ar, $\delta = O(r^q)$, $q > 1$.

Consider the following random walk over small spheres which starts at $x \in G \backslash \Gamma_\delta$. We set $X_0 = x$. If $X_0 \in G \backslash \Gamma_{ar}$ then the next point X_1 is found as a uniformly distributed point on the sphere of radius ar with center at X_0. If $X_0 \in \Gamma_{ar} \backslash \Gamma_\delta$ then X_1 has the uniform distribution on the tangent sphere of radius $\rho(X_0, \partial G)$. Clearly, X_1 can be interpreted in both cases as

$$X_1 = X_0 + aw(\vartheta_1),$$

where ϑ_1 is the first passage time of the process $X_0 + aw(t)$ to the corresponding sphere. We can repeat this starting from X_1 and so on. Let $\varkappa = \varkappa_x$ be the first number at which $X_\varkappa \in \Gamma_\delta$ and set $\vartheta_\kappa = 0$ for $k > \varkappa$ and $X_k = X_\varkappa$ for $k \geq \varkappa$. As a result, we obtain the random walk

$$X_0 = x$$
$$X_1 = X_0 + aw(\vartheta_1)$$
$$\dots\dots\dots\dots\dots\dots$$
$$X_k = X_{k-1} + a(w(\vartheta_1 + \cdots + \vartheta_k) - w(\vartheta_1 + \cdots + \vartheta_{k-1})),\ k = 1, \ldots, \varkappa,$$
$$X_k = X_\varkappa,\ k \geq \varkappa,$$

which stops at a random step $\varkappa$. It is a Markov chain. We emphasize that, fortunately, we need not solve a quite difficult problem of simulating the first passage times ϑ_k to find X_k (see the corresponding discussion in the introduction to Chap. 5).

Let $\mathcal{B}_k = \sigma(X_0, X_1, \ldots, X_k)$, $k = 1, 2, \ldots$, be the sequence of σ-algebras generated by the random walk $X_0, X_1, \ldots, X_k, \ldots$. Presuppose that a method approximating the solution of (4.3)-(4.5) is proposed and the sequences $Y_0, Y_1, \ldots, Y_k, \ldots$, $Z_0, Z_1, \ldots, Z_k, \ldots$ which approximate $Y_{x,y}(\vartheta_1 + \cdots + \vartheta_k)$, $Z_{x,y,z}(\vartheta_1 + \cdots + \vartheta_k)$, $k = 1, 2, \ldots$, respectively, are constructed so that Y_k, Z_k are $\mathcal{B}_k$-measurable and they are stopped at the random step $\varkappa$. Let $\bar{X}_\varkappa$ be the point on the boundary ∂G closest to $X_\varkappa$. Put $\bar{Y}_\varkappa = Y_\varkappa$, $\bar{Z}_\varkappa = Z_\varkappa$. Introduce the function

$$v(x,y,z) = E(\varphi(X_x(\tau))Y_{x,y}(\tau) + Z_{x,y,z}(\tau)). \qquad (4.6)$$

Clearly,
$$v(x,y,z) = u(x)y + z. \qquad (4.7)$$

We are interested in the difference

$$\mathcal{R} = Ev(\bar{X}_\varkappa, \bar{Y}_\varkappa, \bar{Z}_\varkappa) - Ev(X_x(\tau), Y_{x,y}(\tau), Z_{x,y,z}(\tau))$$
$$= Ev(\bar{X}_\varkappa, \bar{Y}_\varkappa, \bar{Z}_\varkappa) - v(x,y,z)$$

since $Ev(\bar{X}_\varkappa, \bar{Y}_\varkappa, \bar{Z}_\varkappa) = E(\varphi(\bar{X}_\varkappa)\bar{Y}_\varkappa + \bar{Z}_\varkappa) = E(\varphi(\bar{X}_\varkappa)Y_\varkappa + Z_\varkappa)$ is taken as an approximation of $v(x,y,z) = u(x)y + z$. We have

$$v(\bar{X}_\varkappa, \bar{Y}_\varkappa, \bar{Z}_\varkappa) - v(x,y,z) = (v(\bar{X}_\varkappa, \bar{Y}_\varkappa, \bar{Z}_\varkappa) - v(X_\varkappa, Y_\varkappa, Z_\varkappa))$$
$$+(v(X_\varkappa, Y_\varkappa, Z_\varkappa) - v(X_{\varkappa-1}, Y_{\varkappa-1}, Z_{\varkappa-1})) + \cdots + (v(X_1, Y_1, Z_1) - v(x,y,z))$$
$$= (u(\bar{X}_\varkappa) - u(X_\varkappa))Y_\varkappa + \sum_{k=1}^{\infty}(v(X_k, Y_k, Z_k) - v(X_{k-1}, Y_{k-1}, Z_{k-1}))\chi_{\varkappa \geq k}$$

and, consequently,

$$\mathcal{R} = E(u(\bar{X}_\varkappa) - u(X_\varkappa))Y_\varkappa$$
$$+ \sum_{k=1}^{\infty} E(v(X_k, Y_k, Z_k) - v(X_{k-1}, Y_{k-1}, Z_{k-1}))\chi_{\varkappa \geq k}. \qquad (4.8)$$

An estimate of $\mathcal{R}$ depends on a bound of the first term and on a one-step approximation which gives bounds for the summands in right-hand side of (4.8). Our aim is to find one-step approximations which do not require simulation of ϑ_k and errors of which can be bounded without using any derivatives or at least without using high-order derivatives of the solution $u(x)$ of the original problem (4.1). The latter is very important for problems with small parameter at higher derivatives because a boundary layer arises in such a situation, and the higher derivatives the larger their values. Such approximations are based on simulation of some conditional expectations like

$$\xi^i = E(\int_0^\vartheta w^i(s)ds/w(\vartheta)), \quad \xi^{ij} = E(\int_0^\vartheta w^i(s)dw^j(s)/w(\vartheta)).$$

Section 6.4.1 is devoted to auxiliary lemmas and to simulation of ξ^i, ξ^{ij}. Various one-step approximations are constructed in Sect. 6.4.2.

The estimate of the first term in (4.8) essentially depends on δ. The average number of steps $E\varkappa$ also depends on δ. A choice of δ is connected with accuracy of a one-step approximation. As usual, $\delta = O(r^k)$ if the order of one-step approximation is equal to $O(r^{k+2})$. The convergence theorems are proved in Sects. 6.4.3 and 6.4.4 which start with theorems on the average number of steps $E\varkappa$ and with other results relevant for evaluation of the sum in (4.8) (see Sect. 6.4.3).

In the case of a small parameter at second derivatives (this case is treated in Sect. 6.5), the system (4.3)-(4.5) becomes a system with small noise and we construct some specific methods for its approximate integration. Another way rests on the fact that in the almost entire domain G except a narrow boundary layer the solution of the Dirichlet problem can be found with high accuracy and simply by analytical tools (this part of the solution is known as an external expansion). Having this in mind, we propose a method of random walk in the narrow layer to find the remaining part of the solution (known as an interior expansion). The effectiveness of this analytic-numerical method is achieved because of small average number of steps for the random paths in the very narrow domain.

6.4.1 Conditional expectation of Ito integrals connected with Wiener process in the ball

Here we will exploit both a probabilistic representation and an explicit form of the solution of the Dirichlet problem for the Poisson equation in the ball $U_r = \{x = (x^1, \ldots, x^d): |x|^2 = (x^1)^2 + \cdots + (x^d)^2 \leq r^2\}$:

$$\frac{1}{2}\Delta u + g(x) = 0, \quad |x| < r, \tag{4.9}$$

$$u\,|_{|x|=r} = \varphi(x). \tag{4.10}$$

In (4.9)-(4.10), $g(x) \in C^1(|x| \leq r)$, $\varphi(x) \in C(|x| = r)$.

6.4 Specific random walks for elliptic equations

The probabilistic representation of the solution of the problem (4.9)-(4.10) has the form

$$u(x) = E\varphi(x + w(\vartheta_x)) + E \int_0^{\vartheta_x} g(x + w(s))ds, \quad (4.11)$$

where $w(t) = (w^1(t), \ldots, w^d(t))^\mathsf{T}$ is a d-dimensional standard Wiener process and ϑ_x is the first passage time of the process $x + w(t)$ to the sphere ∂U_r.

The explicit formula for the solution has the following form [216]:

$$u(x) = \int_{|\xi|=r} P_r(x,\xi)\varphi(\xi)\,dS_\xi + \int_{|\xi|<r} G_r(x,\xi)g(\xi)\,d\xi, \quad (4.12)$$

where P_r is the Poisson kernel:

$$P_r(x,\xi) = \frac{r^2 - |x|^2}{\sigma_d\, r\, |x-\xi|^d} \quad (4.13)$$

and G_r is the Green function which for $d = 2$ is equal to

$$G_r(x,\xi) = \frac{1}{2\pi} \ln \frac{|x|\,|(r/|x|)^2 x - \xi|}{r|x-\xi|}, \quad d = 2, \quad (4.14)$$

and for $d > 2$ is equal to

$$G_r(x,\xi) = \frac{1}{(d-2)\sigma_d} \Big(\frac{1}{|x-\xi|^{d-2}} - \frac{(r/|x|)^{d-2}}{|(r/|x|)^2 x - \xi|^{d-2}}\Big), \quad d > 2. \quad (4.15)$$

In (4.13) and (4.15), σ_d is the area of the unit sphere in $\mathbf{R}^d$: $\sigma_d = 2\pi^{d/2}/\Gamma(d/2)$. Recall that $\sigma_d\, r^{d-1}$ is the area of the sphere ∂U_r and $\sigma_d\, r^d/d$ is the volume of the ball U_r.

Proceeding to simulation of the conditional expectation $E(\int_0^\vartheta w^i(s)ds /w(\vartheta))$ where $\vartheta = \vartheta_0$ is the first passage time of the Wiener process $w(t)$ to the sphere ∂U_r, we assume that

$$E(\int_0^\vartheta w^i(s)ds/w(\vartheta)) = \alpha w^i(\vartheta), \quad i = 1, \ldots, d. \quad (4.16)$$

If (4.16) is true then the constant α can be found from the condition

$$E(\int_0^\vartheta w^i(s)ds - \alpha w^i(\vartheta))^2 \longrightarrow \min_\alpha,$$

i.e.,

$$\alpha = \frac{Ew^i(\vartheta)\int_0^\vartheta w^i(s)ds}{E[w^i(\vartheta)]^2}. \quad (4.17)$$

Lemma 4.1. *For every* $i = 1, \ldots, d$ *the following formulas hold:*

$$E(w^i(\vartheta))^2 = \frac{r^2}{d}, \qquad (4.18)$$

$$Ew^i(\vartheta) \int_0^\vartheta w^i(s)ds = E\int_0^\vartheta (w^i(s))^2 ds = \frac{r^4}{2d(d+2)}, \qquad (4.19)$$

and, consequently, α *from* (4.17) *is equal to*

$$\alpha = \frac{r^2}{2(d+2)}. \qquad (4.20)$$

Proof. The relation (4.18) is evidently follows from the identity $(w^1(\vartheta))^2 + \cdots + (w^d(\vartheta))^2 = r^2$. Further, Ito's formula implies

$$dw^i(t) \int_0^t w^i(s)ds = \int_0^t w^i(s)ds \times dw^i(t) + (w^i(t))^2 dt$$

and, therefore,

$$Ew^i(\vartheta) \int_0^\vartheta w^i(s)ds = E\int_0^\vartheta (w^i(s))^2 ds.$$

It is not difficult to verify that the function $u = r^4 - |x|^4$ is a solution to the problem

$$\frac{1}{2}\Delta u + 2(d+2)|x|^2 = 0, \quad u|_{|x|=r} = 0.$$

Therefore (see (4.11))

$$u(0) = r^4 = 2(d+2) E\int_0^\vartheta \sum_{k=1}^d (w^k(s))^2 ds = 2d(d+2) E\int_0^\vartheta (w^i(s))^2 ds \quad (4.21)$$

that gives (4.19). $\square$

Thus, the hypothesis (4.16) is true. The following theorems are proved in [186].

Theorem 4.2. *For every* $i = 1, \ldots, d$ *the equality*

$$E(\int_0^\vartheta w^i(s)ds/w(\vartheta)) = \frac{r^2}{2(d+2)} w^i(\vartheta) \qquad (4.22)$$

holds

Theorem 4.3. *For every* $i, j = 1, \ldots, d$ *the following formulas hold:*

$$E[\int_0^\vartheta w^i(s)dw^i(s)/w(\vartheta)] = \frac{1}{2}[w^i(\vartheta)]^2 - \frac{r^2}{2d}, \qquad (4.23)$$

$$E[\int_0^\vartheta w^i(s)dw^j(s)/w(\vartheta)] = \frac{1}{2}w^i(\vartheta)w^j(\vartheta), \quad i \neq j. \qquad (4.24)$$

Introduce the functions
$$h_m(x) = E\vartheta_x^m, \quad m = 1, 2, \ldots,$$
where $x \in U_r$ and ϑ_x is the first passage time of the process $x + w(t)$ to the sphere ∂U_r.

As it follows from one of Dynkin's theorems (see [48, Theorem 13.17]), the function $h_m(x)$ is the only solution of the Dirichlet problem
$$\frac{1}{2}\Delta h_1 + 1 = 0, \quad h_1\mid_{\partial U_r} = 0,$$
$$\frac{1}{2}\Delta h_m + m h_{m-1}(x) = 0, \quad h_m\mid_{\partial U_r} = 0, \quad m = 2, 3, \ldots. \quad (4.25)$$

The solution of this problem is obviously a function of the variable $\chi = (x,x)^{1/2} = |x|$, $0 \le \chi \le r$. We denote this function as $q_m(\chi)$. We easily obtain the following boundary value problem for $d > 1$ (we recall that d is a dimension of the Wiener process $w(t)$):

$$\frac{1}{2}q_1'' + \frac{d-1}{2\chi}q_1' + 1 = 0, \quad q_1(0) < \infty, \quad q_1(r) = 0,$$
$$\frac{1}{2}q_m'' + \frac{d-1}{2\chi}q_m' + m q_{m-1}(\chi) = 0, \quad q_m(0) < \infty, \quad q_m(r) = 0. \quad (4.26)$$

We note that if $d = 1$ then (4.25) can be rewritten in the form
$$\frac{1}{2}h_m'' + m h_{m-1}(x) = 0, \quad h_m(-r) = h_m(r) = 0.$$

The equations (4.26) are solvable by quadratures. Their solution has the form
$$q_m(\chi) = \alpha_0 \chi^{2m} + \alpha_1 \chi^{2(m-1)} r^2 + \alpha_2 \chi^{2(m-2)} r^4 + \cdots + \alpha_m r^{2m}.$$

Then, we can sequentially obtain:
$$h_1(x) = \frac{r^2 - |x|^2}{d},$$
$$h_2(x) = \frac{|x|^4}{d(d+2)} - \frac{2r^2|x|^2}{d^2} + \frac{(d+4)r^4}{d^2(d+2)},$$

and so on. In particular
$$E\vartheta = \frac{r^2}{d}, \quad E\vartheta^2 = \frac{d+4}{d^2(d+2)}r^4, \quad D\vartheta = \frac{2}{d^2(d+2)}r^4. \quad (4.27)$$

But such formulas become complicated with growth of m. For example,
$$E\vartheta^3 = \frac{d^2 + 12d + 48}{d^3(d+2)(d+4)}r^6.$$

Therefore, it is useful to obtain some simple bounds for $h_m(x)$ [186].

Lemma 4.4. *The functions $h_m(x)$ are bounded as:*

$$\frac{1}{d^m}(r^2 - |x|^2)^m \leq h_m(x) \leq \frac{m!}{d^m} r^{2m-2}(r^2 - |x|^2), \quad m = 1, 2, \ldots. \quad (4.28)$$

Consequently,

$$\frac{1}{d^m} r^{2m} \leq E\vartheta^m \leq \frac{m!}{d^m} r^{2m} \quad (4.29)$$

and for $\lambda < d/r^2$

$$E\exp(\lambda\vartheta) \leq \frac{d}{d - \lambda r^2}. \quad (4.30)$$

6.4.2 Specific one-step approximations for elliptic equations

In this subsection we construct some one-step approximations of the solution of the system (4.3)-(4.5).

For definiteness, let $X_0 = x \in G \backslash \Gamma_{ar}$. Then, the next point X_1 has the uniform distribution on the sphere of radius ar with center at X_0:

$$X_1 = x + aw(\vartheta), \quad (4.31)$$

where ϑ is the first passage time of the solution $X_x(t)$ of the equation (4.3) to the sphere $\partial U_r(x)$ of radius r with center at x.

Consider the solution $X_x(t)$, $Y_{x,y}(t)$, $Z_{x,y,z}(t)$ of the system (4.3)-(4.5) at the time ϑ: $X_x(\vartheta)$, $Y_{x,y}(\vartheta)$, $Z_{x,y,z}(\vartheta)$. As we have just seen, $X_1 = X_x(\vartheta)$ can be simulated exactly. Our aim is to construct an approximation Y_1, Z_1 of $Y_{x,y}(\vartheta)$, $Z_{x,y,z}(\vartheta)$ so that the difference

$$\begin{aligned}
d &= E(v(X_1, Y_1, Z_1) - v(X_x(\vartheta), Y_{x,y}(\vartheta), Z_{x,y,z}(\vartheta))) \\
&= E(u(X_1)Y_1 + Z_1 - u(X_x(\vartheta))Y_{x,y}(\vartheta) - Z_{x,y,z}(\vartheta)) \\
&= Eu(X_x(\vartheta))(Y_1 - Y_{x,y}(\vartheta)) + E(Z_1 - Z_{x,y,z}(\vartheta)) \quad (4.32)
\end{aligned}$$

is small.

Repeatedly applying Ito's formula like the Wagner-Platen expansion (see Sect. 1.2.2), we obtain

$$\begin{aligned}
Y_{x,y}(\vartheta) &= y + \frac{1}{a} y \sum_{i=1}^{d} b^i(x) w^i(\vartheta) + c(x) y \vartheta \\
&\quad + \frac{1}{a^2} y \sum_{i=1}^{d} \sum_{j=1}^{d} b^i(x) b^j(x) \int_0^\vartheta w^j(t)\, dw^i(t) \\
&\quad + y \sum_{i=1}^{d} \sum_{j=1}^{d} \frac{\partial b^i}{\partial x^j}(x) \int_0^\vartheta w^j(t)\, dw^i(t) + \rho_{11} + \rho_{12} + \rho_{13}, \quad (4.33)
\end{aligned}$$

where

6.4 Specific random walks for elliptic equations

$$\rho_{11} = a \sum_{i=1}^{d} \int_{0}^{\vartheta} \int_{0}^{t} \frac{\partial c}{\partial x^i}(X_x(s)) Y_{x,y}(s) \, dw^i(s) \, dt$$

$$+ \frac{1}{a} \sum_{i=1}^{d} \int_{0}^{\vartheta} \int_{0}^{t} c(X_x(s)) b^i(X_x(s)) Y_{x,y}(s) \, dw^i(s) \, dt$$

$$+ \frac{1}{a} \sum_{i=1}^{d} \int_{0}^{\vartheta} \int_{0}^{t} c(X_x(s)) b^i(X_x(s)) Y_{x,y}(s) \, ds \, dw^i(t)$$

$$+ \sum_{i=1}^{d} \sum_{j=1}^{d} \int_{0}^{\vartheta} \int_{0}^{t} \left(\frac{a}{2} \frac{\partial^2 b^i}{(\partial x^j)^2}(X_x(s)) + \frac{1}{a} \frac{\partial b^i}{\partial x^j}(X_x(s)) b^j(X_x(s)) \right)$$

$$\times Y_{x,y}(s) \, ds \, dw^i(t), \tag{4.34}$$

$$\rho_{12} = \int_{0}^{\vartheta} \int_{0}^{t} \left(c^2(X_x(s)) + \frac{a^2}{2} \sum_{i=1}^{d} \frac{\partial^2 c}{(\partial x^i)^2}(X_x(s)) \right) Y_{x,y}(s) \, ds \, dt$$

$$+ \sum_{i=1}^{d} \int_{0}^{\vartheta} \int_{0}^{t} \frac{\partial c}{\partial x^i}(X_x(s)) b^i(X_x(s)) Y_{x,y}(s) \, ds \, dt, \tag{4.35}$$

and ρ_{13} contains a sum of integrals like

$$I_{i_1,i_2,i_3} = \int_{0}^{\vartheta} \int_{0}^{t} \int_{0}^{s} f^{i_1 i_2 i_3}(X_x(s_1)) Y_{x,y}(s_1) \, dw^{i_1}(s_1) \, dw^{i_2}(s) \, dw^{i_3}(t), \tag{4.36}$$

$$i_1 = 0, \ldots, n, \quad i_2 \neq 0, \quad i_3 \neq 0,$$

where $f^{i_1 i_2 i_3}$ is a finite sum of products and each product has not more than three factors of the form b^i, $\partial b^i/\partial x^j$, $\partial^2 b^i/\partial x^j \partial x^k$, $\partial^3 b^i/\partial x^j \partial x^k \partial x^l$, and c. We underline that $\rho_{13} = 0$ if $b = 0$.

We have for Z:

$$Z_{x,y,z}(\vartheta) = z + g(x) \, y \, \vartheta + \rho_{21} + \rho_{22}, \tag{4.37}$$

where

$$\rho_{21} = a \sum_{i=1}^{d} \int_{0}^{\vartheta} \int_{0}^{t} \frac{\partial g}{\partial x^i}(X_x(s)) Y_{x,y}(s) dw^i(s) dt$$

$$+ \frac{1}{a} \sum_{i=1}^{d} \int_{0}^{\vartheta} \int_{0}^{t} g(X_x(s)) b^i(X_x(s)) Y_{x,y}(s) \, dw^i(s) \, dt \tag{4.38}$$

and

$$\rho_{22} = \sum_{i=1}^{d} \int_0^{\vartheta} \int_0^t \frac{\partial g}{\partial x^i}(X_x(s))\, b^i(X_x(s))\, Y_{x,y}(s)\, ds\, dt$$

$$+ \int_0^{\vartheta} \int_0^t \left(g(X_x(s)) c(X_x(s)) + \frac{a^2}{2} \sum_{i=1}^{d} \frac{\partial^2 g}{(\partial x^i)^2}(X_x(s)) \right) Y_{x,y}(s)\, ds\, dt. \tag{4.39}$$

Let us put

$$Y_1 = y + \frac{1}{a} y \sum_{i=1}^{d} b^i(x) w^i(\vartheta) + c(x) y \frac{r^2}{d}$$

$$+ \frac{1}{2a^2} y \sum_{i=1}^{d} \sum_{j=1}^{d} b^i(x) b^j(x) w^i(\vartheta) w^j(\vartheta)$$

$$- \frac{1}{2a^2} y \frac{r^2}{d} \sum_{i=1}^{d} [b^i(x)]^2 + \frac{1}{2} y \sum_{i=1}^{d} \sum_{j=1}^{d} \frac{\partial b^i}{\partial x^j}(x)\, w^i(\vartheta)\, w^j(\vartheta)$$

$$- \frac{1}{2} y \frac{r^2}{d} \sum_{i=1}^{d} \frac{\partial b^i}{\partial x^i}(x), \tag{4.40}$$

$$Z_1 = z + g(x) y \frac{r^2}{d}. \tag{4.41}$$

We note that $Y_1 > 0$ for a sufficiently small r since we supposed that $y > 0$.
We have

$$d = E\left[u(X_x(\vartheta)) E(Y_1 - Y_{x,y}(\vartheta) \mid w(\vartheta)) \right] + E(Z_1 - Z_{x,y,z}(\vartheta)). \tag{4.42}$$

Then, due to the relation $E(\vartheta \mid w(\vartheta)) = E\vartheta = r^2/d$, Theorem 4.3, and the formulas (4.33) and (4.37), we have

$$d = -E[u(X_x(\vartheta))] E(\rho_{11} + \rho_{12} + \rho_{13} \mid w(\vartheta)) - E(\rho_{21} + \rho_{22})$$
$$= -E[u(X_x(\vartheta))](\rho_{11} + \rho_{12} + \rho_{13}) - E(\rho_{21} + \rho_{22}). \tag{4.43}$$

Let

$$M_0(x) = \max_{\xi \in \bar{U}(x)} |u(\xi)|, \quad M_1(x) = \max_{\xi \in \bar{U}(x),\, 1 \le i \le d} \left| \frac{\partial u}{\partial x^i}(\xi) \right|.$$

By a thorough analysis of the one-step error (4.43), the following theorem is proved in [186].

Theorem 4.5. *The one-step error $d = d(x, y, r)$ of the approximation (4.31), (4.40)-(4.41) is of the form*

$$|d| \le (K_0 M_0(x) + K_1 M_1(x) + K_2)\, y r^4, \tag{4.44}$$

where K_0, K_1, K_2 are constants depending on a, b, c, and g only. That is the degree of smallness of this approximation with respect to r is equal to 4.

Let us consider now the simpler approximation

$$Y_1 = y + \frac{1}{a} y \sum_{i=1}^{d} b^i(x) w^i(\vartheta) + c(x) y \frac{r^2}{d}, \tag{4.45}$$

$$Z_1 = z + g(x) y \frac{r^2}{d}. \tag{4.46}$$

Introduce

$$M_2(x) := \max_{\xi \in \bar{U}(x),\ 1 \leq i,j \leq d} \left| \frac{\partial^2 u}{\partial x^i \partial x^j}(\xi) \right|.$$

We have the following result [186].

Theorem 4.6. *The one-step error $d = d(x, y, r)$ of the approximation (4.31), (4.45)-(4.46) has the form*

$$|d| \leq (K_0 M_0(x) + K_1 M_1(x) + K_2 M_2(x) + K_3)\, y\, r^4. \tag{4.47}$$

Remark 4.7. The degree of smallness of both approximations (4.40)-(4.41) and (4.45)-(4.46) is equal to 4. But the estimate (4.44) does not depend on second derivatives of the function u.

Now consider the Dirichlet problem for the nonhomogeneous Helmholtz equation

$$\frac{1}{2} a^2 \Delta u + c(x) u + g(x) = 0,\ x \in G, \tag{4.48}$$

$$u\,|_{\partial G} = \varphi(x), \tag{4.49}$$

i.e., when $b^i(x) = 0$, $i = 1, \ldots, d$, in (4.1).

In this case we construct two one-step approximation given in the next theorems (see their proofs in [186]).

Theorem 4.8. *Consider the one-step approximation*

$$X_1 = x + a w(\vartheta), \tag{4.50}$$

$$Y_1 = y + c(x) y \frac{r^2}{d} + a \sum_{i=1}^{d} \frac{\partial c}{\partial x^i}(x) y \frac{r^2}{2(d+2)} w^i(\vartheta)$$
$$+ \frac{1}{2} c_1(x) y \frac{4+d}{d^2(2+d)} r^4, \tag{4.51}$$

$$Z_1 = z + g(x) y \frac{r^2}{d} + \frac{1}{2} g_1(x) y \frac{4+d}{d^2(2+d)} r^4, \tag{4.52}$$

where

$$c_1(x) = c^2(x) + \frac{a^2}{2} \sum_{i=1}^{d} \frac{\partial^2 c}{(\partial x^i)^2}(x),$$

$$g_1(x) = g(x) c(x) + \frac{a^2}{2} \sum_{i=1}^{d} \frac{\partial^2 g}{(\partial x^i)^2}(x).$$

The one-step error $d = d(x, y, r)$ of this approximation has the form

$$|d| \leq (K_0 M_0(x) + a^2 K_1 M_1(x) + a^4 K_2 M_2(x) + K_3) \, y \, r^6, \qquad (4.53)$$

where K_0, K_1, K_2, K_3 depend on c and g only.

Theorem 4.9. *The one-step error $d = d(x, y, r)$ of the approximation*

$$X_1 = x + a w(\vartheta), \ Y_1 = y + c(x) y \frac{r^2}{d}, \ Z_1 = z + g(x) y \frac{r^2}{d} \qquad (4.54)$$

has the form

$$|d| \leq [K_0 M_0(x) + a^2 K_1 M_1(x) + K_2] \, y \, r^4. \qquad (4.55)$$

6.4.3 The average number of steps

In this subsection, we consider the question about average characteristics of the number of steps $\varkappa$ of the homogeneous Markov chain X_k (see the beginning of the current section) before it gets into a neighborhood of the boundary ∂G. Then, in the next subsection, we construct a number of algorithms for the Dirichlet problem which are based on the one-step approximations proposed in Sect. 6.4.2. Finally, we prove some convergence theorems using the results of Sects. 6.4.2 and 6.4.3.

In connection with the homogeneous Markov chain X_k from the introduction to Sect. 6.4, we introduce the one-step transition function (cf. Sect. 5.1.2):

$$P(x, B) = P(X_1 \in B \mid X_0 = x),$$

where B is a Borel set belonging to $\bar{G}$. If $x \in G \backslash \Gamma_{ar}$ then $P(x, B)$ is concentrated on the sphere $\partial U_{ar}(x)$ of radius ar with center at x. If $x \in \Gamma_{ar} \backslash \Gamma_\delta$ then $P(x, B)$ is concentrated on the sphere $\partial U_\rho(x)$ of radius $\rho(x, \partial G)$ with center at x. And if $x \in \Gamma_\delta$ then $P(x, B)$ is concentrated at the point x. Define an operation P acting on functions $v(x)$, $x \in \bar{G}$, by the formula

$$Pv(x) = \int_{\bar{G}} P(x, dy) v(y) = Ev(X_1), \ X_0 = x.$$

Consider the boundary value problem in $\bar{G}$ (cf. (5.1.13)-(5.1.14) and (3.14)-(3.15)):

$$Pv(x) - v(x) = -f(x), \ x \in G \backslash \Gamma_\delta, \qquad (4.56)$$

$$v(x) = 0, \ x \in \Gamma_\delta, \tag{4.57}$$

which is connected with the chain X_k.

In (4.56), $f(x)$ is a continuous function defined on the compact set $G\backslash\Gamma_\delta$: $f \in C(G\backslash\Gamma_\delta)$. It is not difficult to prove that there exists a unique solution of the problem (4.56)-(4.57) which is a continuous function on $G\backslash\Gamma_\delta$. This solution is known to be the following function [301] (cf. (5.1.15)):

$$v(x) = E \sum_{k=0}^{\varkappa_x - 1} f(X_k), \ X_0 = x, \tag{4.58}$$

where $\varkappa_x$ relates to the chain starting at x. If $v(x)$ is the solution of the boundary value problem (4.56)-(4.57) with a function $f(x)$ such that $f(x) \geq 1$ for $x \in G\backslash\Gamma_\delta$, then, thanks to (4.58), we have

$$E\varkappa_x \leq v(x). \tag{4.59}$$

Consider the function (cf. Remark 5.1.6):

$$V_1(x) = \begin{cases} A^2 + (\alpha, x) - x^2, & x \in G\backslash\Gamma_\delta, \\ 0, & x \in \Gamma_\delta, \end{cases}$$

where the constant A^2 and the vector α are such that the inequality $A^2 + (\alpha, x) - x^2 \geq 0$ is valid for all $x \in \bar{G}$. The next lemma is proved by arguments similar to those used in the proof of Lemma 5.1.5 and in Remark 5.1.6 (see further details in [186]).

Lemma 4.10. *The inequalities*

$$PV_1(x) - V_1(x) \leq -a^2 r^2, \quad x \in G\backslash\Gamma_{\alpha r}, \tag{4.60}$$

$$PV_1(x) - V_1(x) \leq 0, \quad x \in \Gamma_{\alpha r}\backslash\Gamma_\delta, \tag{4.61}$$

hold.

We also introduce the function

$$V_2(x) = \begin{cases} \ln(\alpha r/\delta) + 1, & x \in G\backslash\Gamma_{\alpha r}, \\ \ln(\rho(x)/\delta) + 1, & x \in \Gamma_{\alpha r}\backslash\Gamma_\delta, \\ 0, & x \in \Gamma_\delta, \end{cases}$$

where $\rho(x) = \rho(x, \partial G)$.

Lemma 4.11. *(see [186]). For a sufficiently small $r > 0$ the following inequalities hold:*

$$PV_2(x) - V_2(x) \leq 0, \quad x \in G\backslash\Gamma_{\alpha r}, \tag{4.62}$$

$$PV_2(x) - V_2(x) \leq -C_d, \quad x \in \Gamma_{\alpha r}\backslash\Gamma_\delta, \tag{4.63}$$

where C_d does not depend on x.

If the domain G is convex, the assumption on smallness of r can be omitted.

Theorem 4.12. *There exist constants B and C such that for any x and sufficiently small $r > 0$:*

$$E\varkappa_x \leq \frac{B}{a^2 r^2} + C \ln \frac{\alpha r}{\delta} . \tag{4.64}$$

If $\delta = O(r^p)$, $p > 1$, then

$$E\varkappa_x \leq \frac{B+1}{a^2 r^2} . \tag{4.65}$$

If G is convex and $r \geq l/2$ where l is a diameter of G, then the random walk over touching spheres is realized and

$$E\varkappa_x \leq C \ln \frac{l}{2\delta} . \tag{4.66}$$

Proof. The inequalities (4.64)-(4.66) follow from Lemmas 4.10 and 4.11 and from (4.59) if we take $v(x)$ of the form:

$$v(x) = \frac{V_1(x)}{a^2 r^2} + \frac{V_2(x)}{C_d} . \tag{4.67}$$

□

Remark 4.13. We emphasize that the number p does not play any essential role for the upper bound of the average number of steps $E\varkappa$.

Lemma 4.14. *Let $q(x) > 0$, $q \in C(G \backslash \Gamma_\delta)$, $f(x) \geq 0$, $f \in C(G \backslash \Gamma_\delta)$, $f(x) = 0$ for $x \in \Gamma_\delta$. Let $z(x)$ be a solution to the boundary value problem*

$$q(x)Pz(x) - z(x) = -f(x) , \quad x \in G \backslash \Gamma_\delta , \tag{4.68}$$

$$z(x) = 0 , \quad x \in \Gamma_\delta . \tag{4.69}$$

Then for $x \in G \backslash \Gamma_\delta$

$$z(x) = f(x) + E \sum_{k=1}^{\varkappa_x - 1} f(X_k) \Pi_{i=0}^{k-1} q(X_i) . \tag{4.70}$$

Proof. We have for $x \in G \backslash \Gamma_\delta$:

$z(x) = f(x) + q(x)Pz(x) = f(x) + q(x)Ez(X_1)$
$= f(x) + q(x)E(f(X_1) + q(X_1)Pz(X_1))$
$= f(x) + q(x)E(\chi_{\varkappa_x > 1} f(X_1)) + q(x)E(\chi_{\varkappa_x > 1} q(X_1)E(z(X_2)/X_1))$
$= f(x) + q(x)E(\chi_{\varkappa_x > 1} f(X_1)) + q(x)E(\chi_{\varkappa_x > 2} q(X_1)z(X_2))$
$= f(x) + q(x)E(\chi_{\varkappa_x > 1} f(X_1)) + q(x)E(\chi_{\varkappa_x > 2} q(X_1)f(X_2))$
$\quad + q(x)E(\chi_{\varkappa_x > 3} q(X_1)q(X_2)z(X_3))$
$= \cdots = f(x) + q(x)E(\chi_{\varkappa_x > 1} f(X_1))$
$\quad + \cdots + q(x)E(\chi_{\varkappa_x > N} q(X_1) \cdots q(X_{N-1})f(X_N))$
$\quad + q(x)E(\chi_{\varkappa_x > N+1} q(X_1) \cdots q(X_N)z(X_{N+1})) .$

Now we tend N to infinity and obtain (4.70). □

Corollary 4.15. *Suppose the conditions of Lemma 4.14 are satisfied. If* $q = \text{const} > 1$ *and* $f(x) = 1$ *for* $x \in G \backslash \Gamma_\delta$ *then*

$$z(x) = E(1 + q + q^2 + \cdots + q^{\varkappa_x - 1}) = \frac{1}{q-1}(Eq^{\varkappa_x} - 1) \ .$$

If $q = \text{const} > 1$ *and* $f(x) \geq c$ *for* $x \in G \backslash \Gamma_\delta$ *then*

$$Eq^{\varkappa_x} < \infty, \quad E(1 + q + \cdots + q^{\varkappa_x - 1}) \leq \frac{1}{c}z(x) \ .$$

Lemma 4.16. *Let* $\delta = O(r^p)$, $p > 1$. *Then there exist constants* $\beta > 0$ *and* $K > 0$ *such that for all sufficiently small* r

$$E \sum_{k=0}^{\varkappa_x - 1} (1 + \beta r^2)^k = O\left(\frac{1}{r^2}\right) , \tag{4.71}$$

$$E(1 + \beta r^2)^{\varkappa_x} < K \ , \tag{4.72}$$

$$P(\varkappa_x \geq k) \leq K(1 - \beta r^2)^k \ . \tag{4.73}$$

Proof. For the function $v(x)$ from (4.67) we have

$$(1 + \beta r^2) P v - (1 + \beta r^2) v \leq -(1 + \beta r^2), \quad x \in G \backslash \Gamma_\delta \ ,$$

or

$$(1 + \beta r^2) P v - v \leq \beta r^2 v - (1 + \beta r^2), \quad x \in G \backslash \Gamma_\delta \ ,$$

$$v = 0, \quad x \in \Gamma_\delta \ .$$

Thus, the function $v(x)$ is a solution to the problem (4.68)-(4.69) with $q(x) = 1 + \beta r^2$ and with $f(x)$ which satisfies the inequality

$$f(x) \geq 1 + \beta r^2 - \beta r^2 v = 1 + \beta r^2 - \frac{\beta}{a^2} V_1(x) - \frac{\beta r^2}{C_d} V_2(x) \ .$$

Clearly, $f(x) \geq 1/2$ for sufficiently small β and r. Using Corollary 4.15, we obtain

$$E \sum_{k=0}^{\varkappa_x - 1} (1 + \beta r^2)^k \leq 2v(x)$$

and, consequently, (4.71) is proved.

The relation (4.71) implies (4.72). The relation (4.73) is obtained from (4.72) by the Chebyshev inequality. □

6.4.4 Numerical algorithms and convergence theorems

Here we construct a number of algorithms for the Dirichlet problem (4.1), which are based on the one-step approximations proposed in Sect. 6.4.2, and we prove their convergence.

We suppose that the domain G, the coefficients $b^i(x)$, $c(x)$, $g(x)$, and the function $\varphi(x)$ in (4.1) satisfy the conditions (i)–(ii) (see them at the beginning of the current section). We recall that Γ_δ is the interior of a δ-neighborhood of the boundary ∂G belonging to $\bar{G}$. Let $U_1 \in \mathbf{R}^d$ be an open unit ball with center at the origin and with the boundary ∂U_1. Let ξ be a point uniformly distributed on the sphere ∂U_1 and $\xi_1, \xi_2, \ldots$ be independent random points distributed as ξ (see Remark 5.1.3 for their modeling).

Using the one-step approximation (4.45)-(4.46), we construct the following algorithm.

Algorithm 4.17 *If $X_k \in G \backslash \Gamma_{ar}$, we set $r_{k+1} = r$. If $X_k \in \Gamma_{ar} \backslash \Gamma_{r^2}$, we set $r_{k+1} = \rho(X_k, \partial G)/a$. And in both cases*

$$X_{k+1} = X_k + ar_{k+1}\xi_{k+1}, \quad X_0 = x, \tag{4.74}$$

$$Y_{k+1} = Y_k[1 + \frac{r_{k+1}}{a}\sum_{i=1}^{d} b^i(X_k)\xi_{k+1}^i + c(X_k)\frac{r_{k+1}^2}{d}], \quad Y_0 = 1, \tag{4.75}$$

$$Z_{k+1} = Z_k + Y_k\, g(X_k)\frac{r_{k+1}^2}{d}, \quad Z_0 = 0. \tag{4.76}$$

Let $\varkappa = \varkappa_x$ be the first number at which $X_\varkappa \in \Gamma_{r^2}$. Then we set $X_k = X_\varkappa$ for $k \geq \varkappa$, i.e., our algorithm is stopped at the random step $\varkappa$. Having obtained $X_\varkappa$, we find the point $\bar{X}_\varkappa \in \partial G$ which is the closest to $X_\varkappa$. We take $(\bar{X}_\varkappa, Y_\varkappa, Z_\varkappa)$ as the final state of the Markov chain.

Using the obtained $\bar{X}_\varkappa, Y_\varkappa, Z_\varkappa$, we evaluate

$$v(\bar{X}_\varkappa, Y_\varkappa, Z_\varkappa) = u(\bar{X}_\varkappa)Y_\varkappa + Z_\varkappa = \varphi(\bar{X}_\varkappa)Y_\varkappa + Z_\varkappa.$$

The solution of the problem (4.1) is approximately equal to

$$u(x) \approx E\left(\varphi(\bar{X}_\varkappa)Y_\varkappa + Z_\varkappa\right) \approx \frac{1}{M}\sum_{m=1}^{M}\left(\varphi(\bar{X}_\varkappa^{(m)})Y_\varkappa^{(m)} + Z_\varkappa^{(m)}\right), \tag{4.77}$$

where $\bar{X}_\varkappa^{(m)}, Y_\varkappa^{(m)}, Z_\varkappa^{(m)}$, $m = 1, \ldots, M$, are independent realizations of the algorithm (4.74)-(4.76). The first approximate equality in (4.77) involves an error due to replacing $X_x(\tau), Y_{x,1}(\tau), Z_{x,1,0}(\tau)$ by $\bar{X}_\varkappa, Y_\varkappa, Z_\varkappa$; in the second approximate equality the error comes from the Monte-Carlo method. The first error is estimated by $O(r^2)$ (see Theorem 4.20 below) and the second one is $O(1/\sqrt{M})$.

Construct the algorithm based on the one-step approximation (4.40)-(4.41).

6.4 Specific random walks for elliptic equations

Algorithm 4.18 *Define r_{k+1} as in Algorithm 4.17. Then*

$$X_{k+1} = X_k + a r_{k+1} \xi_{k+1} , \quad X_0 = x , \tag{4.78}$$

$$Y_{k+1} = Y_k [1 + \frac{r_{k+1}}{a} \sum_{i=1}^{d} b^i(X_k) \xi_{k+1}^i + c(X_k) \frac{r_{k+1}^2}{d}$$
$$+ \gamma(X_k, r_{k+1}, \xi_{k+1})] , \quad Y_0 = 1 , \tag{4.79}$$

$$Z_{k+1} = Z_k + Y_k \, g(X_k) \frac{r_{k+1}^2}{d} , \quad Z_0 = 0 , \tag{4.80}$$

where

$$\gamma(x, r, \xi) = \frac{r^2}{2a^2} \sum_{i=1}^{d} \sum_{j=1}^{d} b^i(x) b^j(x) \xi^i \xi^j - \frac{r^2}{2a^2 d} \sum_{i=1}^{d} [b^i(x)]^2$$
$$+ \frac{r^2}{2} \sum_{i=1}^{d} \sum_{j=1}^{d} \frac{\partial b^i}{\partial x^j}(x) \xi^i \xi^j - \frac{r^2}{2d} \sum_{i=1}^{d} \frac{\partial b^i}{\partial x^i}(x) .$$

The algorithm is stopped at a random step $\varkappa$ when $X_\varkappa \in \Gamma_{r^2}$. Having obtained $X_\varkappa$, we find the point $\bar{X}_\varkappa \in \partial G$ which is the closest to $X_\varkappa$. We take $(\bar{X}_\varkappa, Y_\varkappa, Z_\varkappa)$ as the final state of the Markov chain.

Now consider the Dirichlet problem for the nonhomogeneous Helmholtz equation (4.48)-(4.49). In this case we can also suggest two algorithms. One of them is based on the one-step approximation (4.54). Here we write down the other one which is based on the one-step approximation (4.51)-(4.52).

Algorithm 4.19 *We set $r_{k+1} = r$ if $X_k \in G \backslash \Gamma_{ar}$ and $r_{k+1} = \rho(X_k, \partial G)/a$ if $X_k \in \Gamma_{ar} \backslash \Gamma_{r^4}$. Then*

$$X_{k+1} = X_k + a r_{k+1} \xi_{k+1} , \quad X_0 = x , \tag{4.81}$$

$$Y_{k+1} = Y_k \left[1 + c(X_k) \frac{r_{k+1}^2}{d} + \frac{a \, r_{k+1}^3}{2(d+2)} \sum_{i=1}^{d} \frac{\partial c}{\partial x^i}(X_k) \, \xi_{k+1}^i \right.$$
$$\left. + \frac{4+d}{2d^2(2+d)} c_1(X_k) \, r_{k+1}^4 \right] , \quad Y_0 = 1 , \tag{4.82}$$

$$Z_{k+1} = Z_k + Y_k [g(X_k) \frac{r_{k+1}^2}{d} + \frac{4+d}{2d^2(2+d)} g_1(X_k) \, r_{k+1}^4] , \quad Z_0 = 0 . \tag{4.83}$$

The algorithm is stopped at a random step $\varkappa$ when $X_\varkappa \in \Gamma_{r^4}$. Having obtained $X_\varkappa$, we find the point $\bar{X}_\varkappa \in \partial G$ which is the closest to $X_\varkappa$. We take $(\bar{X}_\varkappa, Y_\varkappa, Z_\varkappa)$ as the final state of the Markov chain.

We note that by Theorem 4.12 the average number of steps for all the algorithms presented here is $O(1/r^2)$.

Proceeding to convergence theorems, let us use the relation (4.8):

$$|E(\varphi(\bar{X}_\varkappa)Y_\varkappa - Z_\varkappa) - u(x)| = |\mathcal{R}| \leq |E(u(\bar{X}_\varkappa) - u(X_\varkappa))Y_\varkappa| + \sum_{k=1}^{\infty} |\tilde{d}_k|,$$

where

$$\tilde{d}_k = E\chi_{\varkappa > k-1}(v(X_k, Y_k, Z_k) - v(X_{k-1}, Y_{k-1}, Z_{k-1})).$$

For brevity, we write $\varkappa$ instead of $\varkappa_x$ everywhere.

Clearly,

$$u(X_{k-1})Y_{k-1} - Z_{k-1} = v(X_{k-1}, Y_{k-1}, Z_{k-1})$$
$$= E(v(X_{X_{k-1}}(\vartheta_k), Y_{X_{k-1}, Y_{k-1}}(\vartheta_k), Z_{X_{k-1}, Y_{k-1}, Z_{k-1}}(\vartheta_k))/\mathcal{B}_{k-1}).$$

Therefore

$$\tilde{d}_k = E\chi_{\varkappa > k-1} d_k,$$

where

$$d_k = E[v(X_k, Y_k, Z_k) \\ - v(X_{X_{k-1}}(\vartheta_k), Y_{X_{k-1}, Y_{k-1}}(\vartheta_k), Z_{X_{k-1}, Y_{k-1}, Z_{k-1}}(\vartheta_k))/\mathcal{B}_{k-1}] \quad (4.84)$$

is a one-step error at the point $(X_{k-1}, Y_{k-1}, Z_{k-1})$. Thus

$$|\mathcal{R}| \leq |E(u(\bar{X}_\varkappa) - u(X_\varkappa))Y_\varkappa| + \sum_{k=1}^{\infty} |E\chi_{\varkappa > k-1} d_k|. \quad (4.85)$$

Theorem 4.20. *Let $c(x) \leq -c_0 < 0$. Then Algorithms 4.17 and 4.18 have the second order of convergence with respect to r, i.e., for all sufficiently small r*

$$|E(\varphi(\bar{X}_\varkappa)Y_\varkappa - Z_\varkappa) - u(x)| \leq Kr^2. \quad (4.86)$$

In addition, the constant K for Algorithm 4.17 *depends on first and second derivatives of the required solution $u(x)$ while this constant for Algorithm* 4.18 *depends on first derivatives only.*

Proof. We restrict ourselves to the proof of Algorithm 4.18. Due to Theorem 4.5, the one-step error d_k from (4.84) satisfies the inequality

$$|d_k| \leq KY_{k-1}r^4, \quad (4.87)$$

where (see (4.44))

$$K_0 M_0(x) + K_1 M_1(x) + K_2 \leq K$$

for all $x \in \bar{G}$.

Since $X_\varkappa \in \Gamma_{r^2}$, we have
$$|u(\bar{X}_\varkappa) - u(X_\varkappa)| \leq Kr^2 .$$

Therefore
$$|\mathcal{R}| \leq Kr^2 EY_\varkappa + Kr^4 \sum_{k=0}^{\infty} E\chi_{\varkappa>k} Y_k . \tag{4.88}$$

Clearly, K does not depend on the second derivatives of $u(x)$ here.

It is easy to see that
$$|\gamma(x,r,\xi)| = O(r^2) , \quad E(\gamma(X_k, r_{k+1}, \xi_{k+1})/\mathcal{B}_k) = 0 . \tag{4.89}$$

Then we have for $k > 0$:

$$E\chi_{\varkappa>k} Y_k \leq E\chi_{\varkappa>k-1} Y_k$$
$$= E[\chi_{\varkappa>k-1} Y_{k-1} \, E(1 + \frac{r_k}{a} \sum_{i=1}^{d} b^i(X_{k-1})\xi_k^i + c(X_{k-1})\frac{r_k^2}{2}$$
$$+\gamma(X_{k-1}, r_k, \xi_k)/\mathcal{B}_{k-1}]$$
$$= E[\chi_{\varkappa>k-1} Y_{k-1}(1 + c(X_{k-1})\frac{r_k^2}{2})] \leq (1 - \frac{c_0}{2}r^2) \, E\chi_{\varkappa>k-1} Y_{k-1}$$
$$\leq (1 - \frac{c_0}{2}r^2)^2 \, E\chi_{\varkappa>k-2} Y_{k-2} \leq \cdots \leq (1 - \frac{c_0}{2}r^2)^k .$$

From here
$$Kr^4 \sum_{k=0}^{\infty} E\chi_{\varkappa>k} Y_k \leq Kr^2 . \tag{4.90}$$

Further,
$$EY_\varkappa = \sum_{k=1}^{\infty} E\chi_{\varkappa=k} Y_k = \sum_{k=1}^{\infty} (E\chi_{\varkappa>k-1} Y_k - E\chi_{\varkappa>k} Y_k)$$
$$= E\chi_{\varkappa>0} Y_1 + \sum_{k=1}^{\infty} E\chi_{\varkappa>k}(Y_{k+1} - Y_k) .$$

But
$$E\chi_{\varkappa>k}(Y_{k+1} - Y_k)$$
$$= E\chi_{\varkappa>k} Y_k[\frac{r_{k+1}}{a} \sum_{i=1}^{d} b^i(X_k)\xi_{k+1}^i + c(X_k)\frac{r_{k+1}^2}{2} + \gamma(X_k, r_{k+1}, \xi_{k+1})]$$
$$= E[\chi_{\varkappa>k} Y_k \, E(\frac{r_{k+1}}{a} \sum_{i=1}^{d} b^i(X_k)\xi_{k+1}^i + c(X_k)\frac{r_{k+1}^2}{2} + \gamma(X_k, r_{k+1}, \xi_{k+1})/\mathcal{B}_k)]$$
$$= E[\chi_{\varkappa>k} Y_k \, c(X_k)\frac{r_{k+1}^2}{2}] < 0 .$$

Hence
$$EY_\varkappa \leq E\chi_{\varkappa>0}Y_1 \leq EY_1 < 1 . \qquad (4.91)$$

The relations (4.88), (4.90), and (4.91) imply (4.86). Theorem 4.20 is proved. □

We analogously prove the following result.

Theorem 4.21. *Let $c(x) \leq 0$. Then Algorithm 4.19 has the fourth order of convergence with respect to r:*

$$|E(\varphi(\bar{X}_\varkappa)Y_\varkappa - Z_\varkappa) - u(x)| \leq Kr^4 . \qquad (4.92)$$

The constant K depends on first and second derivatives of $u(x)$.

Remark 4.22. The more simple algorithm based on the one-step approximation (4.54) has the second order of convergence with respect to r and the constant K depends on first derivatives of $u(x)$ only.

Remark 4.23. In this section we have proposed a number of algorithms for (4.1), i.e., for (3.1) with $a^{ij}(x) = a^2 \delta_{ij}$, where δ_{ij} is the Kronecker delta. All the results obtained here can be carried over to the case of elliptic equation (3.1) with constant coefficients a^{ij}.

6.5 Methods for elliptic equations with small parameter at higher derivatives

Proceeding to numerical investigation of a boundary layer, let us consider the model problem for the nonhomogeneous Helmholtz equation :

$$\frac{1}{2}\varepsilon^2 \Delta u + c(x)u = g(x), \quad x \in U_R , \qquad (5.1)$$

$$u \mid_{\partial U_R} = 0 , \qquad (5.2)$$

where $\varepsilon \ll 1$, $U_R \subset \mathbf{R}^d$ is an open ball of radius R with center at the origin, $c(x)$ and $g(x)$ belong to $C^\infty(\bar{U}_R)$, and $c(x) \leq -c_0 < 0$, $x \in \bar{U}_R$.

A solution $u(x, \varepsilon)$ of this problem has a fluent alteration everywhere in U_R except a small neighborhood of ∂U_R which is called boundary layer and which is narrowed with decreasing ε. The solution $u(x, \varepsilon)$ varies sharply in the boundary layer. It is well known (see [111] and references therein) that the width of the boundary layer for the problem (5.1)-(5.2) is proportional to ε, i.e., the boundary layer has the form $\Gamma_{l\varepsilon}$ (l is a positive number). Moreover, it is known that

$$|u(x,\varepsilon)| \leq K, \quad |\frac{\partial u}{\partial x^i}(x,\varepsilon)| \leq K, \quad |\frac{\partial^2 u}{\partial x^i \partial x^j}(x,\varepsilon)| \leq K, \quad x \in U_R \setminus \Gamma_{l\varepsilon},$$

$$|u(x,\varepsilon)| \leq K, \quad |\frac{\partial u}{\partial x^i}(x,\varepsilon)| \leq \frac{K}{\varepsilon}, \quad |\frac{\partial^2 u}{\partial x^i \partial x^j}(x,\varepsilon)| \leq \frac{K}{\varepsilon^2}, \quad x \in \Gamma_{l\varepsilon} . \qquad (5.3)$$

6.5 Methods for elliptic equations with small parameter

An analytical approach to the problem (5.1)-(5.2) consists in construction of an external asymptotic expansion $V(x,\varepsilon)$ and of an interior asymptotic expansion $W(x,\varepsilon)$ [111]. They give the solution in $U_R\backslash\Gamma_{l\varepsilon}$ and $\Gamma_{l\varepsilon}$, respectively. The external expansion can be written as

$$V(x,\varepsilon) = \sum_{k=0}^{\infty} \varepsilon^{2k} v_k(x),$$

where

$$v_0(x) = \frac{g(x)}{c(x)}, \quad v_k(x) = -\frac{\Delta v_{k-1}(x)}{c(x)}, \quad k \geq 1.$$

The function $V(x,\varepsilon)$ is an asymptotic solution of (5.1)-(5.2) in $U_R\backslash\Gamma_{l\varepsilon}$, i.e., the function

$$V_m(x,\varepsilon) = \sum_{k=0}^{m} \varepsilon^{2k} v_k(x) \tag{5.4}$$

differs from the solution in $U_R\backslash\Gamma_{l\varepsilon}$ by $O(\varepsilon^{2m+2})$.

The interior expansion $W(x,\varepsilon)$ is needed to compensate a discrepancy in the boundary conditions. It turned out that outside the boundary layer $W(x,\varepsilon) = O(\varepsilon^N)$, $\varepsilon \to 0$, for any N. The sum $V + W$ is an asymptotic solution of the problem (5.1)-(5.2). The interior expansion is constructed in a more complicated way than the external one, and it is not given here.

It should be mentioned that the problem (5.1)-(5.2) is one of the simplest in the theory of boundary layer. If, for instance, the condition $c(x) \leq -c_0 < 0$, $x \in \bar{U}_R$, is violated so that the function $c(x)$ may take zero values then analytical investigation of the corresponding problem becomes exceedingly intricate. Therefore, a numerical approach to problems with small parameter at higher derivatives is actual. But it should not be supposed that one can use general numerical methods (for example, the methods from Sects. 6.3 and 6.4) without taking into account the smallness of the parameter at higher derivatives. Principal difficulties lie in the fact that derivatives of the solution in the boundary layer are large and the average number of steps evaluated in Theorem 4.12 is as big as $O(1/\varepsilon^2 r^2)$. Let us analyze these and some other difficulties in the case of the model problem (5.1)-(5.2).

As before, we consider a random walk over spheres with radius εr in $U_R\backslash\Gamma_{\varepsilon r}$ (we have ε instead of a now) and over spheres tangent to ∂U_R in $\Gamma_{\varepsilon r}\backslash\Gamma_\delta$ where δ is sufficiently small (in any case $\delta < \varepsilon r/2$).

Now it is convenient to present the error $\mathcal{R}$ (see (4.85)) in the following form

$$|\mathcal{R}| \leq |E(u(\bar{X}_\varkappa) - u(X_\varkappa))Y_\varkappa| + \sum_{k=1}^{\infty} |E\chi_{\varkappa > k-1} d_k| \leq |E(u(\bar{X}_\varkappa) - u(X_\varkappa))Y_\varkappa|$$

$$+ \sum_{k=1}^{\infty} |E\chi_{\Gamma_{l\varepsilon}\backslash\Gamma_\delta}(X_{k-1})d_k| + \sum_{k=1}^{\infty} |E\chi_{U_R\backslash\Gamma_{l\varepsilon}}(X_{k-1})d_k| \tag{5.5}$$

because
$$\chi_{\varkappa>k-1} = \chi_{\Gamma_{l\varepsilon}\backslash\Gamma_\delta}(X_{k-1}) + \chi_{U_R\backslash\Gamma_{l\varepsilon}}(X_{k-1}).$$

Let the one-step error d_k be bounded by $\delta_0(r,\varepsilon)Y_{k-1}$ in the part $\Gamma_{l\varepsilon}\backslash\Gamma_\delta$ of the boundary layer $\Gamma_{l\varepsilon}$ and by $\delta_1(r,\varepsilon)Y_{k-1}$ outside the boundary layer, i.e., in $U_R\backslash\Gamma_{l\varepsilon}$. We note that the method (4.74)-(4.76) under $b(x) = 0$ and the method (4.81)-(4.83) have $Y_k \leq 1$ for sufficiently small r if $c(x) \leq -c_0 < 0$. It follows from (5.5) that for these methods

$$|\mathcal{R}| \leq |E(u(\bar{X}_\varkappa) - u(X_\varkappa))Y_\varkappa| + \delta_0(r,\varepsilon)E\varkappa_0 + \delta_1(r,\varepsilon)E\varkappa_1, \tag{5.6}$$

where $\varkappa_0$ and $\varkappa_1$ are random numbers of steps inside and outside the boundary layer, respectively. Clearly, $\varkappa_0$ and $\varkappa_1$ depend on x. Due to Theorem 4.12, we have $E\varkappa_1 \leq K/\varepsilon^2 r^2$. Fortunately, as it is shown in the lemma below, $E\varkappa_0 \leq K/r^2$ (see its proof in [186]).

Lemma 5.1. *There exists a constant $K > 0$ such that for any $x \in U_R\backslash\Gamma_\delta$ and sufficiently small ε and r*

$$E\varkappa_0 \leq \frac{K}{r^2}. \tag{5.7}$$

We return to the inequality (5.6). The first term in the right-hand side of (5.6) is bounded by $K\delta/\varepsilon$ according to (5.3) (we note in passing that for the problem (5.1)-(5.2) we need not find $\bar{X}_\varkappa$ as $u(\bar{X}_\varkappa) = 0$). If we choose $\delta = O(r^p)$, then the first term can be done sufficiently small. At the same time, due to Theorem 4.12, the average number of steps depends on p insignificantly and, as before, it is estimated as $O(1/\varepsilon^2 r^2)$. The factor $E\varkappa_0 = O(1/r^2)$ in the second term (Lemma 5.1) is comparatively small, and the other factor $\delta_0(r,\varepsilon)$ depends on behavior of the solution in the boundary layer and it may take large values. But the errors of the methods from the previous section do not contain too higher order derivatives of the solution. Therefore, the second term can also be done small. The third term in (5.6) has very large factor $E\varkappa_1 = O(1/\varepsilon^2 r^2)$ and, consequently, this term can be decreased by means of $\delta_1(r,\varepsilon)$ only. Thus, the principal problem is to construct a sufficiently accurate and effective one-step approximation in the larger domain $U_R\backslash\Gamma_{l\varepsilon}$. Let us take into account that in the case of the problem (5.1)-(5.2) the system (4.3)-(4.5) is a system with small noise:

$$dX = \varepsilon dw(t) \tag{5.8}$$

$$dY = c(X)Y dt \tag{5.9}$$

$$dZ = g(X)Y dt. \tag{5.10}$$

In Chap. 3, we construct specific methods for systems with small noise. The errors of these methods have the form $O(h^p + \varepsilon^k h^q)$, $q < p$, where h is a step with respect to time. The time-step order of such a method is equal to q which is comparatively low and, thanks to this fact, one may reach a certain efficiency. Moreover, according to large p and the factor ε^k at h^q, the method error becomes sufficiently small, and the method reaches high exactness. These ideas can be carried over to approximation under space-discretization as well. We shall construct an efficient one-step approximation in the main domain $U_R\backslash\Gamma_{l\varepsilon}$ with an error of the form $O(r^{2p}+\varepsilon^k r^{2q})$. We recall that the solution $u(x,\varepsilon)$ has a fluent alteration in $U_R\backslash\Gamma_{l\varepsilon}$.

First we analyze the method based on the one-step approximation (4.54). According to (5.3), $M_0(x)$ and $M_1(x)$, $x \in \Gamma_{l\varepsilon}$, in (4.55) are bounded by K and K/ε, respectively. Hence $\delta_0(r,\varepsilon) \leq Kr^4$ (of course, we have to take ε instead of a in (4.55)) and, due to Lemma 5.1, the second term in (5.6) has the acceptable bound $O(r^2)$. Clearly, the third term has the following bound:

$$\delta_1(r,\varepsilon)E\varkappa_1 \leq Kr^4 \frac{K}{\varepsilon^2 r^2} \leq \frac{Kr^2}{\varepsilon^2},$$

and, to obtain an acceptable accuracy, we have to take a very small r. Moreover, this circumstance leads to an increase of the average number of steps.

We analogously get for the method (4.81)-(4.83):

$$\delta_0(r,\varepsilon)E\varkappa_0 \leq Kr^4, \quad \delta_1(r,\varepsilon) \leq Kr^6, \quad \delta_1(r,\varepsilon)E\varkappa_1 \leq \frac{Kr^4}{\varepsilon^2}.$$

This method can be simplified without an essential loss of accuracy. To this end, consider the following algorithm.

Algorithm 5.2 We set $r_{k+1} = r$ if $X_k \in U_R\backslash\Gamma_{\varepsilon r}$ and $r_{k+1} = (R-|X_k|)/\varepsilon$ if $X_k \in \Gamma_{\varepsilon r}\backslash\Gamma_\delta$ for $\delta < \varepsilon r$. Then

$$X_{k+1} = X_k + \varepsilon r_{k+1}\xi_{k+1}, \quad X_0 = x, \tag{5.11}$$

$$Y_{k+1} = Y_k\left[1 + c(X_k)\frac{r_{k+1}^2}{d} + \frac{\varepsilon r_{k+1}^3}{2(d+2)}\sum_{i=1}^d \frac{\partial c}{\partial x^i}(X_k)\xi_{k+1}^i \right.$$
$$\left. + \frac{4+d}{2d^2(2+d)}c^2(X_k)r_{k+1}^4\right], \quad Y_0 = 1, \tag{5.12}$$

$$Z_{k+1} = Z_k + Y_k\left[g(X_k)\frac{r_{k+1}^2}{d} + \frac{4+d}{2d^2(2+d)}c(X_k)g(X_k)r_{k+1}^4\right], \quad Z_0 = 0. \tag{5.13}$$

Let $\varkappa = \varkappa_x$ be the first number at which $X_\varkappa \in \Gamma_\delta$. Then we set $X_k = X_\varkappa$ for $k \geq \varkappa$, i.e., our algorithm is stopped at the random step $\varkappa$. Having obtained $X_\varkappa$, we find the point $\bar{X}_\varkappa \in \partial G$ which is the closest to $X_\varkappa$. We take $(\bar{X}_\varkappa, Y_\varkappa, Z_\varkappa)$ as the final state of the Markov chain.

Note that a choice of δ in this algorithm is specified below depending on the problem considered.

The method (5.11)-(5.13) does not require calculation of the second derivatives $\partial^2 c/(\partial x^i)^2$ and $\partial^2 g/(\partial x^i)^2$ at each step in contrast to the method (4.81)-(4.83).

One can prove that

$$|d| \leq K\left(M_0(x) + \varepsilon^2 M_1(x)\right) y\, r^6 + K\, \varepsilon^4 M_2(x)\, y\, r^4 + K\left(\varepsilon^2 r^4 + r^6\right) y\,.$$

Therefore (see (5.3)),

$$|\delta_i(r,\varepsilon)| \leq K(\varepsilon^2 r^4 + r^6), \quad i = 0, 1\,. \tag{5.14}$$

The error (5.14) is only of the fourth order with respect to r (due to this fact the method (5.11)-(5.13) is quite simple) but at the same time it is sufficiently small due to the factor ε^2. Using (5.14) with regard to $\delta = r^3$, it is not difficult to obtain the following result (we note that now the first term in (5.6) is $O(r^3/\varepsilon) \leq Kr^2 + Kr^4/\varepsilon^2$).

Theorem 5.3. *Let $\delta = r^3$. The error of Algorithm 5.2 is estimated as*

$$|\mathcal{R}| \leq Kr^2 + K\frac{r^4}{\varepsilon^2} \tag{5.15}$$

and the average number of steps for this method is equal to $O(1/\varepsilon^2 r^2)$.

We emphasize that the large average number of steps leads to extraordinary computational costs. At the same time, we can find the solution of the problem (5.1)-(5.2) in $U_R \backslash \Gamma_{l\varepsilon}$ with high accuracy using the truncated external expansion (5.4). We use this fact and construct below an analytic-numerical method.

We set

$$u(x,\varepsilon) \approx V_m(x,\varepsilon)\,, \quad x \in U_R \backslash \Gamma_{l\varepsilon}\,,$$

and, instead of (5.1)-(5.2), we introduce the problem the Helmholtz equation with small parameter:

$$\frac{1}{2}\varepsilon^2 \Delta u + c(x)u = g(x), \quad R - l\varepsilon < |x| < R\,, \tag{5.16}$$

$$u\,|_{|x|=R-l\varepsilon} = V_m(x,\varepsilon), \quad u\,|_{|x|=R} = 0\,. \tag{5.17}$$

Consider the random walk defined by $r < \max(\varepsilon, l\varepsilon)$ and $\delta \ll r$ in the layer $R - l\varepsilon \leq |x| \leq R$: if $R - l\varepsilon \leq |X_k| < R - l\varepsilon + \delta$ or $R - \delta < |X_k| \leq R$, then $X_{k+1} = X_k$; if $R - l\varepsilon + \delta \leq |X_k| < R - l\varepsilon + \varepsilon r$ or $R - \varepsilon r < |X_k| \leq R - \delta$, then r_{k+1} is equal to $(|X_k| - (R - l\varepsilon))/\varepsilon$ or $(R - |X_k|)/\varepsilon$, respectively; if $R - l\varepsilon + \varepsilon r \leq |X_k| \leq R - \varepsilon r$ then $r_{k+1} = r$. In the second and third cases we put

$$X_{k+1} = X_k + \varepsilon r_{k+1} \xi_{k+1}\,. \tag{5.18}$$

It is not difficult to prove the following theorem.

Theorem 5.4. *Let* $\delta = r^5$. *Then the error of Algorithm 5.2 for the problem* (5.16)-(5.17) *is estimated as*

$$|\mathcal{R}| \le K(\varepsilon^2 r^2 + r^4) + K\frac{r^5}{\varepsilon} \qquad (5.19)$$

and the average number of steps is equal to $O(1/r^2)$.

It is clear that the error of the analytic-numerical method for the original problem (5.1)-(5.2) is larger than (5.19) by $O(\varepsilon^{2m+2})$. The proposed analytic-numerical method is very effective: it is more accurate (compare the errors (5.19) and (5.15)) and it has the lesser average number of steps.

Remark 5.5. Undoubtedly, many results obtained here for the model problem (5.1)-(5.2) can be used for more general problems. In particular, they can be carried over to the problem (4.1) with $a = \varepsilon$, $b(x) = 0$, $c(x) \le -c_0 < 0$ without any essential change.

6.6 Methods for the Neumann problem for parabolic equations

In this section we construct some Markov chains with reflection such that the expectation of a certain functional of chain paths is close to the solution of the Neumann problem for parabolic equations. These Markov chains weakly approximate the solution of the system of SDEs which is characteristic for the Neumann problem.

Let G be a bounded domain in $\mathbf{R}^d$ and $Q = [T_0, T_1) \times G$ be a cylinder in $\mathbf{R}^{d+1}$. Consider the parabolic equation:

$$\frac{\partial u}{\partial t} + \frac{1}{2}\sum_{i,j=1}^{d} a^{ij}(t,x)\frac{\partial^2 u}{\partial x^i \partial x^j} + \sum_{i=1}^{d} b^i(t,x)\frac{\partial u}{\partial x^i} + c(t,x)u$$
$$+ g(t,x) = 0, \quad (t,x) \in Q, \qquad (6.1)$$

with the initial condition

$$u(T_1, x) = f(x) \qquad (6.2)$$

and the Neumann boundary condition

$$\frac{\partial u}{\partial \nu} + \varphi(t,x)u = \psi(t,x), \quad t \in [T_0, T_1], \quad x \in \partial G, \qquad (6.3)$$

where ν is the direction of the inner normal to the surface ∂G at a point $x \in \partial G$.

We assume that the coefficients $a^{ij} = a^{ji}$ satisfy the strict ellipticity condition in $\bar{Q}$ and that the surface ∂G and the coefficients of the problem (6.1)-(6.3) are sufficiently smooth.

In what follows, we require the existence of the classical solution of the problem (6.1)-(6.3) [174] possessing continuous in $\bar{Q}$ second derivatives with respect to t and x^i and third derivatives with respect to x^i. This solution exists, for example, in the case of a homogeneous boundary value problem ($\psi(t, x) = 0$) if $g = 0$ in the interior part of some neighborhood of the set $[T_0, T_1) \times \partial G$ and $f = 0$ in the interior part of some neighborhood of ∂G.

The solution of the problem (6.1)-(6.3) has the following probabilistic representation [70, 82, 110]:

$$u(t, x) = E\left[f(X_{t,x}(T_1))Y_{t,x,1}(T_1) + Z_{t,x,1,0}(T_1)\right], \quad (6.4)$$

where $X_{t,x}(s)$, $Y_{t,x,y}(s)$, $Z_{t,x,y,z}(s)$, $s \geq t$, is the solution of the Cauchy problem for the system of SDEs:

$$dX = b(s, X) I_G(X) \, ds + \sigma(s, X) I_G(X) \, dw(s)$$
$$+ \nu(X) I_{\partial G}(X) \, d\mu(s), \quad X(t) = x, \quad (6.5)$$
$$dY = c(s, X) I_G(X) Y \, ds + \varphi(s, X) I_{\partial G}(X) Y \, d\mu(s), \quad Y(t) = y, \quad (6.6)$$
$$dZ = g(s, X) I_G(X) Y \, ds + \psi(s, X) I_{\partial G}(X) Y \, d\mu(s), \quad Z(t) = z. \quad (6.7)$$

In (6.5)-(6.7), $(t, x) \in Q$, $w(s) = (w^1(s), \ldots, w^d(s))^\mathsf{T}$ is a standard Wiener process, $b(s, x)$ is a d-dimensional column-vector composed of the coefficients $b^i(s, x)$, the $d \times d$ matrix $\sigma(s, x)$ is obtained from the formula $\sigma(s, x)\sigma^\mathsf{T}(s, x) = a(s, x)$, $a(s, x) = \{a^{ij}(s, x)\}$, $i, j = 1, \ldots, d$, $\mu(s)$ is the local time of the process X on the boundary ∂G, i.e., it is a scalar increasing process which grows only when $X(s) \in \partial G$, Y and Z are scalars, and $I_A(x)$ is the indicator function of a set A.

Our aim is to construct a Markov chain (t_k, X_k, Y_k, Z_k) which approximate in the weak sense the solution of the system (6.5)-(6.7). Due to the reflection, such a Markov chain is terminated on the upper base of the cylinder $\bar{Q}$. We shall construct this chain analogously to the chains from Sect. 6.1. We require that the conditional expectation of the difference

$$d = u(t_{k+1}, X_{k+1})Y_{k+1} + Z_{k+1} - u(t_k, X_k)Y_k - Z_k \quad (6.8)$$

is small at each step provided X_k, Y_k, Z_k are known (for definiteness, we suppose that all t_k are deterministic).

Let $r > 0$ be a sufficiently small number. If the distance $\rho(X_k, \partial G)$ between the point X_k and the boundary ∂G is greater than r, then given the point (t_k, X_k) we construct the point (t_{k+1}, X_{k+1}) using one of the methods for solving the Dirichlet problem from Sect. 6.1 which one-step order of accuracy is $O(r^4)$.

Now let X_k be a point near the boundary (a boundary point), i.e., $\rho(X_k, \partial G) \leq r$. We suppose that $t_k \leq T_1 - r^2$. Otherwise, we terminate the chain at the point $(t_{k+1}, X_{k+1}, Y_{k+1}, Z_{k+1}) = (T_1, X_k, Y_k, Z_k)$. In this case the error on the last step is $O(r^2)$. For convenience, we denote by (t_0, x_0, y_0, z_0),

$t_0 \le T_1 - r^2$, $\rho(x_0, \partial G) \le r$, the point (t_k, X_k, Y_k, Z_k). Our nearest aim is to construct a point $t_1 = t_0 + r^2 = t_0 + h$, X_1, Y_1, Z_1 such that the expectation Ed of the difference (6.8) is of order $O(r^3)$.

6.6.1 One-step approximation for boundary points

Let x_0^π be the projection of x_0 on ∂G and $r_0 := \rho(x_0, \partial G) = \rho(x_0, x_0^\pi)$, $r_0 \le r$. In a neighborhood of the point x_0 we introduce a new orthogonal coordinate system such that its origin coincides with x_0 and one of the coordinate vectors (for definiteness, the first one) coincides with $\nu(x_0)$. We denote by $\chi = (\chi^1, \ldots, \chi^d)^\mathsf{T}$ the coordinates of the vector $x - x_0$ in the new coordinate system. This transformation of variables has the form

$$\chi = Q(x - x_0), \quad x = x_0 + Q^\mathsf{T} \chi , \qquad (6.9)$$

where $Q = \{q_{ij}\}$ is an orthogonal matrix.

In the new variables the point x_0 has zero coordinates, the point x_0^π has the coordinates $(-r_0, 0, \ldots, 0)$, and the equation (6.1) is written as

$$\frac{\partial u}{\partial t} + \frac{1}{2} \sum_{i,j=1}^{d} a^{ij}(t, \chi) \frac{\partial^2 u}{\partial \chi^i \partial \chi^j} + \sum_{i=1}^{d} \beta^i(t, \chi) \frac{\partial u}{\partial \chi^i} + c(t, \chi) u$$
$$+ g(t, \chi) = 0, \quad (t, x) \in Q, \qquad (6.10)$$

and the boundary condition (6.3) takes the form

$$\frac{\partial u}{\partial \nu} + \varphi(t, \chi) u = \psi(t, \chi), \quad t \in [T_0, T_1], \quad x \in \partial G . \qquad (6.11)$$

In (6.10)-(6.11) we preserve the notation for the functions u, c, g, φ, ψ which depend on t, χ according to (6.9). Further, the coefficients $a^{ij} = \sum_{k,m=1}^{d} q_{ik} a^{km} q_{jm}$ form the matrix $QAQ^\mathsf{T} = \{a^{ij}\}$, and the coefficients $\beta^i = \sum_{k=1}^{d} q_{ik} b^k$ form the vector Qb.

Let us write two relations for the function u which are important for further analysis. For brevity, we will write $(-r_0, 0)$ instead of $(-r_0, 0, \ldots, 0)$. We have

$$u(t_0, -r_0, 0) = u(t_0, 0, 0) + \frac{\partial u}{\partial \chi^1}(t_0, 0, 0) \times (-r_0) + O(r^2) ,$$

$$\frac{\partial u}{\partial \chi^1}(t_0, -r_0, 0) = \frac{\partial u}{\partial \chi^1}(t_0, 0, 0) + \frac{\partial^2 u}{(\partial \chi^1)^2}(t_0, 0, 0) \times (-r_0) + O(r^2) .$$

The condition (6.11) at the point $(t_0, -r_0, 0)$ can be written in the form

$$\frac{\partial u}{\partial \chi^1}(t_0, -r_0, 0) = -\varphi(t_0, -r_0, 0)\, u(t_0, -r_0, 0) + \psi(t_0, -r_0, 0) .$$

The above three relations imply

$$\left(\frac{\partial u}{\partial \chi^1}\right)_0 = -\varphi_0\,(1+\varphi_0 r_0)u_0 + r_0 \left(\frac{\partial^2 u}{(\partial \chi^1)^2}\right)_0$$
$$+(1+\varphi_0 r_0)\psi_0 + O(r^2)\,, \qquad (6.12)$$

where u_0, φ_0, and ψ_0 mean $u(t_0,0,0)$, $\varphi(t_0,-r_0,0)$, and $\psi(t_0,-r_0,0)$, respectively. The symbol $(\cdot)_0$ means that the corresponding function is evaluated at the point $(t_0,0,0)$.

Further, differentiating (6.11) with respect to $\chi^2,\ldots,\chi^d$ at the point $(t_0,-r_0,0)$, we obtain

$$\frac{\partial^2 u}{\partial \chi^1 \partial \chi^j}(t_0,-r_0,0) + \frac{\partial \varphi}{\partial \chi^j}(t_0,-r_0,0)\,u(t_0,-r_0,0)$$
$$+\varphi(t_0,-r_0,0)\frac{\partial u}{\partial \chi^j}(t_0,-r_0,0) = \frac{\partial \psi}{\partial \chi^j}(t_0,-r_0,0),$$

whence it follows that

$$\varphi_0 \left(\frac{\partial u}{\partial \chi^j}\right)_0 = -\varphi'_{j0}\,u_0 - \left(\frac{\partial^2 u}{\partial \chi^1 \partial \chi^j}\right)_0 + \psi'_{j0} + O(r), \qquad (6.13)$$

where φ'_{j0} and ψ'_{j0} are the values of the corresponding derivatives at the point $(t_0,-r_0,0)$.

Let a point (t_0,x_0,y_0,z_0) be such that $\rho(x_0,\partial G) \leq r$, $t_0 \leq T_1 - r^2$. Now we form $(t_1,X_1,Y_1,Z_1) = (t_0 + r^2, x_0 + \Delta X, y_0 + \Delta Y, z_0 + \Delta Z)$ so that the point (t_1,X_1) belongs to $\bar{Q}$, the random variable ΔX is of order $O(r)$, and the equality

$$E\,(u(t_1,X_1)\,Y_1 + Z_1 - u(t_0,x_0)\,y_0 - z_0) = O(r^3) \qquad (6.14)$$

holds. This relation is written in the new coordinates as

$$E\,(u(t_1,\chi)\,Y_1 + Z_1 - u(t_0,x_0)\,y_0 - z_0) = O(r^3), \qquad (6.15)$$

where $\chi = Q\,\Delta X$.

Theorem 6.1. *Let*

$$t_1 = t_0 + r^2, \quad \chi^1 = \chi_0^1 + \chi_1^1\,, \quad \chi_0^1 = \beta_0^1\,r^2, \qquad (6.16)$$

$$\chi_1^1 = \left[\alpha_0^{11}\,r^2 + r_0^2\right]^{1/2} - r_0 - \beta_0^1\,r^2, \qquad (6.17)$$

$$\chi^j = \chi_0^j + \varphi_0 \chi_1^j + \chi_2^j,\quad \chi_0^j = \beta_0^j\,r^2,\quad \chi_1^j = -\alpha_0^{1j}\,r^2,\quad \chi_2^j = \sum_{l=2}^{d} \lambda^{jl}\,\nu_l\,r,$$
$$(6.18)$$

$$j = 2,\ldots,d,$$

6.6 Methods for the Neumann problem for parabolic equations 401

where ν_l are independent random variables taking the values ± 1 with probability $1/2$, the numbers λ^{jl}, $j,l = 2,\ldots,d$, satisfy the equalities

$$\sum_{k=2}^{d} \lambda^{ik}\lambda^{jk} = \alpha^{ij}, \quad i,j = 2,\ldots,d, \qquad (6.19)$$

(i.e., they form the $(d-1) \times (d-1)$-matrix λ satisfying the equality $\lambda\lambda^\mathsf{T} = \{\alpha^{i,j}\}_{i,j=2\div d}$ which results in $E\chi_2^i \chi_2^j = \alpha^{ij} r^2$), and

$$\Delta Y = \left(c_0\, r^2 + \varphi_0\, (1 + \varphi_0\, r_0)\chi_1^1 + \sum_{j=2}^{d} \varphi'_{j0}\, \chi_1^j + \varphi_0^2\, (\chi_1^1)^2 \right) y_0, \qquad (6.20)$$

$$\Delta Y = \left(g_0\, r^2 - \psi_0\, (1 + \varphi_0\, r_0)\chi_1^1 - \sum_{j=2}^{d} \psi'_{j0}\, \chi_1^j - \varphi_0\psi_0\, (\chi_1^1)^2 \right) y_0. \qquad (6.21)$$

Then the relation (6.15) (the relation (6.14) in the old coordinates) holds.

Proof. We have

$$d = u(t_1,\chi)\, Y_1 + Z_1 - u_0\, y_0 - z_0$$

$$= \left[u_0 + \left(\frac{\partial u}{\partial t}\right)_0 r^2 + \left(\frac{\partial u}{\partial \chi^1}\right)_0 \chi^1 + \sum_{j=2}^{d}\left(\frac{\partial u}{\partial \chi^j}\right)_0 \chi^j + \frac{1}{2}\left(\frac{\partial^2 u}{(\partial \chi^1)^2}\right)_0 (\chi^1)^2 \right.$$

$$+ \sum_{j=2}^{d} \left(\frac{\partial^2 u}{\partial \chi^1 \partial \chi^j}\right)_0 \chi^1 \chi^j + \frac{1}{2}\sum_{i,j=2}^{d}\left(\frac{\partial^2 u}{\partial \chi^i \partial \chi^j}\right)_0 \chi^i \chi^j + O(r^3) \bigg] (y_0 + \Delta Y)$$

$$+ \Delta Z - u_0 y_0. \qquad (6.22)$$

Using (6.12), (6.13), (6.17), (6.18), we get

$$\left(\frac{\partial u}{\partial \chi^1}\right)_0 \chi^1 = \left(\frac{\partial u}{\partial \chi^1}\right)_0 \chi_0^1$$

$$+ \left[-\varphi_0\, (1+\varphi_0 r_0) u_0 + r_0 \left(\frac{\partial^2 u}{(\partial \chi^1)^2}\right)_0 + (1+\varphi_0 r_0)\psi_0 \right] \chi_1^1 + O(r^3),$$

$$\sum_{j=2}^{d}\left(\frac{\partial u}{\partial \chi^j}\right)_0 \chi^j = \sum_{j=2}^{d}\left(\frac{\partial u}{\partial \chi^j}\right)_0 (\chi_0^j + \chi_2^j) + \varphi_0 \sum_{j=2}^{d}\left(\frac{\partial u}{\partial \chi^j}\right)_0 \chi_1^j$$

$$= \sum_{j=2}^{d}\left(\frac{\partial u}{\partial \chi^j}\right)_0 (\chi_0^j + \chi_2^j)$$

$$+ \sum_{j=2}^{d}\left(-\varphi'_{j0}\, u_0 - \left(\frac{\partial^2 u}{\partial \chi^1 \partial \chi^j}\right)_0 + \psi'_{j0}\right) \chi_1^j + O(r^3),$$

$$\sum_{j=2}^{d}\left(\frac{\partial^2 u}{\partial \chi^1 \partial \chi^j}\right)_0 \chi^1 \chi^j = \sum_{j=2}^{d}\left(\frac{\partial^2 u}{\partial \chi^1 \partial \chi^j}\right)_0 \chi^1 \chi_2^j + O(r^3),$$

$$\frac{1}{2}\sum_{i,j=2}^{d}\left(\frac{\partial^2 u}{\partial \chi^i \partial \chi^j}\right)_0 \chi^i \chi^j = \frac{1}{2}\sum_{i,j=2}^{d}\left(\frac{\partial^2 u}{\partial \chi^i \partial \chi^j}\right)_0 \chi_2^i \chi_2^j + O(r^3).$$

Substituting these expressions in (6.22) and making some elementary calculations (note that $2r_0\chi_1^1 + (\chi^1)^2 = \alpha_0^{11} r^2 + O(r^3)$), we obtain

$$d = \left[\left(\frac{\partial u}{\partial t}\right)_0 r^2 + \left(\frac{\partial u}{\partial \chi^1}\right)_0 \beta_0^1 r^2 + +\frac{1}{2}\left(\frac{\partial^2 u}{(\partial \chi^1)^2}\right)_0 \alpha_0^{11} r^2 \right.$$

$$+ (-\varphi_0\,(1+\varphi_0 r_0)u_0 + \psi_0(1+\varphi_0 r_0))\chi_1^1 + \sum_{j=2}^{d}\left(\frac{\partial u}{\partial \chi^j}\right)_0 \beta_0^j r^2$$

$$+ \sum_{j=2}^{d}\left(\frac{\partial u}{\partial \chi^j}\right)_0 \chi_2^j + \sum_{j=2}^{d}\left(\frac{\partial^2 u}{\partial \chi^1 \partial \chi^j}\right)_0 \alpha_0^{1j} r^2 + \sum_{j=2}^{d}\alpha_0^{1j}(\varphi_{j0}'\,u_0 - \psi_{j0}')\,r^2$$

$$\left. + \sum_{j=2}^{d}\left(\frac{\partial^2 u}{\partial \chi^1 \partial \chi^j}\right)_0 \chi^1 \chi_2^j + \frac{1}{2}\sum_{i,j=2}^{d}\left(\frac{\partial^2 u}{\partial \chi^i \partial \chi^j}\right)_0 \chi_2^i \chi_2^j + O(r^3)\right] y_0$$

$$+ u_0 \Delta Y + (-\varphi_{j0}'\,u_0 + \psi_{j0}')\chi_1^1 \Delta Y + \sum_{j=2}^{d}\left(\frac{\partial u}{\partial \chi^j}\right)_0 \chi_2^j \Delta Y + \Delta Z + O(r^3).$$

Further, taking into account that $E\chi_2^j = 0$, $j = 2, \ldots, d$, we get

$$Ed = \left[\left(\frac{\partial u}{\partial t}\right)_0 r^2 + \sum_{i=1}^{d}\beta_0^i\left(\frac{\partial u}{\partial \chi^i}\right)_0 r^2 + \frac{1}{2}\sum_{i,j=1}^{d}\alpha_0^{ij}\left(\frac{\partial^2 u}{\partial \chi^i \partial \chi^j}\right)_0 r^2\right.$$

$$+ \left(-\varphi_0\,(1+\varphi_0 r_0)\chi_1^1 + \sum_{j=2}^{d}\alpha_0^{1j}\varphi_{j0}'\,r^2\right)u_0 + \psi_0(1+\varphi_0 r_0)\chi_1^1$$

$$\left. - \sum_{j=2}^{d}\alpha_0^{1j}\psi_{j0}'\,r^2 \right] y_0$$

$$+ u_0\left[c_0 r^2 + \varphi_0\,(1+\varphi_0 r_0)\chi_1^1 - \sum_{j=2}^{d}\alpha_0^{1j}\varphi_{j0}'\,r^2 + \varphi_0^2\left(\chi_1^1\right)^2\right] y_0$$

$$- \varphi_0^2 u_0 \left(\chi_1^1\right)^2 y_0 + \psi_0\left(\chi_1^1\right)^2 \varphi_0 + g_0 r^2 y_0 - \psi_0(1+\varphi_0 r_0)\chi_1^1$$

$$+ \sum_{j=2}^{d}\alpha_0^{1j}\psi_{j0}'\,r^2 y_0 - \varphi_0\psi_0\left(\chi_1^1\right)^2 y_0 + O(r^3).$$

$$= \left[\left(\frac{\partial u}{\partial t}\right)_0 + \sum_{i=1}^{d} \beta_0^i \left(\frac{\partial u}{\partial \chi^i}\right)_0 + \frac{1}{2}\sum_{i,j=1}^{d} \alpha_0^{ij}\left(\frac{\partial^2 u}{\partial \chi^i \partial \chi^j}\right)_0 + c_0 u_0 + g_0\right]$$
$$\times r^2 y_0 + O(r^3) = O(r^3) \ .$$

Theorem 6.1 is proved. □

6.6.2 Convergence theorems

Theorem 6.2. *Suppose that the one-step transition for nonboundary points is made by a method with one-step error $O(r^4)$ whose time step h_k satisfies the inequality $h_k \geq C_0 r^2$ and whose increment ΔX satisfies the inequality $|\Delta X| \leq C_1 r$, where C_0 and C_1 are some positive constants (for instance, one can exploit methods from Sect. 6.1 here). Let the one-step transition for boundary points (t_0, x_0), $t_0 < T_1 - r^2$, be made in accordance with Theorem 6.1. Let the last step $\varkappa$, when $T_1 - r^2 \leq t_{\varkappa-1} < T_1$, result in the transition to the point $(t_\varkappa, X_\varkappa, Y_\varkappa, Z_\varkappa) = (T_1, X_{\varkappa-1}, Y_{\varkappa-1}, Z_{\varkappa-1})$. Then such a method has weak order of accuracy $O(r^2)$, i.e.,*

$$u(t, x) = E\left[f(X_\varkappa)Y_\varkappa + Z_\varkappa\right] + O(r^2), \quad t_0 = t, \quad X_0 = x. \tag{6.23}$$

The average number of steps is equal to $E\varkappa = O(1/r^2)$.

Proof. The proof of this theorem is analogous to the one of Theorems 1.12 and 1.13. The following arguments are used here. It is obvious that the number of steps for nonboundary points does not exceed $(T_1 - T_0)/Cr^2 = O(1/r^2)$. Since the error at each such step is $O(r^4)$, the total error over these steps is $O(r^2)$. It turns out (see Lemma 6.3 below) that the average number of steps for boundary points is estimated by $O(1/r)$. Since the one-step error for boundary points is $O(r^3)$, total error over these steps is $O(r^2)$ as well. Finally, the error on the last step does not exceed $O(r^2)$. Thus, the total error of the method is $O(r^2)$. Theorem 6.2 is proved. □

Now we turn our attention to the formulation and proof of the lemma mentioned above. Along with the surface ∂G, we consider the surface $\partial_l G$ which is concentric to ∂G and belongs to G, and the distance l between ∂G and $\partial_l G$ is fixed and small so that any ball of radius l with a center on $\partial_l G$ entirely belongs to G. The layer between $\partial_l G$ and ∂G is denoted as G_l. Consider a Markov chain (t_k, X_k) generated by one or another method specified in Theorem 6.2 which starts from the point (t, x), i.e., $t_0 = t$ and $X_0 = x$.

Let (t_k, X_k), $k = 0, \ldots, \varkappa$, be a certain path of this chain. Consider all the time moments $t_{k_1}, \ldots, t_{k_q}$ which the path spends in the cylindric layer $\bar{Q}_r = [T_0, T_1 - r^2] \times \bar{G}_r$. The random number q depends on t and x: $q = q(t, x)$ (of course, q also depends on r).

Lemma 6.3. *The average time* $v(t,x) = Eq(t,x)$ *which the chain* (t_k, X_k) *spends in the cylindric layer* $\bar{Q}_r = [T_0, T_1 - r^2] \times \bar{G}_r$ *uniformly in* $(t,x) \in \bar{Q}$ *satisfies the estimate*

$$v(t,x) \leq \frac{C}{r}. \tag{6.24}$$

Proof. We use here the same technique as, e.g., in the proof of Theorem 1.13. In connection with the Markov chain (t_k, X_k), we consider the boundary value problem

$$\begin{aligned} PV - V &= -g(t,x), & (t,x) &\in [T_0, T_1 - r^2] \times \bar{G}, \\ V(t,x) &= 0, & (t,x) &\in (T_1 - r^2, T_1] \times \bar{G}, \end{aligned} \tag{6.25}$$

where P is the one-step transition operator $Pv(t,x) = Ev(t_1, X_1)$, $t_0 = t$, $X_0 = x$. As is known (see, e.g., Sect. 6.1.2), if $V(x)$ is the solution of the boundary value problem (6.25) with a function $g(t,x)$ such that

$$g(t,x) \geq I_{\bar{Q}_r}(t,x), \quad (t,x) \in [T_0, T_1 - r^2] \times \bar{G}, \tag{6.26}$$

then

$$v(t,x) \leq V(t,x). \tag{6.27}$$

Introduce the functions

$$w(x) = \begin{cases} \rho^2(x, \partial_l G), & x \in \bar{G}_l, \\ 0, & x \in \bar{G} \setminus \bar{G}_l, \end{cases}$$

and

$$W(t,x) = \begin{cases} 0, & (t,x) \in (T_1 - r^2, T_1] \times \bar{G}, \\ K(T_1 - t) + w(x), & (t,x) \in [T_0, T_1 - r^2] \times \bar{G}. \end{cases} \tag{6.28}$$

This function satisfies the boundary condition in (6.25), $K \geq 0$ is a constant whose value is specified below. Now we evaluate $PW - W$ (i.e., the function $-g(t,x)$) for various points (t,x).

Let a point (t,x) be such that $T_0 \leq t \leq T_1 - r^2$ and x belongs to the layer between the surfaces $\partial_l G$ and $\partial_r G$, $r \ll l$ and, besides, $\rho(x, \partial_l G) \geq C_1 r$. But virtue of this choice of (t,x), the point $x + \Delta X \in \bar{G}$ (see the conditions of Theorem 6.2). Therefore, $w(x) = \rho^2(x, \partial_l G)$, $w(x + \Delta X) = \rho^2(x + \Delta X, \partial_l G)$. Let p_x be the projection of x on the surface $\partial_l G$. We have $\rho(x + \Delta X, \partial_l G) = \rho(x, \partial_l G) + (\Delta X, (x - p_x)/|x - p_x|) + O(r^2)$, where $(\cdot, \cdot)$ is the scalar product. Hence

$$E\left[w(x + \Delta X) - w(x)\right]$$
$$= E\left[\left(\rho(x, \partial_l G) + \left(\Delta X, \frac{x - p_x}{|x - p_x|}\right)\right)^2 - \rho^2(x, \partial_l G)\right] + O(r^2)$$
$$= 2\rho(x, \partial_l G) E\left(\Delta X, \frac{x - p_x}{|x - p_x|}\right) + E\left(\Delta X, \frac{x - p_x}{|x - p_x|}\right)^2 + O(r^2) = O(r^2)$$

since $|E\Delta X| = O(r^2)$ and $E|\Delta X|^2 = O(r^2)$ for the methods exploited in Theorem 6.2. Therefore

$$PW - W = -K\Delta t + O(r^2) . \tag{6.29}$$

Now let $\rho(x, \partial_l G) < C_1 r$. In this case both $w(x) = O(r^2)$ and $w(x+\Delta X) = O(r^2)$ and, hence, (6.29) holds again.

We obviously have

$$PW - W = -K\Delta t$$

for points x which are inside $\partial_l G$ and for which the distance $\rho(x, \partial_l G) > C_1 r$.

Since $O(r^2)$ in (6.29) satisfies the inequality $|O(r^2)| \leq Cr^2$ uniformly in (t, x), where $C > 0$ is a constant, and since $\Delta t \geq C_0 r^2$, we can find a constant K such that the right-hand side of (6.29) is non-positive. Thus, the function $f(t, x) = -(PW - W)$ is nonnegative.

Finally, we consider a point (t, x) such that $T_0 \leq t \leq T_1 - r^2$, $x \in \bar{G}_r$. In connection with the notation used in Theorem 6.1 it is convenient to use (t_0, x_0) instead of (t, x) here. We have

$$\rho(x_0, \partial_l G) = l - r_0, \quad \rho(x_0 + \Delta X, \partial_l G) = l - r_0 - \chi^1 + O(r^2) .$$

Further,

$$E\left[w(x_0 + \Delta X) - w(x_0)\right] = \left(l - r_0 - \chi^1\right)^2 - (l - r_0)^2 + O(r^2)$$
$$= -2l\chi^1 + O(r^2) .$$

It is not difficult to see that χ^1 has a uniform lower bound proportional to r. Hence, we obtain for such points (t, x):

$$PW(t, x) - W(t, x) = -f(t, x), \quad f(t, x) \geq qr.$$

As a result, we have constructed the function W which is a solution of the problem (6.25) with the nonnegative everywhere function $f(t, x)$ satisfying the inequality $f \geq qr$ in Q_r. Now we consider the function $V(t, x) = W(t, x)/qr$. It is evident that $V(t, x)$ is the solution of the problem (6.25) with the function $g(t, x) = f(t, x)/qr$ which satisfies the inequality (6.26). Therefore, (6.27) holds and $v(t, x)$ satisfies the estimate (6.24). □

The proof of Theorem 6.2 inspires the following ideas. Since the time which the chain spends in $\bar{G}_r$ is $O(1/r)$, it is sufficient for convergence of a method to require that the one-step error in $\bar{G}_r$ is $O(r^2)$ instead of $O(r^3)$. However, in this case the accuracy order of the method is $O(r)$ only, but at this expense we can make an attempt to construct an essentially more simple method.

Theorem 6.4. *Let*

$$t_1 = t_0 + r^2, \qquad (6.30)$$

$$\chi^1 = qr, \qquad (6.31)$$

where q is a positive number,

$$\chi^j = 0, \quad j = 2, \ldots, d, \qquad (6.32)$$

$$\Delta Y = \varphi_0 \, \chi^1 \, y_0 = \varphi_0 \, qr \, y_0, \qquad (6.33)$$

$$\Delta Z = -\psi_0 \, \chi^1 \, y_0 = -\psi_0 \, qr \, y_0. \qquad (6.34)$$

Then the relations (6.14)-(6.15) hold with the replacement of $O(r^3)$ by $O(r^2)$, and the method from Theorem 6.2 with (6.16)-(6.21) replaced by (6.30)-(6.34) has the order of accuracy $O(r)$, i.e., the relation (6.23) holds with the replacement of $O(r^2)$ by $O(r)$.

Proof. We have

$$d = u(t_1, \chi) Y_1 + Z_1 - u_0 y_0 - z_0$$

$$= \left[u_0 + \left(\frac{\partial u}{\partial \chi^1} \right)_0 \chi^1 + O(r^2) \right] (y_0 + \varphi_0 \, \chi^1 \, y_0) - \psi_0 \, \chi^1 \, y_0 - u_0 \, y_0$$

$$= u_0 \, \varphi_0 \, \chi^1 \, y_0 + \left(\frac{\partial u}{\partial \chi^1} \right)_0 \chi^1 \, y_0 - \psi_0 \, \chi^1 \, y_0 + O(r^2).$$

At the same time, it follows from (6.12) that

$$\left(\frac{\partial u}{\partial \chi^1} \right)_0 + \varphi_0 \, u_0 - \psi_0 = O(r),$$

whence $d = O(r^2)$.

The last assertion of this theorem is proved by reference to Lemma 6.3 as we did in Theorem 6.2. Theorem 6.4 is proved. □

Remark 6.5. It is possible to construct a method of order $O(r^2)$ using the simpler formulas (6.30)-(6.34). This can be done as follows. We introduce a layer $\bar{Q}_\delta$ with δ proportional to r^2 instead of the boundary layer $\bar{Q}_r$ used in Theorem 6.2. We use a standard method for nonboundary points and Theorem 6.4 with the replacement of r by δ for boundary points. Here the one-step error is $O(\delta^2)$ and the average number of steps which the chain spends in $\bar{Q}_\delta$ is $O(1/\delta)$. Therefore, the summarized error over boundary points is equal to $O(\delta) = O(r^2)$. And the total error of the method is $O(r^2)$. In this case the number of steps is larger than in the method from Theorem 6.2, of course. But apparently the number of steps will be still of order $O(1/r^2)$.

7 Probabilistic approach to numerical solution of the Cauchy problem for nonlinear parabolic equations

Nonlinear partial differential equations (PDEs) of parabolic type are of great interest both in theoretical and applied aspects. To mention just a few, the Burgers equation, Fisher–Kolmogorov–Petrovskii–Piskunov (FKPP) and Ginzburg–Landau equations, equations with nonlinear diffusion and with blow-up solutions, reaction-diffusion systems are examples of parabolic nonlinear PDEs. They are suggested as mathematical models of problems in many fields such as fluid dynamics, filtration, combustion, biochemistry, dynamics of populations, etc. Nonlinear PDEs are usually not susceptible of analytic solution and mostly investigated by means of numerical methods. Their investigation is presented in many publications in which deterministic approaches are applicable (see, e.g., [42, 91, 143, 259, 269, 291] and references therein). A few authors only exploit probabilistic approaches (see [70, 141, 288] and references therein).

A probabilistic approach to constructing new layer methods for solving nonlinear PDEs of parabolic type is proposed in [188]. It is based on the well-known probabilistic representations of solutions of linear parabolic equations and on the ideas of weak sense numerical integration of SDEs (see Chap. 2). In spite of the probabilistic nature these methods are nevertheless deterministic.

We start this chapter (Sect. 7.1) with the probabilistic approach to the Cauchy problem for linear parabolic equations. The approach cannot be carried over to nonlinear problems on the whole. However, its local version can be generalized to the Cauchy problem for semilinear parabolic equations. This is done in Sect. 7.2, where some layer methods are constructed in the nonlinear case. A practical realization of layer methods requires space discretization and an interpolation. We propose a number of numerical algorithms and prove their convergence. The principal ideas are demonstrated in the one-dimensional case. The multi-dimensional case is shortly discussed in Sect. 7.3. In addition, we show how the results obtained can be extended to reaction-diffusion systems. Some numerical tests are presented in Sect. 7.4.

The second part of this chapter (Sects. 7.5-7.7) deals with application of the probabilistic approach to the Cauchy problem for semilinear parabolic equations with small parameter [206]. Nonlinear parabolic equations with small parameter arise in a variety of applications (see, e.g., [42, 66, 111, 122, 253] and references therein). For instance, they are used in gas dynamics, when one has to take into account small viscosity and small heat conductivity. Some problems of combustion are described by PDEs with small parameter. They also arise as the result of introducing artificial viscosity in systems of first-order hyperbolic equations that is one of the popular approaches to numerical solution of inviscid problems of gas dynamics [219, 249, 303]. The probabilistic representations of the solution to the Cauchy problem for semilinear parabolic equations with small parameter are connected with systems of SDEs with small noise. To construct effective layer methods for this problem, we exploit special weak approximations for SDEs with small noise proposed in Chap. 3.

The probabilistic approach considered in this chapter takes into account a coefficient dependence on the space variables and a relationship between diffusion and advection in an intrinsic manner. In particular, the layer methods allow us to avoid difficulties stemming from essentially changing coefficients and strong advection. Some other probabilistic numerical methods for nonlinear PDEs are available, e.g., in [141, 288].

7.1 Probabilistic approach to linear parabolic equations

Consider the Cauchy problem for linear parabolic equation

$$\frac{\partial u}{\partial t} + \frac{1}{2}\sum_{i,j=1}^{d} a^{ij}(t,x)\frac{\partial^2 u}{\partial x^i \partial x^j} + \sum_{i=1}^{d} b^i(t,x)\frac{\partial u}{\partial x^i} + c(t,x)u + g(t,x) = 0,$$

(1.1)

$$t_0 \leq t < T, \quad x \in \mathbf{R}^d,$$

with the initial condition

$$u(T,x) = \varphi(x).$$

(1.2)

The matrix $a(t,x) = \{a^{ij}(t,x)\}$ is supposed to be symmetric and positive semidefinite.

Let $\sigma(t,x)$ be a matrix obtained from the equation

$$a(t,x) = \sigma(t,x)\sigma^\top(t,x).$$

This equation is solvable with respect to σ (for instance, by a lower triangular matrix) at least for a positively definite a.

The solution to the problem (1.1)-(1.2) has various probabilistic representations (cf. (2.4.23)):

$$u(t,x) = E[\varphi(X_{t,x}(T))\,Y_{t,x,1}(T) + Z_{t,x,1,0}(T)], \quad t \leq T, \ x \in \mathbf{R}^d,$$

(1.3)

where $X_{t,x}(s)$, $Y_{t,x,y}(s)$, $Z_{t,x,y,z}(s)$, $s \geq t$, is the solution of the Cauchy problem for the system of SDEs

7.1 Probabilistic approach to linear parabolic equations

$$dX = b(s, X)ds - \sigma(s, X)\mu(s, X)ds + \sigma(s, X)dw(s), \quad X(t) = x, \quad (1.4)$$
$$dY = c(s, X)Y ds + \mu^\top(s, X)Y dw(s), \quad Y(t) = y, \quad (1.5)$$
$$dZ = g(s, X)Y ds + F^\top(s, X)Y dw(s), \quad Z(t) = z. \quad (1.6)$$

Here $w(s) = (w^1(s), \ldots, w^d(s))^\top$ is a d-dimensional standard Wiener process, $b(s, x)$ is the column-vectors of dimension d composed from the coefficients $b^i(s, x)$, $\mu(s, x)$ and $F(s, x)$ are arbitrary column-vectors of dimension d, Y and Z are scalars. The usual representation (see [48]) can be seen in (1.3)-(1.6) if $\mu = 0$, $F = 0$. The case $F = 0$ rests on Girsanov's theorem. For $F \neq 0$, the representation (1.3) is evidently true as well. We note that the representations (1.3)-(1.6) are the well-known Feynman–Kac formula.

In what follows it is supposed that all the coefficients of (1.1) and (1.4)-(1.6) and the solution of the problem (1.1)-(1.2) (which is supposed to exist and to be unique) are sufficiently smooth and satisfy some conditions of growth under large $|x|$ so that these conditions are sufficient for applying the theory of weak methods (see Chap. 2).

Let us consider a time discretization, for definiteness the equidistant one:

$$T = t_N > t_{N-1} > \cdots > t_0 = t, \quad \frac{T - t_0}{N} = h.$$

Recall that a weak approximation of the system (1.4)-(1.6) consists in construction of the system of stochastic difference equations

$$X_0 = x, \quad X_{m+1} = X_m + A(t_m, X_m, h; \xi_m), \quad (1.7)$$
$$Y_0 = 1, \quad Y_{m+1} = Y_m + \alpha(t_m, X_m, h; \xi_m)Y_m, \quad (1.8)$$
$$Z_0 = 0, \quad Z_{m+1} = Z_m + \beta(t_m, X_m, h; \xi_m)Y_m, \quad m = 0, 1, \ldots, N-1, \quad (1.9)$$

where X_m is a d-dimensional vector, Y_m and Z_m are scalars, ξ_m is a random vector of a certain dimension, A is a d-dimensional vector function, α and β are scalar functions, ξ_m is independent of $X_0, \ldots, X_m$ and $\xi_0, \ldots, \xi_{m-1}$.

Let the system (1.7)-(1.9) be a weak scheme of order p for the system (1.4)-(1.6), i.e., (see Chap. 2)

$$\bar{u}(t_0, x) = \bar{u}(t, x) := E(\varphi(X_N)Y_N + Z_N) = u(t, x) + R_N, \quad (1.10)$$

where

$$|R_N| \leq K(1 + |x|^\varkappa)h^p,$$

and $K > 0$, $\varkappa \geq 0$ are some constants.

Standard numerical methods for PDEs, including the finite difference ones (see, e.g., [241, 245, 258, 273, 304]), can successfully be applied provided the dimension d of the space variable x is comparatively small ($d \leq 3$) while for larger dimensions these numerical procedures become unrealistic due to a huge volume of computations. Fortunately, in many cases functionals only or even individual values of a solution have to be found. A probabilistic approach

7 Probabilistic approach to the Cauchy problem for nonlinear PDEs

for such problems has an essential advantage as long as the problem under consideration can be reduced to solving the corresponding system of ordinary stochastic differential equations.

Let us recall (see details in Chap. 2) that the probabilistic representation (1.3)-(1.6) and its approximation (1.10), (1.7)-(1.9) give an example of the approach which allows to find the individual values $u(t,x)$ of the solution of problem (1.1)-(1.2) even in essentially multi-dimensional $(d>3)$ cases. The value $\bar{u}(t,x)$ is evaluated by applying the Monte Carlo technique:

$$\bar{u}(t,x) \approx \frac{1}{M} \sum_{m=1}^{M} (\varphi(X_N^{(m)}) Y_N^{(m)} + Z_N^{(m)}),$$

where $(X_N^{(m)}, Y_N^{(m)}, Z_N^{(m)})$, $m=1,\ldots,M$, are independent realizations of the process defined by (1.7)-(1.9).

But it should be noted that the probabilistic approach is useful not only in this respect. Here we exploit it to construct some layer methods. To show this, let us consider the Cauchy problem

$$X_k = x, \quad X_{l+1} = X_l + A(t_l, X_l, h; \xi_l) \tag{1.11}$$
$$Y_k = y, \quad Y_{l+1} = Y_l + \alpha(t_l, X_l, h; \xi_l) Y_l \tag{1.12}$$
$$Z_k = z, \quad Z_{l+1} = Z_l + \beta(t_l, X_l, h; \xi_l) Y_l, \tag{1.13}$$
$$l = k, k+1, \ldots, N-1; \quad 0 \leq k \leq N-1,$$

which is related to the system (1.7)-(1.9). Denote by $\bar{X}_{t_k,x}(t_l)$, $\bar{Y}_{t_k,x,y}(t_l)$, $\bar{Z}_{t_k,x,y,z}(t_l)$, $t_l \geq t_k$, the solution of this problem.

Introduce the function (recall that $T=t_N$):

$$\bar{u}(t_k, x, y, z) = E(\varphi(\bar{X}_{t_k,x}(T)) \bar{Y}_{t_k,x,y}(T) + \bar{Z}_{t_k,x,y,z}(T)).$$

Clearly, the function $\bar{u}(t_k, x, y, z)$ can be written as

$$\bar{u}(t_k, x, y, z) = \bar{u}(t_k, x) y + z,$$

where

$$\bar{u}(t_k, x) = E(\varphi(\bar{X}_{t_k,x}(T)) \bar{Y}_{t_k,x,1}(T) + \bar{Z}_{t_k,x,1,0}(T)).$$

Let $t = t_0 \leq t_k < t_l \leq T$. Since

$$\bar{X}_{t_k,x}(T) = \bar{X}_{t_l, \bar{X}_{t_k,x}(t_l)}(T)$$
$$\bar{Y}_{t_k,x,1}(T) = \bar{Y}_{t_l, \bar{X}_{t_k,x}(t_l), \bar{Y}_{t_k,x,1}(t_l)}(T)$$
$$\bar{Z}_{t_k,x,1,0}(T) = \bar{Z}_{t_l, \bar{X}_{t_k,x}(t_l), \bar{Y}_{t_k,x,1}(t_l), \bar{Z}_{t_k,x,1,0}(t_l)}(T),$$

we have

7.1 Probabilistic approach to linear parabolic equations

$$\bar{u}(t_k, x) = EE[\varphi(\bar{X}_{t_l, \bar{X}_{t_k,x}(t_l)}(T))\bar{Y}_{t_l, \bar{X}_{t_k,x}(t_l), \bar{Y}_{t_k,x,1}(t_l)}(T)$$
$$+ \bar{Z}_{t_l, \bar{X}_{t_k,x}(t_l), \bar{Y}_{t_k,x,1}(t_l), \bar{Z}_{t_k,x,1,0}(t_l)}(T) / \bar{X}_{t_k,x}(t_l), \bar{Y}_{t_k,x,1}(t_l), \bar{Z}_{t_k,x,1,0}(t_l)]$$
$$= E(\bar{u}(t_l, \bar{X}_{t_k,x}(t_l))\bar{Y}_{t_k,x,1}(t_l) + \bar{Z}_{t_k,x,1,0}(t_l)), \quad \bar{u}(t_N, x) = \varphi(x). \quad (1.14)$$

Using (1.14) sequentially with $l = k+1$:

$$\bar{u}(t_k, x) = E(\bar{u}(t_{k+1}, \bar{X}_{t_k,x}(t_{k+1}))\bar{Y}_{t_k,x,1}(t_{k+1}) + \bar{Z}_{t_k,x,1,0}(t_{k+1})), \quad (1.15)$$
$$k = N-1, \ldots, 0,$$

one can recurrently find the approximate solution $\bar{u}(t_{N-1}, x)$, $\bar{u}(t_{N-2}, x), \ldots,$ $\bar{u}(t_0, x)$ of the problem (1.1)-(1.2) starting from

$$\bar{u}(t_N, x) = \varphi(x). \quad (1.16)$$

This method is deterministic if we are able to calculate the expectations explicitly (see, e.g., the formulas (1.20) or (1.23) below). To realize (1.15) numerically, it is sufficient to calculate the functions $\bar{u}(t_k, x)$ at some knots x_i applying some kind of interpolation at every layer.

Further, it is more convenient to clarify some additional ideas on simple examples. To this end, let us consider the following one-dimensional ($d=1$) problem for the heat equation:

$$\frac{\partial u}{\partial t} + \frac{\sigma^2}{2}\frac{\partial^2 u}{\partial x^2} = 0, \quad t < 0, \quad x \in \mathbf{R}, \quad u(0, x) = \varphi(x). \quad (1.17)$$

Since $c = 0$, $g = 0$, we omit the equations for Y and Z. We have

$$dX = \sigma dw(s), \quad X(t_0) = x, \quad t_0 < 0. \quad (1.18)$$

Example 1.1. Consider the weak Euler scheme

$$X_{k+1} = X_k + \sigma\sqrt{h}\xi_k, \quad X_0 = x, \quad (1.19)$$

where $P(\xi_k = \pm 1) = 1/2$.

If we set $l = k+1$ in (1.14), we obtain

$$\bar{u}(t_k, x) = E\bar{u}(t_{k+1}, \bar{X}_{t_k,x}(t_{k+1}))$$
$$= \frac{1}{2}\bar{u}(t_{k+1}, x - \sigma\sqrt{h}) + \frac{1}{2}\bar{u}(t_{k+1}, x + \sigma\sqrt{h}),$$
$$\bar{u}(t_N, x) = \varphi(x). \quad (1.20)$$

Here $t_N = 0$, $h = -t_0/N$, $t_k = -h(N-k) = t_{k+1} - h$, $k = N-1, \ldots, 0$. The relation (1.20) is a linear difference equation. The equation (1.19) can be considered as a characteristic one for (1.20), and the formula

$$\bar{u}(t_k, x) = E\varphi(\bar{X}_{t_k,x}(t_N)) \quad (1.21)$$

gives the probabilistic representation of the solution to the equation (1.20). It is well known (see Chap. 2) that this solution is distinguished from the solution of the problem (1.17) by a quantity of order $O(h)$.

It is not difficult to see that evaluation of the values $\bar{u}(t_k, x_i)$ by layers due to the formula (1.20) coincides with the simplest explicit finite difference scheme for solving (1.17) if we set $h_t = h$, $h_x = \sigma\sqrt{h}$ and consider the equidistant space discretization : $x_i = x_0 + i\sigma\sqrt{h}$, $i = 0, \pm 1, \pm 2, \ldots$, x_0 is a point belonging to **R**.

If we need the solution of (1.17) for all points (t_k, x_i), we can use (1.20) to find $\bar{u}(t_k, x_i)$ layerwise. But if we need the solution at a separate point (t_k, x), the formula (1.21) is more convenient. Of course, in the last case the Monte Carlo error arises in addition.

Example 1.2. Now consider a more general scheme than (1.19):

$$X_{k+1} = X_k + \alpha\sqrt{h}\eta_k, \quad X_0 = x, \qquad (1.22)$$

where the constant $\alpha \geq \sigma$, $P(\eta = \pm 1) = \sigma^2/2\alpha^2$, $P(\eta = 0) = 1 - \sigma^2/\alpha^2$. Instead of (1.20), we get

$$\bar{u}(t_k, x) = E\bar{u}(t_{k+1}, \bar{X}_{t_k,x}(t_{k+1})) = (1 - \frac{\sigma^2}{\alpha^2})\bar{u}(t_{k+1}, x)$$
$$+ \frac{\sigma^2}{2\alpha^2}\bar{u}(t_{k+1}, x - \alpha\sqrt{h}) + \frac{\sigma^2}{2\alpha^2}\bar{u}(t_{k+1}, x + \alpha\sqrt{h}),$$
$$\bar{u}(t_N, x) = \varphi(x). \qquad (1.23)$$

Again due to the theory of weak methods for SDEs, the formula (1.21) with $\bar{X}$ from (1.22) approximates the solution of the problem (1.17) with accuracy $O(h)$. The formula (1.21) can be realized either by the Monte Carlo method or layerwise according to (1.23). The layer realization (1.23) is deterministic and coincides (if we choose the corresponding space net) with the following finite difference scheme

$$\frac{\bar{u}(t_k, x_i) - \bar{u}(t_{k+1}, x_i)}{h_t} = \frac{\sigma^2}{2}\frac{\bar{u}(t_{k+1}, x_{i+1}) - 2\bar{u}(t_{k+1}, x_i) + \bar{u}(t_{k+1}, x_{i-1})}{h_x^2},$$
$$\qquad (1.24)$$
$$h_t = h, \quad h_x = \alpha\sqrt{h}.$$

Due to the Lax–Richtmyer equivalence theorem, the method (1.24) (or, what is the same, the method (1.23)) converges with the rate $O(h)$ if $\alpha \geq \sigma$. If $\alpha < \sigma$, the numerical approximation (1.24) is not stable from the point of view of the theory of finite difference methods, and the method (1.24) diverges. We underline that there does not exist any probabilistic scheme of the form (1.11)-(1.13), (1.15) corresponding to (1.24) with $\alpha < \sigma$, i.e., there is no such a bad probabilistic scheme. Convergence theorems for weak methods (in comparison with the theory of finite difference methods) do not contain any conditions on stability of their approximations. The point is that X_{k+1} (and consequently the distribution of X_{k+1} which generalizes the step h_x) of a suitable weak scheme is connected with h_t, X_k, and with the coefficients of the problem in a reasonable way. Thus, methods having the probabilistic nature like (1.11)-(1.13), (1.15) are more adjusted (especially when coefficients of the considered problem are nonconstant) because the suitable choice of h_x is achieved automatically.

Let us note that the methods (1.20) and (1.23) do not need any interpolation because the layer $\bar{u}(t_k, x_i)$ makes use of the previous layer $\bar{u}(t_{k+1}, x)$ at the knots x_j only. But such a property of layer methods under consideration is rather exception than a rule. In conclusion, let us give two other examples.

Example 1.3. Consider the scheme (1.22) with $\alpha = \sigma\sqrt{3}$:

$$X_{k+1} = X_k + \sigma\sqrt{3h}\eta_k , \quad X_0 = x , \tag{1.25}$$

where $P(\eta = \pm 1) = 1/6$, $P(\eta = 0) = 2/3$. Since

$$E\eta = E\eta^3 = 0, \quad E(\sqrt{3}\eta)^2 = 1, \quad E(\sqrt{3}\eta)^4 = 3,$$

this scheme has the second order of accuracy (cf. Sect. 2.2).

We obtain the following finite difference method from (1.25):

$$\bar{u}(t_k, x_i) = \frac{1}{6}(\bar{u}(t_{k+1}, x_{i+1}) + \bar{u}(t_{k+1}, x_{i-1})) + \frac{2}{3}\bar{u}(t_{k+1}, x_i), \tag{1.26}$$

where $x_{i+1} - x_i = \sigma\sqrt{3h}$.

Since the scheme (1.25) is of the second order, the method (1.26) is also of order two, i.e., $|u(t_k, x_i) - \bar{u}(t_k, x_i)| = O(h^2)$. The method (1.26) is known as the finite difference method of excited accuracy.

Example 1.4. Consider another scheme

$$X_{k+1} = X_k + \sigma\sqrt{h}\zeta_k , \quad X_0 = x , \tag{1.27}$$

where $P(\zeta = 0) = p$, $P(\zeta = \pm\alpha) = q$, $P(\zeta = \pm\beta) = r$.

If, for example, $\alpha = 1$, $\beta = \sqrt{6}$, $p = 1/3$, $q = 3/10$, $r = 1/30$, then

$$E\zeta = E\zeta^3 = E\zeta^5 = 0, \quad E\zeta^2 = 1, \quad E\zeta^4 = 3, \quad E\zeta^6 = 15,$$

and the scheme is of order three (cf. Sect. 2.3.3). The corresponding method

$$\bar{u}(t_k, x) = E\bar{u}(t_{k+1}, \bar{X}_{t_k,x}(t_{k+1})) = E\bar{u}(t_{k+1}, x + \sigma\sqrt{h}\zeta_k)$$
$$= \frac{1}{30}\bar{u}(t_{k+1}, x - \sigma\sqrt{6h}) + \frac{3}{10}\bar{u}(t_{k+1}, x - \sigma\sqrt{h}) + \frac{1}{3}\bar{u}(t_{k+1}, x)$$
$$+ \frac{3}{10}\bar{u}(t_{k+1}, x + \sigma\sqrt{h}) + \frac{1}{30}\bar{u}(t_{k+1}, x + \sigma\sqrt{6h}) \tag{1.28}$$

is of order three too. But an interpolation is necessary for numerical realization of (1.28) on some net of knots x_i because of incommensurability of $\sigma\sqrt{h}$ and $(\sigma\sqrt{6h} - \sigma\sqrt{h})$.

We note in passing that, for example, the scheme

$$X_{k+1} = X_k + \sigma\sqrt{h}\nu_k, \quad X_0 = x,$$

where $P(\nu = 0) = 7/18$, $P(\nu = \pm 1) = 1/4$, $P(\nu = \pm 2) = 1/20$, $P(\nu = \pm 3) = 1/180$, also induces a method of order three. Evidently, this method has the form

$$\bar{u}(t_k, x_i) = \frac{1}{180}\bar{u}(t_{k+1}, x_{i-3}) + \frac{1}{20}\bar{u}(t_{k+1}, x_{i-2}) + \frac{1}{4}\bar{u}(t_{k+1}, x_{i-1})$$
$$+ \frac{7}{18}\bar{u}(t_{k+1}, x_i) + \frac{1}{4}\bar{u}(t_{k+1}, x_{i+1}) + \frac{1}{20}\bar{u}(t_{k+1}, x_{i+2})$$
$$+ \frac{1}{180}\bar{u}(t_{k+1}, x_{i+3}), \tag{1.29}$$

where $x_{i+1} - x_i = \sigma\sqrt{h}$.

Remark 1.5. Consider the Cauchy problem for an autonomous linear parabolic equation in its usual form (with positive direction of time θ):

$$\frac{\partial v}{\partial \theta} = \frac{1}{2}\sum_{i,j=1}^{d} a^{ij}(x)\frac{\partial^2 v}{\partial x^i \partial x^j} + \sum_{i=1}^{d} b^i(x)\frac{\partial v}{\partial x^i} + c(x)v + g(x), \quad \theta > 0, \ x \in \mathbf{R}^d,$$
$$\tag{1.30}$$
$$v(0, x) = \varphi(x). \tag{1.31}$$

Changing the variables $t = \theta$, $u(t, x) = v(-t, x)$, we get the Cauchy problem of the form (1.1)-(1.2) for the function $u(t, x)$ where $t < 0$, $x \in \mathbf{R}^d$, $T = 0$, $u(T, x) = \varphi(x)$. In the considered case the system (1.4)-(1.6) is autonomous as well (we suppose the function $\mu(s, x)$ in (1.4)-(1.6) to be independent of s). Therefore (see (1.3))

$$v(\theta, x) = u(-\theta, x) = E(\varphi(X_{-\theta,x}(0))Y_{-\theta,x,1}(0) + Z_{-\theta,x,1,0}(0))$$
$$= E(\varphi(X_{0,x}(\theta))Y_{0,x,1}(\theta) + Z_{0,x,1,0}(\theta)), \quad \theta > 0, \ x \in \mathbf{R}^d,$$

i.e., we can consider the positive direction of time for both the parabolic equation and its characteristic system of SDEs. According to this fact, we can write the following more convenient procedure in place of (1.15), (1.16):

$$\bar{v}(0, x) = \varphi(x),$$
$$\bar{v}(\theta_{k+1}, x) = E(\bar{v}(\theta_k, \bar{X}_{0,x}(h))\bar{Y}_{0,x,1}(h) + \bar{Z}_{0,x,1,0}(h)), \tag{1.32}$$
$$k = 0, \ldots, N-1,$$

where $0 = \theta_0 < \theta_1 < \cdots < \theta_N = \theta$; $h = \theta/N$ (of course, we consider A, α, and β in the scheme (1.11)-(1.13) to be independent of t_m). At the same time, we preferred to follow the general style of our exposition in Examples 1.1-1.4.

7.2 Layer methods for semilinear parabolic equations

For simplicity, we restrict ourselves to the one-dimensional case $d = 1$ in this section.

7.2.1 The construction of layer methods

Consider the Cauchy problem

$$\frac{\partial u}{\partial t} + \frac{1}{2}\sigma^2(t,x,u)\frac{\partial^2 u}{\partial x^2} + b(t,x,u)\frac{\partial u}{\partial x} + g(t,x,u) = 0, \ t_0 \leq t < T, \ x \in \mathbf{R}, \tag{2.1}$$

$$u(T,x) = \varphi(x). \tag{2.2}$$

Let $u = u(t,x)$ be the solution of the problem (2.1)-(2.2), which is supposed to exist, to be unique, to be sufficiently smooth, and to satisfy some conditions of boundedness. One can find many theoretical results on this topic in [91, 143, 259, 269, 291] (see also references therein). If we substitute $u = u(t,x)$ in the coefficients σ^2, b, g, we obtain a linear parabolic equation. We suppose that all the requirements mentioned in the previous section in connection with the equation (1.1) are fulfilled for the obtained linear equation as well. Let us note that in comparison with (1.1) this linear equation does not contain a term linear in u. It is so due to the general form of g in (2.1). Sometimes it may be more preferable to represent $g(t,x,u)$ as $g(t,x,u) = c(t,x)u + g_0(t,x,u)$ (for instance, in the case of small $g_0(t,x,u)$) and to substitute $u = u(t,x)$ in the function g_0 only. Clearly, in this case we obtain another linear equation and another probabilistic representation. For definiteness, we shall consider the case without term linear in u, i.e., we take $c(s,x) \equiv 0$, and we also put $\mu(s,x) \equiv 0$, $F(s,x) \equiv 0$ in the system (1.4)-(1.6).

We have (see (1.3) with $Y \equiv 1$)

$$u(t,x) = E(\varphi(X_{t,x}(T)) + \int_t^T g(s, X_{t,x}(s), u(s, X_{t,x}(s)))ds), \ t \leq T, \ x \in \mathbf{R}, \tag{2.3}$$

where $X_{t,x}(s)$ is the solution of the Cauchy problem for the equation

$$dX = b(s, X, u(s,X))ds + \sigma(s, X, u(s,X))dw(s), \quad X(t) = x.$$

Consider the equidistant time discretization

$$T = t_N > t_{N-1} > \cdots > t_0 = t, \quad \frac{T - t_0}{N} = h.$$

Due to (2.3), we have

$$u(t_k, x) = E(u(t_{k+1}, X_{t_k,x}(t_{k+1})))$$
$$+ \int_{t_k}^{t_{k+1}} g(s, X_{t_k,x}(s), u(s, X_{t_k,x}(s))) ds$$
$$= E(u(t_{k+1}, X_{t_k,x}(t_{k+1})) + Z_{t_k,x,0}(t_{k+1})), \qquad (2.4)$$

where X, Z satisfy the following system

$$dX = b(s, X, u(s, X)) ds + \sigma(s, X, u(s, X)) dw(s), \quad X(t_k) = x, \qquad (2.5)$$

$$dZ = g(s, X, u(s, X)) ds, \quad Z(t_k) = 0. \qquad (2.6)$$

Applying the explicit weak Euler scheme with the simplest simulation of noise to the system (2.5)-(2.6), we get

$$X_{t_k,x}(t_{k+1}) \simeq \bar{X}_{t_k,x}(t_{k+1}) = x + b(t_k, x, u(t_k, x))h + \sigma(t_k, x, u(t_k, x))\sqrt{h}\xi_k, \qquad (2.7)$$

$$Z_{t_k,x,0}(t_{k+1}) \simeq \bar{Z}_{t_k,x,0}(t_{k+1}) = g(t_k, x, u(t_k, x))h, \qquad (2.8)$$

where $\xi_{N-1}, \xi_{N-2}, \ldots, \xi_0$ are i.i.d. random variables which are distributed by the law $P(\xi = \pm 1) = 1/2$.

Using (2.4), we obtain

$$u(t_k, x) \simeq E(u(t_{k+1}, \bar{X}_{t_k,x}(t_{k+1})) + \bar{Z}_{t_k,x,0}(t_{k+1}))$$
$$= \frac{1}{2} u(t_{k+1}, x + b(t_k, x, u(t_k, x))h + \sigma(t_k, x, u(t_k, x))\sqrt{h})$$
$$+ \frac{1}{2} u(t_{k+1}, x + b(t_k, x, u(t_k, x))h - \sigma(t_k, x, u(t_k, x))\sqrt{h})$$
$$+ g(t_k, x, u(t_k, x))h. \qquad (2.9)$$

Following (2.9), one can write for the approximations $\bar{u}(t_k, x)$:

$$\bar{u}(t_N, x) = \varphi(x),$$

$$\bar{u}(t_k, x) = \frac{1}{2} \bar{u}(t_{k+1}, x + b(t_k, x, \bar{u}(t_k, x))h + \sigma(t_k, x, \bar{u}(t_k, x))\sqrt{h})$$
$$+ \frac{1}{2} \bar{u}(t_{k+1}, x + b(t_k, x, \bar{u}(t_k, x))h - \sigma(t_k, x, \bar{u}(t_k, x))\sqrt{h})$$
$$+ g(t_k, x, \bar{u}(t_k, x))h, \quad k = N-1, \ldots, 1, 0. \qquad (2.10)$$

The method (2.10) is an implicit layer method for solution of the Cauchy problem (2.1)-(2.2). This method is deterministic though the probabilistic approach is used for its derivation. Recall that it rests on the explicit Euler scheme.

Now let us use the following implicit scheme instead of (2.7)-(2.8):

$$\bar{X}_{t_k,x}(t_{k+1}) := \bar{X}_{k+1} = x + b(t_{k+1}, \bar{X}_{k+1}, u(t_{k+1}, \bar{X}_{k+1}))h$$
$$+ \sigma(t_{k+1}, \bar{X}_k, u(t_{k+1}, \bar{X}_k))\sqrt{h}\xi_k, \qquad (2.11)$$

$$\bar{Z}_{t_k,x,0}(t_{k+1}) := \bar{Z}_{k+1} = g(t_{k+1}, \bar{X}_{k+1}, u(t_{k+1}, \bar{X}_{k+1}))h, \quad (2.12)$$

where $\xi_{N-1}, \xi_{N-2}, \ldots, \xi_0$ are the same as in (2.7).

Let $\bar{X}_{k+1} = \bar{X}_{k+1}(\xi_k)$ be the solution of (2.11) (recall that the function $u(t_{k+1}, x)$ is supposed to be known). The variable ξ_k gets two different values. Denote by $\bar{X}^1_{k+1}$, $\bar{X}^2_{k+1}$ the corresponding values of $\bar{X}_{k+1}$. Introduce the analogous notation for the two values of $\bar{Z}_{k+1}$. As a result, we obtain the method

$$\bar{u}(t_N, x) = \varphi(x),$$

$$\bar{u}(t_k, x) = \frac{1}{2}(\bar{u}(t_{k+1}, \bar{X}^1_{k+1}) + \bar{Z}^1_{k+1}) + \frac{1}{2}(\bar{u}(t_{k+1}, \bar{X}^2_{k+1}) + \bar{Z}^2_{k+1}). \quad (2.13)$$

It is deterministic just as the method (2.10).

The formula (2.13) is explicit but to find $\bar{X}_{k+1}$ we have to use the implicit scheme (2.11). Therefore, both the method (2.10) and the method (2.13) are implicit.

To find $\bar{u}(t_k, x)$ from (2.10), one can apply the method of simple iteration. If we take $\bar{u}(t_{k+1}, x)$ as a null iteration, we get the following first iteration (we denote this iteration as $\bar{u}(t_k, x)$ again):

$$\bar{u}(t_N, x) = \varphi(x),$$

$$\bar{u}(t_k, x) = \frac{1}{2}\bar{u}(t_{k+1}, x + b(t_k, x, \bar{u}(t_{k+1}, x))h + \sigma(t_k, x, \bar{u}(t_{k+1}, x))\sqrt{h})$$
$$+ \frac{1}{2}\bar{u}(t_{k+1}, x + b(t_k, x, \bar{u}(t_{k+1}, x))h - \sigma(t_k, x, \bar{u}(t_{k+1}, x))\sqrt{h})$$
$$+ g(t_k, x, \bar{u}(t_{k+1}, x))h, \quad k = N-1, \ldots, 1, 0. \quad (2.14)$$

The formula (2.14) gives an explicit method for recurrent layerwise solution of the problem (2.1)-(2.2). We note that if we apply another approximate method to solve (2.10) (for example, taking the second iteration), we obtain another explicit method which can possess better properties than (2.14) (just as in numerical integration of ordinary differential equations).

Analogously, applying the method of simple iteration to (2.11) with x as a null iteration and substituting the obtained first iteration in (2.12) and (2.13), we obtain the following explicit method which slightly differs from (2.14):

$$\bar{u}(t_N, x) = \varphi(x),$$

$$\bar{u}(t_k, x) = \frac{1}{2}\bar{u}(t_{k+1}, \ x + b(t_{k+1}, x, \bar{u}(t_{k+1}, x))h + \sigma(t_{k+1}, x, \bar{u}(t_{k+1}, x))\sqrt{h})$$
$$+ \frac{1}{2}\bar{u}(t_{k+1}, \ x + b(t_{k+1}, x, \bar{u}(t_{k+1}, x))h - \sigma(t_{k+1}, x, \bar{u}(t_{k+1}, x))\sqrt{h})$$
$$+ g(t_{k+1}, x, \bar{u}(t_{k+1}, x))h, \quad k = N-1, \ldots, 1, 0. \quad (2.15)$$

Now we make use of a higher-order method of numerical integration of SDEs in the case of the equation (2.1) with constant σ. Let us apply the second-order (in the weak sense) Runge–Kutta scheme (see Sect. 2.2.2) to the system (2.5)-(2.6) with constant σ. We obtain (instead of (2.7)-(2.8)):

$$X_{t_k,x}(t_{k+1}) \simeq \bar{X}_{t_k,x}(t_{k+1}) = x + \sigma\sqrt{h}\xi_k + \frac{1}{2}b(t_k, x, u(t_k, x))h$$
$$+\frac{1}{2}b(t_{k+1}, x + b(t_k, x, u(t_k, x))h$$
$$+\sigma\sqrt{h}\xi_k, u(t_{k+1}, x + b(t_k, x, u(t_k, x))h + \sigma\sqrt{h}\xi_k))h, \quad (2.16)$$

$$Z_{t_k,x,0}(t_{k+1}) \simeq \bar{Z}_{t_k,x,0}(t_{k+1})$$
$$= \frac{1}{2}g(t_k, x, u(t_k, x))h + \frac{1}{2}g(t_{k+1}, x + b(t_k, x, u(t_k, x))h$$
$$+\sigma\sqrt{h}\xi_k, u(t_{k+1}, x + b(t_k, x, u(t_k, x))h + \sigma\sqrt{h}\xi_k))h, \quad (2.17)$$

where $\xi_{N-1}, \xi_{N-2}, \ldots, \xi_0$ are i.i.d. random variables distributed by the law $P(\xi = 0) = 2/3$, $P(\xi = \pm\sqrt{3}) = 1/6$.

Then, we obtain the following implicit layer method instead of (2.10):

$$\bar{u}(t_N, x) = \varphi(x),$$

$$\bar{u}(t_k, x) = \frac{2}{3}\bar{u}(t_{k+1}, x + \frac{1}{2}\bar{b}h + \frac{1}{2}b(t_{k+1}, x + \bar{b}h, \bar{u}(t_{k+1}, x + \bar{b}h))h)$$
$$+\frac{1}{6}\bar{u}(t_{k+1}, x + \sigma\sqrt{3h} + \frac{1}{2}\bar{b}h$$
$$+\frac{1}{2}b(t_{k+1}, x + \sigma\sqrt{3h} + \bar{b}h, \bar{u}(t_{k+1}, x + \sigma\sqrt{3h} + \bar{b}h))h)$$
$$+\frac{1}{6}\bar{u}(t_{k+1}, x - \sigma\sqrt{3h} + \frac{1}{2}\bar{b}h$$
$$+\frac{1}{2}b(t_{k+1}, x - \sigma\sqrt{3h} + \bar{b}h, \bar{u}(t_{k+1}, x - \sigma\sqrt{3h} + \bar{b}h))h)$$
$$+\frac{1}{2}g(t_k, x, u(t_k, x))h + \frac{1}{3}g(t_{k+1}, x + \bar{b}h, \bar{u}(t_{k+1}, x + \bar{b}h))h$$
$$+\frac{1}{12}g(t_{k+1}, x + \sigma\sqrt{3h} + \bar{b}h, \bar{u}(t_{k+1}, x + \sigma\sqrt{3h} + \bar{b}h))h$$
$$+\frac{1}{12}g(t_{k+1}, x - \sigma\sqrt{3h} + \bar{b}h, \bar{u}(t_{k+1}, x - \sigma\sqrt{3h} + \bar{b}h))h, \quad (2.18)$$

where $\bar{b} = b(t_k, x, \bar{u}(t_k, x))$.

This method has the one-step error of order three. If we take $\bar{u}(t_{k+1}, x)$ as a null iteration, we obtain the first iteration differing from the solution of (2.18) by a quantity of order $O(h^2)$, and only starting from the second iteration we attain the needed exactness. So, the implicit method (2.18) becomes the explicit one of the same order after two simple iterations.

7.2 Layer methods for semilinear parabolic equations   419

Clearly, resting on the ideas leading to the methods obtained, one can construct a lot of new methods using some other probabilistic representations or other methods of numerical integration of SDEs.

7.2.2 Convergence theorem for a layer method

We continue to treat the problem (2.1)-(2.2).

We assume (remind that for simplicity in writing the case $d=1$ is considered):

(i) The coefficients $b(t,x,u)$, $\sigma(t,x,u)$, $g(t,x,u)$ are uniformly bounded:

$$|b| \leq K, \ |\sigma| \leq K, \ |g| \leq K, \ t_0 \leq t \leq T, \ x \in \mathbf{R}, \ u_\circ < u < u^\circ, \qquad (2.19)$$

where $-\infty \leq u_\circ < u^\circ \leq \infty$ are some constants.

(ii) The coefficients $b(t,x,u)$, $\sigma(t,x,u)$, $g(t,x,u)$ uniformly satisfy the Lipschitz condition with respect to x and u:

$$\begin{aligned}
|b(t,x_2,u_2) - b(t,x_1,u_1)| &+ |\sigma(t,x_2,u_2) - \sigma(t,x_1,u_1)| \\
+ |g(t,x_2,u_2) - g(t,x_1,u_1)| & \\
\leq K(|x_2 - x_1| &+ |u_2 - u_1|), \\
t_0 \leq t \leq T, \ x_1, x_2 \in \mathbf{R}, \ u_\circ &< u_1, u_2 < u^\circ.
\end{aligned} \qquad (2.20)$$

(iii) There exists a unique bounded solution $u(t,x)$ of the problem (2.1)-(2.2) such that

$$u_\circ < u_* \leq u(t,x) \leq u^* < u^\circ, \ t_0 \leq t \leq T, \ x \in \mathbf{R}, \qquad (2.21)$$

and there exist the uniformly bounded derivatives:

$$\left|\frac{\partial^m u}{\partial t^i \partial x^l}\right| \leq K, \ i = 0, \ l = 1,2,3,4; \ i = 1, \ l = 0,1,2; \ i = 2, \ l = 0; \qquad (2.22)$$

$$t_0 \leq t \leq T, \ x \in \mathbf{R}.$$

First of all let us evaluate the one-step error of the method (method (2.14)):

$$\bar{u}(t_N, x) = \varphi(x),$$

$$\begin{aligned}
\bar{u}(t_k, x) = &\frac{1}{2}\bar{u}(t_{k+1}, x + b(t_k, x, \bar{u}(t_{k+1}, x))h + \sigma(t_k, x, \bar{u}(t_{k+1}, x))\sqrt{h}) \\
&+ \frac{1}{2}\bar{u}(t_{k+1}, x + b(t_k, x, \bar{u}(t_{k+1}, x))h - \sigma(t_k, x, \bar{u}(t_{k+1}, x))\sqrt{h}) \\
&+ g(t_k, x, \bar{u}(t_{k+1}, x))h, \quad k = N-1, \ldots, 1, 0.
\end{aligned} \qquad (2.23)$$

This error on the k-th layer (on the $(N-k)$-th step) is evidently equal to $v(t_k, x) - u(t_k, x)$, where

$$v(t_k, x) = \frac{1}{2}u(t_{k+1}, x + b(t_k, x, u(t_{k+1}, x))h + \sigma(t_k, x, u(t_{k+1}, x))\sqrt{h})$$
$$+ \frac{1}{2}u(t_{k+1}, x + b(t_k, x, u(t_{k+1}, x))h - \sigma(t_k, x, u(t_k, x))\sqrt{h})$$
$$+ g(t_k, x, u(t_{k+1}, x))h. \qquad (2.24)$$

Lemma 2.1. *Under the assumptions* (i)-(iii) *the one-step error of the method* (2.23) *has the second order of smallness with respect to* h:

$$|v(t_k, x) - u(t_k, x)| \le Ch^2, \qquad (2.25)$$

where C does not depend on x, h, k.

Proof. Expanding the functions $u(t_k + h, x + bh \pm \sigma\sqrt{h})$ at (t_k, x) in powers of h and $bh \pm \sigma\sqrt{h}$ and using the assumptions on boundedness (2.19) and (2.22), we get

$$v(t_k, x) = u(t_k, x) + \frac{\partial u}{\partial t}(t_k, x)h + \frac{\partial u}{\partial x}(t_k, x)bh$$
$$+ \frac{1}{2}\frac{\partial^2 u}{\partial x^2}(t_k, x)\sigma^2 h + gh + O(h^2). \qquad (2.26)$$

In (2.26) the coefficients b, σ^2, g have t_k, x, $u(t_{k+1}, x)$ as their arguments, and

$$|O(h^2)| \le Ch^2, \qquad (2.27)$$

where C does not depend on x, h, k. Now applying the Lipschitz condition (2.20) with respect to the variable u, it is not difficult to obtain

$$v(t_k, x) = u(t_k, x) + \frac{\partial u}{\partial t}(t_k, x)h + \frac{\partial u}{\partial x}(t_k, x)b(t_k, x, u(t_k, x))h$$
$$+ \frac{1}{2}\frac{\partial^2 u}{\partial x^2}(t_k, x)\sigma^2(t_k, x, u(t_k, x))h + g(t_k, x, u(t_k, x))h$$
$$+ O(h^2), \qquad (2.28)$$

where $O(h^2)$ satisfies the relation (2.27) again.

Because $u(t, x)$ is a solution of the equation (2.1), the inequality (2.25) follows from (2.28). □

Theorem 2.2. *Under the assumptions* (i)-(iii) *the method* (2.23) *has the first order of convergence, i.e.,*

$$|\bar{u}(t_k, x) - u(t_k, x)| \le Kh, \qquad (2.29)$$

where K does not depend on x, h, k.

Proof. Denote the error of the method (2.23) on the k-th layer as $R(t_k, x) := \bar{u}(t_k, x) - u(t_k, x)$. Then, we have

7.2 Layer methods for semilinear parabolic equations

$$\bar{u}(t_k, x) = u(t_k, x) + R(t_k, x), \ \bar{u}(t_{k+1}, x) = u(t_{k+1}, x) + R(t_{k+1}, x). \quad (2.30)$$

By (2.23) and (2.30), we get

$$u(t_k, x) + R(t_k, x)$$
$$= \bar{u}(t_k, x) = \frac{1}{2}\bar{u}(t_{k+1}, x + \bar{b}h + \bar{\sigma}\sqrt{h}) + \frac{1}{2}\bar{u}(t_{k+1}, x + \bar{b}h - \bar{\sigma}\sqrt{h}) + \bar{g}h$$
$$= \frac{1}{2}u(t_{k+1}, x + \bar{b}h + \bar{\sigma}\sqrt{h}) + \frac{1}{2}u(t_{k+1}, x + \bar{b}h - \bar{\sigma}\sqrt{h}) + \bar{g}h$$
$$+\frac{1}{2}R(t_{k+1}, x + \bar{b}h + \bar{\sigma}\sqrt{h}) + \frac{1}{2}R(t_{k+1}, x + \bar{b}h - \bar{\sigma}\sqrt{h}), \quad (2.31)$$

where $\bar{b}, \bar{\sigma}, \bar{g}$ are the coefficients $b(t, x, u), \sigma(t, x, u), g(t, x, u)$ calculated at $t = t_k$, $x = x$, $u = \bar{u}(t_{k+1}, x) = u(t_{k+1}, x) + R(t_{k+1}, x)$. For example, $\bar{b} = b(t_k, x, u(t_{k+1}, x) + R(t_{k+1}, x))$.

Here we have to assume for a while that the value $u(t_{k+1}, x) + R(t_{k+1}, x)$ remains in the interval (u_o, u°) (see the conditions (2.19) and (2.20)). Clearly, $R(t_N, x) = 0$, and below we prove recurrently that $R(t_k, x)$ is sufficiently small under a sufficiently small h. Thereupon, thanks to (2.21), this suggestion will be justified.

We have

$$\bar{b} = b(t_k, x, u(t_{k+1}, x) + R(t_{k+1}, x)) = b(t_k, x, u(t_{k+1}, x)) + \Delta b = b + \Delta b,$$

where $b := b(t_k, x, u(t_{k+1}, x))$ and Δb satisfies the inequality (thanks to (2.20))

$$|\Delta b| \leq K|R(t_{k+1}, x)|. \quad (2.32)$$

Analogously,

$$\bar{\sigma} = \sigma + \Delta \sigma, \ |\Delta \sigma| \leq K|R(t_{k+1}, x)|, \ \bar{g} = g + \Delta g, \ |\Delta g| \leq K|R(t_{k+1}, x)|. \quad (2.33)$$

It is not difficult to see that (2.32) and (2.33) imply the equalities

$$u(t_{k+1}, x + \bar{b}h \pm \bar{\sigma}\sqrt{h}) = u(t_{k+1}, x + bh \pm \sigma\sqrt{h})$$
$$+ \frac{\partial u}{\partial x}(t_{k+1}, x + bh)\left(\Delta b h \pm \Delta \sigma \sqrt{h}\right) + \Delta_\pm h, \quad (2.34)$$

where $\Delta_\pm$ satisfy the inequality of the type (2.32). Substituting this in (2.31), we obtain

$$u(t_k, x) + R(t_k, x)$$
$$= \frac{1}{2}u(t_{k+1}, x + bh + \sigma\sqrt{h}) + \frac{1}{2}u(t_{k+1}, x + bh - \sigma\sqrt{h}) + gh$$
$$+ \frac{1}{2}R(t_{k+1}, x + \bar{b}h + \bar{\sigma}\sqrt{h}) + \frac{1}{2}R(t_{k+1}, x + \bar{b}h - \bar{\sigma}\sqrt{h}) + r_k$$
$$= v(t_k, x) + \frac{1}{2}R(t_{k+1}, x + \bar{b}h + \bar{\sigma}\sqrt{h})$$
$$+ \frac{1}{2}R(t_{k+1}, x + \bar{b}h - \bar{\sigma}\sqrt{h}) + r_k, \quad (2.35)$$

where
$$|r_k| \leq K|R(t_{k+1}, x)|\, h\,. \tag{2.36}$$

Finally, using Lemma 2.1, we get
$$R(t_k, x) = \frac{1}{2} R(t_{k+1}, x + \bar{b}h + \bar{\sigma}\sqrt{h}) + \frac{1}{2} R(t_{k+1}, x + \bar{b}h - \bar{\sigma}\sqrt{h})$$
$$+ r_k + O(h^2)\,. \tag{2.37}$$

Now introduce
$$R_k := \sup_{-\infty < x < \infty} |R(t_k, x)|\,. \tag{2.38}$$

It follows from (2.36) and (2.37) (in addition recall that $R(t_N, x) = 0$):
$$R_N = 0,\ R_k \leq R_{k+1} + K R_{k+1} h + C h^2,\ k = N-1, \ldots, 1, 0\,, \tag{2.39}$$

which implies
$$R_k \leq \frac{C}{K} \left(e^{K(T - t_0)} - 1 \right) h,\ k = N, \ldots, 0\,.$$

Theorem 2.2 is proved. □

Remark 2.3. The result (2.29) for the method (2.23) can be justified under some other conditions as well. For instance, it is possible to allow a linear growth of the coefficients b, σ, g under $|x| \to \infty$ instead of the condition (i) if we assume in addition that the derivatives of the solution $u(t, x)$ from (2.22) are not only bounded but some of them go to zero under $|x| \to \infty$ (namely, if the expressions $\left| \dfrac{\partial^m u}{\partial t^i \partial x^l} \right| (1 + |x|^l)$, $i = 0$, $l = 1, 2, 3, 4$; $i = 1$, $l = 1, 2$, are uniformly bounded). Besides, we emphasize that the conditions of Theorem 2.2 are not necessary and the method (2.23) can be applied much broader than it is determined by $(i) - (iii)$. At the same time, the conditions $(i) - (iii)$ are fairly suitable in many situations.

7.2.3 Numerical algorithms

A recursive procedure can be applied to implement the method (2.23). But for large $T - t_0$ and small h such a procedure is computationally too expensive. To avoid recursive calculations and to construct a numerical algorithm, the method (2.23) (just as other layer methods) needs a discretization in the variable x. Consider the equidistant space discretization: $x_j = x_0 + j\alpha h$, $j = 0, \pm 1, \pm 2, \ldots$, $x_0 \in \mathbf{R}$, $\alpha > 0$ is a number, i.e., h_x is taken to be equal to $\alpha h = \alpha h_t$. Using, for example, the linear interpolation, we construct the following algorithm.

7.2 Layer methods for semilinear parabolic equations

Algorithm 2.4 *The algorithm is defined by the following formulas*

$$\bar{u}(t_N, x) = \varphi(x),$$

$$\bar{u}(t_k, x_j)$$
$$= \frac{1}{2}\bar{u}(t_{k+1}, x_j + b(t_{k+1}, x_j, \bar{u}(t_{k+1}, x_j))h + \sigma(t_{k+1}, x_j, \bar{u}(t_{k+1}, x_j))\sqrt{h})$$
$$+ \frac{1}{2}\bar{u}(t_{k+1}, x_j + b(t_{k+1}, x_j, \bar{u}(t_{k+1}, x_j))h - \sigma(t_{k+1}, x_j, \bar{u}(t_{k+1}, x_j))\sqrt{h})$$
$$+ g(t_k, x_j, \bar{u}(t_{k+1}, x_j))h, \quad x_j = x_0 + j\alpha h, \ j = 0, \pm 1, \pm 2, \ldots, \quad (2.40)$$

$$\bar{u}(t_k, x) = \frac{x_{j+1} - x}{\alpha h}\bar{u}(t_k, x_j) + \frac{x - x_j}{\alpha h}\bar{u}(t_k, x_{j+1}), \ x_j \leq x \leq x_{j+1}, \quad (2.41)$$
$$k = N-1, \ldots, 1, 0.$$

Theorem 2.5. *Under the assumptions (i)-(iii) Algorithm 2.4 has the first order of convergence, i.e., the approximation $\bar{u}(t_k, x)$ from the formula (2.41) satisfies the relation*

$$|\bar{u}(t_k, x) - u(t_k, x)| \leq Kh, \quad (2.42)$$

where K does not depend on x, h, k.

Proof. Let us introduce the error of Algorithm 2.4 on the k-th layer

$$R(t_k, x) := \bar{u}(t_k, x) - u(t_k, x)$$

and R_k in accordance with (2.38):

$$R_k := \sup_{-\infty < x < \infty} |R(t_k, x)|.$$

Of course, these new $R(t_k, x)$ and R_k differ from the old ones. Just as earlier, we are able to obtain for the nodes x_j (cf. (2.37)):

$$R(t_k, x_j) = \frac{1}{2}R(t_{k+1}, x_j + \bar{b}h + \bar{\sigma}\sqrt{h}) + \frac{1}{2}R(t_{k+1}, x_j + \bar{b}h - \bar{\sigma}\sqrt{h})$$
$$+ r_k + O(h^2).$$

Hence

$$|R(t_k, x_j)| \leq R_{k+1} + KR_{k+1}h + Ch^2. \quad (2.43)$$

We have

$$u(t_k, x) = \frac{x_{j+1} - x}{\alpha h}u(t_k, x_j) + \frac{x - x_j}{\alpha h}u(t_k, x_{j+1}) + O(h^2), \ x_j \leq x \leq x_{j+1}, \quad (2.44)$$

where the interpolation error $O(h^2)$ satisfies the inequality of the form (2.27).

We get from (2.44) and (2.41):
$$R(t_k, x) = \frac{x_{j+1} - x}{\alpha h} R(t_k, x_j) + \frac{x - x_j}{\alpha h} R(t_k, x_{j+1}) + O(h^2), \quad x_j \leq x \leq x_{j+1},$$
whence due to (2.43) for all x:
$$|R(t_k, x)| \leq R_{k+1} + K R_{k+1} h + C h^2, \qquad (2.45)$$
of course, with another constant C.

The inequality (2.45) implies (2.39). Theorem 2.5 is proved. □

Remark 2.6. Along with the linear interpolation (2.41) it is natural to use the spline approximation of the form
$$\bar{u}(t_k, x) = \sum_{i=-\infty}^{\infty} \bar{u}(t_k, x_i) B\left(\frac{x - i\alpha h}{\alpha h}\right), \quad x_i = x_0 + i\alpha h, \ x \in \mathbf{R}, \qquad (2.46)$$
$$k = N - 1, \ldots, 1, 0,$$
where $B(x)$ is the standard cubic B-spline
$$B(x) = \begin{cases} \frac{2}{3} - x^2 + \frac{1}{2}|x|^3, & |x| \leq 1, \\ \frac{1}{6}(2 - |x|)^3, & 1 \leq |x| \leq 2, \\ 0, & |x| \geq 2. \end{cases}$$

The spline (2.46) is twice continuously differentiable, and because $B(x)$ is locally supported, the series (2.46) does not have more than four nonzero terms for any $x \in \mathbf{R}$.

It is known (see, e.g., [22]) that the spline
$$\Lambda(x) = \sum_{i=-\infty}^{\infty} f(x_i) B\left(\frac{x - i\alpha h}{\alpha h}\right)$$
possesses fairly good approximating and smoothing properties. In particular, if there exists a third derivative of $f(x)$ and it is bounded then there exist constants C_1 and C_2 such that
$$|f(x) - \Lambda(x)| \leq C_1 h^2, \ |f'(x) - \Lambda'(x)| \leq C_2 h, \ x \in \mathbf{R}.$$

And since the sequence $B_i(x) \doteq B\left(\frac{x - i\alpha h}{\alpha h}\right)$ provides a nonnegative partition of unity:
$$\sum_{i=-\infty}^{\infty} B_i(x) = 1, \ B_i(x) \geq 0, \text{ all } i,$$
the proof of Theorem 2.5 can be carried over for the case of the approximation (2.46).

7.2 Layer methods for semilinear parabolic equations

Remark 2.7. To reduce the amount of nodes x_j, it is natural to take advantage of cubic interpolation with step $h_x = \beta\sqrt{h}$ instead of linear interpolation with step $h_x = \alpha h$. Then we obtain the following algorithm:

$$\bar{u}(t_N, x) = \varphi(x),$$

$$\bar{u}(t_k, x_j)$$
$$= \frac{1}{2}\bar{u}(t_{k+1}, x_j + b(t_{k+1}, x_j, \bar{u}(t_{k+1}, x_j))h + \sigma(t_{k+1}, x_j, \bar{u}(t_{k+1}, x_j))\sqrt{h})$$
$$+ \frac{1}{2}\bar{u}(t_{k+1}, x_j + b(t_{k+1}, x_j, \bar{u}(t_{k+1}, x_j))h - \sigma(t_{k+1}, x_j, \bar{u}(t_{k+1}, x_j))\sqrt{h})$$
$$+ g(t_k, x_j, \bar{u}(t_{k+1}, x_j))h, \ x_j = x_0 + j\beta\sqrt{h}, \ j = 0, \pm 1, \pm 2, \ldots, \quad (2.47)$$

$$\bar{u}(t_k, x) = \sum_{i=0}^{3} \Phi_{j,i}(x)\, \bar{u}(t_k, x_{j+i}), \ x_{j+1} \leq x \leq x_{j+2},$$

$$\Phi_{j,i}(x) = \prod_{k=0, k\neq i}^{3} \frac{x - x_{j+k}}{x_{j+i} - x_{j+k}}, \ k = N-1, \ldots, 1, 0. \quad (2.48)$$

It can be shown as earlier that the inequality (2.43) holds for the algorithm (2.47)-(2.48). But in place of (2.44) we get

$$u(t_k, x) = \sum_{i=0}^{3} \Phi_{j,i}(x)\, u(t_k, x_{j+i}) + O(h_x^4), \ x_{j+1} \leq x \leq x_{j+2},$$

and, consequently,

$$R(t_k, x) = \sum_{i=0}^{3} \Phi_{j,i}(x)\, R(t_k, x_{j+i}) + O(h^2), \ x_{j+1} \leq x \leq x_{j+2}. \quad (2.49)$$

Though $\sum_{i=0}^{3} \Phi_{j,i}(x) = 1$ for any x, the sum of the absolute values $\sum_{i=0}^{3} |\Phi_{j,i}(x)|$ can take values greater than one. And instead of the inequality (2.45), we can obtain the following one only:

$$|R(t_k, x)| \leq AR_{k+1} + KR_{k+1}h + Ch^2,$$

where the constant A is greater than one, unfortunately.

Therefore, our proof of Theorem 2.5 cannot be carried over for the case of cubic interpolation. At the same time, a number of numerical experiments show a good quality of the procedure (2.47)-(2.48) (see, e.g., Sects. 7.7, 8.3, and 8.6). It can be explained in the following way. Introduce

$$R_k^* := \sup_{x_j,\ j=0,\pm 1,\pm 2,\ldots,} |R(t_k, x_j)|.$$

Clearly, $R_k^* \leq R_k$. It is natural that the difference between R_k^* and R_k is sufficiently small. Let us suppose that

$$R_k \leq R_k^* + A R_k^* h + B h^2,$$

where A and B are nonnegative constants which do not depend on k and h. Then we get from (2.43):

$$R_k^* \leq R_{k+1}^* + K R_{k+1}^* h + C h^2,$$

whence $R_k^* = O(h)$.

Finally, using (2.49), we obtain that the algorithm (2.47)-(2.48) has the first order of convergence.

Remark 2.8. Consider the Cauchy problem for an autonomous semilinear parabolic equation with positive direction of time t:

$$\frac{\partial u}{\partial t} = \frac{1}{2}\sigma^2(x,u)\frac{\partial^2 u}{\partial x^2} + b(x,u)\frac{\partial u}{\partial x} + g(x,u),\ t > 0,\ x \in \mathbf{R}, \qquad (2.50)$$

$$u(0,x) = \varphi(x). \qquad (2.51)$$

If we substitute a solution $u(t,x)$ of the problem (2.50)-(2.51) in the coefficients σ, b, g, the equation (2.50) becomes nonautonomous and that is why the reasoning of Remark 1.5 cannot be carried over for the problem (2.50)-(2.51). Nevertheless, the following procedure with positive direction of time can be obtained from (2.40)-(2.41):

$$\bar{u}(0,x) = \varphi(x),$$

$$\bar{u}(t_{k+1}, x_j) = \frac{1}{2}\bar{u}(t_k, x_j + b(x_j, \bar{u}(t_k, x_j))h + \sigma(x_j, \bar{u}(t_k, x_j))\sqrt{h})$$
$$+ \frac{1}{2}\bar{u}(t_k, x_j + b(x_j, \bar{u}(t_k, x_j))h - \sigma(x_j, \bar{u}(t_k, x_j))\sqrt{h})$$
$$+ g(x_j, \bar{u}(t_k, x_j))h,$$
$$x_j = x_0 + j\alpha h,\ j = 0, \pm 1, \pm 2, \ldots,\ t_k = kh,\ h = t/N, \qquad (2.52)$$

$$\bar{u}(t_k, x) = \frac{x_{j+1} - x}{\alpha h}\bar{u}(t_k, x_j) + \frac{x - x_j}{\alpha h}\bar{u}(t_k, x_{j+1}),\ x_j \leq x \leq x_{j+1}, \qquad (2.53)$$
$$k = 0, 1, \ldots, N-1.$$

This procedure is used in numerical examples in Sect. 7.4.

Remark 2.9. Clearly, all the methods and algorithms from this section can be considered with variable time and space steps.

7.3 Multi-dimensional case

7.3.1 Multidimensional parabolic equation

Consider the Cauchy problem for multi-dimensional semilinear parabolic equation

$$\frac{\partial u}{\partial t} + \frac{1}{2}\sum_{i,j=1}^{d} a^{ij}(t,x,u)\frac{\partial^2 u}{\partial x^i \partial x^j} + \sum_{i=1}^{d} b^i(t,x,u)\frac{\partial u}{\partial x^i} + g(t,x,u) = 0, \quad (3.1)$$

$$t_0 \leq t < T, \ x \in \mathbf{R}^d,$$

$$u(T,x) = \varphi(x). \quad (3.2)$$

As in Sect. 7.2, we can write the same relations (2.3)-(2.8) here but with the distinction that x, X, and b are d-dimensional vectors, σ is a $d \times d$-dimensional matrix such that $\sigma\sigma^\top = a = \{a^{ij}\}$, and $\xi_{N-1}, \xi_{N-2}, \ldots, \xi_0$ in (2.7) are i.i.d. d-dimensional vectors with i.i.d. components ξ_k^i, $i = 1, \ldots, d$, and each component ξ^i is distributed by the law $P(\xi = \pm 1) = 1/2$.

Using (2.4), we obtain (here we restrict ourselves to the two-dimensional case for simplicity in writing):

$$u(t_k, x) = u(t_k, x^1, x^2)$$
$$\simeq Eu(t_{k+1}, \bar{X}^1_{t_k,x}(t_{k+1}), \bar{X}^2_{t_k,x}(t_{k+1})) + E\bar{Z}_{t_k,x,0}(t_{k+1})$$
$$= \frac{1}{4}u(t_{k+1}, x^1 + b^1 h + \sigma^{11}\sqrt{h} + \sigma^{12}\sqrt{h}, x^2 + b^2 h + \sigma^{21}\sqrt{h} + \sigma^{22}\sqrt{h})$$
$$+ \frac{1}{4}u(t_{k+1}, x^1 + b^1 h + \sigma^{11}\sqrt{h} - \sigma^{12}\sqrt{h}, x^2 + b^2 h + \sigma^{21}\sqrt{h} - \sigma^{22}\sqrt{h})$$
$$+ \frac{1}{4}u(t_{k+1}, x^1 + b^1 h - \sigma^{11}\sqrt{h} + \sigma^{12}\sqrt{h}, x^2 + b^2 h - \sigma^{21}\sqrt{h} + \sigma^{22}\sqrt{h})$$
$$+ \frac{1}{4}u(t_{k+1}, x^1 + b^1 h - \sigma^{11}\sqrt{h} - \sigma^{12}\sqrt{h}, x^2 + b^2 h - \sigma^{21}\sqrt{h} - \sigma^{22}\sqrt{h})$$
$$+ gh, \quad (3.3)$$

where $b^i = b^i(t_k, x^1, x^2, u(t_k, x^1, x^2))$, $\sigma^{ij} = \sigma^{ij}(t_k, x^1, x^2, u(t_k, x^1, x^2))$, $i,j = 1, 2$, $g = g(t_k, x^1, x^2, u(t_k, x^1, x^2))$.

Then a method analogous to (2.23) has the form:

$$\bar{u}(t_N, x^1, x^2) = \varphi(x^1, x^2),$$

$$\bar{u}(t_k, x^1, x^2)$$
$$= \frac{1}{4}\bar{u}(t_{k+1}, x^1 + \bar{b}^1 h + \bar{\sigma}^{11}\sqrt{h} + \bar{\sigma}^{12}\sqrt{h}, x^2 + \bar{b}^2 h + \bar{\sigma}^{21}\sqrt{h} + \bar{\sigma}^{22}\sqrt{h})$$
$$+ \frac{1}{4}\bar{u}(t_{k+1}, x^1 + \bar{b}^1 h + \bar{\sigma}^{11}\sqrt{h} - \bar{\sigma}^{12}\sqrt{h}, x^2 + \bar{b}^2 h + \bar{\sigma}^{21}\sqrt{h} - \bar{\sigma}^{22}\sqrt{h})$$

$$+\frac{1}{4}\bar{u}(t_{k+1}, x^1 + \bar{b}^1 h - \bar{\sigma}^{11}\sqrt{h} + \bar{\sigma}^{12}\sqrt{h}, x^2 + \bar{b}^2 h - \bar{\sigma}^{21}\sqrt{h} + \bar{\sigma}^{22}\sqrt{h})$$
$$+\frac{1}{4}\bar{u}(t_{k+1}, x^1 + \bar{b}^1 h - \bar{\sigma}^{11}\sqrt{h} - \bar{\sigma}^{12}\sqrt{h}, x^2 + \bar{b}^2 h - \bar{\sigma}^{21}\sqrt{h} - \bar{\sigma}^{22}\sqrt{h})$$
$$+\bar{g}h, \qquad (3.4)$$

where $\bar{b}^i = b^i(t_k, x^1, x^2, \bar{u}(t_{k+1}, x^1, x^2))$, $\bar{\sigma}^{ij} = \sigma^{ij}(t_k, x^1, x^2, \bar{u}(t_{k+1}, x^1, x^2))$, $i, j = 1, 2$, $\bar{g} = g(t_k, x^1, x^2, \bar{u}(t_{k+1}, x^1, x^2))$, $k = N-1, \ldots, 1, 0$.

Consider the equidistant space discretization: $x_j^1 = x_0^1 + j\alpha^1 h$, $x_l^2 = x_0^2 + l\alpha^2 h$, $j, l = 0, \pm 1, \pm 2, \ldots$, (x_0^1, x_0^2) is a point belonging to $\mathbf{R}^2$ and $\alpha^1 > 0$, $\alpha^2 > 0$ are some numbers, i.e., h_{x^1}, h_{x^2} are taken to be equal to $\alpha^1 h$, $\alpha^2 h$, respectively. Using the sequential linear interpolation, we construct the following algorithm based on the method (3.4):

$$\bar{u}(t_N, x^1, x^2) = \varphi(x^1, x^2),$$

$$\bar{u}(t_k, x_j^1, x_l^2)$$
$$= \frac{1}{4}\bar{u}(t_{k+1}, x_j^1 + \bar{b}^1 h + \bar{\sigma}^{11}\sqrt{h} + \bar{\sigma}^{12}\sqrt{h}, x_l^2 + \bar{b}^2 h + \bar{\sigma}^{21}\sqrt{h} + \bar{\sigma}^{22}\sqrt{h})$$
$$+\frac{1}{4}\bar{u}(t_{k+1}, x_j^1 + \bar{b}^1 h + \bar{\sigma}^{11}\sqrt{h} - \bar{\sigma}^{12}\sqrt{h}, x_l^2 + \bar{b}^2 h + \bar{\sigma}^{21}\sqrt{h} - \bar{\sigma}^{22}\sqrt{h})$$
$$+\frac{1}{4}\bar{u}(t_{k+1}, x_j^1 + \bar{b}^1 h - \bar{\sigma}^{11}\sqrt{h} + \bar{\sigma}^{12}\sqrt{h}, x_l^2 + \bar{b}^2 h - \bar{\sigma}^{21}\sqrt{h} + \bar{\sigma}^{22}\sqrt{h})$$
$$+\frac{1}{4}\bar{u}(t_{k+1}, x_j^1 + \bar{b}^1 h - \bar{\sigma}^{11}\sqrt{h} - \bar{\sigma}^{12}\sqrt{h}, x_l^2 + \bar{b}^2 h - \bar{\sigma}^{21}\sqrt{h} - \bar{\sigma}^{22}\sqrt{h})$$
$$+\bar{g}h, \qquad (3.5)$$

where all the coefficients $\bar{b}$ and $\bar{\sigma}$ are calculated at $t_k, x_j^1, x_l^2, \bar{u}(t_{k+1}, x_j^1, x_l^2)$,

$$\bar{u}(t_k, x^1, x^2) = \frac{x_{j+1}^1 - x^1}{\alpha^1 h} \frac{x_{l+1}^2 - x^2}{\alpha^2 h} \bar{u}(t_k, x_j^1, x_l^2)$$
$$+ \frac{x_{j+1}^1 - x^1}{\alpha^1 h} \frac{x^2 - x_l^2}{\alpha^2 h} \bar{u}(t_k, x_j^1, x_{l+1}^2)$$
$$+ \frac{x^1 - x_j^1}{\alpha^1 h} \frac{x_{l+1}^2 - x^2}{\alpha^2 h} \bar{u}(t_k, x_{j+1}^1, x_l^2)$$
$$+ \frac{x^1 - x_j^1}{\alpha^1 h} \frac{x^2 - x_l^2}{\alpha^2 h} \bar{u}(t_k, x_{j+1}^1, x_{l+1}^2), \qquad (3.6)$$

$$x_j^1 \le x^1 \le x_{j+1}^1, \ x_l^2 \le x^2 \le x_{l+1}^2, \ (x^1, x^2) \ne (x_i^1, x_m^2),$$
$$i, m = 0, \pm 1, \pm 2, \ldots, \ k = N-1, \ldots, 1, 0.$$

Remark 3.1. The sequential linear interpolation in (3.6) is not linear with respect to both variables x^1 and x^2. The following triangular interpolation of $u(t, x^1, x^2)$

$$\bar{u}(t_k, x^1, x^2) = \left(1 - \frac{x^1 - x_j^1}{\alpha^1 h} - \frac{x^2 - x_l^2}{\alpha^2 h}\right) \bar{u}(t_k, x_j^1, x_l^2)$$
$$+ \frac{x^2 - x_l^2}{\alpha^2 h} \bar{u}(t_k, x_j^1, x_{l+1}^2) + \frac{x^1 - x_j^1}{\alpha^1 h} \bar{u}(t_k, x_{j+1}^1, x_l^2) \quad (3.7)$$

is linear and it has an error of $O(h^2)$ just as the interpolation (3.6). This interpolation is not suitable for the all points (x^1, x^2) from the rectangle $\Pi_{j,l} = \{(x^1, x^2) : x_j^1 \leq x^1 \leq x_{j+1}^1, x_l^2 \leq x^2 \leq x_{l+1}^2\}$ by the same reason as it was mentioned in Remark 2.7. But for the points from the triangle with the corners (x_j^1, x_l^2), (x_j^1, x_{l+1}^2), (x_{j+1}^1, x_l^2), the interpolation (3.7) is suitable because

$$\frac{x^1 - x_j^1}{\alpha^1 h} + \frac{x^2 - x_l^2}{\alpha^2 h} \leq 1 \quad (3.8)$$

for these points.

For the other points of the rectangle $\Pi_{j,l}$, we have

$$\frac{x_{j+1}^1 - x^1}{\alpha^1 h} + \frac{x_{l+1}^2 - x^2}{\alpha^2 h} < 1, \quad (3.9)$$

and we can use the formula

$$\bar{u}(t_k, x^1, x^2) = \left(1 - \frac{x_{j+1}^1 - x^1}{\alpha^1 h} - \frac{x_{l+1}^2 - x^2}{\alpha^2 h}\right) \bar{u}(t_k, x_{j+1}^1, x_{l+1}^2)$$
$$+ \frac{x_{j+1}^1 - x^1}{\alpha^1 h} \bar{u}(t_k, x_j^1, x_{l+1}^2) + \frac{x_{l+1}^2 - x^2}{\alpha^2 h} \bar{u}(t_k, x_{j+1}^1, x_l^2). \quad (3.10)$$

Thus, the formulas (3.7) and (3.10) for $(x^1, x^2) \in \Pi_{j,l}$ satisfy (3.8) and (3.9), respectively, and they give the other suitable rule of interpolation.

Convergence theorems for the method (3.4) and for the algorithm (3.5) with the interpolation (3.6) or (3.7)-(3.10) are analogous to Theorems 2.2 and 2.5.

7.3.2 Probabilistic approach to reaction-diffusion systems

Reaction-diffusion systems have received a great deal of attention, motivated by both their widespread occurrence in models from physics, chemistry, and biology, and by the richness of the structure of their solution sets. Reaction-diffusion systems can give rise to a number of interesting phenomena like, e.g., threshold behavior, multiple steady states and hysteresis, spatial patterns, moving fronts or pulses and oscillations (see, e.g. [91, 269]).

The methods constructed in previous sections can also be applied to the Cauchy problem for systems of reaction-diffusion equations of the form (for simplicity we write them for the one-dimensional x):

$$\frac{\partial u_q}{\partial t} + L_q u_q + g_q(t,x,u) = 0, \ t_0 \le t < T, \ x \in \mathbf{R}, \ q = 1,\ldots,n, \quad (3.11)$$

$$u_q(T,x) = \varphi_q(x), \quad (3.12)$$

where

$$u := (u_1,\ldots,u_n),$$

$$L_q := \frac{1}{2}\sigma_q^2(t,x,u)\frac{\partial^2}{\partial x^2} + b_q(t,x,u)\frac{\partial}{\partial x}.$$

It is not difficult to derive the method which is analogous to (2.23):

$$\bar{u}_q(t_N,x) = \varphi_q(x),$$

$$\bar{u}_q(t_k,x) = \frac{1}{2}\bar{u}_q(t_{k+1}, x + b_q(t_k,x,\bar{u}(t_{k+1},x))h + \sigma_q(t_k,x,\bar{u}(t_{k+1},x))\sqrt{h})$$
$$+\frac{1}{2}\bar{u}_q(t_{k+1}, x + b_q(t_k,x,\bar{u}(t_{k+1},x))h - \sigma_q(t_k,x,\bar{u}(t_{k+1},x))\sqrt{h})$$
$$+g_q(t_k,x,\bar{u}(t_{k+1},x))h, \quad k = N-1,\ldots,1,0, \quad (3.13)$$

and then the corresponding algorithm (see (2.40)-(2.41)).

The system (3.11) is such that the linear system of parabolic equations obtained after substituting $u = u(t,x)$ in the coefficients σ_q, b_q, g_q splits and, therefore, every parabolic equation can be solved separately. In connection with this fact, one can consider n separate simple systems of the type (2.5)-(2.6). Such a receipt is impossible for reaction-diffusion systems containing equations with derivatives of different functions among $u_1,\ldots,u_n$. Consider, for example, the system

$$\frac{\partial u_q}{\partial t} + \frac{1}{2}\sigma^2(t,x,u)\frac{\partial^2 u_q}{\partial x^2} + \sum_{j=1}^{n}\sigma(t,x,u)b_{jq}(t,x,u)\frac{\partial u_j}{\partial x} + g_q(t,x,u) = 0 \quad (3.14)$$

with the conditions (3.12) (we pay attention that σ in (3.14) does not depend on q). In this case one can use the following probabilistic representation [177]:

$$u_q(t_k,x) = E\sum_{l=1}^{n} u_l(t_{k+1}, X_{t_k,x}(t_{k+1}))Y^l_{t_k,x,q}(t_{k+1}) + EZ_{t_k,x,q,0}(t_{k+1}), \quad (3.15)$$

where $X_{t_k,x}(s),\ Y^l_{t_k,x,q}(s),\ Z_{t_k,x,q,0}(s)$ is the solution of the Cauchy problem for the system of SDEs

$$dX = \sigma(s,X,u(s,X))dw(s), \ X(t_k) = x,$$

$$dY^j = \sum_{l=1}^{n} b_{jl}(s,X,u(s,X))Y^l dw(s), \ Y^j(t_k) = \delta_{jq} = \begin{cases} 0, \ j \ne q, \\ 1, \ j = q, \end{cases}$$

$$dZ = \sum_{l=1}^{n} g_l(s, X, u(s, X))Y^l ds, \quad Z(t_k) = 0. \tag{3.16}$$

Now it is not difficult to derive the method which is analogous to (2.23):

$$\bar{u}_q(t_N, x) = \varphi_q(x),$$

$$\bar{u}_q(t_k, x) = \frac{1}{2} \sum_{l=1}^{n} \bar{u}_l(t_{k+1}, x + \sigma(t_k, x, \bar{u}(t_{k+1}, x))\sqrt{h})$$
$$\times (\delta_{lq} + b_{lq}(t_k, x, \bar{u}(t_{k+1}, x))\sqrt{h})$$
$$+ \frac{1}{2} \sum_{l=1}^{n} \bar{u}_l(t_{k+1}, x - \sigma(t_k, x, \bar{u}(t_{k+1}, x))\sqrt{h})$$
$$\times (\delta_{lq} - b_{lq}(t_k, x, \bar{u}(t_{k+1}, x))\sqrt{h})$$
$$+ g_q(t_k, x, \bar{u}(t_{k+1}, x))h, \quad k = N-1, \ldots, 1, 0, \tag{3.17}$$

and then the corresponding algorithm.

Convergence Theorems 2.2 and 2.5 can be carried over to these method and algorithm.

7.4 Numerical examples

Example 4.1. Consider the quasilinear equation with power law nonlinearities (see, e.g., [259]):

$$\frac{\partial u}{\partial t} = \frac{\partial}{\partial x}(u^\alpha \frac{\partial u}{\partial x}) + u^{\alpha+1}, \quad t > 0, \ x \in \mathbf{R}, \tag{4.1}$$

where $\alpha > 0$ is a constant.

The equation (4.1) has blow-up solutions. In particular, it has the following blow-up automodelling solution (see Fig. 4.1 as well):

$$u(t, x) = \begin{cases} (T_0 - t)^{-1/\alpha} \left(\frac{2(\alpha+1)}{\alpha(\alpha+2)} \cos^2 \frac{\pi x}{L}\right)^{1/\alpha}, & |x| < \frac{L}{2}, \\ 0, & |x| \geq \frac{L}{2}, \ 0 < t < T_0, \end{cases} \tag{4.2}$$

where

$$L = \frac{2\pi}{\alpha}(\alpha+1)^{1/2}.$$

The temperature $u(t, x)$ grows infinitely when $t \to T_0$. At the same time, the heat is localized in the interval $(-L/2, L/2)$. The function

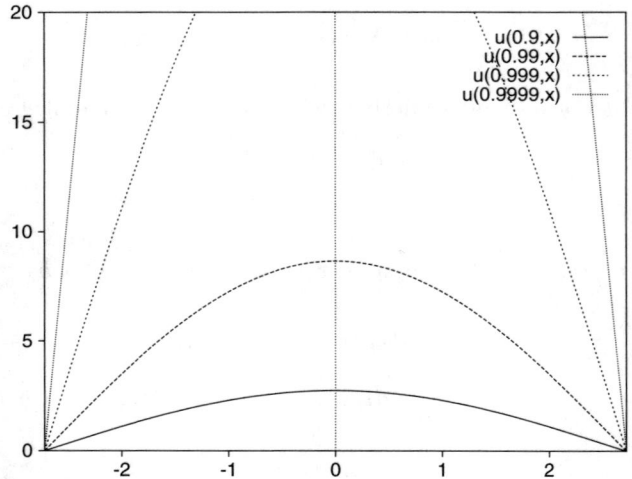

Fig. 4.1. The solution of (4.1)-(4.2) with $\alpha = 2$, $T_0 = 1$.

$$v = \frac{1}{\alpha+1} u^{\alpha+1}$$

satisfies the equation

$$\frac{\partial v}{\partial t} = \frac{1}{2} 2(\alpha+1)^{\alpha/(\alpha+1)} v^{\alpha/(\alpha+1)} \frac{\partial^2 v}{\partial x^2} + (\alpha+1)^{(2\alpha+1)/(\alpha+1)} v^{(2\alpha+1)/(\alpha+1)} \tag{4.3}$$

which is of the form (2.50).

We use the algorithm (2.52)-(2.53) to find the solution of (4.3) with $\alpha = 2$ and with the initial condition

$$v(0,x) = \begin{cases} \frac{\sqrt{3}}{8} \cos^3 \frac{\pi x}{L}, & |x| < \frac{L}{2}, \\ 0, & |x| \geq \frac{L}{2}. \end{cases} \tag{4.4}$$

In this case $T_0 = 1$.

A dependence on h of the following errors

$$err_{\bar{v}} = \max_{x_i} |\bar{v}(t,x_i) - v(t,x_i)|,$$

$$err_{\bar{u}} = \max_{x_i} |\bar{u}(t,x_i) - u(t,x_i)|, \quad \bar{u}(t,x_i) = (3\bar{v}(t,x_i))^{1/3},$$

$$err_{\bar{u}}[-2,2] = \max_{|x_i| \leq 2} |\bar{u}(t,x_i) - u(t,x_i)|$$

is given in Table 4.1 for $t = 0.5$. We take $h_t = h_x = h$ here.

Table 4.1. The absolute errors of the algorithm (2.52)-(2.53) for the Cauchy problem (4.3)-(4.4) at $t = 0.5$.

	$h = 10^{-1}$	$h = 10^{-2}$	$h = 10^{-3}$	$h = 10^{-4}$
$err_{\bar{v}}$	$0.9931 \cdot 10^{-1}$	$1.422 \cdot 10^{-2}$	$1.489 \cdot 10^{-3}$	$1.500 \cdot 10^{-4}$
$err_{\bar{u}}$	$0.9572 \cdot 10^{-1}$	$1.643 \cdot 10^{-2}$	$4.432 \cdot 10^{-3}$	$13.77 \cdot 10^{-4}$
$err_{\bar{u}}[-2, 2]$	$0.7015 \cdot 10^{-1}$	$0.9552 \cdot 10^{-2}$	$1.215 \cdot 10^{-3}$	$1.003 \cdot 10^{-4}$

Table 4.2. The relative errors $\delta(t, h)$ of the algorithm (2.52)-(2.53) for the Cauchy problem (4.3)-(4.4) and the explosion time.

	$h = 10^{-1}$	$h = 10^{-2}$	$h = 10^{-3}$	$h = 10^{-4}$
$t = 0.9$	$3.644 \cdot 10^{-1}$	$1.024 \cdot 10^{-1}$	$1.313 \cdot 10^{-2}$	$1.353 \cdot 10^{-3}$
$t = 0.99$	------	$5.298 \cdot 10^{-1}$	$1.815 \cdot 10^{-1}$	$2.585 \cdot 10^{-2}$
$t = 0.999$	------	------	$6.167 \cdot 10^{-1}$	$2.436 \cdot 10^{-1}$
$t = 0.9999$	------	------	------	$6.704 \cdot 10^{-1}$
t^*	1.5	1.07	1.008	1.0001

Rather large values of $err_{\bar{u}}$ are due to the fact that the values $v(t, x_i)$ for x_i close to the ends of the interval $(-L/2, L/2)$ are very small and, consequently, for such x_i

$$|\bar{u}(t, x_i) - u(t, x_i)| = |(3\bar{v}(t, x_i))^{1/3} - (3v(t, x_i))^{1/3}|$$
$$\simeq 3^{1/3}|\bar{v}(t, x_i) - v(t, x_i)|^{1/3},$$

i.e., $err_{\bar{u}} = O(h^{1/3})$. But the difference $\bar{u}(t, x_i) - u(t, x_i)$ on a subinterval $(-a, a)$, $a < L/2$, behaves as $O(h)$ (see the row $err_{\bar{u}}[-2, 2]$ in Table 4.1).

For times t which are close to the explosion time T_0, the errors $err_{\bar{v}}$ become fairly large (we pay attention that v in our example is proportional to u^3). However, if we are interested in finding the explosion time, it is natural to consider another characteristic when $t \to T_0$. Table 4.2 presents the values

$$\delta(t, h) = \frac{err_{\bar{u}}}{u(t, 0)}$$

and the time t^* at which the values of $\bar{u}$ become larger than 10^4, i.e., this time evaluates the explosion time.

Example 4.2. Consider the one-dimensional Burgers equation

$$\frac{\partial u}{\partial t} = \frac{1}{2}\sigma^2 \frac{\partial^2 u}{\partial x^2} - u\frac{\partial u}{\partial x}, \quad t > 0, \ x \in \mathbf{R}, \quad (4.5)$$

$$u(0, x) = \varphi(x). \quad (4.6)$$

Due to the Cole–Hopf transformation, the solution of this problem can be found explicitly:

$$u(t,x) = \frac{\int_{-\infty}^{\infty} K(t,x,y)\,\varphi(y)\exp\left(-\frac{1}{\sigma^2}\int_0^y \varphi(\xi)d\xi\right) dy}{\int_{-\infty}^{\infty} K(t,x,y)\exp\left(-\frac{1}{\sigma^2}\int_0^y \varphi(\xi)d\xi\right) dy}, \quad (4.7)$$

$$K(t,x,y) = \frac{1}{\sqrt{2\pi\sigma^2 t}}\exp\left(-\frac{(x-y)^2}{2\sigma^2 t}\right). \quad (4.8)$$

If

$$u(0,x) = \psi(x) = \begin{cases} 1, & x < 0, \\ 0, & x \geq 0, \end{cases} \quad (4.9)$$

then

$$u(t,x) = \psi(t,x)$$

$$:= 1 - \frac{\operatorname{erfc}\left(-\frac{x}{\sqrt{2\sigma^2 t}}\right)}{\operatorname{erfc}\left(-\frac{x}{\sqrt{2\sigma^2 t}}\right) + \exp\left(\frac{t-2x}{2\sigma^2}\right)\left[2 - \operatorname{erfc}\left(\frac{t-x}{\sqrt{2\sigma^2 t}}\right)\right]}, \quad (4.10)$$

where

$$\operatorname{erfc}(x) = \frac{2}{\sqrt{\pi}}\int_x^{\infty} \exp(-\xi^2)\,d\xi.$$

For a sufficiently small σ the solution (4.9) is close to the travelling shock wave $\psi(x - t/2)$ with speed $1/2$.

Tables 4.3 and 4.4 give numerical results obtained by applying the algorithm (2.52)-(2.53) with $h_t = h_x = h$ to the Cauchy problem (4.5), (4.9). They present the errors of approximate solution $\bar{u}$ in the discrete Chebyshev norm (the top position) and in the l^1-norm (the lower position):

Table 4.3. Burgers equation. Dependence of the errors $\operatorname{err}_{\bar{u}}^c$ and $\operatorname{err}_{\bar{u}}^l$ on h and σ under fixed $t = 1$.

	$h = 10^{-1}$	$h = 10^{-2}$	$h = 10^{-3}$	$h = 10^{-4}$
$\sigma = 0.05$	> 0.5 0.2516	> 0.5 0.1198	> 0.5 $0.2960 \cdot 10^{-1}$	0.3232 $0.3349 \cdot 10^{-2}$
$\sigma = 0.1$	> 0.5 0.1874	> 0.5 $0.7501 \cdot 10^{-1}$	0.2104 $0.8495 \cdot 10^{-2}$	$0.2198 \cdot 10^{-1}$ $0.8692 \cdot 10^{-3}$
$\sigma = 0.2$	0.4849 0.1057	0.1484 $0.2316 \cdot 10^{-1}$	$0.1582 \cdot 10^{-1}$ $0.2412 \cdot 10^{-2}$	$0.1625 \cdot 10^{-2}$ $0.2485 \cdot 10^{-3}$
$\sigma = 0.5$	$0.7295 \cdot 10^{-1}$ $0.5137 \cdot 10^{-1}$	$0.7704 \cdot 10^{-2}$ $0.5580 \cdot 10^{-2}$	$0.8448 \cdot 10^{-3}$ $0.6035 \cdot 10^{-3}$	$0.9010 \cdot 10^{-4}$ $0.6538 \cdot 10^{-4}$
$\sigma = 1$	$0.1033 \cdot 10^{-1}$ $0.2150 \cdot 10^{-1}$	$0.1151 \cdot 10^{-2}$ $0.2631 \cdot 10^{-2}$	$0.1351 \cdot 10^{-3}$ $0.2769 \cdot 10^{-3}$	$0.1506 \cdot 10^{-4}$ $0.3247 \cdot 10^{-4}$

Table 4.4. Burgers equation. Dependence of the errors $err_{\bar{u}}^c$ and $err_{\bar{u}}^l$ on h and t under fixed $\sigma = 0.5$.

	$h = 10^{-1}$	$h = 10^{-2}$	$h = 10^{-3}$
$t = 0.02$	------	0.1348	$0.4270 \cdot 10^{-1}$
	------	$0.1923 \cdot 10^{-1}$	$0.4703 \cdot 10^{-2}$
$t = 0.1$	0.1427	$0.2216 \cdot 10^{-1}$	$0.1099 \cdot 10^{-2}$
	$0.2850 \cdot 10^{-1}$	$0.6567 \cdot 10^{-2}$	$0.3503 \cdot 10^{-3}$
$t = 0.5$	$0.6368 \cdot 10^{-1}$	$0.5438 \cdot 10^{-2}$	$0.6311 \cdot 10^{-3}$
	$0.3198 \cdot 10^{-1}$	$0.3463 \cdot 10^{-2}$	$0.3824 \cdot 10^{-3}$
$t = 2.5$	0.1147	$0.1296 \cdot 10^{-1}$	$0.1362 \cdot 10^{-2}$
	0.1018	$0.1125 \cdot 10^{-1}$	$0.1181 \cdot 10^{-2}$

$$err_{\bar{u}}^c = \max_{x_i} |\bar{u}(t, x_i) - u(t, x_i)|,$$

$$err_{\bar{u}}^l = \sum_i |\bar{u}(t, x_i) - u(t, x_i)| \times h.$$

These results illustrate the good properties of the algorithm (2.52)-(2.53). Besides, they show wider capabilities of the algorithm than it is ensured by Theorem 2.2 (we have in mind the discontinuity of the function $\psi(x)$). The large values of the errors (especially of $err_{\bar{u}}^c$) for small σ and t can easily be explained: the corresponding solution has very large derivatives with respect to x in these cases. Clearly, the errors can be decreased if we improve the exactness of interpolation, e.g., choosing a smaller h_x. In connection with this example, see numerical experiments in [24, 25] and in Sects. 7.7, 8.3.1, and 8.6.1 as well.

Example 4.3. Consider the asymptotic behavior of some solutions to the problem (4.5)-(4.6). Figure 4.2 shows that the solution of the problem (4.5), (4.9) for large t is close to a wave which preserves its shape and moves with speed $1/2$. Figure 4.3 is related to the solution with the initial data

$$u(0, x) = \begin{cases} 1, & x < -10, \\ 0.75, & -10 < x < 0, \\ 0, & x > 0. \end{cases} \quad (4.11)$$

Comparing these two figures, one can conclude that there exist a limit shape and a limit speed of the waves which are the same for the initial conditions (4.9) and (4.11).

The following two-parameter solution of the Burgers equation (4.5) was found in [24]:

$$u^{a,b}(t, x) = b - a \tanh \frac{a(x - bt)}{\sigma^2}, \quad (4.12)$$

where a and b are constants and (we remind)

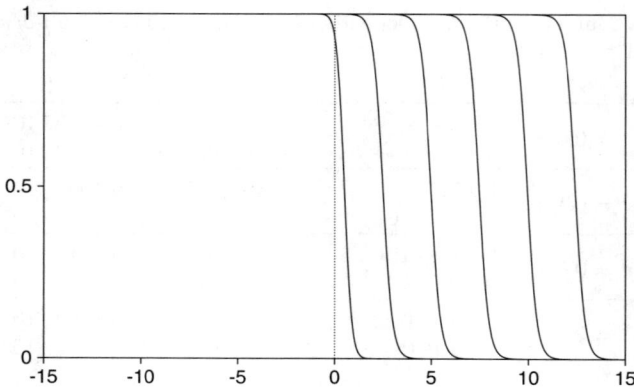

Fig. 4.2. The solution to the problem (4.5), (4.9) at $t = 1$, $t = 5$, $t = 10$, $t = 15$, $t = 20$, $t = 25$; $\sigma = 0.5$.

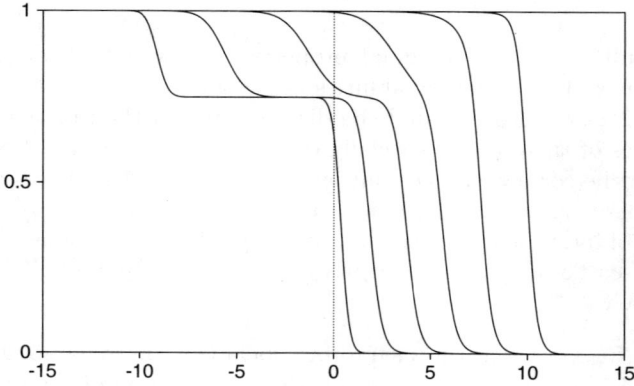

Fig. 4.3. The solution to the problem (4.5), (4.11) at $t = 1$, $t = 5$, $t = 10$, $t = 15$, $t = 20$, $t = 25$; $\sigma = 0.5$.

$$\tanh x = \operatorname{sign}(x) \frac{1 - e^{-2|x|}}{1 + e^{-2|x|}}.$$

Clearly, $u^{a,b}(t, x) = u^{|a|,b}(t, x)$ (therefore, it is sufficient to consider the case $a \geq 0$ only) and

$$b - a \leq u^{a,b}(t, x) \leq b + a, \quad a > 0.$$

Let us also note that the function

$$u_0(x) = -\tanh \frac{x}{\sigma^2}$$

is a stationary solution of the Burgers equation, i.e.,

$$\frac{1}{2}\sigma^2 \frac{d^2 u_0}{dx^2} - u_0 \frac{du_0}{dx} = 0, \quad (4.13)$$

and if some function $u_0(x)$ is a solution of the steady-state Burgers equation (4.13) then the function

$$u(t,x) = b + au_0(a(x-bt))$$

is a solution of the equation (4.5).

The solution $u^{a,b}(t,x)$ is a traveling wave which runs with speed b, i.e., it runs from the left to the right if $b > 0$, it runs conversely if $b < 0$, and it is immovable if $b = 0$. The shape of this wave is determined by the function $u^{a,b}(0,x)$.

We say that the shape of $u(t,x)$ converges to $f(x)$ as t tends to infinity if there exists a function $m(t)$ such that

$$\lim_{t\to\infty} \sup_x |u(t, x + m(t)) - f(x)| = 0.$$

It is not difficult to prove that the shape of $\psi(t,x)$ (see (4.10)) converges to $u^{0.5,0.5}(0,x) = 0.5 - 0.5\tanh(x/2\sigma^2)$. To show this, one should set $m(t) = t/2$ (let us remind that $\text{erfc}(-x) + \text{erfc}\, x = 2$ and therefore $\psi(t, t/2) = 1/2$).

The following assertion can be proved by using the explicit formula (4.7) (see [112] as well).

Proposition 4.4. *Let*

$$u(0,x) = \begin{cases} c, & x < l_0, \\ \lambda(x), & l_0 \leq x \leq l_0 + l, \\ d, & x > l_0 + l, \end{cases} \quad (4.14)$$

where c, d, l_0, l are some constants: $c > d$, $l \geq 0$; $\lambda(x)$ is a measurable function and $d \leq \lambda(x) \leq c$.

Then the shape of $u(t,x)$ converges to $u^{a,b}(0,x)$ with

$$a = \frac{c-d}{2} > 0, \quad b = \frac{c+d}{2}.$$

More precisely:

$$\lim_{t\to\infty} \sup_x |u(t, x + l_0 + bt + \alpha) - u^{a,b}(0,x)| = 0, \quad (4.15)$$

where

$$\alpha = \frac{S - d \times l}{c - d}, \quad S = \int_0^l \lambda(\xi) d\xi.$$

Thus, the limit shape of a solution of the Burgers equation with initial data of the form (4.14) depends on c and d only and for large t it is close to the traveling symmetric wave of the shape $u^{a,b}(0,x)$ with speed b and with center

$$m(t) = l_0 + bt + \alpha. \quad (4.16)$$

Table 4.5. Burgers equation. Dependence of the errors $err_{\bar{u}}$ and $sherr_{\bar{u}}$ on h and t under fixed $\sigma = 0.5$.

	$h = 10^{-1}$	$h = 10^{-2}$	$h = 10^{-3}$
$t = 10$	0.3098	$0.3751 \cdot 10^{-1}$	$0.3829 \cdot 10^{-2}$
	$0.1866 \cdot 10^{-1}$	$0.7637 \cdot 10^{-3}$	$0.2068 \cdot 10^{-3}$
$t = 20$	0.5324	$0.7048 \cdot 10^{-1}$	$0.7153 \cdot 10^{-2}$
	$0.1906 \cdot 10^{-1}$	$0.9203 \cdot 10^{-3}$	$0.1264 \cdot 10^{-3}$
$t = 30$	0.6970	0.1033	$0.1048 \cdot 10^{-1}$
	$0.1900 \cdot 10^{-1}$	$0.9212 \cdot 10^{-3}$	$0.1274 \cdot 10^{-3}$

Table 4.5 gives numerical results obtained by the algorithm (2.52)-(2.53) with $h_t = h_x = h$ applied to the Cauchy problem (4.5), (4.9) with $\sigma = 0.5$. The table presents the usual errors $err_{\bar{u}} = err_{\bar{u}}^c$ of the approximate solution $\bar{u}$ and the distances $sherr_{\bar{u}}$ of $\bar{u}$ from the shape (shape errors) determined by $u^{0.5,0.5}(0,x) = 0.5 - 0.5\tanh(x/2\sigma^2)$. These distances are calculated by the formula
$$sherr_{\bar{u}} = \max_{x_i} |\bar{u}(t, x_i + \bar{m}(t)) - u^{0.5,0.5}(0, x_i)|,$$
where $\bar{m}(t)$ is a root of the equation $\bar{u}(t,x) = 1/2$.

We see that the shape error $sherr_{\bar{u}}$ is stabilized as t tends to infinity, and it tends to zero if h tends to zero. This proves that the solution of the procedure (2.52)-(2.53) has a limit shape which is close to the limit shape of the solution of the problem (4.5), (4.9) for small h.

7.5 Probabilistic approach to semilinear parabolic equations with small parameter

In this section we apply the probabilistic approach described in previous sections to the Cauchy problem for semilinear parabolic equations with small parameter of the form

$$\frac{\partial u}{\partial t} + \frac{\varepsilon^2}{2} \sum_{i,j=1}^{d} a^{ij}(t,x,u) \frac{\partial^2 u}{\partial x^i \partial x^j} + \sum_{i=1}^{d} (b^i(t,x,u) + \varepsilon^2 c^i(t,x,u)) \frac{\partial u}{\partial x^i}$$
$$+ g(t,x,u) = 0, \quad t \in [t_0, T], \ x \in \mathbf{R}^d, \tag{5.1}$$

$$u(T,x) = \varphi(x). \tag{5.2}$$

The probabilistic representations of the solution to this problem are connected with systems of SDEs with small noise. Let the Cauchy problem (5.1)-(5.2) have the unique solution $u = u(t,x)$ which is sufficiently smooth and satisfies some needed conditions of boundedness (see the corresponding theoretical results, e.g., in [143, 269, 291]). If we substitute $u = u(t,x)$ in

7.5 Probabilistic approach to parabolic equations with small parameter

the coefficients of (5.1), we obtain a linear parabolic equation with small parameter. The solution of this linear equation has the following probabilistic representation

$$u(t,x) = E(\varphi(X_{t,x}(T)) + Z_{t,x,0}(T)), \ t \leq T, \ x \in \mathbf{R}^d, \tag{5.3}$$

where $X_{t,x}(s)$, $Z_{t,x,z}(s)$, $s \geq t$, is the solution of the Cauchy problem for the system of SDEs with small noise:

$$dX = (b(s,X,u(s,X)) + \varepsilon^2 c(s,X,u(s,X)))ds$$
$$+\varepsilon\sigma(s,X,u(s,X))dw, \ X(t) = x, \tag{5.4}$$

$$dZ = g(s,X,u(s,X))ds, \ Z(t) = z. \tag{5.5}$$

Here $w(s) = (w^1(s), \ldots, w^d(s))^\top$ is a d-dimensional standard Wiener process, $b(s,x,u)$ and $c(s,x,u)$ are d-dimensional column-vectors compounded from the coefficients $b^i(s,x,u)$ and $c^i(s,x,u)$ of (5.1), $\sigma(s,x,u)$ is a $d \times d$-matrix obtained from the equation $a(s,x,u) = \sigma(s,x,u)\sigma^\top(s,x,u)$, where $a = \{a^{ij}\}$; the equation is solvable with respect to σ (for instance, by a lower triangular matrix) at least in the case of a positively definite a.

Just for systems like (5.4)-(5.5), special weak approximations are proposed in Chap. 3. Applying these special approximations, we get new layer methods intended to solve the Cauchy problem (5.1)-(5.2).

To simplify the notation, we write $u(t,x)$ and $X_{t,x}(s)$ instead of $u(t,x;\varepsilon)$ and $X^\varepsilon_{t,x}(s)$ here and in what follows.

In Sects. 7.5.1-7.5.4, specific implicit and explicit layer methods for the problem (5.1)-(5.2) are proposed. Some theorems on their rates of convergence both in the regular and in the singular cases are given. For simplicity, we deal with the one-dimensional case of the problem (5.1)-(5.2). Extensions to the multi-dimensional case and systems of reaction-diffusion equations can be done as in Sect. 7.3 (see details in [206]). In Sect. 7.6, we propose both two-layer and three-layer methods in a particular case of the problem (5.1)-(5.2). Some results of numerical tests on the Burgers equation with small viscosity and on the generalized FKPP-equation with small parameter are given in Sect. 7.7.

For simplicity, we consider the Cauchy problem (5.1)-(5.2) for $d = 1$:

$$\frac{\partial u}{\partial t} + \frac{\varepsilon^2}{2}\sigma^2(t,x,u)\frac{\partial^2 u}{\partial x^2} + (b(t,x,u) + \varepsilon^2 c(t,x,u))\frac{\partial u}{\partial x} + g(t,x,u) = 0, \tag{5.6}$$
$$t \in [t_0,T], \ x \in \mathbf{R},$$

$$u(T,x) = \varphi(x). \tag{5.7}$$

The probabilistic representation of the solution $u(t,x)$ to this problem has the form (5.3)-(5.5) with $d = 1$.

7.5.1 Implicit layer method and its convergence

Applying the Runge–Kutta scheme (3.5.10) to the system (5.4)-(5.5), we get

$$X_{t_k,x}(t_{k+1}) \simeq \bar{X}_{t_k,x}(t_{k+1}) = x + \frac{1}{2}hb_k$$
$$+ \frac{1}{2}hb(t_{k+1}, x + hb_k, u(t_{k+1}, x + hb_k)) + \varepsilon^2 hc_k + \varepsilon h^{1/2}\sigma_k \xi_k,$$
$$Z_{t_k,x,z}(t_{k+1}) \simeq \bar{Z}_{t_k,x,z}(t_{k+1}) = z + \frac{1}{2}hg_k$$
$$+ \frac{1}{2}hg(t_{k+1}, x + hb_k, u(t_{k+1}, x + hb_k)), \qquad (5.8)$$

where b_k, c_k, σ_k, and g_k are the coefficients b, c, σ, and g calculated at the point $(t_k, x, u(t_k, x))$.

Using a probabilistic representation of the form (2.4), we obtain

$$u(t_k, x)$$
$$\simeq E(u(t_{k+1}, \bar{X}_{t_k,x}(t_{k+1})) + \bar{Z}_{t_k,x,0}(t_{k+1}))$$
$$= \frac{1}{2}u(t_{k+1}, x + h[b_k + b(t_{k+1}, x + hb_k, u(t_{k+1}, x + hb_k))]/2$$
$$+ \varepsilon^2 hc_k + \varepsilon h^{1/2}\sigma_k)$$
$$+ \frac{1}{2}u(t_{k+1}, x + h[b_k + b(t_{k+1}, x + hb_k, u(t_{k+1}, x + hb_k))]/2$$
$$+ \varepsilon^2 hc_k - \varepsilon h^{1/2}\sigma_k)$$
$$+ \frac{1}{2}hg_k + \frac{1}{2}hg(t_{k+1}, x + hb_k, u(t_{k+1}, x + hb_k)).$$

We can approximate $u(t_k, x)$ by $v(t_k, x)$ found from

$$v(t_k, x)$$
$$= \frac{1}{2}u(t_{k+1}, x + h[\tilde{b}_k + b(t_{k+1}, x + h\tilde{b}_k, u(t_{k+1}, x + h\tilde{b}_k))]/2$$
$$+ \varepsilon^2 h\tilde{c}_k + \varepsilon h^{1/2}\tilde{\sigma}_k)$$
$$+ \frac{1}{2}u(t_{k+1}, x + h[\tilde{b}_k + b(t_{k+1}, x + h\tilde{b}_k, u(t_{k+1}, x + h\tilde{b}_k))]/2$$
$$+ \varepsilon^2 h\tilde{c}_k - \varepsilon h^{1/2}\tilde{\sigma}_k)$$
$$+ \frac{1}{2}h\tilde{g}_k + \frac{1}{2}hg(t_{k+1}, x + h\tilde{b}_k, u(t_{k+1}, x + h\tilde{b}_k)), \qquad (5.9)$$

where $\tilde{b}_k$, $\tilde{c}_k$, $\tilde{\sigma}_k$, and $\tilde{g}_k$ are the coefficients b, c, σ, and g calculated at the point $(t_k, x, v(t_k, x))$.

The corresponding implicit layer method has the form

$$\bar{u}(t_N, x) = \varphi(x),$$

7.5 Probabilistic approach to parabolic equations with small parameter

$$\bar{u}(t_k, x)$$
$$= \frac{1}{2}\bar{u}(t_{k+1}, x + h[\bar{b}_k + b(t_{k+1}, x + h\bar{b}_k, \bar{u}(t_{k+1}, x + h\bar{b}_k))]/2$$
$$+\varepsilon^2 h\bar{c}_k + \varepsilon h^{1/2}\bar{\sigma}_k)$$
$$+\frac{1}{2}\bar{u}(t_{k+1}, x + h[\bar{b}_k + b(t_{k+1}, x + h\bar{b}_k, \bar{u}(t_{k+1}, x + h\bar{b}_k))]/2$$
$$+\varepsilon^2 h\bar{c}_k - \varepsilon h^{1/2}\bar{\sigma}_k)$$
$$+\frac{1}{2}h\bar{g}_k + \frac{1}{2}hg(t_{k+1}, x + h\bar{b}_k, \bar{u}(t_{k+1}, x + h\bar{b}_k)), \qquad (5.10)$$

$$k = N-1, \ldots, 0,$$

where $\bar{b}_k$, $\bar{c}_k$, $\bar{\sigma}_k$, and $\bar{g}_k$ are the coefficients b, c, σ, and g calculated at the point $(t_k, x, \bar{u}(t_k, x))$.

Let us assume that
(i) The coefficients $b(t, x, u)$ and $g(t, x, u)$ and their first and second derivatives are continuous and uniformly bounded, the coefficients $c(t, x, u)$ and $\sigma(t, x, u)$ and their first derivatives are continuous and uniformly bounded:

$$\left|\frac{\partial^{i+j+l} b}{\partial t^i \partial x^j \partial u^l}\right| \leq K, \quad \left|\frac{\partial^{i+j+l} g}{\partial t^i \partial x^j \partial u^l}\right| \leq K, \quad 0 \leq i+j+l \leq 2,$$

$$\left|\frac{\partial^{i+j+l} c}{\partial t^i \partial x^j \partial u^l}\right| \leq K, \quad \left|\frac{\partial^{i+j+l} \sigma}{\partial t^i \partial x^j \partial u^l}\right| \leq K, \quad 0 \leq i+j+l \leq 1, \qquad (5.11)$$

$$t_0 \leq t \leq T, \ x \in \mathbf{R}, \ u_\circ < u < u^\circ,$$

where $-\infty \leq u_\circ$, $u^\circ \leq \infty$ are some constants.
(ii) There exists a unique bounded solution $u(t, x)$ of the problem (5.6)-(5.7) such that

$$u_\circ < u_* \leq u(t, x) \leq u^* < u^\circ, \qquad (5.12)$$

where u_*, u^* are some constants, and there exist the uniformly bounded derivatives:

$$\left|\frac{\partial^{i+j} u}{\partial t^i \partial x^j}\right| \leq K, \ i = 0, \ j = 1, 2, 3, 4; \ i = 1, \ j = 0, 1, 2; \qquad (5.13)$$

$$i = 2, \ j = 0, 1; \ i = 3, \ j = 0; \ t_0 \leq t \leq T, \ x \in \mathbf{R}, \ 0 < \varepsilon \leq \varepsilon^*.$$

Below, in Sect. 7.5.3, we consider the singular case, when the condition (5.13) is not fulfilled.

Lemma 5.1. (*see [206]*). *Under the assumptions* (i) - (ii), *the one-step error of the implicit layer method* (5.10) *is estimated by* $O(h^3 + \varepsilon^2 h^2)$, *i.e.,*

$$|v(t_k, x) - u(t_k, x)| \leq C\left(h^3 + \varepsilon^2 h^2\right),$$

where $v(t_k, x)$ *is found from* (5.9), C *does not depend on* h, k, x, ε.

Theorem 5.2. *Under the assumptions (i) - (ii), the global error of the implicit layer method (5.10) is estimated by $O(h^2 + \varepsilon^2 h)$:*

$$|u(t_k, x) - \bar{u}(t_k, x)| \leq K \left(h^2 + \varepsilon^2 h \right),$$

where the constant K does not depend on h, k, x, ε.

A proof of this theorem is analogous to the proof of Theorem 2.2 (see details in [206]).

Remark 5.3. For linear parabolic equations, i.e., when the coefficients of (5.6) do not depend on u, the method (5.10) becomes the explicit one with the global error $O(h^2 + \varepsilon^2 h)$ and can be applied to solving linear parabolic equations with small parameter. Note also that if the dimension d of the linear problem is high ($d > 3$ in practice) and it is enough to find the solution at a few points only, the Monte Carlo technique is preferable.

7.5.2 Explicit layer methods

To implement the implicit method (5.10), one can use the method of simple iteration. If we take $u(t_{k+1}, x)$ as a null iteration, in the case of $b(t, x, u) \neq b(t, x)$ or $g(t, x, u) \neq g(t, x)$, the first iteration provides the one-step error $O(h^2)$ only. One can show that applying the second iteration, we get $O(h^3 + \varepsilon^2 h^2)$ as the one-step error. However it is possible to reach the same one-step accuracy by some modification of the first iteration that reduces the number of recalculations. The explicit layer method obtained on this way has the form (we use the same notation $\bar{u}(t_k, x)$ again)

$$\bar{u}(t_N, x) = \varphi(x),$$

$$\hat{b}_k = b(t_k, x, \bar{u}(t_{k+1}, x)), \ \hat{c}_k = c(t_k, x, \bar{u}(t_{k+1}, x)), \ \hat{\sigma}_k = \sigma(t_k, x, \bar{u}(t_{k+1}, x)),$$

$$\bar{u}^{(1)}(t_k, x) = \bar{u}(t_{k+1}, x + h\hat{b}_k) + hg(t_k, x, \bar{u}(t_{k+1}, x)),$$

$$\bar{u}(t_k, x)$$
$$= \frac{1}{2} \bar{u}(t_{k+1}, x + h[b(t_k, x, \bar{u}^{(1)}(t_k, x)) + b(t_{k+1}, x + h\hat{b}_k, \bar{u}(t_{k+1}, x + h\hat{b}_k))]/2$$
$$+ \varepsilon^2 h \hat{c}_k + \varepsilon h^{1/2} \hat{\sigma}_k)$$
$$+ \frac{1}{2} \bar{u}(t_{k+1}, x + h[b(t_k, x, \bar{u}^{(1)}(t_k, x)) + b(t_{k+1}, x + h\hat{b}_k, \bar{u}(t_{k+1}, x + h\hat{b}_k))]/2$$
$$+ \varepsilon^2 h \hat{c}_k - \varepsilon h^{1/2} \hat{\sigma}_k)$$
$$+ \frac{1}{2} hg(t_k, x, \bar{u}^{(1)}(t_k, x)) + \frac{1}{2} hg(t_{k+1}, x + h\hat{b}_k, \bar{u}(t_{k+1}, x + h\hat{b}_k)), \quad (5.14)$$

$$k = N - 1, \ldots, 0.$$

The following theorem can be proved by the arguments like those used for Theorem 2.2.

7.5 Probabilistic approach to parabolic equations with small parameter   443

Theorem 5.4. *Under the assumptions (i) - (ii), the global error of the explicit layer method (5.14) is estimated by $O(h^2 + \varepsilon^2 h)$:*

$$|u(t_k, x) - \bar{u}(t_k, x)| \leq K\left(h^2 + \varepsilon^2 h\right),$$

where the constant K does not depend on h, k, x, ε.

Remark 5.5. Naturally, we can take other weak approximations (more accurate than the ones we use above) of SDEs with small noise from Chap. 3 to construct the corresponding high-order (with respect to h and ε) methods for the problem (5.6)-(5.7). In Sect. 7.6 we give high-order methods in some particular cases of the equation (5.6).

7.5.3 Singular case

The estimates of errors for the methods proposed above (Theorems 5.2 and 5.4) are obtained provided the bounds of derivatives of the solution to the problems considered are uniform with respect to $x \in \mathbf{R}^d$, $t \in [t_0, T]$, and $0 < \varepsilon \leq \varepsilon^*$ (see (5.13)). This assumption is ensured, e.g., in the following case. Consider the first-order partial differential equation, obtained from (5.6) with $\varepsilon = 0$:

$$\frac{\partial u^0}{\partial t} + b(t, x, u^0)\frac{\partial u^0}{\partial x} + g(t, x, u^0) = 0, \ t \in [t_0, T), \ x \in \mathbf{R}, \qquad (5.15)$$

$$u^0(T, x) = \varphi(x). \qquad (5.16)$$

If the coefficients of the equation (5.15) and the initial condition (5.16) are such that the solution $u^0(t, x)$, $x \in \mathbf{R}$, is sufficiently smooth for $t_0 \leq t \leq T$, then the derivatives of the solution $u(t, x)$ to (5.15)-(5.16) can be uniformly bounded with respect to $0 \leq \varepsilon \leq \varepsilon^*$ for $t \in [t_0, T]$ (see [79, 269]). Note that generally the assumption, that the coefficients of (5.15) and the initial condition $\varphi(x)$ are bounded and smooth functions, is not enough to ensure the regular behavior of $u^0(t, x)$ at any $t < T$ [79].

A lot of physical phenomena (e.g., formation and propagation of shock waves) having singular behavior is described by equations with small parameter. The derivatives of their solutions go to infinity as $\varepsilon \to 0$ and, rigorously speaking, the results of Theorems 5.2 and 5.4 become inapplicable.

After the known change of variables $t = \varepsilon^2 t'$, $x = \varepsilon^2 x'$, the problem (5.6)-(5.7) is rewritten for $v(t', x') := u(\varepsilon^2 t', \varepsilon^2 x')$ in the form

$$\frac{\partial v}{\partial t'} + \frac{1}{2}\sigma^2(\varepsilon^2 t', \varepsilon^2 x', v)\frac{\partial^2 v}{\partial x'^2} + (b(\varepsilon^2 t', \varepsilon^2 x', v) + \varepsilon^2 c(\varepsilon^2 t', \varepsilon^2 x', v))\frac{\partial v}{\partial x'}$$
$$+\varepsilon^2 g(\varepsilon^2 t', \varepsilon^2 x', v) = 0, \ t' \in [t_0/\varepsilon^2, T/\varepsilon^2), \ x' \in \mathbf{R}, \ 0 < \varepsilon \leq \varepsilon^*, \qquad (5.17)$$

$$v(T/\varepsilon^2, x') = \varphi(\varepsilon^2 x'). \qquad (5.18)$$

If the assumptions like (ii) hold for the solution $v(t', x')$ of (5.17)-(5.18) (we observe that the problem (5.17)-(5.18) is considered on long time intervals), then the derivatives of the solution $u(t,x)$ to (5.6)-(5.7) are estimated as:

$$\left|\frac{\partial^{i+j}u}{\partial t^i \partial x^j}\right| \leq \frac{K}{\varepsilon^{2(i+j)}}, \ t \in [t_0, T], \ x \in \mathbf{R}, \ 0 < \varepsilon \leq \varepsilon^*. \tag{5.19}$$

These bounds are natural ones for the problem (5.6)-(5.7) in the singular case.

If one followed the arguments of Lemma 5.1 and Theorem 5.2 in the singular case (i.e., taking the assumptions (i), (5.12), (5.19) instead of (i)-(ii)), the estimate of the form $\frac{K}{C}\frac{h}{\varepsilon^2}(e^{K(T-t_0)/\varepsilon^2} - 1)$ would be obtained for the proposed methods. Due to the big factor $1/\varepsilon^2$ in the exponent, this estimate is meaningless for practical purposes. Our numerical tests (see Sect. 7.7.1) demonstrate that the proposed methods are of essentially better quality than could be predicted by this estimate. Most likely, the methods work fairly well in the singular case because the derivatives are large only in a small domain known as interior layer (see, e.g., [111]) that is attributable to the majority of interesting applications. The further theoretical investigation, namely, obtaining a realistic estimate for the errors of the methods proposed, should rest on a stability analysis and on more extensive properties of the solution considered. A similar problem for finite elements methods has been considered in a few papers (see [57] and references therein).

However, in some particular, but important, singular cases of the problem (5.6)-(5.7) we get reasonable estimates (without $1/\varepsilon^2$ in the exponent) for the errors of the proposed methods by the arguments of Lemma 5.1 and Theorem 5.2. As an example, we give the following two theorems here (see their proofs in the preprint [203]).

Theorem 5.6. *Assume the coefficients b and σ in (5.6) are independent of u. Let the conditions (i) and (5.12) hold, and let the derivatives $|\partial^{i+j}u/\partial t^i \partial x^j|$, $i = 0, \ j = 1, 2, 3, 4;\ i = 1, j = 0, 1, 2;\ i = 2, \ j = 0, 1;\ i = 3, \ j = 0$, satisfy (5.19). Then for a sufficiently small h/ε^2, the global error of the explicit layer method (5.14) is estimated as*

$$|u(t_k, x) - \bar{u}(t_k, x)| \leq \frac{K}{C}\frac{h}{\varepsilon^4}(e^{K(T-t_0)} - 1),$$

where the constants C and K do not depend on h, k, x, ε.

In a lot of applications (e.g., in shock waves) the derivatives are significant only in a small interval (interior layer) $(x_*(t), x^*(t))$ with width $|x_*(t) - x^*(t)| \sim \varepsilon^2$:

$$\left|\frac{\partial^{i+j}u}{\partial t^i \partial x^j}\right| \leq \frac{K}{\varepsilon^{2(i+j)}}, \ t \in [t_0, T], \ x \in (x_*(t), x^*(t)), \ 0 < \varepsilon \leq \varepsilon^*, \tag{5.20}$$

7.5 Probabilistic approach to the Cauchy problem for nonlinear PDEs

and

$$\left|\frac{\partial^{i+j} u}{\partial t^i \partial x^j}\right| \leq K, \ t \in [t_0, T], \ x \notin (x_*(t), x^*(t)), \ 0 < \varepsilon \leq \varepsilon^*,$$

$$\int_{x \notin (x_*(t), x^*(t))} \left|\frac{\partial^{i+j} u}{\partial t^i \partial x^j}\right| dx \leq K, \ i + j \neq 0, \ t \in [t_0, T], \ 0 < \varepsilon \leq \varepsilon^*. \quad (5.21)$$

Theorem 5.7. *Assume the coefficients b and σ in (5.6) are independent of u. Let the conditions (i) and (5.12) hold and the derivatives $|\partial^{i+j} u / \partial t^i \partial x^j|$, $i = 0, \ j = 1, 2, 3, 4; \ i = 1, \ j = 0, 1, 2; \ i = 2, \ j = 0, 1; \ i = 3, \ j = 0$, satisfy (5.20) and (5.21). Then for a sufficiently small h/ε^2, the global error of the explicit layer method (5.14) is estimated in L_1-norm as*

$$\int_{\mathbf{R}} |u(t_k, x) - \bar{u}(t_k, x)| dx \leq \frac{K}{C} \frac{h}{\varepsilon^2} \left(e^{K(T-t_0)} - 1 \right), \quad (5.22)$$

where the constants C and K do not depend on h, k, x, ε.

The analogous theorems for the more simple method (2.23) give the same estimates of its error. However, in our experiments the layer method (5.14) gives better results than (2.23). To show the advantages of the method (5.14) in the singular case theoretically, further investigation is required. Seemingly, a more accurate analysis of the error of the method (5.14) should rest on more extensive properties of the solution $u(t, x)$.

See also Remarks 6.1 and 7.1, where in the singular situation we give reasonable estimates of the errors for some other particular cases of the problem (5.6)-(5.7).

7.5.4 Numerical algorithms based on interpolation

Introduce an equidistant space discretization: $\{x_j = x_0 + jh_x, \ j = 0, \pm 1, \pm 2, \ldots\}$, $x_0 \in \mathbf{R}$, h_x is a sufficiently small positive number. Since it does not lead to any misunderstanding, we use the old notation $\bar{u}$, $\bar{u}^{(1)}$, etc. for new values here.

Theorem 5.8. *Under the assumptions (i)-(ii), the numerical algorithm based on the explicit method (5.14) and on the linear interpolation:*

$$\bar{u}(t_N, x) = \varphi(x),$$

$$\hat{b}_{k,j} = b(t_k, x_j, \bar{u}(t_{k+1}, x_j)), \ \hat{c}_{k,j} = c(t_k, x_j, \bar{u}(t_{k+1}, x_j)),$$
$$\hat{\sigma}_{k,j} = \sigma(t_k, x_j, \bar{u}(t_{k+1}, x_j)),$$

$$\bar{u}^{(1)}(t_k, x_j) = \bar{u}(t_{k+1}, x_j + h\hat{b}_{k,j}) + hg(t_k, x_j, \bar{u}(t_{k+1}, x_j)),$$

$$\bar{u}(t_k, x_j)$$
$$= \frac{1}{2}\bar{u}(t_{k+1}, x_j + h[b(t_k, x_j, \bar{u}^{(1)}(t_k, x_j))$$
$$+ b(t_{k+1}, x_j + h\hat{b}_{k,j}, \bar{u}(t_{k+1}, x_j + h\hat{b}_{k,j}))]/2 + \varepsilon^2 h\hat{c}_{k,j} + \varepsilon h^{1/2}\hat{\sigma}_{k,j})$$
$$+ \frac{1}{2}\bar{u}(t_{k+1}, x_j + h[b(t_k, x_j, \bar{u}^{(1)}(t_k, x_j))$$
$$+ b(t_{k+1}, x_j + h\hat{b}_{k,j}, \bar{u}(t_{k+1}, x_j + h\hat{b}_{k,j}))]/2 + \varepsilon^2 h\hat{c}_{k,j} - \varepsilon h^{1/2}\hat{\sigma}_{k,j})$$
$$+ \frac{1}{2}hg(t_k, x_j, \bar{u}^{(1)}(t_k, x_j))$$
$$+ \frac{1}{2}hg(t_{k+1}, x_j + h\hat{b}_{k,j}, \bar{u}(t_{k+1}, x_j + h\hat{b}_{k,j})),$$

$$\bar{u}(t_k, x) = \frac{x_{j+1} - x}{h_x}\bar{u}(t_k, x_j) + \frac{x - x_j}{h_x}\bar{u}(t_k, x_{j+1}), \quad x_j < x < x_{j+1}, \quad (5.23)$$
$$j = 0, \pm 1, \pm 2, \ldots, \quad k = N-1, \ldots, 0,$$

has global error estimated by $O(h^2 + \varepsilon^2 h)$ if the value of h_x is selected as $h_x = \alpha \min(h^{3/2}, \varepsilon h)$, where α is a positive constant.

This theorem can be proved by the arguments like those used for Theorem 2.5 (see details in [206]). Remarks 2.6-2.9 are applicable here.

7.6 High-order methods for semilinear equation with small constant diffusion and zero advection

Here we restrict ourselves to the case of $d = 1$ for simplicity again. Consider the Cauchy problem for the semilinear heat equation

$$\frac{\partial u}{\partial t} + \frac{\varepsilon^2}{2}\frac{\partial^2 u}{\partial x^2} + g(t, x, u) = 0, \, t \in [t_0, T), \, x \in \mathbf{R}, \quad (6.1)$$

$$u(T, x) = \varphi(x). \quad (6.2)$$

We assume that $g(t, x, u)$ is a uniformly bounded and sufficiently smooth function and conditions like (ii) from Sect. 7.5.1 are fulfilled for the solution $u(t, x)$ to (6.1)-(6.2). Note that to construct high-order methods we need uniform boundedness of the derivatives of $u(t, x)$ with higher orders than in the assumption (5.13). To realize the methods of this section, we can avoid any interpolation choosing a special space discretization. The methods of this section are tested by simulation of the generalized FKPP-equation with a small parameter (see Sect. 7.7.2).

The probabilistic representation of the solution to (6.1)-(6.2) has the form (see (5.3)-(5.5)):

$$u(t, x) = E(\varphi(X_{t,x}(T)) + Z_{t,x,0}(T)), \quad (6.3)$$

7.6 High-order methods for semilinear equation with small parameter   447

where $X_{t,x}(s)$, $Z_{t,x,z}(s)$, $s \geq t$, satisfies the system

$$dX = \varepsilon dw(s), \quad X(t) = x,$$
$$dZ = g(s, X, u(s, X))ds, \quad Z(t) = z. \qquad (6.4)$$

Note that the system (6.4) is a system of differential equations with small *additive* noise.

7.6.1 Two-layer methods

Let us write down the layer methods (2.23) and (5.14) and the second-order method (2.18) in the case of the problem (6.1)-(6.2).

The explicit layer method (2.23) *with error* $O(h)$ *has the form*

$$\bar{u}(t_N, x_j) = \varphi(x_j),$$

$$\bar{u}(t_k, x_j) = \frac{1}{2}\bar{u}(t_{k+1}, x_j + \varepsilon h^{1/2}) + \frac{1}{2}\bar{u}(t_{k+1}, x_j - \varepsilon h^{1/2})$$
$$+ hg(t_{k+1}, x_j, \bar{u}(t_{k+1}, x_j)), \qquad (6.5)$$

$$x_j = x_0 + j\varepsilon h^{1/2}, \ j = 0, \pm 1, \pm 2, \ldots, \ k = N-1, \ldots, 0.$$

Note that it coincides with the well-known finite-difference scheme under the special relation of time and space steps ($h_x = \varepsilon h_t^{1/2}$) in the scheme.

In the case of the problem (6.1)-(6.2) *the explicit layer method* (5.14) *with error* $O(h^2 + \varepsilon^2 h)$ *takes the form*

$$\bar{u}(t_N, x_j) = \varphi(x_j),$$

$$\bar{u}^{(1)}(t_k, x_j) = \bar{u}(t_{k+1}, x_j) + hg(t_k, x_j, \bar{u}(t_{k+1}, x_j)),$$

$$\bar{u}(t_k, x_j) = \frac{1}{2}\bar{u}(t_{k+1}, x_j + \varepsilon h^{1/2}) + \frac{1}{2}\bar{u}(t_{k+1}, x_j - \varepsilon h^{1/2})$$
$$+ \frac{1}{2}h[g(t_k, x_j, \bar{u}^{(1)}(t_k, x_j)) + g(t_{k+1}, x_j, \bar{u}(t_{k+1}, x_j))], \qquad (6.6)$$

$$x_j = x_0 + j\varepsilon h^{1/2}, \ j = 0, \pm 1, \pm 2, \ldots, \ k = N-1, \ldots, 0.$$

In the case of the problem (6.1)-(6.2) *the implicit layer method* (2.18) *with error* $O(h^2)$ *has the form*

$$\bar{u}(t_N, x_j) = \varphi(x_j),$$

$$\bar{u}(t_k, x_j) = \frac{1}{6}\bar{u}(t_{k+1}, x_j + \sqrt{3}\varepsilon h^{1/2}) + \frac{2}{3}\bar{u}(t_{k+1}, x_j) + \frac{1}{6}\bar{u}(t_{k+1}, x_j - \sqrt{3}\varepsilon h^{1/2})$$
$$+ \frac{h}{2}g(t_k, x_j, \bar{u}(t_k, x_j)) + \frac{h}{3}g(t_{k+1}, x_j, \bar{u}(t_{k+1}, x_j))$$
$$+ \frac{h}{12}g(t_{k+1}, x_j + \sqrt{3}\varepsilon h^{1/2}, \bar{u}(t_{k+1}, x_j + \sqrt{3}\varepsilon h^{1/2}))$$
$$+ \frac{h}{12}g(t_{k+1}, x_j - \sqrt{3}\varepsilon h^{1/2}, \bar{u}(t_{k+1}, x_j - \sqrt{3}\varepsilon h^{1/2})), \tag{6.7}$$

$$x_j = x_0 + j\sqrt{3}\varepsilon h^{1/2}, \ j = 0, \pm 1, \pm 2, \ldots, \ k = N-1, N-2, \ldots, 0.$$

To solve the algebraic equations obtained at each step of the method (6.7), one can use the Newton method or the method of simple iteration.

Remark 6.1. In the singular case the natural bounds for derivatives of the solution to (6.1)-(6.2) have the form

$$\left|\frac{\partial^{i+j} u}{\partial t^i \partial x^j}\right| \leq \frac{K}{\varepsilon^j}, \ t \in [t_0, T], \ x \in \mathbf{R}, \ 0 < \varepsilon \leq \varepsilon^*. \tag{6.8}$$

These bounds are obtained using the following change of variables: $t = t'$, $x = \varepsilon x'$ (cf. Sect. 7.5.3).

One can prove under (6.8) that the errors of both methods (6.5) and (6.6) are estimated as

$$|u(t_k, x) - \bar{u}(t_k, x)| \leq Kh,$$

where the constant K does not depend on x, k, h, ε. Nevertheless, the method (6.6) gives better results than (6.5) in our experiments. One can explain this by the fact that the constant K of (6.6) is essentially less than the K of (6.5). Under (6.8) the error of the method (6.7) remains $O(h^2)$.

Note that in Sect. 7.7.2 we present results of testing these methods (instead of (6.5) we use a modification of it) on an equation in which the term g depends on ε and the derivatives of the solution have other bounds than (6.8) (see Remark 7.1 and other details in Sect. 7.7.2).

7.6.2 Three-layer methods

Here we obtain two three-layer methods. We estimate their one-step errors but do not prove their convergence that requires stability analysis of multi-layer methods. We test these methods in our experiments, and they give fairly good results.

To calculate $\bar{u}(t_{k+1}, x)$ by a three-layer method, two previous layers are used. So, to start simulations we should know $\bar{u}(t_N, x)$ and $\bar{u}(t_{N-1}, x)$. To simulate $\bar{u}(t_{N-1}, x)$ one can use, e.g., the two-layer method (6.7) with a sufficiently small step. Below we consider this layer to be known and denote $\psi(x) := \bar{u}(t_{N-1}, x)$.

7.6 High-order methods for semilinear equation with small parameter

Apply the special Runge–Kutta scheme (3.5.18) to approximate (6.4):

$$X_{t_k,x}(t_{k+1}) \simeq \bar{X}_{t_k,x}(t_{k+1}) = x + \varepsilon h^{1/2}\xi_k,$$

$$Z_{t_k,x,z}(t_{k+1}) \simeq \bar{Z}_{t_k,x,z}(t_{k+1}) = z + \frac{h}{6}(g(t_k, x, u(t_k, x))$$
$$+ 2g(t_{k+1/2}, x, u(t_{k+1/2}, x))$$
$$+ 2g(t_{k+1/2}, x + \varepsilon h^{1/2}\xi_k, u(t_{k+1/2}, x + \varepsilon h^{1/2}\xi_k))$$
$$+ g(t_{k+1}, x + \varepsilon h^{1/2}\xi_k, u(t_{k+1}, x + \varepsilon h^{1/2}\xi_k))), \qquad (6.9)$$

where ξ_k are i.i.d. variables with the law $P(\xi = 0) = 2/3$, $P(\xi = \pm\sqrt{3}) = 1/6$.

The implicit method with one-step error $O(h^5 + \varepsilon^2 h^3)$ has the form (to get the method we use the scheme (6.9) with the time step $2h$):

$$\bar{u}(t_N, x_j) = \varphi(x_j), \quad \bar{u}(t_{N-1}, x_j) = \psi(x_j),$$

$$\bar{u}(t_k, x_j) = \frac{1}{6}\bar{u}(t_{k+2}, x_j + \sqrt{6}\varepsilon h^{1/2}) + \frac{2}{3}\bar{u}(t_{k+2}, x_j) + \frac{1}{6}\bar{u}(t_{k+2}, x_j - \sqrt{6}\varepsilon h^{1/2})$$
$$+ \frac{h}{3}g(t_k, x_j, \bar{u}(t_k, x_j)) + \frac{10h}{9}g(t_{k+1}, x_j, \bar{u}(t_{k+1}, x_j))$$
$$+ \frac{h}{9}g(t_{k+1}, x_j + \sqrt{6}\varepsilon h^{1/2}, \bar{u}(t_{k+1}, x_j + \sqrt{6}\varepsilon h^{1/2}))$$
$$+ \frac{h}{9}g(t_{k+1}, x_j - \sqrt{6}\varepsilon h^{1/2}, \bar{u}(t_{k+1}, x_j - \sqrt{6}\varepsilon h^{1/2}))$$
$$+ \frac{h}{18}g(t_{k+2}, x_j + \sqrt{6}\varepsilon h^{1/2}, \bar{u}(t_{k+2}, x_j + \sqrt{6}\varepsilon h^{1/2}))$$
$$+ \frac{2h}{9}g(t_{k+2}, x_j, \bar{u}(t_{k+2}, x_j))$$
$$+ \frac{h}{18}g(t_{k+2}, x_j - \sqrt{6}\varepsilon h^{1/2}, \bar{u}(t_{k+2}, x_j - \sqrt{6}\varepsilon h^{1/2})), \qquad (6.10)$$

$$x_j = x_0 + j\sqrt{6}\varepsilon h^{1/2}, \ j = 0, \pm 1, \pm 2, \ldots, \ k = N-2, N-3, \ldots, 0.$$

Let us look at the stability properties of this method in the simple case when u and g in (6.1) do not depend on x, i.e., apply the method (6.10) to the ordinary differential equation

$$\frac{du}{dt} + g(t, u) = 0, \ t \leq T, \ u(T) = \varphi. \qquad (6.11)$$

Recall (see, e.g., [97]) that a linear n-step method for (6.11)

$$\alpha_n u_k + \alpha_{n-1} u_{k+1} + \cdots + \alpha_0 u_{k+n} = h(\beta_n g_k + \cdots + \beta_0 g_{k+n}),$$
$$g_i = g(t_i, u_i), \ \alpha_n \neq 0, \ |\alpha_0| + |\beta_0| > 0,$$

is zero-stable (D-stable) if the generating polynomial

$$\alpha_n \lambda^n + \alpha_{n-1}\lambda^{n-1} + \cdots + \alpha_0 = 0 \qquad (6.12)$$

satisfies the root condition: the roots of (6.12) lie on or within the unit circle, and the roots on the unit circle are simple.

In the case of (6.11) the method (6.10) coincides with the Milne two-step method which is of the order $O(h^4)$ and is zero-stable. Its generating polynomial has two roots: 1 and -1. As is known [97], the root -1 can be dangerous for some differential equations. The method (6.10) has unstable behavior in our numerical tests on the generalized FKPP-equation with a small parameter (7.8)-(7.10) (Sect. 7.7.2). One can see that the method (6.10) does not preserve the property $u \leq 1$ of the problem (7.8)-(7.10) that leads to an unstable behavior of the approximate solutions. We modify the method (6.10) in the experiments: if $\bar{u}(t_k, x_j) > 1$, we put $\bar{u}(t_k, x_j) = 1$. Since locally, in a single step, the arising difference $0 < \bar{u}(t_k, x_j) - 1$ is not greater than the one-step error of this method, this modification does not change the one-step accuracy order of the method. The modified method turned out to work fairly well if applied to the generalized FKPP-equation. However, the modification is based on the knowledge of the properties of the solution, and it may be difficult to find such a modification for another problem. Fortunately, we are able to approximate the system (6.4) by another weak scheme and obtain a method for (6.1)-(6.2) with better stability properties in the sense considered above but with the one-step error of lower order (see the method (6.13) below). Let us note that it is possible to reach both the same one-step accuracy $O(h^5 + \varepsilon^2 h^3)$ and the better stability properties by a four-layer method.

Approximate the system (6.4) by the special scheme with one-step order $O(h^4 + \varepsilon^2 h^3)$:

$$X_{t_k, x}(t_{k+1}) \simeq \bar{X}_{t_k, x}(t_{k+1}) = x + \varepsilon h^{1/2} \xi_k,$$

$$Z_{t_k, x, z}(t_{k+1}) \simeq \bar{Z}_{t_k, x, z}(t_{k+1}) = z + \frac{h}{12}[5g(t_k, x, u(t_k, x))$$
$$+ g(t_{k+1}, x, u(t_{k+1}, x))]$$
$$+ 7g(t_{k+1}, x + \varepsilon h^{1/2}\xi_k, u(t_{k+1}, x + \varepsilon h^{1/2}\xi_k))$$
$$- g(t_{k+2}, x + \varepsilon h^{1/2}\xi_k, u(t_{k+2}, x + \varepsilon h^{1/2}\xi_k))),$$

where ξ_k are i.i.d. variables with the law $P(\xi = 0) = 2/3$, $P(\xi = \pm\sqrt{3}) = 1/6$.

The three-layer implicit method with one-step error $O(h^4 + \varepsilon^2 h^3)$ has the form

$$\bar{u}(t_N, x_j) = \varphi(x_j), \quad \bar{u}(t_{N-1}, x_j) = \psi(x_j),$$

$$\bar{u}(t_k, x_j) = \frac{1}{6}\bar{u}(t_{k+1}, x_j + \sqrt{3}\varepsilon h^{1/2}) + \frac{2}{3}\bar{u}(t_{k+1}, x_j) + \frac{1}{6}\bar{u}(t_{k+1}, x_j - \sqrt{3}\varepsilon h^{1/2})$$
$$+ \frac{5h}{12}g(t_k, x_j, \bar{u}(t_k, x_j)) + \frac{17h}{36}g(t_{k+1}, x_j, \bar{u}(t_{k+1}, x_j))$$
$$+ \frac{7h}{72}g(t_{k+1}, x_j + \sqrt{3}\varepsilon h^{1/2}, \bar{u}(t_{k+1}, x_j + \sqrt{3}\varepsilon h^{1/2}))$$

$$+\frac{7h}{72}g(t_{k+1}, x_j - \sqrt{3}\varepsilon h^{1/2}, \bar{u}(t_{k+1}, x_j - \sqrt{3}\varepsilon h^{1/2}))$$
$$-\frac{h}{72}g(t_{k+2}, x_j + \sqrt{3}\varepsilon h^{1/2}, \bar{u}(t_{k+2}, x_j + \sqrt{3}\varepsilon h^{1/2}))$$
$$-\frac{h}{18}g(t_{k+2}, x_j, \bar{u}(t_{k+2}, x_j))$$
$$-\frac{h}{72}g(t_{k+2}, x_j - \sqrt{3}\varepsilon h^{1/2}, \bar{u}(t_{k+2}, x_j - \sqrt{3}\varepsilon h^{1/2})), \qquad (6.13)$$

$$x_j = x_0 + j\sqrt{3}\varepsilon h^{1/2}, \quad j = 0, \pm 1, \pm 2, \ldots, \quad k = N-2, N-3, \ldots, 0.$$

For (6.11) this method coincides with one of the implicit two-step Adams methods of order $O(h^3)$, and the roots of its generating polynomial are 1 and 0. One can expect that in the case of the problem (6.1)-(6.2) the method (6.13) also possesses better stability properties than (6.10). In our numerical tests on the generalized FKPP-equation with small parameter (Sect. 7.7.2) the method (6.13) has stable behavior.

To solve the algebraic equations, obtained at each step of the methods (6.10) and (6.13), one can use the Newton method or the method of simple iteration.

Remark 6.2. The methods of this section can be extended for problems of a higher dimension or for systems of reaction-diffusion equations. Additionally using some other weak approximations to SDEs with small additive noise, new layer methods can be constructed. For instance, three- and four-layer methods with the one-step error $O(h^5 + \varepsilon^2 h^2)$ can be obtained. It is also not difficult to get an implicit four-layer method with the one-step error $O(h^5 + \varepsilon^2 h^3)$ for (6.1)-(6.2) possessing good stability properties in the above sense, or an explicit four-layer method with the one-step error $O(h^4 + \varepsilon^2 h^3)$, and so on.

7.7 Numerical tests

We note in Remark 2.7 that the algorithms rested on *cubic interpolation* give quite good results. We use the advantage of the cubic interpolation in our numerical tests on the Burgers equation. Let us recall that a sufficiently smooth function $f(x)$, $x \in \mathbf{R}$, can be interpolated by cubic interpolation as

$$f(x) \simeq \bar{f}(x) = \sum_{i=0}^{3} \Phi_{j,i}(x) f(x_{j+i}), \quad x_{j+1} < x < x_{j+2},$$

$$\Phi_{j,i}(x) = \prod_{m=0, m\neq i}^{3} \frac{x - x_{j+m}}{x_{j+i} - x_{j+m}}, \qquad (7.1)$$

where $x_j = x_0 + j \times h_x$, $x_0 \in \mathbf{R}$, $j = 0, \pm 1, \pm 2, \ldots$, h_x is a positive number.

The error of the cubic interpolation (7.1) is estimated by

$$|\bar{f}(x) - f(x)| \le \frac{3}{128} \max_{x_j < x < x_{j+3}} \left|\frac{\partial^4 u}{\partial x^4}\right| \times h_x^4, \quad x_{j+1} < x < x_{j+2}.$$

Recall (see Theorem 5.8) that the algorithm (5.23), based on the layer method (5.14) and on linear interpolation, has the error estimated by $O(h^2 + \varepsilon^2 h)$ provided $h_x = \min(h^{3/2}, \varepsilon h)$. One can expect that under the assumptions (i)-(ii) from Sect. 7.5.1 the algorithm based on the layer method (5.14) and on the cubic interpolation (7.1) can achieve the same accuracy $O(h^2 + \varepsilon^2 h)$ with h_x taken equal to $\min(h^{3/4}, \sqrt{\varepsilon h})$ only. Our numerical tests on the Burgers equation support this supposition.

7.7.1 The Burgers equation with small viscosity

The one-dimensional Burgers equation with small viscosity has the form

$$\frac{\partial u}{\partial t} = \frac{\varepsilon^2}{2} \frac{\partial^2 u}{\partial x^2} - u \frac{\partial u}{\partial x}, \quad t > 0, \ x \in \mathbf{R}, \tag{7.2}$$

$$u(0, x) = \varphi(x). \tag{7.3}$$

By means of the Cole–Hopf transformation, one can find the explicit solution of the problem (7.2)-(7.3):

$$u(t, x) = \frac{\int_{-\infty}^{\infty} K(t, x, y) \varphi(y) \exp\left(-\frac{1}{\varepsilon^2} \int_0^y \varphi(\xi) d\xi\right) dy}{\int_{-\infty}^{\infty} K(t, x, y) \exp\left(-\frac{1}{\varepsilon^2} \int_0^y \varphi(\xi) d\xi\right) dy},$$

$$K(t, x, y) = \frac{1}{\sqrt{2\pi\varepsilon^2 t}} \exp\left(-\frac{(x-y)^2}{2\varepsilon^2 t}\right).$$

Let us take the initial condition $\varphi(x)$ of the form

$$\varphi(x) = \begin{cases} c, & x < l_0, \\ \lambda(x), & l_0 \le x \le l_0 + l, \\ d, & x > l_0 + l, \end{cases} \tag{7.4}$$

where c, d, l_0, l are some numbers, $c > d$, $l \ge 0$; $\lambda(x)$ is a bounded measurable function, and $d \le \lambda(x) \le c$.

Recall some theoretical facts concerning the problem (7.2)-(7.3), (7.4) (see details, e.g., in [111, 269]). The solution $u(t, x)$ to (7.2)-(7.3), (7.4) is uniformly bounded:

$$d \le u(t, x) \le c, \ x \in \mathbf{R}, \ 0 \le t, \ 0 \le \varepsilon \le \varepsilon^*.$$

Let the initial condition $\varphi(x)$ be a sufficiently smooth function. Introduce the time moment T such that the solution of the hyperbolic problem obtained from (7.2)-(7.3), (7.4) for $\varepsilon = 0$ is smooth at $t < T$ and discontinuous at $t \geq T$. The solution $u(t,x)$ to (7.2)-(7.3), (7.4) is regular for $t \leq t_* < T$:

$$\left|\frac{\partial^{i+j}u}{\partial t^i \partial x^j}(t,x)\right| \leq K, \ x \in \mathbf{R}, \ 0 \leq t \leq t_*, \ 0 \leq \varepsilon \leq \varepsilon^*.$$

If $t \geq T$ then the solution is singular in an interval $(x_*(t), x^*(t))$ with width $|x^*(t) - x_*(t)| \sim \varepsilon^2$:

$$\left|\frac{\partial^{i+j}u}{\partial t^i \partial x^j}(t,x)\right| \leq \frac{K}{\varepsilon^{2(i+j)}}, \ x \in (x_*(t), x^*(t)), \ t \geq T, \ 0 < \varepsilon \leq \varepsilon^*,$$

$$\left|\frac{\partial^{i+j}u}{\partial t^i \partial x^j}(t,x)\right| \leq K, \ x \notin (x_*(t), x^*(t)), \ t \geq T, \ 0 < \varepsilon \leq \varepsilon^*.$$

In our experiments we take $\lambda(x)$ equal to

$$\lambda(x) = a - b\sin\frac{\pi x}{\mu}, \ \mu > 0, \ b > 0,$$

$$c = a + b, \ d = a - b, \ l = \mu, \ l_0 = -\frac{\mu}{2}. \quad (7.5)$$

For this $\lambda(x)$ the moment T can easily be found: $T = \mu/(\pi b)$.

We compare the behavior of two algorithms. The first one is based on the layer method (5.14) with the cubic interpolation (7.1). In the case of the problem (7.2)-(7.3) it has the form

$$\bar{u}(0,x) = \varphi(x),$$

$$\bar{u}(t_{k+1}, x_j) = \frac{1}{2}\bar{u}(t_k, x_j - h\bar{u}(t_k, x_j - h\bar{u}(t_k, x_j)) + \varepsilon h^{1/2})$$
$$+ \frac{1}{2}\bar{u}(t_k, x_j - h\bar{u}(t_k, x_j - h\bar{u}(t_k, x_j)) - \varepsilon h^{1/2}),$$

$$\bar{u}(t_k, x) = \sum_{i=0}^{3} \Phi_{j,i}(x)\, \bar{u}(t_k, x_{j+i}), \ x_{j+1} < x < x_{j+2},$$

$$\Phi_{j,i}(x) = \prod_{m=0, m \neq i}^{3} \frac{x - x_{j+m}}{x_{j+i} - x_{j+m}}, \quad (7.6)$$

$$j = 0, \pm 1, \pm 2, \ldots, \ k = 0, \ldots, N-1,$$

where $x_j = x_0 + j \times h_x$.

The second algorithm is based on the layer method (2.23) and on the cubic interpolation (7.1):

$$\bar{u}(0,x) = \varphi(x),$$

7 Probabilistic approach to the Cauchy problem for nonlinear PDEs

$$\bar{u}(t_{k+1}, x_j) = \frac{1}{2}\bar{u}(t_k, x_j - h\bar{u}(t_k, x_j) + \varepsilon h^{1/2})$$
$$+ \frac{1}{2}\bar{u}(t_k, x_j - h\bar{u}(t_k, x_j) - \varepsilon h^{1/2}),$$

$$\bar{u}(t_k, x) = \sum_{i=0}^{3} \Phi_{j,i}(x)\,\bar{u}(t_k, x_{j+i}), \quad x_{j+1} < x < x_{j+2},$$

$$\Phi_{j,i}(x) = \prod_{m=0, m\neq i}^{3} \frac{x - x_{j+m}}{x_{j+i} - x_{j+m}}, \tag{7.7}$$

$$j = 0, \pm 1, \pm 2, \ldots, \quad k = 0, \ldots, N-1.$$

Table 7.1 gives the results of simulation of the problem (7.2)-(7.3) with $\varphi(x)$ from (7.4), (7.5) in the case of the regular solution. In this case the assumptions (i)-(ii) from Sect. 7.5.1 are fulfilled, and the algorithm (7.6) has the error estimated by $O(h^2 + \varepsilon^2 h)$ and the algorithm (7.7) has the error estimated by $O(h)$. The value of h_x is taken equal to $h^{3/4}$. We present the errors of the approximate solutions $\bar{u}$ in the discrete Chebyshev norm and in the l^1-norm:

$$err^c(t) = \max_{x_i} |\bar{u}(t, x_i) - u(t, x_i)|,$$

$$err^l(t) = \sum_i |\bar{u}(t, x_i) - u(t, x_i)| \times h_x.$$

One can infer from Table 7.1 that the proposed special algorithm (7.6) with error $O(h^2 + \varepsilon^2 h)$ requires less computational effort than the algorithm (7.7) with error $O(h)$ and that the experimental data conform to the orders of accuracy of the algorithms given by the theoretical results.

Table 7.1. The Burgers equation (regular solution). Dependence of the errors $err^c(t_*)$ and $err^l(t_*)$ on h and ε for the algorithms (7.6) and (7.7) when $a = b = 0.5$, $\mu = 8$, and $t_* = 4$ ($T \approx 5.09$).

		algorithm (7.6)		algorithm (7.7)	
ε	h	$err^c(t_*)$	$err^l(t_*)$	$err^c(t_*)$	$err^l(t_*)$
	0.3	$0.1351 \cdot 10^{-1}$	$0.1531 \cdot 10^{-1}$	0.1130	0.1397
0.3	0.1	$0.2146 \cdot 10^{-2}$	$0.3347 \cdot 10^{-2}$	$0.3978 \cdot 10^{-1}$	$0.4628 \cdot 10^{-1}$
	0.01	$0.2295 \cdot 10^{-3}$	$0.3874 \cdot 10^{-3}$	$0.4221 \cdot 10^{-2}$	$0.4799 \cdot 10^{-2}$
	0.001	$0.2265 \cdot 10^{-4}$	$0.3947 \cdot 10^{-4}$	$0.4244 \cdot 10^{-3}$	$0.4814 \cdot 10^{-3}$
	0.3	$0.2325 \cdot 10^{-1}$	$0.2051 \cdot 10^{-1}$	0.1539	0.1519
	0.1	$0.4255 \cdot 10^{-2}$	$0.2287 \cdot 10^{-2}$	$0.6084 \cdot 10^{-1}$	$0.5007 \cdot 10^{-1}$
0.1	0.03	$0.3489 \cdot 10^{-3}$	$0.2396 \cdot 10^{-3}$	$0.2029 \cdot 10^{-1}$	$0.1553 \cdot 10^{-1}$
	0.01	$0.4444 \cdot 10^{-4}$	$0.5442 \cdot 10^{-4}$	$0.6751 \cdot 10^{-2}$	$0.5169 \cdot 10^{-2}$
	0.001	$0.5529 \cdot 10^{-5}$	$0.6374 \cdot 10^{-5}$	$0.6806 \cdot 10^{-3}$	$0.5189 \cdot 10^{-3}$

Table 7.2. The Burgers equation (singular solution). The errors $err^c(t)$ and $err^l(t)$ for $t = 8$ ($T \approx 5.09$). Other parameter values are the same as in Table 7.1. The time steps h and h_* are used when $t \leq t_*$ and $t > t_*$, respectively.

ε	h	h_*	algorithm (7.6)		algorithm (7.7)	
			$err^c(t)$	$err^l(t)$	$err^c(t)$	$err^l(t)$
0.3	0.1	0.01	$0.6322 \cdot 10^{-2}$	$0.2713 \cdot 10^{-2}$	0.1693	$0.6555 \cdot 10^{-1}$
	0.01	0.001	$0.4036 \cdot 10^{-3}$	$0.2482 \cdot 10^{-3}$	$0.1771 \cdot 10^{-1}$	$0.6782 \cdot 10^{-2}$
	0.001	0.0001	$0.5977 \cdot 10^{-4}$	$0.3760 \cdot 10^{-4}$	$0.1776 \cdot 10^{-2}$	$0.6931 \cdot 10^{-3}$
0.1	0.1	0.001	$0.5553 \cdot 10^{-1}$	$0.2351 \cdot 10^{-2}$	> 0.5	$0.6594 \cdot 10^{-1}$
	0.03	0.001	$0.1219 \cdot 10^{-1}$	$0.4699 \cdot 10^{-3}$	> 0.5	$0.3189 \cdot 10^{-1}$
	0.03	0.0001	$0.3955 \cdot 10^{-2}$	$0.1718 \cdot 10^{-3}$	0.4029	$0.1716 \cdot 10^{-1}$
	0.01	0.0001	$0.7047 \cdot 10^{-3}$	$0.3007 \cdot 10^{-4}$	0.1687	$0.6828 \cdot 10^{-2}$
	0.001	0.0001	$0.4139 \cdot 10^{-3}$	$0.2312 \cdot 10^{-4}$	$0.5461 \cdot 10^{-1}$	$0.2185 \cdot 10^{-2}$

To find the solution $u(t,x)$ to the problem (7.2)-(7.3), (7.4) for $t > T$, when the solution is singular, we realize the following numerical procedure: we simulate the problem by the algorithms (7.6) and (7.7) with a sufficiently big time step h and with $h_x = h^{3/4}$ up to the time moment $t_* < T$; then we change the time step h to a smaller one h_*, take $h_x = h_*$, and continue the simulations.

Table 7.2 gives the results of simulation of the problem (7.2)-(7.3) with $\varphi(x)$ from (7.4), (7.5) when $t > T$. One can see that in the singular case the behavior of the algorithm (7.6) is also better than the behavior of (7.7).

In connection with this example, see the numerical experiments in [25] and in Sect. 7.4 as well.

7.7.2 The generalized FKPP-equation with a small parameter

Consider the problem

$$\frac{\partial u}{\partial t} = \frac{\varepsilon^2}{2} \frac{\partial^2 u}{\partial x^2} + g(x,u;\varepsilon), \quad t > 0, \; x \in \mathbf{R}, \tag{7.8}$$

$$u(0,x) = \chi_-(x) = \begin{cases} 1, & x < 0 \\ 1/2, & x = 0 \\ 0, & x > 0, \end{cases} \tag{7.9}$$

and take

$$g(x,u;\varepsilon) = \frac{1}{\varepsilon^2} c(x)\, u(1-u),$$

$$c(x) = c + \frac{a}{\pi} \arctan \alpha(x-b). \tag{7.10}$$

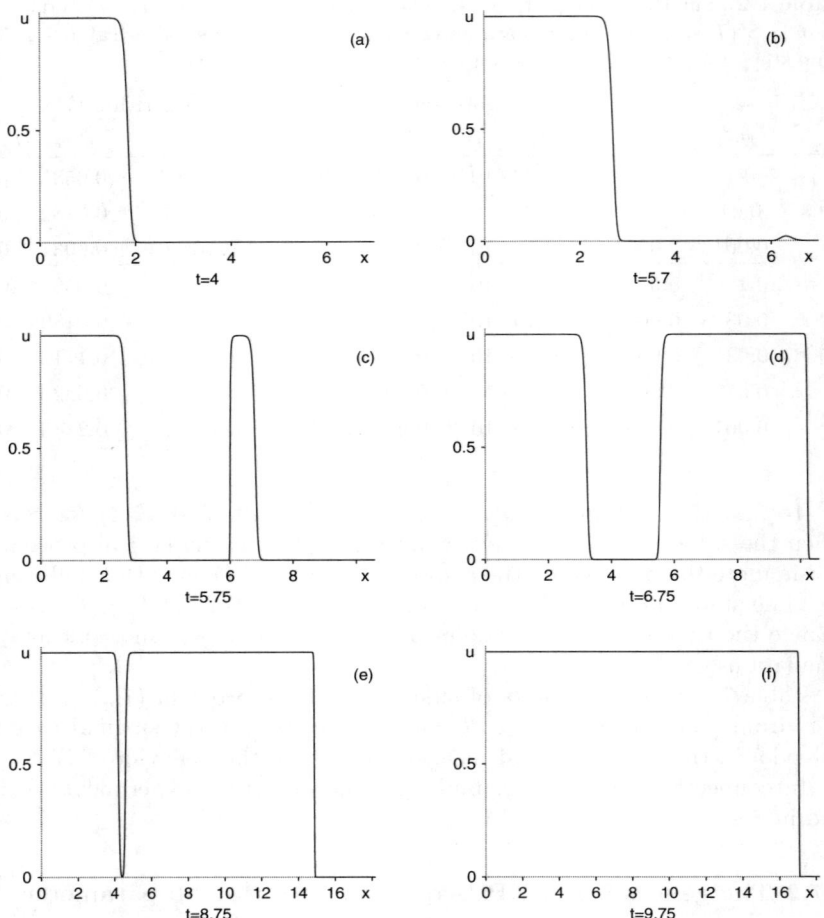

Fig. 7.1. The FKPP-equation. Evolution of the solution $u(t,x)$ to the problem (7.8)-(7.10) for $\varepsilon = 0.1$, $c = 1.125$, $a = 2$, $b = 6$, $\alpha = 150$ simulated by the method (6.13) with $h = 0.0001$.

Here $\varepsilon > 0$ is a small parameter, $\alpha > 0$ is a big number, c, a, and b are positive constants, and $a/2 < c < 3a/2$.

The problem (7.8)-(7.10) is a generalization of the FKPP-equation. The theoretical results for this problem obtained in [70] give the following. For $t < T_0 \approx \dfrac{b\sqrt{2a}}{c+0.5a}$, the wave propagates to the right of the domain $G_0 = \{x < 0\}$ with the velocity $\sqrt{2c-a}$, "taking no notice" of the fact that after $x = b$ the coefficient $c(x)$ takes a larger value $c + a/2$. But at the time T_0, a new "source" arises at the point $x = b$, away from which the front starts propagating in both directions: to the left with the velocity close to $\sqrt{2c-a}$ and to the right with the velocity close to $\sqrt{2c+a}$. Figure 7.1, obtained in our numerical experiments, demonstrates this phenomenon.

In the experiments we observe that for our values of the parameters ($\varepsilon = 0.1$, see the caption to Fig. 7.1)

$$\min_{-\infty<x<b} \bar{u}(5.75,x) \approx 10^{-71}$$

while there is already the new front at $x = b$ (see Fig. 7.1c). So, the "channel" through which the new "source" is initialized is very narrow. This fact has to be taken into account for realizing numerical procedures on a computer. For instance, when $\varepsilon = 0.04$

$$\min_{-\infty<x<b} \bar{u}(5.75,x) \approx 10^{-439},$$

which is less than the smallest positive number ($\sim 10^{-308}$) realizable by many compilers. To observe the phenomenon in this case, one has to compose a special numerical procedure or use a special compiler.

The following simple calculations (which, of course, do not pretend on an exhaustive explanation) clarify such a high sensitivity. We see from Fig. 7.2 that for time $t \in [4.8, 5.8]$ and $\varepsilon = 0.2$ the arising new source is such that the second derivative $\partial^2 u/\partial x^2$ for, e.g., $x \geq 4$ is not too large and therefore the equation (7.8) can be approximately replaced by

$$\frac{du}{dt} = \frac{1}{\varepsilon^2}c(x)u(1-u), \quad \varepsilon = 0.2, \quad x \geq 4. \tag{7.11}$$

Then

$$u(t,x) = \frac{u(t_0,x)e^{c(x)(t-t_0)/\varepsilon^2}}{1 + u(t_0,x)(e^{c(x)(t-t_0)/\varepsilon^2} - 1)}. \tag{7.12}$$

Let $x = 6.4$, $t_0 = 4.8$, $t = 5.8$. In this case $c(x) \approx 2.125$ and $u(5.8, 6.4) \approx 0.6$ (see Fig. 7.2). Then by (7.12) we get

$$u(4.8, 6.4) \approx 1.3 \times 10^{-23}. \tag{7.13}$$

At the same time, if such a small source arises at a point $x < b$ where $c(x) \approx 1.125$, we obtain the following very small value $u(5.8,x)|_{x<b} \approx 2.1 \times 10^{-11}$, i.e., u grows at $x = 6.4$ much faster than at $x < b$.

An additional confirmation of the high sensitiveness of the considered model is given, for instance, by the following experiment. If we put $u(0,x) = \mu$ for $x > 0$ with a small positive μ, e.g. 10^{-15}, in the initial condition (7.9) and take the other parameters as in Fig. 7.1, the new "source" arises at a moment $t \ll 1$.

Here we compare five numerical methods: the methods (6.6), (6.7), (6.10), (6.13) given in Sect. 7.6 (of course, we take their versions adapted to problems

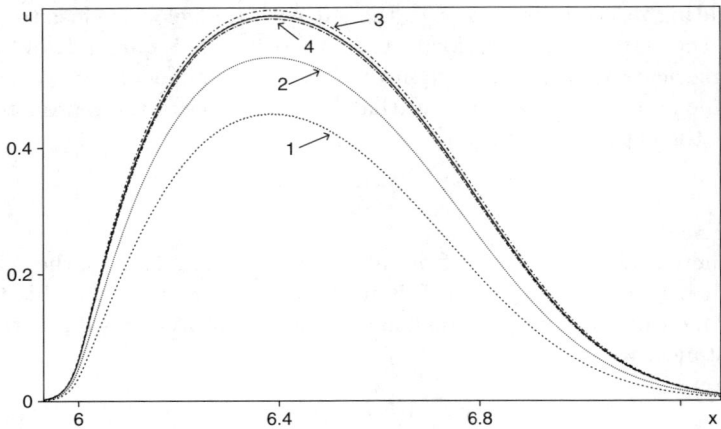

Fig. 7.2. The FKPP-equation (new source appearance). Comparison of the methods. The solid curve is simulated by (6.7) and (6.13) with $h = 0.0001$, and it visually coincides with the exact solution. The curves 1, 2 are simulated by (7.14) and (6.6) with $h = 0.0001$; the curves 3, 4 by (6.7) and (6.13) with $h = 0.001$. Here $\varepsilon = 0.2$, $t = 5.8$, and the other parameter values are the same as in Fig. 7.1.

with positive direction of time, cf. Remark 2.8) and the first-order method written below. The results of the numerical tests are given on Figs. 7.2 and 7.3.

The first-order method:

$$\bar{u}(0, x_j) = \chi_-(x_j),$$

$$\bar{u}(t_{k+1}, x_j) = \frac{1}{2}\bar{u}(t_k, x_j + \varepsilon h^{1/2}) + \frac{1}{2}\bar{u}(t_k, x_j - \varepsilon h^{1/2})$$
$$+ \frac{h}{2}(g(x_{j-1}, \bar{u}(t_k, x_{j-1})) + g(x_{j+1}, \bar{u}(t_k, x_{j+1}))), \quad (7.14)$$

$$x_j = x_0 + j\varepsilon h^{1/2}, \ j = 0, \pm 1, \ldots, \ k = 0, \ldots, N-1.$$

It can be checked that for a sufficiently small h this method preserves the monotonicity property of the solution. The first-order method (6.5), which has $hg(x_j, \bar{u}(t_k, x_j))$ instead of $h(g(x_{j-1}, \bar{u}(t_k, x_{j-1})) + g(x_{j+1}, \bar{u}(t_k, x_{j+1})))/2$, does not preserve the monotonicity property and has unstable behavior for the problem considered.

The algebraic equations arising in the implementation of the methods (6.7), (6.10), and (6.13) at each step are quadratic ones and are solved exactly. The results of testing of the three-layer method (6.10) are discussed in Sect. 7.6.2.

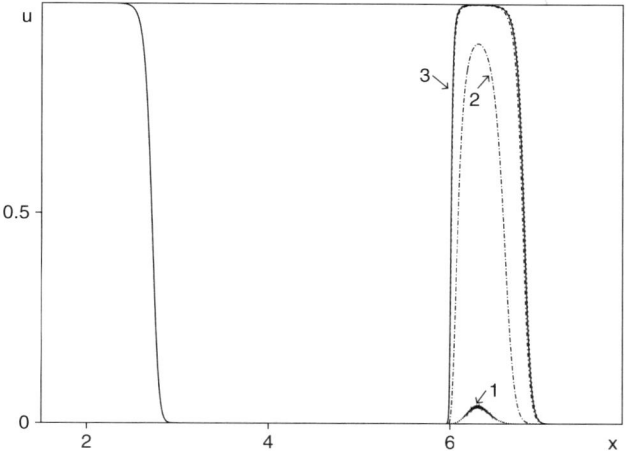

Fig. 7.3. The FKPP-equation. Comparison of the methods. The curve 3 is simulated by (6.7) and (6.13) with $h = 0.0005$ and $h = 0.0001$, and it visually coincides with the exact solution. The curves 1 and 2 are simulated by (7.14) and (6.6) with $h = 0.0001$. Here $t = 5.75$, and the other parameter values are the same as in Fig. 7.1.

Remark 7.1. Derivatives of the solution to (7.8)-(7.10) can be estimated as

$$\left|\frac{\partial^{i+j}u}{\partial t^i \partial x^j}\right| \leq \frac{K}{\varepsilon^{2(i+j)}}, \quad t \in [t_0, T], \ x \in \mathbf{R}, \ 0 < \varepsilon \leq \varepsilon^*. \tag{7.15}$$

One can prove that under (7.15) and a sufficiently small h/ε^2 the errors of both methods (6.5) and (6.6) are estimated as

$$|u(t_k, x) - \bar{u}(t_k, x)| \leq K\frac{h}{\varepsilon^4},$$

and the error of the method (6.7) is estimated as

$$|u(t_k, x) - \bar{u}(t_k, x)| \leq K\frac{h^2}{\varepsilon^6},$$

where the constant K does not depend on x, k, h, ε.

8 Numerical solution of the nonlinear Dirichlet and Neumann problems based on the probabilistic approach

In this chapter we apply the probabilistic approach of Chap. 7 to nonlinear problems with Dirichlet and Neumann boundary conditions [210,212]. We recall that the probabilistic approach is based on making use of the well-known probabilistic representations of solutions to linear parabolic equations and the ideas of weak sense numerical integration of SDEs. Despite their probabilistic nature these methods are nevertheless deterministic. The probabilistic approach takes into account a coefficient dependence on the space variables and a relationship between diffusion and advection in an intrinsic manner. In particular, the layer methods allow us to avoid difficulties stemming from essentially changing coefficients and strong advection.

Sections 8.1-8.3 are devoted to the nonlinear Dirichlet problems while Sects. 8.4-8.6 deal with the nonlinear Neumann problems.

8.1 Layer methods for the Dirichlet problem for semilinear parabolic equations

Let G be a bounded domain in $\mathbf{R}^d$, $Q = [t_0, T) \times G$ be a cylinder in $\mathbf{R}^{d+1}$, $\Gamma = \overline{Q} \setminus Q$. The set Γ is a part of the boundary of the cylinder Q consisting of the upper base and the lateral surface. Consider the Dirichlet problem for the semilinear parabolic equation

$$\frac{\partial u}{\partial t} + \frac{1}{2}\sum_{i,j=1}^{d} a^{ij}(t,x,u)\frac{\partial^2 u}{\partial x^i \partial x^j} + \sum_{i=1}^{d} b^i(t,x,u)\frac{\partial u}{\partial x^i}$$
$$+g(t,x,u) = 0, \ (t,x) \in Q, \quad (1.1)$$

$$u(t,x)|_\Gamma = \varphi(t,x). \quad (1.2)$$

The form of equation (1.1) is relevant to a probabilistic approach, i.e., the equation is considered under $t < T$, and the "initial" conditions are prescribed at $t = T$. Using the well-known probabilistic representation of the solution to (1.1)-(1.2) (see [48, 70]), we get

$$u(t,x) = E(\varphi(\tau, X_{t,x}(\tau)) + Z_{t,x,0}(\tau)). \quad (1.3)$$

462   8 Numerical solution of the nonlinear Dirichlet and Neumann problems

In (1.3), $X_{t,x}(s)$, $Z_{t,x,z}(s)$, $(t,x) \in Q$, $s \geq t$, is the solution of the Cauchy problem for the Ito system of SDEs

$$dX = b(s, X, u(s, X))\, ds + \sigma(s, X, u(s, X))\, dw(s)\,, \quad X(t) = x\,,$$

$$dZ = g(s, X, u(s, X))\, ds\,, \quad Z(t) = z\,, \qquad (1.4)$$

where $w(s) = (w^1(s), \ldots, w^d(s))^\top$ is a standard Wiener process, $b(s,x,u) = (b^1(s,x,u), \ldots, b^d(s,x,u))^\top$ is a column vector, the matrix $\sigma = \sigma(s,x,u)$ is obtained from the equation

$$\sigma \sigma^\top = a\,, \quad \sigma = \{\sigma^{ij}(s,x,u)\}\,, \quad a = \{a^{ij}(s,x,u)\}\,,$$

and $\tau = \tau_{t,x}$ is the first exit time of the trajectory $(s, X_{t,x}(s))$ from the domain Q.

If the equation (1.1) is linear, the system (1.4) does not contain the unknown function $u(s,x)$ and therefore one can use weak approximation schemes for solving (1.4) with the Monte Carlo realization of representation (1.3). Some constructive schemes are presented in Chap. 6. The procedures of Chap. 6 together with the Monte Carlo approach allow us to find a value $u(t,x)$ at a single point even when the domain G has high dimension.

Of course, the nonlinear case is much more complicated. But we are aimed to construct layer methods and due to this fact it becomes possible to use a one-step (local) version of the representation (1.3) (see formula (1.7) below). Introduce a time discretization, for definiteness the equidistant one:

$$T = t_N > t_{N-1} > \cdots > t_0, \quad h := \frac{T - t_0}{N}.$$

The methods proposed in this chapter give an approximation $\bar{u}(t_k, x)$ of the solution $u(t_k, x)$, $k = N, \ldots, 0$, $x \in \overline{G}$, i.e., step by step everywhere in the domain $\overline{G}$. It is feasible if the dimension of the domain G is comparatively small ($d \leq 3$). To construct the layer methods, we exploit the ideas of weak sense numerical integration of SDEs in bounded domain and obtain some approximate relations on the basis of (1.7), (1.4). The relations allow us to express $\bar{u}(t_k, x)$, $k = N-1, \ldots, 0$, recurrently in terms of $\bar{u}(t_{k+1}, x)$. Despite their probabilistic nature these methods turn out to be deterministic as in the previous chapter.

In Sects. 8.1.1-8.1.3, we derive layer methods relying on the numerical integration of SDEs and prove their convergence using both deterministic and probabilistic type arguments. To realize layer methods in practice, we need a discretization in the variable x with some kind of interpolation at every step to turn an applied method into an algorithm. Such numerical algorithms are constructed in Sect. 8.1.4. A majority of ideas can be demonstrated at $d = 1$, and we restrict ourselves to this case in the current section. The case $d \geq 2$ is shortly discussed in Sect. 8.2. Numerical tests are presented in Sect. 8.3 where, in particular, we compare the proposed layer methods with the well-known finite-difference schemes.

8.1.1 Construction of a layer method of first order

The boundary value problem (1.1)-(1.2) in the one-dimensional case has the form

$$\frac{\partial u}{\partial t} + \frac{1}{2}\sigma^2(t,x,u)\frac{\partial^2 u}{\partial x^2} + b(t,x,u)\frac{\partial u}{\partial x} + g(t,x,u) = 0, \quad (t,x) \in Q, \quad (1.5)$$

$$u(t,x)|_\Gamma = \varphi(t,x). \quad (1.6)$$

In this case Q is the partly open rectangle: $Q = [t_0, T) \times (\alpha, \beta)$, and Γ consists of the upper base $\{T\} \times [\alpha, \beta]$ and two vertical intervals: $[t_0, T) \times \{\alpha\}$ and $[t_0, T) \times \{\beta\}$. We assume that $\sigma(t, x, u) \geq \sigma_* > 0$ for $(t, x) \in \bar{Q}$, $-\infty < u < \infty$.

Let $u = u(t, x)$ be the solution to problem (1.5)-(1.6), which is supposed to exist, to be unique, and to be sufficiently smooth. One can find many theoretical results on this topic in [91, 143, 259, 269, 291] (see also references therein).

Analogously to (1.3), we have

$$u(t_k, x) = E(u(\vartheta_{t_k,x}, X_{t_k,x}(\vartheta_{t_k,x})) + Z_{t_k,x,0}(\vartheta_{t_k,x})), \quad (1.7)$$

where $\vartheta_{t_k,x} = \vartheta_{t_k,x}(t_{k+1}) := \tau_{t_k,x} \wedge t_{k+1}$, and X, Z satisfy system (1.4).

Let us suppose for a while that it is possible to extend the coefficients of the equation (1.5) so that the new equation has a solution $u(t, x)$ on $[t_0, T) \times \mathbf{R}$ which is an extension of the solution to the boundary value problem (1.5)-(1.6). Then, instead of (1.7), we obtain (we suppose the layer $u(t_{k+1}, x)$ to be known)

$$u(t_k, x) = E[u(t_{k+1}, X_{t_k,x}(t_{k+1})) + Z_{t_k,x,0}(t_{k+1})]. \quad (1.8)$$

Approximation for internal points. Applying the explicit weak Euler scheme with the simplest simulation of noise to system (1.4), we get

$$\bar{X}_{t_k,x}(t_{k+1}) = x + h\,b(t_k, x, u(t_k, x)) + \sqrt{h}\,\sigma(t_k, x, u(t_k, x))\,\xi, \quad (1.9)$$

$$\bar{Z}_{t_k,x,0}(t_{k+1}) = hg(t_k, x, u(t_k, x)), \quad (1.10)$$

where the ξ is distributed by the law: $P(\xi = \pm 1) = 1/2$.

Using (1.8), we get to within $O(h^2)$:

$$\begin{aligned}
u(t_k, x) &\simeq E[u(t_{k+1}, \bar{X}_{t_k,x}(t_{k+1})) + \bar{Z}_{t_k,x,0}(t_{k+1})] \\
&= \frac{1}{2}u(t_{k+1}, x + hb(t_k, x, u(t_k, x)) - \sqrt{h}\sigma(t_k, x, u(t_k, x))) \\
&\quad + \frac{1}{2}u(t_{k+1}, x + hb(t_k, x, u(t_k, x)) + \sqrt{h}\sigma(t_k, x, u(t_k, x))) \\
&\quad + hg(t_k, x, u(t_k, x)). \quad (1.11)
\end{aligned}$$

464   8 Numerical solution of the nonlinear Dirichlet and Neumann problems

Now we can obtain an implicit relation for an approximation of $u(t_k, x)$. Applying the method of simple iteration to the implicit relation and taking $u(t_{k+1}, x)$ as a null iteration, we get the following explicit one-step approximation $v(t_k, x)$ of $u(t_k, x)$:

$$v(t_k, x) = \frac{1}{2}u(t_{k+1}, x+hb_k-\sqrt{h}\sigma_k) + \frac{1}{2}u(t_{k+1}, x+hb_k+\sqrt{h}\sigma_k) + hg_k, \quad (1.12)$$

where b_k, σ_k, g_k are the coefficients $b(t, x, u)$, $\sigma(t, x, u)$, $g(t, x, u)$ calculated at the point $(t_k, x, u(t_{k+1}, x))$.

But in reality we know the layer $u(t_{k+1}, x)$ for $\alpha \leq x \leq \beta$ only. At the same time the argument $x + hb_k - \sqrt{h}\sigma_k$ for x close to α is less than α and the argument $x + hb_k + \sqrt{h}\sigma_k$ for x close to β is more than β. Thus, we need to extend the layer $u(t_{k+1}, x)$ in a constructive manner.

Approximation for points near the boundary. Using the explicit weak Euler scheme for the initial point (t, α) with $t_k \leq t \leq t_{k+1}$, we put (cf. (1.9)-(1.10)):

$$\bar{X}_{t,\alpha}(t_{k+1}) = x + b(t, \alpha, u(t, \alpha)) \times (t_{k+1} - t) + \sigma(t, \alpha, u(t, \alpha))\sqrt{t_{k+1} - t}\,\xi\,,$$

$$\bar{Z}_{t,\alpha,0}(t_{k+1}) = g(t, \alpha, u(t, \alpha)) \times (t_{k+1} - t)\,. \quad (1.13)$$

Analogously, we define $\bar{X}_{t,\beta}(t_{k+1})$, $\bar{Z}_{t,\beta,0}(t_{k+1})$.

We have (see (1.11) and (1.13)) for $t_k \leq t \leq t_{k+1}$:

$$u(t, \alpha) \simeq E[u(t_{k+1}, \bar{X}_{t,\alpha}(t_{k+1})) + \bar{Z}_{t,\alpha,0}(t_{k+1})]$$

$$= \frac{1}{2}u(t_{k+1}, \alpha + b(t, \alpha, u(t, \alpha)) \times (t_{k+1} - t) - \sigma(t, \alpha, u(t, \alpha))\sqrt{t_{k+1} - t})$$

$$+ \frac{1}{2}u(t_{k+1}, \alpha + b(t, \alpha, u(t, \alpha)) \times (t_{k+1} - t) + \sigma(t, \alpha, u(t, \alpha))\sqrt{t_{k+1} - t})$$

$$+ g(t, \alpha, u(t, \alpha)) \times (t_{k+1} - t)\,. \quad (1.14)$$

If we replace (recall that $u(t, \alpha) = \varphi(t, \alpha)$ due to (1.6)) the argument $(t, \alpha, u(t, \alpha)) = (t, \alpha, \varphi(t, \alpha))$ by $(t_k, \alpha, \varphi(t_{k+1}, \alpha))$, the right-hand side of (1.14) is changed by a quantity of order $O(h^2)$. Since the approximation in (1.14) is also of order $O(h^2)$, we get

$$\frac{1}{2}u(t_{k+1}, \alpha + b(t_k, \alpha, \varphi(t_{k+1}, \alpha)) \times (t_{k+1} - t)$$
$$-\sigma(t_k, \alpha, \varphi(t_{k+1}, \alpha))\sqrt{t_{k+1} - t})$$
$$= \varphi(t, \alpha) - \frac{1}{2}u(t_{k+1}, \alpha + b(t_k, \alpha, \varphi(t_{k+1}, \alpha)) \times (t_{k+1} - t)$$
$$+\sigma(t_k, \alpha, \varphi(t_{k+1}, \alpha))\sqrt{t_{k+1} - t})$$
$$-g(t_k, \alpha, \varphi(t_{k+1}, \alpha)) \times (t_{k+1} - t) + O(h^2)\,. \quad (1.15)$$

8.1 Layer methods for the Dirichlet problem

Introduce
$$\alpha_0 := \alpha + hb(t_k, \alpha, \varphi(t_{k+1}, \alpha)) - \sqrt{h}\sigma(t_k, \alpha, \varphi(t_{k+1}, \alpha)).$$
Clearly $\alpha_0 < \alpha$ and
$$\alpha_0 \leq \alpha + b(t_k, \alpha, \varphi(t_{k+1}, \alpha)) \times (t_{k+1} - t)$$
$$-\sigma(t_k, \alpha, \varphi(t_{k+1}, \alpha))\sqrt{t_{k+1} - t} \leq \alpha$$
for $t_k \leq t \leq t_{k+1}$ under a sufficiently small h.

Analogously
$$\frac{1}{2}u(t_{k+1}, \beta + b(t_k, \beta, \varphi(t_{k+1}, \beta)) \times (t_{k+1} - t)$$
$$+\sigma(t_k, \beta, \varphi(t_{k+1}, \beta))\sqrt{t_{k+1} - t})$$
$$= \varphi(t, \beta) - \frac{1}{2}u(t_{k+1}, \beta + b(t_k, \beta, \varphi(t_{k+1}, \beta)) \times (t_{k+1} - t)$$
$$-\sigma(t_k, \beta, \varphi(t_{k+1}, \beta))\sqrt{t_{k+1} - t})$$
$$-g(t_k, \beta, \varphi(t_{k+1}, \beta)) \times (t_{k+1} - t) + O(h^2),$$
$$\beta_0 := \beta + hb(t_k, \beta, \varphi(t_{k+1}, \beta)) + \sqrt{h}\sigma(t_k, \beta, \varphi(t_{k+1}, \beta)). \quad (1.16)$$

The relations (1.15)-(1.16) give the desired extension of the function $u(t_{k+1}, x)$ on the interval $[\alpha_0, \beta_0]$.

Let us return to the formula (1.12) now. The arguments $x + hb_k - \sqrt{h}\sigma_k$ and $x + hb_k + \sqrt{h}\sigma_k$ are monotone increasing functions in $x \in [\alpha, \beta]$ for a sufficiently small h. Their values belong to $[\alpha_0, \beta_0]$, and $x + hb_k + \sqrt{h}\sigma_k$ is always (for $x \in [\alpha, \beta]$) more than α while $x + hb_k - \sqrt{h}\sigma_k$ is less than β. Let $x + hb_k - \sqrt{h}\sigma_k < \alpha$ (clearly it is possible for x close to α). Due to the above, there exists a unique root $\gamma_k(x)$, $0 < \gamma_k(x) \leq 1$, of the quadratic equation
$$\alpha + \gamma_k h\, b(t_k, \alpha, \varphi(t_{k+1}, \alpha)) - \sqrt{\gamma_k h}\sigma(t_k, \alpha, \varphi(t_{k+1}, \alpha))$$
$$= x + hb_k - \sqrt{h}\sigma_k. \quad (1.17)$$

Analogously, if $x + hb_k + \sqrt{h}\sigma_k > \beta$, then there exists a unique root $\delta_k(x)$, $0 < \delta_k(x) \leq 1$, of the quadratic equation
$$\beta + \delta_k h\, b(t_k, \beta, \varphi(t_{k+1}, \beta)) + \sqrt{\delta_k h}\sigma(t_k, \beta, \varphi(t_{k+1}, \beta))$$
$$= x + hb_k + \sqrt{h}\sigma_k. \quad (1.18)$$

If, for instance, $x + hb_k - \sqrt{h}\sigma_k < \alpha$, then one can replace the value $u(t_{k+1}, x + hb_k - \sqrt{h}\sigma_k)/2$ in (1.12) by the value due to (1.17) and (1.15):
$$\frac{1}{2}u(t_{k+1}, x + hb_k - \sqrt{h}\sigma_k)$$
$$= \frac{1}{2}u(t_{k+1}, \alpha + \gamma_k hb(t_k, \alpha, \varphi(t_{k+1}, \alpha)) - \sqrt{\gamma_k h}\sigma(t_k, \alpha, \varphi(t_{k+1}, \alpha)))$$
$$\approx \varphi(t_{k+1-\gamma_k}, \alpha)$$
$$-\frac{1}{2}\gamma_k h\, u(t_{k+1}, \alpha + b(t_k, \alpha, \varphi(t_{k+1}, \alpha)) + \sqrt{\gamma_k h}\sigma(t_k, \alpha, \varphi(t_{k+1}, \alpha)))$$
$$-\gamma_k h\, g(t_k, \alpha, \varphi(t_{k+1}, \alpha)),$$
where $t_{k+1-\gamma_k} = t_k + h(1 - \gamma_k)$.

8 Numerical solution of the nonlinear Dirichlet and Neumann problems

The layer method. As a result, we obtain the following one-step approximation $v(t_k, x)$ for $u(t_k, x)$:

$$v(t_k, x) = \frac{1}{2}u(t_{k+1}, x + hb_k - \sqrt{h}\sigma_k) + \frac{1}{2}u(t_{k+1}, x + hb_k + \sqrt{h}\sigma_k)$$
$$+ hg_k, \text{ if } x + hb_k \pm \sqrt{h}\sigma_k \in [\alpha, \beta];$$

$$v(t_k, x) = \varphi(t_{k+1-\gamma_k}, \alpha) - \gamma_k hg(t_k, \alpha, \varphi(t_{k+1}, \alpha))$$
$$-\frac{1}{2}u(t_{k+1}, \alpha + \gamma_k hb(t_k, \alpha, \varphi(t_{k+1}, \alpha)) + \sqrt{\gamma_k h}\sigma(t_k, \alpha, \varphi(t_{k+1}, \alpha)))$$
$$+\frac{1}{2}u(t_{k+1}, x + hb_k + \sqrt{h}\sigma_k) + hg_k, \text{ if } x + hb_k - \sqrt{h}\sigma_k < \alpha;$$

$$v(t_k, x) = \frac{1}{2}u(t_{k+1}, x + hb_k - \sqrt{h}\sigma_k) + \varphi(t_{k+1-\delta_k}, \beta)$$
$$-\delta_k h\, g(t_k, \beta, \varphi(t_{k+1}, \beta)) - \frac{1}{2}u(t_{k+1}, \beta + \delta_k h\, b(t_k, \beta, \varphi(t_{k+1}, \beta)))$$
$$-\sqrt{\delta_k h}\sigma(t_k, \beta, \varphi(t_{k+1}, \beta))) + hg_k, \text{ if } x + hb_k + \sqrt{h}\sigma_k > \beta, \quad (1.19)$$
$$k = N-1, \ldots, 1, 0,$$

where b_k, σ_k, g_k are the coefficients $b(t, x, u)$, $\sigma(t, x, u)$, $g(t, x, u)$ calculated at the point $(t_k, x, u(t_{k+1}, x))$ and γ_k, δ_k are the corresponding roots of the equations (1.17) and (1.18).

Thus *the layer method* acquires the form

$$\bar{u}(t_N, x) = \varphi(t_N, x), \quad x \in [\alpha, \beta],$$

$$\bar{u}(t_k, x) = \frac{1}{2}\bar{u}(t_{k+1}, x + h\bar{b}_k - \sqrt{h}\bar{\sigma}_k) + \frac{1}{2}\bar{u}(t_{k+1}, x + h\bar{b}_k + \sqrt{h}\bar{\sigma}_k) + h\bar{g}_k,$$
$$\text{if } x + h\bar{b}_k \pm \sqrt{h}\bar{\sigma}_k \in [\alpha, \beta];$$

$$\bar{u}(t_k, x) = \varphi(t_{k+1-\bar{\gamma}_k}, \alpha) - \bar{\gamma}_k hg(t_k, \alpha, \varphi(t_{k+1}, \alpha))$$
$$-\frac{1}{2}\bar{u}(t_{k+1}, \alpha + \bar{\gamma}_k hb(t_k, \alpha, \varphi(t_{k+1}, \alpha)) + \sqrt{\bar{\gamma}_k h}\sigma(t_k, \alpha, \varphi(t_{k+1}, \alpha)))$$
$$+\frac{1}{2}\bar{u}(t_{k+1}, x + h\bar{b}_k + \sqrt{h}\bar{\sigma}_k) + h\bar{g}_k, \text{ if } x + h\bar{b}_k - \sqrt{h}\bar{\sigma}_k < \alpha;$$

$$\bar{u}(t_k, x) = \frac{1}{2}\bar{u}(t_{k+1}, x + h\bar{b}_k - \sqrt{h}\bar{\sigma}_k) + \varphi(t_{k+1-\bar{\delta}_k}, \beta)$$
$$-\bar{\delta}_k hg(t_k, \beta, \varphi(t_{k+1}, \beta)) - \frac{1}{2}\bar{u}(t_{k+1}, \beta + \bar{\delta}_k hb(t_k, \beta, \varphi(t_{k+1}, \beta)))$$
$$-\sqrt{\bar{\delta}_k h}\sigma(t_k, \beta, \varphi(t_{k+1}, \beta))) + h\bar{g}_k, \text{ if } x + h\bar{b}_k + \sqrt{h}\bar{\sigma}_k > \beta; \quad (1.20)$$

8.1 Layer methods for the Dirichlet problem   467

$$k = N-1, \ldots, 1, 0,$$

where $\bar{b}_k$, $\bar{\sigma}_k$, $\bar{g}_k$ are the coefficients $b(t, x, u)$, $\sigma(t, x, u)$, $g(t, x, u)$ calculated at the point $(t_k, x, \bar{u}(t_{k+1}, x))$ and $\bar{\gamma}_k$, $\bar{\delta}_k$ are the corresponding roots of the equations (1.17) and (1.18) with the right-hand sides $x + h\bar{b}_k - \sqrt{h}\bar{\sigma}_k$ and $x + h\bar{b}_k + \sqrt{h}\bar{\sigma}_k$.

The method (1.20) is an explicit layer method for solving the Dirichlet problem (1.5)-(1.6). This method is deterministic, even though it is constructed by a probabilistic approach. The method is of the first order of convergence with respect to h (see Theorem 1.3 below).

Remark 1.1. Let us briefly discuss some differences between the layer methods obtained here and the well-known finite-difference methods. Finite-difference methods also allow us to express an approximate solution on the layer $t = t_k$ recurrently in terms of the solution on the layer $t = t_{k+1}$. For their construction, both the time step Δt and the space step Δx are used. Moreover, the knots of the layer $t = t_{k+1}$ used to evaluate $\bar{u}(t_k, x_j)$ are definitely prescribed. In our methods we use the time step h only, and the points from the layer $t = t_{k+1}$ to evaluate $\bar{u}(t_k, x)$ arise automatically. A location of these points depends on the coefficients of the problem considered and on the weak scheme chosen. As a result, the probabilistic approach takes into account a coefficient dependence on the space variables and a relationship between diffusion and advection in an intrinsic manner. In particular, the layer methods allow us to avoid difficulties stemming from essentially changing coefficients and strong advection. We should also note that the probabilistic approach gives a natural way to derive a lot of various new methods.

8.1.2 Convergence theorem

We make the following assumptions.

(i) There exists a unique solution $u(t, x)$ of problem (1.5)-(1.6) such that

$$u_\circ < u_* \leq u(t, x) \leq u^* < u^\circ, \ t_0 \leq t \leq T, \ x \in [\alpha, \beta], \qquad (1.21)$$

where $u_\circ$, u_*, u^*, u° are some constants, and there exist uniformly bounded derivatives:

$$\left| \frac{\partial^{i+j} u}{\partial t^i \partial x^j} \right| \leq K, \ i = 0, \ j = 1, 2, 3, 4; \ i = 1, \ j = 0, 1, 2; \ i = 2, \ j = 0; \qquad (1.22)$$

$$t_0 \leq t \leq T, \ x \in [\alpha, \beta].$$

(ii) The coefficients $b(t, x, u)$, $\sigma(t, x, u)$, $g(t, x, u)$ and their first and second derivatives in x and u are uniformly bounded:

$$\left| \frac{\partial^{i+j} b}{\partial x^i \partial u^j} \right| \leq K, \ \left| \frac{\partial^{i+j} \sigma}{\partial x^i \partial u^j} \right| \leq K, \ \left| \frac{\partial^{i+j} g}{\partial x^i \partial u^j} \right| \leq K, \ 0 \leq i+j \leq 2, \qquad (1.23)$$

$$t_0 \le t \le T, \ x \in [\alpha, \beta], \ u_\circ < u < u^\circ.$$

The lemma on the one-step error $\rho(t_k, x)$ of method (1.20) and the global convergence theorem are proved analogously to Lemma 7.2.1 and Theorem 7.2.2, respectively (see the complete proofs in [210]).

Lemma 1.2. *Under the assumptions (i) and (ii), the one-step error $\rho(t_k, x)$ of the method (1.20) has the second order of smallness with respect to h, i.e.,*

$$|\rho(t_k, x)| = |v(t_k, x) - u(t_k, x)| \le Ch^2,$$

where $v(t_k, x)$ is defined by (1.19), C does not depend on h, k, x.

Theorem 1.3. *Under the assumptions (i) and (ii), the method (1.20) has the first order of convergence with respect to h, i.e.,*

$$|\bar{u}(t_k, x) - u(t_k, x)| \le Kh,$$

where K does not depend on h, k, x.

8.1.3 A layer method with a simpler approximation near the boundary

Without exploiting the idea used above of involving the points outside the interval $[\alpha, \beta]$ while constructing a layer method, it is possible to get a layer method that is simpler but with a larger one-step error near the boundary than (1.20) (see Lemma 1.4 below). Let us note that in spite of the greater one-step boundary error the global error of this method will be $O(h)$ again (see Theorem 1.5). Here we approximate the solution $u(t_k, x)$, when the point x is close to α (or β), using values of the solution at a point $(t_{k+\lambda_k}, \alpha)$ with some $\lambda_k \in (0, 1)$ (or at a point $(t_{k+\mu_k}, \beta)$ with some $\mu_k \in (0, 1)$) and at the point $(t_{k+1}, x + h\bar{b}_k + \sqrt{h}\bar{\sigma}_k)$ (or $(t_{k+1}, x + h\bar{b}_k - \sqrt{h}\bar{\sigma}_k)$) with some (positive) weights. These two weights may be interpreted as probabilities of reaching and not reaching of α (or β). The method obtained on this way has the form

$$\bar{u}(t_N, x) = \varphi(t_N, x), \ x \in [\alpha, \beta],$$

$$\bar{u}(t_k, x) = \frac{1}{2}\bar{u}(t_{k+1}, x + h\bar{b}_k - \sqrt{h}\bar{\sigma}_k) + \frac{1}{2}\bar{u}(t_{k+1}, x + h\bar{b}_k + \sqrt{h}\bar{\sigma}_k) + h\bar{g}_k,$$
$$\text{if } x + h\bar{b}_k \pm \sqrt{h}\bar{\sigma}_k \in [\alpha, \beta];$$

$$\bar{u}(t_k, x) = \frac{1}{1+\sqrt{\bar{\lambda}_k}}\varphi(t_{k+\bar{\lambda}_k}, \alpha) + \frac{\sqrt{\bar{\lambda}_k}}{1+\sqrt{\bar{\lambda}_k}}\bar{u}(t_{k+1}, x + h\bar{b}_k + \sqrt{h}\bar{\sigma}_k)$$
$$+ \sqrt{\bar{\lambda}_k}h\bar{g}_k, \text{ if } x + h\bar{b}_k - \sqrt{h}\bar{\sigma}_k < \alpha;$$

8.1 Layer methods for the Dirichlet problem

$$\bar{u}(t_k, x) = \frac{1}{1+\sqrt{\bar{\mu}_k}} \varphi(t_{k+\bar{\mu}_k}, \beta) + \frac{\sqrt{\bar{\mu}_k}}{1+\sqrt{\bar{\mu}_k}} \bar{u}(t_{k+1}, x + h\bar{b}_k - \sqrt{h}\bar{\sigma}_k)$$
$$+ \sqrt{\bar{\mu}_k} h \bar{g}_k, \text{ if } x + h\bar{b}_k + \sqrt{h}\bar{\sigma}_k > \beta; \quad k = N-1, \ldots, 1, 0, \quad (1.24)$$

where $\bar{b}_k$, $\bar{\sigma}_k$, $\bar{g}_k$ are the coefficients $b(t, x, u)$, $\sigma(t, x, u)$, $g(t, x, u)$ calculated at the point $(t_k, x, \bar{u}(t_{k+1}, x))$ and $0 < \bar{\lambda}_k, \bar{\mu}_k < 1$ are roots of the quadratic equations (it is not difficult to verify that the roots exist and are unique)

$$\alpha = x + \bar{\lambda}_k h \bar{b}_k - \sqrt{\bar{\lambda}_k h}\, \bar{\sigma}_k, \ \beta = x + \bar{\mu}_k h \bar{b}_k + \sqrt{\bar{\mu}_k h}\, \bar{\sigma}_k.$$

This method involves one value of the function $\varphi(t, x)$ and one value of the approximate solution $\bar{u}(t_{k+1}, y)$ on the previous layer in contrast to the method (1.20) which requires evaluating one value of the function $\varphi(t, x)$ and two values of the approximate solution $\bar{u}(t_{k+1}, y)$ on the previous layer. The method (1.24) is based on Algorithm 6.2.9 for linear Dirichlet problems.

Lemma 1.4. *Under the assumptions* (i) *and* (ii), *the one-step error* $\rho(t_k, x)$ *of the method* (1.24) *is estimated as*

$$|\rho(t_k, x)| \leq Ch^2 \text{ if } x + hb_k \pm \sqrt{h}\sigma_k \in [\alpha, \beta];$$

$$|\rho(t_k, x)| \leq Ch^{3/2} \text{ if } x + hb_k - \sqrt{h}\sigma_k < \alpha \text{ or } x + hb_k + \sqrt{h}\sigma_k > \beta.$$

The proof is analogous to that of Lemma 7.2.1 (see also the proof of Lemma 1.2 in [210]). The following convergence theorem for the method (1.24) takes place.

Theorem 1.5. *Under the assumptions* (i) *and* (ii), *the method* (1.24) *has the global error estimated as*

$$|\bar{u}(t_k, x) - u(t_k, x)| \leq Kh, \quad (1.25)$$

where K does not depend on h, k, x.

Proof. If we followed the deterministic way of proving Theorem 7.2.2 and Theorem 1.3, we would get that the global error of method (1.24) is $O(\sqrt{h})$. To prove the estimate (1.25), we exploit ideas of proving convergence theorems for probabilistic methods solving linear boundary value problems from Chap. 6 (cf. Theorems 6.2.11 and 6.2.4).

To this end, in connection with the layer method (1.24), we introduce the Markov chain (ϑ_i, X_i), $i \geq k$, $(\vartheta_k, X_k) = (t_k, x) \in \bar{Q}$, which stops on Γ at a random moment $\varkappa \leq N$. For $(\vartheta_i, X_i) \notin \Gamma$, we define

$$X_{i+1}^{\pm} := X_i + h\bar{b}_i \pm h^{1/2}\bar{\sigma}_i.$$

If $X_{i+1}^{\pm} \in [\alpha, \beta]$ then $\vartheta_{i+1} = \vartheta_i + h$ and X_{i+1} takes values X_{i+1}^{-} or X_{i+1}^{+} with

$$P\{(\vartheta_{i+1}, X_{i+1}) = (\vartheta_i + h, X_{i+1}^-)\}$$
$$= P\{(\vartheta_{i+1}, X_{i+1}) = (\vartheta_i + h, X_{i+1}^+)\} = \frac{1}{2};$$

if $X_{i+1}^- < \alpha$ then

$$P\{(\vartheta_{i+1}, X_{i+1}) = (\vartheta_i + \bar{\lambda}_i h, \alpha)\} = \frac{1}{1+\sqrt{\bar{\lambda}_i}},$$

$$P\{(\vartheta_{i+1}, X_{i+1}) = (\vartheta_i + h, X_{i+1}^+)\} = \frac{\sqrt{\bar{\lambda}_i}}{1+\sqrt{\bar{\lambda}_i}};$$

if $X_{i+1}^+ > \beta$ then

$$P\{(\vartheta_{i+1}, X_{i+1}) = (\vartheta_i + h, X_{i+1}^-)\} = \frac{\sqrt{\bar{\mu}_i}}{1+\sqrt{\bar{\mu}_i}},$$

$$P\{(\vartheta_{i+1}, X_{i+1}) = (\vartheta_i + \bar{\mu}_i h, \beta)\} = \frac{1}{1+\sqrt{\bar{\mu}_i}}.$$

Here $\bar{b}_i = b(\vartheta_i, X_i, \bar{u}(\vartheta_i + h, X_i))$, $\bar{\sigma}_i = \sigma(\vartheta_i, X_i, \bar{u}(\vartheta_i + h, X_i))$, $\bar{u}(t_k, x)$ is considered to be known from (1.24), and $0 < \bar{\lambda}_i, \bar{\mu}_i < 1$ are roots of the quadratic equations

$$\alpha = X_i + \bar{\lambda}_i h \bar{b}_i - \bar{\sigma}_i \sqrt{\bar{\lambda}_i h}, \quad \beta = X_i + \bar{\mu}_i h \bar{b}_i + \sqrt{\bar{\mu}_i h} \bar{\sigma}_i. \tag{1.26}$$

If $(\vartheta_i, X_i) \in \Gamma$, the Markov chain stops and $\varkappa = i$. Let us note that ϑ_i coincides with t_i except, may be, the last moment $\varkappa$.

Now introduce the random sequence Z_i, $Z_k = 0$:

if $X_{i+1}^\pm \in [\alpha, \beta]$, then $Z_{i+1} = Z_i + h\bar{g}_i$;

if $X_{i+1}^- < \alpha$, then $Z_{i+1} = Z_i + \sqrt{\bar{\lambda}_i} h \bar{g}_i$;

if $X_{i+1}^+ > \beta$, then $Z_{i+1} = Z_i + \sqrt{\bar{\mu}_i} h \bar{g}_i$; $i = k, \ldots, \varkappa - 1$,

where $\bar{g}_i = g(\vartheta_i, X_i, \bar{u}(\vartheta_i + h, X_i))$.

Define the boundary layer $\partial \Gamma \in \overline{Q}$: for all the points $(t_k, x) \in \overline{Q} \setminus \partial \Gamma$ both points $x + h\bar{b}_k \pm h^{1/2} \bar{\sigma}_k$ belong to $[\alpha, \beta]$. Clearly, for the points $(t_k, x) \in \partial \Gamma$ either $x + h\bar{b}_k - h^{1/2} \bar{\sigma}_k \notin [\alpha, \beta]$ or $x + h\bar{b}_k + h^{1/2} \bar{\sigma}_k \notin [\alpha, \beta]$.

It is not difficult to show that the mean of the number of steps $\nu(t_k, x)$, which the Markov chain (ϑ_i, X_i), $i = k, \ldots, \varkappa$, $\vartheta_k = t_k$, $X_k = x$, spends in the layer $\partial \Gamma$ is estimated as

$$E\nu(t_k, x) \leq H, \tag{1.27}$$

where H does not depend on h, k, x (cf. Lemma 6.2.2).

8.1 Layer methods for the Dirichlet problem

One can see that

$$\bar{u}(t_k, x) = E\left[\bar{u}(\vartheta_\varkappa, X_\varkappa) + Z_\varkappa\right] = E\left[\varphi(\vartheta_\varkappa, X_\varkappa) + Z_\varkappa\right] = E\left[u(\vartheta_\varkappa, X_\varkappa) + Z_\varkappa\right].$$

We have $R(t_N, x) := \bar{u}(t_N, x) - u(t_N, x) = 0$ and for $k = N-1, \ldots, 0$:

$$R(t_k, x)$$

$$:= \bar{u}(t_k, x) - u(t_k, x) = E \sum_{i=k}^{\varkappa-1} [u(\vartheta_{i+1}, X_{i+1}) - u(\vartheta_i, X_i) + Z_{i+1} - Z_i]$$

$$= E \sum_{i=k}^{\varkappa-1} I_{\overline{Q}\setminus\partial\Gamma}(\vartheta_i, X_i) [u(\vartheta_{i+1}, X_{i+1}) - u(\vartheta_i, X_i) + Z_{i+1} - Z_i]$$

$$+ E \sum_{i=k}^{\varkappa-1} I_{\partial\Gamma}(\vartheta_i, X_i) [u(\vartheta_{i+1}, X_{i+1}) - u(\vartheta_i, X_i) + Z_{i+1} - Z_i]$$

$$= \sum_{i=k}^{N-1} EE(\chi_{\varkappa>i} I_{\overline{Q}\setminus\partial\Gamma}(\vartheta_i, X_i)[u(\vartheta_{i+1}, X_{i+1}) - u(\vartheta_i, X_i)$$

$$+ Z_{i+1} - Z_i]/X_i, Z_i) + \sum_{i=k}^{N-1} EE(\chi_{\varkappa>i} I_{\partial\Gamma}(\vartheta_i, X_i)$$

$$\times [u(\vartheta_{i+1}, X_{i+1}) - u(\vartheta_i, X_i) + Z_{i+1} - Z_i]/X_i, Z_i)$$

$$= \sum_{i=k}^{N-1} E(\chi_{\varkappa>i} I_{\overline{Q}\setminus\partial\Gamma}(\vartheta_i, X_i)$$

$$\times E\left[u(\vartheta_{i+1}, X_{i+1}) - u(\vartheta_i, X_i) + Z_{i+1} - Z_i/X_i, Z_i\right])$$

$$+ \sum_{i=k}^{N-1} E(\chi_{\varkappa>i} I_{\partial\Gamma}(\vartheta_i, X_i)$$

$$\times E\left[u(\vartheta_{i+1}, X_{i+1}) - u(\vartheta_i, X_i) + Z_{i+1} - Z_i/X_i, Z_i\right]). \tag{1.28}$$

In (1.28) and below we use the standard properties of conditional mathematical expectations taking into account that the indicator functions $\chi_{\varkappa>i}$, $I_{\overline{Q}\setminus\partial\Gamma}(\vartheta_i, X_i)$, and $I_{\partial\Gamma}(\vartheta_i, X_i)$ are measurable with respect to X_i.

To calculate the conditional expectations in (1.28), we exploit a lemma from [82, Section 10]. In our case the lemma allows to evaluate a conditional expectation as the ordinary mathematical expectation under fixed values of the random variables X_i, Z_i.

We get for $(\vartheta_i, X_i) \in \overline{Q} \setminus \partial\Gamma$:

$$A_i := E\left[u(\vartheta_{i+1}, X_{i+1}) - u(\vartheta_i, X_i) + Z_{i+1} - Z_i/X_i, Z_i\right]$$

$$= \frac{1}{2} u(t_{i+1}, X_i + h\bar{b}_i - h^{1/2}\bar{\sigma}_i) + \frac{1}{2} u(t_{i+1}, X_i + h\bar{b}_i + h^{1/2}\bar{\sigma}_i)$$

$$- u(t_i, X_i) + h\bar{g}_i. \tag{1.29}$$

We expand the terms of (1.29) at the point (t_i, X_i):

$$u(t_{i+1}, X_i + h\bar{b}_i \pm \sqrt{h}\bar{\sigma}_i)$$
$$= u(t_i, X_i) + \frac{\partial u}{\partial t}h + (h\bar{b}_i \pm \sqrt{h}\bar{\sigma}_i)\frac{\partial u}{\partial x}$$
$$+ \frac{\bar{\sigma}_i^2}{2}\frac{\partial^2 u}{\partial x^2}h \pm \bar{b}_i\bar{\sigma}_i\frac{\partial^2 u}{\partial x^2}h^{3/2} \pm \bar{\sigma}_i\frac{\partial^2 u}{\partial t \partial x}h^{3/2} \pm \frac{\bar{\sigma}_i^3}{6}\frac{\partial^3 u}{\partial x^3}h^{3/2} + O(h^2). \quad (1.30)$$

Then, attracting relations

$$\bar{b}_i = b(t_i, X_i, \bar{u}(t_{i+1}, X_i)) = b(t_k, X_i, u(t_{i+1}, X_i) + R(t_{k+1}, X_i))$$
$$= b(t_k, X_i, u(t_{k+1}, X_i)) + \Delta b = b(t_k, X_i, u(t_k, X_i)) + \Delta b + O(h),$$

$$\bar{\sigma}_i = \sigma(t_i, X_i, u(t_i, X_i)) + \Delta\sigma + O(h),$$
$$\bar{\sigma}_i^2 = \sigma^2(t_i, X_i, u(t_i, X_i)) + \Delta\sigma^2 + O(h),$$
$$\bar{g}_i = g(t_i, X_i, u(t_i, X_i)) + \Delta g + O(h), \quad (1.31)$$

where

$$|\Delta b|, |\Delta\sigma|, |\Delta\sigma^2|, |\Delta g| \leq K |R(t_{i+1}, X_i)|,$$

and taking into account that $u(t,x)$ is the solution of the problem (1.5)-(1.6), we obtain

$$A_i = r_i + O(h^2), \quad (1.32)$$

where

$$|r_i| \leq Kh|R(\vartheta_i + h, X_i)|.$$

Now let $(\vartheta_i, X_i) \in \partial\Gamma$ be such that X_i is close to α. We have

$$B_i := E\left[u(\vartheta_{i+1}, X_{i+1}) - u(\vartheta_i, X_i) + Z_{i+1} - Z_i / X_i, Z_i\right]$$
$$= \frac{1}{1+\sqrt{\bar{\lambda}_i}}\varphi(t_{i+\bar{\lambda}_i}, \alpha) + \frac{\sqrt{\bar{\lambda}_i}}{1+\sqrt{\bar{\lambda}_i}}u(t_{i+1}, X_i + h\bar{b}_i + h^{1/2}\bar{\sigma}_i)$$
$$- u(t_i, X_i) + \sqrt{\bar{\lambda}_i}h\bar{g}_i$$
$$= \frac{1}{1+\sqrt{\bar{\lambda}_i}}u(t_{i+\bar{\lambda}_i}, \alpha) + \frac{\sqrt{\bar{\lambda}_i}}{1+\sqrt{\bar{\lambda}_i}}u(t_{i+1}, X_i + h\bar{b}_i + h^{1/2}\bar{\sigma}_i)$$
$$- u(t_i, X_i) + \sqrt{\bar{\lambda}_i}h\bar{g}_i. \quad (1.33)$$

We expand the terms of (1.33) at the point (t_i, X_i). Then, using (1.26), attracting relations like (1.31), and taking into account that $u(t,x)$ is the solution of the problem (1.5)-(1.6), we get

$$B_i = \bar{r}_i + O(h^{3/2}), \quad (1.34)$$

where

$$|\bar{r}_i| \leq Kh|R(\vartheta_i + h, X_i)|.$$

An analogous relation can be obtained for $(\vartheta_i, X_i) \in \partial \Gamma$ with X_i being close to β.

Substituting (1.32) and (1.34) in (1.28), we obtain

$$R(t_k, x) = E \sum_{i=k}^{\varkappa-1} I_{\overline{Q} \setminus \partial \Gamma}(\vartheta_i, X_i) \left[r_i + O(h^2) \right]$$

$$+ E \sum_{i=k}^{\varkappa-1} I_{\partial \Gamma}(\vartheta_i, X_i) \left[\bar{r}_i + O(h^{3/2}) \right]. \quad (1.35)$$

Due to (1.27),

$$\left| E \sum_{i=k}^{\varkappa-1} I_{\partial \Gamma}(\vartheta_i, X_i) O(h^{3/2}) \right| \leq CHh^{3/2}.$$

Then, from (1.35), we obtain for $R_k := \max_{x \in [\alpha, \beta]} |R(t_k, x)|$ that

$$R_k \leq Kh \sum_{i=k}^{N-1} R_{i+1} + Ch. \quad (1.36)$$

Introduce $\varepsilon_k := Kh \sum_{i=k}^{N-1} R_{i+1} + Ch$, $k = N-1, \ldots, 0$. Due to (1.36), $R_k \leq \varepsilon_k$. Consequently, $\varepsilon_k = KhR_{k+1} + \varepsilon_{k+1} \leq (1 + Kh)\varepsilon_{k+1}$, $k = N-2, \ldots, 0$. Then (since $\varepsilon_{N-1} = Ch$)

$$R_k \leq \varepsilon_k \leq Ce^{K(T-t_0)} h, \ k = N, \ldots, 0.$$

Theorem 1.5 is proved. □

Remark 1.6. The assertions of Lemma 1.4 and Theorem 1.5 are also valid if we take weaker assumptions on the coefficients than (ii), namely:

$$|b| \leq K, \ |\sigma| \leq K, \ |g| \leq K,$$

$$|b(t, x_2, u_2) - b(t, x_1, u_1)| + |\sigma(t, x_2, u_2) - \sigma(t, x_1, u_1)|$$
$$+ |g(t, x_2, u_2) - g(t, x_1, u_1)|$$
$$\leq K(|x_2 - x_1| + |u_2 - u_1|), \ t_0 \leq t \leq T, \ x \in [\alpha, \beta], \ u_\circ < u < u^\circ.$$

Remark 1.7. It follows from the proof of Theorem 1.5 that to construct a first-order method we can use an approximation of $u(t_k, x)$ which one-step error near the boundary (i.e., when $(t_k, x) \in \partial \Gamma$) is estimated as $O(h)$ only (cf. Lemma 1.4). For instance, we can approximate the solution $u(t_k, x)$, when x is close to α, by values of the solution at the point (t_{k+1}, α) and at a point $(t_{k+1}, \hat{x}_k) \in \overline{Q} \setminus \partial \Gamma$ (e.g., one can take $\hat{x}_k = \alpha + h^{1/2} \max \sigma + h \max |b|$, where the maxima are taken over $(t, x) \in \overline{Q}$, $u \in [u_\circ, u^\circ]$) with the weights

$p = \dfrac{\hat{x}_k - x}{\hat{x}_k - \alpha}$ and $q = 1 - p$ respectively. Analogously, we can approximate $u(t_k, x)$ when x is close to β. Making use of this approximation for $(t_k, x) \in \partial \Gamma$ and the Euler approximation for $(t_k, x) \in \overline{Q} \backslash \partial \Gamma$, we will get the new layer method with the first order of convergence (see also Sect. 6.2.1, where such a construction is used in Algorithm 6.2.1 for solving linear Dirichlet problems by the Monte Carlo approach). This layer method is slightly easier than (1.24) and its generalization to the multi-dimensional case is also easier. But we prefer to present the method (1.24) and its generalization (see Sect. 8.2) because, first, we would like to have more methods, and second, methods with a more accurate approximation near the boundary give more accurate results even if they have the same order of convergence as methods less accurate near the boundary (see, e.g., numerical tests in Sect. 8.3). Besides, the methods discussed in this remark are practically of the same complexity.

Remark 1.8. We can also conclude from the proof of Theorem 1.5 that if we use an approximation of $u(t_k, x)$ which one-step error is $O(h^{3/2})$ for $(t_k, x) \in \partial \Gamma$ (as it is for (1.24)) and is at least $O(h^{5/2})$ for $(t_k, x) \in \overline{Q} \backslash \partial \Gamma$, we will obtain the new layer method with the global error $O(h^{3/2})$. For instance, in the case of constant σ it is possible to get such a method by attracting the second-order weak Runge–Kutta scheme (2.2.20) from Sect. 2.2.2 instead of the weak Euler scheme (1.9) used for construction of (1.24) (see also Sect. 7.2).

8.1.4 Numerical algorithms and their convergence

To become a numerical algorithm, the method (1.20) (just as other layer methods) needs a discretization in the variable x. Consider an equidistant space discretization with a space step h_x (recall that the notation for time step is h): $x_j = \alpha + j h_x$, $j = 0, 1, 2, \ldots, M$, $h_x = (\beta - \alpha)/M$. Using, for example, linear interpolation, we construct the following algorithm (we denote it as $\bar{u}(t_k, x)$ again, since this does not cause any confusion).

Algorithm 1.9 *The algorithm is defined by the following formulas*

$$\bar{u}(t_N, x) = \varphi(t_N, x), \ x \in [\alpha, \beta],$$

$$\bar{u}(t_k, x_j) = \frac{1}{2}\bar{u}(t_{k+1}, x_j + h\bar{b}_{k,j} - \sqrt{h}\bar{\sigma}_{k,j}) + \frac{1}{2}\bar{u}(t_{k+1}, x_j + h\bar{b}_{k,j} + \sqrt{h}\bar{\sigma}_{k,j})$$
$$+ h\bar{g}_{k,j}, \ \text{if} \ x_j + h\bar{b}_{k,j} \pm \sqrt{h}\bar{\sigma}_{k,j} \in [\alpha, \beta] ;$$

$$\bar{u}(t_k, x_j) = \varphi(t_{k+1-\bar{\gamma}_{k,j}}, \alpha) - \bar{\gamma}_{k,j} h g(t_k, \alpha, \varphi(t_{k+1}, \alpha))$$
$$- \frac{1}{2}\bar{u}(t_{k+1}, \alpha + \bar{\gamma}_{k,j} h b(t_k, \alpha, \varphi(t_{k+1}, \alpha)) + \sqrt{\bar{\gamma}_{k,j} h} \sigma(t_k, \alpha, \varphi(t_{k+1}, \alpha)))$$
$$+ \frac{1}{2}\bar{u}(t_{k+1}, x_j + h\bar{b}_{k,j} + \sqrt{h}\bar{\sigma}_{k,j}) + h\bar{g}_{k,j}, \ \text{if} \ x_j + h\bar{b}_{k,j} - \sqrt{h}\bar{\sigma}_{k,j} < \alpha;$$

8.1 Layer methods for the Dirichlet problem   475

$$\bar{u}(t_k, x_j) = \frac{1}{2}\bar{u}(t_{k+1}, x_j + h\bar{b}_{k,j} - \sqrt{h}\bar{\sigma}_{k,j}) + \varphi(t_{k+1-\bar{\delta}_{k,j}}, \beta)$$
$$-\bar{\delta}_{k,j}hg(t_k, \beta, \varphi(t_{k+1}, \beta))$$
$$-\frac{1}{2}\bar{u}(t_{k+1}, \beta + \bar{\delta}_{k,j}hb(t_k, \beta, \varphi(t_{k+1}, \beta))) - \sqrt{\bar{\delta}_{k,j}}h\sigma(t_k, \beta, \varphi(t_{k+1}, \beta)))$$
$$+h\bar{g}_{k,j}, \text{ if } x_j + h\bar{b}_{k,j} + \sqrt{h}\bar{\sigma}_{k,j} > \beta; \quad (1.37)$$
$$j = 1, 2, \ldots, M-1,$$
$$\bar{u}(t_k, x) = \frac{x_{j+1} - x}{h_x}\bar{u}(t_k, x_j) + \frac{x - x_j}{h_x}\bar{u}(t_k, x_{j+1}), \; x_j < x < x_{j+1}, \quad (1.38)$$
$$j = 0, 1, 2, \ldots, M-1, \; k = N-1, \ldots, 1, 0,$$

where $\bar{b}_{k,j}$, $\bar{\sigma}_{k,j}$, $\bar{g}_{k,j}$ are the coefficients $b(t, x, u)$, $\sigma(t, x, u)$, $g(t, x, u)$ calculated at the point $(t_k, x_j, \bar{u}(t_{k+1}, x_j))$ and $0 < \bar{\gamma}_{k,j}, \bar{\delta}_{k,j} \leq 1$ are roots of the equations (1.17) and (1.18) with the right-hand sides $x_j + h\bar{b}_{k,j} - \sqrt{h}\bar{\sigma}_{k,j}$ and $x_j + h\bar{b}_{k,j} + \sqrt{h}\bar{\sigma}_{k,j}$, respectively.

Theorem 1.10. *If the value of h_x is taken equal to $\varkappa h$, $\varkappa$ is a positive constant, then under the assumptions (i) and (ii) Algorithm 1.9 has the first order of convergence, i.e., the approximation $\bar{u}(t_k, x)$ from (1.37)-(1.38) satisfies the relation*

$$|\bar{u}(t_k, x) - u(t_k, x)| \leq Kh, \quad (1.39)$$

where K does not depend on x, h, k.

The proof of Theorem 1.10 differs only little from the proof of Theorem 7.2.5 and is therefore omitted.

Remark 1.11. Using probabilistic arguments, it is possible to prove that the algorithm based on the method (1.24) and linear interpolation has the global error $O(h)$ for $h_x = \varkappa h$.

Remark 1.12. As it was mentioned in Remark 7.2.7, it is natural to consider cubic interpolation instead of the linear one for constructing numerical algorithms. The use of cubic interpolation allows us to take the space step $h_x = \varkappa\sqrt{h}$ (in contrast to $h_x = \varkappa h$ for linear interpolation) and, thus, to reduce the volume of computations. Moreover, if we use cubic interpolation, we can avoid special formulas near the boundary choosing some appropriate $\varkappa$ (indeed, we can take, e.g., $\varkappa = 2 \max_{t \in [t_0, T], x \in \overline{G}, u \in [u_\circ, u^\circ]} \sigma(t, x, u)$, then for a sufficiently small h the points $x_j + h\bar{b}_{k,j} \pm \sqrt{h}\bar{\sigma}_{k,j}, \; j = 1, 2, \ldots, M-1$, always belong to $[\alpha, \beta]$). Unfortunately, we have not succeeded in proving a convergence theorem in the case of cubic interpolation. The way of proving Theorem 1.10 gives us some restriction on the type of interpolation procedure which can be used for constructing the numerical algorithm (see also Chap. 7). The restriction is such that the sum of the absolute values of the coefficients staying at $\bar{u}(t_k, \cdot)$ in the interpolation procedure must not be greater than one. Linear interpolation and B-splines of the order $O(h_x^2)$ satisfy the restriction. However, cubic interpolation of the order $O(h_x^4)$ does not satisfy it. In Sect. 8.3 we test an algorithm based on cubic interpolation. The tests give fairly good results. See also some theoretical explanations and numerical tests in Chap. 7.

Remark 1.13. Clearly, the algorithms can be considered with variable time steps and space steps. An algorithm with variable space steps is used in our numerical tests (see Sect. 8.3.1).

8.2 Extension to the multi-dimensional Dirichlet problem

In this section we generalize the layer method (1.24) to the multi-dimensional case $(d > 1)$ (see also Algorithm 6.2.9 for linear Dirichlet problems). A generalization of the layer method (1.20) to the multi-dimensional case is complicated, and it is not considered here.

As has been mentioned before, layer methods are feasible if the dimension d of the domain G is not more than 3. This is why, we restrict ourselves here to the cases $d = 2$ and $d = 3$. We remark only that it is not difficult to generalize the layer method (1.24) for an arbitrary d.

Consider the case $d = 2$. Introduce the notation ${}_iX_{k+1} := ({}_iX^1_{k+1}, {}_iX^2_{k+1})$,

$$\begin{aligned}
{}_iX^1_{k+1} &= x^1 + h\bar{b}_k^1 + \sqrt{h}\bar{\sigma}_k^{11}{}_i\xi^1 + \sqrt{h}\bar{\sigma}_k^{12}{}_i\xi^2, \\
{}_iX^2_{k+1} &= x^2 + \bar{b}_k^2 h + \sqrt{h}\bar{\sigma}_k^{21}{}_i\xi^1 + \sqrt{h}\bar{\sigma}_k^{22}{}_i\xi^2, \\
i &= 1, 2, 3, 4, \quad x = (x^1, x^2) \in G \subset \mathbf{R}^2,
\end{aligned}$$

where ${}_1\xi = (-1, -1)$, ${}_2\xi = (-1, 1)$, ${}_3\xi = -{}_1\xi$, ${}_4\xi = -{}_2\xi$ and $\bar{b}_k = (\bar{b}_k^1, \bar{b}_k^2)$, $\bar{\sigma}_k = \{\bar{\sigma}_k^{jl}\}$ are the coefficients $b(t, x, u)$, $\sigma(t, x, u)$ calculated at the point $(t_k, x, \bar{u}(t_{k+1}, x))$.

If the point $x = (x^1, x^2) \in G$ is sufficiently far from the boundary ∂G (more precisely, if the points ${}_iX_{k+1}$, $i = 1, 2, 3, 4$, belong to $\overline{G}$), the layer method has the form (cf. (7.3.4)):

$$\bar{u}(t_k, x^1, x^2) = \sum_{i=1}^{4} \frac{1}{4}\bar{u}(t_{k+1}, {}_iX^1_{k+1}, {}_iX^2_{k+1}) + h\bar{g}_k, \qquad (2.1)$$

where $\bar{g}_k$ is the coefficient $g(t, x, u)$ calculated at the point $(t_k, x, \bar{u}(t_{k+1}, x))$.

If the point $x = (x^1, x^2) \in G$ is close to the boundary ∂G, then some of the points ${}_iX_{k+1} = ({}_iX^1_{k+1}, {}_iX^2_{k+1})$, $i = 1, 2, 3, 4$, may be outside the domain $\overline{G}$. Let us connect the point x with the points ${}_{i^*}X_{k+1}$, which are outside $\overline{G}$, by the curves $\psi_{i^*}(\lambda) = (\psi_{i^*}^1(\lambda), \psi_{i^*}^2(\lambda))$:

8.2 Extension to the multi-dimensional Dirichlet problem

$$\psi_{i*}^1(\lambda) = x^1 + \lambda h \bar{b}_k^1 + \sqrt{\lambda h} \bar{\sigma}_k^{11}{}_{i*}\xi^1 + \sqrt{\lambda h} \bar{\sigma}_k^{12}{}_{i*}\xi^2,$$

$$\psi_{i*}^2(\lambda) = x^2 + \lambda h \bar{b}_k^2 + \sqrt{\lambda h} \bar{\sigma}_k^{21}{}_{i*}\xi^1 + \sqrt{\lambda h} \bar{\sigma}_k^{22}{}_{i*}\xi^2, \ 0 \leq \lambda \leq 1.$$

The boundary ∂G is assumed to be sufficiently smooth. For a sufficiently small h there is a unique value of $\lambda = {}_{i*}\bar{\lambda}_k$, $0 < {}_{i*}\bar{\lambda}_k < 1$, such that the point ${}_{i*}\eta_k = ({}_{i*}\eta_k^1, {}_{i*}\eta_k^2)$, where

$${}_{i*}\eta_k^1 = x^1 + {}_{i*}\bar{\lambda}_k h \bar{b}_k^1 + \sqrt{{}_{i*}\bar{\lambda}_k h} \bar{\sigma}_k^{11}{}_{i*}\xi^1 + \sqrt{{}_{i*}\bar{\lambda}_k h} \bar{\sigma}_k^{12}{}_{i*}\xi^2,$$

$${}_{i*}\eta_k^2 = x^2 + {}_{i*}\bar{\lambda}_k h \bar{b}_k^2 + \sqrt{{}_{i*}\bar{\lambda}_k h} \bar{\sigma}_k^{21}{}_{i*}\xi^1 + \sqrt{{}_{i*}\bar{\lambda}_k h} \bar{\sigma}_k^{22}{}_{i*}\xi^2,$$

belongs to the boundary ∂G.

Put ${}_j\bar{\lambda}_k = 1$ and ${}_j\eta_k = {}_jX_{k+1}$ for the points ${}_jX_{k+1}$ belonging to $\overline{G}$. Then the layer method takes the form

$$\bar{u}(t_k, x^1, x^2)$$

$$= \frac{\sqrt{{}_2\bar{\lambda}_k \cdot {}_3\bar{\lambda}_k \cdot {}_4\bar{\lambda}_k}}{(\sqrt{{}_1\bar{\lambda}_k} + \sqrt{{}_3\bar{\lambda}_k})(\sqrt{{}_1\bar{\lambda}_k \cdot {}_3\bar{\lambda}_k} + \sqrt{{}_2\bar{\lambda}_k \cdot {}_4\bar{\lambda}_k})} \bar{u}(t_{k+{}_1\bar{\lambda}_k}, {}_1\eta_k^1, {}_1\eta_k^2)$$

$$+ \frac{\sqrt{{}_1\bar{\lambda}_k \cdot {}_3\bar{\lambda}_k \cdot {}_4\bar{\lambda}_k}}{(\sqrt{{}_2\bar{\lambda}_k} + \sqrt{{}_4\bar{\lambda}_k})(\sqrt{{}_1\bar{\lambda}_k \cdot {}_3\bar{\lambda}_k} + \sqrt{{}_2\bar{\lambda}_k \cdot {}_4\bar{\lambda}_k})} \bar{u}(t_{k+{}_2\bar{\lambda}_k}, {}_2\eta_k^1, {}_2\eta_k^2)$$

$$+ \frac{\sqrt{{}_1\bar{\lambda}_k \cdot {}_2\bar{\lambda}_k \cdot {}_4\bar{\lambda}_k}}{(\sqrt{{}_1\bar{\lambda}_k} + \sqrt{{}_3\bar{\lambda}_k})(\sqrt{{}_1\bar{\lambda}_k \cdot {}_3\bar{\lambda}_k} + \sqrt{{}_2\bar{\lambda}_k \cdot {}_4\bar{\lambda}_k})} \bar{u}(t_{k+{}_3\bar{\lambda}_k}, {}_3\eta_k^1, {}_3\eta_k^2)$$

$$+ \frac{\sqrt{{}_1\bar{\lambda}_k \cdot {}_2\bar{\lambda}_k \cdot {}_3\bar{\lambda}_k}}{(\sqrt{{}_2\bar{\lambda}_k} + \sqrt{{}_4\bar{\lambda}_k})(\sqrt{{}_1\bar{\lambda}_k \cdot {}_3\bar{\lambda}_k} + \sqrt{{}_2\bar{\lambda}_k \cdot {}_4\bar{\lambda}_k})} \bar{u}(t_{k+{}_4\bar{\lambda}_k}, {}_4\eta_k^1, {}_4\eta_k^2)$$

$$+ h\bar{g}_k \frac{2\sqrt{{}_1\bar{\lambda}_k \cdot {}_2\bar{\lambda}_k \cdot {}_3\bar{\lambda}_k \cdot {}_4\bar{\lambda}_k}}{\sqrt{{}_1\bar{\lambda}_k \cdot {}_3\bar{\lambda}_k} + \sqrt{{}_2\bar{\lambda}_k \cdot {}_4\bar{\lambda}_k}}. \tag{2.2}$$

Recall that if ${}_i\eta_k = ({}_i\eta_k^1, {}_i\eta_k^2) \in \partial G$ then $\bar{u}(t_{k+{}_i\bar{\lambda}_k}, {}_i\eta_k^1, {}_i\eta_k^2) = \varphi(t_{k+{}_i\bar{\lambda}_k}, {}_i\eta_k^1, {}_i\eta_k^2)$ (see (1.2)).

The errors of the one-step approximations corresponding to (2.1) and (2.2) are $O(h^2)$ and $O(h^{3/2})$, respectively. By probabilistic arguments (see the proof of Theorem 1.5), the global error of the layer method (2.1)-(2.2) is estimated by $O(h)$.

Now consider the case $d = 3$. Introduce the notation ${}_iX_{k+1} = ({}_iX_{k+1}^1, {}_iX_{k+1}^2, {}_iX_{k+1}^3)$, $i = 1, 2, \ldots, 8$, where

$${}_iX_{k+1}^j := x^j + h\bar{b}_k^j + \sqrt{h}\bar{\sigma}_k^{j1}{}_i\xi^1 + \sqrt{h}\bar{\sigma}_k^{j2}{}_i\xi^2 + \sqrt{h}\bar{\sigma}_k^{j3}{}_i\xi^3, \ j = 1, 2, 3,$$

$$x = (x^1, x^2, x^3) \in G \subset \mathbf{R}^3.$$

Here $\bar{b}_k = \{\bar{b}_k^j\}$ and $\bar{\sigma}_k = \{\bar{\sigma}_k^{jl}\}$ are the coefficients $b(t, x, u)$ and $\sigma(t, x, u)$ calculated at the point $(t_k, x, \bar{u}(t_k, x))$ and ${}_i\xi = ({}_i\xi^1, {}_i\xi^2, {}_i\xi^3)$, $i = 1, \ldots, 8$, are the following vectors:

$$_1\xi = (-1,-1,-1),\ _2\xi = (-1,-1,1),\ _3\xi = (-1,1,-1),\ _4\xi = (1,-1,-1),$$

$$_{i+4}\xi = -\,_i\xi,\ i = 1,2,3,4.$$

If the points $_iX_{k+1}$, $i = 1, 2, \ldots, 8$, belong to $\overline{G}$, the layer method has the form

$$\bar{u}(t_k, x) = \sum_{i=1}^{8} \frac{1}{8} \bar{u}(t_{k+1},\,_iX_{k+1}) + h\bar{g}_k, \qquad (2.3)$$

where $\bar{g}_k$ is the coefficient $g(t, x, u)$ calculated at the point $(t_k, x, \bar{u}(t_{k+1}, x))$.

If some points $_{i^*}X_{k+1} \notin \overline{G}$, we connect the point x with the points $_{i^*}X_{k+1}$ by the curves $\psi_{i^*}(\lambda) = (\psi^1_{i^*}(\lambda), \psi^2_{i^*}(\lambda), \psi^3_{i^*}(\lambda))$,

$$\psi^j_{i^*}(\lambda) = x^j + \lambda h \bar{b}^j_k + \sqrt{\lambda h}\,\bar{\sigma}^{j1}_k\,_{i^*}\xi^1 + \sqrt{\lambda h}\,\bar{\sigma}^{j2}_k\,_{i^*}\xi^2 + \sqrt{\lambda h}\,\bar{\sigma}^{j3}_k\,_{i^*}\xi^3,$$

$$j = 1, 2, 3,\ 0 \leq \lambda \leq 1.$$

Due to the smoothness of the boundary ∂G, for a sufficiently small h there is a unique value of $\lambda = \,_{i^*}\bar{\lambda}_k$, $0 < \,_{i^*}\bar{\lambda}_k < 1$, such that the point $_{i^*}\eta_k = (_{i^*}\eta^1_k,\,_{i^*}\eta^2_k,\,_{i^*}\eta^3_k)$, where

$$_{i^*}\eta^j_k = x^j + _{i^*}\bar{\lambda}_k h \bar{b}^j_k + \sqrt{_{i^*}\bar{\lambda}_k h}\,\bar{\sigma}^{j1}_k\,_{i^*}\xi^1 + \sqrt{_{i^*}\bar{\lambda}_k h}\,\bar{\sigma}^{j2}_k\,_{i^*}\xi^2 + \sqrt{_{i^*}\bar{\lambda}_k h}\,\bar{\sigma}^{j3}_k\,_{i^*}\xi^3,$$

$$j = 1, 2, 3,$$

belongs to the boundary ∂G.

Put $_j\bar{\lambda}_k = 1$ and $_j\eta_k = \,_jX_{k+1}$ for the points $_jX_{k+1}$ belonging to $\overline{G}$. Then the layer method takes the form

$$\bar{u}(t_k, x) = \sum_{i=1}^{4} \frac{\gamma_k}{\sqrt{_i\bar{\lambda}_k} + \sqrt{_{i+4}\bar{\lambda}_k}} \left(\frac{1}{\sqrt{_i\bar{\lambda}_k}} \bar{u}(t_{k+\,_i\bar{\lambda}_k},\,_i\eta_k) \right.$$
$$\left. + \frac{1}{\sqrt{_{i+4}\bar{\lambda}_k}} \bar{u}(t_{k+\,_{i+4}\bar{\lambda}_k},\,_{i+4}\eta_k) \right) + 4\gamma_k h \bar{g}_k, \qquad (2.4)$$

where

$$\gamma_k = \left(\sum_{i=1}^{4} \frac{1}{\sqrt{_i\bar{\lambda}_k \cdot\,_{i+4}\bar{\lambda}_k}} \right)^{-1}.$$

To construct the corresponding numerical algorithms, we attract linear interpolation as in the previous section (see also Sect. 7.3). For example, consider the case $d = 2$. To this end, put the domain $\overline{G}$ into a rectangle Π with corners (x^1_0, x^2_0), $(x^1_0, x^2_{M_2})$, $(x^1_{M_1}, x^2_0)$, $(x^1_{M_1}, x^2_{M_2})$ and introduce the equidistant space discretization of the rectangle Π:

$$\Delta_{M_1, M_2} := \{(x^1_j, x^2_l) : x^1_j = x^1_0 + jh_{x^1},\ x^2_l = x^2_0 + lh_{x^2},$$
$$j = 0, \ldots, M_1,\ l = 0, \ldots, M_2\},$$

$$h_{x^1} = \frac{x^1_{M_1} - x^1_0}{M_1}, \ h_{x^2} = \frac{x^2_{M_1} - x^2_0}{M_2}.$$

The values of $\bar{u}(t_k, x^1_j, x^2_l)$ at the nodes of $\Delta_{M_1,M_2} \cap \overline{G}$ are found in accordance with (2.1)-(2.2). Let $(x^1, x^2) \in \overline{G}$ and $x^1_j \leq x^1 \leq x^1_{j+1}$, $x^2_l \leq x^2 \leq x^2_{l+1}$. If all the nodes (x^1_j, x^2_l), (x^1_j, x^2_{l+1}), (x^1_{j+1}, x^2_l), $(x^1_{j+1}, x^2_{l+1}) \in \overline{G}$, the value of $\bar{u}(t_k, x^1, x^2)$ is evaluated as

$$\begin{aligned}\bar{u}(t_k, x^1, x^2) &= \frac{x^1_{j+1} - x^1}{h_{x^1}} \frac{x^2_{l+1} - x^2}{h_{x^2}} \bar{u}(t_k, x^1_j, x^2_l) \\ &+ \frac{x^1_{j+1} - x^1}{h_{x^1}} \frac{x^2 - x^2_l}{h_{x^2}} \bar{u}(t_k, x^1_j, x^2_{l+1}) \\ &+ \frac{x^1 - x^1_j}{h_{x^1}} \frac{x^2_{l+1} - x^2}{h_{x^2}} \bar{u}(t_k, x^1_{j+1}, x^2_l) \\ &+ \frac{x^1 - x^1_j}{h_{x^1}} \frac{x^2 - x^2_l}{h_{x^2}} \bar{u}(t_k, x^1_{j+1}, x^2_{l+1}). \end{aligned} \quad (2.5)$$

If the point $x = (x^1, x^2)$: $x^1_j \leq x^1 \leq x^1_{j+1}$, $x^2_l \leq x^2 \leq x^2_{l+1}$ is such that some of the nodes (x^1_j, x^2_l), (x^1_j, x^2_{l+1}), (x^1_{j+1}, x^2_l), (x^1_{j+1}, x^2_{l+1}) do not belong to $\overline{G}$, then we use some points on the boundary ∂G (due to (1.2) we know values of $u(t, x)$ for $x \in \partial G$) to find $\bar{u}(t_k, x^1, x^2)$ by linear interpolation.

If we take $h_{x^i} = \varkappa^i h$, $i = 1, 2$, $\varkappa^1, \varkappa^2 > 0$ are positive constants, the error of the proposed algorithm is estimated as $O(h)$. This algorithm was used in [297] to simulate FKPP-equation with random advection.

8.3 Numerical tests of layer methods for the Dirichlet problems

In the previous sections we deal with semilinear parabolic equations with negative direction of time t: the equations are considered under $t < T$ and the "initial" conditions are given at $t = T$. This form of equations is suitable for the probabilistic approach which we use to construct numerical methods. Of course, the proposed methods are adaptable to semilinear parabolic equations with positive direction of time, and this adaptation is particularly easy in the autonomous case (cf. Remark 7.2.8). In our numerical tests we use algorithms with positive direction of time (see, e.g., (3.15)-(3.16) below).

8.3.1 The Burgers equation

Consider the Dirichlet problem for the one-dimensional Burgers equation:

$$\frac{\partial u}{\partial t} = \frac{\varepsilon^2}{2} \frac{\partial^2 u}{\partial x^2} - u \frac{\partial u}{\partial x}, \ t > 0, \ x \in (-1, 1), \quad (3.1)$$

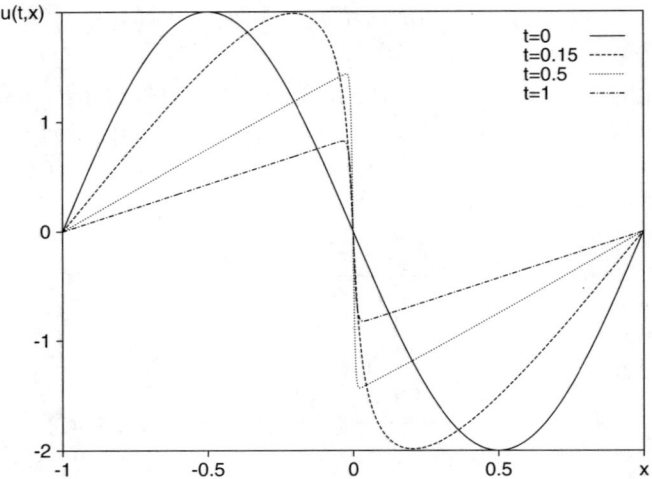

Fig. 3.1. A typical solution $u(t,x)$ of the problem (3.1)-(3.3) for $\varepsilon = 0.1$, $A = 2$ and various time moments.

$$u(0,x) = -A\sin\pi x, \quad x \in [-1,1], \tag{3.2}$$

$$u(t,\pm 1) = 0, \quad t > 0. \tag{3.3}$$

This problem was used for testing various numerical methods in, e.g., [1, 14, 65] (see also references therein). By means of the Cole–Hopf transformation, one can find the explicit solution of the problem (3.1)-(3.3) in the form

$$u(t,x) = -A \frac{\int_{-\infty}^{\infty} \sin\pi(x-y) \exp\left(-\frac{A}{\pi\varepsilon^2}\cos\pi(x-y) - \frac{y^2}{2\varepsilon^2 t}\right) dy}{\int_{-\infty}^{\infty} \exp\left(-\frac{A}{\pi\varepsilon^2}\cos\pi(x-y) - \frac{y^2}{2\varepsilon^2 t}\right) dy} \tag{3.4}$$

or

$$u(t,x) = \frac{\pi\varepsilon^2}{2} \frac{\sum_{n=1}^{\infty} na_n \exp\left(-\frac{1}{8}\varepsilon^2\pi^2 n^2 t\right) \sin\frac{1}{2}\pi n(x+1)}{\frac{1}{2}a_0 + \sum_{n=1}^{\infty} a_n \exp\left(-\frac{1}{8}\varepsilon^2\pi^2 n^2 t\right) \cos\frac{1}{2}\pi n(x+1)} \tag{3.5}$$

with

$$a_n = \int_{-1}^{1} \exp\left(-\frac{A}{\pi\varepsilon^2}\cos\pi x\right) \cos\frac{1}{2}\pi n(x+1)\, dx.$$

We simulate the problem (3.1)-(3.3) on relatively small time intervals $[0,T]$, where the formula (3.4) is more convenient. For a small ε, there is a thin

8.3 Numerical tests of layer methods for the Dirichlet problems

internal layer with width $\sim \varepsilon^2$, where the solution to (3.1)-(3.3) has singular behavior (see, e.g., [111] and references therein). Derivatives of the solution go to infinity as $\varepsilon \to 0$. A typical behavior of the solution is demonstrated on Fig. 3.1.

Here we test three algorithms. The first one is algorithm (1.37)-(1.38). The second one is the algorithm based on layer method (1.24) and linear interpolation. In these two algorithms (they both use linear interpolation), we take the space step h_x being equal to the time step h. The third algorithm is based on cubic interpolation (see Remark 1.12). In the case of the problem (3.1)-(3.3) it has the form

$$\bar{u}(0,x) = -A\sin \pi x, \ x \in [-1,1],$$

$$\bar{u}(t_{k+1}, x_0) = \bar{u}(t_{k+1}, -1) = 0,$$

$$\bar{u}(t_{k+1}, x_M) = \bar{u}(t_{k+1}, 1) = 0,$$

$$\bar{u}(t_{k+1}, x_j) = \frac{1}{2}\bar{u}(t_k, x_j - h\bar{u}(t_k, x_j) - \varepsilon h^{1/2}) + \frac{1}{2}\bar{u}(t_k, x_j - h\bar{u}(t_k, x_j) + \varepsilon h^{1/2}),$$

$$j = 1, \ldots, M-1,$$

$$\bar{u}(t_k, x) = \sum_{i=0}^{3} \Phi_{j,i}(x)\bar{u}(t_k, x_{j+i}), \ x_j < x < x_{j+3},$$

$$\Phi_{j,i}(x) = \prod_{m=0, m\neq i}^{3} \frac{x - x_{j+m}}{x_{j+i} - x_{j+m}}, \tag{3.6}$$

$$k = 0, \ldots, N-1.$$

Here we use a nonequidistant discretization of the interval $[-1,1]$. We take $h_x := x_{j+1} - x_j = \varepsilon\sqrt{h}$ in $[-0.1, 0.1]$ and $h_x = \sqrt{h}$ outside $[-0.1, 0.1]$. Such a choice of h_x is dictated by the fact that if h is comparatively large (e.g., $h = \varepsilon^2$), the equidistant discretization with $h_x = \sqrt{h}$ has not more than one node in the thin internal layer. For a sufficiently small h, it is possible to use the equidistant discretization for cubic interpolation as well.

Since $\varepsilon \ll 1$, the points $x_j - h\bar{u}(t_k, x_j) \pm \varepsilon h^{1/2}$, $j = 1, \ldots, M-1$, belong to the interval $(-1, 1)$. Thus, we avoid using special formulas near the boundary in (3.6) (see Remark 1.12 as well).

Table 3.1 gives numerical results obtained by using the algorithms (1.37)-(1.38) and (3.6). The algorithm based on the layer method (1.24) and linear interpolation gives results practically identical to the ones for (1.37)-(1.38). We present the errors of the approximate solutions $\bar{u}$ in the discrete Chebyshev norm and in the l^1-norm:

$$err^c(t,h) = \max_{x_i} |\bar{u}(t, x_i) - u(t, x_i)|,$$

$$err^l(t,h) = \sum_i |\bar{u}(t, x_i) - u(t, x_i)| \times h_x.$$

Table 3.1. The Burgers equation. Dependence of the errors $err^c(t,h)$ and $err^l(t,h)$ in h for algorithms (1.37)-(1.38) and (3.6) when $t = 0.5$, $\varepsilon = 0.1$, and $A = 2$.

h	algorithm (1.37)-(1.38)		algorithm (3.6)	
	$err^c(t,h)$	$err^l(t,h)$	$err^c(t,h)$	$err^l(t,h)$
0.01	$1.239 \cdot 10^{-1}$	$3.035 \cdot 10^{-2}$	$1.854 \cdot 10^{-1}$	$3.081 \cdot 10^{-2}$
0.0016	$4.574 \cdot 10^{-2}$	$5.311 \cdot 10^{-3}$	$5.855 \cdot 10^{-2}$	$5.481 \cdot 10^{-3}$
0.0001	$2.673 \cdot 10^{-3}$	$3.288 \cdot 10^{-4}$	$3.737 \cdot 10^{-3}$	$3.466 \cdot 10^{-4}$
0.000016	$4.261 \cdot 10^{-4}$	$5.259 \cdot 10^{-5}$	$5.919 \cdot 10^{-4}$	$5.527 \cdot 10^{-5}$

The algorithms based on linear interpolation require both larger volume of computations per time layer and larger amount of memory than the algorithm (3.6) based on cubic interpolation. For instance, in the considered case the algorithm (1.37)-(1.38) with $h = 0.0001$ needs 2×10^4 computations of $\bar{u}(t_k, x)$ per layer $t = t_k$ and to store an array of 2×10^4 elements, and the algorithm (3.6) with the same step $h = 0.0001$ requires only 380 computations of $\bar{u}(t_k, x)$ per layer and an array of 380 elements (see also Remark 1.12).

8.3.2 Comparison analysis

In this section we give some comparison analysis of the layer methods proposed in this chapter and the well-known finite-difference schemes (see also Remark 1.1). We use (3.1)-(3.3) as a test problem again. Here we compare the algorithm (3.6) with two explicit finite-difference schemes (3.7) and (3.8) of order $O(\Delta t, \Delta x^2)$, where Δt is a time step and Δx is a space step. These finite-difference schemes are used for simulation of the Burgers equation in [1, 66].

The method of differences forward in time and central differences in space (the FTCS scheme) applied to the divergent form of the Burgers equation is written as

$$\bar{u}(0, x) = -A \sin \pi x, \quad x \in [-1, 1],$$

$$\bar{u}(t_{k+1}, x_0) = \bar{u}(t_{k+1}, -1) = 0, \quad \bar{u}(t_{k+1}, x_M) = \bar{u}(t_{k+1}, 1) = 0,$$

$$\bar{u}(t_{k+1}, x_j) = \bar{u}(t_k, x_j) - \frac{\Delta t}{4\Delta x}(\bar{u}^2(t_k, x_{j+1}) - \bar{u}^2(t_k, x_{j-1}))$$
$$+ \frac{\varepsilon^2}{2}\frac{\Delta t}{\Delta x^2}(\bar{u}(t_k, x_{j+1}) - 2\bar{u}(t_k, x_j) + \bar{u}(t_k, x_{j-1})), \quad (3.7)$$

$$j = 1, \ldots, M-1, \quad k = 0, \ldots, N-1,$$

where the step of a time discretization $\Delta t := T/N$ and $t_k = k \times \Delta t$ and the step of space discretization $\Delta x := 2/M$ and $x_j = -1 + j \times \Delta x$.

In the case of the problem (3.1)-(3.3) the Brailovskaya scheme has the form (see [1])

8.3 Numerical tests of layer methods for the Dirichlet problems

Table 3.2. The Burgers equation. The relative errors $\delta^l(t,h)$ (*top position*) and $\delta^c(t,h)$ (*lower position*) of algorithm (3.6) and finite-diference schemes (3.7) and (3.8) are given for $t = 0.08$, $\varepsilon = 0.1$, $h = \Delta t = 0.0016$, and various A.

A	algorithm (3.6)	scheme (3.7)	scheme (3.8)
5	$7.79 \cdot 10^{-3}$	$2.22 \cdot 10^{-2}$	$2.01 \cdot 10^{-2}$
	$5.28 \cdot 10^{-2}$	$2.05 \cdot 10^{-1}$	$1.66 \cdot 10^{-1}$
		oscillations	*oscillations*
10	$1.87 \cdot 10^{-2}$	$\gg 100$	$2.35 \cdot 10^{-2}$
	$9.96 \cdot 10^{-1}$	$\gg 100$	$6.94 \cdot 10^{-2}$
			oscillations
15	$2.70 \cdot 10^{-2}$		$3.64 \cdot 10^{-1}$
	$9.84 \cdot 10^{-1}$	*overflow*	$3.42 \cdot 10^{0}$
			big oscillations

Table 3.3. The Burgers equation. The relative errors $\delta^l(t,h)$ (*top position*) and $\delta^c(t,h)$ (*lower position*) of algorithm (3.6) and finite-diference schemes (3.7) and (3.8) are given for $h = \Delta t = 0.0001$, the other parameters are as in Table 3.2.

A	algorithm (3.6)	scheme (3.7)	scheme (3.8)
5	$1.26 \cdot 10^{-3}$	$8.88 \cdot 10^{-4}$	$7.81 \cdot 10^{-4}$
	$4.68 \cdot 10^{-2}$	$1.55 \cdot 10^{-2}$	$1.10 \cdot 10^{-2}$
10	$1.24 \cdot 10^{-3}$	$3.56 \cdot 10^{-3}$	$3.58 \cdot 10^{-3}$
	$9.25 \cdot 10^{-2}$	$1.54 \cdot 10^{-1}$	$1.54 \cdot 10^{-1}$
		oscillations	*oscillations*
15	$1.91 \cdot 10^{-3}$	$5.07 \cdot 10^{-3}$	$5.11 \cdot 10^{-3}$
	$1.99 \cdot 10^{-1}$	$1.81 \cdot 10^{-1}$	$1.84 \cdot 10^{-1}$
		oscillations	*oscillations*

$$\bar{u}(0,x) = -A\sin\pi x, \ x \in [-1,1],$$

$$\bar{u}(t_{k+1}, x_0) = \hat{u}(t_{k+1}, x_0) = 0, \ \bar{u}(t_{k+1}, x_M) = \hat{u}(t_{k+1}, x_M) = 0,$$

$$\hat{u}(t_{k+1}, x_j) = \bar{u}(t_k, x_j) - \frac{\Delta t}{4\Delta x}(\bar{u}^2(t_k, x_{j+1}) - \bar{u}^2(t_k, x_{j-1}))$$
$$+ \frac{\varepsilon^2}{2}\frac{\Delta t}{\Delta x^2}(\bar{u}(t_k, x_{j+1}) - 2\bar{u}(t_k, x_j) + \bar{u}(t_k, x_{j-1})),$$

$$\bar{u}(t_{k+1}, x_j) = \bar{u}(t_k, x_j) - \frac{\Delta t}{4\Delta x}(\hat{u}^2(t_k, x_{j+1}) - \hat{u}^2(t_k, x_{j-1}))$$
$$+ \frac{\varepsilon^2}{2}\frac{\Delta t}{\Delta x^2}(\bar{u}(t_k, x_{j+1}) - 2\bar{u}(t_k, x_j) + \bar{u}(t_k, x_{j-1})), \quad (3.8)$$

$$j = 1, \ldots, M-1, \quad k = 0, \ldots, N-1.$$

The space step Δx in the finite-difference schemes (3.7) and (3.8) is selected as $\Delta x = \varkappa \times \varepsilon \sqrt{\Delta t}$. The results of Tables 3.2 and 3.3 correspond to $\varkappa = 4$.

As in Sect. 8.3.1, we realize the algorithm (3.6) using a nonequidistant discretization of the interval $[-1, 1]$. For the time step $h = 0.0016$ (Table 3.2), we take $h_x := x_{j+1} - x_j = \varepsilon \sqrt{h}$ in $[-0.1, 0.1]$ and $h_x = 2\sqrt{h}$ outside $[-0.1, 0.1]$. And for $h = 0.0001$ (Table 3.3) we choose $h_x = \varepsilon \sqrt{h}$ in $[-0.02, 0.02]$ and $h_x = 2\sqrt{h}$ outside $[-0.02, 0.02]$ (see also the explanations in Sect. 8.3.1).

Tables 3.2 and 3.3 present the relative errors $\delta^c(t, h)$ and $\delta^l(t, h)$. The error $\delta^c(t, h)$ is equal to

$$\delta^c(t, h) = \frac{\max_{x_i} |\bar{u}(t, x_i) - u(t, x_i)|}{\max_{x_i} |u(t, x_i)|}$$

for all three methods. The error $\delta^l(t, h)$ is equal to

$$\delta^l(t, h) = \frac{1}{\max_{x_i} |u(t, x_i)|} \sum_i |\bar{u}(t, x_i) - u(t, x_i)| \times h_x$$

for the algorithm (3.6) while for the schemes (3.7) and (3.8) it is given by

$$\delta^l(t, h) = \frac{1}{\max_{x_i} |u(t, x_i)|} \sum_i |\bar{u}(t, x_i) - u(t, x_i)| \times \Delta x.$$

The comment *"oscillations"* means that the numerical solution has oscillations in a neighborhood of $x = 0$. An illustration of such oscillations is given in Fig. 3.2. The comment *"overflow"* indicates that overflow error occurs during simulation.

Let us observe that if we take $\varkappa = 2$ in order to improve accuracy of the results obtained by the Brailovskaya scheme (3.8), e.g., for $\Delta t = 0.0016$, $A = 10$ (see Table 3.2 and Fig. 3.2), the numerical solution becomes more unstable and overflow error occurs. If we take $\varkappa = 8$ in this case, the errors and amplitude of oscillations become greater than for $\varkappa = 4$.

We can conclude from the results presented in Tables 3.2 and 3.3 and in Fig. 3.2 that the algorithm (3.6) based on the layer method demonstrates a more stable behavior than the finite-difference schemes when the parameter A is sufficiently large. In the considered problem (3.1)-(3.3) large values of A lead, in particular, to large advection in a neighborhood of $x = 0$. Our experiments confirm that the layer methods allow us to avoid difficulties stemming from strong advection (see Remark 1.1). It should also be mentioned that the algorithms based on layer methods require more CPU time than finite-difference schemes. For example, in the case of parameters as in Table 3.3 to solve (3.1)-(3.3) by the algorithm (3.6) we need ≈ 2 sec while the schemes (3.7) and (3.8) require ≈ 0.3 and ≈ 0.4 sec correspondingly. But the algorithm

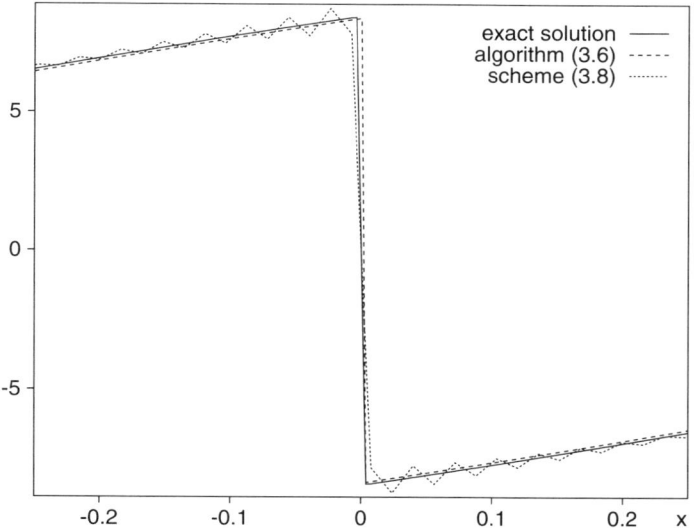

Fig. 3.2. Solution of the problem (3.1)-(3.3) for $A = 10$, the other parameters are as in Table 3.2.

(3.6) gives us quite appropriate results with the greater step $h = 0.0016$ (see Table 3.2 and Fig. 3.2) and in this case it requires ≈ 0.06 sec. Simulations were made on PC with Intel Pentium 233 MHz processor using Borland C compiler.

8.3.3 Quasilinear equation with power law nonlinearities

Consider the Dirichlet problem for quasilinear parabolic equation with power law nonlinearities [258, 259]:

$$\frac{\partial u}{\partial t} = \frac{1}{2}\frac{\partial}{\partial x}\left(u^q \frac{\partial u}{\partial x}\right), \quad t \in (0,1), \quad x > 0, \quad q > 0, \tag{3.9}$$

with the initial condition

$$u(0,x) = (1 - x/L)^{2/q}, \quad x \in [0, L],$$
$$u(0,x) = 0, \quad x > L, \tag{3.10}$$

and the boundary regime

$$u(t,0) = (1-t)^{-1/q}, \quad t \in [0,1), \tag{3.11}$$

where $L = \sqrt{(q+2)/q}$.

The exact solution of this problem has the form [258, 259]:

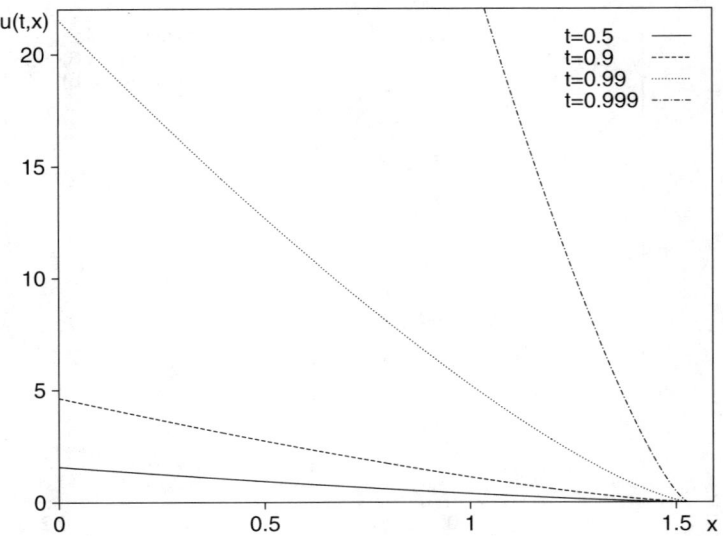

Fig. 3.3. A typical solution $u(t,x)$ of problem (3.9)-(3.11) for $q = 1.5$ and various time moments.

$$u(t,x) = \left(\frac{1 - x/L}{\sqrt{1-t}}\right)^{2/q} \text{ for } x \in [0, L]$$

and

$$u(t,x) = 0 \text{ for } x > L.$$

The temperature $u(t,x)$ grows infinitely as $t \to 1$. At the same time the heat remains localized in the interval $[0, L]$. Figure 3.3 presents a typical behavior of the solution to (3.9)-(3.11).

The equation (3.9) is not of the form (1.5). The function

$$v = u^{q+1}$$

satisfies the problem

$$\frac{\partial v}{\partial t} = \frac{1}{2} v^{q/(q+1)} \frac{\partial^2 v}{\partial x^2}, \; t \in (0,1), \; x > 0, \tag{3.12}$$

$$v(0,x) = (1 - x/L)^{2(q+1)/q}, \; x \in [0, L],$$

$$v(0,x) = 0, \; x > L, \tag{3.13}$$

$$v(t,0) = (1-t)^{-(q+1)/q}, \; t \in [0,1). \tag{3.14}$$

The equation (3.12) has the form (1.5).

We simulate the solution of (3.12)-(3.14) by two algorithms: the algorithm (1.37)-(1.38) and the algorithm based on the layer method (1.24) and linear

8.3 Numerical tests of layer methods for the Dirichlet problems

interpolation. The last one in the case of the problem (3.12)-(3.14) has the form

$$\bar{v}(0,x) = \begin{cases} (1-x/L)^{2(q+1)/q}, & x \in [0,L], \\ 0, & x \in (L,\infty), \end{cases}$$

$$\bar{v}(t_{k+1},x_j) = \frac{1}{2}\bar{v}(t_k, x_j - \sqrt{h}\,(\bar{v}(t_k,x_j))^{q/2(q+1)})$$
$$+ \frac{1}{2}\bar{v}(t_k, x_j + \sqrt{h}\,(\bar{v}(t_k,x_j))^{q/2(q+1)}),$$
$$\text{if } x_j - \sqrt{h}(\bar{v}(t_k,x_j))^{q/2(q+1)} \geq 0;$$

$$\bar{v}(t_{k+1},x_j) = \frac{1}{1+\sqrt{\bar{\lambda}_k}}(1 - t_{k+1-\bar{\lambda}_k})^{-(q+1)/q}$$
$$+ \frac{\sqrt{\bar{\lambda}_k}}{1+\sqrt{\bar{\lambda}_k}}\bar{v}(t_k, x_j + \sqrt{h}(\bar{v}(t_k,x_j))^{q/2(q+1)}),$$

$$\bar{\lambda}_k = \left(\frac{x_j}{(\bar{v}(t_k,x_j))^{q/2(q+1)} \cdot \sqrt{h}}\right)^2, \text{ if } x_j - (\bar{v}(t_k,x_j))^{q/2(q+1)} \cdot \sqrt{h} < 0; \quad (3.15)$$

$$\bar{v}(t_{k+1},x) = \frac{x_{j+1}-x}{h_x}\bar{v}(t_{k+1},x_j) + \frac{x-x_j}{h_x}\bar{v}(t_{k+1},x_{j+1}), \; x_j \leq x \leq x_{j+1},$$
(3.16)
$$j = 0, 1, 2, \ldots, \quad k = 1, \ldots, N,$$

where $x_j = j \times h_x$, $t_k = k \times h$.

In our tests we take $h_x = h$. Tables 3.4 and 3.5 give numerical results obtained by using the algorithm (3.15)-(3.16). The algorithm (1.37)-(1.38) gives similar results and they are omitted here. Table 3.4 presents the errors

$$err_{\bar{v}}(t,h) := \max_j |\bar{v}(t,x_j) - v(t,x_j)|,$$

$$err_{\bar{u}}(t,h) := \max_j |\bar{u}(t,x_j) - u(t,x_j)|, \; \bar{u}(t,x_j) = (\bar{v}(t,x_j))^{1/(q+1)}.$$

For times t which are close to the explosion time $t = 1$, the functions $u(t,x)$ and $v(t,x)$ take large values and the absolute errors become fairly large. In Table 3.5 we present the relative error

$$\delta(t,h) := \frac{err_{\bar{u}}(t,h)}{u(t,0)}$$

at times close to the explosion.

In the experiments, the tested algorithms converge as $O(h)$ which is in complete agreement with our theoretical results.

Table 3.4. Quasilinear equation with power law nonlinearities. Dependence of errors $err_{\bar{v}}(t,h)$ *(top position)* and $err_{\bar{u}}(t,h)$ *(lower position)* in h and t for the algorithm (3.15)-(3.16) under $q = 1.5$.

	$h = 10^{-1}$	$h = 10^{-2}$	$h = 10^{-3}$	$h = 10^{-4}$
$t = 0.5$	$0.8664 \cdot 10^{-1}$	$0.8786 \cdot 10^{-2}$	$0.9705 \cdot 10^{-3}$	$1.018 \cdot 10^{-4}$
	$0.3542 \cdot 10^{-1}$	$0.7693 \cdot 10^{-2}$	$1.685 \cdot 10^{-3}$	$3.622 \cdot 10^{-4}$
$t = 0.9$	> 5	$8.094 \cdot 10^{-1}$	$8.265 \cdot 10^{-2}$	$8.817 \cdot 10^{-3}$
	$5.910 \cdot 10^{-1}$	$8.109 \cdot 10^{-2}$	$8.656 \cdot 10^{-3}$	$8.918 \cdot 10^{-4}$

Table 3.5. Quasilinear equation with power law nonlinearities. Dependence of the relative error $\delta(t,h)$ in h and t for the algorithm (3.15)-(3.16) under $q = 1.5$.

	$h = 10^{-1}$	$h = 10^{-2}$	$h = 10^{-3}$	$h = 10^{-4}$
$t = 0.9$	$1.273 \cdot 10^{-1}$	$1.747 \cdot 10^{-2}$	$1.865 \cdot 10^{-3}$	$1.921 \cdot 10^{-4}$
$t = 0.99$	—	$1.392 \cdot 10^{-1}$	$1.789 \cdot 10^{-2}$	$1.913 \cdot 10^{-3}$
$t = 0.999$	—	—	$1.398 \cdot 10^{-1}$	$1.801 \cdot 10^{-2}$
$t = 0.9999$	—	—	—	$1.400 \cdot 10^{-1}$

8.4 Layer methods for the Neumann problem for semilinear parabolic equations

Let G be a bounded domain in $\mathbf{R}^d$, $Q = [t_0, T) \times G$ be a cylinder in $\mathbf{R}^{d+1}$, $\Gamma = \overline{Q} \setminus Q$. The set Γ is a part of the boundary of the cylinder Q consisting of the upper base and the lateral surface. Consider the Neumann problem for the semilinear parabolic equation

$$\frac{\partial u}{\partial t} + \frac{1}{2}\sum_{i,j=1}^{d} a^{ij}(t,x,u)\frac{\partial^2 u}{\partial x^i \partial x^j} + \sum_{i=1}^{d} b^i(t,x,u)\frac{\partial u}{\partial x^i} + g(t,x,u) = 0, \quad (4.1)$$
$$(t,x) \in Q,$$

with the initial condition

$$u(T,x) = \varphi(x) \qquad (4.2)$$

and the boundary condition

$$\frac{\partial u}{\partial \nu} = \psi(t,x,u), \ t \in [t_0, T], \ x \in \partial G, \qquad (4.3)$$

where ν is the direction of the internal normal to the boundary ∂G at the point $x \in \partial G$.

Using the well known probabilistic representation of the solution to (4.1)-(4.3) (see [70, 82]), we get

8.4 Layer methods forf the Neumann problem   489

$$u(t,x) = E(\varphi(X_{t,x}(T)) + Z_{t,x,0}(T)), \quad (4.4)$$

where $X_{t,x}(s)$, $Z_{t,x,z}(s)$, $t_0 \leq t < T$, $s \geq t$, $x \in \overline{G}$, is a solution of the Cauchy problem to the Ito system of SDEs

$$dX = b(s, X, u(s,X))I_G(X)ds + \sigma(s, X, u(s,X))I_G(X)dw(s)$$
$$+\nu(X)I_{\partial G}(X)d\mu(s), \quad X(t) = x,$$

$$dZ = g(s, X, u(s,X))I_G(X)ds + \psi(s, X, u(s,X))I_{\partial G}(X)d\mu(s),$$
$$Z(t) = z. \quad (4.5)$$

Here $w(s) = (w^1(s), \ldots, w^d(s))^\top$ is a standard Wiener process, $b(s,x,u) = (b^1(s,x,u), \ldots, b^d(s,x,u))^\top$ is a column vector, the matrix $\sigma = \sigma(s,x,u)$ is obtained from the equation

$$\sigma\sigma^\top = a, \ \sigma = \{\sigma^{ij}(s,x,u)\}, \ a = \{a^{ij}(s,x,u)\}, \ i,j = 1,\ldots,d,$$

$\mu(s)$ is a local time of the process X on ∂G, and $I_A(x)$ is the indicator of a set A.

Layer methods proposed in this section for the problem (4.1)-(4.3) are based on weak schemes for integration of SDEs with reflection from Chap. 6.

In Sect. 8.4.1, two layer methods for the nonlinear Neumann problem are constructed. Some convergence theorems are given in Sect. 8.4.2. To realize a layer method in practice, a discretization in the variable x with interpolation at every step is needed to turn the method into an algorithm. Such numerical algorithms are given in Sect. 8.4.3. We restrict ourselves to the one-dimensional ($d = 1$) case in this sections. The case $d \geq 2$ is discussed in Sect. 8.5. Numerical tests are presented in Sect. 8.6. Their results are in complete agreement with theoretical ones. In particular, it was demonstrated in numerical tests that layer methods may be preferable to finite-difference ones in the case of strong advection.

8.4.1 Construction of layer methods

The Neumann boundary value problem in the one-dimensional case has the form

$$\frac{\partial u}{\partial t} + \frac{1}{2}\sigma^2(t,x,u)\frac{\partial^2 u}{\partial x^2} + b(t,x,u)\frac{\partial u}{\partial x} + g(t,x,u) = 0, \quad (4.6)$$
$$t_0 \leq t < T, \ \alpha < x < \beta,$$

$$u(T,x) = \varphi(x), \ \alpha \leq x \leq \beta; \quad (4.7)$$

$$\frac{\partial u}{\partial x}(t,\alpha) = \psi_1(t,u(t,\alpha)), \ \frac{\partial u}{\partial x}(t,\beta) = \psi_2(t,u(t,\beta)), \ t_0 \leq t \leq T. \quad (4.8)$$

In this case Q is a partly open rectangle: $Q = [t_0, T) \times (\alpha, \beta)$, and Γ consists of the upper base $\{T\} \times [\alpha, \beta]$ and two vertical intervals: $[t_0, T) \times \{\alpha\}$ and $[t_0, T) \times \{\beta\}$. We assume that $\sigma(t, x, u) \geq \sigma_* > 0$ for $(t, x) \in \overline{Q}$, $-\infty < u < \infty$.

Let $u = u(t, x)$ be a solution of the problem (4.6)-(4.8) which is supposed to exist, to be unique, and to be sufficiently smooth. Theoretical results on this topic are available in [143, 291] (see also references therein).

Analogously to (4.4), we have the local representation

$$u(t_k, x) = E(u(t_{k+1}, X_{t_k,x}(t_{k+1})) + Z_{t_k,x,0}(t_{k+1})), \tag{4.9}$$

where $X_{t,x}(s)$, $Z_{t,x,z}(s)$, $t_0 \leq t < T$, $s \geq t$, $x \in [\alpha, \beta]$, satisfy (4.5).

Method of order $O(h)$. Applying a slightly modified weak scheme with one-step boundary order $O(h^{3/2})$ from Sect. 6.6 to the system (4.5), it is not difficult to obtain

$$X_{t_k,x}(t_{k+1}) \simeq \bar{X}_{t_k,x}(t_{k+1}) = x + h\tilde{b}_k + h^{1/2}\tilde{\sigma}_k \xi_k,$$
$$Z_{t_k,x,z}(t_{k+1}) \simeq \bar{Z}_{t_k,x,z}(t_{k+1}) = z + h\tilde{g}_k, \quad \text{if } x + h\tilde{b}_k \pm h^{1/2}\tilde{\sigma}_k \in [\alpha, \beta];$$

$$\bar{X}_{t_k,x}(t_{k+1}) = x + (\alpha - x) + \sqrt{h\tilde{\sigma}_k^2 + (\alpha - x)^2},$$
$$\bar{Z}_{t_k,x,z}(t_{k+1}) = z + h\tilde{g}_k - \psi_1(t_k, u(t_k, \alpha))$$
$$\times (\alpha - x - h\tilde{b}_k + \sqrt{h\tilde{\sigma}_k^2 + (\alpha - x)^2}), \quad \text{if } x + h\tilde{b}_k - h^{1/2}\tilde{\sigma}_k < \alpha;$$

$$\bar{X}_{t_k,x}(t_{k+1}) = x + (\beta - x) - \sqrt{h\tilde{\sigma}_k^2 + (\beta - x)^2},$$
$$\bar{Z}_{t_k,x,z}(t_{k+1}) = z + h\tilde{g}_k - \psi_2(t_k, u(t_k, \beta))$$
$$\times (\beta - x - h\tilde{b}_k - \sqrt{h\tilde{\sigma}_k^2 + (\beta - x)^2}), \quad \text{if } x + h\tilde{b}_k + h^{1/2}\tilde{\sigma}_k > \beta. \tag{4.10}$$

Here $\tilde{b}_k$, $\tilde{\sigma}_k$, $\tilde{g}_k$ are the coefficients $b(t, x, u)$, $\sigma(t, x, u)$, $g(t, x, u)$ calculated at the point $(t_k, x, u(t_k, x))$ and $\xi_{N-1}, \xi_{N-2}, \ldots, \xi_0$ are i.i.d. random variables with the law $P(\xi = \pm 1) = 1/2$.

One can see that using the approximation (4.10) and the representation (4.9), we get an implicit one-step approximation for $u(t_k, x)$. Applying the method of simple iteration to this implicit approximation with $u(t_{k+1}, x)$ as a null iteration, we come to the explicit one-step approximation $v(t_k, x)$ of $u(t_k, x)$:

$$v(t_k, x) = \frac{1}{2} u(t_{k+1}, x + hb_k - h^{1/2}\sigma_k) + \frac{1}{2} u(t_{k+1}, x + hb_k + h^{1/2}\sigma_k)$$
$$+ hg_k, \quad \text{if } x + hb_k \pm h^{1/2}\sigma_k \in [\alpha, \beta];$$

8.4 Layer methods forf the Neumann problem

$$v(t_k, x) = u(t_{k+1}, \alpha + \sqrt{h\sigma_k^2 + (\alpha - x)^2})$$
$$-\psi_1(t_{k+1}, u(t_{k+1}, \alpha)) \times (\alpha - x - hb_k + \sqrt{h\sigma_k^2 + (\alpha - x)^2}) + hg_k,$$
$$\text{if } x + hb_k - h^{1/2}\sigma_k < \alpha;$$

$$v(t_k, x) = u(t_{k+1}, \beta - \sqrt{h\sigma_k^2 + (\beta - x)^2})$$
$$-\psi_2(t_{k+1}, u(t_{k+1}, \beta)) \times (\beta - x - hb_k - \sqrt{h\sigma_k^2 + (\beta - x)^2}) + hg_k,$$
$$\text{if } x + hb_k + h^{1/2}\sigma_k > \beta; \qquad (4.11)$$
$$k = N-1, \ldots, 1, 0,$$

where b_k, σ_k, g_k are the coefficients b, σ, g calculated at the point $(t_k, x, u(t_{k+1}, x))$. Let us observe that within the limits of the considered accuracy it is possible to take t_{k+1} instead of t_k. That is why, one can take, for instance, $\psi_1(t_{k+1}, u(t_{k+1}, \alpha))$ instead of $\psi_1(t_k, u(t_{k+1}, \alpha))$ in (4.11).

The corresponding explicit layer method for solving the Neumann problem (4.6)-(4.8) has the form

$$\bar{u}(t_N, x) = \varphi(t_N, x), \quad x \in [\alpha, \beta],$$

$$\bar{u}(t_k, x) = \frac{1}{2}\bar{u}(t_{k+1}, x + h\bar{b}_k - h^{1/2}\bar{\sigma}_k) + \frac{1}{2}\bar{u}(t_{k+1}, x + h\bar{b}_k + h^{1/2}\bar{\sigma}_k)$$
$$+ h\bar{g}_k, \text{ if } x + h\bar{b}_k \pm h^{1/2}\bar{\sigma}_k \in [\alpha, \beta];$$

$$\bar{u}(t_k, x) = \bar{u}(t_{k+1}, \alpha + \sqrt{h\bar{\sigma}_k^2 + (\alpha - x)^2})$$
$$-\psi_1(t_{k+1}, \bar{u}(t_{k+1}, \alpha)) \times (\alpha - x - h\bar{b}_k + \sqrt{h\bar{\sigma}_k^2 + (\alpha - x)^2}) + h\bar{g}_k,$$
$$\text{if } x + h\bar{b}_k - h^{1/2}\bar{\sigma}_k < \alpha;$$

$$\bar{u}(t_k, x) = \bar{u}(t_{k+1}, \beta - \sqrt{h\bar{\sigma}_k^2 + (\beta - x)^2})$$
$$-\psi_2(t_{k+1}, \bar{u}(t_{k+1}, \beta)) \times (\beta - x - h\bar{b}_k - \sqrt{h\bar{\sigma}_k^2 + (\beta - x)^2}) + h\bar{g}_k,$$
$$\text{if } x + h\bar{b}_k + h^{1/2}\bar{\sigma}_k > \beta; \qquad (4.12)$$
$$k = N-1, \ldots, 1, 0,$$

where $\bar{b}_k = \bar{b}_k(x) = b(t_k, x, \bar{u}(t_{k+1}, x))$, $\bar{\sigma}_k = \bar{\sigma}_k(x) = \sigma(t_k, x, \bar{u}(t_{k+1}, x))$, $\bar{g}_k = \bar{g}_k(x) = g(t_k, x, \bar{u}(t_{k+1}, x))$.

This layer method has the one-step error near the boundary estimated by $O(h^{3/2})$ and for internal points estimated by $O(h^2)$ (see Lemma 4.1 in the next subsection). We prove that its order of convergence is $O(h)$ when the boundary condition does not depend on the solution (see Theorem 4.3). Apparently, this is so in the general case as well (see Remark 4.4).

Another method with the same one-step error is given in Sect. 8.4.4.

8 Numerical solution of the nonlinear Dirichlet and Neumann problems

Method of order $O(\sqrt{h})$. Applying the weak scheme with one-step boundary order $O(h)$ from Sect. 6.6 to the system (4.5), it is not difficult to obtain

$$X_{t_k,x}(t_{k+1}) \simeq \bar{X}_{t_k,x}(t_{k+1}) = x + h\tilde{b}_k + h^{1/2}\tilde{\sigma}_k\xi_k,$$
$$Z_{t_k,x,z}(t_{k+1}) \simeq \bar{Z}_{t_k,x,z}(t_{k+1}) = z + h\tilde{g}_k, \text{ if } x + h\tilde{b}_k \pm h^{1/2}\tilde{\sigma}_k \in [\alpha,\beta];$$

$$\bar{X}_{t_k,x}(t_{k+1}) = x + qh^{1/2}, \ \bar{Z}_{t_k,x,z}(t_{k+1}) = z - \psi_1(t_k, u(t_k,\alpha))qh^{1/2},$$
$$\text{if } x + h\tilde{b}_k - h^{1/2}\tilde{\sigma}_k < \alpha;$$

$$\bar{X}_{t_k,x}(t_{k+1}) = x - qh^{1/2}, \ \bar{Z}_{t_k,x,z}(t_{k+1}) = z + \psi_2(t_k, u(t_k,\beta))qh^{1/2},$$
$$\text{if } x + h\tilde{b}_k + h^{1/2}\tilde{\sigma}_k > \beta. \tag{4.13}$$

Here $\tilde{b}_k, \tilde{\sigma}_k, \tilde{g}_k$ are the coefficients $b(t,x,u), \sigma(t,x,u), g(t,x,u)$ calculated at the point $(t_k, x, u(t_k,x))$, $\xi_{N-1}, \xi_{N-2}, \ldots, \xi_0$ are i.i.d. random variables with the law $P(\xi = \pm 1) = 1/2$, and q is a positive number (see Remark 4.7 below, where a discussion on choosing q is given).

As before, we obtain the following explicit one-step approximation $v(t_k, x)$ of $u(t_k, x)$:

$$v(t_k, x) = \frac{1}{2}u(t_{k+1}, x + hb_k - h^{1/2}\sigma_k) + \frac{1}{2}u(t_{k+1}, x + hb_k + h^{1/2}\sigma_k) + hg_k,$$
$$\text{if } x + hb_k \pm h^{1/2}\sigma_k \in [\alpha,\beta];$$

$$v(t_k, x) = u(t_{k+1}, x + qh^{1/2}) - \psi_1(t_{k+1}, u(t_{k+1}, \alpha))qh^{1/2},$$
$$\text{if } x + hb_k - h^{1/2}\sigma_k < \alpha;$$

$$v(t_k, x) = u(t_{k+1}, x - qh^{1/2}) + \psi_2(t_{k+1}, u(t_{k+1}, \beta))qh^{1/2},$$
$$\text{if } x + hb_k + h^{1/2}\sigma_k > \beta; \tag{4.14}$$

$$k = N-1, \ldots, 1, 0.$$

The corresponding explicit layer method for solving the Neumann problem (4.6)-(4.8) has the form

$$\bar{u}(t_N, x) = \varphi(t_N, x), \ x \in [\alpha, \beta],$$

$$\bar{u}(t_k, x) = \frac{1}{2}\bar{u}(t_{k+1}, x + h\tilde{b}_k - h^{1/2}\tilde{\sigma}_k) + \frac{1}{2}\bar{u}(t_{k+1}, x + h\tilde{b}_k + h^{1/2}\tilde{\sigma}_k)$$
$$+ h\tilde{g}_k, \text{ if } x + h\tilde{b}_k \pm h^{1/2}\tilde{\sigma}_k \in [\alpha,\beta];$$

$$\bar{u}(t_k, x) = \bar{u}(t_{k+1}, x + qh^{1/2}) - \psi_1(t_{k+1}, \bar{u}(t_{k+1}, \alpha))qh^{1/2},$$
$$\text{if } x + h\bar{b}_k - h^{1/2}\bar{\sigma}_k < \alpha;$$

$$\bar{u}(t_k, x) = \bar{u}(t_{k+1}, x - qh^{1/2}) + \psi_2(t_{k+1}, \bar{u}(t_{k+1}, \beta))qh^{1/2},$$
$$\text{if } x + h\bar{b}_k + h^{1/2}\bar{\sigma}_k > \beta; \qquad (4.15)$$

$$k = N - 1, \ldots, 1, 0,$$

where $\bar{b}_k = \bar{b}_k(x) = b(t_k, x, \bar{u}(t_{k+1}, x))$, $\bar{\sigma}_k = \bar{\sigma}_k(x) = \sigma(t_k, x, \bar{u}(t_{k+1}, x))$, $\bar{g}_k = \bar{g}_k(x) = g(t_k, x, \bar{u}(t_{k+1}, x))$.

This layer method is simpler but less accurate than (4.12). Its one-step error near the boundary is $O(h)$ and for internal points is $O(h^2)$ (see Lemma 4.5 in the next subsection). We prove that its order of convergence is $O(h^{1/2})$ when the boundary condition does not depend on the solution (see Theorem 4.6). Apparently, this is so in the general case as well.

A method of the same convergence order is proposed for the linear Neumann problem in [41]. This method is extended to the nonlinear problem in Sect. 8.4.4.

8.4.2 Convergence theorems

We make the following assumptions.

(i) There exists a unique solution $u(t, x)$ of the problem (4.6)-(4.8) such that

$$-\infty \leq u_\circ < u_* \leq u(t, x) \leq u^* < u^\circ \leq \infty, \ t_0 \leq t \leq T, \ x \in [\alpha, \beta], \quad (4.16)$$

where $u_\circ$, u_*, u^*, u° are some constants, and there exist the uniformly bounded derivatives:

$$\left|\frac{\partial^{i+j} u}{\partial t^i \partial x^j}\right| \leq K, \ i = 0, \ j = 1, 2, 3, 4; \ i = 1, \ j = 0, 1, 2; \ i = 2, \ j = 0; \quad (4.17)$$

$$t_0 \leq t \leq T, \ x \in [\alpha, \beta].$$

(ii) The coefficients $b(t, x, u)$, $\sigma(t, x, u)$, $g(t, x, u)$ are uniformly bounded and uniformly satisfy the Lipschitz condition with respect to x and u:

$$|b| \leq K, \ |\sigma| \leq K, \ |g| \leq K,$$
$$|b(t, x_2, u_2) - b(t, x_1, u_1)| + |\sigma(t, x_2, u_2) - \sigma(t, x_1, u_1)|$$
$$+ |g(t, x_2, u_2) - g(t, x_1, u_1)| \leq K(|x_2 - x_1| + |u_2 - u_1|), \quad (4.18)$$
$$t_0 \leq t \leq T, \ x \in [\alpha, \beta], \ u_\circ < u < u^\circ.$$

Lemma 4.1. *Under the assumptions* (i) *and* (ii), *the one-step error* $\rho(t_k, x)$ *of the method* (4.12) *is estimated as*

$$|\rho(t_k, x)| = |v(t_k, x) - u(t_k, x)| \leq Ch^2, \quad x + hb_k \pm h^{1/2}\sigma_k \in [\alpha, \beta]; \quad (4.19)$$

$$|\rho(t_k, x)| = |v(t_k, x) - u(t_k, x)| \leq Ch^{3/2}, \quad (4.20)$$
$$x + hb_k - h^{1/2}\sigma_k < \alpha \text{ or } x + hb_k + h^{1/2}\sigma_k > \beta,$$

where $v(t_k, x)$ *is the corresponding one-step approximation, C does not depend on h, k, x.*

Proof. If both points $x + hb_k \pm \sqrt{h}\sigma_k$ belong to $[\alpha, \beta]$, (4.19) follows directly from Lemma 7.2.1.

Let us consider the case when the point $x + hb_k - h^{1/2}\sigma_k < \alpha$. Due to (4.11), we get

$$v(t_k, x) = u(t_{k+1}, x + \Delta X^\alpha) - \psi_1(t_{k+1}, u(t_{k+1}, \alpha)) \times (\Delta X^\alpha - hb_k) + hg_k, \quad (4.21)$$

where

$$\Delta X^\alpha := \alpha - x + \sqrt{h\sigma_k^2 + (\alpha - x)^2}.$$

It is clear that

$$|\alpha - x| \leq Ch^{1/2}, \quad |\Delta X^\alpha| \leq Ch^{1/2}. \quad (4.22)$$

Taking into account that $\psi_1(t_{k+1}, u(t_{k+1}, \alpha)) = u'_x(t_{k+1}, \alpha)$ (see (4.8)), then expanding the functions $u(t_{k+1}, x + \Delta X^\alpha)$ and $u'_x(t_{k+1}, x + (\alpha - x))$ at the point (t_k, x), and using the assumptions (i), (ii), and the inequalities (4.22), we get

$$v(t_k, x) = u + \frac{\partial u}{\partial t}h + \frac{\partial u}{\partial x}\Delta X^\alpha + \frac{1}{2}\frac{\partial^2 u}{\partial x^2}(\Delta X^\alpha)^2$$
$$- \frac{\partial u}{\partial x}(\Delta X_\alpha - hb_k) - \frac{\partial^2 u}{\partial x^2}(\alpha - x)\Delta X^\alpha + g_k h + O(h^{3/2})$$
$$= u + h(\frac{\partial u}{\partial t} + b_k \frac{\partial u}{\partial x} + g_k) + \frac{1}{2}\frac{\partial^2 u}{\partial x^2}\Delta X^\alpha(\Delta X^\alpha - 2(\alpha - x))$$
$$+ O(h^{3/2}), \quad (4.23)$$

where the function u and its derivatives are calculated at the point (t_k, x). The expression $\Delta X^\alpha(\Delta X^\alpha - 2(\alpha - x))$ is equal to $h\sigma_k^2$.

Due to the assumptions (i) and (ii), we obtain

$$b_k = b(t_k, x, u(t_{k+1}, x)) = \tilde{b}_k + O(h),$$
$$\sigma_k^2 = \tilde{\sigma}_k^2 + O(h), \quad g_k = \tilde{g}_k + O(h),$$

where $\tilde{b}_k$, $\tilde{\sigma}_k$, $\tilde{g}_k$ are calculated at the point $(t_k, x, u(t_k, x))$.

Then we get from (4.23) that

$$v(t_k, x) = u + h\left(\frac{\partial u}{\partial t} + b\frac{\partial u}{\partial x} + \frac{\sigma^2}{2}\frac{\partial^2 u}{\partial x^2} + g\right) + O(h^{3/2}). \quad (4.24)$$

Since $u(t, x)$ is the solution of problem (4.6)-(4.8), the relation (4.24) implies

$$v(t_k, x) = u(t_k, x) + O(h^{3/2}).$$

The case $x + h\check{b}_k + h^{1/2}\check{\sigma}_k > \beta$ can be considered analogously. □

To prove the theorem on global convergence of the method (4.12), we need some auxiliary constructions. Let us introduce the random sequence X_i, Z_i:

$$X_k = x, \ Z_k = 0,$$

$$X_{i+1} = X_i + h\check{b}_i + h^{1/2}\check{\sigma}_i \xi_i, \ Z_{i+1} = Z_i + h\check{g}_i,$$
$$\text{if } X_i + h\check{b}_i \pm h^{1/2}\check{\sigma}_i \in [\alpha, \beta];$$

$$X_{i+1} = X_i + \Delta X_i^\alpha, \ Z_{i+1} = Z_i + h\check{g}_i - \psi_1(t_{i+1}) \times (\Delta X_i^\alpha - h\check{b}_i),$$
$$\Delta X_i^\alpha := (\alpha - X_i) + \sqrt{h\check{\sigma}_i^2 + (\alpha - X_i)^2},$$
$$\text{if } X_i + h\check{b}_i - h^{1/2}\check{\sigma}_i < \alpha;$$

$$X_{i+1} = X_i + \Delta X_i^\beta, \ Z_{i+1} = Z_i + h\check{g}_i - \psi_2(t_{i+1}) \times (\Delta X_i^\beta - h\check{b}_i),$$
$$\Delta X_i^\beta := (\beta - X_i) - \sqrt{h\check{\sigma}_i^2 + (\beta - X_i)^2},$$
$$\text{if } X_i + h\check{b}_i + h^{1/2}\check{\sigma}_i > \beta; \quad (4.25)$$
$$i = k, \ldots, N-1, \ k \geq 0.$$

Here ξ_i are i.i.d. random variables with the law $P(\xi = \pm 1) = 1/2$ and $\check{b}_i = \bar{b}_i(X_i) = b(t_i, X_i, \bar{u}(t_{i+1}, X_i))$, $\check{\sigma}_i = \bar{\sigma}_i(X_i) = \sigma(t_i, X_i, \bar{u}(t_{i+1}, X_i))$, $\check{g}_i = \bar{g}_i(X_i) = g(t_i, X_i, \bar{u}(t_{i+1}, X_i))$. Let us note that the function $\bar{u}(t_i, x)$, $i = 0, \ldots, N$, $x \in [\alpha, \beta]$, is uniquely defined by (4.12). Evidently, the sequence (t_i, X_i) is a Markov chain.

Introduce the boundary layer $\partial \Gamma \in \overline{Q}$: for all the points $(t_k, x) \in \overline{Q} \setminus \partial \Gamma$, both points $x + h\bar{b}_k(x) \pm h^{1/2}\bar{\sigma}_k(x)$ belong to $[\alpha, \beta]$, and for the points $(t_k, x) \in \partial \Gamma$, either $x + h\bar{b}_k(x) - h^{1/2}\bar{\sigma}_k(x) \notin [\alpha, \beta]$ or $x + h\bar{b}_k(x) + h^{1/2}\bar{\sigma}_k(x) \notin [\alpha, \beta]$.

Lemma 4.2. *Under the assumptions (i) and (ii), the mean number of steps $\varkappa(t_k, x)$ which the Markov chain (t_i, X_i), $i = k, \ldots, N$, $k \geq 0$, $X_k = x$, spends in the layer $\partial \Gamma$, is estimated as*

$$E\varkappa(t_k, x) \leq \frac{C}{\sqrt{h}},$$

where C does not depend on h, k, x.

The proof of Lemma 4.2 differs only little from the proof of Lemma 6.6.3 and is therefore omitted.

Theorem 4.3. *Let the Neumann problem for the equation (4.6) with condition (4.7) have the following boundary conditions*

$$\frac{\partial u}{\partial x}(t,\alpha) = \psi_1(t), \ \frac{\partial u}{\partial x}(t,\beta) = \psi_2(t), \ t_0 \leq t \leq T. \tag{4.26}$$

Under the assumptions (i) and (ii), the method (4.12) has the first order of convergence with respect to h, i.e.,

$$|\bar{u}(t_k, x) - u(t_k, x)| \leq Kh,$$

where K does not depend on h, k, x.

To prove this theorem, we exploit ideas of proving convergence theorems for probabilistic methods from Chap. 6. This proof is analogous to the proof of Theorem 1.5 (see details in [212]).

Remark 4.4. Most likely, the conclusion of Theorem 4.3 is true under the boundary conditions (4.8). We have not succeeded in proving such a general theorem but we can prove it in the case of the linear boundary conditions

$$\frac{\partial u}{\partial x}(t,\alpha) = \varphi_1(t)u(t,\alpha) + \psi_1(t),$$
$$\frac{\partial u}{\partial x}(t,\beta) = \varphi_2(t)u(t,\beta) + \psi_2(t), \ t_0 \leq t \leq T. \tag{4.27}$$

Moreover, numerical experiments confirm the just mentioned conjecture (see Sect. 8.6.1).

It turns out that the method (4.15) in the case (4.26) (and in the case (4.27) as well) is convergent with order $O(h^{1/2})$. As above, this fact is apparently true for the case of general boundary conditions.

Let us formulate the corresponding results. First we note that the method (4.15) generates a Markov chain (t_i, X_i) for which Lemma 4.2 takes place.

Lemma 4.5. *Under the assumptions (i) and (ii), the one-step error $\rho(t_k, x)$ of the method (4.15) is estimated as*

$$|\rho(t_k, x)| = |v(t_k, x) - u(t_k, x)| \leq Ch^2, \ x + hb_k \pm h^{1/2}\sigma_k \in [\alpha, \beta];$$
$$|\rho(t_k, x)| = |v(t_k, x) - u(t_k, x)| \leq Ch,$$
$$x + hb_k - h^{1/2}\sigma_k < \alpha \ or \ x + hb_k + h^{1/2}\sigma_k > \beta,$$

where $v(t_k, x)$ is defined by (4.14), C does not depend on h, k, x.

Theorem 4.6. *Under the assumptions (i) and (ii), the method (4.15) for the Neumann problem (4.6)-(4.7), (4.26) is of order $O(h^{1/2})$, i.e.,*

$$|\bar{u}(t_k,x) - u(t_k,x)| \leq K h^{1/2}, \tag{4.28}$$

where K does not depend on h, k, x.

Proofs of Lemma 4.5 and Theorem 4.6 are similar to those of Lemma 4.1 and Theorem 4.3.

Remark 4.7. The layer method (4.15) has the parameter q, which, in principle, may be any positive number. Naturally, the value of q affects the method accuracy: K in (4.28) depends on q. By an extended analysis of the one-step boundary error and of the mean number of steps of the corresponding Markov chain in the boundary layer $\partial \Gamma$, we get

$$K \leq C_1 \left(\frac{1}{q} \max_{(t,x)\in \bar{Q}} \left|\frac{\partial u}{\partial t}\right| + \frac{q}{2} \max_{(t,x)\in \bar{Q}} \left|\frac{\partial^2 u}{\partial x^2}\right| \right) + C_2,$$

where C_i, $i=1,2$, do not depend on h, k, x, and q.

Evidently, both large and small values of q are not appropriate. If we know estimates of derivatives of the solution to the considered problem, it is not difficult to find an appropriate q. But generally the choice of q requires a special consideration.

Let $b(t,x,u) \equiv 0$ and $g(t,x,u) \equiv 0$. In this case the one-step boundary error $\rho(t_k,x)$ of the method (4.15) near α is evaluated as

$$\rho(t_k,x) = \frac{1}{2}\frac{\partial^2 u}{\partial x^2}(t_k,x) \times \left(q^2 h + 2(x-\alpha)qh^{1/2} - h\sigma_k^2 \right) + O(h^{3/2}),$$

$$x - h^{1/2}\sigma_k < \alpha,$$

and analogously near β. Taking $qh^{1/2} = \alpha - x + \sqrt{h\sigma_k^2 + (\alpha-x)^2}$, we obtain $\rho(t_k,x) = O(h^{3/2})$. Substitution of this q (depending on k and x) in (4.15) gives us a method with convergence order $O(h)$, which coincides with the method (4.12). Such an analysis also suggests that it is preferable to take $q \approx \sigma$.

8.4.3 Numerical algorithms

Consider the equidistant space discretization with space step h_x (recall that the notation for time step is h): $x_j = \alpha + jh_x$, $j = 0, 1, 2, \ldots, M$, $h_x = (\beta-\alpha)/M$.

Using linear interpolation, we construct the following algorithm on the basis of the method (4.12) (we denote it as $\bar{u}(t_k,x)$ again, since this should not cause any confusion).

Algorithm 4.8 *The algorithm is defined by the following formulas*

$$\bar{u}(t_N, x) = \varphi(t_N, x), \ x \in [\alpha, \beta],$$

$$\bar{u}(t_k, x_j) = \frac{1}{2}\bar{u}(t_{k+1}, x_j + h\bar{b}_{k,j} - h^{1/2}\bar{\sigma}_{k,j})$$
$$+ \frac{1}{2}\bar{u}(t_{k+1}, x_j + h\bar{b}_{k,j} + h^{1/2}\bar{\sigma}_{k,j}) + h\bar{g}_{k,j},$$
$$\text{if } x_j + h\bar{b}_{k,j} \pm h^{1/2}\bar{\sigma}_{k,j} \in [\alpha, \beta];$$

$$\bar{u}(t_k, x_j) = \bar{u}\left(t_{k+1}, \alpha + \sqrt{h\bar{\sigma}_{k,j}^2 + (\alpha - x_j)^2}\right)$$
$$-\psi_1(t_{k+1}, \bar{u}(t_{k+1}, \alpha)) \times \left(\alpha - x_j - h\bar{b}_{k,j} + \sqrt{h\bar{\sigma}_{k,j}^2 + (\alpha - x_j)^2}\right) + h\bar{g}_{k,j},$$
$$\text{if } x_j + h\bar{b}_{k,j} - h^{1/2}\bar{\sigma}_{k,j} < \alpha;$$

$$\bar{u}(t_k, x_j) = \bar{u}\left(t_{k+1}, \beta - \sqrt{h\bar{\sigma}_{k,j}^2 + (\beta - x_j)^2}\right)$$
$$-\psi_2(t_{k+1}, \bar{u}(t_{k+1}, \beta)) \times \left(\beta - x_j - h\bar{b}_{k,j} - \sqrt{h\bar{\sigma}_{k,j}^2 + (\beta - x_j)^2}\right) + h\bar{g}_{k,j},$$
$$\text{if } x_j + h\bar{b}_{k,j} + h^{1/2}\bar{\sigma}_{k,j} > \beta; \ j = 1, 2, \ldots, M-1, \quad (4.29)$$

$$\bar{u}(t_k, x) = \frac{x_{j+1} - x}{h_x}\bar{u}(t_k, x_j) + \frac{x - x_j}{h_x}\bar{u}(t_k, x_{j+1}), \ x_j < x < x_{j+1}, \quad (4.30)$$
$$j = 0, 1, 2, \ldots, M-1, \ k = N-1, \ldots, 1, 0,$$

where $\bar{b}_{k,j}, \bar{\sigma}_{k,j}, \bar{g}_{k,j}$ are the coefficients b, σ, g calculated at the point $(t_k, x_j, \bar{u}(t_{k+1}, x_j))$.

Using probabilistic arguments, we prove the convergence theorem (see details in [212]).

Theorem 4.9. *Consider the problem* (4.6)-(4.7), (4.26). *If h_x is taken equal to $\varkappa h$, $\varkappa$ is a positive constant, then under the assumptions (i) and (ii) Algorithm 4.8 has the first order of convergence, i.e., the approximation $\bar{u}(t_k, x)$ from the formulas* (4.29)-(4.30) *satisfies the relation*

$$|\bar{u}(t_k, x) - u(t_k, x)| \leq Kh,$$

where K does not depend on x, h, k.

Remark 4.10. We pay attention that the factor $\varkappa$ in Theorem 4.9 is arbitrary. To choose h_x, we do not need any stability criteria in comparison with finite-difference schemes, on which the Lax–Richtmyer equivalence theorem imposes a requirement on relation between the time step Δt and the space step Δx (see Example 7.1.2 for a detailed discussion). At the same time, accuracy of the algorithm (4.29)-(4.30) depends on $\varkappa$. In practice, a choice of $\varkappa$ is connected with interpolation properties of the solution.

Remarks 1.12 and 1.13 are valid here.

On the basis of linear interpolation and the layer method (4.15), we get the following algorithm.

Algorithm 4.11 *The algorithm is defined by the following formulas*

$$\bar{u}(t_N, x) = \varphi(t_N, x), \ x \in [\alpha, \beta],$$

$$\bar{u}(t_k, x_j) = \frac{1}{2}\bar{u}(t_{k+1}, x_j + h\bar{b}_{k,j} - h^{1/2}\bar{\sigma}_{k,j}) + \frac{1}{2}\bar{u}(t_{k+1}, x_j + h\bar{b}_{k,j} + h^{1/2}\bar{\sigma}_{k,j})$$
$$+ h\bar{g}_{k,j}, \ \text{if} \ x_j + h\bar{b}_{k,j} \pm h^{1/2}\bar{\sigma}_{k,j} \in [\alpha, \beta] \, ;$$

$$\bar{u}(t_k, x_j) = \bar{u}(t_{k+1}, x_j + q\sqrt{h}) - \psi_1(t_{k+1}, \bar{u}(t_{k+1}, \alpha)) \times qh^{1/2},$$
$$\text{if} \ x_j + h\bar{b}_{k,j} - h^{1/2}\bar{\sigma}_{k,j} < \alpha;$$

$$\bar{u}(t_k, x_j) = \bar{u}(t_{k+1}, x_j - q\sqrt{h}) + \psi_2(t_{k+1}, \bar{u}(t_{k+1}, \beta)) \times qh^{1/2},$$
$$\text{if} \ x_j + h\bar{b}_{k,j} + h^{1/2}\bar{\sigma}_{k,j} > \beta; \ j = 1, 2, \ldots, M-1, \quad (4.31)$$

$$\bar{u}(t_k, x) = \frac{x_{j+1} - x}{h_x}\bar{u}(t_k, x_j) + \frac{x - x_j}{h_x}\bar{u}(t_k, x_{j+1}), \ x_j < x < x_{j+1}, \quad (4.32)$$
$$j = 0, 1, 2, \ldots, M-1, \ k = N-1, \ldots, 1, 0,$$

where $\bar{b}_{k,j}$, $\bar{\sigma}_{k,j}$, $\bar{g}_{k,j}$ are the coefficients b, σ, g calculated at the point $(t_k, x_j, \bar{u}(t_{k+1}, x_j))$.

Theorem 4.12. *Consider the problem (4.6)-(4.7), (4.26). If h_x is taken equal to $\varkappa h^{3/4}$, $\varkappa$ is a positive constant, then under the assumptions (i) and (ii) Algorithm 4.11 has order of convergence $O(\sqrt{h})$, i.e., the approximation $\bar{u}(t_k, x)$ from the formulas (4.31)-(4.32) satisfies the relation*

$$|\bar{u}(t_k, x) - u(t_k, x)| \leq K\sqrt{h},$$

where K does not depend on x, h, k.

8.4.4 Some other layer methods

In this subsection two additional methods in the case of $d = 1$ are given.

Using the concept of fictitious knots, we obtain the following method (see details in the preprint [208]):

$$\bar{u}(t_N, x) = \varphi(x), \ x \in [\alpha, \beta],$$

$$\bar{u}(t_k,x) = \frac{1}{2}\bar{u}(t_{k+1},x+h\bar{b}_k-h^{1/2}\bar{\sigma}_k) + \frac{1}{2}\bar{u}(t_{k+1},x+h\bar{b}_k+h^{1/2}\bar{\sigma}_k)$$
$$+h\bar{g}_k, \text{ if } x+h\bar{b}_k\pm h^{1/2}\bar{\sigma}_k \in [\alpha,\beta],$$

$$\bar{u}(t_k,x) = \frac{1}{2}\bar{u}(t_{k+1},2\alpha-x-h\bar{b}_k+h^{1/2}\bar{\sigma}_k)$$
$$-\psi_1(t_{k+1},\bar{u}(t_{k+1},\alpha))\times(\alpha-x-h\bar{b}_k+h^{1/2}\bar{\sigma}_k)$$
$$+\frac{1}{2}\bar{u}(t_{k+1},x+h\bar{b}_k+h^{1/2}\bar{\sigma}_k)+h\bar{g}_k, \text{ if } x+h\bar{b}_k-h^{1/2}\bar{\sigma}_k<\alpha,$$

$$\bar{u}(t_k,x) = \frac{1}{2}\bar{u}(t_{k+1},x+h\bar{b}_k-h^{1/2}\bar{\sigma}_k) + \frac{1}{2}\bar{u}(t_{k+1},2\beta-x-h\bar{b}_k-h^{1/2}\bar{\sigma}_k)$$
$$+\psi_2(t_{k+1},\bar{u}(t_{k+1},\beta))\times(x+h\bar{b}_k+h^{1/2}\bar{\sigma}_k-\beta)+h\bar{g}_k,$$
$$\text{if } x+h\bar{b}_k+h^{1/2}\bar{\sigma}_k>\beta; \qquad (4.33)$$

$$k=N-1,\ldots,1,0,$$

where $\bar{b}_k$, $\bar{\sigma}_k$, $\bar{g}_k$ are the coefficients b, σ, g calculated at the point $(t_k,x,\bar{u}(t_{k+1},x))$.

The method (4.33) is an explicit layer method for solving the Neumann problem (4.6)-(4.8). We prove that its one-step error near the boundary is $O(h^{3/2})$ and for internal points is $O(h^2)$. Apparently, this method has order of convergence $O(h)$.

The method (4.33) is more complicated than the method (4.12). At the same time, it demonstrates more accurate results than (4.12) in our numerical tests (see Sect. 8.6.1).

The method (4.15) is an extension of the method of order $O(h^{1/2})$ from Sect. 6.6 to the nonlinear case. In [41] another method of order $O(h^{1/2})$ for linear Neumann problems is proposed. Its extension to the nonlinear Neumann problem (4.6)-(4.8) has the form

$$\bar{u}(t_N,x) = \varphi(t_N,x), \ x \in [\alpha,\beta],$$

$$\bar{u}(t_k,x) = \frac{1}{2}\bar{u}(t_{k+1},x+h\bar{b}_k-h^{1/2}\bar{\sigma}_k) + \frac{1}{2}\bar{u}(t_{k+1},x+h\bar{b}_k+h^{1/2}\bar{\sigma}_k)$$
$$+h\bar{g}_k, \text{ if } x+h\bar{b}_k\pm h^{1/2}\bar{\sigma}_k \in [\alpha,\beta];$$

$$\bar{u}(t_k,x) = \frac{1}{2}\bar{u}(t_{k+1},\alpha) + \frac{1}{2}\bar{u}(t_{k+1},x+h\bar{b}_k+h^{1/2}\bar{\sigma}_k)$$
$$-\frac{1}{2}\psi_1(t_{k+1},\bar{u}(t_{k+1},\alpha))(\alpha-x-h\bar{b}_k+h^{1/2}\bar{\sigma}_k)+h\bar{g}_k,$$
$$\text{if } x+h\bar{b}_k-h^{1/2}\bar{\sigma}_k<\alpha;$$

$$\bar{u}(t_k, x) = \frac{1}{2}\bar{u}(t_{k+1}, x + h\bar{b}_k - h^{1/2}\bar{\sigma}_k) + \frac{1}{2}\bar{u}(t_{k+1}, \beta)$$
$$-\frac{1}{2}\psi_2(t_{k+1}, \bar{u}(t_{k+1}, \beta))(\beta - x - h\bar{b}_k - h^{1/2}\bar{\sigma}_k) + h\bar{g}_k,$$
$$\text{if } x + h\bar{b}_k + h^{1/2}\bar{\sigma}_k > \beta; \qquad (4.34)$$
$$k = N-1, \ldots, 1, 0,$$

where $\bar{b}_k = \bar{b}_k(x) = b(t_k, x, \bar{u}(t_{k+1}, x))$, $\bar{\sigma}_k = \bar{\sigma}_k(x) = \sigma(t_k, x, \bar{u}(t_{k+1}, x))$, $\bar{g}_k = \bar{g}_k(x) = g(t_k, x, \bar{u}(t_{k+1}, x))$.

We prove that the one-step error of this method near the boundary is $O(h)$ and for internal points is $O(h^2)$. Apparently, this layer method has order of convergence $O(h^{1/2})$. It is more complicated near the boundary than (4.15). At the same time, the method (4.34) demonstrates more accurate results than (4.15) in our numerical tests (see Sect. 8.6.1).

Algorithms based on linear interpolation and on the layer methods of this section can be written as in Sect. 8.4.3.

8.5 Extension to the multi-dimensional Neumann problem

It is not difficult to generalize the layer methods given above for an arbitrary dimension d. For instance, let us extend the method (4.15) to the case of $d = 2$. Recall that in this case σ is a 2×2-matrix satisfying the relation $\sigma\sigma^\top = a$.

Introduce the notation $_iX_{k+1} := (_iX^1_{k+1}, {}_iX^2_{k+1})$,

$$_iX^1_{k+1} = x^1 + h\bar{b}^1_k + \sqrt{h}\bar{\sigma}^{11}_k \, _i\xi^1 + \sqrt{h}\bar{\sigma}^{12}_k \, _i\xi^2,$$
$$_iX^2_{k+1} = x^2 + h\bar{b}^2_k + \sqrt{h}\bar{\sigma}^{21}_k \, _i\xi^1 + \sqrt{h}\bar{\sigma}^{22}_k \, _i\xi^2,$$
$$i = 1, 2, 3, 4, \ x = (x^1, x^2) \in \overline{G} \subset \mathbf{R}^2,$$

where $_1\xi = (-1, -1)$, $_2\xi = (-1, 1)$, $_3\xi = -_1\xi$, $_4\xi = -_2\xi$, and $\bar{b}_k = (\bar{b}^1_k, \bar{b}^2_k)$ and $\bar{\sigma}_k = \{\bar{\sigma}^{jl}_k\}$ are the coefficients $b(t, x, u)$, $\sigma(t, x, u)$ calculated at the point $(t_k, x, \bar{u}(t_{k+1}, x))$.

If the point $x = (x^1, x^2) \in G$ is sufficiently far from the boundary ∂G (more precisely, if the points $_iX_{k+1}$, $i = 1, 2, 3, 4$, belong to $\overline{G}$), the layer method has the form (cf. (7.3.4) and (2.1)):

$$\bar{u}(t_k, x^1, x^2) = \sum_{i=1}^{4} \frac{1}{4}\bar{u}(t_{k+1}, {}_iX^1_{k+1}, {}_iX^2_{k+1}) + h\bar{g}_k, \qquad (5.1)$$

where $\bar{g}_k$ is the coefficient $g(t, x, u)$ calculated at the point $(t_k, x, \bar{u}(t_{k+1}, x))$.

If the point $x = (x^1, x^2) \in \overline{G}$ is close or belongs to the boundary ∂G, then some of the points $_iX_{k+1} = (_iX^1_{k+1}, {}_iX^2_{k+1})$, $i = 1, 2, 3, 4$, may be outside the domain $\overline{G}$. In this case let us consider the projection $\bar{x}$ of the point x on ∂G. Let $\nu = (\nu^1, \nu^2)$ be the unit vector of the internal normal at the point $\bar{x}$. Clearly, if $x \neq \bar{x}$, $\nu = (x - \bar{x})/|x - \bar{x}|$. Then we put

$$\bar{u}(t_k, x^1, x^2) = \bar{u}(t_{k+1}, x + qh^{1/2}\nu) - \psi(t_{k+1}, \bar{x}, \bar{u}(t_{k+1}, \bar{x})) \times qh^{1/2}. \quad (5.2)$$

Thus, we obtain the method (5.1)-(5.2): the rule (5.1) is for points $x = (x^1, x^2) \in G$ such that all the corresponding points $_iX = (_iX^1, {}_iX^2)$, $i = 1, 2, 3, 4$, belong to $\overline{G}$, and the rule (5.2) is applied otherwise. The error of the one-step approximation corresponding to (5.1) is of order $O(h^2)$ and that corresponding to (5.2) is of order $O(h)$. If the function ψ does not depend on u, we can prove that the layer method (5.1)-(5.2) has the global error estimated by $O(h^{1/2})$. These assertions can be checked directly without requiring new ideas than those in Sect. 8.4.2.

To construct the corresponding numerical algorithms, we use linear interpolation as in the previous section. If the domain G is a rectangle Π with the corners (x^1_0, x^2_0), $(x^1_0, x^2_{M_2})$, $(x^1_{M_1}, x^2_0)$, $(x^1_{M_1}, x^2_{M_2})$, we introduce the equidistant space discretization:

$$\Delta_{M_1, M_2} := \{(x^1_j, x^2_l) : x^1_j = x^1_0 + jh_{x^1},\ x^2_l = x^2_0 + lh_{x^2},$$
$$j = 0, \ldots, M_1,\ l = 0, \ldots, M_2\},$$

$$h_{x^1} = \frac{x^1_{M_1} - x^1_0}{M_1},\ h_{x^2} = \frac{x^2_{M_2} - x^2_0}{M_2}.$$

The values of $\bar{u}(t_k, x^1_j, x^2_l)$ at the nodes of Δ_{M_1, M_2} are found in accordance with (5.1)-(5.2). Let $x^1_j < x^1 < x^1_{j+1}$, $x^2_l < x^2 < x^2_{l+1}$. Then the value of $\bar{u}(t_k, x^1, x^2)$ is evaluated as

$$\bar{u}(t_k, x^1, x^2) = \frac{x^1_{j+1} - x^1}{h_{x^1}} \frac{x^2_{l+1} - x^2}{h_{x^2}} \bar{u}(t_k, x^1_j, x^2_l)$$
$$+ \frac{x^1_{j+1} - x^1}{h_{x^1}} \frac{x^2 - x^2_l}{h_{x^2}} \bar{u}(t_k, x^1_j, x^2_{l+1})$$
$$+ \frac{x^1 - x^1_j}{h_{x^1}} \frac{x^2_{l+1} - x^2}{h_{x^2}} \bar{u}(t_k, x^1_{j+1}, x^2_l)$$
$$+ \frac{x^1 - x^1_j}{h_{x^1}} \frac{x^2 - x^2_l}{h_{x^2}} \bar{u}(t_k, x^1_{j+1}, x^2_{l+1}). \quad (5.3)$$

If the function ψ does not depend on u, we can prove that taking $h_{x^i} = \varkappa^i h^{3/4}$, $i = 1, 2$, $\varkappa^1, \varkappa^2 > 0$, the error of this algorithm is estimated as $O(h^{1/2})$.

The case of an arbitrary domain G requires a special consideration. For instance, for a sufficiently wide class of domains G, it is possible to find one-to-one mapping of G onto a domain G' with a rectangular grid (see, e.g., [66] and references therein). Then we can use the above algorithm in G' and map the results onto G.

Remark 5.1. Combining methods from Sects. 8.1-8.2 and from Sects. 8.4-8.5, we can solve mixed boundary value problems, i.e., when we have the Dirichlet condition on a part of the boundary ∂G and the Neumann condition on the rest of ∂G.

8.6 Numerical tests for the Neumann problem

8.6.1 Comparison of various layer methods

Consider the Neumann problem for the one-dimensional Burgers equation:

$$\frac{\partial u}{\partial t} = \frac{\sigma^2}{2} \frac{\partial^2 u}{\partial x^2} - u \frac{\partial u}{\partial x}, \quad t > 0, \ x \in (-4, 4), \tag{6.1}$$

$$u(0, x) = -\frac{\sigma^2 \sinh x}{\cosh x + A}, \quad x \in [-4, 4], \tag{6.2}$$

$$\frac{\partial u}{\partial x}(t, \pm 4) = -\sigma^2 \frac{1 + A \exp(-\sigma^2 t/2) \cosh 4}{[\cosh 4 + A \exp(-\sigma^2 t/2)]^2}, \quad t \geq 0. \tag{6.3}$$

Here A is a positive constant.

The exact solution to this problem has the form (see [18])

$$u(t, x) = -\frac{\sigma^2 \sinh x}{\cosh x + A \exp(-\sigma^2 t/2)}.$$

In the tests we use cubic interpolation (cf. Remark 1.12)

$$\bar{u}(t_k, x) = \sum_{i=0}^{3} \Phi_{j,i}(x) \, \bar{u}(t_k, x_{j+i}), \quad x_j < x < x_{j+3},$$

$$\Phi_{j,i}(x) = \prod_{m=0, m \neq i}^{3} \frac{x - x_{j+m}}{x_{j+i} - x_{j+m}}. \tag{6.4}$$

Here we test the following five algorithms: (i) the algorithm (4.29)-(4.30) (Algorithm 4.8), (ii) the algorithm based on layer method (4.12) and cubic interpolation (6.4) (algorithm (4.29), (6.4)), (iii) the algorithm based on layer method (4.33) and cubic interpolation (6.4), (iv) the algorithm based on layer method (4.15) and cubic interpolation (6.4) (algorithm (4.31), (6.4)), and (v) the algorithm based on layer method (4.34) and cubic interpolation (6.4). We

504   8 Numerical solution of the nonlinear Dirichlet and Neumann problems

Table 6.1. Dependence of the errors $err^c(t)$ (*lower position*) and $err^l(t)$ (*top position*) in h when $t = 2$, $\sigma = 1.5$, and $A = 2$.

h	algorithm (4.29)-(4.30)	algorithm (4.29), (6.4)	algorithm (4.33), (6.4)	algorithm (4.31), (6.4)	algorithm (4.34), (6.4)
0.16	$5.216 \cdot 10^{-1}$	$7.434 \cdot 10^{-1}$	$5.967 \cdot 10^{-2}$	> 1	$7.333 \cdot 10^{-1}$
	$8.509 \cdot 10^{-2}$	$1.177 \cdot 10^{-1}$	$1.380 \cdot 10^{-2}$	$3.328 \cdot 10^{-1}$	$1.098 \cdot 10^{-1}$
0.01	$3.170 \cdot 10^{-2}$	$1.888 \cdot 10^{-2}$	$3.867 \cdot 10^{-3}$	$3.722 \cdot 10^{-1}$	$1.346 \cdot 10^{-1}$
	$5.748 \cdot 10^{-3}$	$3.737 \cdot 10^{-3}$	$1.224 \cdot 10^{-3}$	$6.161 \cdot 10^{-2}$	$2.192 \cdot 10^{-2}$
0.0016	$4.479 \cdot 10^{-3}$	$3.835 \cdot 10^{-3}$	$7.124 \cdot 10^{-4}$	$1.653 \cdot 10^{-1}$	$4.909 \cdot 10^{-2}$
	$8.149 \cdot 10^{-4}$	$7.444 \cdot 10^{-4}$	$2.127 \cdot 10^{-4}$	$2.750 \cdot 10^{-2}$	$8.172 \cdot 10^{-3}$
0.0001	$2.387 \cdot 10^{-4}$	$2.711 \cdot 10^{-4}$	$4.639 \cdot 10^{-5}$	$4.378 \cdot 10^{-2}$	$1.168 \cdot 10^{-2}$
	$4.479 \cdot 10^{-5}$	$5.213 \cdot 10^{-5}$	$1.357 \cdot 10^{-5}$	$7.307 \cdot 10^{-3}$	$1.968 \cdot 10^{-3}$

take the space step $h_x = h$ for linear interpolation and $h_x = \sqrt{h}$ for cubic interpolation. The parameter q of algorithm (4.31), (6.4) is taken equal to 1.

Table 6.1 and Figure 6.1 give numerical results obtained by these algorithms. In the table the errors of the approximate solutions $\bar{u}$ are presented in the discrete Chebyshev norm (lower position) and in the l^1-norm (top position):

$$err^c(t) = \max_{x_i} |\bar{u}(t, x_i) - u(t, x_i)|,$$
$$err^l(t) = \sum_i |\bar{u}(t, x_i) - u(t, x_i)| \times h_x. \qquad (6.5)$$

In the experiments the algorithm (4.31), (6.4) and the algorithm (4.34), (6.4) converge as $O(h^{1/2})$, the other algorithms converge as $O(h)$. We note that the algorithm (4.34), (6.4) gives more accurate results than the algorithm (4.31), (6.4), and the algorithm (4.33), (6.4) is more accurate than the algorithms (4.29). The algorithms (4.29)-(4.30) and (4.29), (6.4) demonstrate almost the same accuracy. But the algorithm (4.29)-(4.30) (as well as other algorithms based on linear interpolation) requires both larger volume of computations per time layer and larger amount of memory than the algorithm (4.29), (6.4) based on cubic interpolation (see also Remark 1.12 and numerical tests in Sect. 8.3).

Further, the boundary condition (6.3) can be rewritten in the form

$$\frac{\partial u}{\partial x}(t, \pm 4) = u(t, \pm 4)(\frac{1}{\sigma^2} u(t, \pm 4) - 1) - \frac{\sigma^2 \exp(\pm 4)}{\cosh 4 + A \exp(-\sigma^2 t/2)}, \quad t \geq 0. \quad (6.6)$$

In order to provide an experimental verification of the conjecture from Remark 4.4 we apply the algorithm (4.29), (6.4) to (6.1) with the initial condition (6.2) and the nonlinear boundary condition (6.6). Taking the same

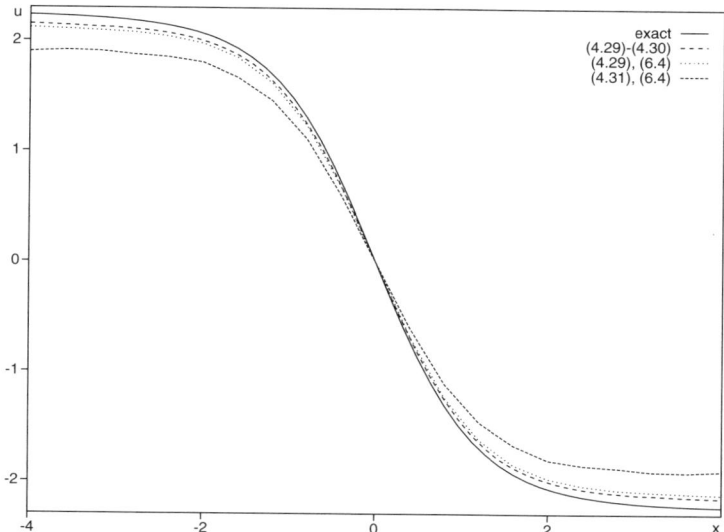

Fig. 6.1. Solution of the problem (6.1)-(6.3). Here $h = 0.16$, the other parameters are as in Table 6.1.

values of parameters as in Table 6.1, we obtain, in particular, that for $h = 0.01$ the error $err^c(2) = 1.103 \times 10^{-3}$ and for $h = 0.0001$ the error $err^c(2) = 1.138 \times 10^{-5}$ that experimentally confirms the conjecture.

In Appendix A.5 the algorithm (4.29), (6.4) is implemented for (6.1)-(6.3). By the program from Listing A.12 the results presented in the second column of Table 6.1 and also in Fig. 6.1 were obtained.

8.6.2 A comparison analysis of layer methods and finite-difference schemes

Here the test problem is the Burgers equation

$$\frac{\partial u}{\partial t} = \frac{\sigma^2}{2}\frac{\partial^2 u}{\partial x^2} - u\frac{\partial u}{\partial x}, \quad t > 0, \ x \in (-2, 8), \tag{6.7}$$

with the following initial and boundary conditions

$$u(0, x) = \varphi(x) := \begin{cases} a, & x \in [-2, 0), \\ (a+b)/2, & x = 0, \\ b, & x \in (0, 8], \end{cases} \tag{6.8}$$

$$\frac{\partial u}{\partial x}(t, x) = \psi(t, x), \ t > 0, \ x \in \{-2, 8\}, \tag{6.9}$$

where

$$\psi(t,x) = -\frac{(a-b)^2 J_1(t,x) J_2(t,x)}{\sigma^2 \times (J_1(t,x) + J_2(t,x))^2}$$
$$+ \sqrt{\frac{2}{\pi \sigma^2 t}} \exp\left(\frac{-x^2}{2\sigma^2 t}\right) \frac{b-a}{J_1(t,x) + J_2(t,x)},$$

and
$$J_1(t,x) = \exp\left(\frac{a(at-2x)}{2\sigma^2}\right) \operatorname{erfc}\left(\frac{x-at}{\sqrt{2\sigma^2 t}}\right),$$
$$J_2(t,x) = \exp\left(\frac{b(bt-2x)}{2\sigma^2}\right) \operatorname{erfc}\left(\frac{bt-x}{\sqrt{2\sigma^2 t}}\right).$$

The exact solution of this problem is
$$u(t,x) = \frac{a J_1(t,x) + b J_2(t,x)}{J_1(t,x) + J_2(t,x)}.$$

We compare the algorithm (4.29), (6.4) with the method of differences forward in time and central differences in space applied to the divergent form of the Burgers equation. This finite-difference scheme in application to the problem (6.7)-(6.9) is written as

$$\bar{u}(0,x) = \varphi(x), \; x \in [-2, 8],$$
$$\bar{u}(t_{k+1}, x_{-1}) = \bar{u}(t_{k+1}, x_1) - 2\Delta x \times \psi(t_{k+1}, x_0),$$
$$\bar{u}(t_{k+1}, x_{M+1}) = \bar{u}(t_{k+1}, x_{M-1}) + 2\Delta x \times \psi(t_{k+1}, x_M),$$

$$\bar{u}(t_{k+1}, x_j) = \bar{u}(t_k, x_j) - \frac{\Delta t}{4\Delta x}(\bar{u}^2(t_k, x_{j+1}) - \bar{u}^2(t_k, x_{j-1}))$$
$$+ \frac{\sigma^2}{2} \frac{\Delta t}{\Delta x^2}(\bar{u}(t_k, x_{j+1}) - 2\bar{u}(t_k, x_j) + \bar{u}(t_k, x_{j-1})), \quad (6.10)$$
$$j = 0, \ldots, M, \; k = 0, \ldots, N-1,$$

where the step of time discretization $\Delta t := T/N$ and $t_k = k \times \Delta t$ and the step of space discretization $\Delta x := 10/M$ and $x_j = -2 + j \times \Delta x$.

The explicit scheme (6.10) is of order $O(\Delta t, \Delta x^2)$. It is used for simulation of the Burgers equation in [1, 66].

In the experiments we take the space step $h_x = \sqrt{h}$ for the algorithm (4.29), (6.4) and the space step $\Delta x = \sqrt{\Delta t}$ for the finite-difference scheme (6.10).

Table 6.2 presents the errors in the discrete Chebyshev norm (lower position in the table) and in the l^1-norm (top position) (see (6.5)). The comment "overflow" indicates that overflow error occurs during simulation. The comment "oscillations" means that the numerical solution has oscillations (see Fig. 6.2). We see that the algorithm (4.29), (6.4) demonstrates a more stable behavior than the finite-difference scheme (6.10). In the test problem

8.6 Numerical tests for the Neumann problem

Table 6.2. Comparison analysis. Dependence of the errors $err^c(t)$ (*lower position*) and $err^l(t)$ (*top position*) in h when $t = 0.6$, $\sigma = 0.4$, $a = 11$, and $b = 9$.

h	0.01	0.0016	0.0004	0.0001
algorithm	4.859×10^{-1}	1.031×10^{-1}	2.659×10^{-2}	6.792×10^{-3}
(4.29), (6.4)	1.208	3.425×10^{-1}	8.980×10^{-2}	2.308×10^{-2}
		2.531	5.057×10^{-2}	1.234×10^{-2}
scheme (6.10)	overflow	6.375	1.261×10^{-1}	2.766×10^{-2}
		oscillations		

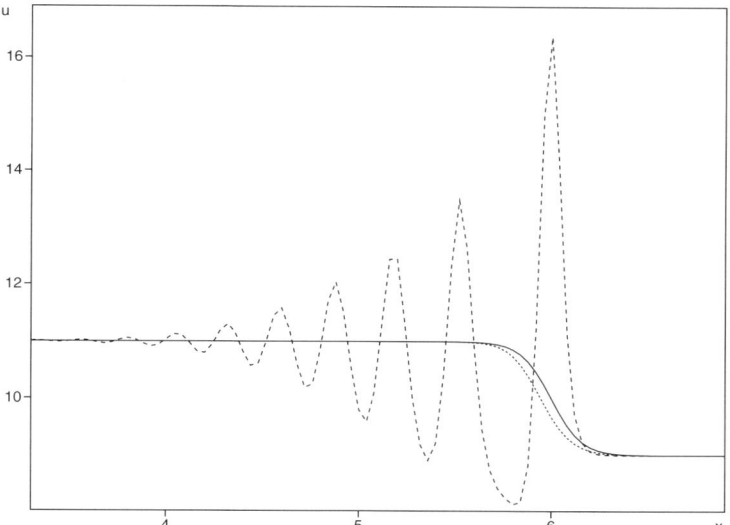

Fig. 6.2. Solution of the problem (6.7)-(6.9). *Solid line* is the exact solution, *dotted line* is the algorithm (4.29), (6.4), and *dashed line* is the scheme (6.10). Here $h = 0.0016$, the other parameters are as in Table 6.2.

(6.7)-(6.9) large values of a, b lead, in particular, to large advection in a neighborhood of the front. These experiments confirm that the layer methods allow us to avoid difficulties stemming from strong advection (see also comparison analysis in Sect. 8.3.2). We note that the algorithms based on layer methods require more CPU time than finite-difference schemes if you use the same step h. For example, in the case of parameters as in Table 6.2 to solve (6.7)-(6.9) by the algorithm (4.29), (6.4) with $h = 0.0004$ we need ≈ 5 sec while the scheme (6.10) requires ≈ 2.8 sec. But the algorithm (4.29), (6.4) gives us quite appropriate results with the greater step $h = 0.0016$ (see Table 6.2 and Fig. 6.2) and in this case it requires ≈ 0.7 sec. Simulations were made on PC with Intel Pentium 233 MHz processor using Borland C compiler.

9 Application of stochastic numerics to models with stochastic resonance and to Brownian ratchets

In this chapter we consider such models of stochastic dynamics as systems with stochastic resonance and stochastic ratchets. The term "stochastic resonance" (SR) is used in connection with effects attributable to the interaction between a periodic applied force and noise in nonlinear systems. SR was first considered in the context of a model concerning climate dynamics [19, 20, 224]. Then SR has been observed in a large variety of systems including lasers, noise-driven electronic circuits, superconducting quantum interference devices, chemical reactions, etc. As a survey on SR, one can use the reviews [3, 77]. One of the remarkable properties of systems with SR is the existence of regular oscillations under a certain set of parameters of the system. In Sect. 9.1 we give constructive sufficient conditions for regular oscillations in systems with stochastic resonance [207]. Using these conditions, we propose a numerical procedure for indicating domains of parameters corresponding to the regular oscillations and consider regular oscillations in various systems with SR.

Stochastic ratchets are defined as systems which are able to produce a directed current through the rectification of noise although on average no macroscopic force is acting (see, e.g., the reviews [116, 246] and references therein). Much interest in these simple nonequilibrium models is stimulated by their potential relevance to protein motors, transport in noncentrosymmetric materials, and novel particle pumps and separation techniques. Analytical and numerical studies of noise-induced directed transport mainly deal with evaluating the mean velocity. Having values of the mean velocity only, it is impossible to describe the noise-induced transport in detail. For example, it may happen that the mean velocity is small but the trajectories travel far in both positive and negative directions. On the other hand, we can observe large mean velocity both in the case of unidirectional transport (when there is practically no movement in one of the directions) and in the opposite case when the trajectories travel far in both directions. In Sect. 9.2 we are interested in the noise-induced unidirectional transport [209]. One of motivations for consideration of unidirectional transport may be possible biological applications of the Brownian ratchets to modeling molecular motors [54, 116, 161, 246, 293]. As is known (see, e.g., [116, 293]), molecular motors are microscopic objects that unidirectionally move along one-dimensional periodic structures and the problem of explaining this unidirectionality belongs to a larger class of such problems involving rectifying processes at a small scale. In Sect. 9.2, we give constructive conditions for existence of unidirectional transport for systems with state-dependent noise and for forced thermal ratchets. Using them, domains of parameters corresponding to unidirectional transport are identified.

9.1 Noise-induced regular oscillations in systems with stochastic resonance

A typical system, for which the SR phenomenon is observed, has the form of the Ito equation

$$dX = a(X)dt + b(t)dt + \sigma(t,X)dw(t), \tag{1.1}$$

where b and σ are periodic in t, $w(t)$ is a standard Wiener process. For instance, the system

$$dX = (\alpha X - X^3)dt + A\sin\omega t\, dt + \sigma dw(t) \tag{1.2}$$

has the form (1.1). The following system in the sense of Stratonovich [12]

$$dX = (\alpha - X - 2c\frac{X}{1+X^2})dt + A\sin\omega t\, dt + \sigma\frac{X}{1+X^2}\circ dw(t) \tag{1.3}$$

can be presented in the form (1.1) as well.

Let us investigate the conditions of arising regular oscillations for a specific system which is similar to the equation (1.2):

$$dX = (X - X^3)dt + A\chi(t;\theta)dt + \sigma dw(t), \tag{1.4}$$

where $\chi(t;\theta)$ is the following θ-periodic function:

$$\chi(t;\theta) = \begin{cases} 1, & 0 \le t < \theta/2, \\ -1, & \theta/2 \le t < \theta. \end{cases} \tag{1.5}$$

Thus, θ is a period and A is an amplitude of the constraining oscillations.

For clarity of exposition, we give an explanation of the mechanism of arising regular oscillations. In the absence of noise ($\sigma = 0$) and periodic forcing ($A = 0$) equation (1.4) has the stationary points $x = -1$, $x = 0$, $x = 1$. The points $x = -1$ and $x = 1$ are stable and $x = 0$ is unstable (see Fig. 1.1(a), where $f(x)$ is the right side of the equation (1.4) for $\sigma = 0$, $A = 0$). For $\sigma = 0$ and not large $A > 0$, the stationary points are displaced as shown in Fig. 1.1(b) during the first half-period and as in Fig. 1.1(c) during the second half-period. Clearly, for $\sigma = 0$ a point from a neighborhood of $x = -1$ (from the left well) cannot get into a neighborhood of $x = 1$ (into the right well) and vice versa. Such transitions become possible for $\sigma \ne 0$.

9.1 Noise-induced regular oscillations in systems with stochastic resonance

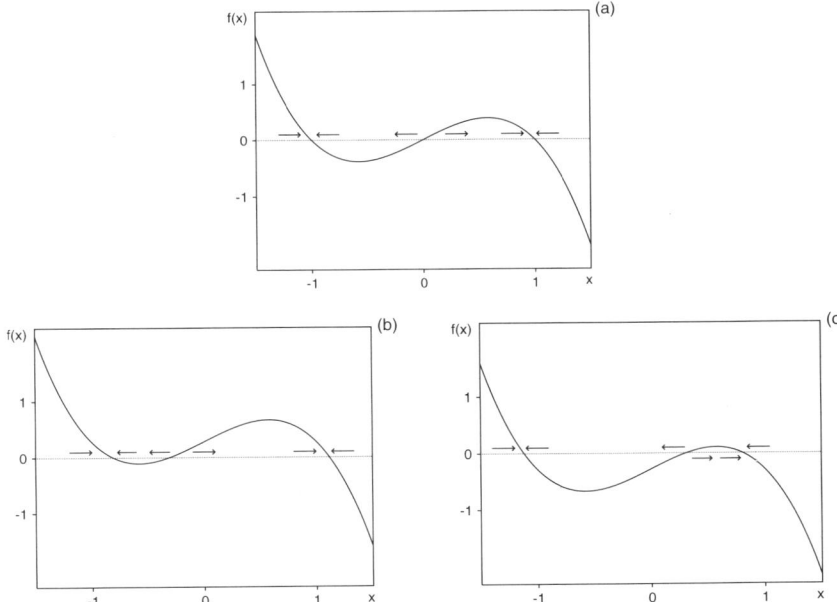

Fig. 1.1. The function $f(x) = x - x^3 + A$ for (a) $A = 0$, (b) $A = 0.28$, and (c) $A = -0.28$.

Regular transitions (oscillations) arise if a point from the left well attains the point $x = 1$ with probability close to 1 at a random time $\tau < \theta/2$ and after that it remains in the right well with probability close to 1 during the time $\theta/2 - \tau$. Indeed, the system acts by virtue of Fig. 1.1(c) after the half-period and, due to the symmetry, the situation repeats with changing the left well for the right one.

In Sect. 9.1.1 we investigate two probabilities in conjunction with Fig. 1.1(b): the probability of attainability of the point $x = 1$ from $x = -1$ for a time less than $\theta/2$ (which can be considered as the probability of getting into the right well from the left one) and the probability of unattainability of the point $x = 0$ from $x = 1$ during the first half-period $\theta/2$. It is clear that the closeness of the product p of these probabilities to 1 is a sufficient condition for the presence of regular oscillations. We observe that at the same time some fluctuations of such a regular behavior are unavoidable: it always remains a positive probability of unattainability from the left well into the right one, sometimes more than two transitions may occur during one period and so on. In other words, we assign a probability p_{ro} to the very phenomenon of regular oscillations and the above-mentioned product p bounds this probability from below. The magnitude of p is found by numerical solution of two boundary value problems of parabolic type. As a result, given a level of probability, a domain of parameters can be found such that the probability p_{ro} is above this level.

The approach proposed here and the approaches based on Kramers' theory of diffusion over a potential barrier are compared in Sect. 9.1.2. Some other measures of SR are also shortly discussed.

Implementation of the proposed approach for various systems with SR is considered in Sects. 9.1.3-9.1.5. In Sect. 9.1.3 we consider a system of the form (1.4) with σ depending on t, X (or on X only). Due to the multiplicative noise, one can essentially extend the domain of system parameters guaranteeing regular oscillations. In particular, we succeed to get high-frequency regular oscillations. Section 9.1.4 deals with large-amplitude regular oscillations in a monostable system. In Sect. 9.1.5 a system of two coupled oscillators is considered. An increase of coupling leads to shift of the domain of parameters corresponding to the regular oscillations.

9.1.1 Sufficient conditions for regular oscillations

The main conception. Let $X_{s,x}(t)$ be the solution of (1.4) which starts from the point x at the moment s. If $s = 0$, we write $X_x(t)$ instead of $X_{0,x}(t)$. It is known [19] that for suitable A, θ, σ a point from a neighborhood of the point $x = -1$ gets into a neighborhood of the point $x = 1$ during the first half-period $\theta/2$ with the probability close to 1 and remains there up to the end of the half-period. The same takes place in the time interval $[\theta/2, \theta]$ in the reverse order. Then all the events are repeated.

Let us underline: under the regular oscillations we understand a behavior of the solution $X(t)$ such that $X_{-1}(t)$ reaches $x = 1$ at a random time moment τ less than $\theta/2$ and $X_{\tau,1}(t)$ remains greater than zero during the rest $\theta/2 - \tau$ of the half-period. We emphasize that the transitions occur at random time moments, i.e., the phase at which the transitions occur is random.

As it has been explained at the beginning of this section, acceptable sufficient conditions for the regular oscillations are the following ones: the probability $p_{-1,1} = p_{-1,1}(A, \theta, \sigma) := P(X_{-1}(t) < 1, \ 0 \le t \le \theta/2)$ has to be small, and the probability $p_{1,0} = p_{1,0}(A, \theta, \sigma) := P(X_1(t) > 0, \ 0 \le t \le \theta/2)$ has to be close to 1. The oscillations will occur with the probability p_{ro} which exceeds the product

$$p = p(A, \theta, \sigma) := q_{-1,1}(A, \theta, \sigma) \cdot p_{1,0}(A, \theta, \sigma), \tag{1.6}$$

where $q_{-1,1} = 1 - p_{-1,1}$. So, we conclude that *the closeness of $p = p(A, \theta, \sigma)$ to 1 is a sufficient condition of regular oscillations*. Hereafter this condition is referred to as (RO). Thus, in practice, we can indicate domains of parameters, for which the regular oscillations take place with the probability exceeding the given level, by evaluating the product p.

Remark 1.1. The condition of closeness of the probability $q_{-1,1}(A, \theta, \sigma)$ to 1 is necessary, but closeness of $p_{1,0}(A, \theta, \sigma)$ to 1 is not necessary for the regular

9.1 Noise-induced regular oscillations in systems with stochastic resonance 513

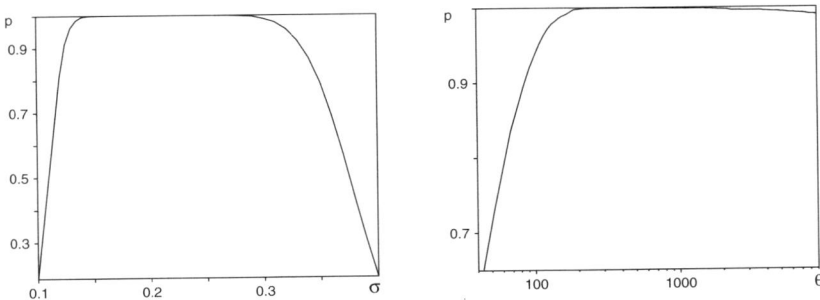

Fig. 1.2. Dependence of the product $p(A, \theta, \sigma)$ in σ for $A = 0.28$, $\theta = 10^4/3$ (*left*) and in θ for $A = 0.28$, $\sigma = 0.29$ (*right*); θ in the logarithmic scale.

oscillations. Indeed, $X_{-1}(t)$ reaches $x = 1$ after some time $\tau > 0$ and in fact we need that $X_{\tau,1}(t)$ remains in the neighborhood of $x = 1$ during a time less than $\theta/2$. It is not difficult to get

$$p_{ro}(A, \theta, \sigma) = \int_0^{\theta/2} p_1'(t; A, \sigma) \cdot p_2(\theta/2 - t; A, \sigma) dt, \quad p_1' = \frac{dp_1}{dt},$$

where $p_1(t; A, \sigma) := P(\tau_{-1}(1) \le t)$, $p_2(t; A, \sigma) := P(\tau_1(0) > t)$, and $\tau_{-1}(1)$ and $\tau_1(0)$ are the first-passage times of $X_{-1}(t)$ to $x = 1$ and of $X_1(t)$ to $x = 0$, respectively. Note that $p_1(\theta/2; A, \sigma) = q_{-1,1}(A, \theta, \sigma)$ and $p_2(\theta/2; A, \sigma) = p_{1,0}(A, \theta, \sigma)$. One can see that $p(A, \theta, \sigma) \le p_{ro}(A, \theta, \sigma) \le q_{-1,1}(A, \theta, \sigma)$. Further, if $p_{1,0} \approx 1$ then $p \approx q_{-1,1}$ and $p_{ro} \approx p$. Closeness of p_{ro} to 1 gives the necessary and sufficient condition of the regular oscillations. But this condition is less constructive than the given above sufficient condition (RO). Besides, the product p from (1.6) approximates p_{ro} quite accurate for a wide set of parameters according to our numerical experiments. For example, if we put the curves of both p_{ro} and p on Fig. 1.2, they coincide visually. At the same time we should note that for some sets of parameters p_{ro} and p can be not so close, e.g., for $A = 0.28$, $\theta = 30$, and $\sigma = 0.6$ we have $p_{ro} = 0.67$ and $p = 0.53$.

Evaluation of the product p. Our urgent aim is to evaluate $p(A, \theta, \sigma)$. Introduce the functions

$$u(s, x) = u(s, x; A, \theta, \sigma) := 1 - P(X_{s,x}(t) < 1, \ s \le t \le \frac{\theta}{2}),$$

$$0 \le s \le \frac{\theta}{2}, \ x \le 1,$$

and

$$v(s, x) = v(s, x; A, \theta, \sigma) := P(X_{s,x}(t) > 0, \ s \le t \le \frac{\theta}{2}), \ 0 \le s \le \frac{\theta}{2}, \ x \ge 0.$$

We get

$$1 - p_{-1,1} = u(0, -1), \quad p_{1,0} = v(0, 1), \quad p = u(0, -1) \cdot v(0, 1).$$

The function $u(s, x)$ satisfies the following mixed problem

$$\frac{\partial u}{\partial s} + \frac{1}{2}\sigma^2 \frac{\partial^2 u}{\partial x^2} + (x - x^3 + A)\frac{\partial u}{\partial x} = 0, \quad 0 \leq s < \frac{\theta}{2}, \quad x < 1, \quad (1.7)$$

with the initial and boundary conditions:

$$u(\frac{\theta}{2}, x) = 0, \quad x < 1, \quad u(s, 1) = 1, \quad 0 \leq s \leq \frac{\theta}{2}. \quad (1.8)$$

The solution of (1.7)-(1.8) has the probabilistic representation

$$u(s, x) = E\varphi\left(X_{s,x}(\tau_{s,x} \wedge \frac{\theta}{2})\right), \quad (1.9)$$

where $\tau_{s,x}$ is the first (random) moment at which $X_{s,x}(t) = 1$ and

$$\varphi(x) = \begin{cases} 0, & x < 1, \\ 1, & x = 1. \end{cases}$$

We analogously get that the function $v(s, x)$ satisfies the mixed problem

$$\frac{\partial v}{\partial s} + \frac{1}{2}\sigma^2 \frac{\partial^2 v}{\partial x^2} + (x - x^3 + A)\frac{\partial v}{\partial x} = 0, \quad 0 \leq s < \frac{\theta}{2}, \quad x > 0, \quad (1.10)$$

$$v(\frac{\theta}{2}, x) = 1, \quad x > 0, \quad v(s, 0) = 0, \quad 0 \leq s \leq \frac{\theta}{2}. \quad (1.11)$$

The solution of (1.10)-(1.11) has the probabilistic representation

$$u(s, x) = E\psi\left(X_{s,x}(\tau_{s,x} \wedge \frac{\theta}{2})\right), \quad (1.12)$$

where $\tau_{s,x}$ is the first (random) moment at which $X_{s,x}(t) = 0$ and

$$\psi(x) = \begin{cases} 1, & x > 0, \\ 0, & x = 0. \end{cases}$$

One can prove that the function $u(s, x; A, \theta, \sigma)$ is increasing and the function $v(s, x; A, \theta, \sigma)$ is decreasing with respect to θ. And typically there is a fairly wide range of $\theta \in (\theta_*(A, \sigma), \theta^*(A, \sigma))$, where under fixed A and σ the product $p(A, \theta, \sigma)$ is close to its maximum. Analogously, the product $p(A, \theta, \sigma)$ is close to its maximum in a range of the noise intensity $\sigma \in (\sigma_*(A, \theta), \sigma^*(A, \theta))$. We take the amplitude A of the constraining oscillations less than $A^* = 2\sqrt{3}/9$ so that the system (1.4) with $A < A^*$ and $\sigma = 0$ has three stationary points for each of the half-periods. Evidently, the

9.1 Noise-induced regular oscillations in systems with stochastic resonance

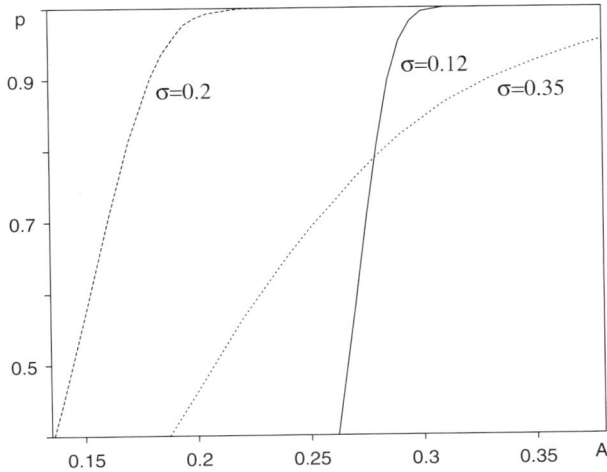

Fig. 1.3. Dependence of the product $p(A, \theta, \sigma)$ in A for $\theta = 10^4/3$ and various σ.

product $p(A, \theta, \sigma)$ is an increasing function with respect to A under fixed θ and σ.

As is known and has already been mentioned, there is a domain of parameters such that the regular oscillations are observed. Due to the sufficient condition of regular oscillations (RO), we are able to get estimates of this domain in terms of the product $p(A, \theta, \sigma)$.

To find the probabilities $q_{-1,1}$ and $p_{1,0}$, we have to solve the problems (1.7)-(1.8) and (1.10)-(1.11) numerically. In a number of tests we have used both finite-difference schemes and probabilistic methods from Chap. 6 and have seen ourselves that they give coincident results. Due to the fact that the absolute value of the term $x - x^3$ becomes large at $|x| \gg 1$, there are difficulties in implementation of finite-difference schemes for solving the boundary value problems (1.7)-(1.8) and (1.10)-(1.11). The difficulties do not arise in simulating the problems by the probabilistic methods. Moreover, we need in the individual values $u(0, -1)$, $v(0, 1)$ only and in such a case the probabilistic approach with the Monte Carlo technique is most relevant. That is why we mainly use the probabilistic methods of Chap. 6 in our experiments and attract finite-difference ones from time to time to control the obtained results.

Numerical results. Figures 1.2 and 1.3 show typical behavior of the product $p(A, \theta, \sigma)$. The remarkable feature is that there is a range of parameters where the product p is close to 1 that corresponds to the regular oscillations. Given the level of the product p, the domains of parameters are indicated on Fig. 1.4. Let us emphasize that the range of parameters for the regular oscillations is fairly large. For some fixed A and σ, the system (1.4) can be

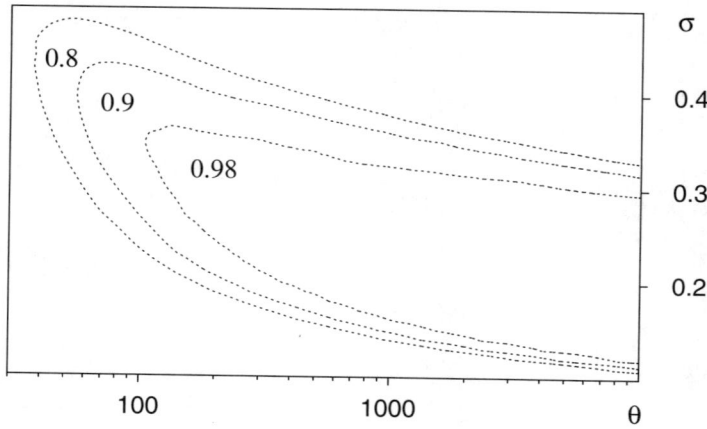

Fig. 1.4. Level curves of the product $p(A, \theta, \sigma)$ in the plane (θ, σ) for $A = 0.28$; θ in the logarithmic scale.

turned to a regular behavior (p becomes close to 1) by choosing the period θ (cf. Fig. 1.2 (right), Fig. 1.4, and the properties of the functions u and v listed above). We should also mention that the product p can have maximum in θ for fixed A and σ which is far from 1, and, evidently, in this case the system is far from the regular behavior. Influence of the amplitude A on the effect of synchronization in systems with SR was considered, e.g. in [3] (see also references therein). It was stated that increase of A leads to an extension of the range of σ (for fixed θ), where the regular oscillations occur. This follows from our analysis as well (see Fig. 1.3 and the properties of the functions u and v). Let us also observe that in the case of the model (1.4) the regular oscillations are realized for sufficiently large periods θ only. This corresponds to the common knowledge on SR [3,77]. To get high-frequency regular oscillations (i.e., for rather small θ), we involve into consideration models with specific multiplicative noises in Sect. 9.1.3.

Figure 1.5 presents typical sample trajectories of the solution to (1.4). We take values of parameters corresponding to Fig. 1.2. For the parameters A, θ, σ such that the product $p(A, \theta, \sigma)$ is close to 1, i.e., the sufficient condition (RO) takes place, we observe the regular oscillations (see Fig. 1.5(b)). A sample trajectory in the case when $q_{-1,1} \approx 0.8$ and $p_{1,0} \approx 1$, i.e., when the necessary condition does not fulfill, is given on Fig. 1.5(a). One can see that transitions between two wells during $\theta/2$ occur with the probability close to 0.8. Figure 1.5(c) demonstrates a typical trajectory in the case of $q_{-1,1} \approx 1$ and $p_{1,0} \approx 0.8$. After reaching $x = 1$ ($x = -1$), the trajectory remains in the corresponding well during the rest of the half-period with the probability close to 0.8. To simulate trajectories, we use the mean-square Euler method.

9.1 Noise-induced regular oscillations in systems with stochastic resonance

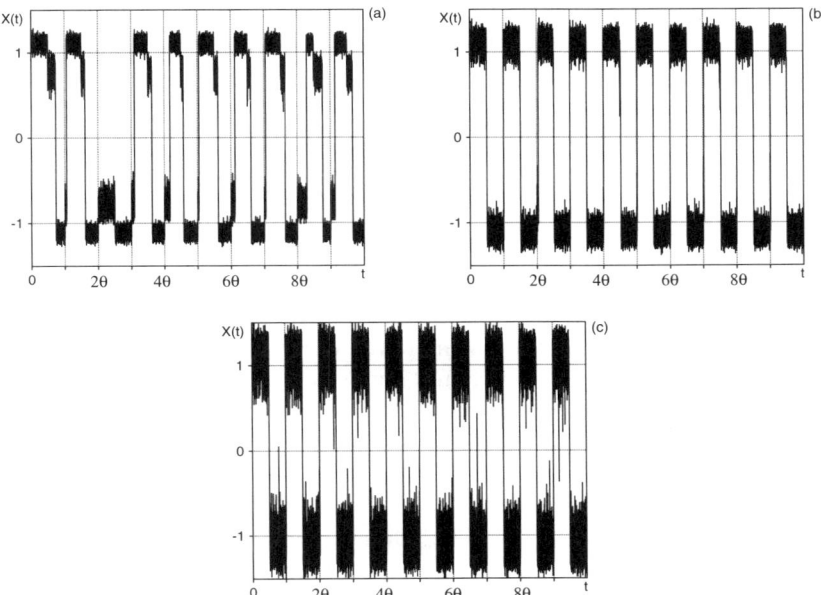

Fig. 1.5. Sample trajectories of the solution to (1.4) for $A = 0.28$, $\theta = 10^4/3$, and various σ: (a) $\sigma = 0.12$, (b) $\sigma = 0.2$, (c) $\sigma = 0.35$.

Remark 1.2. We also implement the approach for the model with sinusoidal forcing (1.2). Some numerical experiments are performed. In particular, they approve the fact that the domain of parameters corresponding to the regular oscillations in the case of sinusoidal forcing is narrower than in the case of periodic rectangular pulses forcing.

9.1.2 Comparison with the approach based on Kramers' theory of diffusion over a potential barrier

Let us consider system (1.4) when it acts by virtue of Fig. 1.1(b), i.e.,

$$dX = a(X)dt + \sigma dw(t), \ a(x) = x - x^3 + A, \ A < 2\sqrt{3}/9.$$

Evaluate some mean characteristics of $\tau_{-1}(1)$ (the first-passage time of $X_{-1}(t)$ to $x = 1$) and $\tau_1(0)$ (the first-passage time of $X_1(t)$ to $x = 0$).

The mean value $E\tau_{-1}(1)$ can be found in the following way. Consider the boundary value problem

$$\frac{1}{2}\sigma^2 \psi'' + a(x)\psi' + 1 = 0, \ \psi(C; C) = 0, \ \psi(1; C) = 0, \ C < -1,$$

for the function $\psi(x; C)$, where C is a parameter. It is known [48, 82] that $\psi(-1; C)$ is equal to the mean value of the first-exit time of the process

$X_{-1}(t)$ from the interval $(C, 1)$. Clearly, $E\tau_{-1}(1) = \lim_{C \to -\infty} \psi(-1; C)$. In addition, one can prove that $E\tau_{-1}(1) = \Psi(-1)$, where $\Psi(x)$ is a solution to the problem

$$\frac{1}{2}\sigma^2 \Psi'' + a(x)\Psi' + 1 = 0, \ \Psi'(-\infty) = 0, \ \Psi(1) = 0. \tag{1.13}$$

The second moment $E\tau_{-1}^2(1)$ is equal to $\Psi_1(-1)$, where $\Psi_1(x)$ is a solution to the problem

$$\frac{1}{2}\sigma^2 \Psi_1'' + a(x)\Psi_1' + 2\Psi(x) = 0, \ \Psi_1'(-\infty) = 0, \ \Psi_1(1) = 0,$$

and $\Psi(x)$ is the solution of problem (1.13) [48,82]. The mean $E\tau_1(0)$ and the second moment $E\tau_1^2(0)$ can be found analogously.

The approach of [19, 20] based on Kramers' theory employs the following conditions as sufficient ones for the existence of regular oscillations (from the principal point of view our exposition here only slightly differs from [19,20]): (i) $E\tau_{-1}(1) << \theta/2$, (ii) $E\tau_1(0) >> \theta/2$, (iii) $(D\tau_{-1}(1))^{1/2} = [E\tau_{-1}^2(1) - (E\tau_{-1}(1))^2]^{1/2} << \theta/2$. These conditions are fairly constructive because all the magnitudes $E\tau_{-1}(1)$, $E\tau_1(0)$, and $E\tau_{-1}^2(1)$ can be found by quadratures. Moreover, for small σ they can be expressed by exponential Kramers formulas.

In the previous subsection we propose an alternative approach, which is sufficiently constructive as well and possesses more generality. Besides, one can get the more exhaustive answers using the sufficient condition (RO) from Sect. 9.1.1 in comparison to the conditions of [19,20] which are only qualitative in nature. Let us emphasize once more that the probability in Sect. 9.1.1 is assigned to the very phenomenon of regular oscillations. The universality and utility of the proposed approach are demonstrated in the next subsections. At the same time, Kramers-like approaches are analytically tractable in some limit cases while our approach is numerical. We should also mention some other measures commonly used for SR (signal-to-noise ratio, response amplitude, waiting time distributions) [3, 77]. As is known [36, 77], a maximum of signal-to-noise ratio does not directly reflect the regular oscillations (synchronization between the hopping and driving) in systems with SR. Such characteristics of SR as the response amplitude at the frequency of the periodic signal and waiting time distributions reflect the phenomenon of regular oscillations. Generally, waiting time distributions are less effective in computational sense than the measure p introduced in Sect. 9.1.1. And the p gives more information on the regular oscillations than the response amplitude.

9.1.3 High-frequency regular oscillations in systems with multiplicative noise

In the case of system (1.4) it is impossible to get high-frequency regular oscillations. Indeed, if we decrease the period length θ, we should increase the

9.1 Noise-induced regular oscillations in systems with stochastic resonance

noise level σ to preserve the level of the probability $q_{-1,1}$ of escape from the metastable state to absolutely stable state. But the probability $p_{1,0}$ of return from the absolutely stable state to the metastable one decreases with an increase of σ. Therefore, the product p becomes low and the regular oscillations disappear. In this subsection we consider some specific systems with multiplicative noise such that the probability $p_{1,0}$ is always equal to 1 and due to this fact we are able to obtain the high-frequency regular oscillations.

Consider the model with multiplicative time-dependent noise

$$dX = (X - X^3)dt + A\chi(t;\theta)dt + \sigma\gamma(t, X;\theta)dw(t), \qquad (1.14)$$

where $\chi(t;\theta)$ is the θ-periodic function defined in (1.5) and $\gamma(t, x; \theta)$ is the following θ-periodic function

$$\gamma(t, x; \theta) = \begin{cases} 1, & 0 \leq t < \theta/2, \ x < 1, \\ 0, & 0 \leq t < \theta/2, \ x \geq 1, \\ 1, & \theta/2 \leq t < \theta, \ x > -1, \\ 0, & \theta/2 \leq t < \theta, \ x \leq -1. \end{cases}$$

Periodically modulated noise is not uncommon and it arises, for example, at the output of any amplifier whose gain varies periodically in time. And a system with diffusion coefficient which depends on its state plays an important role in a number of physical systems (see, e.g. [106, 248, 272] and references therein).

It is evident that $p_{1,0} = 1$ in the case of (1.14) and, consequently, the necessary and sufficient condition of regular oscillations consists in the closeness of the probability $q_{-1,1}(A, \theta, \sigma)$ to 1. This probability can be close to 1 even for a fairly small θ (i.e., for high frequency $\omega = 2\pi/\theta$) and for very small A under an appropriate value of σ. Thus, it is possible to organize the high-frequency regular oscillations in system (1.14) with small periodic forcing. Moreover, in the case of the model (1.14) the regular oscillations can be obtained under zero A.

Figure 1.6 demonstrates level lines of $q_{-1,1}(A, \theta, \sigma)$ in the plane (θ, σ) for $A = 0.28$. A typical trajectory with the high-frequency oscillations is given on Fig. 1.7. Let us observe that rather long excursions of trajectories (up to $x = \pm 5$ or even more) are possible. Consider, for instance, a trajectory $X_{s,1}(t)$, $0 \leq s < \theta/2$, $t \geq s$. When $t < \theta/2$, we have $\chi(t;\theta) > 0$ and the noise $\sigma\gamma(t, x;\theta) = \sigma$ for $x < 1$ and $\sigma\gamma(t, x;\theta) = 0$ for $x \geq 1$. Consequently, $X_{s,1}(t) \geq 1$, $s \leq t < \theta/2$, and the noise is switched off for this $X_{s,1}(t)$. At $t = \theta/2$ the noise is switched on for all $x > -1$ and affects the trajectory $X_{s,1}(t)$ which can fluctuate then in both directions, in particular, up to a large positive x. Such large fluctuations occur often because to achieve the high-frequency regular oscillations, we take sufficiently large values of the noise intensity σ.

Now consider the model with multiplicative time-independent noise

$$dX = (X - X^3)dt + A\chi(t;\theta)dt + \sigma\gamma(X)dw(t), \qquad (1.15)$$

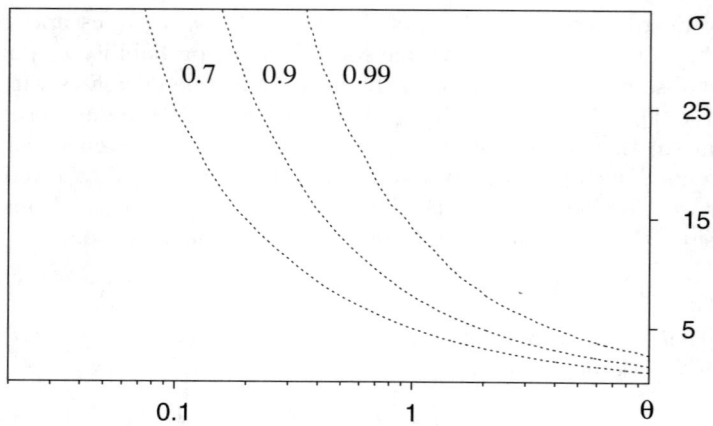

Fig. 1.6. Level curves of the probability $q_{-1,1}(A, \theta, \sigma)$ in the plane (θ, σ) for $A = 0.28$; θ in the logarithmic scale.

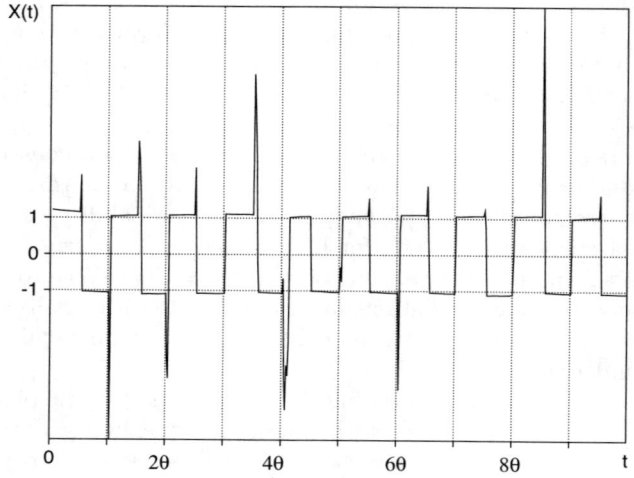

Fig. 1.7. Sample trajectory of the solution to (1.14) for $A = 0.28$, $\theta = 2\pi/\omega \approx 0.524$ ($\omega = 12$), $\sigma = 35$.

where

$$\gamma(x) = \begin{cases} 1, & -1 < x < 1, \\ 0, & \text{otherwise.} \end{cases} \tag{1.16}$$

Let the solution $X(t)$ of (1.15) start from $x = -1$. During the time $[0, \theta/2)$ the drift in the system (1.15) corresponds to Fig. 1.1(b). Clearly, the probability of attainability of the point $x = 1$ for the time less than $\theta/2$ is not less than

9.1 Noise-induced regular oscillations in systems with stochastic resonance

$q_{-1,1}$ in the model (1.4). After reaching the point $x = 1$ at a random moment, the trajectory moves deterministically in positive direction to a point $X(\theta/2) > 1$. Then the drift in (1.15) becomes corresponding to Fig. 1.1(c) and the trajectory changes its movement direction. The trajectory comes back to the point $x = 1$ at a moment $\theta/2 + \tau$, where τ is random. It remains the time $\theta/2 - \tau$ for the trajectory to reach the point $x = -1$. The random moment τ is less than a quantity s^* which can be evaluated in the following way. Let the solution $X(t)$ of the equation

$$X' = X - X^3 + A\chi(t;\theta)$$

start from $x = 1$. Then the trajectory $X(t)$ moves in positive direction up to $t = \theta/2$, when the trajectory changes its movement direction, and comes back to the point $x = 1$ at the instance $t^* \in (\theta/2, \theta)$. The value of the desired s^* is equal to $t^* - \theta/2$.

Introduce the probability

$$p^*_{-1,1} = p^*_{1,-1}(A,\theta,\sigma) := P(X_{-1}(t) < 1,\ 0 \le t \le \theta/2 - s^*).$$

The sufficient condition of regular oscillations of the solution to (1.15) consists in the closeness of the probability $q^*_{-1,1} = 1 - p^*_{1,-1}$ to 1. By the same arguments as in Sect. 9.1.1, it is not difficult to see that $q^*_{-1,1} = u(0,-1)$, where $u(s,x)$ is the solution to the following mixed problem

$$\frac{\partial u}{\partial s} + \frac{\sigma^2}{2}\gamma(x)\frac{\partial^2 u}{\partial^2 x} + (x - x^3 + A)\frac{\partial u}{\partial x} = 0,\ 0 \le s < \theta/2 - s^*,\ x < 1, \quad (1.17)$$

$$u(\theta/2 - s^*, x) = 0,\ x < 1,\quad u(s,1) = 1,\ 0 \le s \le \theta/2 - s^*. \quad (1.18)$$

The regular oscillations in the case of (1.15) are observed under a more wide set of parameters than for the equation (1.4) however under a more restricted set of parameters than for the equation (1.14). Figure 1.8 shows a typical trajectory of the solution to (1.15) under values of parameters such that they do not ensure the regular oscillations in the case of the model (1.4).

Remark 1.3. Let us discuss simulation of SDEs (1.14) and (1.16). If a model has discontinuous in time and continuous in space coefficients, there are no serious problems in its simulation. Despite the diffusion coefficient in the equation (1.14) is discontinuous in t and x, principal difficulties do not arise as well. It is so because any trajectory of (1.14) feels the discontinuity of the diffusion coefficient in x not more than once during the half-period $\theta/2$. As to the equation (1.16), discontinuity in x of the diffusion coefficient leads to some problems in its numerical simulation. Indeed, if $X(t) \ge -1$ at a moment $t \in [n\theta, (n+1/2)\theta)$, $n = 0,1,2,\ldots$, then $X(s) > -1$ for all $s \in (t, (n+1/2)\theta)$ with probability 1. But due to the discretization error, the mean-square Euler approximation $\bar{X}_k$ of $X(t_k)$ violates this property and can become less than -1. As a result, it gives a too distorted image of the real

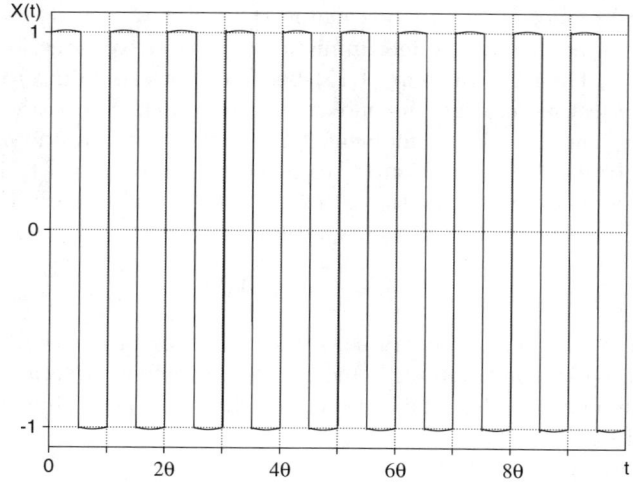

Fig. 1.8. Sample trajectory of the solution to (1.15) for $A = 0.02$, $\theta = 2\pi/\omega \approx 62.83$ ($\omega = 0.1$), $\sigma = 3$.

behavior. To overcome this difficulty, we propose a modified approximation $\tilde{X}_k$ (agreeing with the above-mentioned property of trajectories):

$$\begin{aligned}
&\tilde{X}_0 = x,\ t_0 = 0,\\
&\hat{X}_{k+1} = \tilde{X}_k + h \times (\tilde{X}_k - \tilde{X}_k^3) + hA\chi(t_k;\theta) + h^{1/2}\sigma\xi_k;\\
&\text{if } \tilde{X}_k < -1 \text{ or } \tilde{X}_k > 1\ ,\ \text{then } \tilde{X}_{k+1} = \hat{X}_{k+1};\\
&\text{if } -1 \leq \tilde{X}_k \leq 1 \text{ and } \chi(t_k;\theta) > 0,\ \text{then } \tilde{X}_{k+1} = \max(-1, \hat{X}_{k+1});\\
&\text{if } -1 \leq \tilde{X}_k \leq 1 \text{ and } \chi(t_k;\theta) < 0,\ \text{then } \tilde{X}_{k+1} = \min(\hat{X}_{k+1}, 1);\\
&k = 0, 1, 2, \ldots\ .
\end{aligned} \quad (1.19)$$

Here h is a step of the time discretization, ξ_k are independent normally distributed random variables with zero mean and unit variance.

We also take the continuous $\gamma(x) = \arctan(\alpha(x-1)) + \arctan(-\alpha(x+1))$ instead of the function (1.16) and integrate (1.15) with this $\gamma(x)$ for a large α by the mean-square Euler method with a sufficiently small time step. Simulations with the continuous $\gamma(x)$ and simulations due to the method (1.19) give similar results.

The same difficulties arise in numerical solution of the boundary value problem (1.17)-(1.18). They can be overcome in a similar manner.

Remark 1.4. Using the approach proposed in Sect. 9.1.1, one can obtain a sufficient condition for the regular oscillations in system (1.3) which has an asymmetrical bistable potential (see [12]). Let the system (1.3) have two

9.1 Noise-induced regular oscillations in systems with stochastic resonance

stable points x_- and x_+, $x_- < x_+$, and one unstable x_u, $x_- < x_u < x_+$, in the absence of periodic forcing and noise. To give a sufficient condition for the regular oscillations in the asymmetrical case, four probabilities have to be considered: the probability q_{x_-,x_+} with which the trajectory starting from $x = x_-$ reaches the point $x = x_+$ during the first half-period of the periodic forcing (i.e., when the periodic forcing is positive); the probability p_{x_+,x_u} of unattainability of the point $x = x_u$ during the first half-period by the trajectory starting from $x = x_+$; the probability q_{x_+,x_-} with which the trajectory starting from $x = x_+$ reaches the point $x = x_-$ during the second half-period of the periodic forcing; the probability p_{x_-,x_u} of unattainability of the point $x = x_u$ during the second half-period by the trajectory starting from $x = x_-$. Due to the asymmetry, $q_{x_-,x_+} \neq q_{x_+,x_-}$ and $p_{x_+,x_u} \neq p_{x_-,x_u}$. In this situation, the sufficient condition of regular oscillations consists in the closeness of the products $q_{x_-,x_+} \cdot p_{x_+,x_u}$ and $q_{x_+,x_-} \cdot p_{x_-,x_u}$ to 1. One can easily write down boundary value problems for these probabilities.

It follows from the analysis given in [12] that the model (1.3) can operate in the regime of high-frequency regular oscillations. Probably, this becomes possible owing to the specific type of noise which acts, in essence, in bounded range of states x only (cf. the models (1.14) and (1.15) which also have bounded-type noise).

9.1.4 Large-amplitude regular oscillations in monostable system

Consider the stochastic differential equation

$$dX = b(X)dt + A\chi(t;\theta)dt + \sigma dw, \tag{1.20}$$

where $\chi(t;\theta)$ is the θ-periodic function from (1.5) and

$$b(x) = \begin{cases} -2\alpha(x+\beta), & x < -\beta, \\ -\gamma\pi \sin \pi x, & |x| < 1, \\ 0, & 1 \leq |x| \leq \beta, \\ -2\alpha(x-\beta), & x > \beta, \end{cases} \tag{1.21}$$

$\alpha, \beta, \gamma > 0$ are some constants. See graphics of $b(x)$ and its potential $F(x) = -\int b(x)dx$ on Fig. 1.9.

Under $A = 0$, $\sigma = 0$ the solution to the equation (1.20) has the unique globally stable point $x = 0$. If $\sigma = 0$ and A is not large, the equation (1.20) has a θ-periodic solution with the amplitude less than 1 (see Fig. 1.10(a)). After adding the noise of a certain level, the system does not exhibit regular oscillations (see Fig. 1.10(b)). But an increase of the noise intensity leads to regular oscillations with large amplitude, approximately equal to β (see Fig. 1.10(c)).

To find a set of parameters under which the regular oscillations with large amplitude are observed, one can use the approach of Sect. 9.1.1 again. Introduce the probabilities

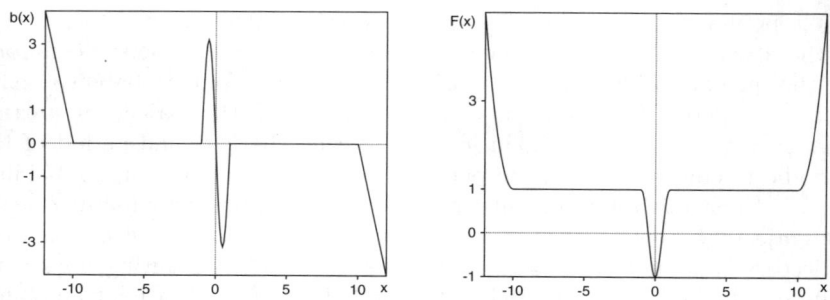

Fig. 1.9. The function $b(x)$ (see (1.21)) and the potential $F(x)$ under $\alpha = 1$, $\beta = 10$, $\gamma = 1$.

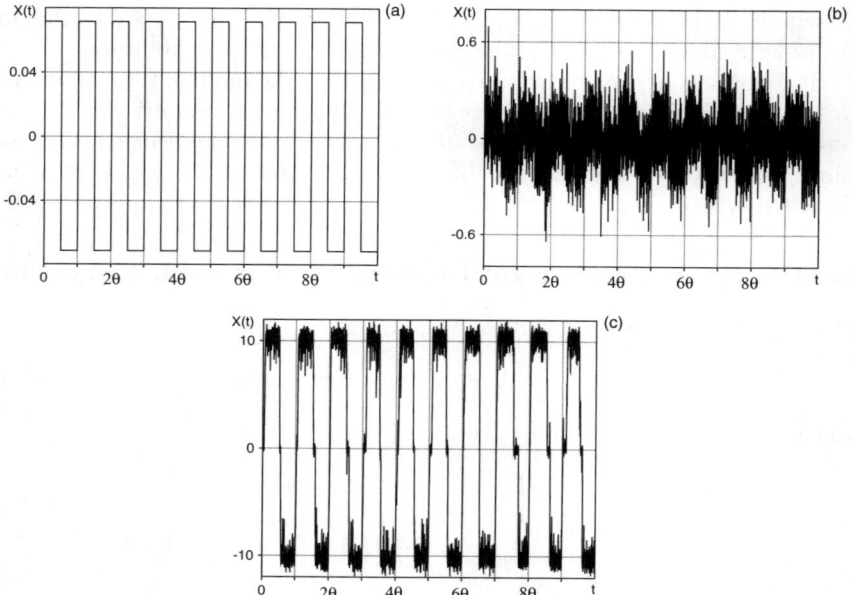

Fig. 1.10. Sample trajectories of the solution to (1.20) under $\alpha = 1$, $\beta = 10$, $\gamma = 1$, $A = 0.7$, $\theta \approx 628.32$ ($\omega = 0.01$), and various σ: (a) $\sigma = 0$, (b) $\sigma = 0.55$, (c) $\sigma = 1$.

$$q_{-\beta,\beta} := 1 - P(X_{-\beta}(t) < \beta,\ 0 \leq t \leq \theta/2),$$
$$p_{\beta,1} := P(X_{\beta}(t) > 1,\ 0 \leq t \leq \theta/2).$$

Then the sufficient condition for the regular oscillations with the amplitude β consists in the closeness of the product $q_{-\beta,\beta} \cdot p_{\beta,1}$ to 1. It is not difficult to write the boundary value problems for calculating these probabilities just as in Sect. 9.1.1.

9.1.5 Regular oscillations in a system of two coupled oscillators

In this subsection we apply the proposed above approach to the system of two mutually coupled bistable overdamped oscillators

$$dX^1 = (\alpha_1 X^1 - (X^1)^3)dt + c \cdot (X^2 - X^1)dt + A\chi(t;\theta)dt + \sigma dw_1(t)$$
$$dX^2 = (\alpha_2 X^2 - (X^2)^3)dt + c \cdot (X^1 - X^2)dt + A\chi(t;\theta)dt + \sigma dw_2(t), \quad (1.22)$$

where $w_1(t)$ and $w_2(t)$ are independent standard Wiener processes, the function $\chi(t;\theta)$ is defined in (1.5), the coefficients α_1 and α_2, the strength of coupling c, and the noise intensity σ are some non-negative constants.

SR in a system similar to (1.22) was considered in [220]. The authors of [220] came to the conclusion that the maximum of the signal-to-noise ratio taken over noise intensity is a nonmonotonous function of coupling. An array of coupled stochastic oscillators is numerically investigated in [296] using methods of Chap. 3.

Here we are interested in the regular oscillations. Introduce the notation: $x_- = (-1, -1)$, $x_+ = (1, 1)$, $x_u = (0, 0)$ are the points belonging to $\mathbf{R}^2$, $X_x(t)$ is the solution of the system (1.22) which starts from the point $x \in \mathbf{R}^2$ at the zero instant,

$$p_{x_-, x_+} = p_{x_-, x_+}(A, \theta, \sigma, c)$$
$$:= P(X_{x_-}(t) \in \mathbf{R}^2 \backslash \{x^1 > 1, x^2 > 1\}, 0 \le t \le \theta/2),$$
$$p_{x_+, x_u} = p_{x_+, x_u}(A, \theta, \sigma, c) := P(X_{x_+}(t) \in \{x^1 > 0, x^2 > 0\}, 0 \le t \le \theta/2),$$

and $q_{x_-, x_+} := 1 - p_{x_-, x_+}$.

A sufficient condition of regular oscillations consists in the closeness of the product $p(A, \theta, \sigma, c) := q_{x_-, x_+} \cdot p_{x_+, x_u}$ to 1. It is not difficult to write down the boundary value problems for calculating the probabilities q_{x_-, x_+} and p_{x_+, x_u} analogously to (1.7)-(1.8) and (1.10)-(1.11). Solving these problems numerically, we find the product p which bounds the probability of regular oscillations from below. Figures 1.11, 1.12, and 1.13 present results of our calculations of p. One can see that an increase of coupling leads to shift of the domain of parameters corresponding to regular oscillations. The domain is shifted to the range of larger noise intensities (see Figs. 1.12 and 1.13). An increase of the coupling can both decrease and increase the product p depending on the taken A, σ, θ (see Figs. 1.11).

Figures 1.14 and 1.15 show typical trajectories of the first oscillator under various collections of the parameters. Figure 1.14 demonstrates that for some fixed A, σ, θ disappearance of regular oscillations for the system (1.22) can result from an increase of coupling c. And vice versa, Figure 1.15 presents a situation when an increase of coupling leads to arising regular oscillations.

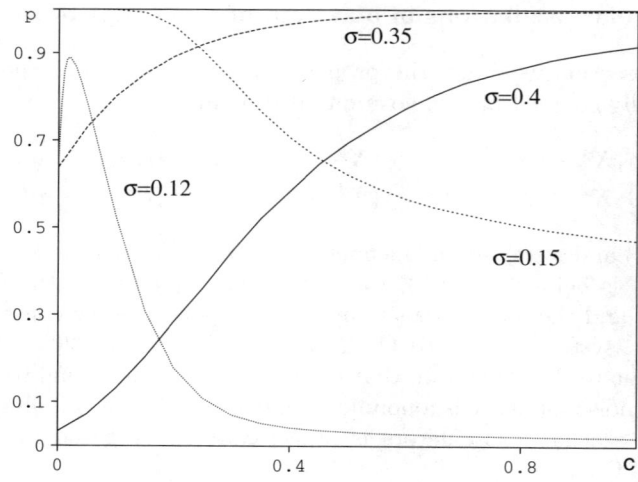

Fig. 1.11. Two coupled oscillators. Dependence of the product $p(A, \sigma, \theta, c)$ in c under $A = 0.28$, $\theta = 10^4/3$, $\alpha_1 = \alpha_2 = 1$, and various σ.

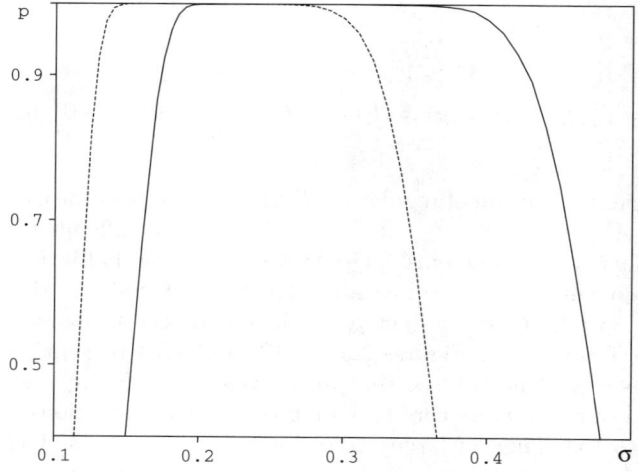

Fig. 1.12. Two coupled oscillators. Dependence of the product $p(A, \sigma, \theta, c)$ in σ under $A = 0.28$, $\theta = 10^4/3$, $\alpha_1 = \alpha_2 = 1$, and $c = 0$ (*dashed line*), $c = 2$ (*solid line*).

9.2 Noise-induced unidirectional transport

In this section we consider the noise-induced unidirectional transport in Brownian ratchets [209]. In Sect. 9.2.1, we study the detailed structure of

9.2 Noise-induced unidirectional transport   527

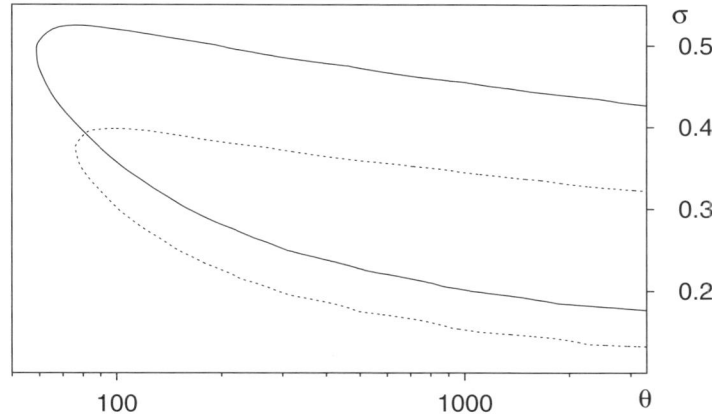

Fig. 1.13. Two coupled oscillators. Curves of the level 0.9 of the product $p(A, \sigma, \theta, c)$ in the plane (θ, σ) under $A = 0.28$ and $c = 0$ (*dashed line*), $c = 2$ (*solid line*); θ in the logarithmic scale.

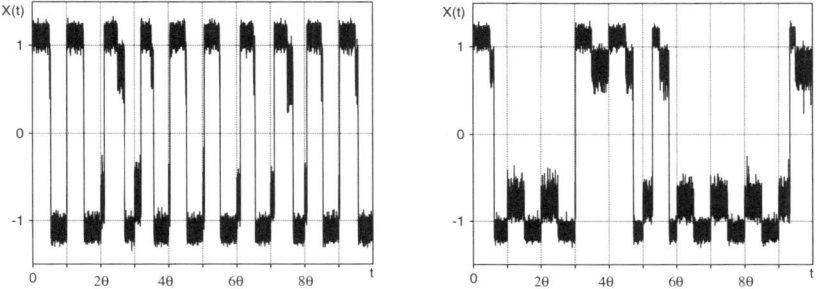

Fig. 1.14. Two coupled oscillators. Sample trajectories of the first oscillator under $A = 0.28$, $\theta = 10^4/3$, $\sigma = 0.15$, $\alpha_1 = \alpha_2 = 1$, and $c = 0$ (*left*), $c = 2$ (*right*).

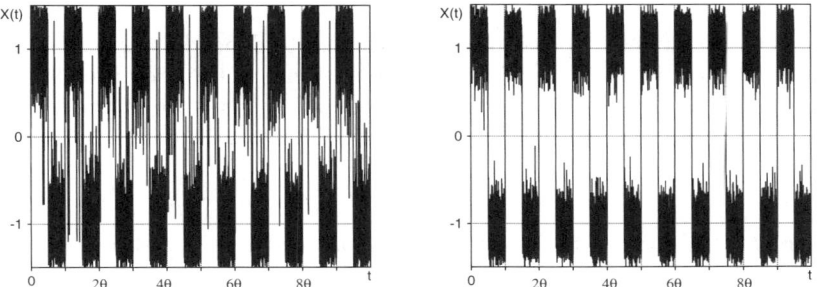

Fig. 1.15. Two coupled oscillators. Sample trajectories of the first oscillator under $\sigma = 0.45$ and $c = 0$ (*left*), $c = 3$ (*right*). Other parameter values are as in Fig. 1.14.

transport in systems with state-dependent noise [32, 119] and obtain an analytical condition for unidirectional transport. We consider the probability q that the trajectory $X_x(t)$, $X(0) = x$, first reaches the right end of the interval $(x - L,\ x + L)$, L being a period of the ratchet potential. This probability is found analytically by solving the corresponding boundary value problem for a second-order ordinary differential equation. The condition for the unidirectional transport to be in the positive direction is closeness of the probability q to 1.

In Sect. 9.2.2, we consider forced thermal ratchets [13, 116, 161, 205] (see (2.13) below). To propose a condition for unidirectional transport, we have to consider two probabilities: the probability $Q_{0,x}^{<}$ that the trajectory $X_{0,x}(t)$, $X(0) = x$, escapes from $(x - L,\ x + L)$ through the right end during the first half-period $[0, T/2)$ of the periodic forcing with a period T, and the probability $P_{T/2,x}^{>}$ that $X_{T/2,x}(t)$, $X(T/2) = x$, does not escape from $(x - L,\ x + L)$ through the left end during the second half-period $[T/2, T)$. The condition for the unidirectional transport to be in the positive direction consists in closeness of the product $Q_{0,x}^{<} \cdot P_{T/2,x}^{>}$ to 1. For large T, we get a qualitative condition of the unidirectionality attracting results of Sect. 9.2.1. But in a general case we should evaluate the probabilities $Q_{0,x}^{<}$, $P_{T/2,x}^{>}$ by solving two boundary value problems for linear parabolic equations. As a result, we propose an effective (numerical) tool for identifying domains of parameters, where the noise-induced unidirectional transport is realized. We restrict ourselves to consideration of classical ratchet models but the technique is universal and can be applied to various systems with the noise-induced transport.

We present some results of numerical experiments which, in particular, demonstrate (see, e.g., Fig. 2.3) that noise-induced unidirectional transport is observed for a sufficiently wide range of parameters. We also note that there is no unidirectional transport for both sufficiently small and large noise intensities.

9.2.1 Systems with state-dependent diffusion

It was shown in [32, 119] (see also [205]) that state-dependent diffusion can induce transport in a system which is at equilibrium in the presence of thermal noise only. Here we are interested in the detailed structure of the transport in such systems. We consider the stochastic differential equation in the sense of Ito:

$$dX = f(X)dt + \sigma(X)dw(t), \qquad (2.1)$$

where $f(x)$ and $\sigma(x)$ are L-periodic functions and $w(t)$ is a standard Wiener process.

Introduce the process $\Phi(t) = X(t) \pmod{L}$ on a circle of radius $L/2\pi$. The process $\Phi(t)$ is continuous on the circle. Due to the periodicity of f and σ, we can write (2.1) in the form

9.2 Noise-induced unidirectional transport

$$dX = f(\Phi)dt + \sigma(\Phi)dw(t). \tag{2.2}$$

Under sufficiently wide assumptions (e.g., $\sigma(x) \neq 0$, $x \in R$), $\Phi(t)$ is an ergodic process (see, e.g., [124]). Its invariant density $p(\varphi)$, $0 \leq \varphi \leq L$, is L-periodic and satisfies the stationary Fokker–Planck equation

$$\frac{1}{2}\frac{\partial^2}{\partial \varphi^2}(\sigma^2 p) - \frac{\partial}{\partial \varphi}(fp) = 0, \quad p(0) = p(L), \quad \int_0^L p(\varphi)d\varphi = 1.$$

Solving this problem, we get

$$p(\varphi) = \frac{Cr(\varphi)}{\sigma^2(\varphi)}[r(L)\int_\varphi^L r^{-1}(\vartheta)d\vartheta + \int_0^\varphi r^{-1}(\vartheta)d\vartheta], \tag{2.3}$$

where

$$r(\varphi) = \exp(\rho(\varphi)), \quad \rho(\varphi) = 2\int_0^\varphi \frac{f(\xi)}{\sigma^2(\xi)}d\xi,$$

and $C > 0$ is found from the condition of normalization.

Let $EX(0) < \infty$. Due to the ergodicity of $\Phi(t)$, we have for the mean velocity $\bar{v}$ of $X(t)$:

$$\bar{v} := \lim_{t\to\infty}\frac{EX(t)}{t} = \lim_{t\to\infty}\frac{EX(0)}{t} + \lim_{t\to\infty}\frac{1}{t}\int_0^t Ef(\Phi(s))ds$$

$$= \int_0^L f(\varphi)p(\varphi)d\varphi = \frac{LC}{2}[e^\rho - 1], \quad \rho := \rho(L) = 2\int_0^L \frac{f(\xi)}{\sigma^2(\xi)}d\xi. \tag{2.4}$$

We see that the number ρ gives an integral of the ratio of the drift $f(x)$ to diffusivity $\sigma^2(x)/2$ over the ratchet period L. The sign of $\bar{v}$ depends on the sign of $e^\rho - 1$ only. Evidently, the necessary and sufficient condition for zero mean velocity consists in the equality $\rho = \rho(L) = 0$ (see also [32, 119, 205]).

For instance, if $\sigma \equiv \text{const}$ and the potential

$$V(x) = -\int f(x)dx$$

is an L-periodic function (e.g., a ratchet potential), we get the well-known fact of thermodynamics [62] that $\bar{v} = 0$. At the same time, for some L-periodic potential $V(x)$ and L-periodic state-dependent $\sigma(x)$ the number ρ can be non-zero, i.e., $\bar{v} \neq 0$, and a noise-induced transport can be observed [32, 119]. Our immediate aim is to find sufficient conditions for *unidirectional* transport.

Let $X_{s,x}(t)$ be the solution of (2.2) which starts from the point x at the time s. If $s = 0$, we write $X_x(t)$ instead of $X_{0,x}(t)$. Consider an interval $(x - mL, x + nL)$, where m and n are positive integers. The trajectory $X_x(t)$ reaches one of the points $x - mL$, $x + nL$ in a finite (random) time τ with probability 1. Here τ is the first exit time of $X_x(t)$ from the interval $(x - mL, x + nL)$. Introduce the notation:

$$p_{m,n} := P(X_x(\tau) = x - mL), \quad q_{m,n} := P(X_x(\tau) = x + nL).$$

Theorem 2.1. *The following expressions are valid:*

$$p_{m,n} = \frac{e^{n\rho} - 1}{e^{(m+n)\rho} - 1}, \quad q_{m,n} = \frac{e^{n\rho}(e^{m\rho} - 1)}{e^{(m+n)\rho} - 1}. \quad (2.5)$$

In particular,

$$p := p_{1,1} = \frac{1}{1 + e^{\rho}}, \quad q := q_{1,1} = \frac{1}{1 + e^{-\rho}}. \quad (2.6)$$

To prove the theorem, we use the following arguments. Let τ_y, $x - mL < y < x + nL$, be the first exit time of trajectory $X_y(t)$ from the interval $(x - mL, x + nL)$ (clearly $\tau = \tau_x$), and let φ be a function defined on the set consisting of two points: $x - mL$ and $x + nL$. It is well known [48, 82] that the function

$$u(y) = u(y; x) := E\varphi(X_y(\tau_y))$$

satisfies the boundary value problem

$$\frac{1}{2}\sigma^2(y)\frac{d^2u}{dy^2} + f(y)\frac{du}{dy} = 0, \quad x - mL < y < x + nL, \quad (2.7)$$

$$u(x - mL) = \varphi(x - mL), \quad u(x + nL) = \varphi(x + nL). \quad (2.8)$$

If $\varphi(x - mL) = 1$, $\varphi(x + nL) = 0$, then $u(y; x) = P(X_y(\tau_y) = x - mL)$ and $p_{m,n} = u(x; x)$. Solving the problem (2.7)-(2.8) and transforming the obtained expression for $u(x; x)$, we arrive at (2.5).

Remark 2.2. The formulas (2.5) remain true with the same ρ if we consider a SDE in the sense of Stratonovich:

$$dX = f(X)dt + \sigma(X) \circ dw(t).$$

It is equivalent to the Ito equation

$$dX = f(X)dt + \frac{1}{2}\sigma(X)\frac{d\sigma}{dx}(X)dt + \sigma(X)dw(t).$$

We have

$$\rho_{str}(\varphi) = 2 \int_0^{\varphi} \frac{1}{\sigma^2(\xi)}\left(f(\xi) + \frac{1}{2}\sigma(\xi)\frac{d\sigma}{dx}(\xi)\right)d\xi = \rho_{ito}(\varphi) + \ln\frac{\sigma(\varphi)}{\sigma(0)}.$$

Due to the periodicity of $\sigma(\varphi)$, we obtain $\rho_{str} = \rho_{str}(L) = \rho_{ito}(L) = \rho$.

We remark that the probabilities $p_{m,n}$ and $q_{m,n}$ do not depend on x. Let $\tau_0 = \tau$ be the first exit time of $X_{x_0}(t)$ from the interval $(x_0 - L, x_0 + L)$ and $x_1 := X_{x_0}(\tau_0)$. Then, $x_1 = x_0 - L$ with the probability p and $x_1 = x_0 + L$ with the probability $q = 1 - p$. Let τ_1 be the first exit time of $X_{x_1}(t)$ from the interval $(x_1 - L, x_1 + L)$. Clearly, the conditional probabilities

9.2 Noise-induced unidirectional transport   531

$P(X_{x_0}(\tau_0 + \tau_1) = X_{x_1}(\tau_1) = x_1 \mp L \mid X_{x_0}(\tau_0) = x_1)$ are equal to p and q, respectively. If we continue, we obtain the random sequences $\tau_0, \tau_1, \ldots, \tau_k, \ldots$ and x_0, $x_1 = X_{x_0}(\tau_0)$, ..., $x_k = X_{x_0}(\tau_0 + \cdots + \tau_{k-1})$, ... Using these, we can get a concise description of the evolution of $X_{x_0}(t)$. Let us consider in brief the main features of these sequences.

The sequence τ_k. The sequence consists of independent identically distributed (i.i.d.) random variables with distribution of τ. We observe that all the basic probabilistic characteristics of the random variable τ can be found by solving deterministic differential equations. For instance, the probability $P(\tau < s, X_x(\tau) = x - L)$ can be evaluated by solving a mixed problem for a backward Kolmogorov equation and the characteristic function or Laplace transform for τ can be evaluated by solving a boundary value problem for an ordinary differential equation (see, e.g., [48]). It turns out that such important characteristics as $E\tau$ and $D\tau$ can be found by quadratures. Namely $E\tau = u(0)$, where $u(x)$ is a solution of the following boundary value problem [48, 82]:

$$\frac{1}{2}\sigma^2(x)u'' + f(x)u' = -1, \ u(-L) = u(L) = 0. \tag{2.9}$$

We get

$$u(x) = -G(x) + C_1 \int_0^x \exp\left(-2\int_0^\xi \frac{f(s)}{\sigma^2(s)}ds\right) d\xi + C_2,$$

where

$$G(x) = \int_0^x \left[\frac{2}{\sigma^2(\xi)} \int_\xi^x \exp\left(2\int_\eta^\xi \frac{f(s)}{\sigma^2(s)}ds\right) d\eta\right] d\xi,$$

and the constants C_1, C_2 have to be found from (2.9). Thus

$$E\tau = u(0) = C_2 = \frac{e^\rho G(L) + G(-L)}{1 + e^\rho}. \tag{2.10}$$

The second moment $E\tau^2$ can also be expressed in quadratures: $E\tau^2 = v(0)$, where $v(x)$ is a solution to the boundary value problem [48]

$$\frac{1}{2}\sigma^2(x)v'' + f(x)v' = -2u(x), \ v(-L) = v(L) = 0.$$

We do not write the corresponding explicit expression for $v(x)$ because of its bulky form.

Clearly, behavior of the sum $\tau_0 + \cdots + \tau_N$ as $N \to \infty$ is governed by the law of large numbers and by the central limit theorem.

The sequence x_k. It is not difficult to see that $x_{k+1} = x_k + \xi_k$, where ξ_k are i.i.d. random variables. Any ξ is a Bernoulli random variable taking values $-L$ and L with probabilities $P(\xi = -L) = p = 1/(1+e^\rho)$, $P(\xi = L) = q = 1/(1+e^{-\rho})$. The sequence x_k can be considered as a trajectory of the random walk. The theory of such a random walk is well developed (see, e.g., [59]).

Unidirectional transport. Let $\rho > 0$. Then the transport is positive. If the probability p is small, then the retrograde steps are infrequent and in such a case it is natural to consider the transport as a unidirectional one. An acceptable condition for the noise-induced unidirectional transport to be to the right is closeness of the probability p to zero. The probability p is smaller, when ρ is larger (see (2.6)). So, we have obtained the very simple characteristic ρ of transport unidirectionality for the model (2.1). Analogously, for $\rho < 0$ the transport is negative and an acceptable condition for the noise-induced unidirectional transport to be to the left is closeness of the probability q to zero.

Note that there is no definite relation between the mean velocity $\bar{v}$ (see formula (2.4)) and the characteristic of unidirectionality ρ because for large (small) ρ the constant C in (2.4) can be small (large). The following relation between ρ, $\bar{v}$, and $E\tau$ holds:

$$\bar{v} = \frac{L}{E\tau} \frac{e^\rho - 1}{e^\rho + 1}. \tag{2.11}$$

A heuristic proof of (2.11) is as follows. Consider the sequence $x_0, x_1, \ldots, x_N$ defined above. We have $P(x_{i+1} = x_i - L) = p = 1/(1+e^\rho)$ and $P(x_{i+1} = x_i + L) = q = 1/(1+e^{-\rho})$ (see (2.6)). Then the mean advance of the sequence for N steps $E(x_N - x_0)$ is equal to $NL \times (1/(1+e^{-\rho}) - 1/(1+e^\rho))$. Taking into account that the mean time of the N steps is equal to $NE\tau$, we come to (2.11). The rigorous proof consists in direct verification of the formula (2.11). This is possible due to the known expressions for $E\tau$ and the constant C.

Remark 2.3. Consider the segment $X_{x_k}(t)$, $\tau_{k-1} \leq t \leq \tau_k$, of the trajectory $X_{x_0}(t)$. Let $x_{k+1} = X_{x_k}(\tau_k) = x_k + L$, i.e., the considered trajectory shifts to the right at the $(k+1)^{th}$ step. In such a situation, we can assert only that the trajectory does not jump back over the distance $2L$. However, the trajectory can jump back over the distance close to $2L$ during the time (τ_{k-1}, τ_k). Indeed, the trajectory may come up close to $x_k + L$, then turn back and come up close to $x_k - L$, and finally reach $x_k + L$.

Example 2.4. To illustrate the results of this section, we take the coefficients of (2.1) in the form [32]:

$$f(x) = f_0 \sin(2\pi x), \quad \sigma(x) = \frac{\sigma_0}{\sqrt{1 - \alpha \cos(2\pi x + \phi)}} \tag{2.12}$$

with $f_0, \sigma_0 > 0$, $0 < \alpha < 1$. In this case $\rho = \dfrac{\alpha f_0}{\sigma_0^2} \sin \phi$. Note that if $\phi = k\pi$ with integer k, then ρ is equal to 0 and there is no transport.

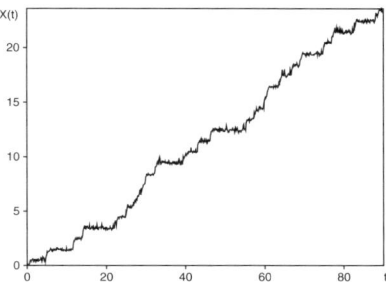 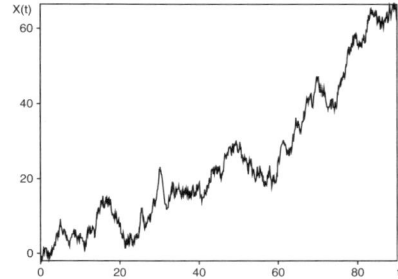

Fig. 2.1. Systems with state-dependent noise. Sample trajectories of the solution to (2.1) with the coefficients of (2.12) for $f_0 = 1$, $\alpha = 0.9$, $\phi = \pi/2$, and $\sigma_0 = 0.44$ (*left*) and $\sigma_0 = 3$ (*right*).

Figure 2.1 gives typical trajectories of the solution $X(t)$ to (2.1) with the coefficients of (2.12). Figure 2.1 (left) corresponds to the regime of the unidirectional transport: there is practically no movement to the left ($\rho \doteq 4.65$ and $p = 0.0095$). If we increase the noise intensity, the mean velocity of the transport increases but the transport becomes non-unidirectional (see Fig. 2.1 (right), the corresponding $\rho = 0.1$ and $p = 0.475$). Let us note that if we decrease the noise intensity, the mean velocity decreases, and for a sufficiently small noise the transport is negligibly small and not interesting from the point of view of possible applications.

As for an experimental verification of effects in systems with state-dependent noise, let us refer the reader to, for instance, the paper of Büttiker [32], where possible physical experiments are discussed.

9.2.2 Forced thermal ratchets

In this subsection we consider a periodically forced thermal ratchet of the form [161] (see also [13, 116, 205, 246]):

$$dX = f(X)dt + A\chi(t;T)dt + \sigma dw(t), \tag{2.13}$$

where $f(x) = -dV(x)/dx$, $V(x)$ is an L-periodic ratchet potential $V(x) = V(x+L)$, $x \in R$, possessing no reflection symmetry (there is no ϕ such that $V(x+\phi) = V(-x+\phi)$ for all $x \in (0, L/2)$), A, T, and σ are some positive constants,

$$\chi(t;T) = \begin{cases} 1, & 0 \leq t < T/2, \\ -1, & T/2 \leq t < T, \end{cases}$$

and $\chi(t;T)$ is T-periodical.

As is known, forced thermal ratchets exhibit the noise-induced transport. Here we study conditions, when the transport is unidirectional.

A qualitative condition of unidirectionality. In connection with (2.13), consider two SDEs

$$dX^+ = f(X^+)dt + Adt + \sigma dw(t), \tag{2.14}$$

$$dX^- = f(X^-)dt - Adt + \sigma dw(t). \tag{2.15}$$

Just as we get (2.4), it is possible to find expressions for the mean velocities (see, e.g., [205]): $\bar{v}^\pm = \lim_{t\to\infty} EX^\pm(t)/t$. The asymmetry of the ratchet potential $V(x)$ can result in $\bar{v}^+ \neq -\bar{v}^-$. If the period T of $\chi(t;T)$ is sufficiently large, the mean velocity $\bar{v}$ of $X(t)$ can approximately be evaluated by $\bar{v} \doteq (\bar{v}^+ + \bar{v}^-)/2$ (see [13, 116, 161, 205, 246]).

Analogously to p, q, and τ of the previous section, introduce the probabilities p^+ and q^+ (p^- and q^-) and the random time τ^+ (τ^-) for the process $X^+(t)$ ($X^-(t)$):

$$p^\pm := P(X_x^\pm(\tau^\pm) = x - L), \ q^\pm := P(X_x^\pm(\tau^\pm) = x + L),$$

and $\tau^\pm$ are the first exit times of $X_x^\pm(t)$ from the interval $(x - L, x + L)$. Because $\rho^\pm := 2\int_0^L \frac{f(x) \pm A}{\sigma^2}dx = \pm\frac{2AL}{\sigma^2}$, we get from Theorem 2.1: $p^+ = q^-$ and $q^+ = p^-$.

Let T be so large that $E\tau^\pm << T/2$. Evidently, in this case the transport cannot be unidirectional. Indeed, the condition of closeness of q^+ to 1 is necessary for unidirectionality of the transport for the model (2.13). So, the transport is positively unidirectional for the first half period and negatively unidirectional for the second half period. Consequently, it is not unidirectional as a whole.

Consider another case. If $v^+ > v^-$, then $E\tau^+ < E\tau^-$ (see (2.11)). Now let T be such that $E\tau^+ << T/2$, $E\tau^- >> T/2$, and, as before, q^+ be close to 1. Clearly, then one can expect the unidirectional transport in (2.13) in positive direction. Analogously, if T is such that $E\tau^- << T/2$, $E\tau^+ >> T/2$, and p^- is close to 1, one can expect the unidirectional transport to be in the negative direction. These qualitative sufficient conditions of the unidirectional transport are fairly constructive because the magnitudes $E\tau^\pm$ can be found by quadratures (see (2.10)).

Example 2.5. Consider the model (2.13) with a simple ratchet potential [161]

$$V(x) = \begin{cases} \dfrac{h}{l}x, \ 0 \leq x < l, \\ \dfrac{h}{L-l}(L-x), \ l \leq x < L. \end{cases} \tag{2.16}$$

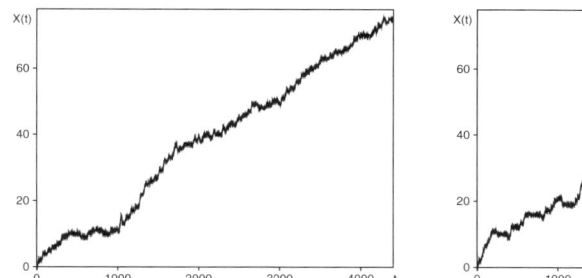
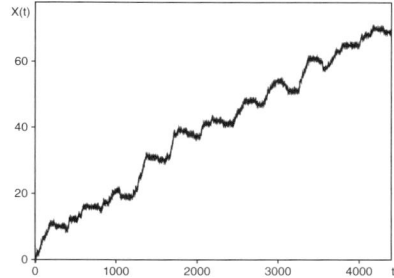

Fig. 2.2. Forced thermal ratchets. Sample trajectories of the solution to (2.13) with the potential of (2.16) for $L = 1$, $l = 0.75$, $h = 1/6$, $A = 0.2$, $\sigma = 0.2$, and $T = 60$ (*left*) and $T = 400$ (*right*).

From (2.10) we get

$$E\tau^+ = \frac{\sigma^2}{2} \frac{L^2 h^2}{(Al-h)^2(A(L-l)+h)^2}$$

$$\times \frac{\exp\left(-\frac{2}{\sigma^2}(A(L-l)+h)\right) + \exp\left(-\frac{2}{\sigma^2}(Al-h)\right) - \exp\left(-\frac{2AL}{\sigma^2}\right) - 1}{\exp\left(-\frac{2AL}{\sigma^2}\right) + 1}$$

$$+ \left(\frac{l^2}{Al-h} + \frac{(L-l)^2}{A(L-l)+h}\right) \frac{1 - \exp\left(-\frac{2AL}{\sigma^2}\right)}{1 + \exp\left(-\frac{2AL}{\sigma^2}\right)}. \qquad (2.17)$$

The value $E\tau^-$ is obtained by substituting $-A$ in (2.17) instead of A. Note that the indeterminacy in (2.17) at $Al = h$ (or in the corresponding formula for $E\tau^-$ at $A(L-l) = h$) can be evaluated. For instance,

$$E\tau^+ = \frac{l^2}{\sigma^2} + \frac{L^2 - l^2}{AL} \frac{1 - \exp\left(-\frac{2AL}{\sigma^2}\right)}{1 + \exp\left(-\frac{2AL}{\sigma^2}\right)} \quad \text{if } Al = h.$$

Let us take $l = \frac{3}{4}L$, $h = \frac{10}{9}Al = \frac{5}{6}AL$, $\frac{2AL}{\sigma^2} \gg 1$. Then

$$E\tau^+ \doteq \frac{7200\sigma^2}{169A^2}\left[\exp\left(\frac{AL}{6\sigma^2}\right) - 1\right] - \frac{87L}{13A},$$

$$E\tau^- \doteq \frac{7200\sigma^2}{17\,689A^2}\left[\exp\left(\frac{7AL}{6\sigma^2}\right) - 1\right] - \frac{33L}{133A}.$$

If, for example, $\frac{2AL}{\sigma^2} = 10$, $L = 1$, $A = 0.2$ (and, consequently, $\sigma = 0.2$), then $E\tau^+ \approx 22$ and $E\tau^- \approx 137$. Figure 2.2 demonstrates sample trajectories of the solution $X(t)$ of (2.13) with the potential $V(x)$ from (2.16) for the parameters given above. Figure 2.2 (left) corresponds to $T/2 = 30$ which is less than $E\tau^-$ and greater than $E\tau^+$. In this case the retrograde steps are infrequent, and we can consider the transport as unidirectional. When we take T such that $E\tau^- \ll T/2$ (see Fig. 2.2 (right)), the retrograde steps become quite frequent.

Let us note that by the qualitative conditions it is quite difficult to indicate parameters corresponding to the unidirectional transport, in particular, due to a large variance of τ^-. Besides, these conditions do not give us a measure of unidirectionality. To obtain a more detailed description of noise-induced transport, we need to consider other characteristics.

A general condition of unidirectionality. Now our aim is to give an appropriate characteristic of unidirectional transport in forced thermal ratchets and to propose a universal (numerical) tool for identifying domains of parameters, where unidirectional transport is realized. Here we use the technique of Sect. 9.1.

Denote by $X_{s,x}(t)$, $t \geq s$, the solution of (2.13) starting at the moment s from the point x. Introduce the probabilities:
the probability $P_{0,x}^>$ (the probability $P_{T/2,x}^>$) that the trajectory $X_{0,x}(t)$ (the trajectory $X_{T/2,x}(t)$) shifts in negative direction not more than on L during the first half-period (the second half-period) of the periodic forcing

$$P_{0,x}^> := P(X_{0,x}(t) > x - L,\ 0 \leq t \leq T/2),$$

$$P_{T/2,x}^> := P(X_{T/2,x}(t) > x - L,\ T/2 \leq t \leq T);$$

the probability $P_{0,x}^<$ (the probability $P_{T/2,x}^<$) that the trajectory $X_{0,x}(t)$ (the trajectory $X_{T/2,x}(t)$) shifts in positive direction not more than on L during the first half-period (the second half-period) of the periodic forcing

$$P_{0,x}^< := P(X_{0,x}(t) < x + L,\ 0 \leq t \leq T/2),$$

$$P_{T/2,x}^< := P(X_{T/2,x}(t) < x + L,\ T/2 \leq t \leq T).$$

Let

$$Q_{0,x}^> := 1 - P_{0,x}^>,\ Q_{T/2,x}^> := 1 - P_{T/2,x}^>,$$
$$Q_{0,x}^< := 1 - P_{0,x}^<,\ Q_{T/2,x}^< := 1 - P_{T/2,x}^<.$$

For example, $Q_{T/2,x}^<$ is the probability for $X_{T/2,x}(t)$ to reach the level $x + L$ at least once during the second half-period.

It is not difficult to see that for $A > 0$

$$Q_{T/2,x}^> > Q_{0,x}^> \text{ and } Q_{0,x}^< > Q_{T/2,x}^<. \tag{2.18}$$

Therefore, if both $P^>_{T/2,x}$ is close to one (i.e., $Q^>_{T/2,x}$ is close to zero) and $Q^<_{0,x}$ is close to one for all $x \in [0, L]$, then during each period of the periodic forcing the trajectory $X(t)$ moves in positive direction and does not move in negative direction with probability close to 1. Analogously, if both $P^<_{0,x}$ is close to one (i.e., $Q^<_{0,x}$ is close to zero) and $Q^>_{T/2,x}$ is close to one, then we have the unidirectional transport to the left with probability close to 1.

So, *closeness of one of the following products*

$$\Pi^+ = \Pi^+_x(A, T, \sigma) := P^>_{T/2,x} \times Q^<_{0,x}, \quad \Pi^- = \Pi^-_x(A, T, \sigma) := P^<_{0,x} \times Q^>_{T/2,x}$$

to 1 *for all* x *is a sufficient condition for the unidirectional transport with high probability.*

Let us note that the phenomenon of unidirectionality cannot be observed with probability 1 because there is always a non-zero probability that the trajectory moves far in either direction.

For definiteness, below we are interested in the transport in positive direction, i.e., when Π^+ is close to 1. The further analysis essentially rests on the possibility to evaluate the probability $P^>_{T/2,x}$ (and $Q^<_{0,x}$) in a constructive way. To this end, we introduce the function

$$u(s, y) = u_{x-L}(s, y) := P(X^-_{s,y}(t) > x - L, \ s \le t \le T/2),$$

$$0 \le s \le T/2, \ y \ge x - L,$$

where $X^-_{s,x}(t)$ is a solution to (2.15). Since the distribution of $X_{T/2,x}(t)$, $T/2 \le t \le T$, coincides with the distribution of $X^-_{0,x}(t)$, $0 \le t \le T/2$, one can see that

$$P^>_{T/2,x} = u_{x-L}(0, x).$$

The function $\tilde{u}(s, y) := P(X^-_{s,y}(t) > x - L)$ solves the corresponding Cauchy problem for the backward Kolmogorov equation (2.19). It is well known that the function $u(s, y)$ satisfies the boundary value problem in the half-band for the same equation:

$$\frac{\partial u}{\partial s} + \frac{\sigma^2}{2}\frac{\partial^2 u}{\partial y^2} + (f(y) - A)\frac{\partial u}{\partial y} = 0, \ 0 \le s < T/2, \ y > x - L, \quad (2.19)$$

$$u(T/2, y) = 1, \ y > x - L; \ u(s, x - L) = 0, \ 0 \le s \le T/2. \quad (2.20)$$

The solution to the problem (2.19)-(2.20) has the following probabilistic representation:

$$u(s, y) = u_{x-L}(s, y) = E\varphi(\tau_{s,y}(x - L), X^-_{s,y}(\tau_{s,y}(x - L))), \quad (2.21)$$

where $(\tau_{s,y}(x - L), X^-_{s,y}(\tau_{s,y}(x - L)))$ is the first exit point of the space-time diffusion $(t, X^-_{s,y}(t))$, $t > s$, from the domain $[0, T/2] \times (x - L, +\infty)$ and

$$\varphi(s, y) = \begin{cases} 1, & s = T/2, \ y > x - L, \\ 0, & 0 \le s \le T/2, \ y = x - L. \end{cases}$$

The probability $Q_{0,x}^<$ can be evaluated analogously. We obtain that $P_{0,x}^< = 1 - Q_{0,x}^<$ is equal to

$$P_{0,x}^< = v_{x+L}(0, x),$$

where $v(s, y) = v_{x+L}(s, y)$ is the solution of the boundary value problem

$$\frac{\partial v}{\partial s} + \frac{\sigma^2}{2} \frac{\partial^2 v}{\partial y^2} + (f(y) + A) \frac{\partial v}{\partial y} = 0, \ 0 \le s < T/2, \ y < x + L, \quad (2.22)$$

$$v(T/2, y) = 1, \ y < x + L; \ v(s, x + L) = 0, \ 0 \le s \le T/2. \quad (2.23)$$

The solution of this problem has the following probabilistic representation:

$$v(s, y) = v_{x+L}(s, y) = E\psi(\tau_{s,y}(x+L), X_{s,y}^+(\tau_{s,y}(x+L))), \quad (2.24)$$

where $(\tau_{s,y}(x+L), X_{s,y}^+(\tau_{s,y}(x+L)))$ is the first exit point of the space-time diffusion $(t, X_{s,y}^+(t)), t > s$, from the domain $[0, T/2) \times (-\infty, x+L)$ and

$$\psi(s, y) = \begin{cases} 1, \ s = T/2, \ y < x + L, \\ 0, \ 0 \le s \le T/2, \ y = x + L. \end{cases}$$

As a result, we have obtained the following theorem.

Theorem 2.6. *A sufficient condition for the positive unidirectional transport in the model (2.13) consists in closeness of the product $\Pi^+ = u(0, x) \times (1 - v(0, x))$ to 1 for all $0 \le x < L$. Here $u(0, x), v(0, x)$ are values of the functions $u(s, y), v(s, y)$ at $(s, y) = (0, x)$, where the functions are solutions of the boundary value problems (2.19)-(2.20) and (2.22)-(2.23). The individual values $u(0, x)$ and $v(0, x)$ can be found as probabilistic representations (2.21) and (2.24) for fixed $(s, y) = (0, x)$. An analogous assertion is true for negative unidirectional transport.*

Remark 2.7. Similarly, we can state boundary value problems for evaluating the probabilities $P(X_{T/2,x}(t) > x - L^-, \ T/2 \le t \le T)$ and $P(X_{0,x}(t) < x + L^+, \ 0 \le t \le T/2)$ with $L^- = mL$, $L^+ = nL$, m, n are positive integers. These probabilities can be used for a detailed description of the transport much as the probabilities $P_{T/2,x}^>$ and $P_{0,x}^<$ are employed above.

It is possible to prove that $u_{x-L}(0, x; A, T, \sigma)$ is decreasing with respect to A, T, and σ, and $v_{x-L}(0, x; A, T, \sigma)$ is increasing with respect to A, T, and σ. Then the product Π^+ has a maximum in T for fixed A and σ. Analogously, the product Π^+ has a maximum in σ for fixed A and T. The remarkable feature of the phenomenon considered here is that for some ratchet potentials $V(x)$ the product Π^+ is close to 1 for a sufficiently wide range of parameters A, T, σ. In particular, this is confirmed in our tests (see Fig. 2.3). We take the amplitude A of the periodic forcing less than A^* so that there is no transport in the system (2.13) for $A < A^*$ and $\sigma = 0$.

9.2 Noise-induced unidirectional transport 539

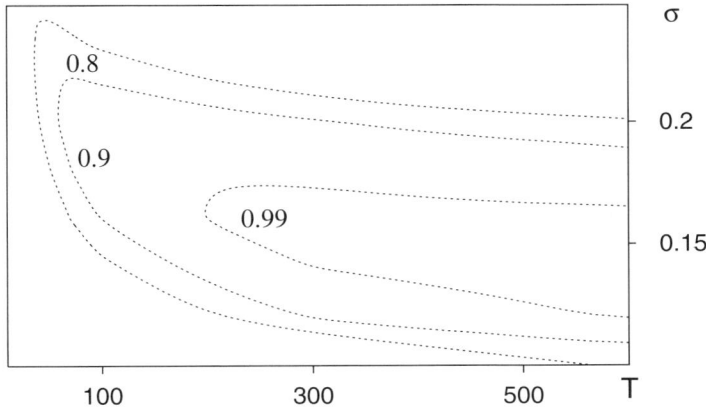

Fig. 2.3. Forced thermal ratchets. Level curves of the product Π^+ in the plane (T,σ) for $A = 0.6$ and the potential $V(x)$ of (2.25) with $L = 1$.

Remark 2.8. If the probability $P^>_{T/2,x}$ is close to 1, there is practically no transport in a negative direction, and if, in addition, $E\tau^+ \ll T/2$, then the transport is positively unidirectional. So, closeness of the probability $P^>_{T/2,x}$ to 1 and $E\tau^+ \ll T/2$ give us the other condition for unidirectional transport in a positive direction. This condition is less general than the one of Theorem 2.6 but it is substantially easier to evaluate $E\tau^+$ than $Q^<_{0,x}$. We note in passing that in this case the mean shift $\bar{\Delta}$ of $X(t)$ during the single period T of the periodic forcing is approximately estimated as $\bar{\Delta} \doteq \bar{v}^+ T/2$.

Numerical results. In a general case to find domains of parameters corresponding to unidirectional transport, one should solve the linear boundary value problems (2.19)-(2.20) and (2.22)-(2.23) numerically. We perform some numerical experiments. We take the following ratchet potential $V(x)$

$$V(x) = -\frac{L}{2\pi}\left(\sin\frac{2\pi x}{L} + \frac{1}{4}\sin\frac{4\pi x}{L}\right), \quad L > 0, \tag{2.25}$$

that is used for some tests in, e.g., [13, 205].

Figure 2.3 gives level curves of the product Π^+. In accordance with our tests the probabilities $P^>_{T/2,x}$ and $Q^<_{0,x}$ depend only weakly on x, and for definiteness we take x in the presented tests such that the potential $V(x)$ has a local minimum at this point. One can see that the domain of parameters corresponding to the noise-induced unidirectional transport is sufficiently large. Let us remark that there is no unidirectional transport for both sufficiently large and small noise intensities. Figure 2.4 demonstrates typical trajectories of the solution $X(t)$ to (2.13) with the potential of (2.25). Figure 2.4 (left) corresponds to the regime of the unidirectional transport. One can see that

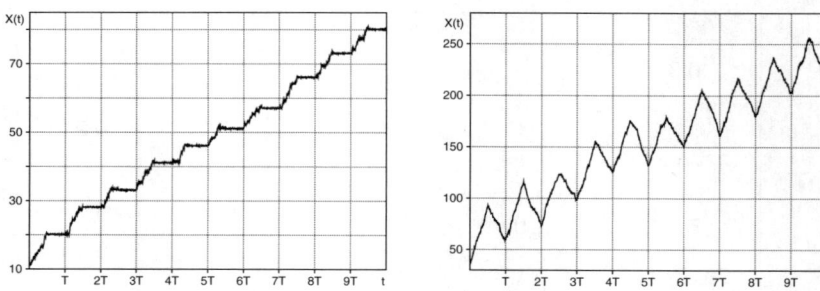

Fig. 2.4. Forced thermal ratchets. Sample trajectories of the solution to (2.13) with the potential $V(x)$ of (2.25) for $L=1$, $A=0.6$, $T=400$, and $\sigma=0.15$ (*left*) and $\sigma=0.4$ (*right*).

during the first half-period of the periodic forcing the trajectory moves to the right on a distance of $5-10$ periods of the potential $V(x)$. At the same time, during the second-half period of the periodic forcing the trajectory increment is practically equal to zero. If we increase the noise intensity, the mean shift $\bar{\Delta}$ during the single period of the periodic forcing increases but the transport becomes non-unidirectional (see Fig. 2.4 (right)).

Let us note that the approach proposed above is universal and can be extended to more complicated models.

In our tests we use both finite-difference schemes and probability methods of Chap. 6 to solve the problems (2.19)-(2.20) and (2.22)-(2.23). Let us observe that we need the individual values $u_{x-L}(0,x)$ and $v_{x+L}(0,x)$ only and in such a case the probabilistic approach with the Monte Carlo technique is most relevant.

A Appendix:
Practical guidance to implementation of the stochastic numerical methods

An overwhelming majority of the methods proposed in this book are brought to numerical algorithms. Then it only remains to write a computer program, which is usually not complicated, and to use the method in practice. Nevertheless, we give some illustrations in this Appendix how our methods can be implemented. A successful realization of a method depends on our ability to choose a step of numerical integration to obtain a solution of SDEs with a required accuracy. Further, when the Monte Carlo technique is used, the supplementary problem of choosing a number of independent realizations arises as well. These problems are discussed in Sects. A.1 and A.2 which deal with realizations of mean-square and weak methods.

Algorithms and, consequently the programs, are slightly more complicated in the case of simulating SDEs in bounded domains. Implementation issues for such algorithms are considered in Sects. A.3 and A.4. Section A.5 is devoted to realization of our algorithms for nonlinear PDEs. Section A.6 contains some supplementary procedures used in programs from the previous sections.

All the examples in this Appendix are written in ANSI C. The purpose of these examples is to give an illustration of implementation of our algorithms. The programs are written in a rather free manner and should not be considered as a software product. At the same time, they are in a working condition. Also see implementation of numerical methods for SDEs in [132].

A.1 Mean-square methods

This section deals with implementation of mean-square methods presented in Chap. 1 and Sects. 3.1-3.3, 4.2-4.5, and 4.7. Practical realization of these methods is, to a large extent, similar to implementation of numerical methods from deterministic numerical analysis in the case of the initial value problem for deterministic ordinary differential equations. The only difference is that stochastic methods require a source of randomness that is discussed in Sect. 2.6. Program examples realizing standard mean-square methods like those from Chap. 1 are given, for instance, in [132]. Here, to clarify the matter, we present the implementation of methods which were used in our experiments with the model for synchrotron oscillations of particles in storage rings (4.4.5) from Sect. 4.4.2:

$$dP = -\omega^2 \sin(Q)dt - \sigma_1 \cos(Q)dw_1 - \sigma_2 \sin(Q)dw_2,$$

$$dQ = Pdt. \tag{1.1}$$

The program `hmmd.c` given in Listing A.1 corresponds to the realization of the Euler method (4.4.6):

$$P_{k+1} = P_k - h\omega^2 \sin(Q_k) - h^{1/2}(\sigma_1 \cos(Q_k)\Delta_k w_1 + \sigma_2 \sin(Q_k)\Delta_k w_2),$$
$$Q_{k+1} = Q_k + hP_k, \tag{1.2}$$

and the explicit symplectic method (4.4.7):

$$\mathcal{Q} = Q_k + hP_k,$$
$$P_{k+1} = P_k - h\omega^2 \sin(\mathcal{Q}) - h^{1/2}(\sigma_1 \cos(\mathcal{Q})\Delta_k w_1 + \sigma_2 \sin(\mathcal{Q})\Delta_k w_2),$$
$$Q_{k+1} = \mathcal{Q}. \tag{1.3}$$

Both methods are of first mean-square order.

Mean-square methods are usually used for direct simulations of trajectories. To check an accuracy of simulations or, in other words, to select an appropriate time step, it is advisable to simulate a stochastic system by a method with different time steps and then compare the results. We can be satisfied with our choice h_* of the time step if by further decrease of the step we get trajectories of the system, which are close enough (in terms of the experiment under consideration) to the ones obtained with the step h_*. Of course, this comparison is meaningful only if the simulations are done along the *same* driving sample paths of the Wiener processes involved. Suppose we are implementing a method with a uniform time step that uses an information about trajectories of Wiener processes in the form of their increments only (e.g., an Euler-type mean-square method). Further, assume that we would like to run the corresponding program twice with time steps h and lh, where l is a positive integer. First, we should pass the same seed (in the below program it is the parameter *iseed*) for the RNG in both runs. Second, we need to ensure that the increments of the Wiener processes $\Delta w(lh) = w(t+lh) - w(t)$ on an interval $[t, t+lh]$ are the same in both runs, i.e., if in the run with the step h we simulate the increments $w(t+jh) - w(t+(j-1)h)$, $j = 1, \ldots, l$, on each $[t, t+lh]$, then in the run with the step lh the increments $\Delta w(lh)$ should be evaluated as

$$\Delta w(lh) = \sum_{j=1}^{l} [w(t+jh) - w(t+(j-1)h)].$$

The meaning of the parameter l is the same in both this discussion and the program `hmmd.c` given in Listing A.1. For example, if we would like to simulate trajectories by this program along the same driving sample paths of the Wiener processes using the time steps $h = 0.1$ and $h = 0.01$, we should run it with $l = 10$ for $h = 0.1$ and $l = 1$ for $h = 0.01$. In the described approach we coarsen Wiener sample paths from the discretization with a smaller time step h to the one with the larger step lh. Let us also note that it is possible to use the Brownian bridge to refine Wiener sample paths from the discretization with a larger time step to a discretization with a smaller step.

Listing A.1. Program hmmd.c (mean-square symplectic and Euler methods for the model for synchrotron oscillations).

```
/* Program hmmd.c: mean-square symplectic and Euler methods
   for the model for synchrotron oscillations.
   The program uses the external procedures rng_lcgs.c and
   rng_gau.c */
#include <stdio.h>
#include <math.h>
#include <stochast.h>  /*the header file, see Miscellaneous*/
#define OF    10L    /* regulates the amount of output data*/

main (void)
{
 long i,iseed,j,l;
 double a,a2h,fl,g1,g2,h,h12,p0,q0,p_e,q_e,p_s,q_s;
 double si1,si2,sih12a,sih12b,t,tk,x;
 FILE *out;

 /* input of parameters */
 printf("omega=");  scanf("%lf",&a);    /* omega  */
 printf("sigma1="); scanf("%lf",&si1); /* sigma_1 */
 printf("sigma2="); scanf("%lf",&si2); /* sigma_2 */
 printf("tk=");     scanf("%lf",&tk);   /* time interval*/
 printf("P_0=");    scanf("%lf",&p0);  /*initial value for P*/
 printf("Q_0=");    scanf("%lf",&q0);  /*initial value for Q*/
 printf("h=");      scanf("%lf",&h);    /* time step */
 printf("seed=");   scanf("%ld",&iseed); /* seed for the RNG
                rng_lcgs.c, it can be any negative integer*/
 printf("l=");      scanf("%ld",&l);    /*to ensure the same
                                         sample paths*/
 /* Open output file */
 if((out = fopen("hmmd.trc","w")) == NULL)
  {fprintf(stderr,"Cannot open output file.\n"); return(-1);}
 /* comment in the output file */
 fprintf(out,"#########################################\n");
 fprintf(out,"# model for synchrotron oscillations \n");
 fprintf(out,"# Input data:                \n");
 fprintf(out,"# omega=%f sigma1=%f sigma2=%f tk=%f \n",a,si1,
         si2,tk);
 fprintf(out,"# P_0=%f Q_0=%f h=%f seed=%ld l=%ld\n#\n",
         p0,q0,h,iseed,l);
```

```
/* auxiliary variables */
h12=sqrt(h); sih12a=h12*si1; sih12b=h12*si2; fl=sqrt(l); a2h=a*a*h;
/* initializing variables for the time loop */
p_e=p0; q_e=q0; p_s=p0; q_s=q0; t=0.; i=1;
fprintf(out,"%lf   %lf %lf   %lf %lf\n",t,p_e,q_e,p_s,q_s);

while(t<tk)  /* time loop */
{
 /* simulation of the Wiener increments using*/
    g1=0.; g2=0.; /* initialization */
  for(j=1; j<=l; j++){ g1=g1+rng_gau(&iseed);
                        g2=g2+rng_gau(&iseed); }
  g1/=fl; g2/=fl; /* increments */

  /* m-sq Euler method*/
   x=p_e;
   p_e=p_e-a2h*sin(q_e)-sih12a*cos(q_e)*g1-sih12b*sin(q_e)*g2;
   q_e+=h*x;

  /* first-order explicit symplectic method */
   q_s+=h*p_s;
   p_s=p_s-a2h*sin(q_s)-sih12a*cos(q_s)*g1-sih12b*sin(q_s)*g2;

   t+=h; /* current time */
   if(((i % OF) == 0))
     fprintf(out,"%lf   %lf %lf   %lf %lf\n",t,p_e,q_e,p_s,q_s);
   ++i; /* counter for the output */
 }  /*end of the time loop*/
 fclose(out); return(0);
} /* end of the main program */
```

To generate Gaussian random numbers in the above program, we use the procedure rng_gau.c from Listing A.16, which uses the RNG rng_lcgs.c (see Listing A.14) as a source of uniformly distributed random variables. The output file hmmd.trc of the program hmmd.c is ready for plotting graphs of sample trajectories of $P(t)$, $Q(t)$ in, e.g., *Gnuplot*. Using this program with time steps $h = 0.02$ and $h = 0.002$, sample trajectories plotted on Fig. 4.4.2 (see Sect. 4.4.2) were obtained. Figure 4.4.2 clearly demonstrates that the Euler method (1.2) is unacceptable for simulation of the solution to (1.1) on a long time interval while the symplectic method (1.3) produces quite accurate results despite both methods have the same mean-square order of accuracy.

A.2 Weak methods and the Monte Carlo technique

Weak methods are sufficient for evaluation of mean values and solving problems of mathematical physics by the Monte Carlo technique, and they are simpler than mean-square ones. In this section we consider implementation of weak methods applicable to the Cauchy problem for parabolic PDEs, and we are concerned in realization of the methods from Chap. 2 and Sects. 3.4-3.8, 4.6, and 4.8. Implementation of weak methods for boundary value problems is discussed in Sect. A.4.

Weak methods are usually used together with the Monte Carlo technique, i.e., their implementation includes running a large number of independent trajectories, making final averaging, and estimating the Monte Carlo error. We note that simulation of a large number of independent trajectories requires RNGs with long periods (see Sects. 2.6.1 and 2.6.2). Further, Monte Carlo simulations are well suited to parallel computers. A common way of parallelizing Monte Carlo simulations is to run identical procedures but with different random number sequences on the various processors. A communication between the processors is needed only to start the simulation and to make final averaging and output. Then the use of p independent processors should reduce computational costs of the simulation in p times (see further discussion in Sect. 2.6.4).

By weak methods we obtain weak approximations $\bar{X}(T)$ of solutions $X(T)$ of SDEs, which we use to evaluate averages $Ef(X(T)) \approx Ef(\bar{X}(T))$. The difference $Ef(X(T)) - Ef(\bar{X}(T))$ is the error of numerical integration which depends on the choice of the time step h. First we evaluate $Ef(\bar{X}(T))$, where $\bar{X}(T)$ is simulated with a tentative step h.

To evaluate $Ef(\bar{X}(T))$ in practice, we need to apply the Monte Carlo technique. As a result, in addition to the error of numerical integration, there is also the Monte Carlo error:

$$Ef(\bar{X}(T)) = \frac{1}{M} \sum_{m=1}^{M} f(\bar{X}^{(m)}(T)) + R_{mc}, \qquad (2.1)$$

where M is the number of independent realizations $\bar{X}^{(m)}(T)$ of $\bar{X}(T)$. The Monte Carlo error R_{mc} has zero bias and its variance equals to

$$Var(R_{mc}) = \frac{Var\, f(\bar{X}(T))}{M}, \qquad (2.2)$$

i.e., the simulated $\hat{f} := \frac{1}{M} \sum_{m=1}^{M} f(\bar{X}^{(m)}(T))$ belongs to the confidence interval:

$$\hat{f} \in (Ef(\bar{X}(T)) - c\sqrt{Var(R_{mc})}, Ef(\bar{X}(T)) + c\sqrt{Var(R_{mc})}) \qquad (2.3)$$

with the fiducial probability, for example, 0.997 for $c = 3$ and 0.95 for $c = 2$. Since dependence of $Var\,(R_{mc})$ on h is usually not essential and this variance can be evaluated in numerical simulations together with $\hat{f}$, we can control the Monte Carlo error and decrease it by choosing an appropriate number M of independent realizations. Note that variance reduction techniques (see

Sects. 2.4 and 3.7) can be used for reducing the variance $Var\, f(\bar{X}(T))$ and, consequently, the Monte Carlo error.

If $\hat{f}$ is considered as an estimate for $Ef(\bar{X}(T))$, then its bias is equal to zero. But we need $\hat{f}$ as an estimate for $Ef(X(T))$. In this case its bias is equal to the error of numerical integration and it depends on h. The choice of time step h_* of an implemented method is appropriate if by further decrease of the time step, we obtain the result which is close enough to the one obtained with h_*. Further, the Talay–Tubaro extrapolation method from Sects. 2.2.3 and 3.6 can be used to estimate errors of numerical integration (see practical application of the extrapolation method in, e.g. Sect. 3.8).

Here, as an example, we consider realization of quasi-symplectic weak methods from Sect. 4.8. The program qhn.c given in Listing A.2 corresponds to the simulation of the oscillator with cubic restoring force and additive noise (4.8.13):

$$\ddot{Q} = Q - Q^3 - \nu\dot{Q} + \sigma\dot{w} \tag{2.4}$$

by the weak implicit quasi-symplectic method (4.8.15):

$$\bar{P}_I = P_k + h\left(\frac{\bar{Q}_I + Q_k}{2} - \frac{(\bar{Q}_I + Q_k)^3}{8}\right) + h^{1/2}\sigma\xi_k,$$

$$\bar{Q}_I = Q_k + h(\bar{P}_I + P_k)/2,$$

$$P_{k+1} = (1 - \nu h)\bar{P}_I, \quad Q_{k+1} = \bar{Q}_I, \tag{2.5}$$

and the weak implicit Euler method (4.8.16):

$$P_{k+1} = P_k + h\left(Q_{k+1} - Q_{k+1}^3 - \nu P_{k+1}\right) + h^{1/2}\sigma\xi_k$$

$$Q_{k+1} = Q_k + hP_{k+1}. \tag{2.6}$$

Both methods are of first weak order and require simulation of random variables ξ_k distributed according to the law $P(\xi = \pm 1) = 1/2$. To simulate them, we use the procedure rng_disc.c (see Listing A.17 in Appendix A.6). As a source of uniform random numbers for this procedure, we use the additive lagged Fibonacci generator $F(1279, 418)$ (see the procedure rng_fiba.c from Listing A.15), which seed table is initialized using the generator rng_lcgs.c (see Listing A.14 in Appendix A.6). Note that a simpler method of generating discrete random variables discussed in Remark 2.6.3 can be used here as well.

In the program qhn.c we simulate $EQ^2(t_i)$. The arrays av_e[i] (for the Euler method) and av_s[i] (for the quasi-symplectic) are used for $\sum_{m=1}^{M}\left[\bar{Q}^{(m)}(t_i)\right]^2$ while the arrays vav_e[i] and vav_s[i] are used for $\sum_{m=1}^{M}\left[\bar{Q}^{(m)}(t_i)\right]^4$. At the end of the program, we obtain

$$\hat{Q}_i^2 = \frac{1}{M}\sum_{m=1}^{M}\left[\bar{Q}^{(m)}(t_i))\right]^2$$

and also the corresponding sample variances. Under the assumption that the sample variances are close to $Var(\bar{Q}^2(t_i))$ and taken $c = 2$, we obtain the confidence intervals for $\hat{Q}_i^2$. The output file qhn.trc of this program is ready for plotting graphs in, e.g., *Gnuplot*. The first column in qhn.trc corresponds to time; the second and fourth columns give $E\left(Q(t)\right)^2$ obtained by the Euler method and the quasi-symplectic method, respectively. The third and fifth columns correspond to the half-length of the confidence intervals, i.e., to $2\sqrt{Var(R_{mc})}$, for the data in the second and fourth columns, respectively.

Since we implement here implicit schemes, we need a method for solving nonlinear algebraic equations, which are cubic equations in our case. For this purpose, we are using Newton's method, and just for simplicity we apply it directly to the cubic equation instead of using a universal procedure.

Listing A.2. Program qhn.c (weak symplectic and Euler methods for the cubic oscillator with additive noise).

```
/* program qhn.c: weak implicit quasi-symplectic and Euler methods
   for the cubic oscillator with additive noise. The program uses
      rng_fiba.c, rng_lcgs.c, and rng_disc.c */
#include <malloc.h>
#include <stdio.h>
#include <math.h>
#include <stochast.h>
/*constants for Newton's method*/
#define NT_ACC 0.00000001  /*accuracy */
#define MAXIN  30          /* max number of iterations*/
#define FAB 1279 /*for lagged Fibonacci RNG F(1279,418)*/
double fab[FAB];
double necubic(); /* Newton's method for cubic equation*/

main (void)
{
 int i,n;
 long iseed,j,m;
 FILE *out;
 double p_e,q_e,p_s,q_s,p0,q0;
 double g,h,h12,h2,k0,k1,k1e,k2,k3e,k3s,nh,nu,nuh,nuhp;
 double t,tk,si,sih12,upe,ups,dupe,dups,x;
 double *av_e,*av_s,*vav_e,*vav_s;

 /* Input of parameters */
 printf("nu=");    scanf("%lf",&nu); /* nu */
 printf("noise="); scanf("%lf",&si); /* noise */
 printf("tk=");    scanf("%lf",&tk); /* time interval */
 printf("P_0=");   scanf("%lf",&p0); /*initial value for P*/
```

548   A Appendix: Practical guidance to implementation

```
        printf("Q_0=");   scanf("%lf",&q0); /*initial value for Q*/
        printf("h=");     scanf("%lf",&h);  /* time step */
        printf("seed="); scanf("%ld",&iseed); /* seed for RNG */
        printf("m=");     scanf("%ld",&m);  /*number of independent
                                               realizations */
        n=tk/h+1;  /* number of time steps */
     /* memory allocation for the vectors for averages */
        i=n*sizeof(*av_e);
        if (!(av_e=(double *) malloc((size_t) i)))
           {printf("No memory - 1\n"); return(-1);}
        if (!(av_s=(double *) malloc((size_t) i)))
           {printf("No memory - 2\n"); return(-2);}
        if (!(vav_e=(double *) malloc((size_t) i)))
           {printf("No memory - 3\n"); return(-3);}
        if (!(vav_s=(double *) malloc((size_t) i)))
           {printf("No memory - 4\n"); return(-4);}
     /* and their initialization */
        for(i=0;i<n;i++)
           {av_e[i]=0.;av_s[i]=0.;vav_e[i]=0.;vav_s[i]=0.;}

     /* comment in the file with output */
        if((out=fopen("qhn.trc","w")) == NULL)
           {fprintf(stderr,"Cannot open output file.\n");return(1);}
        fprintf(out,"##########################################\n");
        fprintf(out,"# cubic oscillator with additive noise \n");
        fprintf(out,"# Input data: \n");
        fprintf(out,"#nu=%f noise=%f tk=%f h=%f m=%ld\n",nu,si,tk,h,m);
        fprintf(out,"# P(0)=%f Q(0)=%f seed=%ld\n#\n",p0,q0,iseed);
     /* auxiliary variables */
        h12=sqrt(h); sih12=h12*si; nh=nu*h; h2=h*h; nuh=1.-nh;
        nuhp=1.+nh; k1e=1.-h2/nuhp; k3e=h2/nuhp; k3s=h2/16.;
     /* initialization of F(1279,418) */
        for(i=0;i<=FAB-1;i++)fab[i]=rng_lcgs(&iseed);

        for(j=0;j<m;j++)    /* realization loop */
        {
        /* initialization of variables for the time loop */
          q_e=q0; p_e=p0; q_s=q0; p_s=p0; t=0.; i=1;

          while(t<=tk)   /* time loop */
          { g=rng_disc(fab); /* the dicrete random variable */

          /* implicit weak Euler method */
            k0=-(q_e+h*(p_e+sih12*g)/nuhp);
```

```
    q_e=necubic(q_e,k0,k1e,0.,k3e); /*Newton*/
    p_e=(p_e+h*q_e*(1.-q_e*q_e)+sih12*g)/nuhp;

    /* weak implicit quasi-symplectic method */
    k0=-(q_s+h*(p_s+h*q_s*(1.-q_s*q_s/4.)/4.+sih12*g/2.));
    k1=1.-h2/4+3.*h2*q_s*q_s/16.; k2=3.*h2*q_s/16.;
    x=q_s;
    q_s=necubic(q_s,k0,k1,k2,k3s); /*Newton*/
    p_s=nuh*(p_s+h*(q_s+x)*(1.-(q_s+x)*(q_s+x)/4.)/2.+sih12*g);

    /* for averaging */
    x=q_e*q_e; av_e[i]=av_e[i]+x; vav_e[i]=vav_e[i]+x*x;
    x=q_s*q_s; av_s[i]=av_s[i]+x; vav_s[i]=vav_s[i]+x*x;

    t+=h; ++i; /*current time & index in the vectors*/
  } /* end of time loop */
 } /* end of realization loop */

 /* averaging, MC error & final output */
 x=m; t=0.;
 for(i=1;i<n;i++)
 { t+=h;
  upe=av_e[i]/x; dupe=2.*sqrt(fabs(vav_e[i]/x-upe*upe)/x);
  ups=av_s[i]/x; dups=2.*sqrt(fabs(vav_s[i]/x-ups*ups)/x);
  fprintf(out,"%lf  %lf  %lf  %lf  %lf  \n",t,upe,dupe,ups,dups);
 }
 fclose(out); free(av_e); free(av_s); free(vav_e); free(vav_s);
 return(0); /* end of the main program */
}

double necubic(double x,double k0,double k1,double k2,double k3)
{ /* solving cubic equation by Newton's method */
 double incr=1.;
 int i=0;

 while(fabs(incr)>NT_ACC && i<MAXIN)
 {incr=(k0+x*(k1+x*(k2+k3*x)))/(k1+x*(2.*k2+3.*k3*x));
  x-=incr; ++i;
 }
 if(i == MAXIN)printf("Max number of iterations exceeded!");
 return(x); /* end of procedure necubic() */
}
```
Using this program, we compared behavior of the two methods on long time intervals that is presented on Fig. 4.8.2 in Sect. 4.8.3. We see that even

550   A Appendix: Practical guidance to implementation

for such a small step as $h = 0.01$ the implicit Euler method tends to a wrong limit with increasing t, while the quasi-symplectic method gives quite accurate results, e.g., for $h = 0.25$.

A.3 Algorithms for bounded diffusions

This section deals with implementation of the algorithms for simulation of space and space-time bounded diffusions from Chap. 5. These mean-square methods ensure that the proposed approximations of solutions of the corresponding SDEs belong to a bounded domain. The "ordinary" mean-square methods, which implementation is discussed in Sect. A.1, ensure smallness of time increments at each step, but space increments can take arbitrary large values with some probability. To approximate SDEs in a bounded domain, we have to control space increments at each step in a way such that the constructed approximation belongs to the bounded domain. Of course, this cannot be achieved by "ordinary" mean-square methods, and thus special methods from Chap. 5 are required.

Here we restrict ourselves to the algorithms from Sect. 5.4, by which we can approximate SDEs in space-time bounded domains. Implementation of these algorithms rests on procedures for distributions which are needed to simulate the exit point of the space-time Brownian motion from a space-time parallelepiped (Sect. 5.3). The procedure witau_di.c (see Listing A.3) gives the distribution $\mathcal{P}(t)$ for the first-passage time τ of the one-dimensional standard Wiener process $W(t)$ to the boundary of the interval $[-1, 1]$ found in Lemma 5.3.1 and the procedure witau_de.c (see Listing A.4) gives the density $\mathcal{P}'(t)$ of this distribution. More precisely, in these procedures we simulate the approximate distribution $\bar{\mathcal{P}}(t)$ and density $\bar{\mathcal{P}}'(t)$ using the formulas (5.3.32)-(5.3.33).

Listing A.3. Procedure witau_di.c (simulation of the distribution $\bar{\mathcal{P}}(t)$).

```
/*distribution P(\tau) of exit time of Wiener process from (-1,1)*/
/* it is assumed that erfc() is available */
#include <math.h>
#include <stochast.h>
#define SWI 2./M_PI /* switch between two formulas */
#define MPI2_8 M_PI*M_PI/8.
double witau_di(double t)
{ double ptau, qt;
  if(t<SWI)           /* which formula to use */
  { qt=sqrt(2.*t);
   ptau=2.*(erfc(1./qt)-erfc(3./qt)+erfc(5./qt));
  }
  else
```

```
{ qt=MPI2_8*t;
  ptau=1.-4.*(exp(-qt)-exp(-9.*qt)/3.+exp(-25.*qt)/5.)/M_PI;
}
return(ptau);
}
#undef SWI
#undef MPI2_8
```

Listing A.4. Procedure witau_de.c (simulation of the density $\bar{\mathcal{P}}'(t)$).

```
/*distribution density for exit time
   of Wiener process from (-1,1)*/
#include <math.h>
#include <stochast.h>
#define SWI 2./M_PI /* switch between two formulas */
#define MPI2_8 M_PI*M_PI/8.
#define MPI2 2.*M_PI
double witau_de(double t)
{ double ptau, qt;
 if(t<SWI)
 { qt=1/(2.*t); /* which formula to use */
  ptau=2.*(exp(-qt)-3.*exp(-9.*qt)+5.*exp(-25.*qt))/(t*
     sqrt(MPI2*t));
 }
 else
 { qt=MPI2_8*t;
  ptau=M_PI*(exp(-qt)-3.*exp(-9*qt)+5.*exp(-25*qt))/2.;
 }
 return(ptau);
}
#undef SWI
#undef MPI2_8
#undef MPI2
```

In practice to simulate the random variable τ with the distribution $\mathcal{P}(t)$, we first generate a uniform random variable γ (e.g., by the RNG rng_lcgs.c from Listing A.14) and then solve the nonlinear equation (cf. (5.3.34)):

$$\bar{\mathcal{P}}(\tau) = \gamma \,. \tag{3.1}$$

For this purpose, we use Newton's method together with the bisection method in the procedure witau.c (see Listing A.5). The input for this procedure is γ and the output is τ. We note that the iteration procedure can be implemented in a more sophisticated way than it is done in witau.c.

Listing A.5. Procedure witau.c (evaluation of the inverse function of $\bar{\mathcal{P}}(t)$).

```c
/* inverse of distribution P(\tau) of the exit time
   of Wiener process from (-1,1) */
#include <stdio.h>
#include <math.h>
#include <stochast.h>
#define NT_ACC 0.00000000001 /* accuracy */
#define MAXIN 500 /* max number of iterations*/
double witau(double g)
{ int i;
  double de,delta,di,dii,tau;
  /* first approximation */
  i=(int)(g/0.1);
  switch(i)
  { case 0: tau=0.2; break;
    case 1: tau=0.3; break;
    case 2: tau=0.4; break;
    case 3: tau=0.5; break;
    case 4: tau=0.8; break;
    case 5: tau=0.9; break;
    case 6: tau=1.; break;
    case 7: tau=1.1; break;
    case 8: tau=1.3; break;
    case 9: if(g<0.95) tau=1.5; else tau=1.7; break;
    default: tau=1.7;
  } /* end of first approximation */
  di=witau_di(tau)-g;
  for (i=0;i<MAXIN;i++) /* loop for Newton's iterations */
  {
    de=witau_de(tau); delta=di/de;
    if ((fabs(delta) < NT_ACC) || (fabs(di)<NT_ACC)) return(tau);
    tau-=delta;  dii=witau_di(tau)-g;
    if(di*dii>0.)di=dii; /* if necessary then */
    else {tau+=delta/2.; di=witau_di(tau)-g;} /* bisection mehtod */
  }
  printf("Max number of iterations exceeded in WITAU %lf
           %lf\n",g,delta);
  return(tau);
}
#undef NT_ACC
#undef MAXIN
```

The procedure wiq_di.c (see Listing A.6) gives the conditional probability

$$\mathcal{Q}(\beta;t) = P(W(t) < \beta / \ |W(s)| < 1, \ 0 < s < t) \, ,$$

found in Lemma 5.3.2 and the procedure wiq_de.c (see Listing A.7) gives the corresponding conditional density. Again, in these procedures we simulate the approximate distribution $\bar{\mathcal{Q}}(\beta;t)$ and density $\bar{\mathcal{Q}}'(\beta;t)$ which are obtained by truncating the series (5.3.35) and (5.3.36) (see also the discussion after the proof of Lemma 5.3.2).

Listing A.6. Procedure wiq_di.c (simulation of the distribution $\bar{\mathcal{Q}}(\beta;t)$).

```
/*conditional distribution function Q(\beta;t) */
#include <math.h>
#include <stochast.h>
#define SWI 2./M_PI /* switch between two formulas */
#define MPI2_8 M_PI*M_PI/8.
#define MPI2 2.*M_PI
#ifndef M_PI_2
 #define M_PI_2 M_PI/2.
#endif
double wiq_di(double x, double t)
{double pq, qt;
 if(t<SWI) /* which formula to use */
 { qt=sqrt(2.*t);
   pq=0.5+0.5*(1.-erfc(x/qt)-erfc((2.-x)/qt)+erfc((2.+x)/qt)
       +erfc((4.-x)/qt)-erfc((4.+x)/qt)-erfc((6.-x)/qt)
       +erfc((6.+x)/qt)+erfc((8.-x)/qt))/(1.-witau_di(t));
 }
 else
 { qt=MPI2_8*t;
   pq=0.5+0.5*(sin(M_PI_2*x)+sin(3.*M_PI_2*x)*exp(-8*qt)/3.
       +sin(5.*M_PI_2*x)*exp(-24*qt)/5.)/(1.-exp(-8*qt)/3.
       +exp(-24*qt)/5.);
 }
 return(pq);
}
```

Listing A.7. Procedure wiq_de.c (simulation of the density $\bar{\mathcal{Q}}'(\beta;t)$).

```
/*conditional density function Q(\beta;t) */
#include <math.h>
#include <stochast.h>
#define SWI 2./M_PI /* switch between two formulas */
#define MPI2_8 M_PI*M_PI/8.
#define MPI2 2.*M_PI
#ifndef M_PI_2
 #define M_PI_2 M_PI/2.
#endif
#ifndef M_PI_4
```

```
#define M_PI_4 M_PI/4.
#endif
double wiq_de(double x, double t)
{double pq,qt;
  if(t<SWI) /* which formula to use */
  { qt=2.*t;
    pq=(exp(-x*x/qt)-exp(-(2.-x)*(2.-x)/qt)-exp(-(2.+x)*(2.+x)/qt)
       +exp(-(4.-x)*(4.-x)/qt)+exp(-(4.+x)*(4+x)/qt)
       -exp(-(6.-x)*(6.-x)/qt)-exp(-(6.+x)*(6.+x)/qt)
       +exp(-(8.-x)*(8.-x)/qt))/(sqrt(2.*M_PI*t)*(1.-witau_di(t)));
  }
  else
  { qt=MPI2_8*t;
    pq=M_PI_4*(cos(M_PI_2*x)+cos(3.*M_PI_2*x)*exp(-8*qt)
       +cos(5.*M_PI_2*x)*exp(-24*qt)+cos(7.*M_PI_2*x)*exp(-48*qt))
       /(1.-exp(-8*qt)/3.+exp(-24*qt)/5.);
  }
  return(pq);
}
```

To simulate the random variable β with the conditional distribution $\mathcal{Q}(\beta;t)$ for a given t, we generate a uniform random variable γ (e.g., by the RNG rng_lcgs.c) and then solve the nonlinear equation

$$\gamma = \bar{\mathcal{Q}}(\beta;t).$$

For this purpose, we use Newton's method together with the bisection method in the procedure wiq.c from Listing A.8.

Listing A.8. Procedure wiq.c (evaluation of the inverse function of $\bar{\mathcal{Q}}(\beta;t)$).

```
/* finding inverse of conditional distribution Q(\beta, t) */
#include <stdio.h>
#include <math.h>
#include <stochast.h>
#define NT_ACC 0.00000000001 /* accuracy */
#define MAXIN 500 /* max number of iterations*/
double wiq(double g, double t)
{int i;
 double de,delta,di,dii,beta;
 beta=0.; /* first approximation */
 di=wiq_di(beta,t)-g;
 for (i=0;i<MAXIN;i++) /* loop for Newton's iterations */
 { de=wiq_de(beta,t); delta=di/de;
   if ((fabs(delta) < NT_ACC) || (fabs(di)<NT_ACC)) return(beta);
   beta-=delta; dii=wiq_di(beta,t)-g;
```

```
        if(di*dii>0.)di=dii;
        else {beta+=delta/2.; di=wiq_di(beta,t)-g;} /* bisection method */
    }
    printf("Max number of iterations exceeded in WIQ %lf
        %lf %lf\n",g,t,beta);
    return(beta);
}
#undef NT_ACC
#undef MAXIN
```

The program biharm.c given in Listing A.9 corresponds to Example 5.5.2 from Sect. 5.5. This example concerns with application of the random walk over touching space squares to solving the following two-dimensional problem in the domain $G = \{x = (x_1, x_2) : |x_1| < 1, \ |x_2| < 1\}$:

$$\frac{1}{4}\Delta^2 u = 1, \ x \in G, \tag{3.2}$$

$$u \mid_{\partial G} = \varphi(x), \ \varphi(x_1, \pm 1) = \frac{1 + x_1^4}{12}, \ \varphi(\pm 1, x_2) = \frac{1 + x_2^4}{12},$$

$$\frac{1}{2}\Delta u \mid_{\partial G} = \psi(x), \ \psi(x_1, \pm 1) = \frac{1 + x_1^2}{2}, \ \psi(\pm 1, x_2) = \frac{1 + x_2^2}{2}. \tag{3.3}$$

Introducing the function $v = \frac{1}{2}\Delta u$, we obtain the system of elliptic equations

$$\frac{1}{2}\Delta u - v = 0, \ x \in G, \ u \mid_{\partial G} = \varphi(x) \tag{3.4}$$

$$\frac{1}{2}\Delta v = 1, \ x \in G, \ v \mid_{\partial G} = \psi(x). \tag{3.5}$$

Its exact solution is

$$u(x) = \frac{x_1^4 + x_2^4}{12}, \ v(x) = \frac{x_1^2 + x_2^2}{2}.$$

For the system (3.4)-(3.5), we have the following probabilistic representation

$$u(x) = E\varphi(x + w(\tau)) - E[\tau\psi(x + w(\tau))] + \frac{1}{2}E\tau^2,$$
$$v(x) = E\psi(x + w(\tau)) - E\tau, \tag{3.6}$$

where τ is the first exit time of the process $x + w(s)$ from the domain G.

To simulate the point $(\tau, x + w(\tau))$, we use the random walk over touching space squares, which is terminated in a δ-neighborhood of the boundary ∂G belonging to G. Recall that we are able to exactly simulate both the exit point and the exit time of the Wiener process from a square in accordance with Theorem 5.3.7 (note that here $d = 2$). Then the algorithm gives the result which is free from an error of numerical integration (except an error arising from the substitution $\bar{\mathcal{P}}(t)$ and $\bar{\mathcal{Q}}(\beta;t)$ for $\mathcal{P}(t)$ and $\mathcal{Q}(\beta;t)$, respectively, and an error due to approximation of the boundary conditions depending on δ but these errors are negligibly small). To evaluate the expectations from (3.6), we exploit the Monte Carlo technique and obtain the estimator $(\hat{u},\hat{v})$ for (u,v) which bias is equal to zero. The Monte Carlo error is evaluated in the same way as in Appendix A.2 (cf. (2.1)-(2.2), the constant c is taken equal to 2 here).

Listing A.9. Program biharm.c (solving (3.4)-(3.5) by the random walk over touching space squares).

```
/* Space-time random walk - touching squares for biharm eq.
   The program uses the external procedures witau_di.c, witau_de.c,
   witau.c, wiq_di.c, wiq_de.c, wiq.c, and rng_lcgs.c, and a
   procedure erfc() for evaluating the function erfc(x). */
#include <stdio.h>
#include <math.h>
#include <stochast.h>
#define NT_A 0.00001 /* size of boundary layer */
#define NCR 10000 /* max number of random walk steps */
main (void)
{
  int nu;
  long iseed, j, m, n, ns;
  double u_e,v_e,us,dus,vs,dvs,x,y,x0,y0;
  double ax,bx,ay,by,axb,bxb,ayb,byb,rg;
  double fi,psi,g,r,w,t,tau,ttau,xx,yy;
  /* Input of parameters */
  printf("x_0="); scanf("%lf",&x0); /* the point at which */
  printf("y_0="); scanf("%lf",&y0); /*solution is evaluated*/
  printf("seed="); scanf("%ld",&iseed); /* seed for RNG */
  printf("m="); scanf("%ld",&m); /* number of independent
                                    realizations */
  /* size of the domain G */
   ax=-1.; bx=1.; ay=-1.; by=1.; rg=2.;
  /* size of G-boundary layer */
   axb=ax+NT_A; bxb=bx-NT_A; ayb=ay+NT_A; byb=by-NT_A;
  /* initialization of variables for realization loop */
   ns=0; us=0; dus=0; vs=0; dvs=0;
  for(j=0;j<m;j++) /* realization loop */
  { /* initialization of variables for random walk loop */
    t=0; x=x0; y=y0; n=0;
    while( ( x>axb && x<bxb ) && ( y>ayb && y<byb ) && n<NCR)
    { /* random walk loop */
      /* 1. construction of touching square */
      r=rg;
```

A.3 Algorithms for bounded diffusions 557

```
    if(x-ax<r)r=x-ax; if(y-ay<r)r=y-ay; if(bx-x<r)r=bx-x;
    if(by-y<r)r=by-y;
    /* 2. simulating tau */
    g=1.-sqrt(rng_lcgs(&iseed)); /* 1-\gamma^{1/2} */
    ttau=witau(g); /* \tau from (-1,1) */
    tau=ttau*r*r; /* \tau from (-r,r) */
    /* 3. simulating W */
    g=rng_lcgs(&iseed);
    w=wiq(g,ttau); /* W(ttau) */
    /* on which edge of the square */
    g=rng_lcgs(&iseed);
    if(g<0.5){y+=w*r; if(g<0.25){nu=1;x-=r;} else{nu=2;x+=r;}}
    else{x+=w*r; if(g<0.75){nu=3; y-=r;} else{nu=4; y+=r;} }
    /* current number of steps & current time */
    ++n; t+=tau;
   } /* end of random walk loop: boundary layer is reached or
        the number of steps exceed NCR */
  /* collection of data for averaging */
   ns+=n; /* number of steps*/
   xx=x*x; yy=y*y;
   switch(nu) /* boundary conditions */
   {
     case 1: fi=(1.+yy*yy)/12.; psi=(1.+yy)/2.; break; /*(-1,y)*/
     case 2: fi=(1.+yy*yy)/12.; psi=(1.+yy)/2.; break; /*(1,y) */
     case 3: fi=(1.+xx*xx)/12.; psi=(1.+xx)/2.; break; /*(x,-1)*/
     case 4: fi=(1.+xx*xx)/12.; psi=(1.+xx)/2.; break; /*(x,1) */
   }
   xx=fi-t*psi+t*t/2; us+=xx; dus+=xx*xx;
   xx=psi-t; vs+=xx; dvs+=xx*xx;
  } /* end of realization loop */
/*exact solution */
 u_e=(x0*x0*x0*x0+y0*y0*y0*y0)/12.; v_e=(x0*x0+y0*y0)/2.;
/* final averaging & Monte Carlo error & output */
 xx=m; r=ns; r/=xx;
 us/=xx; dus=2.*sqrt(fabs(dus/xx-us*us)/xx);
 vs/=xx; dvs=2.*sqrt(fabs(dvs/xx-vs*vs)/xx);
 printf("At x=%f y=%f, exact solution u=%f, v=%f \n",x0,y0,u_e,v_e);
 printf("Simulation: u=%f+-%f, v=%f+-%f \n",us,dus,vs,dvs);
 printf(" Average number of steps =%f \n",r);
 } /* end of the main program */
```

Using this program, the results presented in Table 5.5.2 were obtained.

A.4 Random walks for linear boundary value problems

In Appendix A.2 we discussed weak numerical methods for SDEs which are suitable for solving the Cauchy problem for linear parabolic equations. The previous section deals with mean-square approximations of SDEs in bounded domains proposed in Chap. 5, and, as we saw, they can be applied for solving boundary value problems. However, since solutions of boundary value problems for parabolic and elliptic equations can be represented as expectations of solutions of the corresponding systems of SDEs in bounded domains, one can apply far more simple weak approximations which should be subject to limitations related to nonexit from the bounded domains. Such weak methods are constructed in Chap. 6 and in this section we consider their implementation.

Here we restrict ourselves to a practical realization of the simplest random walk for the Dirichlet problem for parabolic equations from Sect. 6.2 (Algorithm 6.2.1). We consider the Dirichlet problem for the heat equation from Sect. 6.2.4:

$$\frac{\partial u}{\partial t} = \frac{1}{2}(1.21 - x_2^2 - x_3^2)\frac{\partial^2 u}{\partial x_1^2} + \frac{1}{2}\frac{\partial^2 u}{\partial x_2^2} + \frac{1}{2}\frac{\partial^2 u}{\partial x_3^2} + 6(1 - 0.5e^{-t})$$
$$\times(x_1^2(1.21 - x_2^2 - x_3^2) + x_2^2) + 0.5e^{-t}(1.21 - x_1^4 - x_2^4), \quad (4.1)$$
$$t \in (0, T], \quad x \in U_1,$$

$$u(0, x) = \frac{1}{2}(1.21 - x_1^4 - x_2^4), \quad x \in \bar{U}_1,$$
$$u(t, x) = (1.21 - x_1^4 - x_2^4)(1 - 0.5e^{-t}), \quad t \in [0, T], \quad x \in \partial U_1, \quad (4.2)$$

where $U_1 \subset \mathbf{R}^3$ is a unit ball with center at the origin. This problem has the exact solution:

$$u(t, x) = (1.21 - x_1^4 - x_2^4)(1 - 0.5e^{-t}).$$

Applying Algorithm 6.2.1 together with the Monte Carlo method to the problem (4.1)-(4.2) (see Listing A.10), we obtain the estimate $\bar{u}(t, x)$ for the solution $u(t, x)$ of this problem. The bias of $\bar{u}(t, x)$ is equal to the error of numerical integration, which is of order $O(h)$. The Monte Carlo error is evaluated in the same way as in Appendix A.2 (cf. (2.1)-(2.2)). The discussion from Appendix A.2 on a proper choice of the step of numerical integration and the number of independent realizations in the Monte Carlo simulation is valid here as well.

As a source of uniform random numbers in the program `rwp.c`, we use the additive lagged Fibonacci generator $F(1279, 418)$, which is realized as the procedure `rng_fiba.c` and which seed table is initialized by the generator `rng_lcgs.c` (see Appendix A.6).

Listing A.10. Program rwp.c (solving the parabolic problem (4.1)-(4.2) by Algorithm 6.2.1).

```c
/* Simulation of parabolic problem by the simplest random walk.
   The program uses rng_fiba.c, rng_lcgs.c, and rng_disc.c */
#include <stdio.h>
#include <math.h>
#include <stochast.h>
#define FAB 1279 /*for lagged Fibonacci RNG F(1279,418)*/
double fab[FAB];
double exact_so(); /* exact solution of the problem */
main (void)
{
 int l;
 long i,iseed,m;
 double ex,h,h12,g1,g2,g3,lambda,r,rl;
 double si,t,tp,u,du,kap,kappa,dkappa,uu;
 double x1,x2,x3,x10,x20,x30,x1p,x2p,x3p,xx1,xx2,z;
 /* Input of parameters */
 printf("x10="); scanf("%lf",&x10); /* coordinates of */
 printf("x20="); scanf("%lf",&x20); /* the point at which */
 printf("x30="); scanf("%lf",&x30); /* solution to be found*/
 printf("tp="); scanf("%lf",&tp); /* time */
 printf("step="); scanf("%lf",&h); /* time step */
 printf("seed="); scanf("%ld",&iseed); /* seed for RNG -
                                         any negative integer*/
 printf("m="); scanf("%ld",&m); /* number of realizations */
 /* initialization of F(1279,418) */
  for(i=0;i<=FAB-1;i++)fab[i]=rng_lcgs(&iseed);
 h12=sqrt(h); /* auxiliary variable */
 lambda=2.; /*the parameter \lambda for the boundary layer*/
 /* initialization of variables for realization loop */
  u=0.; kappa=0.; du=0.; dkappa=0.;
 for(i=0;i<m;i++) /* realization loop */
 {
  /* initialization of variables for time loop */
   x1=x10; x2=x20; x3=x30; z=0.; t=0.; kap=0.;
  while(t<tp) /* time loop */
  {
   si=sqrt(1.21-x2*x2-x3*x3); /* auxiliary variable */
   /* the point which could be closest to the boundary */
    if(x1 > 0) x1p=x1+h12*si; else x1p=x1-h12*si;
    if(x2 > 0) x2p=x2+h12; else x2p=x2-h12;
    if(x3 > 0) x3p=x3+h12; else x3p=x3-h12;
    if(x1p*x1p+x2p*x2p+x3p*x3p > 1.) /*out of the domain?*/
```

```
{ /* point in the boundary layer */
r=sqrt(x1*x1+x2*x2+x3*x3); /* distance from x to 0 */
rl=r-lambda*h12;
g1=rng_fiba(fab);
if(g1<h12*lambda/(1-rl)) /* stop on the boundary? */
{x1/=r; x2/=r; x3/=r; break;}
else /* back into the internal domain */
{ x1=rl*x1/r; x2=rl*x2/r; x3=rl*x3/r; }
}
/* point in the internal domain */
g1=rng_disc(fab); g2=rng_disc(fab); g3=rng_disc(fab);
xx1=x1*x1; xx2=x2*x2; ex=0.5*exp(-(tp-t));
z=z+h*(6.*(1.-ex)*(xx1*(1.21-xx2-x3*x3)+xx2)
  +ex*(1.21-xx1*xx1-xx2*xx2));
x1+=si*h12*g1; x2+=h12*g2; x3+=h12*g3;
++kap; /* number of steps */    t+=h; /* current time */
} /*end of time loop */
/* for averaging */
uu=exact_so(tp-t,x1,x2)+z; /* value on the boundary */
u+=uu; du+=uu*uu; kappa+=kap; dkappa+=kap*kap;
} /* end of realization loop */
/* final averaging& MC error&output */
uu=m; u/=uu; du=2.*sqrt(fabs(du/uu-u*u)/uu);
kappa/=uu; dkappa=2.*sqrt(fabs(dkappa/uu-kappa*kappa)/uu);
printf("Solution: exact %lf, approx %lf +- %lf \n",
 exact_so(tp,x10,x20),u,du);
printf("mean number of steps %lf +- %lf \n",kappa,dkappa);
} /* end of the main program */

double exact_so(double t, double x1, double x2)
{ /* exact solution of the problem, also gives initial and
     boundary values */
 return((1.21-x1*x1*x1*x1-x2*x2*x2*x2)*(1.-0.5*exp(-t)));
}
```

Using this program, the results presented in the upper part of Table 6.2.1 were obtained. The program was run with the parameter $seed = -7$, the other parameters are given in Table 6.2.1.

A.5 Nonlinear PDEs

In this section we consider implementation of algorithms from Chaps. 7 and 8, where layer methods for semilinear parabolic equations are proposed. The layer methods are based on a probabilistic approach which takes into account a coefficient dependence on the space variables and a relationship between diffusion and advection in an intrinsic manner. In particular, the layer methods allow us to avoid difficulties stemming from essentially changing coefficients and strong advection. Despite their probabilistic nature these methods are nevertheless deterministic, and we do not need any source of randomness for their realization. Implementation of layer methods requires space discretization and an interpolation. In Chaps. 7 and 8 convergence theorems are proved for algorithms using linear interpolation. To reduce the amount of nodes in space discretization, it is natural to take advantage of cubic interpolation.

Here we restrict ourselves to the example from Sect. 8.6.1. We consider application of the algorithm based on layer method (8.4.12) and cubic interpolation (8.6.4) (algorithm (8.4.29), (8.6.4)) to the Neumann problem for the one-dimensional Burgers equation:

$$\frac{\partial u}{\partial t} = \frac{\sigma^2}{2}\frac{\partial^2 u}{\partial x^2} - u\frac{\partial u}{\partial x}, \quad t > 0, \ x \in (-4, 4), \tag{5.1}$$

$$u(0, x) = -\frac{\sigma^2 \sinh x}{\cosh x + A}, \quad x \in [-4, 4], \tag{5.2}$$

$$\frac{\partial u}{\partial x}(t, \pm 4) = -\sigma^2 \frac{1 + A\exp(-\sigma^2 t/2)\cosh 4}{[\cosh 4 + A\exp(-\sigma^2 t/2)]^2}, \quad t \geq 0. \tag{5.3}$$

Here A is a positive constant. The exact solution to this problem has the form

$$u(t, x) = -\frac{\sigma^2 \sinh x}{\cosh x + A\exp(-\sigma^2 t/2)}.$$

The algorithm (8.4.29), (8.6.4) for (5.1)-(5.3) is written as

$$\bar{u}(0, x) = -\frac{\sigma^2 \sinh x}{\cosh x + A}, \quad x \in [-4, 4],$$

$$\bar{u}(t_{k+1}, x_j) = \frac{1}{2}\bar{u}(t_k, x_j - h\bar{u}(t_k, x_j) - h^{1/2}\sigma)$$
$$+ \frac{1}{2}\bar{u}(t_k, x_j - h\bar{u}(t_k, x_j) + h^{1/2}\sigma),$$
$$\text{if } x_j - h\bar{u}(t_k, x_j) \pm h^{1/2}\sigma \in [-4, 4];$$

$$\bar{u}(t_{k+1}, x_j) = \bar{u}\left(t_k, -4 + \sqrt{h\sigma^2 + (4 + x_j)^2}\right)$$
$$+ \sigma^2 \frac{1 + A\exp(-\sigma^2 t_k/2)\cosh 4}{[\cosh 4 + A\exp(-\sigma^2 t_k/2)]^2}$$
$$\times \left(-4 - x_j + h\bar{u}(t_k, x_j) + \sqrt{h\sigma^2 + (4 + x_j)^2}\right),$$
$$\text{if } x_j - h\bar{u}(t_k, x_j) - h^{1/2}\sigma < -4;$$

$$\bar{u}(t_{k+1}, x_j) = \bar{u}\left(t_k, 4 - \sqrt{h\sigma^2 + (4 - x_j)^2}\right)$$

$$+\sigma^2 \frac{1 + A\exp(-\sigma^2 t_k/2)\cosh 4}{[\cosh 4 + A\exp(-\sigma^2 t_k/2)]^2}$$

$$\times \left(4 - x_j + h\bar{u}(t_k, x_j) - \sqrt{h\sigma^2 + (4 - x_j)^2}\right),$$

$$\text{if } x_j - h\bar{u}(t_k, x_j) + h^{1/2}\sigma > 4; \ j = 1, 2, \ldots, M - 1, \tag{5.4}$$

$$\bar{u}(t_k, x) = \sum_{i=0}^{3} \Phi_{l,i}(x)\,\bar{u}(t_k, x_{l+i}), \quad x_l < x < x_{l+3},$$

$$\Phi_{l,i}(x) = \prod_{m=0, m \neq i}^{3} \frac{x - x_{l+m}}{x_{l+i} - x_{l+m}}, \tag{5.5}$$

$$k = 0, 1, \ldots, N - 1.$$

Here h is a time step, N is a number of nodes in the equidistant time discretization of an interval $[0, T]$, and M is a number of nodes in the equidistant space discretization of the interval $[-4, 4]$ with the space step $h_x = \sqrt{h}$. This algorithm is realized as the program burgneum.c (see Listing A.12). It uses the procedure cint.c (see Listing A.11) implementing the cubic interpolation (5.5) in the form suitable for our purposes.

Listing A.11. Procedure cint.c (cubic interpolation).

```
/* cubic interpolation */
/* hx - step of discretization, x0 - left end of the interval,
   m - number of nodes, u[] - array with data */
#include <stdio.h>
double cint(double u[], double x, double hx, double x0, long m)
{
  long j;
  double d,r,x1,x2,x3;

  if(m<4){printf("error 1 in cint\n"); return(-55);}
  if(x<=x0){printf("error 2 in cint\n"); return(u[0]);}
  if(x>=x0+hx*(m-1)){printf("error 3 in cint\n");return(u[m-1]);}

  d=x-x0; j=(long) (d/hx); /* x0+j*hx<=x<=(j+1)*hx+x0 */
  if(j>0)--j; if(j==(m-3))--j;
  d-=hx*j; /* xx0=x-x_j */
  x1=d-hx; x2=x1-hx; x3=x2-hx;
  r=(x1*x2*(d*u[j+3]-x3*u[j])/3.-d*x3*(x1*u[j+2]-x2*u[j+1]))
   /(2.*hx*hx*hx);
  return(r);
}
```

A.5 Nonlinear PDEs   563

Listing A.12. Program burgneum.c (solving the Neumann problem (5.1)-(5.3) by the algorithm (5.4), (5.5)).

```c
/* The program burgneum.c: layer method of order O(h) with cubic
   interpolation for solving the Neumann problem for Burgers
   equation. The program uses the external procedure cint.c. */
#include <malloc.h>
#include <stdio.h>
#include <math.h>
#include <stochast.h>
double burg_exn(), burg_bc();
main(void)
{
 long i,m;
 double a,d,h,h12,hx,si,si2h,t,tk,err,merr,ue;
 double aq,bv,x,x01,x0k,y;
 double *v, *w;
 FILE *out;
 /*input of parameters*/
 printf("sigma="); scanf("%lf",&si); /* diffusion */
 printf("A="); scanf("%lf",&a); /* a */
 printf("tk="); scanf("%lf",&tk); /* time interval */
 printf("h="); scanf("%lf",&h); /* time step */
 if ((out = fopen("burgneum.trc", "w"))== NULL)
 { fprintf(stderr, "Cannot open output file.\n");
   return(1); }
 /* output comment in file with data */
 fprintf(out,"##########################################\n");
 fprintf(out,"#layer method of order O(h) with cubic interp.\n");
 fprintf(out,"# Neumann problem for Burgers equation \n");
 fprintf(out,"# sigma=%lf h=%lf tk=%lf A=%lf \n#\n",si,h,tk,a);

 h12=si*sqrt(h); si2h=si*si*h; /* auxiliary variables */
 hx=sqrt(h); /* step of space discretization */
 d=8.; /* length of the space interval */
 x01=-4.; /* left end of the space interval */
 m=(long)(ceil(d/hx))+1;/*number of nodes in space discretization*/
 x0k=x01+(m-1)*hx; /* right end of the space interval,
                       should be 4 */
 /* memory allocation for the vectors for u(t,x) */
 i=m*sizeof(*v);
 if (!(v=(double *) malloc((size_t) i)))
 {printf("No memory - 1\n"); return(-1);}
 if (!(w=(double *) malloc((size_t) i)))
 {printf("No memory - 2\n"); return(-2);}
```

```
/* initial condition */
x=x01;
for(i=0;i<m;i++){v[i]=burg_exn(0.,x,si,a);x+=hx;}
t=0;
while(t < tk) /* time loop */
{
 x=x01;
 for(i=0;i < m;i++)
 { /* space loop */
  y=x-h*v[i];
  if(y+h12<=x0k && y-h12>=x01) /* internal point */
   w[i]=0.5*(cint(v,y+h12,hx,x01,m)+cint(v,y-h12,hx,x01,m));
  else /* boundary zone point */
  {
   if(y-h12 < x01) /* left boundary zone */
   {
    aq=x01+sqrt(si2h+(x-x01)*(x-x01));
    bv=burg_bc(t,x01,si,a); /*left boundary value */
    w[i]=cint(v,aq,hx,x01,m)-bv*(aq-y);
   }
   if(y+h12 > x0k) /* right boundary zone */
   {
    aq=x0k-sqrt(si2h+(x-x0k)*(x-x0k));
    bv=burg_bc(t,x0k,si,a); /* right boundary value */
    w[i]=cint(v,aq,hx,x01,m)-bv*(aq-y);
   }
  } /* end of boundary zone */
  x+=hx;
 } /* end space loop */
 for(i=0;i<m;i++)v[i]=w[i];
 t+=h; /* current time */
} /* end time loop */
/* errors & output */
merr=0.; err=0.; x=x01;
for(i=0;i<m;i++)
{
 ue=burg_exn(t,x,si,a); /* exact solution */
 y=fabs(ue-v[i]);   /* error at this node */
 if(y>merr)merr=y; /* error in Chebyshev norm */
 err+=y;
 fprintf(out,"%lf %lf %lf %lf \n",x,ue,v[i],y);
 x+=hx;
}
```

```
    err*=hx;  /* error in l^1 norm */
    fprintf(out,"# err^l=%e err^c=%e \n",err,merr);
    fclose(out); free(v); free(w);
    return(0);
  } /* end of main */

  double burg_exn(double t, double x, double si, double a)
  { /* exact solution of the problem */
    return(-si*si*sinh(x)/(cosh(x)+a*exp(-si*si*t/2.)));
  }

  double burg_bc(double t, double x, double si, double a)
  { /* boundary condition */
    double ex,si2;
    si2=si*si; ex=a*exp(-si2*t/2.);
    return(-si2*(1+ex*cosh(4.))/((cosh(4.)+ex)*(cosh(4.)+ex)));
  }
```

Using this program, we obtained the numerical results presented in the second column of Table 8.6.1 and also in Fig. 8.6.1.

A.6 Miscellaneous

In this section we are giving the header file stochast.h and portable random number generators used in programs from the previous sections of the Appendix. We start with the header file.

Listing A.13. The header file stochast.h.

```
  #ifndef _MY_H_
  #define _MY_H_
  #ifndef M_PI
   #define M_PI 3.14159265358979323846
  #endif
  double rng_disc(); /* random number generators */
  double rng_fiba();
  double rng_gau();
  double rng_lcgs();
  double witau();      /* distributions for exit points */
  double witau_de(); /* of Wiener process */
  double witau_di();
  double wiq();
  double wiq_de();
  double wiq_di();
  double cint(); /* cubic interpolation */
  #endif /* _MY_H_ */
```

The most popular and well studied uniform RNGs are linear congruential generators. A linear congruential generators (LCG) is defined by the recurrence relation:
$$x_n = (a\, x_{n-1} + c)\,\mathrm{mod}\, m, \qquad (6.1)$$
where the multiplier a and the modulus m are positive integers, the additive constant c is a non-negative integer. The normalized sequence
$$u_n = x_n/m \qquad (6.2)$$
is used as a source of random numbers in $[0,1]$. We denote by $\mathrm{LCG}(m,a,c)$ an LCG generator with the parameters m, a, c. See further details in Sect. 2.6.1.

Computation of the right-hand side of (6.1) cannot be directly implemented in a high-level language, since for a practically valuable LCG the product $a \times (m-1)$ exceeds the maximum value for a 32-bit integer. To avoid this problem, Schrage's method [29] is usually used. This method is based on an approximate factorization of m and operates as follows. Let $a^2 \leq m$ and $q := [m/a]$, $r := m \bmod a$ with square brackets denoting integer part, i.e., $m = aq + r$. Then
$$ax \bmod m = (a(x \bmod q) - [x/q]r) \bmod m\ .$$
If the result is negative, add m.

To get generators with larger periods as well as to improve quality of RNGs, it is possible to combine the outputs x_n^1 and x_n^2 of $\mathrm{LCG}(m_1, a_1, c_1)$ and $\mathrm{LCG}(m_2, a_2, c_2)$, respectively, to create a new random sequence:
$$y_n = (\,x_n^1 + x_n^2)\,\mathrm{mod}\,m_1. \qquad (6.3)$$
In practice to avoid an overflow in computing the intermediate value in (6.3), one can subtract x_n^1 and x_n^2 rather than add, and then add $m_1 - 1$ if the result is non-positive. It is known [145] that the combined generator based on the addition of LCG(2147483563, 40014, 0) and LCG(2147483399, 40692, 0) has quite good properties. This generator can be further improved by shuffling [239].

As a source of uniform random numbers in some of our examples, we use the procedure rng_lcgs.c from Listing A.14, which is a realization of the combined generator based on the addition of LCG(2147483563, 40014, 0) and LCG(2147483399, 40692, 0), with added shuffling.

The following shuffling algorithm is used here [135, 239]. Let x_n^1 and x_n^2 be sequences obtained by $\mathrm{LCG}(m_1, a_1, c_1)$ and $\mathrm{LCG}(m_2, a_2, c_2)$, respectively. Introduce an auxiliary vector V of dimension N (in the procedure rng_lcgs.c it is the array $ishuf[\,]$ and $N = 32$). In the first call of the procedure the vector V is initialized with N values of x^1-sequence and a variable Z (in the procedure it is irv) is set equal to $V[0]$. Then in each call (denote the call number by n): 1) compute the integer $j = [NZ/m_1]$ (i.e., $0 \leq j < N$ is a

random value determined by Z); 2) combine the sequences x_n^1 and x_n^2 as (cf. (6.3)):
$$Z = (V[j] + x_n^2) \bmod m_1;$$
3) put $V[j] = x_{N+n}^1$. The obtained Z is the output of the combined generator and it is also used to compute j in the next call.

We note that the procedure rng_lcgs.c is essentially based on the generator $ran2()$ from [239].

In the first call of rng_lcgs() the parameter *$iseed1$ should be set equal to a negative integer. After that *$iseed1$ should not be changed between calls of this procedure.

Listing A.14. Procedure rng_lcgs.c (uniform (0,1) RNG which is a combination of two LCGs with shuffling).

```
/* uniform (0,1) RNG based on ran2() from [239].
   It is a L'Ecuyer combination of two linear congruential sequences
   with shuffling. In the first call *iseed1=any negative integer;
   after that do not change it between calls. */
#define RNMX (1.0-1.2e-12)
double rng_lcgs(long *iseed1)
{
 int j;
 long k;
 static long iseed2=293451679, irv=0, ishuf[32];
 static double factor=1./2147483563.;
 double rv;
 if (*iseed1 <= 0) /*initialization - only in the first call*/
 {
   if((*iseed1) == 0) *iseed1=1;
   else *iseed1=-(*iseed1);
   iseed2=(*iseed1);
   for (j=39;j>=0;j--) /*initialization of shuffling table*/
   { k=(*iseed1)/53668;
     *iseed1=40014*(*iseed1-k*53668)-k*12211;
     if (*iseed1 < 0) *iseed1+=2147483563;
     if (j < 32) ishuf[j] = *iseed1;
   }
   irv=ishuf[0];
 } /* end of initialization */
 /* first sequence, compute by Schrage's method */
 k=(*iseed1)/53668; *iseed1=40014*(*iseed1-k*53668)-k*12211;
 if (*iseed1 < 0) *iseed1+=2147483563;
 /* second sequence, compute by Schrage's method */
 k=iseed2/52774; iseed2=40692*(iseed2-k*52774)-k*3791;
```

```
            if (iseed2 < 0) iseed2+=2147483399;
         /* shuffling & combination */
            j=irv/67108862; irv=ishuf[j]-iseed2; ishuf[j] = *iseed1;
            if (irv < 1) irv+=2147483562;
            if ((rv=factor*irv) > RNMX) return(RNMX); else return(rv);
         }
         #undef RNMX
```

As a source of uniform random numbers, we also use the additive lagged Fibonacci generator $F(1279, 418)$. We recall that an additive lagged Fibonacci generator is defined by (cf. (2.6.3)):

$$x_n = (x_{n-p} \pm x_{n-q}) \bmod m, \qquad (6.4)$$

where p and q are the lags, $p > q$. The addition (or subtraction) is done modulo any large integer m or with $m = 1$ when x_i are presented as floating point numbers in the interval $[0, 1)$. This method requires storing the p previous values of x_i in an array called a lag table. We denote the generator (6.4) by $F(p, q, m)$ for $m > 1$ and $F(p, q)$ for $m = 1$. See further details in Sect. 2.6.1.

The generator rng_fiba.c from Listing A.15 is the additive lagged Fibonacci generator $F(1279, 418)$ which uses the subtraction operation modulo $m = 1$ and in which x_i are floating point numbers in the interval $[0, 1)$. The lag table (the array $fab[]$ in rng_fiba.c) should be initialized before using this Fibonacci generator. The initialization can be done by, e.g., a linear congruental generator like rng_lcgs.c.

Listing A.15. Procedure rng_fiba.c (the additive lagged Fibonacci generator $F(1279, 418)$).

```
/* additive lagged Fibonacci generator F(1279,418) */
/* the procedure is based on the RANMAR lagged Fibonacci generator
   of Marsaglia, Zaman and Tsang and on fibadd.c by P. Coddington */
#define P 1278
#define Q 417
double rng_fiba(double fab[])
{
  static int i=P+1, j=Q+1;

  if(--i < 0)i=P; if(--j < 0)j=P;
  fab[i]-=fab[j];
  if(fab[i] < 0.)++fab[i];
  return(fab[i]);
}
```

In Listing A.16 we give a realization of the rejection polar method (Marsaglia method) for generation of Gaussian random numbers (2.6.11) (see further details in Sect. 2.6.3). Let us recall the corresponding algorithm. First, generate two independent uniform $(-1, 1)$ random numbers U_1 and U_2 and set $R^2 = U_1^2 + U_2^2$. Second, if $R^2 \geq 1$ then repeat the first step, otherwise

$$X_1 = U_1\sqrt{-2\log R^2/R^2}, \quad X_2 = U_2\sqrt{-2\log R^2/R^2}. \quad (6.5)$$

The obtained X_1 and X_2 are independent normally distributed $\mathcal{N}(0, 1)$ random numbers.

To generate uniform random numbers U, we exploit here the procedure rng_lcgs.c from Listing A.14. The procedure rng_gau.c from Listing A.16 was used in a majority of experiments in the book, where the source of Gaussian random numbers was needed.

Listing A.16. Procedure rng_gau.c (the rejection polar method for generation of Gaussian random numbers).

```
/* Gaussian RNG by the rejection polar method */
/* it is based on gasdev() from [239]*/
/* rng_lcgs.c is used as a source of uniform r.v. */
#include <math.h>
#include <stochast.h>
double rng_gau(long *iseed)
{
 static int isw=0;
 static double x1;
 double c,r,u1,u2;
 if (isw == 0) /* has x1 been used? */
 {
  do
  {
   u1=2.*rng_lcgs(iseed)-1.; u2=2.*rng_lcgs(iseed)-1.;
   r=u1*u1+u2*u2;
  } while (r >= 1.);
  c=sqrt(-2.*log(r)/r);
  x1=u1*c; isw=1; /*x1 will be used in the next call*/
  return(u2*c);
 }
 else
 {
  isw=0;
  return(x1);
 }
}
```

To simulate random variables distributed according to the law $P(\xi = \pm 1) = 1/2$, we used the procedure rng_disc.c.

Listing A.17. Procedure rng_disc.c for simulating the discrete random variable ξ with distribution $P(\xi = \pm 1) = 1/2$.

```
/* returns -1 and +1 with probability 1/2,
   uses additive lagged Fibonacci generator F(1279,418) */
#include <stochast.h>
double rng_disc(double fab[])
{
  if(rng_fiba(fab)<0.5)return(-1.); else return(1.);
}
```

Note that the simpler method of generating discrete random variables discussed in Remark 2.6.3 can be used instead of rng_disc.c.

References

1. Anderson, D.A., Tannehill, J.C., Pletcher, R.H. (1984): Computational fluid mechanics and heat transfer. Hemisphere, New York
2. Anderson, S.L. (1990): Random number generators on vector supercomputers and other advanced architectures. SIAM Review, **32**, 221–251
3. Anishchenko, V.S., Neiman, A.B., Moss, F., Schimansky-Geier, L. (1999): Stochastic resonance: noise enhanced order. Uspechi Fizicheskich Nauk, **169**, 7–38
4. Arnold, L. (1998): Random dynamical systems. Springer, Berlin
5. Arnold, L., Crauel, H., Eckmann, J.-P. (eds) (1991): Lyapunov exponents. Lecture Notes in Math. 1486, Springer
6. Arnold, L., Wihstutz, V. (eds) (1986): Lyapunov exponents. Lecture Notes in Math. 1186, Springer
7. Arnold, V.I. (1989): Mathematical methods of classical mechanics. Springer
8. Auslender, E.I., Milstein, G.N. (1982): Asymptotic expansions of the Lyapunov index for linear stochastic systems with small noises. Prikl. Matem. Mekhan., **46**, 358–365
9. Bally, V. (1997): Approximation scheme for solutions of BSDE, 177-191. In: El Karoui, N., Mazliak, L. (eds.) Backward stochastic differential equations. Pitman, London
10. Bally, V., Pages, G., Printems, J. (2001): A stochastic quantization method for nonlinear problems. Monte Carlo Methods Appl., **7**, 21-34
11. Bally, V., Talay, D. (1996): The law of the Euler scheme for stochastic differential equations I. Convergence rate of the distribution function. Probab. Theory Relat. Fields, **104**, 43-60
12. Bartussek, R., Hänggi, P., Jung, P. (1994): Stochastic resonance in optical bistable systems. Phys. Rev. E, **49**, 3930–3939
13. Bartussek, R., Hänggi, P., Kissner, J.G. (1994): Periodically rocked thermal ratchets. Europh. Lett., **28**, 459–464
14. Basdevant, C., Deville, M., Haldenwang, P., Lacroix, J.M., Onazzani, J., Peyret, R., Orlandi, P., Patera, A.T. (1986): Spectral and finite difference solutions of the Burgers equations. Comp. Fluids, **14**, 23–41
15. Baxendale, P. (1987): Moment stability and large deviations for linear stochastic differential equations. In: Ikeda, N. (ed.) Proc. Taniguchi Symp. on Probab. Meth. in Math. Physics. Katata and Kyoto, 1985 , Kinokuniya, Tokyo, 31–54
16. Belopolskaya, Ya., Milstein, G.N. (2003): An approximation method for Navier-Stokes equations based on probabilistic approach. Statistics&Probability Letters, **64**, 201–211

17. Bensoussan, A., Glowinski, R., Răşcanu, A. (1992): Approximation of some stochastic differential equations by the splitting up method. Appl. Math. Optim., **25**, 81–106
18. Benton, E.R., Platzman, G.W. (1972): A table of solutions of the one-dimensional Burgers equation. Quart. Appl. Math. **30**, 195–212
19. Benzi, R., Parisi, G., Suttera, A., Vulpiani, A. (1982): Stochastic resonance in climatic change. Tellus, **34**, 10–16
20. Benzi, R., Suttera, A., Vulpiani, A. (1981): The mechanism of stochastic resonance. J. Phys. A, **14** L453–L457
21. Bismut, J.-M. (1981): Mécanique aléatoire. Lecture Notes in Mathematics, **866**, Springer
22. de Boor, C. (1978): A practical guide to splines. Springer
23. Borodin, A.N., Salminen, P. (1996): Handbook of Brownian motion - facts and formulae. Birkhäuser, Basel
24. Bossy, M., Fezoui, L., Piperno S. (1997): Comparison of a stochastic particle method and a finite volume deterministic method applied to Burgers equation. Monte Carlo Methods Appl., **3**, 113–140
25. Bossy, M., Talay, D. (1997): A stochastic particle method for the McKean-Vlasov and the Burgers equation. Math. Comp., **66**, 157–192
26. Bouleau, N., Lépingle, D. (1994): Numerical methods for stochastic processes. John Wiley&Sons, New York
27. Box, G.E.P., Muller, M.E. (1958): A note on the generation of random normal deviates. Ann. Math. Stat., **29**, 610–611
28. Boyle, P., Broadie, M., Glasserman, P. (1997): Monte Carlo methods for security pricing. Journal of Economic Dynamics and Control, **21**, 1267-1321
29. Bratley, P., Fox, B.L., Schrage, E.L. (1983) A guide to simulation. Springer
30. Brent, R.P. (1992): Uniform random number generators for supercomputers. In Proceedings Fifth Australian Supercomputer Conference, Melbourne, 1992, 95–104
31. Broadie, M., Glasserman, P. (1996): Estimating security price derivatives using simulation. Manag. Sci., **42**, 269-285
32. Büttiker, M. (1987): Transport as a consequence of state-dependent diffusion. Z. Phys. B, **68**, 161–167
33. Channel, P.J., Scovel, C. (1990): Symplectic integration of Hamiltonian systems. Nonlinearity, **3**, 231–259
34. Chevance, D. (1997): Numerical methods for backward stochastic differential equations. In: Rogers, L.C.G., Talay, D. (eds.) Numerical methods in finance, Cambridge University Press, Cambridge
35. Chitashvili, R.J., Lazrieva, N.L. (1981): Strong solutions of stochastic differential equations with boundary conditions. Stochastics, **5**, 225–309.
36. Choi, M.H., Fox, R.F., Jung, P. (1998): Quantifying stochastic resonance in bistable systems: Response vs residence-time distribution functions. Phys. Rev. E, **57**, 6335–6344
37. Chorin, A.J. (1975): Accurate evaluation of Wiener integrals. Math. Comp. **27**, 1–15
38. Clark, J.M.C., Cameron, R.T. (1980): The maximum rate of convergence of discrete approximations for stochastic differential equations. Lect. Notes Control and Inform. Sci., **25**, 162–171
39. Cyganowski, S., Kloeden, P., Ombach, J. (2002): From elementary probability to stochastic differential equations with MAPLE. Universitext. Springer

40. Coddington, P.D. (1996): Random number generators for parallel computers. The NHSE Review 2 http://nhse.cs.rice.edu/NHSEreview/RNG/PRNGreview.ps
41. Costantini, C., Pacchiarotti, B., Sartoretto, F. (1998): Numerical approximation for functionals of reflecting diffusion processes. SIAM J. Appl. Math., **58**, 73–102
42. Danilov, V.G., Maslov, V.P., Volosov, K.A. (1995): Mathematical modelling of heat and mass transfer processes. Kluwer Academic Publishers, Dordrecht (engl. transl. from Russian, 1987)
43. Davie, A.M., Gaines, J.G. (2001): Convergence of numerical schemes for the solution of parabolic stochastic partial differential equations. Math. Comp., **70**, 121–134
44. De Matteis, A., Pagnutti, S. (1990) Parallelization of random number generators and long-range correlations. Parallel Comp., **15**, 155–164
45. Douglas, J., Ma, J., Protter, P. (1996): Numerical methods for forward-backward stochastic differential equations. Ann. Appl. Probab. **6**, 940-968
46. Duffie, D., Glynn, P. (1995): Efficient Monte Carlo simulation of security prices. Ann. Appl. Prob., **5**, no. 4, 897-905
47. Dupuis, P., Ishii, H. (1991): On Lipschitz continuity of the solution mapping to the Skorokhod problem, with applications. Stoch. Stoch. Rep., **35**, 31–62
48. Dynkin, E.B. (1965): Markov processes. Springer
49. Dynkin, E.B., Yushkevich, A.A. (1969): Markov processes: theorems and problems. Plenum, New York
50. Egorov, A.D., Sobolevsky, P.I., Yanovich, L.A. (1993): Functional integrals: approximative evaluation and applications. Kluwer Academic Publishers, Dordrecht
51. El Karoui, N., Mazliak, L. (eds.) (1997): Backward stochastic differential equations. Pitman, London
52. El Karoui, N., Peng, S., Quenez, M.S. (1997): Backward stochastic differential equations in finance. Math. Finance, **7**, 1–71
53. Elepov, B.S., Kronberg, A.A., Mikhailov, G.A., Sabelfeld, K.K. (1980): Solution of boundary value problems by the Monte Carlo method. Nauka, Novosibirsk (In Russian)
54. Elston, T.C., Peskin, C.S. (2000): The role of protein flexibility in molecular motor function: coupled diffusion in a tilted periodic potential. SIAM J. Appl. Math., **60**, 842–867
55. Elworthy, K.D. (1982): Stochastic differential equations on manifolds. Cambridge Univ. Press, Cambridge
56. Entacher, K. (1998): Bad subsequences of well-known linear congruential pseudorandom number generators. ACM Trans. Model. Comp. Simul., **7**, 61–70
57. Eriksson, K., Johnson, C. (1995): Adaptive finite element methods for parabolic problems IV: Nonlinear problems. SIAM J. Num. Anal., **32**, 1729–1749
58. Ermakov, S.M., Nekrutkin, V.V., Sipin, A.S. (1989): Random processes for classical equations of mathematical physics. Kluwer Academic Publishers
59. Feller, W. (1958): An introduction to probability theory and its applications. John Willey & Sons, Vol. 1
60. Feng, K., Shang, Z.-J. (1995): Volume-preserving algorithms for source-free dynamical systems. Numer. Math., **71**, 451–463

61. Feynman, R.P., Hibbs, A.R. (1965): Quantum mechanics and path integrals, McGraw-Hill, N.Y.
62. Feynman, R.P., Leighton, R.B., Sands, M. (1966): The Feynman lectures on physics. Addison-Wesley, Vol. 1, Chap. 46
63. Fisher, P., Platen, E. (1999): Applications of the balanced method to stochastic differential equations in filtering. Monte Carlo Methods and Applications, **5**, 19-38
64. Fleming, W.H., Rishel, R.W. (1975): Deterministic and stochastic optimal control. Springer
65. Fletcher, C.A.J. (1984): Computational Galerkin methods. Springer
66. Fletcher, C.A.J. (1991): Computational techniques for fluid dynamics. Volumes I, II, Springer
67. Fournié, E., Lasry, J.-M., Lebuchoux, J., Lions, P.-L., Touzi, N. (1999): Applications of Malliavin calculus to Monte Carlo methods in finance. Finance and Stochastics, **3**, 391–412
68. Fournié, E., Lasry, J.-M., Lebuchoux, J., Lions, P.-L. (2001): Applications of Malliavin calculus to Monte Carlo methods in finance. II. Finance and Stochastics, **5**, 201–236
69. Fox, R.F. (1991): Second-order algorithm for the numerical integration of colored-noise problems. Phys.Rev. A, **43**, 2649–2654
70. Freidlin, M.I. (1985): Functional integration and partial differential equations. Princeton Univ. Press, Princeton
71. Freidlin, M.I. (1996): Markov processes and differential equations: asymptotic problems. Birkhäuser, Basel
72. Freidlin, M.I., Wentzell, A.D. (1984): Random perturbations of dynamical systems. Springer
73. Fridman, A. (1964): Partial differential equations of parabolic type. Prentice-Hall, Englewood Cliffs, NJ
74. Gaines, J.G. (1995): Numerical experiments with S(P)DE's. In: Etheridge, A. (Ed.) Stochastic parial differential equations. Cambridge Univ. Press, Cambridge
75. Gaines, J.G., Lyons, T.J. (1994): Random generation of stochastic area integrals. SIAM J. Appl. Math., **54**, no. 4, 1132-1146
76. Gaines, J.G., Lyons, T.J. (1997): Variable step size control in the numerical solution of stochastic differential equations. SIAM J. Appl. Math., **57**, 1455–1484
77. Gammaitoni, L., Hänggi, P., Jung, P., Marchesoni, F. (1998): Stochastic resonance. Rev. Mod. Phys., **70**, 223–287
78. Gardiner, C.W. (1983): Handbook of stochastic methods for physics, chemistry and natural sciences. Springer
79. Gelfand, I.M. (1959): Some problems in the theory of quasi-linear equations. Uspehi Mat. Nauk, **14**, 87–158
80. Gelfand, I.M., Yaglom, A.M. (1960): Integration in functional spaces and its application in quantum physics. J. Math. Phys., **1**, 48–69 (engl. transl. from Russian, Uspechi Mat. Nauk, **11** (1956), 77–114)
81. Gentle, J.E. (1998): Random number generation and Monte Carlo methods. Springer
82. Gichman, I.I., Skorochod, A.V. (1968): Stochastic differential equations. Naukova Dumka, Kiev

83. Gichman, I.I., Skorochod, A.V. (1982): Stochastic differential equations and their applications. Naukova Dumka, Kiev
84. Gladyshev, S.A., Milstein, G.N. (1984): The Runge–Kutta method for calculation of Wiener integrals of functionals of exponential type. Zh. Vychisl. Mat. i Mat. Fiz., **24**, 1136–1149
85. Glasserman, P. (2003): Monte Carlo methods in financial engineering. Springer
86. Gornostyrev, Yu.N., Katsnelson, M.I., Trefilov, A.V., Tretjakov, S.V. (1996): Stochastic approach to simulation of lattice vibrations in strongly anharmonic crystals: Anomalous frequency dependence of the dynamic structure factor. Phys. Rev. B, **54**, 3286–3294
87. Graham, C., Kurtz, Th. G., Méléard, S., Protter, Ph. E., Pulvirenti, M., Talay, D. (1996): Probabilistic models for nonlinear partial differential equations. Lectures given at the 1st Session and Summer School held in Montecatini Terme, May 22–30, 1995. Edited by D. Talay and L. Tubaro. Lecture Notes in Mathematics, 1627, Springer
88. Graham, R. (1982): Hopf bifurcation with fluctuating control parameter. Phys.Rev. A, **25**, 3234–3258
89. Grecksch, W., Kloeden, P. E. (1996): Time-discretised Galerkin approximations of parabolic stochastic PDEs. Bull. Austral. Math. Soc., **54**, 79–85
90. Greenside, H.S., Helfand, E. (1981): Numerical integration of stochastic differential equations. Bell Syst. Techn. J., **60**, 1927–1940
91. Grindrod, P. (1996): The theory and applications of reaction-diffusion equations: patterns and waves. Clarendon Press, Oxford
92. Gyöngy, I. (1998): A note on Euler's approximations. Potential Analysis, **8**, 205-216
93. Gyöngy, I. (1999): Lattice approximations for stochastic quasi-linear parabolic partial differential equations driven by space-time white noise II. Potential Anal., **11**, 1–37
94. Gyöngy, I., Krylov, N. (1996): Existence of strong solutions for Ito's stochastic equations via approximations. Probab. Theory Relat. Fields, **105**, 143–158
95. Gyöngy, I., Nualart, D. (1997): Implicit scheme for stochastic parabolic partial differential equations driven by space-time white noise. Potential Anal., **7**, 725–757
96. Hairer, E., Lubich, C., Wanner, G. (2002): Geometric numerical integration: structure presesrving algorithms for ordinary differential equations. Springer
97. Hairer, E., Nørsett, S.P., Wanner, G. (1993): Solving ordinary differential equations. I. Nonstiff problems. Springer
98. Hairer, E., Wanner, G. (1996): Solving ordinary differential equations. II. Stiff and differential–algebraic problems. Springer
99. Hall, G., Watt, J.M. (1976): Modern numerical methods for ordinary differential equations. Clarendon Press, Oxford
100. Hellekalek, P. (1998): Good random number generators are (not so) easy to find, Math. Comp. Simul., **46**, 485–505
101. Higham, D.J., Kloeden, P.E. (2002): MAPLE and MATLAB for stochastic differential equations in finance. In Programming languages and systems in computational economics and finance (S.S. Nielsen, ed.), Kluwer Academic Publishers
102. Higham, D.J., Mao, X., Stuart, A. (2003) Strong convergence of Euler-type methods for nonlinear stochastic differential equations. SIAM J. Num. Anal., **40**, 1041–1063.

103. Hille, E., Phillips, R.S. (1957): Functional analysis and semigroups. AMS, Providence
104. Hofmann, N., Müller-Gronbach, T., Ritter, K. (2000): Optimal approximation of stochastic differential equations by adaptive step-size control. Math. Comp., **69**, 1017–1034
105. Hofmann, N., Mathé, P. (1997): On quasi-Monte Carlo simulation of stochastic differential equations. Math. Comp., **66**, 573–589
106. Horsthemke, W., Lefever, R. (1984): Noise-induced transitions: theory and applications in physics, chemistry and biology. Springer
107. Hoshino, S., Ichida, K. (1971): Solution of partial differential equations by a modified random walk. Numer. Math., **18**, 61–72
108. Hu, Y. (1996): Semi-implicit Euler-Maruyama scheme for stiff stochastic equations, in Stochastic Analysis and Related Topics V: The Silvri Workshop, Progr. Probab. 38, H. Koerezlioglu,ed., Birkhauser, Boston, 183–202
109. Hu, Y., Watanabe, S. (1996): Donsker delta functions and approximations of heat kernels by the time discretization method. J. Math. Kyoto Univ., **36**, 494-518
110. Ikeda, N., Watanabe, S. (1981): Stochastic differential equations and diffusion processes. North-Holland Publ. Comp., Amsterdam
111. Il'in, A.M. (1989): Matching of asymptotic expansions of solutions of boundary value problems. Nauka, Moscow
112. Il'in, A.M., Oleinik, O.A. (1960): Asymptotic behavior of solutions of the Cauchy problem for some quasi-linear equations for large values of time. Mat. Sbornik, **51**, 191–216
113. Ito, K., McKean, H.P. (1965): Diffusion processes and their sample paths. Springer
114. Izaguirre, J.A., Catarello, D.P., Wozniak, J.M., Skeel, R.D. (2001): Langevin stabilization of molecular dynamics. J. Chem. Phys., **114**, 2090–2098
115. Jacod, J., Protter, P. (1998): Asymptotic error distributions for the Euler method for stochastic differential equations. Ann. Prob., **26**, 267–307
116. Jülicher, F., Ajdari, A., Prost, J. (1997): Modeling molecular motors. Rev. Mod. Phys., **69**, 1269–1281
117. Kalashnikov, V.V., Rachev, S.T. (1988): Mathematical methods for construction of stochastic queuing Models. Nauka, Moscow (English translation by Pacific Grove, California, 1990)
118. Kalman, R.E., Falb, P.L., Arbib, M.A. (1969): Topics in mathematical control theory. McGraw-Hill, New York
119. van Kampen, N. (1988): Relative stability in nonuniform temperature, IBM J. Res. Dev., **32**, 107–111
120. Karatzas, I., Shreve S.E. (1988): Brownian motion and stochastic calculus, Springer
121. Ketter,C., Reimann, P., Hänggi, P., Müller, F. (2000): Drift ratchet. Phys. Rev. E, **61**, 312–323
122. Kevorkian, J., Cole, J.D. (1996): Multiple scale and singular perturbation methods. Springer
123. Khasminskii, R.Z. (1960): Probabilistic representation of solutions of some differential equations. In the book: Trydi VI Vses. Sovechaniya on Th. Prob. and Math. Stat., Vilnius, 177–183

124. Khasminskii, R.Z. (1980): Stability of differential equations under random perturbations of their parameters. Sijthoff&Noordhoff (tranlsated from Russian, 1969)
125. Khasminskii, R.Z., Nevelson, M.B. (1976): Stochastic approximation and recursive estimation, AMS (translated from Russian, 1972)
126. Kinderman, A.J., Monahan, J.F. (1977): Computer generation of random variables using the ratio of uniform deviates. ACM Trans. Math. Soft., **3**, 257–260
127. Kitsul, P.I. (1970): On continuously-discrete filtering of Markov processes of diffusion type. Avtomat. i Telemekh. **31**, no.11, 29-37 (In Russian)
128. Klauder, J.R., Petersen, W.P. (1985): Numerical integration of multiplicative-noise stochastic differential equations. SIAM J. Num. Anal., **22**, 1153–1166
129. Kleinert, H. (1995): Path integrals in quantum mechanics, statistics and polymer physics, World Scientific
130. Kloeden, P.E. (2002): The systematic derivation of higher order numerical methods for stochastic differential equations. Milan J. Math., **70**, 187–207
131. Kloeden, P.E., Platen, E. (1992): Numerical solution of stochastic differential equations. Springer
132. Kloeden, P.E., Platen, E., Schurz, H. (1994): Numerical solution of SDE through computer experiments. Springer
133. Kloeden, P.E., Platen, E., Wright, W. (1992): The approximation of multiple stochastic integrals. Stoch. Anal. Appl., **10**, 431–441
134. Kloeden, P. E., Shott, S.(2001): Linear-implicit strong schemes for Ito-Galerkin approximations of stochastic PDEs. J. Appl. Math. Stochastic Anal., **14**, 47–53
135. Knuth, D.E. (1981): The art of computer programming Vol. 2: Seminumerical methods, Addison-Wesley, Reading, Mass
136. Kohatsu-Higa, A. (1997): High order Ito-Taylor approximations to heat kernels. J. Math. Kyoto Univ., **37**, 129–150
137. Kohatsu-Higa, A. (2001): Weak approximations: a Malliavin calculus approach. Math. of Comp., **70**, 135–175
138. Krylov, N.V. (1977): Controllable processes of diffusion type. Nauka, Moscow
139. Kubo, R., Toda, M., Hashitsume, N. (1985): Statistical physics. Vol. II, Springer
140. Kunita, H. (1990): Stochastic flows and stochastic differential equations. Cambridge University Press, Cambridge
141. Kushner, H.J. (1977): Probability methods for approximations in stochastic control and for elliptic Equations. Academic Press, New York
142. Kuznetsov, D.F. (2001): Numerical integration of stochastic differential equations. University of S.-Peterburg, S.-Petersburg
143. Ladyzhenskaya, O.A., Solonnikov, V.A., Ural'ceva, N.N. (1988): Linear and quasilinear equations of parabolic type. Amer. Math. Soc., Providence, R.I.
144. Landa, P. S. (1998): Noise-induced transport of Brownian particles with consideration for their mass. Phys. Rev. E, **58**, 1325–1333
145. L'Ecuyer, P. (1988): Efficient and portable combined random number generators. Comm. ACM, **31**, 1019–1024
146. L'Ecuyer, P. (1996): Maximally equidistributed combined Tausworthe generators. Math. Comp., **65**, 203–213
147. L'Ecuyer, P. (1998): Random number generation, in Handbook of simulation, Ed.: J. Banks, Wiley, Chapter 4, 93–137

148. L'Ecuyer, P. (1999): Tables of linear congruential generators of different sizes and good lattice structure. Math. Comp., **68**, 249–260
149. L'Ecuyer, P. (1999): Tables of maximally-equidistributed combined LFSR generators. Math. Comp., **68**, 261–269
150. L'Ecuyer, P. (1999): Good parameter sets for combined multiple recursive random number generators. Oper. Research, **47**, 159–164
151. L'Ecuyer, P., Andres, T.H. (1997): A random number generator based on the combination of four LCGs. Math. Comp. Simul., **44**, pp. 99–107
152. L'Ecuyer, P., Simard, R., Chen, E.J., Kelton, W.D. (2002): An object-oriented random number package with many long streams and substreams. Oper. Res., **50**, 1073–1075 (with accompanying online companion 31 p.)
153. Lehmer, D.H. (1951): Mathematical methods in large-scale computing units. Annals Comput. Lab. Harvard Univ., **26**, 141–146
154. Lepingle, D. (1993): Un schéma d'Euler pour équations différentielles stochastiques réfléchies. C.R. Acad. Sci. Paris Série I, **316**, 601–605
155. Leva, J.L. (1992): A fast normal random number generator. ACM Trans. Math. Soft., **18**, 449–453; Algorithm 712: A normal random number generator. ibid, 454–455.
156. Lipster, R.S., Shiryaev, A.N. (1974): Statistics of random processes. Nauka, Moscow
157. Liu, Y. (1993): Numerical approaches to stochastic differential equations with boundary conditions. Thesis, Purdue University
158. Ma, J., Protter, P., Martin, J.S., Torres, S. (2002): Numerical methods for backward stochastic differential equations. Ann. Appl. Prob., **12**, 302–316
159. Mackevicius, V. (1994): Second order weak approximations for Stratonovich stochastic differential equations. Liet. Mat. Rink. **34**, 226–247
160. Mackevicius, V. (1996): Extrapolation of approximations of solutions of stochastic differential equations. Probability theory and mathematical statistics (Tokyo, 1995), World Sci. Publishing, River Edge, NJ, 1996, 276–297
161. Magnasco, M.O. (1993): Forced thermal ratchets. Phys. Rev. Lett., **71**, 1477–1481
162. Mannela, R., Palleschi, V. (1989): Fast and precise algorithm for computer simulation of stochastic differential equations. Phys.Rev. A, **40**, 3381–3386
163. Marsaglia, G. (1962): Improving the polar method for generating a pair of random variables. Boeing Sci. Res. Lab. report D1-82-0203
164. Marsaglia, G.A. (1985): A current view of random number generators, in Computational Science and Statistics, Sixteenth Symposium on the Interface, ed. L. Balliard, Elsevier, Amsterdam, 3–10
165. Marsaglia, G., Tsay, L.-H (1985): Matrices and the structure of random number sequences. Linear Alg. Appl., **67**, 147–156
166. Marsaglia, G., Zaman, A., Marsaglia, J.C.W. (1994): Rapid evaluation of the inverse normal distribution function. Statistics and Probability Letters, **19**, 259–266
167. Marujama, G. (1955): Continuous Markov processes and stochastic equations. Rend. Mat. Circ. Palermo, Ser. 2, **4**, 48–90
168. Mascagni, M. (1999): Some methods of parallel pseudorandom number generation. In: Algorithms for parallel processing, Eds. Heath M.T., Ranade A., Schreiber R.S., Springer, 277–288

169. Mascagni, M., Cuccaro, S.A., Pryor, D.V., Robinson, M.L. (1995): A fast, high-quality, and reproducible lagged-Fibonacci pseudorandom number generator. J. Comp. Phys., **15**, 211–219
170. Mattingly, J.C., Stuart, A.M. (2002): Geometric ergodicity of some hypoelliptic diffusions for particle motions. Markov Processes Relat. Fields, **8**, 199–214
171. Mattingly, J.C., Stuart, A.M., Higham, D.J. (2002): Ergodicity for SDEs and approximations: Locally Lipschitz vector fields and degenerate noise. Stoch. Proc. Appl., **101**, 185–232
172. Menaldi, J.-L. (1983): Stochastic variational inequality for reflected diffusion. Indiana Univ. Math. J., **32**, 733–744
173. Meyn, S.P., Tweedie, R.L. (1992): Markov chains and stochastic stability. Springer
174. Mikhailov, V.P. (1976): Partial differential equations. Nauka, Moscow
175. Milstein, G.N. (1974): Approximate integration of stochastic differential equations. Theor. Prob. Appl., **19**, 583–588
176. Milstein, G.N. (1978): A method with second order accuracy for the integration of stochastic differential equations. Theor. Prob. Appl., **23**, 414–419
177. Milstein, G.N. (1978): Probabilistic solution of linear systems of elliptic and parabolic equations. Theor. Prob. Appl., **23**, 851–855
178. Milstein, G.N. (1985): Weak approximation of solutions of systems of stochastic differential equations. Theor. Prob. Appl., **30**, 706–721
179. Milstein, G.N. (1987): A theorem on the order of convergence of mean-square approximations of solutions of systems of stochastic differential equations. Theor. Prob. Appl., **32**, 809–811
180. Milstein, G.N. (1988): Numerical integration of stochastic differential equations. Ural State University, Sverdlovsk. Engl. transl. by Kluwer Academic Publishers, Mathematics and its Applications, v.313, 1995
181. Milstein, G.N. (1995): The solving of boundary value problems by numerical integration of stochastic equations. Mathem. Comp. Simul., **38**, 77–85
182. Milstein, G.N. (1995): Solution of the first boundary value problem for equations of parabolic type by means of the integration of stochastic differential equations. Theor. Probab. Appl., **40**, 556–563
183. Milstein, G.N. (1996): Application of the numerical integration of stochastic equations for the solution of boundary value problems with Neumann boundary conditions. Theor. Prob. Appl., **41**, 170–177
184. Milstein, G.N. (1996): The simulation of phase trajectories of a diffusion process in a bounded domain. Stochastics and Stochastics Reports, **56**, 103–125
185. Milstein, G.N. (1996): Evaluation of moment Lyapunov exponents for second order stochastic systems. Random&Comp. Dynam., **4**, 301–315
186. Milstein, G.N. (1997): Weak approximation of a diffusion process in a bounded domain. Stoch. Stoch. Rep., **62**, 147–200
187. Milstein, G.N. (1998): On the mean-square approximation of a diffusion process in a bounded domain. Stoch. Stoch. Rep., **64**, 211–233
188. Milstein, G.N. (2002): The probability approach to numerical solution of nonlinear parabolic equations. Num. Meth. PDE, **18**, 490–522
189. Milstein, G.N., Platen, E., Schurz, H. (1998): Balanced implicit methods for stiff stochastic systems. SIAM J. Num. Anal., **35**, 1010–1019

190. Milstein, G.N., P'yanzin, S.A. (1985): Digital modelling of the Kalman-Bucy filter and an optimal filter in quantized arrival of information. Avtomat. i Telemekh., **1**, 59–68
191. Milstein, G.N., P'yanzin, S.A. (1987): Regularization and digital modeling of a Kalman–Bucy filter for systems with degenerate noise in the observations. Avtomat. i Telemekh., **11**, 80–92
192. Milstein, G.N., Reiß, O., Schoenmakers, J.G.M. (2004): A new Monte Carlo method for American options. Intern. J. Theor. Appl. Finance (in print)
193. Milstein, G.N., Repin, Yu.M., Tretyakov, M.V. (2001): Symplectic methods for Hamiltonian systems with additive noise. Preprint No. 640, Weierstraß-Institut für Angewandte Analysis und Stochastik, Berlin
194. Milstein, G.N., Repin, Yu.M., Tretyakov, M.V. (2002): Symplectic integration of Hamiltonian systems with additive noise. SIAM J. Num. Anal., **39**, 2066–2088
195. Milstein, G.N., Repin, Yu.M., Tretyakov, M.V. (2003): Numerical methods for stochastic systems preserving symplectic structure. SIAM J. Num. Anal., **40**, 1583-1604
196. Milstein, G.N., Rybkina, N.F. (1993): An algorithm for random walks over small ellipsoids for solving the general Dirichlet problem. J. Comp. Math. Math. Phys., **33**, 631–647
197. Milstein, G.N., Schoenmakers, J.G.M. (2002): Numerical construction of a hedging strategy against the multi-asset European claim. Stoch. Stoch. Rep., **73**, 125–157
198. Milstein, G.N., Schoenmakers, J.G.M., Spokoiny, V. (2004): Transition density estimation for stochastic differential equations via forward-reverse representations. Bernoulli **10**, No 2
199. Milstein, G.N., Tret'yakov, M.V. (1994): Numerical solution of differential equations with colored noise. J. Stat. Phys., **77**, 691–715
200. Milstein, G.N., Tretyakov, M.V. (1994): Weak approximation for stochastic differential equations with small noises. Preprint No. 123, Weierstraß-Institut für Angewandte Analysis und Stochastik, Berlin
201. Milstein, G.N., Tretyakov, M.V. (1997): Mean-square numerical methods for stochastic differential equations with small noise. SIAM J. Sci. Comp., **18**, 1067–1087
202. Milstein, G.N., Tretyakov, M.V. (1997): Numerical methods in the weak sense for stochastic differential equations with small noise. SIAM J. Num. Anal., **34**, 2142–2167
203. Milstein, G.N., Tretyakov, M.V. (1998): Numerical methods for nonlinear parabolic equations with small parameter based on probability approach. Preprint No. 396, Weierstraß-Institut für Angewandte Analysis und Stochastik, Berlin
204. Milstein, G.N., Tretyakov, M.V. (1999): Simulation of a space-time bounded diffusion. Ann. Appl. Probab., **9**, 732–779
205. Milstein, G.N., Tretyakov, M.V. (1999): Mean velocity of noise-induced transport in the limit of weak periodic forcing. J. Phys. A, **32**, 5795–5805
206. Milstein, G.N., Tretyakov, M.V. (2000): Numerical algorithms for semilinear parabolic equations with small parameter based on weak approximation of stochastic differential equations. Math. Comp., **60**, 237–267
207. Milstein, G.N., Tretyakov, M.V. (2000): Numerical analysis of noise-induced regular oscillations. Physica D, **140**, 244–256

208. Milstein, G.N., Tretyakov, M.V. (2000): Numerical solution of the Neumann problems for nonlinear parabolic equations by probability approach. Preprint No. 589, Weierstraß-Institut für Angewandte Analysis und Stochastik, Berlin
209. Milstein, G.N., Tretyakov, M.V. (2001): Noice-induced unidirectional transport. Stochastics and Dynamics, **1**, 361–375
210. Milstein, G.N., Tretyakov, M.V. (2001): Numerical solution of the Dirichlet problem for nonlinear parabolic equations by a probabilistic approach. IMA J. Num. Anal., **21**, 887–917
211. Milstein, G.N., Tretyakov, M.V. (2002): The simplest random walks for the Dirichlet problem. Theor. Prob. Appl., **47**, 39–58
212. Milstein, G.N., Tretyakov, M.V. (2002) A probabilistic approach to the solution of the Neumann problem for nonlinear parabolic equations. IMA J. Num. Anal., **22**, 599-622
213. Milstein, G.N., Tretyakov, M.V. (2003): Numerical methods for Langevin type equations based on symplectic integrators. IMA J. Num. Anal., **23**, 593-626
214. Milstein, G.N., Tretyakov, M.V. (2003): Numerical analysis of Monte Carlo finite difference evaluation of Greeks. Preprint No. 808, Weierstraß-Institut für Angewandte Analysis und Stochastik, Berlin (submitted)
215. Milstein, G.N., Tretyakov, M.V. (2004): Evaluation of conditional Wiener integrals by numerical integration of stochastic differential equations. J. Comp. Phys. (in print)
216. Miranda, C. (1970): Partial differential equations of elliptic type. Springer
217. Moss, F., McClintock, P.V.E. (eds) (1989): Noise in nonlinear dynamical systems, Vol. I,II,III, Cambridge Univ. Press, Cambridge
218. Müller, M.E. (1956): Some continuous Monte Carlo methods for the Dirichlet problem. Ann. Math. Statist., **27**, 569–589
219. von Neumann, J., Richtmyer, R. (1950): A method for the numerical calculation of hydrodynamic shocks. J. Appl. Phys., **21**, 232–257
220. Neiman, A., Schimansky-Geier, L. (1995): Stochastic resonance in two coupled bistable systems. Phys. Lett. A, **197,** 379–386
221. Newton, N.J. (1986): An asymptotically efficient difference formula for solving stochastic differential equations. Stochastics, **19**, 175–206
222. Newton, N.J. (1994): Variance reduction for simulated diffusions. SIAM J. Appl. Math., **54**, 1780–1805
223. Newton, N.J. (1997): Continuous-time Monte Carlo methods and variance reduction. In: Numerical methods in finance, ed. by Rogers, L.C.G. and Talay, D. Cambridge Univ. Press, Cambridge, 22–42
224. Nicolis, C. (1982): Stochastic aspects of climatic transitions – response to a periodic forcing. Tellus, **34**, 1–9
225. Niederreiter, H. (1992): Random number generation and quasi-Monte Carlo methods, SIAM, Philadelphia
226. Nikitin, N.N., Razevig, V.D. (1978): Methods of numeical modeling of stochastic differential equations and estimates of their errors. Zh. Vychisl. Mat. i Mat. Fiz., **18**, 106–117
227. Ostrem, K. (1973): Introduction to the stochastic theory of equations. Mir, Moscow (In Russian)
228. Öttinger, H.C. (1996): Stochastic processes in polymeric fluids: tools and examples for developing simulation algorithms. Springer
229. Paley, R.E.A.C., Wiener, N. (1934): Fourier transforms in the complex domain. AMS, New York.

230. Pardoux, E., Peng, S. (1992): Backward stochastic differential equations and quasilinear parabolic partial differential equations. In: Rozovskii, B.L., Sowers, R.B. (eds.), Stochastic partial differential equations and their applications, 200-217. Lecture Notes in Control and Inform. Sc., **176**, Springer
231. Pardoux, E., Talay, D. (1985): Discretization and simulation of stochastic differential equations. Acta Appl. Math., **3**, 23–47
232. Petersen, W.P. (1994): Some experiments on numerical simulations of stochastic differential equations and a new algorithm. J. Comp. Phys., **113**, 75–81
233. Petersen, W.P. (1998): A general implicit splitting for stabilizing numerical simulations of Ito stochastic differential equations. SIAM J. Num. Anal., **35**, 1439–1451
234. Pettersson, R. (1995): Approximations for stochastic differential equations with reflecting convex boundaries. Stochastic Processes and their Applications, **59**, 295–308
235. Pettersson, R. (1997): Penalization schemes for reflecting stochastic differential equations. Bernoulli, **3**, no. 4, 403–414
236. Platen, E. (1981): An approximation method for a class of Ito processes. Lit. Mat. Sb., **21**, no. 1, 121–133
237. Platen, E. (1999): An introduction to numerical methods for stochastic differential equations. Acta Numerica, **8**, 197-246
238. Platen, E., Wagner, W. (1982): On a Taylor formula for a class of Ito processes. Prob. Math. Statist., **3**, 37–51
239. Press, W.H., Teukolsky, S.A., Vetterling, W.T., Flannery, B.P. (1992): Numerical Recipes in C: The art of scientific computing. Cambridge Univ. Press, Cambridge
240. Protter, P., Talay, D. (1997): The Euler scheme for Levy driven stochastic differential equations. Ann. Probab., **25**, 393–423
241. Quarteroni, A., Valli, A. (1994): Numerical approximation of partial differential equations. Springer
242. Quispel, G.R.W. (1995): Volume-preserving integrators. Phys. Lett. A, **206**, 26–30
243. Rakitskii, Yu.V., Ustinov, S.M., Chernorutskii, I.G. (1979): Numerical methods for solving stiff systems. Nauka, Moscow
244. Rao, N.J., Borwankar, J.D., Ramkrishna, D. (1974): Numerical solution of Ito integral equations. SIAM J. Control, **12**, 124–139
245. Richtmyer, R.D., Morton, K.W. (1967): Difference methods for initial-value problems. Interscience, New York
246. Reimann, P. (2002): Brownian motors: noisy transport far from equilibrium. Phys. Rep., **361**, 57–265
247. Ripoll, M., Ernst, M.H., Español, , P. (2001): Large scale and mesoscopic hydrodynamics for dissipative particle dynamics. J. Chem.Phys., **115**, 7271–7284
248. Risken, H. (1984): The Fokker-Planck equation. Springer
249. Roache, P.J. (1976): Computational fluid dynamics. Hermosa, Albuquerque, N.M.
250. Roepstorff, G. (1994): Path integral approach to quantum physics, Springer
251. Rogers, L.C.G., Talay, D. (eds) (1997): Numerical methods in finance, Cambridge University Press, Cambridge
252. Rogers, L.C.G., Williams, D. (1987): Diffusions, Markov processes, and martingales. Volume 2, Ito calculus. John Willey&Sons Ltd.

253. Roos, H.-G., Stynes, M., Tobiska, L. (1996): Numerical methods for singularity perturbed differential equations: convection-diffusion and flow Problems. Springer
254. Ross, S.M. (1997): Simulation. Academic Press, N.Y.
255. Rumelin, W. (1982): Numerical treatment of stochastic differential equations. SIAM J. Num. Anal., **19**, 604–613
256. Ryden, T., Wiktorsson M. (2001): On the simulation of iterated Ito inegrals. Stoch. Proc. Appl., **91**, 151–168
257. Sabelfeld, K.K. (1991): Monte Carlo methods in boundary value problems. Springer
258. Samarskii, A.A. (1977): Theory of difference schemes. Nauka, Moscow
259. Samarskii, A.A., Galaktionov, V.A., Kurdyumov, S.P., Mikhailov, A.P. (1995): Blow-up in quasilinear parabolic equations. Walter de Gruyter, Berlin (engl. transl. from Russian, 1987)
260. Sanz-Serna, J.M. (1992): Symplectic integrators for Hamiltonian problems: an overview. Acta Numerica, **1**, 243–286
261. Sanz-Serna, J.M., Calvo, M.P. (1994): Numerical Hamiltonian problems. Chapman&Hall
262. Sawford, B.L. (2001): Turbulence relative dispersion. Annu. Rev. Fluid Mech., **33**, 289–317
263. Schenzle, A., Brand, H. (1979): Multiplicative stochastic processes in statistical physics. Phys. Rev. A, **20**, 1628–1647
264. Schimansky-Geier, L., Pöschel, Th. (eds) (1997): Stochastic dynamics. Lecture Notes in Physics, **484**, Springer
265. Seeßelberg, M., Breuer, H.P., Mais, H., Petruccione,F., Honerkamp, J. (1994): Simulation of one-dimensional noisy Hamiltonian systems and their application to particle storage rings. Z. Phys. C, **62**, 62–73
266. Skeel, R. (1999): Integration schemes for molecular dynamics and related applications. In: "Graduate student's guide to numerical analysis'98" edited by M. Ainsworth, J. Levesley, M. Marletta, Springer, 118–176
267. Slominski, L. (1994): On approximation of solutions of multidimensional SDE's with reflecting boundary conditions. Stochastic Processes and their Applications, **50**, 197–219
268. Slominski, L. (2001): Euler's approximations of solutions of SDEs with reflecting boundary. Stochastic Processes and their Applications, **94**, 317–337
269. Smoller, J. (1983): Shock waves and reaction-diffusion equations. Springer
270. Soize, C. (1994): The Fokker-Planck equation for stochastic dynamical systems and its explicit steady state solutions. World Scientific Publishing Co. Pte. Ltd., London
271. Strang, G. (1968): On the construction and comparison of difference schemes. SIAM J. Num. Anal., **5**, 506–517
272. Stratonovich, R.L. (1967): Topics in the theory of random noise. Gordon and Breach, New York
273. Strikwerda, J.C. (1989): Finite difference schemes and partial differential equations. Wadsworth & Brooks/ PacificGrove, California
274. Stuart, A.M., Humphries, A.R. (1996): Dynamical systems and numerical analysis. Cambridge Univ. Press, Cambridge
275. Suris, Yu.B. (1989): The canonicity of mappings generated by Runge–Kutta type methods when integrating the systems $\ddot{x} = -\partial U/\partial x$. U.S.S.R. Comp. Maths. Math. Phys., **29**, 138–144

276. Suris, Yu.B. (1989): On the canonicity of mappings that can be generalized by methods of Runge–Kutta type for integrating systems $\ddot{x} = -\partial U/\partial x$. Zh. Vychisl. Mat. i Mat. Fiz., **29**, 202–211
277. Suris, Yu.B. (1990): Hamiltonian methods of Runge–Kutta type and their variational interpretation. Math. Model., **2**, 78–87
278. Suris, Yu.B. (1991): Preservation of integral invariants in the numerical solution of systems $\ddot{x} = \dot{x} + f(x)$. Zh. Vychisl. Mat. i Mat. Fiz., **31**, 52–63
279. Suris, Yu.B. (1996): Partitioned Runge–Kutta methods as phase volume preserving integrators. Phys. Lett. A, **220**, 63–69
280. Talay, D. (1983): Résolution trajectorillée et analyse numérique des équations différentielles stochastiques. Stochastics, **9**, 275–306
281. Talay, D. (1983): How to discretize stochastic differential equations. Springer Lect. Notes in Math., **972**, 276–292
282. Talay, D. (1984): Efficient numerical schemes for the approximation of expectations of functionals of the solution of an SDE and applications, Springer Lecture Notes in Control and Inform. Sc., **61**, 294–313
283. Talay, D. (1990): Second-order discretization schemes for stochastic differential systems for the computation of the invariant law. Stoch. Stoch. Rep., **29**, 13–36
284. Talay, D. (1991): Approximation of upper Lyapunov exponents of bilinear stochastic differential equations. SIAM J. Num. Anal., **28**, 1141–1164
285. Talay, D. (1999): Approximation of the invariant probability measure of stochastic Hamiltonian dissipative systems with non-globally Lipschitz coefficients. In Bouc, R., Soize, C. (ed) Progress in Stochastic Structural Dynamics, Vol. 152, Publication du L.M.A.-CNRS
286. Talay, D. (2002): Stochastic Hamiltonian systems: exponential convergence to the invariant measure, and discretization by the implicit Euler scheme. Markov Processes Relat. Fields, **8**, no. 2, 163-198
287. Talay, D., Tubaro, L. (1990): Expansion of the global error for numerical schemes solving stochastic differential equations. Stoch. Anal. Appl., **8**, 483–509
288. Talay, D., Tubaro, L. (eds) (1996): Probabilistic models for nonlinear partial differential equations. Springer Lecture Notes in Math., **1627**
289. Talay, D., Zheng, Z. (2003): Quantiles of the Euler scheme for diffusion processes and financial applications. Math. Fin., **13**, 187–199
290. Tausworthe, R.C. (1965): Random numbers generated by linear recurrence modulo two. Math. Comp., **19**, 201–209
291. Taylor, M.E. (1996): Partial differential equations III: Nonlinear equations. Springer
292. Tezuka, S. (1995): Uniform random numbers: Theory and practice. Kluwer Academic Publishers
293. Thomas, N., Thornhill, R.A. (1998): The physics of biological molecular motors. J. Phys. D, **31**, 253–266
294. Thomson, D.J. (1987): Criteria for the selection of stochastic models of particle trajectories in turbulent flows. J. Fluid. Mech., **180**, 529–556
295. Tocino, A., Vigo-Aguiar, J. (2002): Weak second order conditions for stochastic Runge–Kutta methods. SIAM J. Sci. Comp., **24**, no. 2, 507-523
296. Tretyakov, M.V. (1998): Numerical technique for studying stochastic resonanse. Phys. Rev. E, **57**, 4789–4794

297. Tretyakov, M.V., Fedotov, S. (2001): On FKPP equation with Gaussian shear advection. Physica D, **159**, 191–202
298. Tretyakov, M.V., Tretyakov, S.V. (1994): Numerical integration of Hamiltonian systems with external noise. Phys. Lett. A, **194**, 371–374
299. Tropper, M.M. (1977): Ergodic properties and quasideterministic properties of finite-dimensional stochastic systems. J. Stat. Phys., **17**, 491–509
300. Venttsel, A.D., Gladyshev, S.A., Milstein, G.N. (1984): Piecewise constant approximation for Monte Carlo calculation of Wiener integrals. Theor. Prob. Appl., **29**, 715–722
301. Venttsel, A.D. (1996): A course in the theory of random processes. Nauka, Moscow
302. Veretennikov, A. Yu. (2003): On large deviations for approximations of SDEs. Probab. Theory Related Fields, **125**, 135–152
303. Vorozhtsov, E.V., Yanenko, N.N. (1990): Methods for the localization of singularities in numerical solutions of gas dynamic problems. Springer (engl. transl. from Russian, 1985)
304. Vreugdenhil, C.B., Koren, B. (eds) (1993): Numerical methods for advection-diffusion problems. Notes on Numerical Fluid Mechanics, **45**, Wiesbaden, Vieweg
305. Wagner, W. (1988): Monte Carlo evaluation of functionals of solutions of stochastic differential equations. Variance reduction and numerical examples. Stoch. Anal. Appl., 447–468
306. Wagner, W. (1988): Unbiased multi-step estimators for the Monte Carlo evaluation of certain functional integrals. J. Comp. Physics, **79**, 336–352
307. Wagner, W. (1989): Unbiased Monte Carlo estimators for functionals of weak solutions of stochastic differential equations. Stoch. Stoch. Rep., **28**, 1-20
308. Wagner, W., Platen, E. (1978): Approximation of Ito integral equations, Preprint ZIMM. Akad. der Wiss. der DDR, Berlin
309. Watson, E.J. (1962): Primitive polynomials (mod 2). Math. Comp, **16**, 368–369
310. Wiktorsson, M. (2001): Joint characteristic function and simultaneous simulation of iterated Ito integrals for multiple independent Brownian motions. Ann. Appl. Prob., **11**, 470–487
311. Yanenko, N.N. (1971): The method of fractional steps: the solution of problems of mathematical physics in several variables. Springer
312. Yoo, H. (2000): Semi-discretzation of stochastic partial differential equations on R^1 by a finite-difference method. Math. Comp., **69**, 653–666

Index

Absolute stability (A-stability), 69
Additive noise, 60, 112, 184, 199, 224, 250, 255, 262, 324
Algorithms
- based on time-step control for Dirichlet problems, 348, 371, 388
- for Cauchy problem for semilinear parabolic equation, 422, 445
- for diffusion in a space bounded domain, 291
- for diffusion in a space-time bounded domain, 322
- for Dirichlet problem for elliptic equation with small parameter, 395
- for Dirichlet problem for Helmholtz equation, 389
- for Dirichlet problem for semilinear parabolic equation, 474
- for exit point from a space-time parallelepiped, 316
- for exit point to lateral surface of cylinder with cubic base, 315, 556
- for Neumann problem for linear parabolic equation, 403, 405
- for Neumann problem for semilinear parabolic equation, 497, 499, 502
- of order 1/2 for parabolic Dirichlet problem, 359
- of order 3/2 for parabolic Dirichlet problem, 361
- simplest, for elliptic Dirichlet problem, 367
- simplest, for parabolic Dirichlet problem, 355
Approximation
- diffusion with reflection, 337
- exit point, 301, 325
- in bounded domains, 283, 317, 339, 550
- mean-square, 1, 3, 283, 550
- weak, 86, 100, 339
Average number of steps, 292, 322, 351, 356, 367, 384, 404, 470, 495

Balanced method, 33
Boundary layer, 392
Brownian bridge, 137, 543
Brownian motion, 54
- simulation, 316
- space-time, 306, 328
Brownian paths, 131
Burgers equation, 433
- Dirichlet problem, 479, 482
- Neumann problem, 503, 505, 561
- with small viscosity, 452

Cauchy problem
- for linear parabolic equation, 85
- for SDEs, 85
- for semilinear parabolic equations, 408, 415, 426, 427
- for semilinear parabolic equations with small parameter, 438
- for the first-order linear PDE, 83
Class of functions $\mathbf{F}$, 89
Colored noise, 77, 121, 200, 212
Commutative case, 28, 45, 183
Conditional expectations, 376
Conditional probability, 311, 552
Conditional Wiener integrals, 137

Deviation, 2
- mean, 10
- mean-square, 10
Difference method, 75

Diffusion
- Kramers' theory, 517
- reflected, 337
- space bounded, 283, 306
- space-time bounded, 285, 318
- with drift, 302, 328

Dirichlet problem
- for Helmholtz equation, 383
- for linear elliptic equation, 84, 365, 374, 392
- for linear parabolic equation, 129, 306, 329, 339
- for semilinear parabolic equations, 461, 476

Discretization
- space, 284, 422, 428, 474
- space-time, 285
- time, 4, 133

Distribution
- for one-dimensional Wiener process, 308
- Gaussian, 36, 39, 138, 166
- uniform on the surface of the sphere, 290, 341, 371

Drift-implicit methods, 29

Equation
- backward Kolmogorov, 85, 128, 306, 537
- Bellman, 124
- biharmonic, 330
- Burgers, 433, 452, 479, 503, 505, 561
- elliptic, 84, 332, 365, 372, 374, 555
- elliptic with small parameter, 392
- first order PDE, 443
- FKPP, 455, 479
- Fokker–Planck, 529
- heat, 85, 330, 364, 411, 558
- Helmholtz, 383
- Helmholtz, with small parameter, 392, 396
- Langevin, 212
- multi-dimensional semilinear parabolic, 427, 476, 501
- parabolic, 85, 129, 339, 408
- Poisson, 376
- quasilinear with power law nonlinearities, 431, 485
- semilinear heat, 446
- semilinear parabolic, 415, 461, 488
- semilinear parabolic with small parameter, 438
- stochastic Hamiltonian, 211
- with blow-up solutions, 431

Error
- expansion, 105, 202, 358
- Monte Carlo, 85, 123, 203, 207
- of numerical integration, 85

Euler method
- mean-square
- -- convergence, 14
- -- drift-implicit, 30, 38
- -- explicit, 1, 12, 17, 242
- -- for systems with additive noise, 14
- -- split-step backward, 37
- -- variable step-size, 17
- weak, 86, 344, 411

Exit point, 283, 308
- simulation, 301, 316, 325, 335

Expansion
- external asymptotic, 393
- interior asymptotic, 393
- Taylor, of mathematical expectation, 98
- Taylor, of solutions of ordinary differential equations, 18
- Taylor-type for SDEs, 19
- the global errors of weak methods, 105, 202, 359
- Wagner–Platen, 19

Explicit methods
- mean-square
- -- commutative case, 28
- -- Euler, 12
- -- for systems with additive noise, 60
- -- for systems with colored noise, 78
- -- for systems with multiplicative noise, 23
- -- for systems with small noise, 180
- -- Runge–Kutta, 71
- -- symplectic, 221, 231, 235
- -- Taylor-type, 23
- weak
- -- for systems with additive noise, 113, 121
- -- for systems with colored noise, 121

– – for systems with multiplicative noise, 103
– – for systems with small additive noise, 200
– – for systems with small noise, 191, 193, 196
– – symplectic, 254, 256
Extrapolation method, 105, 202, 358

Feynman path integral, 137
Feynman–Kac formula, 85, 128, 409
Finite-difference schemes, 412, 467, 482, 506, 540
First passage time, 283, 287, 290, 302, 307, 308, 313, 375, 380, 550
Fisher–Kolmogorov–Petrovskii–Piskunov (FKPP) equation, 455, 479
Fokker–Planck equation, 529
Formula
– Cameron–Martin, 153
– Feynman–Kac, 85, 128, 409
– Ito, 19, 84
Fully implicit methods, 38
– convergence theorem, 41, 44
– derivative-free, 45
– general construction, 43
– midpoint, 45, 218
– symplectic, 216

Gaussian distribution, 36, 39, 54, 138, 166, 216
– joint, 46
Girsanov's theorem, 409
Girsanov's transformation, 124, 128, 204, 370

Hamiltonian systems, 211
– separable, 220, 231, 234, 254
– with additive noise, 224
Heat equation, 85, 330, 364, 411, 558
– semilinear, 446

Implementation
– algorithms for bounded diffusions, 550
– layer methods, 561
– mean-square methods, 541
– quasi-symplectic methods, 546

– random walks for linear boundary value problems, 558
– RNGs, 566
– weak methods, 544
Implicit methods
– mean-square
– – for systems with additive noise, 65
– – for systems with colored noise, 81
– – for systems with multiplicative noise, 29, 33, 43
– – for systems with small noise, 182
– – symplectic, 216, 225
– weak
– – for conditional Wiener integrals, 145
– – for systems with colored noise, 123
– – for systems with multiplicative noise, 110
– – for systems with small noise, 193
– – symplectic, 251
Increments
– of approximation of solution, 92
– of solution, 92
– of Wiener process, 2, 6
– space, 283
– time, 283
Inequality
– Chebyshev, 295
– Doob, 289
– Hölder, 289
Integral
– Feynman path, 137
– Ito, 20, 21, 45, 54, 92, 377
– Wiener, 130, 135, 137
Interpolation
– B-spline, 424
– cubic, 425, 451, 476, 481, 503
– linear, 11, 354, 422, 428, 445, 474, 497, 502
– triangular, 428
Ito formula, 19, 84
Ito integral, 20, 172
– conditional expectation, 377
– modeling, 45
– – by Fourier method, 54
– – by rectangle method, 52
– – by trapezium method, 53
– – single noise, 45

- order of smallness, 21
- properties, 21, 92
Ito SDEs, 3, 171, 216

Kubo oscillator, 237, 257

Langevin equations, 212, 262
Lax–Richtmyer equivalence theorem, 412, 498
Layer methods, 410
- comparison analysis, 467, 482, 505
- construction, 415, 463, 489
- convergence, 419, 442, 467, 469, 493
- explicit, 417, 442, 447, 467, 469, 476, 490, 499, 502
- high order, 447
- implementation, 561
- implicit, 416, 440, 448
- three-layer, 448
- two-layer, 447
Liouvillian methods, 248
Liouvillian stochastic systems, 246
Lipschitz condition, 419
- globally, 3, 88
- nonglobally, 16, 37
- one-sided, 16, 37
Lyapunov exponent, 33, 186, 206, 333

Markov chain, 286, 291, 296, 323, 349, 355, 360, 375, 398, 469, 495
- boundary value problem, 292, 351, 368, 384, 404
- generator, 292
- one-step transition function, 291, 351, 368, 404
Markov moments, 84, 285, 287, 296, 322
Martingales, 130, 299
Mean number of steps, 292, 322, 351, 356, 367, 384, 404, 470, 495
Mean velocity, 532
Method of characteristics, 83
Methods, mean-square
- balanced, 33
- derivative-free fully implicit, 45
- drift-implicit, 29
- drift-implicit Euler, 30, 38
- explicit Euler, 1, 12
-- convergence, 14

-- for systems with additive noise, 14
- explicit Taylor-type, 23
- first order, 1
- for systems with colored noise
-- explicit, order 2, 78, 80
-- explicit, order 5/2, 79
-- implicit, 81
-- trapezoidal, 82
- for systems with small additive noise, 184
- for systems with small noise, 177
-- $O(h + \cdots)$, 179
-- $O(h^2 + \cdots)$, 179
-- $O(h^4 + \cdots)$, 180
-- implicit, 182
-- Runge–Kutta, 180
- fully implicit, 38
- implementation, 541
- Liouvillian, 248
- midpoint, 45, 218
- of order 3/2 for systems with additive noise, 60
-- explicit, 62
-- implicit, 65
- optimal, 15, 243
- quasi-symplectic, 263, 270
-- first order, 264, 271
-- second order, 265
-- third order, 268
- Runge–Kutta, 71
- split-step backward Euler, 37
- Störmer–Verlet, 235
- symplectic, 221, 544
-- explicit, 221, 231, 235
-- implicit, 216, 225
- two-step difference, 75
- variable step-size, 17
Methods, weak
- Euler, 86
- for systems with additive noise
-- Runge–Kutta, 113
-- second order, 113
-- third order, 121, 413
- for systems with colored noise
-- implicit, 123
-- Runge–Kutta, 121
- for systems with small additive noise, 200

- for systems with small noise
-- Runge–Kutta, 196
-- Taylor-type, 191, 193, 207
- for Wiener integrals, 133, 136, 138, 145
- implementation, 544
- implicit, 110
- midpoint, 252
- quasi-symplectic, 273, 546
- second order
-- explicit, 103, 413
-- Runge–Kutta, 104
- symplectic
-- explicit, 254, 256
-- implicit, 251
Monte Carlo methods, 85, 329, 364, 373, 410, 515
- error, 123, 203, 207
- variance reduction, 123, 157, 159, 205
- Wiener integrals, 131, 152

Neumann problem
- for linear parabolic equation, 397
- for semilinear parabolic equations, 488, 501
Noise
- additive, 9, 14, 60, 184, 199, 224, 250, 255, 324
- colored, 77, 121, 200, 212
- commutative, 28, 45, 183
- multiplicative, 9, 33
- scalar, 1
- small, 171, 212, 394, 438, 447
Noise-induced regular oscillations, 510
- high-frequency, 519
- in system of two coupled oscillators, 525
- large-amplitude, 523
- sufficient condition, 512
Noise-induced unidirectional transport, 528, 532, 534
- forced thermal ratchets, 533
- sufficient condition, 537
- systems with state-dependent diffusion, 528
Nonlinear PDEs, 407
Normal distribution, 36, 39, 166, 216

One-step approximation

- for boundary points, 399
- for solution of ODE, 19
- mean-square, 4
-- drift-implicit, 30
-- for systems with small noise, 173
-- Taylor-type, 25
- weak, 87
-- in bounded domains, 342, 344, 380
-- of third order, 92
Operators L, Λ_r, 20, 60, 84, 88, 172, 200
Order of accuracy
- mean-square, 1
- strong, 5
- weak, 86
Order of smallness of Ito integrals, 21
Ornstein–Uhlenbeck process, 77
Oscillators
- Kubo, 237, 257
- linear with additive noise, 240
- two coupled bistable overdamped, 525
- Van der Pol, 271
- with cubic restoring force under external random excitation, 280, 546
- with linear damping under external random excitation, 276

Parallel random number generators (PRNGs), 168
Partitioned Runge–Kutta (PRK) methods, 223, 254
Phase flow, 211
Phase volume, 246
Probabilistic representations
- Cauchy problem for linear system of parabolic equations, 430
- elliptic Dirichlet problem, 84, 366, 374
- parabolic Cauchy problem, 85, 128, 408
- parabolic Dirichlet problem, 129, 307, 340
- parabolic Neumann problem, 398
- semilinear parabolic equations, 415, 462, 489
Probability space, 3
Problem
- Cauchy, 83, 85, 408, 415

- Dirichlet, 84, 129, 306, 339, 365, 374, 461
- mixed, 503, 514
- Neumann, 397, 488
- nonlinear, 415, 461, 488

Process
- ergodic, 529
- Ornstein–Uhlenbeck, 77
- Wiener, 3, 308, 377

Quantization, 285
Quasi-symplectic methods, 263
- implementation, 546
- mean-square, 270
-- first order, 264, 271
-- second order, 265
-- third order, 268
- weak, 273, 546

Random number generators (RNGs), 159
- discrete, 165, 546
- Gaussian, 166, 544, 568
- parallel (PRNGs), 168
- uniform, 160
-- additive lagged Fibonacci, 161, 546, 568
-- linear congruential (LCGs), 160, 566
-- linear feedback shift register (LFSRs), 162
-- Tausworthe, 162

Random variables
- complicated, 86
- discrete, 166, 546, 570
- for second-order weak methods, 96
- for third-order weak methods, 120
- Gaussian, 36, 39, 138, 166, 216
- truncated, 3, 39, 41, 216
- uniform, 160, 566

Random walk, 349
- for Dirichlet problems, 359
- over boundaries of small space-time parallelepipeds, 285
- over small spheres, 285, 374
- simplest, for Dirichlet problems, 353, 367
- specific, for elliptic equations, 374

Ratchet

- forced thermal, 533
- potential, 533, 534, 539
- with state-dependent diffusion, 528

Reaction-diffusion systems, 429
Richardson–Runge extrapolation, 105
Runge–Kutta methods
- mean-square
-- derivative-free, 74
-- explicit, 71
-- for systems with colored noise, 80
-- for systems with small additive noise, 185
-- for systems with small noise, 180
-- implicit, 75
-- partitioned (PRK), 223
-- symplectic, 226
- weak
-- for systems with additive noise, 113
-- for systems with colored noise, 121
-- for systems with small noise, 196
-- for Wiener integrals, 136, 138, 145, 147
-- second order, 104
-- symplectic, 255

Sample trajectories, 238, 516, 542
SDEs
- Hamiltonian, 211, 220
- in space bounded domain, 283, 286, 302
- in space-time bounded domain, 285, 318
- Ito, 1
- Langevin, 212, 262
- linear, 15, 27, 186, 206, 333
- Liouvillian, 246
- stiff, 33, 66
- Stratonovich, 9, 44, 183, 188, 199, 211, 237, 530
- with additive noise, 14, 60, 112, 199, 250
- with colored noise, 77, 121, 212
- with normal reflection at the boundary, 337
- with small noise, 171, 212, 394, 438
- with stochastic resonance, 510

Semilinear parabolic equations
- Cauchy problem, 415, 426, 427
- Dirichlet problem, 461

- multi-dimensional, 427, 476, 501
- Neumann problem, 488, 561
- probabilistic representation, 415, 462, 489
- reaction-diffusion, 429
- singular case, 443
- with small diffusion, 438, 446

Simple iteration, 42, 417, 442
Splitting technique, 262
Stability, 33, 186, 277, 278, 333, 413
- absolute, 69
- index, 334
- zero-stable (D-stable), 449

Steps
- average number, 292, 322, 367, 404
- of time discretization, 4
- random, 296, 345
- variable, 17, 426

Stochastic model
- for synchrotron oscillations of particles in storage rings, 239, 259, 541
- of laser, 188

Stochastic ratchets, 528
- forced thermal, 533
- with state-dependent diffusion, 528

Stochastic resonance (SR), 509
Strict ellipticity condition, 285, 286, 293, 306, 365

Symplectic methods
- implementation, 543
- mean-square, 216, 218
- weak, 251

Symplectic structure, 211, 215, 227, 233

Talay–Tubaro extrapolation method, 105, 202

Taylor expansion
- of mathematical expectation, 98
- of semigroups, 99
- of solutions of ordinary differential equations, 18
- of solutions of SDEs, 19

Taylor-type methods
- mean-square, 23
-- explicit for systems with additive noise, 62
-- explicit for systems with colored noise, 78
-- for systems with small additive noise, 184
-- for systems with small noise, 177
-- implicit for systems with additive noise, 65
- weak, 103
-- explicit for systems with additive noise, 121
-- for systems with small noise, 191, 193, 207

Theorem
- convergence for balanced method, 35
- convergence for fully implicit methods, 41
- convergence for layer methods, 419
- error of weak schemes in the case of small noise, 190
- extrapolation method, 106
- Girsanov, 124, 204, 340
- Lax–Richtmyer equivalence, 412, 498
- mean-square error in the case of small noise, 176
- mean-square order – fundamental theorem, 4
- preservation of symplectic structure, 215, 227, 233
- variance reduction, 125, 127
- weak order – main theorem, 100

Transformation
- Cole–Hopf, 433, 452, 480
- Girsanov, 124, 128, 204

Variance reduction, 123
- combining method, 128
- for boundary value problems, 129, 373
- method of control variates, 126, 157
- method of important sampling, 123
- systems with small noise, 203

Wagner–Platen expansion, 2, 19
Weak approximations, 86, 339, 409, 558
Weak methods
- Euler, 86
- explicit second order, 103
- for systems with additive noise, 113, 121
- for systems with colored noise, 121, 123

- for systems with small additive noise, 200
- for systems with small noise
- – Runge–Kutta, 196
- – Taylor-type, 191, 193
- for the Dirichlet problem for elliptic equations, 366, 370, 393
- for the Dirichlet problem for parabolic equations, 340, 353
- for the Neumann problem for parabolic equations, 398
- for Wiener integrals, 133
- implementation, 545, 558
- implicit, 110
- Runge–Kutta, 104, 113, 136
- symplectic, 251, 254, 256

Wiener construction of Brownian motion, 54

Wiener integrals, 130
- conditional, 137
- fourth-order Runge–Kutta method, 136, 138
- implicit Runge–Kutta method, 145, 147
- midpoint method, 133
- of functionals of exponential type, 135
- of functionals of integral type, 131
- trapezium method, 133

Wiener measure, 131
- conditional, 137

Wiener process, 3, 308, 377
- exit point from a cube, 315, 556
- exit time from a cube, 313, 550

Printing: Saladruck, Berlin
Binding: Stein+Lehmann, Berlin

Scientific Computation

A Computational Method in Plasma Physics
F. Bauer, O. Betancourt, P. Garabechan

Implementation of Finite Element Methods for Navier-Stokes Equations
F. Thomasset

Finite-Different Techniques for Vectorized Fluid Dynamics Calculations
Edited by D. Book

Unsteady Viscous Flows
D. P. Telionis

Computational Methods for Fluid Flow
R. Peyret, T. D. Taylor

Computational Methods in Bifurcation Theory and Dissipative Structures
M. Kubicek, M. Marek

Optimal Shape Design for Elliptic Systems
O. Pironneau

The Method of Differential Approximation
Yu. I. Shokin

Computational Galerkin Methods
C. A. J. Fletcher

Numerical Methods for Nonlinear Variational Problems
R. Glowinski

Numerical Methods in Fluid Dynamics
Second Edition   M. Holt

Computer Studies of Phase Transitions and Critical Phenomena   O. G. Mouritsen

Finite Element Methods in Linear Ideal Magnetohydrodynamics
R. Gruber, J. Rappaz

Numerical Simulation of Plasmas
Y. N. Dnestrovskii, D. P. Kostomarov

Computational Methods for Kinetic Models of Magnetically Confined Plasmas
J. Killeen, G. D. Kerbel, M. C. McCoy, A. A. Mirin

Spectral Methods in Fluid Dynamics
Second Edition   C. Canuto, M. Y. Hussaini, A. Quarteroni, T. A. Zang

Computational Techniques for Fluid Dynamics 1 Fundamental and General Techniques   Second Edition
C. A. J. Fletcher

Computational Techniques for Fluid Dynamics 2 Specific Techniques for Different Flow Categories   Second Edition
C. A. J. Fletcher

Methods for the Localization of Singularities in Numerical Solutions of Gas Dynamics Problems
E. V. Vorozhtsov, N. N. Yanenko

Classical Orthogonal Polynomials of a Discrete Variable
A. F. Nikiforov, S. K. Suslov, V. B. Uvarov

Flux Coordinates and Magnetic Filed Structure: A Guide to a Fundamental Tool of Plasma Theory
W. D. D'haeseleer, W. N. G. Hitchon, J. D. Callen, J. L. Shohet

Monte Carlo Methods in Boundary Value Problems
K. K. Sabelfeld

The Least-Squares Finite Element Method Theory and Applications in Computational Fluid Dynamics and Electromagnetics
Bo-nan Jiang

Computer Simulation of Dynamic Phenomena
M. L. Wilkins

Grid Generation Methods
V. D. Liseikin

Radiation in Enclosures
A. Mbiock, R. Weber

Large Eddy Simulation for Incompressible Flows An Introduction   Second Edition
P. Sagaut

Higher-Order Numerical Methods for Transient Wave Equations
G. C. Cohen

Fundamentals of Computational Fluid Dynamics
H. Lomax, T. H. Pulliam, D. W. Zingg

The Hybrid Multiscale Simulation Technology An Introduction with Application to Astrophysical and Laboratory Plasmas   A. S. Lipatov

Computational Aerodynamics and Fluid Dynamics An Introduction   J.-J. Chattot

springeronline.com

Scientific Computation

Nonclassical Thermoelastic Problems in Nonlinear Dynamics of Shells Applications of the Bubnov–Galerkin and Finite Difference Numerical Methods
J. Awrejcewicz, V. A. Krys'ko

A Computational Differential Geometry Approach to Grid Generation V. D. Liseikin

Stochastic Numerics for Mathematical Physics G. N. Milstein, M. V. Tretyakov

springeronline.com